W9-ARQ-036

Glencoe

EARTH SCIENCE

GLENCOE

McGraw-Hill

New York, New York   Columbus, Ohio   Woodland Hills, California   Peoria, Illinois

A GLENCOE PROGRAM

Glencoe Earth Science

Student Edition
Teacher Wraparound Edition
Study Guide SE and TE
Reinforcement SE and TE
Enrichment SE and TE
Concept Mapping
Critical Thinking/Problem Solving
Activity Worksheets
Chapter Review
Chapter Review Software
Lab Manual SE and TE
Science Integration Activities
Cross-Curricular Integration
Science and Society Integration

Transparency Packages:
- Teaching Transparencies
- Section Focus Transparencies
- Science Integration Transparencies

The Glencoe Science Professional Development Series
- Performance Assessment in the Science Classroom
- Lab and Safety Skills in the Science Classroom
- Cooperative Learning in the Science Classroom
- Alternate Assessment in the Science Classroom
- Exploring Environmental Issues

Technology Integration
Multicultural Connections
Performance Assessment
Assessment
Lesson Plans
Spanish Resources
MindJogger Videoquizzes and Teacher Guide
English/Spanish Audiocassettes
CD-ROM Multimedia System
Interactive Videodisc Program
Computer Testbank—DOS and Macintosh Versions

Cover Photograph: The sunset on the Grand Canyon in Arizona was photographed by Jack Dykinga.

Glencoe/McGraw-Hill
A Division of The McGraw-Hill Companies

Send all inquiries to:
Glencoe/McGraw-Hill
8787 Orion Place
Columbus, OH 43240

ISBN 0-02-827808-9
Printed in the United States of America.
9 10 11 12 027/043 04 03 02

Authors

Ralph Feather, Jr. is a teacher of geology, astronomy, Earth science, and integrated science, and serves as Science Department Chair in the Derry Area School District in Derry, Pennsylvania. Mr. Feather has 25 years of teaching experience in secondary science. He holds a B.S. in Geology and an M.Ed. in Geoscience from Indiana University of Pennsylvania. Mr. Feather is currently completing a Ph.D. in "Writing Across the Curriculum to Learn Science" at the University of Pittsburgh. Mr. Feather received the 1991 Presidential Award for Excellence in Science Teaching and the 1991 Geological Society of America Award for Excellence in Earth Science Teaching. He is a member of the National Association for Research in Science Teaching, Geological Society of America, National Association of Geology Teachers, National Earth Science Teacher's Association, and the Planetary Society. Mr. Feather is coauthor of *Science Interactions, Merrill Earth Science,* and *Science Connections* from Glencoe Publishing Company.

Susan Leach Snyder is a teacher of Earth science and serves as Science Department Chair at Jones Middle School, Upper Arlington School District, Columbus, Ohio. Ms. Snyder received a B.S. in Comprehensive Science from Miami University, Oxford, Ohio, and an M.S. in Entomology from the University of Hawaii. She has 23 years of teaching experience. Ms. Snyder, in addition to receiving Exemplary Earth Science and Career Awareness in Science Teaching Team awards from NSTA, has received Outstanding Teacher awards from the National Association of Geology Teachers, the National Marine Educators Association, and the Geological Society of America. She has been a state recipient of the Presidential Award for Excellence in Science and Math Teaching from NSTA, the Ohio Teacher of the Year, and one of four finalists for the National Teacher of the year. Ms. Snyder is a coauthor of *Science Interactions* and *Merrill Earth Science* from Glencoe Publishing Company.

Contributing Writers

Linda Barr
Freelance Writer
Westerville, Ohio

Dan Blaustein
Science Teacher and Author
Evanston, Illinois

Pam Bliss
Freelance Science Writer
Grayslake, Illinois

Mary Dylewski
Freelance Science Writer
Houston, Texas

Nancy Ross-Flanigan
Syndicated Science Reporter
Detroit, Michigan

Helen Frensch
Freelance Writer
Santa Barbara, California

Steve Glazer
Freelance Writer
Champaign, Illinois

Rebecca Johnson
Freelance Science Writer and Author
Sioux Falls, South Dakota

Devi Mathieu
Freelance Science Writer
Sebastopol, California

Patricia West
Freelance Writer
Oakland, California

Consultants

Earth Science

Michael Bikerman, Ph.D
Associate Professor of Geology
University of Pittsburgh
Pittsburgh, Pennsylvania

Anthony E. D'Agostino
President
T D Geoscience
Midland, Texas

Richard Duschl
Professor of Science Education
Vanderbilt University
Nashville, Tennessee

Jay Stanley Hobgood, Ph.D.
Associate Professor of Geography
The Ohio State University
Columbus, Ohio

Larry A. Lebofsky, Ph.D.
Senior Research Scientist
Luna & Planetary Laboratory
University of Arizona
Tucson, Arizona

John Misock
Director of Consumer Health Services
Wyoming Department of Agriculture
Cheyenne, Wyoming

Multicultural

Karen Muir, Ph.D.
Lead Instructor
Department of Social and Behavioral Sciences
Columbus State Community College
Columbus, OH

James B. Phipps, Ph.D.
Instructor
Grays Harbor College
Aberdeen, Washington

Safety

Robert Joseph Tatz, Ph.D.
Instructional Lab Supervisor
The Ohio State University
Columbus, Ohio

Reviewers

Beth A. Beckwith
Kenwood Trail Junior High School
Lakeville, Minnesota

David Burch
Eastern Green County Schools
Bloomfield, Indiana

Sister Carmela Anne
Gesu School
Philadelphia, Pennsylvania

Betty Crowder
Rochester Community School
Rochester, Michigan

Susan Engledow
Clay Junior High School
Carmel, Indiana

Dr. Timothy Folkomer
Penn Wood East Junior High School
Yeadon, Pennsylvania

John A. Fradiska, Jr.
Thomas Johnson Middle School
Frederick, Maryland

Michael Goodrich
Lake Oswego High School
Lake Oswego, Oregon

Patricia Kenzig
St. John Bosco
Parma Heights, Ohio

Martha McIlveene
LaFayette Middle School
LaFayette, Georgia

Kathleen McMichael
Escambia High School
Pensacola, Florida

Arren L. Ostroff
Reynolds Junior High School
Lancaster, Pennsylvania

David Rhoads
Junction City High School
Junction City, Kansas

John Rooke
Pittsford Central School
Pittsford, New York

Richard Wisemiller
Clearwater Discovery School
Clearwater, Florida

Contents

Earth Materials 2

Contents

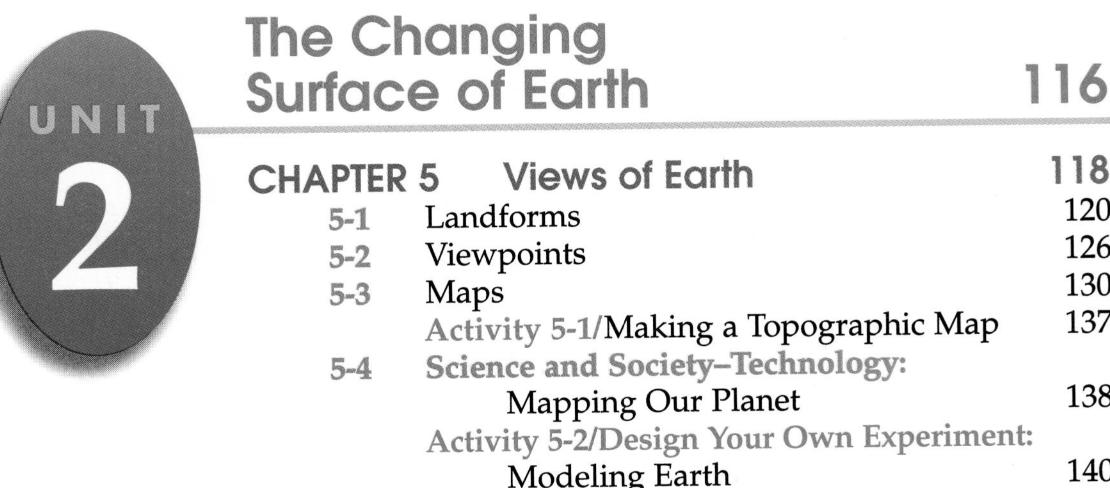

Contents

Contents

UNIT 3

Earth's Internal Processes 232

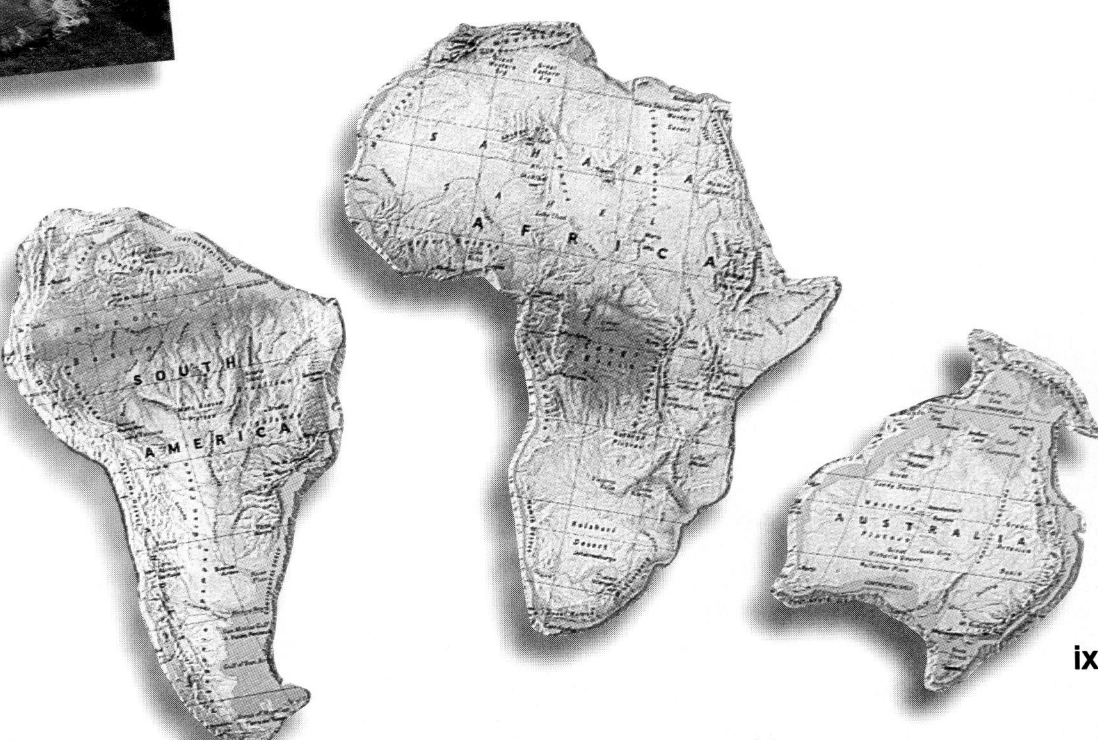

Contents

Contents

Contents

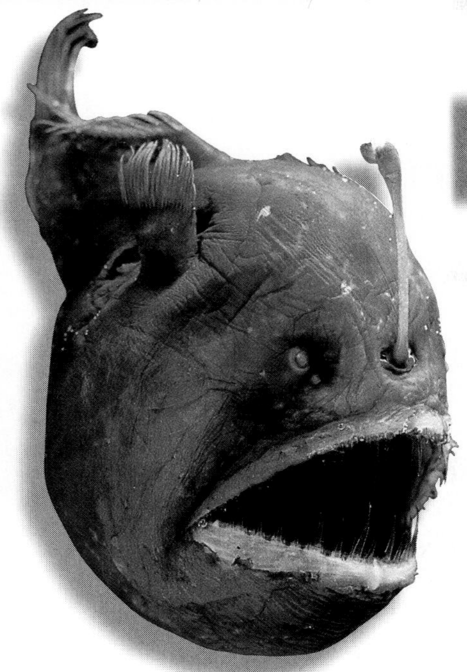

Contents

UNIT 6 You and the Environment 528

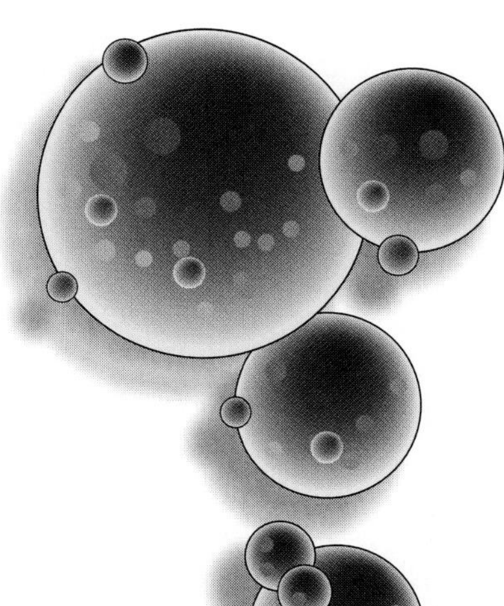

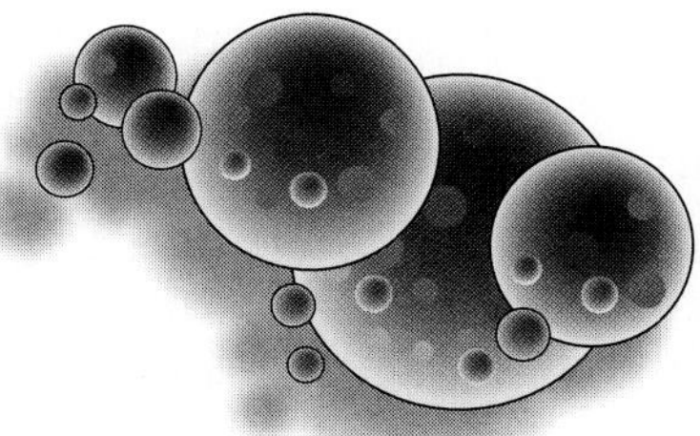

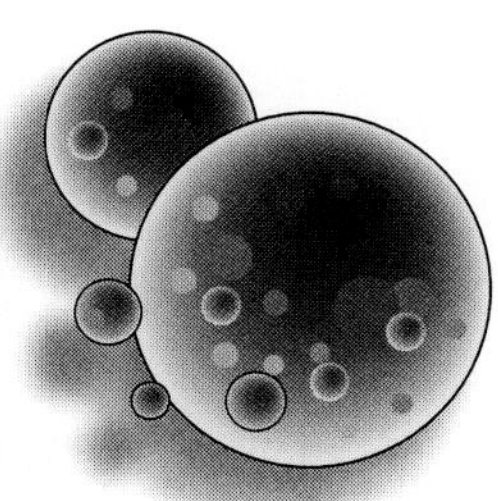

Contents

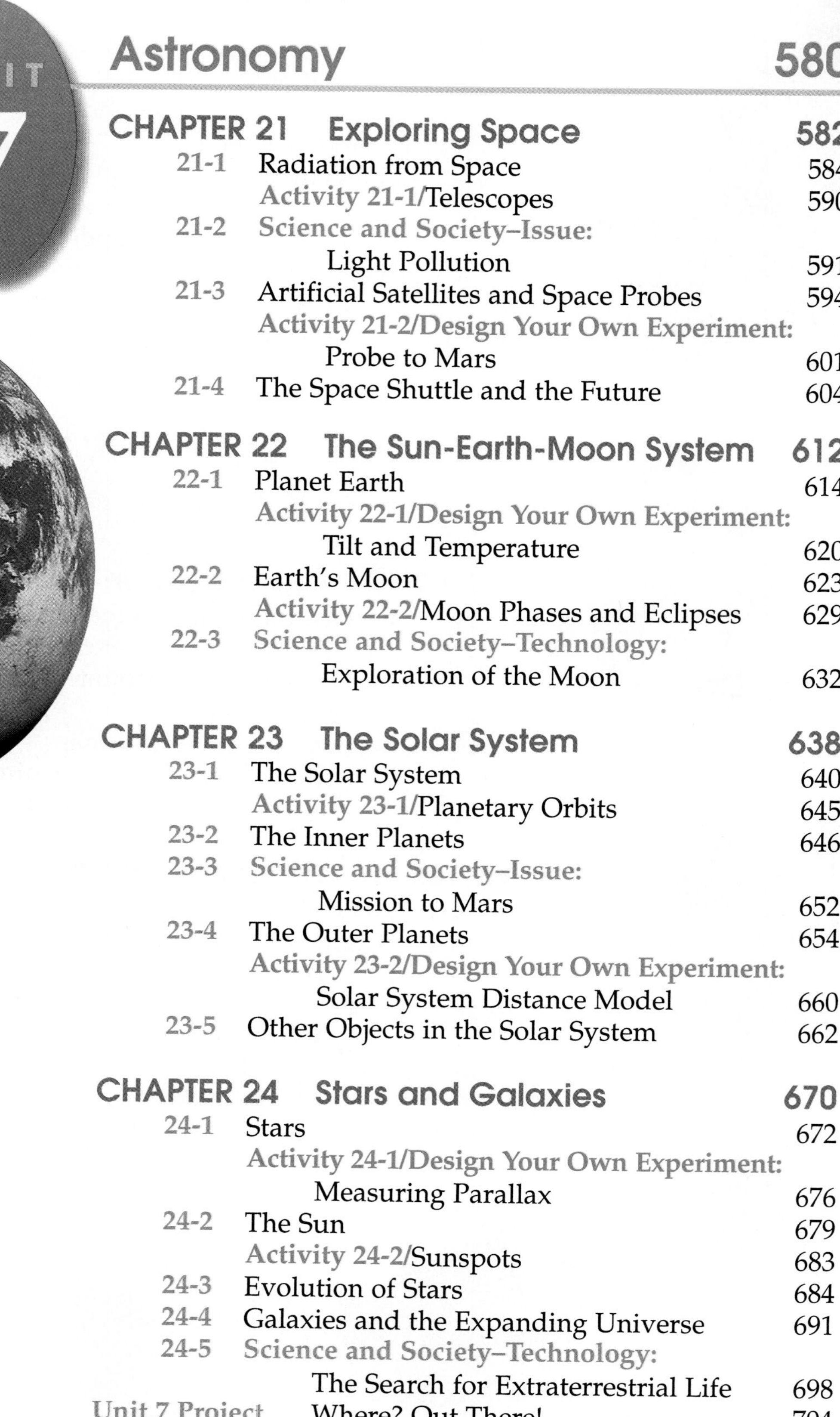

Astronomy 580

Contents

Appendices

Activities

Activities

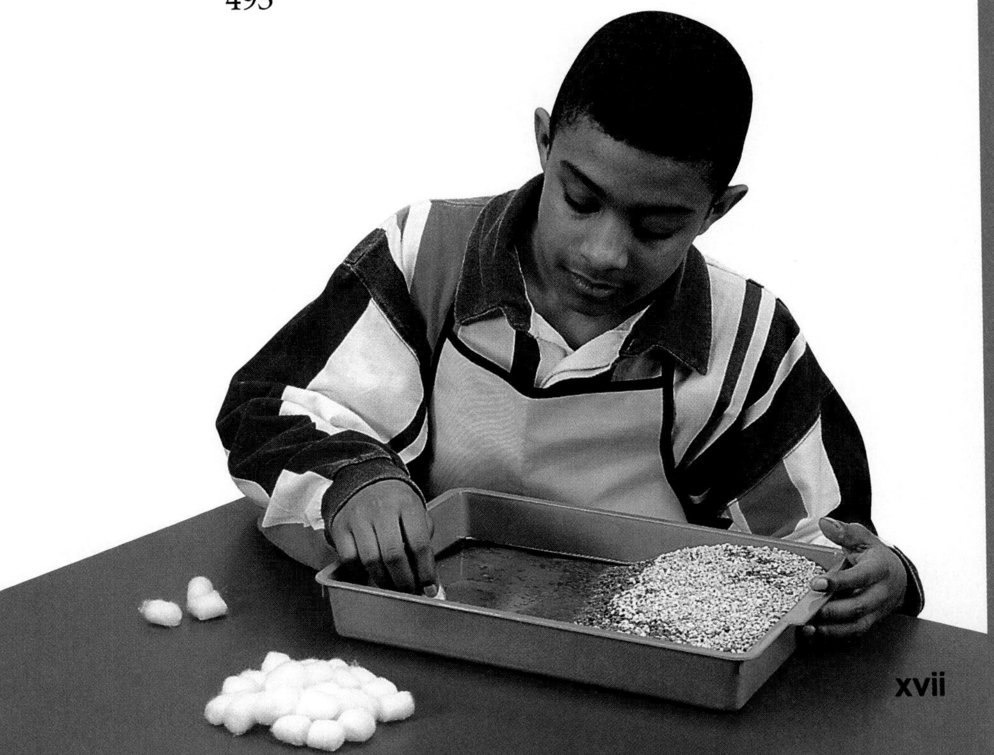

Explore Activities

MiniLABS

MiniLABS

Problem Solving

Using Technology

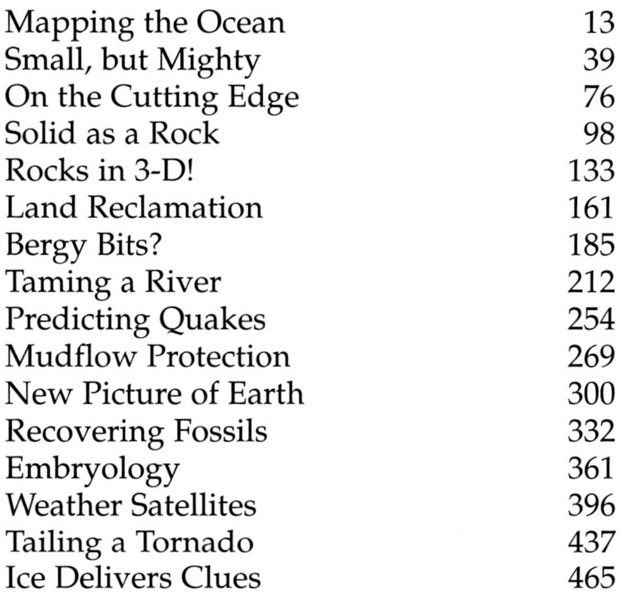

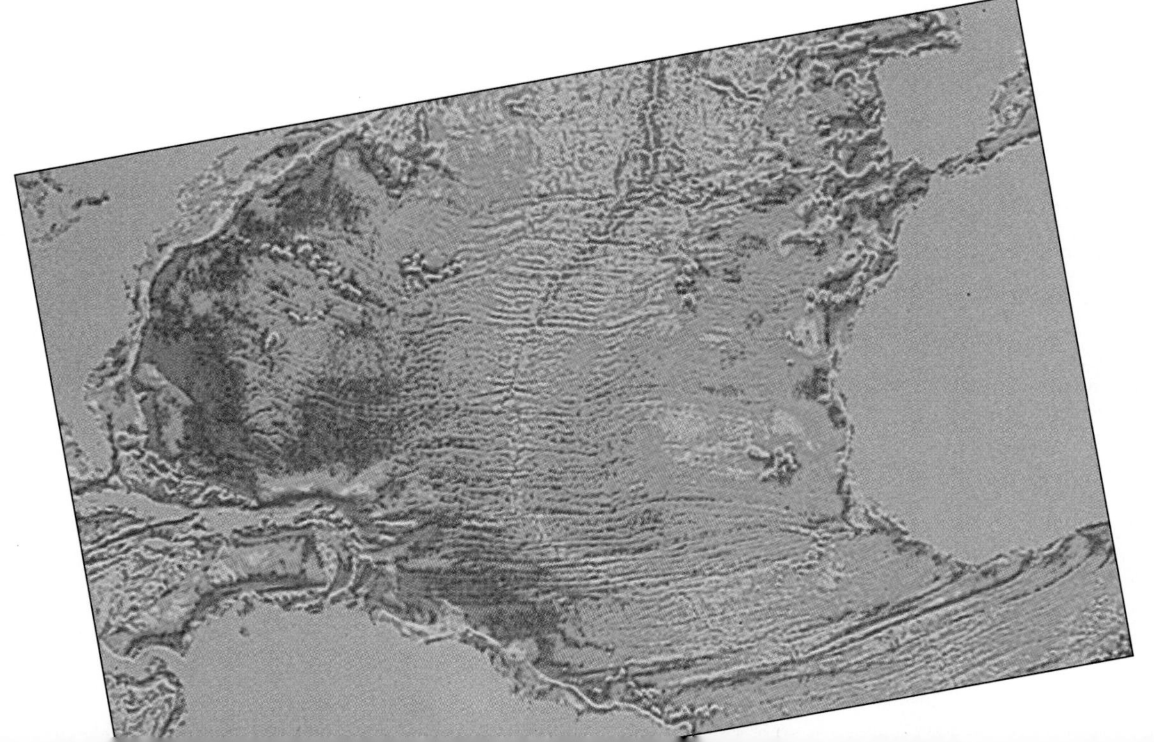

Skill Builders

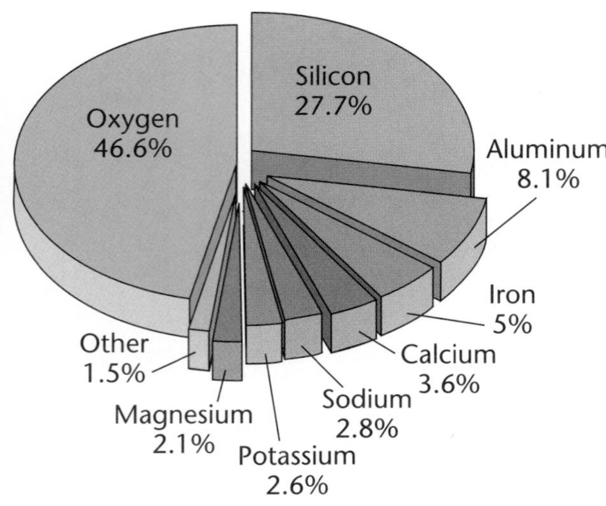

GRAPHICS

ORGANIZING INFORMATION

Element	Atomic No.	Mass No.
Fluorine	9	19
Lithium	3	7
Carbon	6	12
Nitrogen	7	14
Beryllium	4	9
Boron	5	11
Oxygen	8	16
Neon	10	20

THINKING CRITICALLY

DESIGNING AN EXPERIMENT

People and Science

Science Connections

Science and Art

Science and History

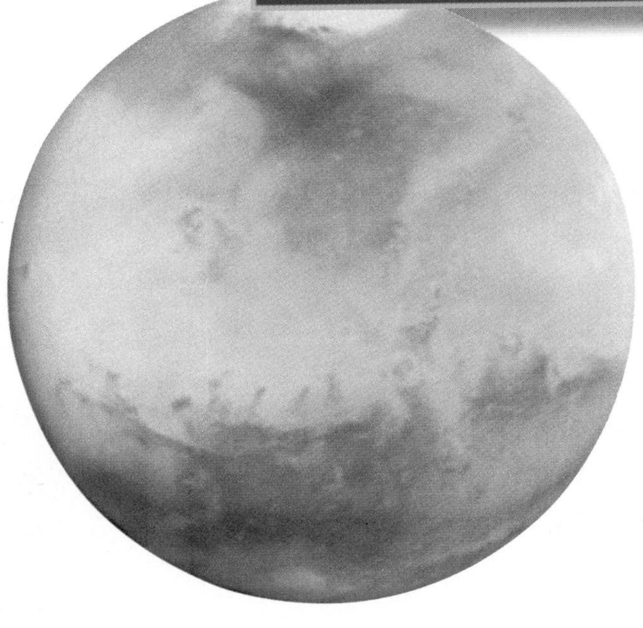

Science and Literature

UNIT 1

Unit Contents

Earth Materials

What's Happening Here?

Earth science is the study of planet Earth and its place in space. Think about that. Where would you begin to study Earth? Would you break apart a mineral and magnify it a thousand times, like the sulfur crystals shown here, to unlock the secrets of its structure? Or would you gaze at the skies to learn more about our planet? Earth scientists do both. They study Earth close up and from a distance, using technology such as microscopes and space satellite cameras. In this unit, you will learn more about Earth science and the materials, such as minerals, that make up Earth.

Science Journal

In your Science Journal, write a paragraph explaining how the microscope and the space satellite both contribute to the study of Earth science.

Previewing the Chapter

Chapter 1

The Nature of Science

The climber in this photo is using the results of many scientific experiments to safely enjoy the thrill of rock climbing. Experiments have led to the development of strong, stretchable ropes that can protect a climber in the event of a fall. The gear used to attach the rope to the rock is the result of years of experimenting with light, strong metal alloys. The rubber on the climber's shoes has been found to be the best compound for providing traction and durability on rock.

EXPLORE ACTIVITY

What is the best shoe design for rock climbing?

1. Make a list of the characteristics you would expect a good shoe to have for rock climbing.
2. Observe the soles of various kinds of shoes. Examine hiking boots, tennis shoes, soccer cleats, baseball cleats, and other shoes.
3. Based on the characteristics you chose, predict which shoes would be the best for climbing steep rocks, or design a rock-climbing shoe that has the best features of several shoes.

Observe: In your Science Journal, discuss some ways you could test the traction of different shoes on rock. Carry out some of these experiments.

Previewing Science Skills

- In the **Skill Builders**, you will **map concepts**, **compare and contrast**, and **measure in SI**.
- In the **Activity**, you will **design an experiment**, **hypothesize**, and **analyze data**.
- In the **MiniLABS**, you will investigate specific areas of study in Earth science, and learn about SI conversions.

1•1 What Is Earth Science?

Science Words

science
Earth science
geology
meteorology
astronomy
oceanography

Objectives

- Differentiate among the following Earth sciences: geology, meteorology, astronomy, and oceanography.
- Identify the topics you'll be studying this year in Earth science.

Science

Science is all around you. It is such a common part of your life that you probably take it for granted. Have you ever wondered why there are seasons, why volcanoes erupt, or whether life exists on other planets? Do you wonder how people find the answers to these and other questions?

They use science. Science means "having knowledge." **Science** is a process of observing and studying things in our world. Many of these observations can't be explained easily and therefore present problems. Science involves trying to solve these problems. Science is a process that enables you to understand your world. Every time you try to find out how and why things look and act the way they do, you are a scientist. For example, if you wonder and try to figure out why some sunsets are more colorful than others, you are doing science. **Figure 1-1** shows just a few of the many worlds you can explore using science.

Collecting Scientific Knowledge

Scientific knowledge has accumulated since people first began observing the world around them. At first, people had only their senses to rely on for their observations. Early astronomers studied the night sky with just their eyes because they didn't

Figure 1-1

Earth science includes the study of climate, volcanoes, space, and much more.

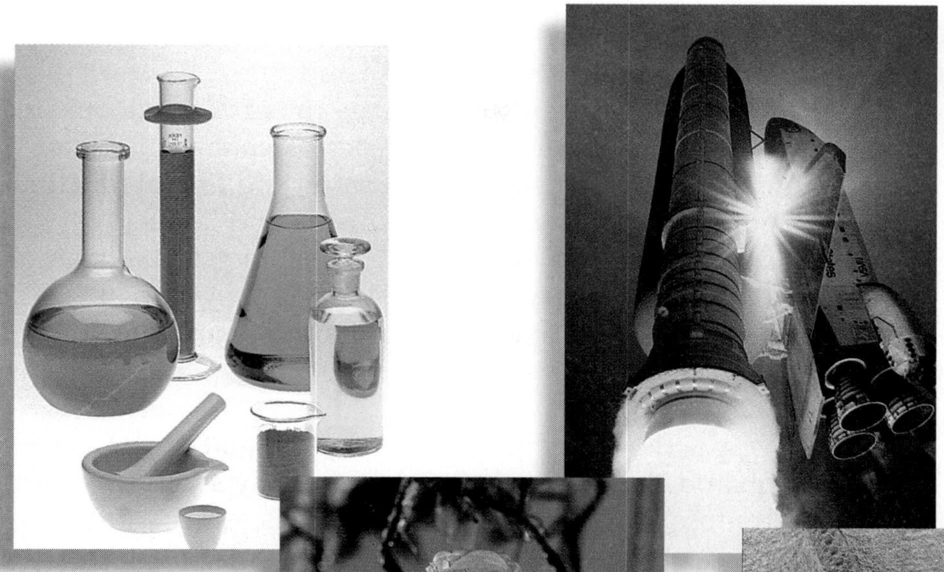

PHYSICS
Basic concepts of physics are involved in the motions of Earth and the moon and in the formation of stars.

Figure 1-2
The major sciences

CHEMISTRY
You'll learn basic concepts of matter when you study rocks and minerals.

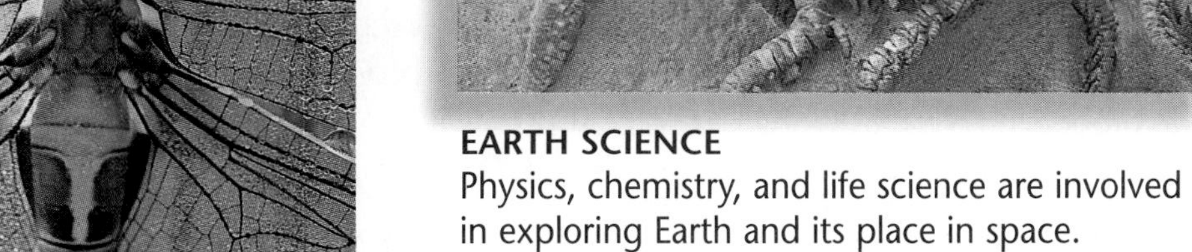

LIFE SCIENCE
Organisms have an important role in Earth's history and the environment.

EARTH SCIENCE
Physics, chemistry, and life science are involved in exploring Earth and its place in space.

have telescopes yet. They acquired knowledge slowly. Today, we have instruments such as microscopes that magnify small objects, satellites that take photographs of Earth and other planets, telescopes that probe the depths of space, and computers that store and analyze information. Today we are learning more information, we are learning it faster, and there are more new inventions and discoveries than ever before.

The Major Sciences

Science can be applied to just about anything, and hundreds of special subject areas fall within the broad scope of "science." But basically, science can be divided into four general areas: chemistry, physics, life science, and Earth science. These general topics do overlap. For example, in Earth science, chemistry, physics, and life science are studied as they relate to Earth. In **Figure 1-2** you can see what these different sciences are about and how they are connected to each other.

CONNECT TO

LIFE SCIENCE

When geologists study fossils, they learn a lot about what life was like in the past. Limestone is a rock that often contains fossils of marine organisms. ***Research*** in the library to find out the names of some organisms that you might find as fossils in limestone.

MiniLAB

How many Earth sciences are there?

Although the four major areas of Earth science are geology, meteorology, astronomy, and oceanography, each of these is composed of subtopic areas of its own.

Procedure

1. Look at the list of specific areas of study below.
2. Identify in which of the four major areas of Earth science they belong.

Analysis

What would a scientist working in each of these fields study?

hydrology	seismology
volcanology	petrology
mineralogy	geomorphology
paleontology	geochemistry
stratigraphy	crystallography

Earth Science

Earth science is the study of Earth and space. It is the study of such things as the transfer of energy in Earth's atmosphere; the evolution of landforms; patterns of change that cause weather; the scale and structure of stars; and the interactions that occur among the water, atmosphere, and land.

Just as science can be divided into the four general areas of Earth science, chemistry, physics, and life science, Earth science in this book is divided into four specific areas of study.

Geology

Geology is the study of Earth, its matter, and the processes that form and change Earth. Some of the things you'll explore are volcanoes, earthquakes, maps, fossils, mountains, and land use. Geologists search for oil, study volcanoes, identify rocks and minerals, study fossils and glaciers, and determine how mountains form.

Meteorology

Meteorology is the study of weather and the forces and processes that cause it. You'll learn about storm patterns, climates, and what factors cause our daily weather. A meteorologist is a scientist who studies weather patterns in order to predict daily weather.

Astronomy

Astronomy is the study of objects in space, including stars, planets, and comets. Before telescopes, like the radio telescope shown in **Figure 1-3,** were invented, this branch of Earth

Figure 1-3

Radio telescopes are used to collect information from beyond our solar system.

science mainly dealt with descriptions of the positions of the stars and planets. Using powerful telescopes, we can observe amazing sights, like that in **Figure 1-4.** Today, astronomers who study space objects seek evidence about the beginning of the universe. The study of astronomy helps scientists understand Earth's origin.

Figure 1-4

Using powerful new telescopes, we can extend our observations deeper into space.

Oceanography

Oceanography is the study of Earth's oceans. Oceanographers conduct research on the physical and chemical properties of ocean water. Oceanographers also study the processes that occur within oceans and the effects humans have on these processes.

As you study these topics, imagine how all this information was collected over the years. People just like you had questions about what they observed, and they used science to find the answers. In section 1-3, you'll learn ways that you too can find answers to questions and solutions to problems.

Section Wrap-up

Review

1. Compare and contrast meteorology and oceanography.
2. List three topics you'll study this year in Earth science.
3. **Think Critically:** The following paragraph summarizes one idea of how the dinosaurs may have died. Explain how this paragraph relates to Earth science.

   Evidence suggests that a large object from outer space may have crashed into Earth's crust. Dust from this collision blocked out the sun's light. Earth became colder, killing some plants and animals.

Skill Builder

Concept Mapping

Make a network tree concept map that shows which of the following topics are studied in a particular topic of Earth science. Use the following terms: *Earth science, geology, waves, currents, astronomy, oceanography, stars, volcanoes, planets, meteorology, fossils, weather, climate.* If you need help, refer to Concept Mapping in the **Skill Handbook.**

Science Journal

Look at the table of contents in this book and name ten careers in Earth science. Make a list of these careers in your Science Journal. Check the career resources in the library or guidance office for help.

SCIENCE & SOCIETY

1•2 TECHNOLOGY: Applying Science

Science Words

technology

Objectives

- List ways technology helps you.
- Discuss ways that the use of technology can be harmful.

Technology and You

The study of science doesn't just add to our understanding of our natural surroundings, it also allows us to make discoveries that help us. **Technology** is the use of scientific discoveries. Everywhere you look, you can see ways that science and technology have shaped your world.

Technology has produced such diverse and important things as robots that check underwater oil rigs for leaks and manufacture products, as shown in **Figure 1-5,** new fibers to make air-supported roofs such as those used in some sports arenas, and calculators and computers that process information.

Technology Is Transferable

The interesting thing about technology is that it is transferable. That means that it can be applied to new situations. For example, all of the types of technology that are mentioned above were developed for use in outer space. Scientists developed robotic parts, new fibers, and microminiaturized instruments for spacecraft and satellites. After these materials were developed, many were modified for use here on Earth. Technology that was developed by the military, like radar and sonar, has applications in astronomy, meteorology, geology, and oceanography today.

Science Journal

Pretend you can travel back in time to the year 1800. Consider all of the advancements that have been made in science since that time. In your Science Journal, write a dialogue between you and someone living at that time as you try to explain what a modern device such as a television set, microwave oven, computer, or car is like.

Figure 1-5

Robots are commonly used in many industries. They are also used to explore other planets.

The Effects of Technology

Let's further examine technology that helps us. Computers do everything from helping us predict weather to monitoring earthquakes. Robotics and computers enable us to make underwater submersibles to study the ocean floor and search for life in space. We use satellites to monitor pollution problems, locate hurricanes, and even track endangered species.

We have the technology to clear forests and build cities, slow down erosion along rivers, and even make it rain in some regions. Humans can change our surroundings on a large scale to meet our needs. We have even developed the technology to spend time in space, as in **Figure 1-6.**

Technology Can Cause Problems

Not all of the changes created by technology are good. For example, many types of technology create water, soil, and air pollution. Technology can cause other problems too. When forests are cleared to build cities, soil erosion increases, and weather patterns change. Although technology sometimes creates new jobs for people, some jobs are lost. In some cases, skills that people have become obsolete. In other instances, robots and computers can be used more cheaply than people.

Figure 1-6

Technology has enabled humans to travel beyond Earth.

Section Wrap-up

Review

1. How can technology be helpful?
2. How can technology cause problems?

Explore the Technology

The Panama Canal allows ships to travel between the Atlantic and Pacific Oceans without having to pass around South America. Because the level of the Atlantic is higher than the Pacific's, the canal was built with a system of locks to prevent flooding of lowlands in Panama. The locks also prevent species of organisms from traveling from one ocean to the other. We have the technological know-how to construct a wider and deeper canal that would permit larger modern ships to get through. It would be built without locks. What should be considered before building the new canal?

SCIENCE & SOCIETY

1•3 Solving Problems

Science Words

scientific method
hypothesis
variable
control
theory
law

Objectives

- Describe some problem-solving strategies.
- List steps commonly used in the scientific method.
- Distinguish among hypotheses, theories, and laws.

Problem-Solving Strategies

Soccer practice, dinner, homework, chores, watching your favorite television program . . . how will you squeeze them all in tonight? Is this a problem you are facing? There are many methods you can use to find solutions to problems. These are called strategies. Let's look at one strategy you can use to solve the problem of how to squeeze so many activities into one evening.

One Approach

To solve any problem, you need to have a strategy. Identifying the problem is the first step of any strategy. Next, you need to collect information about the problem. You need to know the basic facts of when soccer practice begins and ends, how much homework you have, what chores need to be done, and when the TV program is on. After you have determined these things, you might try writing out a time schedule. First, write in the activities that have fixed times. Then, fill in each of the other activities. You may have to try different arrangements before you find the solution that you think is the best.

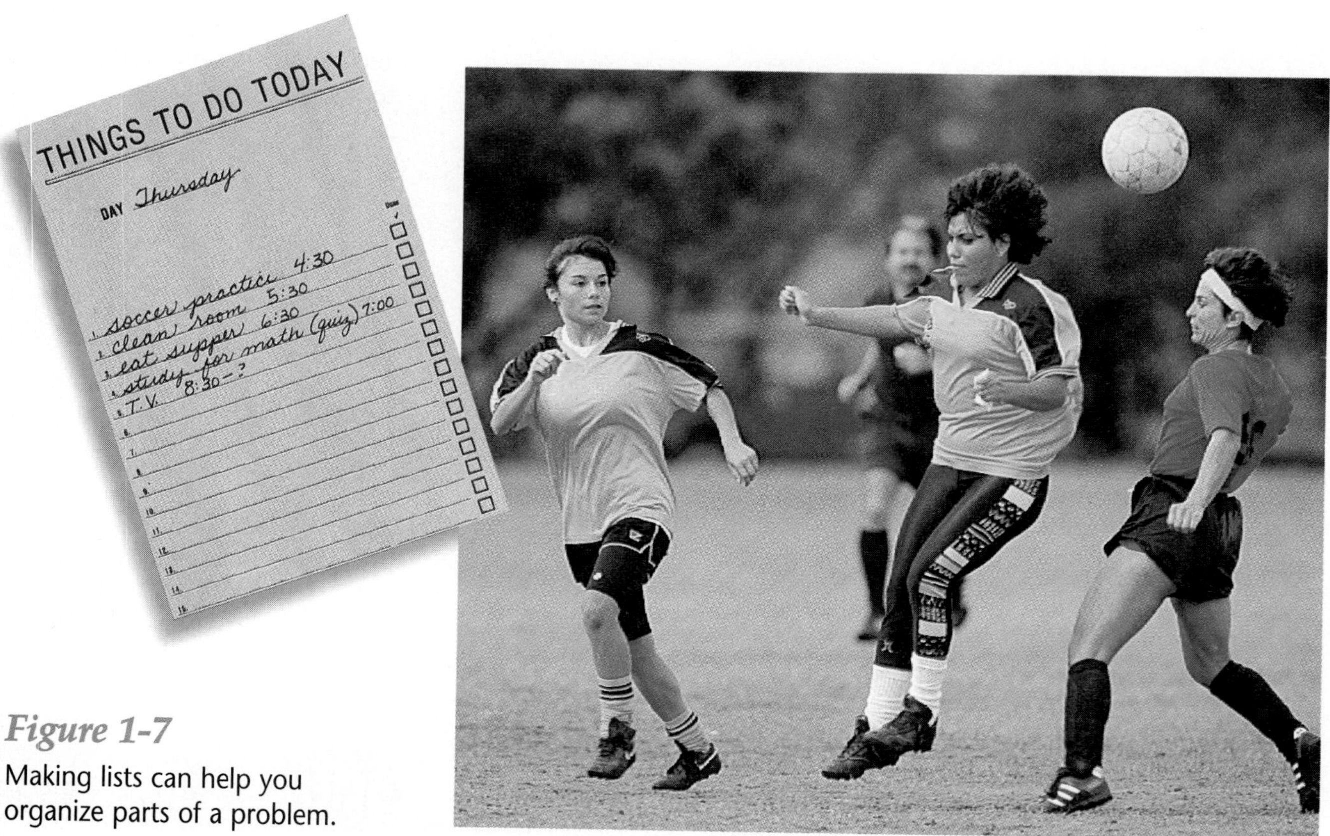

Figure 1-7
Making lists can help you organize parts of a problem.

Other Strategies

There are other ways to solve problems. You might try the strategy of eliminating possibilities. You could do this by trying options until you find the one that works. This method is also known as trial and error.

Sometimes, patterns can help you make predictions about a problem. Organizing data in a table or drawing will help you recognize patterns. For example, suppose three of your friends became sick because of something they ate. You could use a table of what each had recently eaten to look for patterns. Similar food items would be good candidates for causing their illness.

Another strategy is to solve a simpler, related problem, or to make a model, drawing, or graph to help you visualize the problem. If your first strategy does not work, keep trying different strategies.

USING TECHNOLOGY

Mapping the Ocean

When you can't observe something directly, how can you study it? Oceanographers want to study structures on the ocean floor in detail, but it's too expensive and dangerous to survey the entire ocean bottom in person. Thus, some researchers at the Scripps Institute of Oceanography and the National Oceanic and Atmospheric Administration developed a way of using satellites to produce a detailed image of the global seafloor.

A Bumpy Ocean Surface

The images are possible thanks to technology that allows satellites high above Earth to measure the height of the ocean surface to within a centimeter. This gives scientists a detailed map of the shape of the ocean surface, with broad bumps and dips. The dips and bumps are caused by the gravitational attraction between the water and seafloor features. A 2000-meter undersea volcano, for example, produces a bump on the ocean surface that is about 2 meters high. By analyzing the map of the ocean surface, scientists were able to infer the shape of the seafloor.

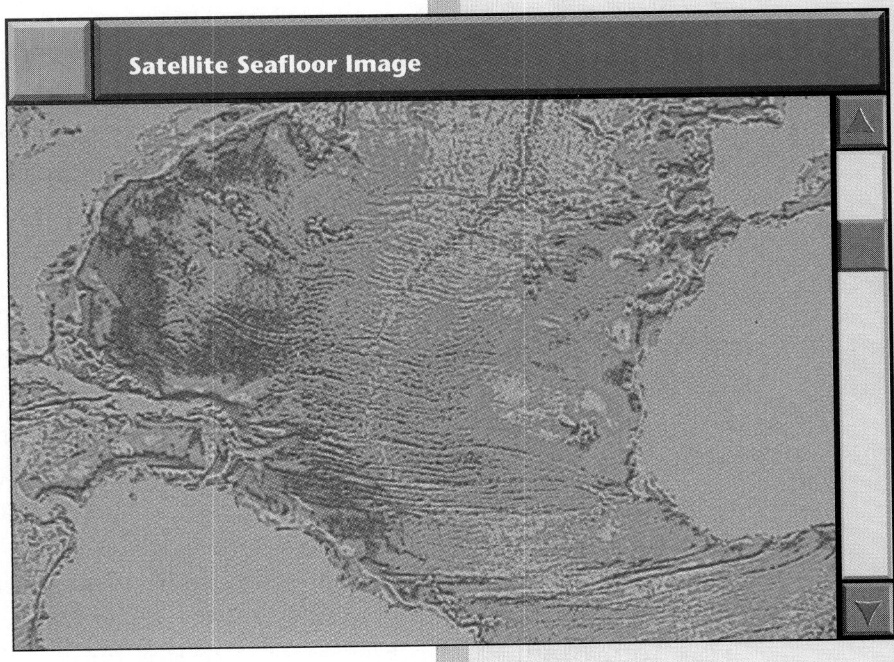

Satellite Seafloor Image

Think Critically:

What do you think might be some limitations of this method of mapping?

Critical Thinking

Imagine yourself mowing the lawn. Suddenly, you come to the edge of a slope, and you discover a large hole where the soil has sunk several feet. The hole wasn't there last week when you mowed the grass. What caused the hole? Where did all the soil go? **Figure 1-8** shows a similar case in which more than just part of a lawn sank down a slope.

After thinking a bit, you remember that there was a severe thunderstorm a few days ago. Perhaps the rain had something to do with the formation of the hole. Maybe the water running down the slope carried away the soil. You may not be aware of it, but you are using critical-thinking skills.

Figure 1-8
This mudslide caused serious damage.

Critical-Thinking Skills

Critical thinking is a process that uses certain skills to solve problems. For example, you *identified the problem* by mentally comparing the slope with how it was the last time you saw it. You *separated important information from unimportant information* when you considered factors that may have affected the slope during the past week. Important factors you may have noted include that it happened only along the slope in the yard, and not on its flatter parts. You also remembered the unusually heavy rain a few days ago. Unimportant factors include the wind that affected all parts of the yard equally, or the sunshine that the entire yard was exposed to. Then you went one step further and analyzed your conclusion that heavy rain could carry away soil on a slope. By performing these three critical-thinking skills, you may have solved the problem.

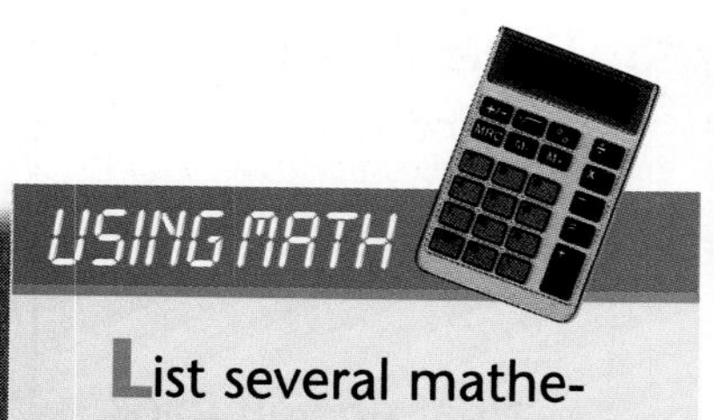

List several mathematical concepts that you might use as you study Earth Science.

Improving Your Critical-Thinking Skills

This book uses an activity called "Flex Your Brain" to help you think about and examine your thinking. "Flex Your Brain" is a way to keep your thinking on track when you are investigating a topic. It takes you through steps of exploration from what you already know and believe, to new conclusions and awareness. Then, it encourages you to review and talk about the steps you took.

"Flex Your Brain" and other features of this book will help you improve your critical-thinking skills. You will get your first chance to "Flex Your Brain" on the next page.

You start with what you know about a topic and move on to new conclusions and new awareness. You end by reviewing and discussing the steps you took.

Flex Your Brain

1 **Topic:** ______________________

2 **What do I already know?**
1. ______
2. ______
3. ______
4. ______
5. ______

3 **Q:** Ask a question ______________________

4 **A:** Guess an answer ______________________

5 **How sure am I? (circle one)**

Not sure				Very sure
1	2	3	4	5

6 **How can I find out?**
1. ______
2. ______
3. ______
4. ______
5. ______

7 **EXPLORE**

8 **Do I think differently?** yes / no

9 **What do I know now?**
1. ______
2. ______
3. ______
4. ______
5. ______

10 **SHARE**
1. ______
2. ______
3. ______

1 Fill in the topic.

2 Jot down what you already know about the topic.

3 Using what you already know (step 2), form a question about the topic. Are you unsure about one of the items you listed? Do you want to know more? Do you want to know what, how, or why? Write down your question.

4 Guess an answer to your question. In the next few steps, you will be exploring the reasonableness of your answer. Write down your guess.

5 Circle the number in the box that matches how sure you are of your answer in step 4. This is your chance to rate your confidence in what you've done so far and, later, to see how your level of sureness affects your thinking.

6 How can you find out more about your topic? You might want to read a book, ask an expert, or do an experiment. Write down ways you can find out more.

7 Make a plan to explore your answer. Use the resources you listed in step 6. Then, carry out your plan.

8 Now that you've explored, go back to your answer in step 4. Would you answer differently?

9 Considering what you learned in your exploration, answer your question again, adding new things you've learned. You may completely change your answer.

10 It's important to be able to talk about thinking. Choose three people to tell about how you arrived at your response in every step. For example, don't just read what you wrote down in step 2. Try to share how you thought of those things.

Problem Solving

Experimenting

Suppose it is your job to pick out new uniforms for your school band. Other students have expressed their concern about how hot uniforms are during football games and parades. You suspect that some colors of uniforms will be warmer than others, so you decide to design an experiment to determine which color absorbs the least amount of heat.

First, obtain fabric samples of each color. Then, cut the samples to the same size and fold them into pockets. Place a thermometer inside each pocket and place them in the sun. After recording the beginning temperature of each sample, and the temperature at the end of ten minutes, analyze your data.

Solve the Problem

1. What is your hypothesis in this experiment?
2. How did you test your hypothesis?

Think Critically:

Suppose the results of your experiment are as follows. Was your hypothesis supported? Explain.

Color	Beginning Temperature	Temperature After 10 Minutes
Red	24°C	26°C
Black	24°C	28°C
Blue	24°C	27°C
White	24°C	25°C
Green	24°C	27°C

Using Scientific Methods

If you want to stop soil from washing away on a slope in your yard, what can you do? Scientists use a series of planned steps, called a **scientific method,** to solve problems. A basic scientific method is listed in **Table 1-1.** It's important to note that scientists don't always follow these exact steps or do the steps in this order. However, most scientists follow some type of step-by-step method. How can a scientific method be used to solve the slope problem?

Steps in a Scientific Method

You've already done the first step by identifying the problem. The next step is to gather facts that you think may have a bearing on the problem. Then make a hypothesis about how you might stop the soil from washing away. A **hypothesis** is a prediction about a problem that can be tested. You might hypothesize that a wall built at the base of the slope will act as a barrier to the soil. Now you're ready to test your hypothesis.

Table 1-1

A Scientific Method
1. Determine the problem.
2. Make a hypothesis.
3. Test your hypothesis.
4. Analyze the results.
5. Draw conclusions.

You might build a wall at the base of the slope and use a hose to wet the soil. If the soil washes away, your hypothesis was not supported and therefore needs to be revised. You could conclude that building a wall like you did does not work to prevent the soil from washing away. You might make a new hypothesis and start the problem-solving process over again.

Variables and Control

An experiment should test only one variable at a time. A **variable** is a changeable factor in an experiment. Building the wall was the variable you tested. Other variables you might decide to test one at a time are planting different types of vegetation on the slope, planting more vegetation on the slope, as in **Figure 1-9,** replacing the soil with another type, or changing the angle of the slope.

A **control** is a standard for comparison. The slope before the wall was built was the control in your experiment. By comparing the control with the other trials, you could tell whether the wall stopped the problem.

CONNECT TO

LIFE SCIENCE

You have learned that oceanographers use sound to study things they cannot see. Biologists use ultrasound, sound above the range that humans can hear, to study cells and tissues they can't see. Suppose you are a biologist testing the effects of ultrasound on cells. ***Identify*** the control in this experiment.

Figure 1-9 Planting trees and other plants on a slope is one way to reduce erosion.

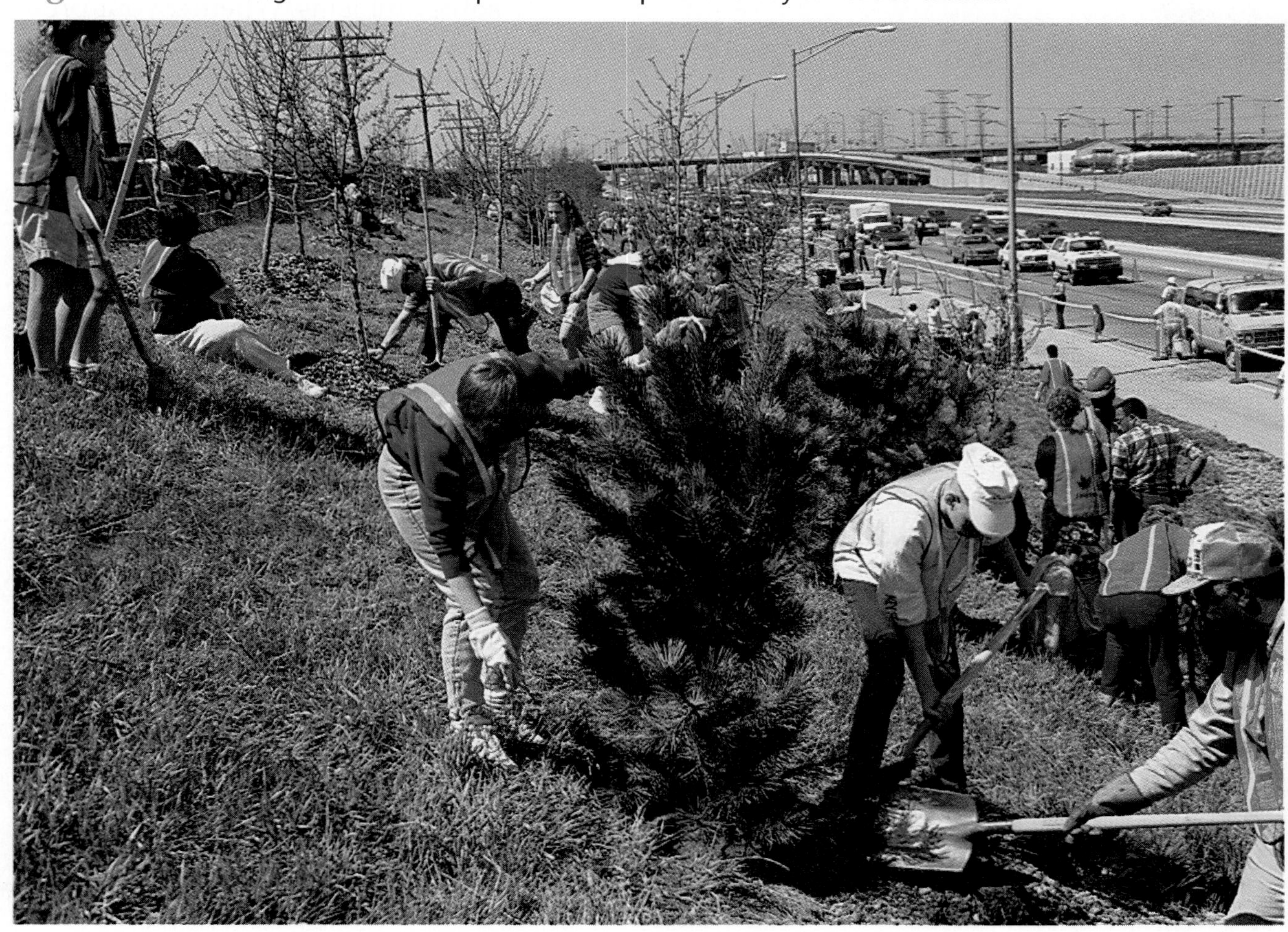

Figure 1-10

Hypotheses are tested many times before they are accepted as theories.

Theories and Laws

Scientists are constantly testing hypotheses. When new data gathered over a long period of time support a hypothesis, scientists become convinced that the hypothesis is correct. They use such hypotheses to form theories. An explanation backed by results obtained from repeated tests or experiments is a **theory.**

A scientific **law** is a rule of nature that describes the behavior of something in nature. Generally, laws predict or describe what will happen in a given situation but don't explain why. An example of a law is Newton's first law of motion. It states that an object continues in motion, or at rest, until it's acted upon by an outside force. You can use scientific methods to help you solve problems. Whether your problem is deciding what route you should follow home or wondering why a rock looks the way it does, a step-by-step method will lead you toward a reasonable solution.

Section Wrap-up

Review

1. What are some strategies you can use to solve problems?
2. What are the steps in a scientific method?
3. **Think Critically:** Imagine that you feel the school building shake, and you think an earthquake was the cause. Is that a hypothesis or a theory?

Skill Builder

Comparing and Contrasting

Compare and contrast a scientific variable with a control. If you need help, refer to Comparing and Contrasting in the **Skill Handbook.**

Using Computers

Word Processing Use your word processing skills to design an experiment to test how changing the type of soil on a slope affects soil loss. Be sure to include your hypothesis and explain how you would test it.

Measurement and Safety

1•4

Science Words

International System of Units (SI)
mass
weight
gravity

Objectives

- List the SI units for the following measurements: length, mass, weight, area, volume, density, and temperature.
- Differentiate between the terms *mass* and *weight* and the terms *area* and *volume.*
- State three lab safety rules.

Measurement

How could you measure the size of the floor of a cave without a ruler or measuring tape? You might count your steps across the cave floor. You could then say that the cave is 25 steps by 30 steps. But this step measurement wouldn't mean the same thing to your friends because their steps would be different from yours. Because of this problem, there are standard units used for measurement.

Today, the measuring system used by most people around the world is the **International System of Units (SI).** SI is a modern version of the metric system. SI is based on a decimal system that uses the number 10 as the base unit.

Length

The standard unit in SI for length is the meter. It's about the length of a guitar. A decimeter is one-tenth of a meter. A centimeter is one one-hundredth of a meter. And a millimeter is one one-thousandth of a meter. A common unit for longer distances is the kilometer. A kilometer is 1000 times greater than a meter. Refer to Appendixes A and B for further explanation of SI and English/SI conversions.

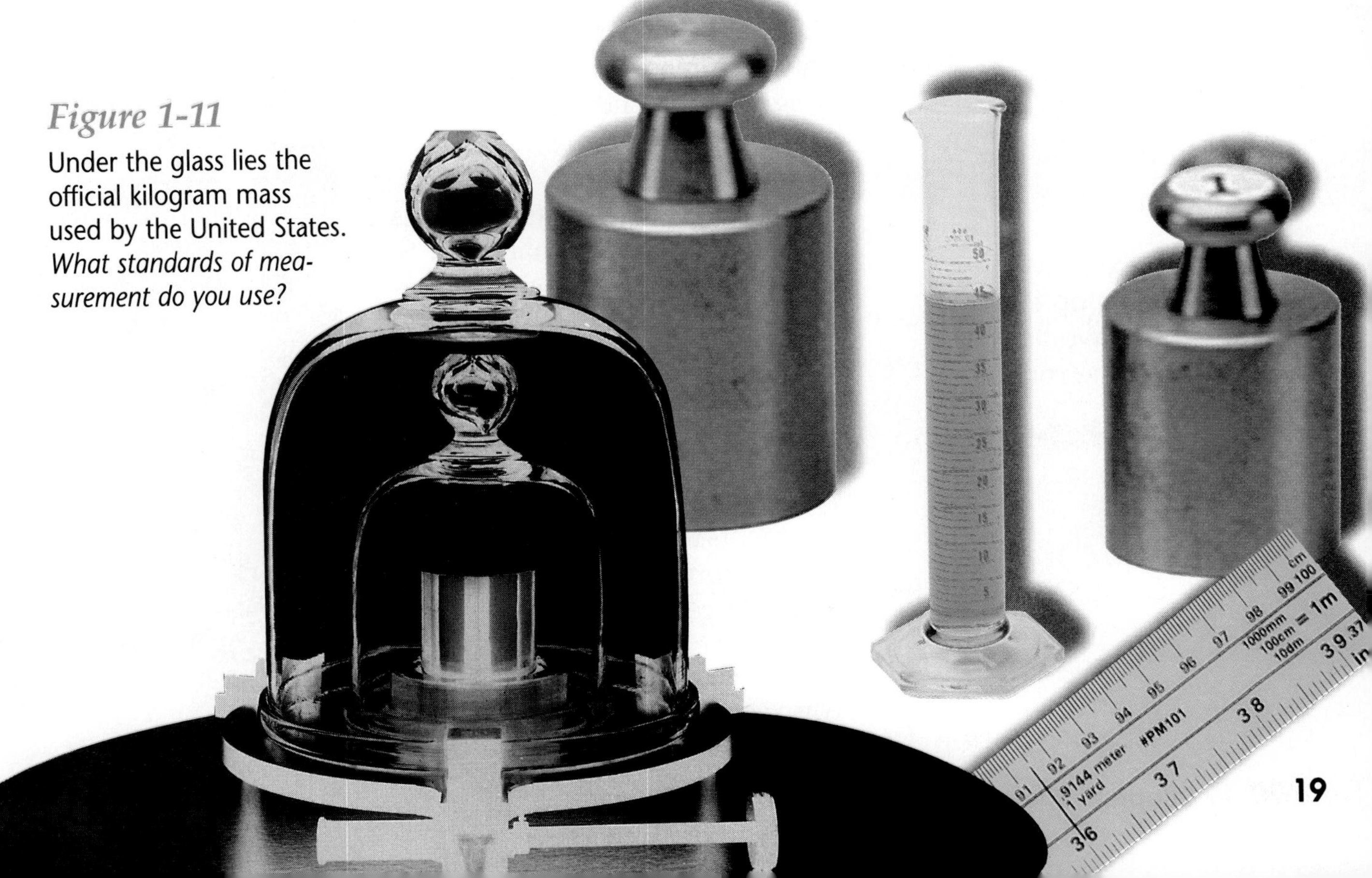

Figure 1-11
Under the glass lies the official kilogram mass used by the United States. *What standards of measurement do you use?*

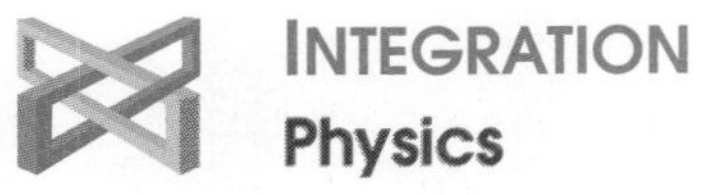

Mass

Mass is a measure of the amount of matter in an object. Mass depends on the number and kinds of atoms that make up an object. Mass does not change with gravitational force. The mass of an object is determined by comparing it with known masses using a balance.

The standard unit of measure for mass is a kilogram. The mass of one bagel is about 57 grams. One gram equals 1000 milligrams, so what would be the mass of one bagel in milligrams? It would be 57 000 milligrams.

Weight

Weight is a measure of gravitational force on a mass. **Gravity** is an attractive force that exists between all objects. If you could weigh yourself on the moon, you would weigh one-sixth the amount you weigh on Earth. This is because the moon's gravitational force is one-sixth that of Earth's.

The standard unit for weight is a newton, named after Sir Isaac Newton, who was the first person to describe gravity. A medium-sized apple weighs about 1 newton.

If an object weighs 30 newtons on Earth, how much would it weigh on the moon?

Area

Some measurements, such as area, require a combination of SI units. Area is the amount of surface included within a set of boundaries. Let's say you want to know the area of a field. First, you'd measure its length and width with a meterstick, and then you'd multiply these two measurements to find the area. In SI, area is expressed in units such as square centimeters (cm^2).

Figure 1-12

When you weigh yourself, you are measuring the force of gravity.

Volume

Figure 1-13

The volume of this aquarium can be found by multiplying its length, width, and height.

Volume is a measure of how much space an object occupies, so if you wanted to know the volume of an aquarium, such as the one in **Figure 1-13,** you'd need to know its length, width, and depth. Then you'd multiply these three measurements to find the volume. The cubic meter (m^3) is the basic unit of volume in SI for solids, but liquid volumes are often measured in liters (L) and milliliters (mL). Liquid volume measurements are made using graduated cylinders and beakers. Because one milliliter of a liquid will just fill a container with a 1 cm^3 volume, milliliters can be expressed as cubic centimeters. For example, an aquarium full of water is 11 356 mL or 11 356 cm^3.

USING MATH

A wood block has dimensions of 2 cm, 1.5 cm, and 3 cm. What is its volume in cubic centimeters?

Density

Density is a measure of the amount of matter that occupies a particular space. It's determined by dividing the mass of an object by its volume.

$$\textbf{density} = \frac{\textbf{mass}}{\textbf{volume}} \qquad D = \frac{m}{v}$$

An SI unit that is often used to express density is grams per cubic centimeter (g/cm^3). How might you express the density of a liquid? We often use grams per milliliter (g/mL).

Air 0.001 g/cm^3
Wood 0.710 g/cm^3
Corn oil 0.925 g/cm^3
Water 1.00 g/cm^3
Plastic 1.17 g/cm^3
Glycerol 1.26 g/cm^3
Rubber 1.34 g/cm^3
Corn syrup 1.38 g/cm^3
Steel 7.81 g/cm^3
Mercury 13.6 g/cm^3

Figure 1-14

Materials of lower density float on other materials of greater density. This woman can easily float in the more dense Dead Sea.

Activity 1-1

Design Your Own Experiment

Rock Density

You've learned that density is the relationship of the mass of an object to its volume. It's pretty easy to find the density of liquids and regularly shaped solids, but how could you find the density of a rock? It's a solid, but it doesn't have a regular geometric shape. How could you find its volume? One way is to measure the displacement of water. When an object is placed in water, it pushes away the water that used to occupy that space. Using this method, the volume of a solid object can be determined.

PREPARATION

Problem

What are two methods you could use to determine the density of a rock?

Form a Hypothesis

Based on what you know about the formula for density, write a hypothesis that explains two methods you could use with the materials listed below to find the density of a rock.

Objectives

- Design an experiment to use two different methods to find the density of a rock.
- Analyze the data that you collect.

Possible Materials 

- graduated cylinder or beaker
- water
- rock
- graph paper (mm or cm grid)
- balance
- goggles
- metric ruler

PLAN THE EXPERIMENT

1. As a group, agree upon and write out your hypothesis statement.
2. As a group, list the steps you need to take to test your hypothesis. Be specific, describing exactly what you will do at each step.
3. Make a list of the materials that you will need to complete each portion of your experiment.
4. Design a data table in your Science Journal so that it will be ready to use as your group collects data.

Check the Plan

1. Read over your entire experiment to make sure that all steps are in logical order.
2. Do you have to run any tests more than one time?
3. *Make sure your teacher approves your plan before you begin.*

DO THE EXPERIMENT

1. Carry out the experiment as planned.
2. While the experiment is going on, write down any observations that you make and complete the data table in your Science Journal.

Analyze and Apply

1. How did you **measure** the volume of the rock in each of your two methods?
2. How did you **measure** the mass of the rock?
3. **Compare** the density figures you got using your two different methods.
4. Which method do you think was the more accurate of the methods you tried? Why?
5. **Compare** your methods with those used by other groups.
6. Of all the methods that people in your class tried, which do you think is the best method to use to **measure** the density of irregularly shaped solids? Why?

Go Further

Suppose you want to determine the density of a very large rock, one that is too big to fit into a beaker. What would you do?

MiniLAB

How do you convert Fahrenheit temperatures to Celsius?

Temperatures are often reported for public weather information in the non-SI units of the Fahrenheit scale, rather than the Celsius scale.

Procedure

1. To convert from °F to °C, subtract 32 from the Fahrenheit temperature and divide by 1.8.
2. To convert from °C to °F, multiply the Celsius temperature by 1.8 and add 32.

Analysis

1. Convert 85°F to °C.
2. Convert -40°C to °F.

Temperature

Temperature is a measure of how hot or cold something is. As you probably know, temperature is measured with a thermometer. What you probably didn't know is that the SI unit for temperature is a kelvin. On the Kelvin scale, absolute zero is 0, the coldest temperature.

Figure 1-15

Temperature affects our daily activities.

The symbol for kelvin is K. Instead of using kelvin thermometers, many scientists use Celsius thermometers. The symbol for a Celsius degree is °C. The Celsius temperature scale is based on the freezing and boiling points of water. The freezing point of pure water is 0°C, and the boiling point is 100°C. A comfortable room temperature is 21°C, and the average human body temperature is about 37°C. You can use these temperatures as reference points when you measure other Celsius temperatures. Now, suppose you wanted to change Celsius temperatures to SI. You'd simply add 273.16 to the degrees Celsius to find the number of kelvins.

degrees Celsius + 273.16 = kelvin

Making accurate measurements in SI is an important part of any experiment. If you don't make accurate measurements, your results and conclusions are invalid.

Figure 1-16

Thermal gloves, goggles, an apron, and safe technique help to protect this student while she heats a solution in a flask.

Safety

Some of the laboratory activities in this book that you'll complete will require you to handle potentially hazardous materials. When performing these activities, safe practices and methods must be used. Scientific equipment and chemicals need to be handled safely and properly, as **Figure 1-16** shows. The safety rules that follow will help you protect yourself and others from injury and will make you aware of possible hazards.

Safety Rules

1. Before beginning any lab, understand the safety symbols shown in Appendix D.
2. Wear goggles and a safety apron whenever an investigation involves heating, pouring, or using chemicals.
3. Always slant test tubes away from yourself and others when heating them. Keep all materials away from open flames. Tie back long hair and loose clothing.
4. Never eat or drink in the lab, and never use laboratory glassware as food or drink containers. Never inhale chemicals, and don't taste any substance or draw any material into a tube with your mouth.
5. Know what to do in case of fire. Also, know the location and proper use of the fire extinguisher, safety shower, fire blanket, first aid kit, and fire alarm.
6. Report any accident or injury to your teacher.
7. When cleaning up, dispose of chemicals and other materials as directed by your teacher, and always wash your hands thoroughly after working in the lab.

Figure 1-17
Laboratories should have proper safety equipment available.

Section Wrap-up

Review

1. Explain the differences between mass and weight and between area and volume.
2. When should you use safety goggles? Refer to Appendix D.
3. **Think Critically:** Which SI units would you use to measure the amount of salt water in an aquarium, the mass of a rock, and the density of a rock?

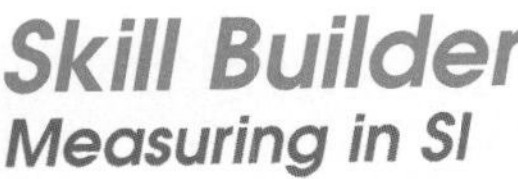

Skill Builder
Measuring in SI

Use your knowledge of SI units to answer the following questions. If you need help, refer to Measuring in SI in the **Skill Handbook.**

1. How many milligrams are in one gram?
2. How many meters are in a kilometer?
3. How many centimeters are in a meter?

USING MATH

If 30 milliliters of fresh water have a mass of 30 grams, what is the density of water? If 100 milliliters of seawater have a mass of 125 grams, what is the density of the seawater?

Science Journal

Locate a road cut or some other rock outcrop or interesting geological feature. Observe the feature, then use what you know about rocks and Earth processes to describe and illustrate the history of the feature.

The Present Is the Key to the Past

Standing on the rim of the Grand Canyon is an incredible experience. There aren't many places where you can look down into a 1.6 km-deep gouge in Earth's crust. The Grand Canyon isn't just a tourist attraction, though. It's also a unique opportunity to observe a large chunk of Earth's geological history. The thick, colorful layers of rocks that can be observed along the canyon walls hold clues to the processes that formed and changed them.

Once Upon a Time, Over a Billion Years Ago . . .

At the very bottom of this awesome gorge lie the oldest rocks, which formed over 1.7 billion years ago. Some of these rocks formed when ancient volcanoes in northern Arizona erupted, while others formed when magma cooled deep within Earth.

Later, quiet seas covered much of the canyon area. When the seas retreated, the minerals and other bits of debris in the water formed layers of sandstones, limestones, and shales. Forces within Earth tilted and folded these layers. Then, over hundreds of millions of years, more layers of shales, limestones, and sandstones formed over this area. The youngest rocks in the canyon are about 225 million years old.

Geologists have proposed many ideas as to how the Grand Canyon formed. Some thought a single, catastrophic event formed the gorge, while others hypothesized that it was a remnant of a once-molten planet. The most accepted hypothesis, however, is that the Colorado River slowly carved the canyon during the past 15 to 20 million years.

Chapter 1 Review

Summary

1-1: What Is Earth Science?

1. Geology is the study of Earth and the processes that form and change it. Meteorology is the study of weather and the forces that cause it. Astronomy is the study of objects in space. Oceanography is the study of Earth's oceans.

1-2: Science and Society: Applying Science

1. Technology has made possible various things you use every day, such as calculators and computers.
2. Technology not only contributes to but also helps solve problems such as pollution and erosion.

1-3: Solving Problems

1. Problem-solving strategies include identifying the problem; collecting data about the problem; eliminating possibilities; using trial and error; solving a simpler, related problem; and making a model or drawing.
2. Commonly used steps in scientific methods include determining the problem, making a hypothesis, testing, analyzing results, and drawing conclusions.
3. A hypothesis is a prediction about a problem that can be tested. Hypotheses may be used to form theories, which are explanations backed by results obtained from repeated tests or experiments. A law is a rule of nature.

1-4: Measurement and Safety

1. In SI, the unit for length is the meter; mass, the kilogram; weight, the newton; area, square centimeters; volume, cubic meters; and density, grams per cubic centimeter.
2. Mass is the amount of matter in an object. Weight is the measure of the force of gravity on an object. Area is the amount of surface in a set of boundaries. Volume is a measure of how much space an object occupies.
3. Lab safety includes understanding how to do the activity, using caution while working with flames, never eating or drinking in the lab, using care with all substances, and reporting any accidents to your teacher.

Key Science Words

a. astronomy
b. control
c. geology
d. gravity
e. Earth science
f. hypothesis
g. International System of Units (SI)
h. law
i. mass
j. meteorology
k. oceanography
l. science
m. scientific method
n. technology
o. theory
p. variable
q. weight

Reviewing Vocabulary

Match each phrase with the correct term from the list of Key Science Words.

1. having knowledge by observing and studying things around you
2. the study of objects in space
3. the use of scientific discoveries
4. a prediction that can be tested
5. a factor in an experiment that changes
6. a scientific rule of nature

Chapter 1 Review

7. a modern version of the metric system
8. the amount of matter in an object
9. a measure of gravitational force
10. the study of weather conditions

Checking Concepts

Choose the word or phrase that completes the sentence or answers the question.

1. The word *science* means to _______.
   a. have knowledge   c. study
   b. observe   d. solve
2. _______ is the study of organisms.
   a. Chemistry   c. Geology
   b. Life Science   d. Physics
3. Oceanographers study Earth's _______.
   a. place in space   c. weather
   b. oceans   d. glaciers
4. ______ involves the study of stars.
   a. Chemistry   c. Astronomy
   b. Physics   d. Geology
5. The volume of a box is best measured in _______.
   a. cubic centimeters   c. newtons
   b. grams   d. centimeters
6. A _______ is a standard used for comparison in an experiment.
   a. variable   c. control
   b. theory   d. law
7. The volume of a beaker of water is best measured in _______.
   a. meters   c. milliliters
   b. centimeters   d. degrees Celsius
8. A balance is used to measure _______.
   a. mass   c. volume
   b. weight   d. density
9. ______ is measured with a thermometer.
   a. Length   c. Temperature
   b. Area   d. Volume
10. Which of these is *not* a lab safety rule?
    a. Slant test tubes away from people when heating them.
    b. Know where the fire extinguisher is.
    c. Wash your hands after working in the lab.
    d. Taste substances to identify them.

Understanding Concepts

Answer the following questions in your Science Journal using complete sentences.

11. Why would a scientist studying a volcano also need some knowledge of physics and chemistry?
12. How have advances in technology affected our planet?
13. How does a theory differ from a hypothesis?
14. Compare and contrast mass and weight.
15. How could you determine the volume of a cardboard box given that its area is 100 m^2?

Thinking Critically

16. Susan has decided to become an oceanographer. In addition to taking life science courses, she must take geology, physics, and chemistry courses. Explain why.
17. Describe how you use technology to do your homework.
18. Suppose you had two plants—a cactus and a palm. You planted them both in potting soil and watered them daily. After a week, the cactus was dead. What problem-solving strategies could you use to find out why the cactus died?
19. Are the steps of a scientific method always followed in the order given on page 16? Explain.
20. The moon's gravitational force is one-sixth that of our planet. How would your mass differ on the moon?

Developing Skills

If you need help, refer to the **Skill Handbook.**

21. Concept Mapping: Make a concept map using the following terms and phrases: *newton, cubic meter, length, mass, volume, kilogram.*

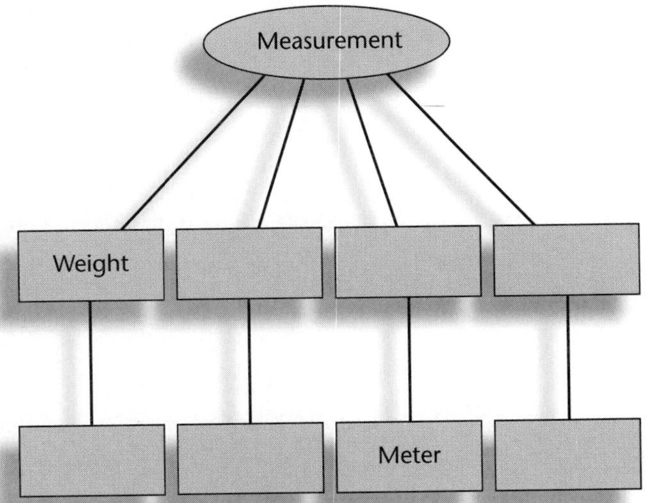

22. Using Variables, Constants, and Controls: Suppose you wanted to know whether crystals grow better in cold or warm water. How would you set up a simple experiment to test this?

23. Comparing and Contrasting: Contrast science and technology.

24. Sequencing: Ann wants to see how much rain falls in her city during March. Sequence the following steps in the most logical order that she needs to follow to solve her problem.

- **a.** Collect rain.
- **b.** Make a rain gauge.
- **c.** Measure the amount of rainfall each day.
- **d.** Research how much rain usually falls during March in her city.
- **e.** Hypothesize how much rain will fall.
- **f.** Graph her results.

25. Measuring in SI: Describe how you would calculate the volume of air inside a balloon like the one in the photograph.

Performance Assessment

1. Carrying Out a Strategy and Collecting Data: Perform an activity to study what happens to air as it is heated. Partially fill a balloon with air. Then tape the balloon to a ring stand with a heating lamp above it. Record the temperature near the balloon, the size of the balloon, and the time at set intervals. Use SI units. Record your data in a table.

2. Designing an Experiment: Design an experiment to determine the effects of water on small plants. Use three identical plants. Record the amount of water used and plant growth using SI units.

3. Forming a Hypothesis: Find a rock that is about the size of your fist, and determine its density. Hypothesize whether the size of a rock affects its density. Test your hypothesis by using a hammer to split the rock into pieces. Determine the density of one of the pieces. Was your hypothesis supported? Explain.

Previewing the Chapter

Chapter 2

Matter and Its Changes

You've probably walked by various bodies of water from time to time. You may have seen mist rising from a lake on a cool autumn morning or noticed ice forming on the lake's surface in the winter. If so, you have observed water as a solid, a liquid, and a gas.

EXPLORE ACTIVITY

Change water from a liquid to a gaseous state.  

1. Pour one cup of water into a glass container.
2. Mark the level of water in the container with a permanent marker.
3. Empty the container into a teapot.
4. With the help of an adult, heat the water in the teapot until it boils.
5. Carefully pour the water in the teapot into the measuring container.

Observe: What did you see coming out of the teapot? What do you notice about the amount of water that remains? In your Science Journal, explain what you think has happened to the water.

Previewing Science Skills

- In the **Skill Builders,** you will **outline, map concepts,** and **make and use tables.**
- In the **Activities,** you will **observe** and **collect and organize data.**
- In the **MiniLABs,** you will **classify, observe,** and **infer.**

2•1 Atoms

Science Words

matter
atom
element
proton
neutron
electron
mass number
atomic number
isotope

Objectives

- Identify matter as anything that has mass and takes up space.
- Describe the internal structure of an atom.
- Explain why isotopes of the same element have the same atomic number but different mass numbers.

The Building Blocks of Matter

What do this book, the air you breathe, and the food you eat all have in common? The book, the air, and the food are all matter. **Matter** is anything that takes up space and has mass.

Look at **Figure 2-1.** You can't always see matter as clearly as you see the sun. For example, you can't see air. But matter, in its various forms, is all around you. Air is a colorless gas, water is a transparent liquid, and rocks are colorful solids. Why do the characteristics of one form of matter differ from the characteristics of another? The answer is because matter is made up of particles called **atoms.** The structure of those atoms determines the characteristics of the matter you observe.

Atoms

All matter is composed of "building blocks." The structure of these building blocks determines the structure of the matter you observe. Think about when you were younger and

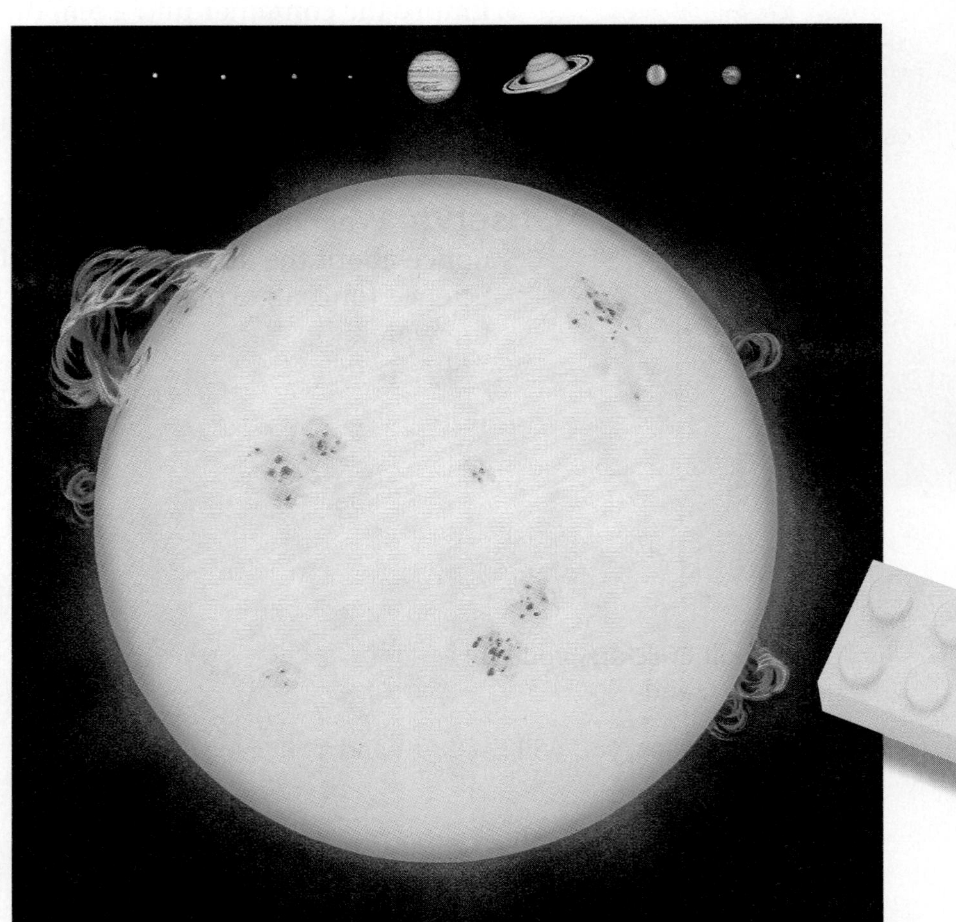

Figure 2-1

More than 99 percent of the matter in our solar system is contained in the sun. *How do we define matter?*

played with snap-together blocks like those in **Figure 2-2.** You could snap the blocks together in many ways to build cars, ships, or buildings. Matter is put together in a similar way. The building blocks of matter are atoms. The arrangement and types of atoms give matter its properties.

Elements

Atoms combine, like the blocks snapping together, to form many different types of matter. Your body has only a few different types of these atoms in it, but they have combined in many different ways to form the matter that composes your body. Other forms of matter contain only one type of atom. Such substances are **elements.** Let's take a look at the structure of an element.

Suppose you have a copper wire. What kind of atoms are in the wire? Because copper is an element, it's made up of only copper atoms. Look at **Table 2-1.** It shows some minerals containing elements and their uses. Appendix F of your book is a table of the known elements called the periodic table.

Figure 2-2

Like atoms, the same few blocks can combine in many different ways. *How could this model help explain the structure of atoms?*

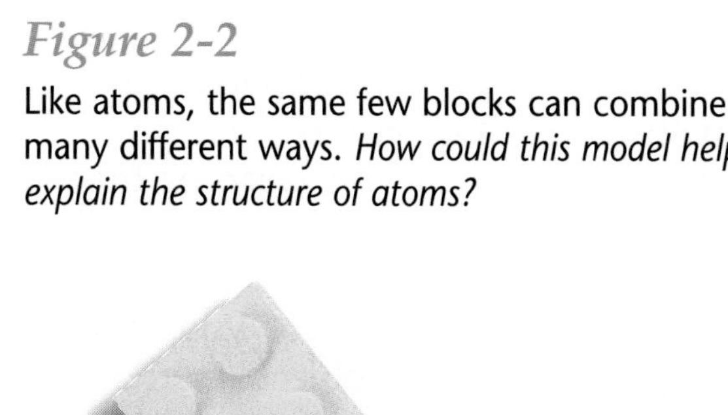

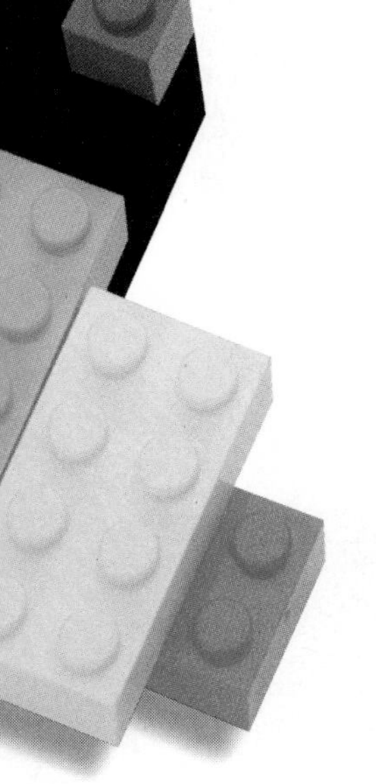

Table 2-1

Some Common Uses of Elements			

Silver	**Bauxite ore** (Contains the element aluminum.)	**Iron ore** (Contains the element iron.)	**Quartz** (Contains the elements silicon and oxygen.)

Jewelry	Soft drink cans	Vehicle	Computer chips

Models of the Structure of Atoms

You already know that atoms are very small. Atoms are far too small to be seen, even with a microscope. How can you study something this small? How can you determine the internal structure of an atom?

When substances are too large or too small to handle or directly observe, models are often used to take their place. Have you ever worked with a model airplane like the one shown in **Figure 2-3A?** If so, your model was a small version of a large object. In the case of atoms, the opposite is true. A large model is made of a very small object. We construct drawings, sculptures, and mental pictures of the internal structure of atoms. These models are based on information we've gathered by observing the ways atoms react when in contact with other atoms or with light.

Protons and Neutrons

Let's construct a mental model of the internal structure of an atom. Three basic particles make up an atom—protons, neutrons, and electrons. Protons and neutrons are located in the center of an atom and make up its nucleus. **Protons** are particles that have a positive electrical charge. **Neutrons** are particles that have no electrical charge. The nucleus, therefore, has a positive charge because there is no negative charge to counter the positive charge from the protons. This positive electrical charge of the nucleus is balanced by the electrons that surround the nucleus.

Electrons

Electrons are negatively charged particles that move around the nucleus. There is one electron for each proton. The hive in **Figure 2-3B** represents the nucleus of the atom. The bees flying around the hive in all directions are like the electrons circling the nucleus. You can't determine exactly where each bee is, but each is usually close to the hive. **Figure 2-3C** shows electrons existing as a negatively charged electron cloud. This cloud completely surrounds the nucleus of the atom. Electrons can be anywhere within the cloud, but evidence suggests that they are located near the nucleus most of the time. They are very much like a swarm of bees flying around its hive.

Figure 2-3

Models help us learn about materials that are too large to handle or too small to directly observe.

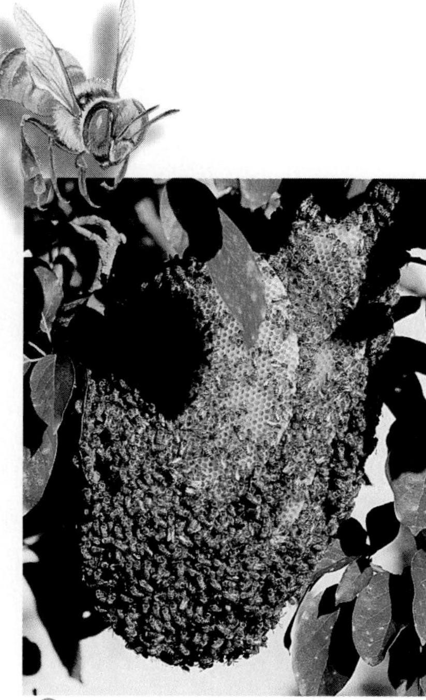

B Bees swarm around their hive but seldom stray far from home. *How is a beehive like a nucleus?*

A This model airplane is a small version of a large object. With atoms, the opposite is true. Scientists construct a large model of a very small atom.

C This model of a helium atom shows two protons and two neutrons in the nucleus, and two electrons in the electron cloud.

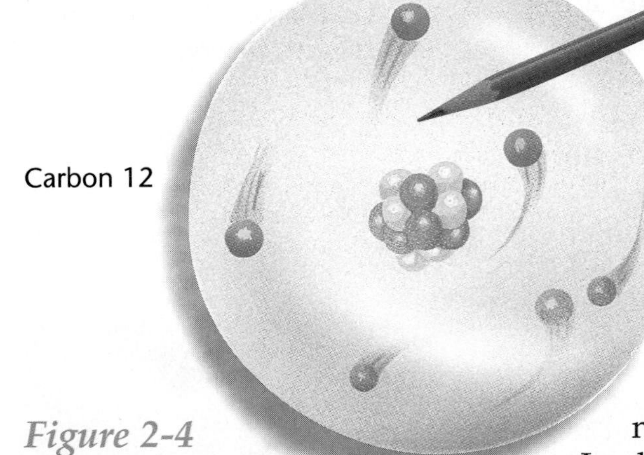

Figure 2-4

The carbon in your pencil lead is mostly carbon-12. Its mass number is 12 and its atomic number is 6. *What is the mass number and atomic number of carbon-14?*

Mass Numbers and Atomic Numbers

Just as an atom has a characteristic number of protons, neutrons, and electrons, it also has a characteristic mass. The **mass number** of an atom is equal to the number of protons and neutrons making up its nucleus. Just as you have more mass when you are carrying many books, an atom has more mass if it contains more particles. Thus, oxygen, which contains eight protons and eight neutrons, has a mass number of 16. Carbon has a mass number of 12. Look at **Figure 2-4.** If carbon has six protons, how many neutrons does it have?

Electrons aren't counted when we compute an atom's mass number. This is because electrons aren't massive enough to be included in the calculations.

Atomic Number

Another property of an atom that's convenient to know is its atomic number. An atom's **atomic number** equals the number of protons in its nucleus. This number also equals the number of electrons contained in its electron cloud. All atoms of a specific element have the same atomic number. For example, all atoms of iron have an atomic number of 26. How many protons does an atom of iron contain? Whether it's in the metal of a car fender or in the nails in a bookcase, an iron atom has 26 protons.

Isotopes

If the number of protons in an atom is changed, a new element is formed. The number of neutrons can be changed, however, without changing the element. All that happens is that the atom's mass changes. Atoms of the same element that have different numbers of neutrons in their nuclei are called **isotopes. Table 2-2** lists isotopes of some common elements. Note that the number of protons remains the same for each element, but the number of neutrons changes.

Some isotopes are used for medical purposes. Others provide us with a way to determine the age of ancient objects. Geologists use these isotopes to date fossils and layers of rock. Archaeologists use them to determine the age of artifacts, such as the mummified body shown in **Figure 2-5.** You will be learning more about isotopes and some of their uses in Chapter 12.

CONNECT TO

LIFE SCIENCE

Some isotopes of elements are radioactive. Physicians can introduce these isotopes into a patient's circulatory system. The low-level radiation they emit allows them to be tracked as they move throughout the patient's body. ***Explain*** how this would be helpful in diagnosing a disease.

Table 2-2

Isotopes			
Isotope	**Number of Protons**	**Number of Neutrons**	**Number of Electrons**
Hydrogen-1	1	0	1
Hydrogen-2	1	1	1
Hydrogen-3	1	2	1
Carbon-12	6	6	6
Carbon-14	6	8	6
Uranium-234	92	142	92
Uranium-235	92	143	92
Uranium-238	92	146	92

Figure 2-5

Some isotopes help archaeologists determine the age of clothing, wood, bones, and other structures. *How does this help us understand ancient civilizations?*

Our model of an atom allows us to predict how a particular element will react when it's in contact with another element. As you continue to investigate matter in this chapter, you will explore how atoms of different elements combine to form the matter around you.

Section Wrap-up

Review

1. How does the air you breathe fit the definition of matter?
2. Explain why the overall electrical charge of an atom is neutral.
3. **Think Critically:** Oxygen-16 and oxygen-17 are two different isotopes of oxygen. The numbers 16 and 17 represent their mass numbers. How many protons and neutrons are in each isotope?

Skill Builder
Comparing and Contrasting

Review the material in Section 2-1. How do atoms and elements differ? How are they the same? If you need help, refer to Comparing and Contrasting in the **Skill Handbook**.

USING MATH

The mass number of nitrogen is 14. Find its atomic number in the table shown in Appendix F. Then determine the number of neutrons in its nucleus.

2•2 Combinations of Atoms

Science Words

molecule
compound
chemical property
ion
mixture
solution

Objectives

- Describe several ways atoms combine to form compounds.
- Compare and contrast compounds and mixtures.

How Atoms Combine

Suppose you've just eaten a snack such as an apple or an orange. It's unlikely that you were concerned about the elements in the snack. Yet anything you eat and anything you touch has some combination of the same few elements in it. The sugar in the apple or orange contains the elements carbon, oxygen, and hydrogen. Some of the air around you contains carbon dioxide, a combination of carbon and oxygen.

In **Figure 2-6,** both the oranges in the tree and the carbon dioxide in the air contain oxygen. Yet in one case oxygen is in a colorless, gaseous form. In the other case it's part of a structure that's hard and colorful. They are different because the atoms of elements combine chemically to form new substances called compounds. When they do, the properties of the individual elements change. Let's see how this happens.

Molecules and Compounds

One way that atoms combine is by sharing the electrons in the outermost portion of their electron clouds. The combined atoms form a **molecule.** For example, two atoms of hydrogen can share electrons with one atom of oxygen to form one molecule of water. Water is a compound composed of molecules. A **compound** is a type of matter that has properties different from the properties of each of the elements in it.

The properties of hydrogen and oxygen are changed when they combine to form water. Under normal conditions on Earth, you will find the elements oxygen and hydrogen only as gases.

Figure 2-6

The air surrounding this tree contains some of the same elements as the fruit hanging from its branches. *How can the same elements, made from the same type of atoms, help make two materials that are so different?*

Figure 2-7

A molecule of water has chemical properties that are different from those of hydrogen and oxygen atoms.

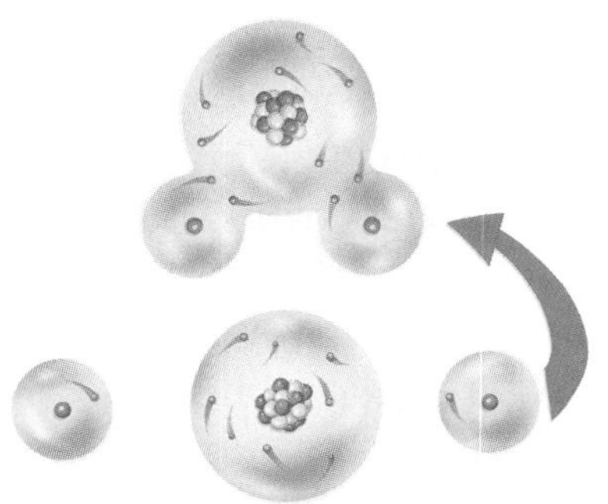

When hydrogen and oxygen combine to form a compound, the new substance has chemical properties that are different from those of the elements in it, as shown in **Figure 2-7.**

Chemical Properties

Chemical properties describe how one substance changes when it reacts with other substances. For example, the chemical properties of iron cause it to change to rust when it reacts with water and oxygen.

USING TECHNOLOGY

Small, but Mighty

Atoms are extremely small. Many are hundreds of billionths of centimeters in diameter. Yet scientists have discovered how to move individual atoms in much the same way you may have used a magnet to move small game pieces around a plastic-covered board. Using a scanning tunneling electron microscope (STEM), physicists drag a fine-tipped needle over the material's surface. When the desired position is reached, the needle is raised and the atom "drops" into this new position.

One at a Time

This discovery may lead to the building of molecules one atom at a time. New drugs that would cure or eliminate certain illnesses could be constructed using this method. Even smaller electrical circuits could be made for everything from watches to computers. The illustration shows xenon atoms arranged to spell out "IBM."

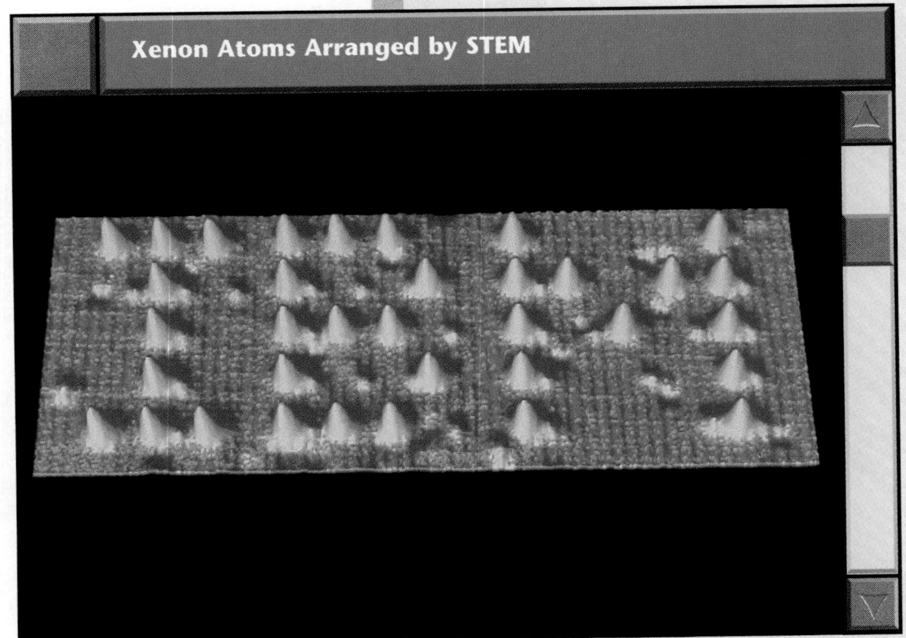

Xenon Atoms Arranged by STEM

Think Critically: How could STEMs be used to permanently store information and what would be the potential impact of this use?

INTEGRATION
Chemistry

MiniLAB

What are some different forms of matter?

Determine the differences between mixtures, compounds, and elements.

Procedure

1. Make a chart with columns titled *Mixtures*, *Compounds*, and *Elements*.
2. Classify each of the items listed below into the proper column on your chart.
3. Experiment to make a solution using two or more of the items listed: air, sand, hydrogen, muddy water, sugar, ice, salt water, water, salt, oxygen, copper.

Analysis

1. How does a solution differ from other mixtures?
2. How does an element differ from a compound?

Ions

You know that atoms combine by sharing electrons from the outer portion of their electron cloud. But atoms also combine because they've become positively or negatively charged.

As you discovered earlier, atoms are usually neutral—they have no overall electrical charge. Under certain conditions, however, atoms can lose or gain electrons. When an atom loses electrons, it has more protons than electrons so the atom is positively charged. When an atom gains electrons, it has more electrons than protons so the atom is negatively charged. Electrically charged atoms are called **ions.**

Ions are attracted to each other when they have opposite charges. Oppositely charged ions join to form electrically neutral compounds. Table salt forms in this way. A sodium (Na) atom loses an electron and becomes a positively charged ion. Then it comes close to a negatively charged chlorine (Cl) ion. They are attracted to each other and form the compound $NaCl$. This is the compound that you use on your french fries or popcorn.

Mixtures

Your backpack probably looks similar to the one shown in **Figure 2-8.** If you look inside, you will see an example of a **mixture.** Many different objects are mixed together, but each retains its own properties. Your math book isn't any different whether it's beside your ruler or beside your history book.

Figure 2-8
Your backpack contains a variety of objects that together form a mixture.

Solutions

Salt water is an example of a kind of mixture called a solution. When one substance of a mixture is dissolved in another substance, a **solution** is formed. In the case of salt water, the properties of the salt molecules aren't changed just because they're mixed in with the water. The salt molecules are separated from each other by other molecules within the water. Therefore, the salt has dissolved in the water. Another property of a solution is that it is the same throughout. One part of a solution is the same as all other parts. To make the salt water a solution, we stir it. Stirring the solution spreads the salt molecules evenly throughout.

Separating Components of Mixtures and Compounds

The components of a mixture can be separated by physical means. You can sit at your desk and pick out the separate items in your backpack. You can let the water evaporate and the salt will remain. But is it possible to separate the components of a compound in a similar way?

Suppose you take sugar and try to separate its carbon atoms from its hydrogen and oxygen atoms. How can you do it? It's much more difficult than separating the components of a mixture. The only way is to separate the carbon, hydrogen, and oxygen atoms of each sugar molecule. This is an example of a chemical change. A chemical change converts one substance into one or more new substances with different chemical properties.

A Full Cup of Tea

You come home from school on a cold, rainy day. You normally drink milk after school, but, given the weather, today you decide to have a cup of hot tea. You brew the tea and pour yourself a generous cup. "Too generous," you note with a frown, for you have filled the cup to its rim. "How will I be able to add sugar to my tea?" Very carefully, you place first one and then two teaspoons of sugar in the cup. You expect it to overflow, but it doesn't.

Solve the Problem:

1. **Why didn't the tea overflow?**
2. **What happened to the two teaspoons of sugar as you gently stirred the cup of tea?**

Think Critically:

Try the same thing, but this time allow the cup of tea to cool down and then try to place two teaspoons of sugar in the totally full cup.

1. Did you observe the same results?
2. Was it as easy to dissolve the sugar into the cold tea?
3. How can you explain any differences you might have observed?

Figure 2-9

The ocean is a mixture of many different forms of matter. Ocean water is a solution.

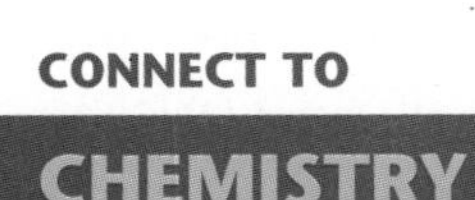

CONNECT TO

CHEMISTRY

When one substance dissolves in another, some of the properties of the dissolving substance change. When salt dissolves in water, ***decide*** if its chemical or physical properties change.

Exploring Matter

Sweetened tea, air, salt water, and the contents of your backpack are all examples of mixtures. The plants and coral shown in **Figure 2-9** are also a mixture. In each case, the materials within the mixture retain their individual properties. The materials themselves are made of compounds. The atoms of these compounds lost their individual chemical properties when they combined. As you continue to explore matter, use the mental models you've developed for atoms, elements, molecules, compounds, and mixtures.

Section Wrap-up

Science Journal

Write a paragraph in your Science Journal explaining why the water molecule shown in **Figure 2-7** on page 39 tends to be positively charged on one side and negatively charged on the other. In other words, explain why it is referred to as a polar molecule.

Review

1. How do atoms or ions combine to form molecules?
2. Why is sweetened tea considered a mixture rather than a compound?
3. **Think Critically:** How can you determine if salt water is a solution or a compound?

Skill Builder

Concept Mapping

Make an events chain map using the terms *mixtures, atoms, molecules, compounds, electrons, protons,* and *neutrons.* If you need help, refer to Concept Mapping in the **Skill Handbook**.

Activity 2-1

Scales of Measurement

How would you describe some of the objects in your classroom? Perhaps the bulletin board is the size of the top half of a door. This description is not very useful. Wouldn't it be better to measure the actual size of the bulletin board? Measuring physical properties in a laboratory experiment will help you make better observations.

Problem

How are physical properties of objects measured in SI?

Materials 

- balance (beam)
- graduated cylinder (100 mL or larger)
- metersticks (2)
- thermometers (3)
- stick or dowel
- rock sample
- string
- globe
- water

Procedure

1. Begin at any station and determine the measurement requested. Record your observations in a data table and list sources of error.
   a. Use a balance to determine the mass, to the nearest 0.1 g, of the rock sample.
   b. Use a graduated cylinder to determine the volume, to the nearest 0.5 mL, of the water.
   c. Use 3 thermometers to measure the average temperature, to the nearest 0.5°C, at a certain location in the room.
   d. Use a meterstick to measure the length, to the nearest 0.1 cm, of the stick or dowel.
   e. Use a meterstick and string to measure the circumference of the globe. Be accurate to the nearest 0.1 cm.
2. Proceed to the other four stations as directed by your teacher.

Analyze

1. **Compare** your measurements with those who used the same objects. Review the values provided by your teacher. How do the values you obtained compare with those provided by your teacher and those of other students?
2. **Calculate** your percentage of error in each case. Use this formula.

$$\frac{\text{your value} - \text{teacher's value}}{\text{teacher's value}} \times 100 = \text{\% of error}$$

Conclude and Apply

3. **Decide** what percentage of error will be acceptable. Generally, being within 5% to 7% of the correct value is considered good. If your values exceed 10% error, what could you do to improve your results and reduce error? What was the most common source of error?

Data and Observations

Station	Sample	Value of Measurement	Causes of Error
a	____	Mass = ____ g	
b	____	Volume = ____ mL	
c	____ (location)	Average temp. = ____ °C	
d	____	Length = ____ cm	
e	____	Circumference = ____ cm	

2•3 Matter

Science Words

physical property
density

Objectives

- Distinguish between chemical and physical properties.
- Contrast the four states of matter.

Physical Properties of Matter

So far in this chapter, you've been investigating chemical properties of matter—the properties that describe how one substance changes into another substance. But you can observe other properties of matter. The properties that you can observe without changing a substance into a new substance are **physical properties.**

What are some physical properties of your clothing? If you say your jeans are blue, soft, and about 80 cm long like the ones in **Figure 2-10,** you've described some of their physical properties. You can observe these without changing the material in your jeans into a new substance.

Density

One physical property that you will use to describe matter is density. **Density** is a measure of the mass of an object divided by its volume. Generally, this measurement is given

Figure 2-10

Blue, soft, room for two legs—these are the physical properties of a pair of blue jeans. *In your Science Journal, list the physical properties of five items in the classroom.*

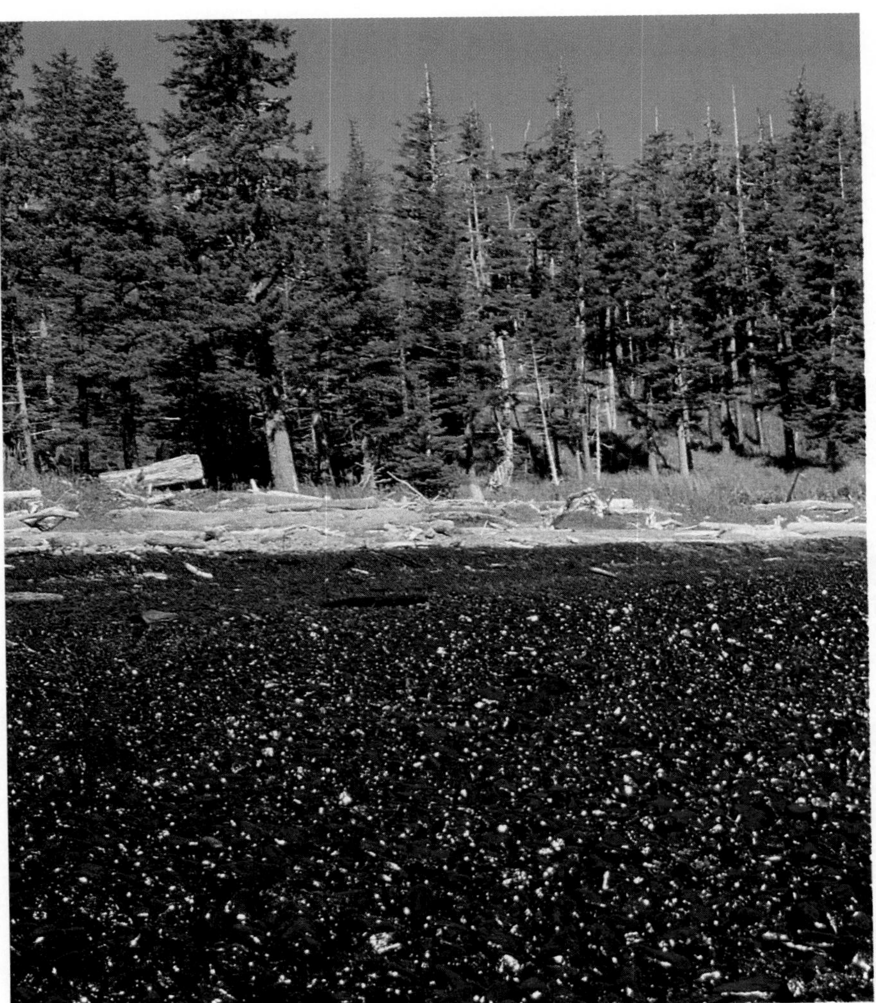

Figure 2-11

One physical property of oil is its density. *How does the density of oil compare with the density of water?*

in grams per cubic centimeter (g/cm^3). For example, the average density of liquid water is 1 g/cm^3. So 1 cm^3 of water has a mass of 1 g.

Suppose you have a small cube and need to find its density. First, measure its mass and volume. Suppose its volume is 2 cm^3 and its mass is 8 g. Use the formula $D = \frac{M}{V}$.

$$D = \frac{M}{V}$$

$$D = \frac{8\ g}{2\ cm^3}$$

$$D = \frac{4\ g}{cm^3} \quad \text{or} \quad 4\ g/cm^3$$

An object that's more dense than water will sink, whereas one that's less dense will float. You've heard about the oil spills off the coast of the United States. Why does this oil, shown in **Figure 2-11,** float on the surface of the water and wash up on the beaches?

You can find the volume of a rectangular solid by multiplying its length, width, and height. A pine board has dimensions of 4.05 cm, 8.85 cm, and 164 cm. Determine the volume and density of the board if its mass is 2580 g.

States of Matter

Think back to breakfast this morning. You may have had solid toast, liquid milk or juice, and of course you breathed air, which is a gas. If you have seen a lightning bolt during a summer storm, you have seen matter in its plasma state. On Earth matter occurs in four physical states. These four states are solid, liquid, gas, and plasma. What causes the differences among these four states of matter?

Figure 2-12

The attraction between atoms or molecules and their rate of movement determines whether matter occurs in a solid, liquid, gas, or plasma state.

Solids

The reason some matter is solid is that its atoms or molecules are in a fixed position relative to each other. The individual atoms or molecules of the turquoise shown in **Figure 2-12A** may vibrate, but they don't switch positions with each other. You can make a mental model of this.

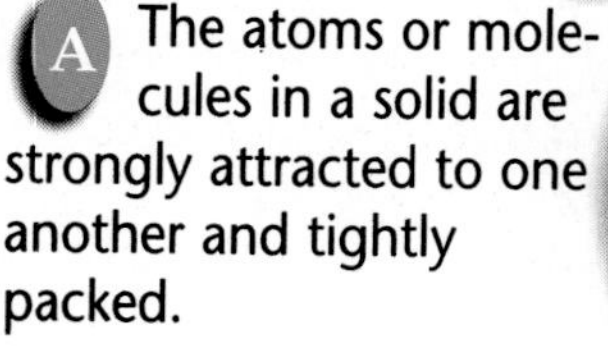

A The atoms or molecules in a solid are strongly attracted to one another and tightly packed.

Suppose you have a puzzle with its many pieces in place. The pieces are packed so tightly that no one piece can switch positions with another piece. But the pieces can move a little. For example, you can twist the whole puzzle a few millimeters without breaking it apart. If it's on a table, you can shake the table and the puzzle's individual pieces will vibrate. But the pieces of the puzzle are held together even though they move.

The puzzle pieces in our model represent atoms or molecules of a substance in a solid state. Such atoms or molecules are strongly attracted to each other and resist being separated.

Liquids

Look at Figure **2-12B.** Atoms or molecules in a liquid are also strongly attracted to each other, but they aren't as strongly attracted as they are in a solid. Atoms or molecules remain close to one another in a liquid but are free to change positions with each other. This allows liquids to flow.

When you sit down to breakfast, you may have several liquids at the table. You may have milk in a glass and syrup on your pancakes. Both are

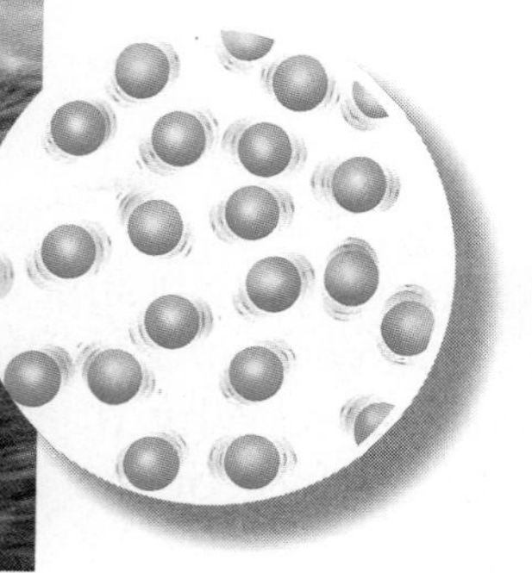

B The atoms or molecules in a liquid are attracted to each other, but not as strongly as those of a solid. They are free to move over and around each other.

substances in the liquid state, even though one flows more freely than the other.

A liquid flows as it takes the shape of the container it's placed in, but it resists changes in volume. You can pour orange juice into a short, wide glass and it will match the shape of the glass. You can then pour the same juice into a tall, skinny glass and it will flow until it matches the shape of its new container. It does so because its molecules move over and around each other.

Gases

Gases behave the way they do because their atoms or molecules have very little attractive force on each other. This causes them to move freely and independently. Air fresheners work because of this property. If an air freshener is placed in a corner, it isn't long before molecules from the air freshener have spread throughout the room. Gases fill the entire container they are placed in. Note the hot air balloon shown in **Figure 2-12C**. The atoms or molecules move apart until they're evenly spaced throughout the balloon.

C The atoms or molecules of a gas are not strongly attracted to each other. *What happened to the gas in this hot air balloon?*

Plasma

What's the most common state of matter? So far we've investigated matter in the solid, liquid, and gaseous states. But most of the matter in the universe is in the plasma state. Matter in this state is composed of ions and electrons. Many of the electrons normally in the electron cloud have escaped and are outside of the ion's electron cloud.

Stars are composed of matter in the plasma state. Plasma also exists in the magnetic field near Jupiter. On Earth, plasma is found in lightning bolts, as shown in **Figure 2-12D.**

D Plasma consists of ions and electrons moving freely.

People and Science

Dr. Samuel B. Mukasa, Isotope Geochemist

On the Job

Q Dr. Mukasa, please explain how you use isotopes in your work.

A Isotopes can be used for tracing the sources of magma (molten rock beneath Earth's surface) and mineral deposits from very deep in Earth. Dating rocks also helps in putting Earth's history together—figuring out when the continents broke up and when mountains formed. Isotopes can be used to date when rocks actually formed.

Q Do you ever work with scientists in other fields?

A Yes, I do. To give you an example, in studying the breakup of the western edge of the old supercontinent Gondwanaland into Antarctica and New Zealand, there are several things you can look for in the rocks. Rocks involved in the breakup zone get very deformed—almost like a car wreck in which two cars slam into each other. The mangled up rocks have patterns of structures in them that are wonderful telltale signs about exactly what the process of breakup was. I work with structural geologists who can interpret these patterns.

Personal Insights

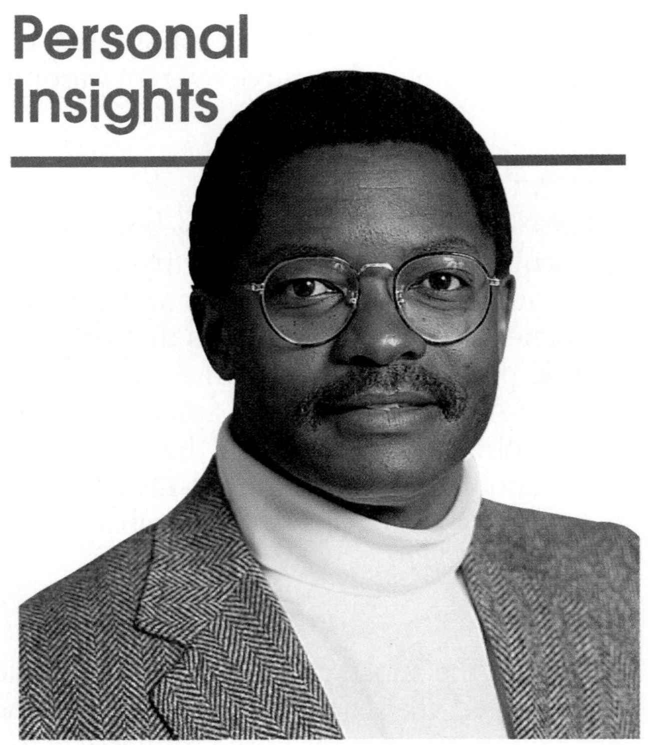

Q Have you had an interest in geology since you were young?

A I grew up in East Africa, and for weekend activities, we used to climb volcanoes. Realizing that Earth was alive and that volcanoes were erupting, I began to wonder, "Why is this happening here and not over there?"

Career Connection

Find the address for the Geological Society of America (ask for help from your school librarian or at the reference desk of your local library). Write for the brochure, "Future Employment Opportunities in the Geological Sciences." When you receive the brochure, go through it, and make a list of promising careers.

- **Oceanographer**
- **Geologist**

Changing the State of Matter

Matter is changed from a liquid to a solid at its freezing point and from a liquid to a gas at its boiling point. You're probably familiar with the freezing and boiling points of water. Water changes from a liquid to a solid at its freezing point of 0°C. It boils at 100°C. Water is the only substance that occurs naturally on Earth as a solid, liquid, and gas.

Other substances don't naturally exist in these three states on Earth. Their boiling and freezing points are above or below the temperatures we experience. Temperatures and conditions needed for matter to exist as plasma are even less common on Earth.

The attraction between atoms or molecules and their rate of movement are two factors that determine the state of matter. When you melt ice, you increase the rate of movement of its molecules. They are then able to move apart. Adding thermal energy to the ice causes this change. As **Figure 2-13** shows, solid metal can be converted into liquid when thermal energy is added.

Changes in state can also occur because of increases or decreases in pressure. You can demonstrate this by applying pressure to an ice cube. It will change to liquid water even though no thermal energy was added.

MiniLAB

What happens when water freezes?

Procedure 

1. Pour water into a graduated cylinder or small beaker so that it is about half-filled.
2. Mark the level of the water with a small piece of masking tape.
3. Place the uncovered container of water in the freezer until all of the water turns to ice.

Analysis

1. How does the level of the ice in the container compare with the original level of the water?
2. What happened to the physical properties of the water when it changed to ice?

Figure 2-13

A solid metal can be converted to liquid by adding thermal energy to its molecules. *What is happening to the molecules during this conversion?*

Science Journal

In your Science Journal, list and classify common solids, liquids, and gases that you use every day.

Changes in Physical Properties

Chemical properties of matter don't change when the matter changes state. But some of its physical properties change. For example, the density of water changes as it changes state. In which state is water the most dense? You may be tempted to say "when it's a solid," but think about this. Although most materials are more dense in their solid state than in their liquid state, ice will float in liquid water; therefore, ice is less dense than liquid water. This is because water molecules move farther apart and the water expands as it freezes.

Some physical properties of substances don't change when they change state. For example, water is colorless and transparent in each of its states.

USING MATH

Twenty cm^3 of water have a mass of 20 g. Its density can be represented by the fraction 20/20. Suppose you freeze the water. Is the density of the ice represented by a proper fraction or an improper fraction? Explain your reasoning.

Classifying Matter

Chemical and physical properties allow us to identify and classify matter. One way to classify matter is by its state. Matter in one state can often be changed to another state by adding or removing thermal energy. Look at **Figure 2-15.** Changes in thermal energy may explain why there is no liquid water on the planet Mars, yet photographs of its surface show what appear to be ancient water-carved valleys. Scientists

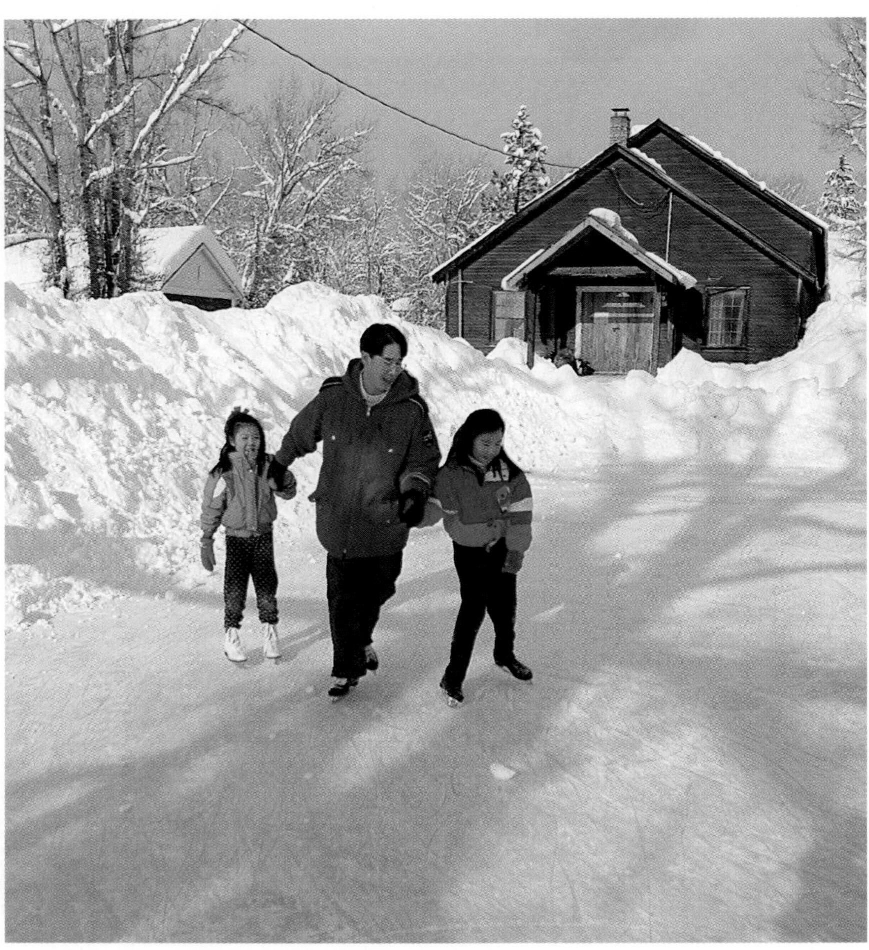

Figure 2-14

If ice were more dense than water, lakes would freeze solid from the bottom up. *What effect would this have on the fish?*

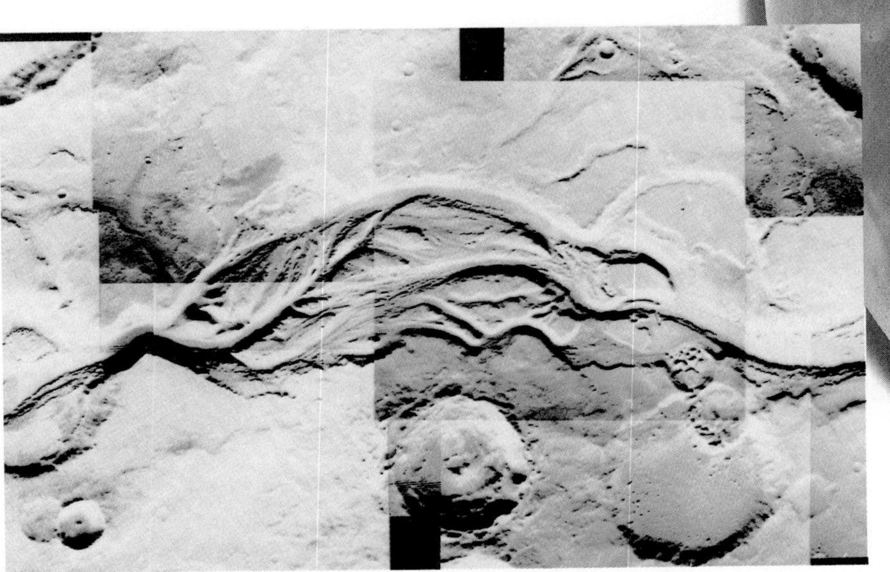

believe that much of the liquid water soaked into the ground and froze, forming permafrost. Some of the water may have frozen to form the polar ice caps. When matter changes state in this way, it retains its chemical properties while some of its physical properties change.

Figure 2-15
This valley on the planet Mars was possibly carved by water.

Section Wrap-up

Review

1. Compare and contrast the movement of water molecules when water is in a solid, liquid, and gaseous state.
2. The planet Mars, as shown in **Figure 2-15,** has what appear to be ancient water-carved channels, yet there's no liquid water on Mars. Explain what may have happened to Mars' liquid water.
3. **Think Critically:** Suppose you blow up a balloon and then place it in a freezer. Later, you find that the balloon has shrunk and has liquid in it. Explain what has happened.

Skill Builder

Making and Using Tables

Jonathon had a glass of milk, an orange, celery sticks, and a peanut butter and jelly sandwich for lunch. Make a table listing the state of matter of each item of his lunch. Also list at least one physical property of each item. If you need help, refer to Making and Using Tables in the **Skill Handbook.**

Using Computers

Graphing Research the melting and boiling points in degrees Celsius of several compounds, including water. Use your computer to make a line graph showing the temperatures at which these several compounds change state from solid to liquid to gas.

Activity 2-2

Design Your Own Experiment

Determining Density

What has a greater density—a rock or a sponge? Does cork have a greater density than clay? There are two factors that contribute to the density of an object. Think back to the equation you learned earlier in this chapter.

PREPARATION

Problem

What processes can be used to determine the densities of several objects?

Form a Hypothesis

State a hypothesis about what process you can use to measure and compare the densities of several materials.

Objectives

- List some ways that the density of an object can be measured.
- Design your own experiment that compares the densities of several materials.

Possible Materials

- pan balance
- beaker (100 mL)
- graduated cylinder (250 mL)
- water
- sponge
- piece of quartz
- piece of clay
- small wood block
- small metal block
- small cork
- rock
- ruler

Safety Precautions

Be wary of sharp edges on some of the materials and take care not to break the beaker or graduated cylinder.

PLAN THE EXPERIMENT

1. As a group, agree upon and write out the hypothesis statement.
2. As a group, list the steps that you need to take to test your hypothesis. Be specific, describing exactly what you will do at each step. List your materials.
3. Using the equation that density = mass/volume, as a group, devise a method of determining the mass and volume of each material to be tested.
4. Design a data table in your Science Journal so that it is ready to use as your group collects data.

Check the Plan

1. Read over your entire experiment to make sure that all steps are in a logical order.
2. Should you run the process more than once for any of the materials?
3. Identify any constants, variables, and controls of the experiment.
4. *Make sure your teacher approves your plan before you proceed.*

DO THE EXPERIMENT

1. Carry out the experiment as planned.
2. While the experiment is going on, write down any observations that you make and complete the data table in your Science Journal.

Analyze and Apply

1. Do you **observe** anything about the way more-dense objects feel compared with less-dense objects of the same size?
2. What happens when a cork is placed in water?
3. Based on this, would you **hypothesize** that a cork is more dense, the same density, or less dense than water?
4. Without measuring the density of an object that floats, **conclude** how you know that it has a density of less than 1.0 g/cm^3.

Go Further

How would the data you collected about the density of the clay be affected if you were to break the clay into two smaller pieces and repeat the experiment? Try it and see if there is any effect on the data.

SCIENCE & SOCIETY

ISSUE:

2•4 Energy from Atoms

Nuclear Energy

Science Words

nuclear energy
fission
nuclear waste

Objectives

- Describe how nuclear energy is made.
- Explain advantages and disadvantages of the Yucca Mountain nuclear waste storage site.

Most electricity in the United States is generated in power plants that use fossil fuels. However, there are alternate sources of energy. **Nuclear energy** is an alternate energy source produced from atomic reactions. When the nucleus of a heavy element is split, lighter elements are formed and energy is released.

Fission

Fission is the splitting of nuclei of atoms in heavy elements such as uranium. The most commonly used fuel in fission power plants is the uranium-235 isotope. It occurs in ore in some sandstones in the Rocky Mountains. After the ore is mined, the uranium is concentrated and then placed in long metal pipes called fuel rods. The fuel rods sit in a pool of cooling water within a large chamber called a nuclear reactor, shown in **Figure 2-16.** Neutrons are fired into the fuel, which begins a chain reaction that in turn releases heat. This heat is used to boil water to make steam. The steam drives a turbine, which turns a generator, producing electricity.

Figure 2-16
Most of the high-level radioactive waste in the United States comes from spent fuel rods used in nuclear power plants.

2 Points of View

Storing Nuclear Waste in Yucca Mountain

One major problem with nuclear energy is the waste material that is produced. **Nuclear waste** from power plants is highly radioactive. Some of this waste will remain radioactive for more than ten-thousand years. Nuclear waste must be safely stored in order to keep it from entering the environment. Yucca Mountain in Nevada has been chosen as a possible storage site for this nuclear waste. Geologists working for the U.S. Department of Energy have concerns about three major hazards: volcanoes, earthquakes, and groundwater. They believe the Yucca Mountain area, shown in **Figure 2-17,** is a good site because the area is remote, very little rain falls, the water table is far below the proposed storage facility, and the land is already owned by the U.S. government. They also believe that any earthquakes that may occur in the area will affect only surface buildings, not the deeply buried storage area.

Figure 2-17

The U.S. Department of Energy's Office of Civilian Radioactive Waste is currently studying the Yucca Mountain site in order to determine whether the site is safe for storage of nuclear waste. The study will continue through the year 2001 and cost approximately $5 billion.

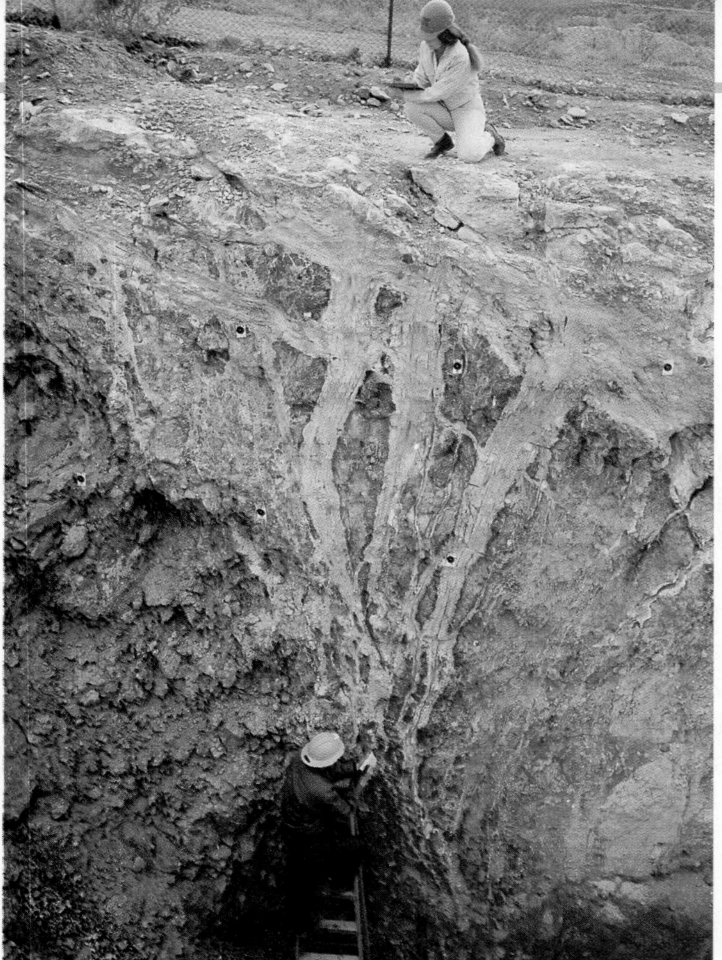

Figure 2-18

Geologists believe that the cream-colored veins of calcite in this deposit were formed either by rainwater percolating down from the surface or by groundwater rising upward from below. *Why is it so important for scientists to determine the cause of this deposit of limestone?*

Search for Other Locations

Not everyone agrees that Yucca Mountain is a safe location for storage of nuclear waste. As shown in **Figure 2-18,** a distinctive deposit of limestone has formed at one location, Trench 14. Geologists for the Department of Energy believe the deposit formed as very small amounts of rainwater percolated downward toward the water table. Geologists working for the state of Nevada believe the Trench 14 limestone was formed by an upwelling of groundwater. This could lead to a release of toxic waste into the environment. Critics of the site are also concerned about the many earthquake faults that exist in the area. They are concerned that an earthquake could cause movement along faults running through the storage facility. This, they believe, could disrupt waste containers and produce cracks through which contaminated groundwater might flow into the environment. Even without an earthquake, critics are concerned about the water. Rainwater could percolate downward or groundwater could rise up. Either way, corrosion of the containers could be accelerated, exposing nuclear waste.

Section Wrap-up

Review

1. What is the source of most of the nuclear waste that needs to be stored in a safe location?
2. List one form of evidence in favor of and one opposed to the use of Yucca Mountain as a nuclear waste storage site.

Explore the Issue

Should the Department of Energy abandon the study of the Yucca Mountain nuclear storage site? If yes, what other arrangements might be made in order to cope with the nuclear waste? If no, how would you alleviate the concerns voiced about the dangers of volcanoes, earthquakes, and groundwater contamination at the Yucca Mountain site?

Chapter 2

Review

Summary

2-1: Atoms

1. Matter is anything that has mass and takes up space. Atoms are the building blocks of matter.
2. Protons and neutrons make up the nucleus of an atom. Protons have a positive charge whereas neutrons have no charge. Electrons surround the nucleus, forming an electron cloud. Electrons are negatively charged.
3. Isotopes are atoms of the same element. Isotopes have the same atomic number, but differ in their number of neutrons.

2-2: Combinations of Atoms

1. A compound is a substance made of two or more elements. The properties of a compound differ from the chemical and physical properties of the elements of which it is composed.
2. Atoms join to form molecules—the building blocks of compounds. A mixture is a substance in which each of the components retains its own properties.

2-3: Matter

1. Physical properties can be observed and measured without causing a chemical change in a substance. Chemical properties can be observed only when one substance reacts with another substance.
2. Atoms or molecules in a solid are in fixed positions relative to one another. In a liquid, the atoms or molecules are close together but are freer to change positions. Atoms or molecules in a gas have very little attractive force on one another. Plasma is composed of ions and electrons.

2-4: Science and Society: Energy from Atoms

1. Nuclear energy is produced from atomic reactions.
2. Use of the Yucca Mountain nuclear waste storage site has both supporters and critics.

Key Science Words

a. atom
b. atomic number
c. chemical property
d. compound
e. density
f. electron
g element
h. fission
i. ion
j. isotope
k. mass number
l. matter
m. mixture
n. molecule
o. neutron
p. nuclear energy
q. nuclear waste
r. physical property
s. proton
t. solution

Reviewing Vocabulary

Match each phrase with the correct term from the list of Key Science Words.

1. building block of matter
2. particle with no electric charge
3. splitting of nuclei of atoms in heavy elements
4. anything that takes up space and has mass
5. a solution is one type
6. moves around the nucleus of an atom
7. composed of only one type of atom
8. mass divided by volume
9. two or more atoms combine to form this building block of compounds
10. atom of the same element but with different number of neutrons

Chapter 2 Review

Checking Concepts

Choose the word or phrase that completes the sentence.

1. _______ contain only one type of atom.
   a. Plasmas c. Elements
   b. Mixtures d. Solids
2. A(n) _______ has a positive charge.
   a. electron c. neutron
   b. proton d. plasma
3. In an atom, the _______ form a cloud around the nucleus.
   a. electrons c. neutrons
   b. protons d. positively charged particles
4. Carbon has a mass number of 12. Thus, it has _______ protons and _______ neutrons.
   a. 6, 6 c. 6, 12
   b. 12, 12 d. 12, 6
5. On Earth, molecular oxygen is usually a _______.
   a. solid c. liquid
   b. gas d. plasma
6. An isotope of carbon is _______.
   a. boron-12 c. carbon-14
   b. nitrogen-12 d. hydrogen-2
7. Electrically charged atoms are _______.
   a. molecules c. isotopes
   b. solutions d. ions
8. The color of your clothes is a(n) _______.
   a. chemical property
   b. physical property
   c. isotope property
   d. molecular property
9. A rock with a volume of 4.0 cm^3 and a density of 3.0 g/cm^3 has a mass of _______.
   a. 0.75 g c. 12.0 g
   b. 3.0 g d. 4.0 g
10. Water changes state at _______.
    a. 0°C and 100°C c. 0°C and 32°C
    b. 32°C and 100°C d. 32°C and 212°C

Understanding Concepts

Answer the following questions in your Science Journal using complete sentences.

11. Compare and contrast protons, electrons, and neutrons.
12. What is the density of 25 mL of cooking oil if its mass is 23 grams?
13. How do compounds and mixtures differ?
14. If an atom has no charge, what is true about its protons and electrons?
15. What process releases energy from an atom? Explain.

Thinking Critically

16. Would isotopes of the same element have the same number of electrons? Explain.
17. Two chlorine ions are both negatively charged. Will they combine to form a molecule? Why or why not?
18. You pour cooking oil into a glass of water. You briefly stir the materials in the glass. Does your glass contain a mixture? Does it contain a solution?
19. When oxygen combines with iron in the presence of water, rust is formed. Is this a chemical or physical property of iron?
20. Why is nuclear waste considered so dangerous?

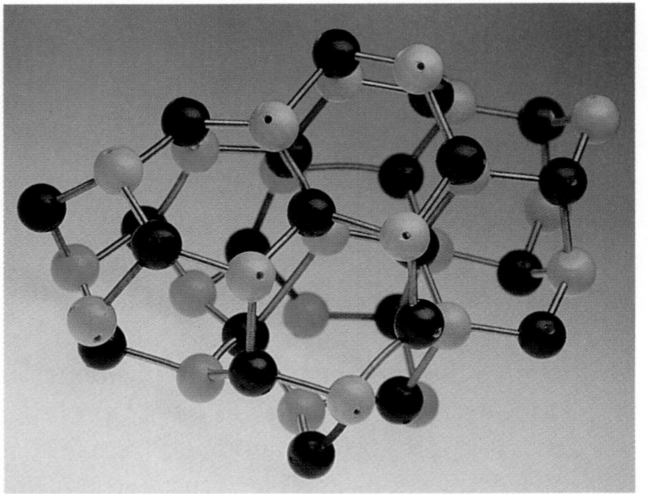

Chapter 2 Review

Developing Skills

If you need help, refer to the **Skill Handbook.**

21. **Classifying:** Use Appendix F to classify the following substances as elements or compounds: iron, aluminum, carbon dioxide, gold, water, and sugar.

Element	Atomic No.	Mass No.
Fluorine	9	19
Lithium	3	7
Carbon	6	12
Nitrogen	7	14
Beryllium	4	9
Boron	5	11
Oxygen	8	16
Neon	10	20

22. **Making and Using Graphs:** Use the data above to make a line graph of increasing mass number and atomic number of each element. What is the relationship between mass number and atomic number?
23. **Hypothesizing:** You put a bottle full of water in the freezer to cool it quickly. You forgot about it, though, and found a broken glass when you went to get it. What hypothesis can you make about water as it changes from a liquid to a solid?
24. **Observing and Inferring:** Your brother is drinking a dark-colored liquid from a clear glass. As you watch from across the room, you think to yourself that the cola is probably refreshing. Is thinking that the liquid is cola an observation or an inference?
25. **Concept Mapping:** Finish the network tree below to illustrate the three main parts of an atom.

Performance Assessment

1. **Model:** Use different-sized and different-colored foam balls and wooden sticks to make models of the atomic structure of the first ten elements in the periodic table.
2. **Making and Using a Classification System:** Classify all the items in your refrigerator according to physical state. Then list at least two chemical and two physical properties of each item. Observe or test the properties you list.
3. **Model:** Research how the model of an atom has evolved with time. Construct models of the early theories of the atom.

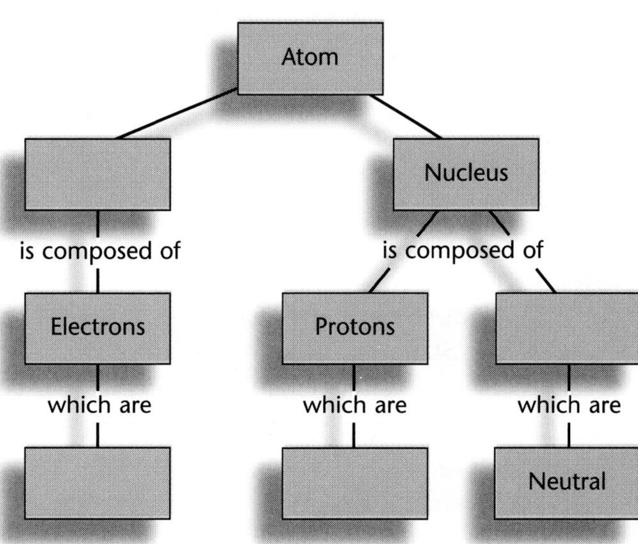

Previewing the Chapter

Minerals

Minerals are used for many things. They are used in jewelry, pencil lead, powders, and in bicycle frames. Table salt, for example, is the mineral halite. The salt flat shown at left is composed of the evaporate minerals halite and gypsum. Evaporate minerals form when water evaporates, leaving salt and other substances behind.

EXPLORE ACTIVITY

Make a saltwater solution.

1. Fill a cup or beaker about half full of warm tap water.
2. Add about a teaspoon of table salt to the water.
3. Stir the water until all the salt has dissolved. Continue to add salt and stir until no more salt will dissolve.
4. Gently pour the water into a pan, and place the pan on a shelf where it will not be disturbed.
5. Check on the pan twice each day for two or three days.

Observe: Keep a log in your Science Journal describing what has happened to the saltwater solution at the end of three days.

Previewing Science Skills

- In the **Skill Builders**, you will classify, graph, and map concepts.
- In the **Activities**, you will observe, collect, and organize data.
- In the **MiniLABs**, you will experiment and compare.

3•1 Minerals

Science Words

mineral
crystal
magma
silicate

Objectives

- List five characteristics all minerals share.
- Give examples of two ways that minerals form.

What is a mineral?

Have you ever used minerals? **Figure 3-1** shows a few of the ways we use minerals every day. A **mineral** is a naturally occurring, inorganic solid with a definite structure and composition. Although 4000 different minerals are found on Earth, they all share five characteristics.

Inorganic and Naturally Occurring Minerals

First, all minerals are formed by natural processes. Rock salt, diamonds, and graphite are minerals because they formed naturally. You'll investigate more about these processes later in this lesson.

Second, minerals are inorganic. They aren't alive, never were alive, and are not made by life processes. Coal is made of carbon from living things. Although geologists do not classify coal as a mineral, some people do. Miners, for example, generally classify anything taken from the ground that has commercial value as a "mineral resource."

Crystalline Solids

The third characteristic that minerals share is that they are all solids. Remember that all solids have a definite volume and shape. A gas such as air and a liquid such as water aren't minerals because neither has definite shape.

Fourth, every mineral is an element or compound with a chemical composition unique to that mineral. For example, rock salt's composition gives it a distinctive flavor.

Finally, the atoms in a mineral are arranged in a pattern that is repeated over and over again. Graphite's arrangement of atoms makes it feel soft and slippery. Opal, on the other hand, is classified as a "mineraloid" because its atoms are not arranged in a definite structure.

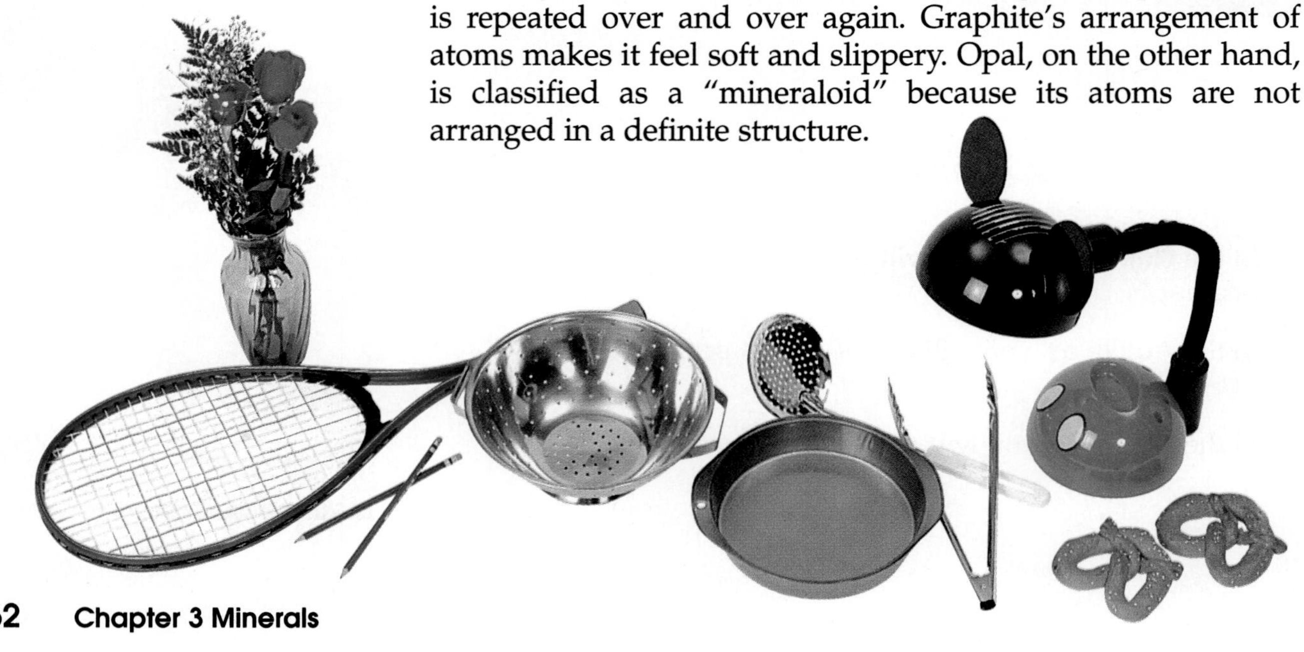

Figure 3-1

Shown here are just a few examples of how we use minerals. *In your Science Journal, list the five characteristics that all minerals share.*

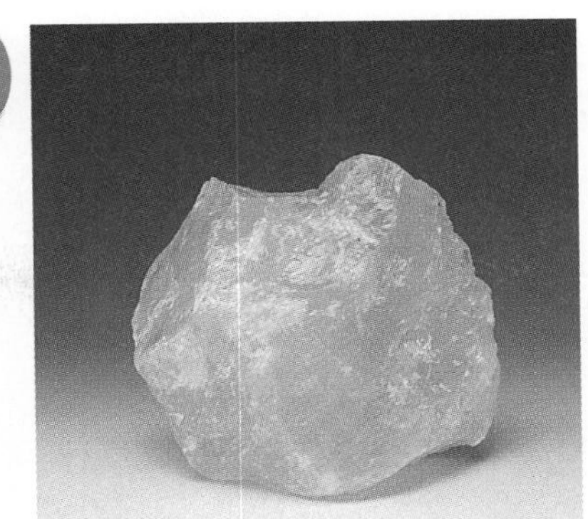

Figure 3-2

Some mineral specimens, such as smoky quartz, have flat surfaces and sharp edges, showing crystal structure on the outside (A). Even if a mineral such as rose quartz doesn't show its crystal structure on the outside, its atoms are still arranged in a crystal structure (B).

The Structure of Minerals

Did you know that each little grain of salt in a salt shaker is a cube? The atoms contained in each grain of salt are grouped in such a way that they form a cube. These cubes are crystals. A **crystal** is a solid in which the atoms are arranged in repeating patterns. **Table 3-1** shows examples of the six major crystal systems.

Crystals

Not all mineral crystals have smooth surfaces and sharp edges like the quartz crystal in **Figure 3-2A.** The quartz in **Figure 3-2B** has atoms arranged in repeating patterns, but you can't see the crystal structure on the outside of the mineral. This is because the quartz in **Figure 3-2B** developed in a tight space, while the quartz in **Figure 3-2A** developed freely in an open space. The hexagonal crystal structure of this quartz is obvious. See **Figure 3-3** for an example of a cubic crystal system.

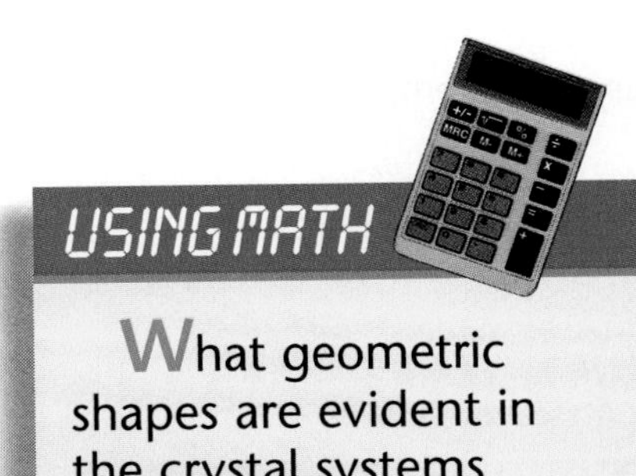

USING MATH

What geometric shapes are evident in the crystal systems shown below?

Table 3-1

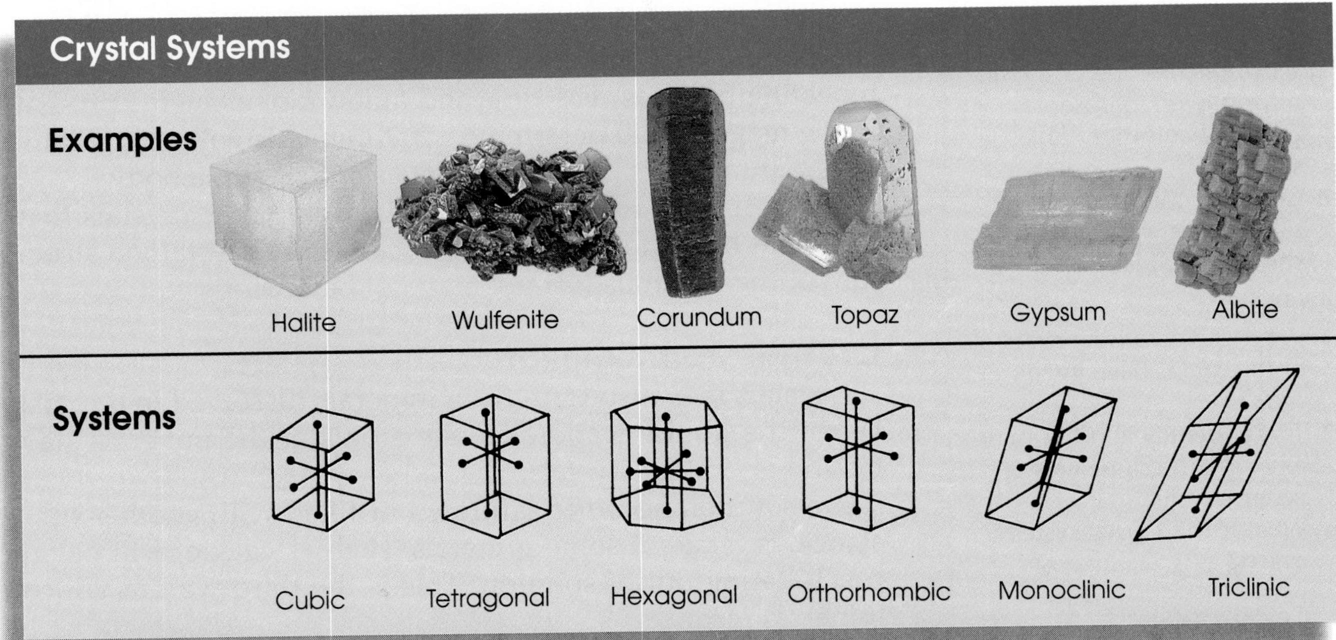

Figure 3-3
Atoms of the mineral galena arrange themselves in a cubic crystal system.

MiniLAB

What do salt crystals look like up close?

Procedure

1. Use a microscope or a magnifying glass to observe grains of common table salt.
2. Compare the shape of the salt crystals with those shown in **Table 3-1.**

Analysis

1. What characteristics do all of the grains have in common?
2. Try to identify another mineral with the same crystal system as salt.
3. What is this crystal system called?

How Minerals Form

There are three ways that minerals form. One way is from the cooling of hot melted rock material, called **magma.** As magma cools, its atoms lose energy, move closer together, and begin to combine into compounds. Molecules of the different compounds that are forming begin to arrange themselves into repeating patterns. The type and amount of elements present in the magma help determine what minerals will form. Many different minerals can form from a single magma body.

When molten rock material cools rapidly, the crystals that form can be quite small. In such cases, the crystalline structure of the minerals formed may not be obvious.

Crystals from Solution

Crystals may also form from minerals dissolved in liquids. When the liquid evaporates, the atoms in the minerals stay behind and form crystals.

If a solution becomes saturated or filled with another substance, crystals of some minerals, such as calcite, will begin precipitating out of solution. This is the third way in which minerals form.

Crystal Formation

So far in this chapter, you've learned about mineral crystals and how they form. You've even observed crystals of table salt form when your saltwater solution evaporated in the Explore Activity. In this activity, you'll have a chance to observe this process further and to learn about another method of mineral formation.

Problem

How do minerals form from solution?

Materials   

- salt solution
- sugar solution
- large test tube
- toothpick
- cotton string
- hand lens
- shallow pans (2)
- test-tube rack
- cardboard
- table salt
- granulated sugar
- hot plate
- thermal mitt

Procedure

1. Gently mix separate solutions of salt in water and sugar in water. Keep stirring the solutions as you add salt or sugar to the water. Stop mixing when no more salt or sugar will dissolve into the solution.
2. Pour the sugar solution into one of the shallow pans. Use the hot plate to gently heat the solution.
3. Place the test tube in the test-tube rack. Using a thermal mitt to protect your hand, pour some of the hot sugar solution into the test tube. **CAUTION:** *The liquid is hot. Do not touch the test tube without protecting your hands.*
4. Tie the thread to one end of the toothpick. Place the toothpick across the opening of the test tube so the thread is in the sugar solution.
5. Cover the test tube with a piece of cardboard and place the rack containing the test tube in a location where it will not be disturbed.
6. Pour a thin layer of the salt solution into the second shallow pan.
7. Place the pan in a location where it will not be disturbed.
8. Leave both the test tube and the shallow pan undisturbed for at least one week.
9. At the end of one week, examine each solution and see whether crystals have formed. Use a hand lens to observe the crystals.

Analyze

1. **Compare** and **contrast** the crystals that formed from the salt and sugar solutions with each other. How do they compare with samples of table salt and sugar?
2. What happened to the salt water in the shallow pan?
3. Did this same process occur in the test tube? **Explain.**

Conclude and Apply

4. What caused the formation of crystals in the test tube? What caused the formation of crystals in the shallow pan?
5. Sugar is harvested from plants. Is sugar a mineral? **Explain.**
6. **Relate** the formation of crystals in this activity to mineral crystal formation.

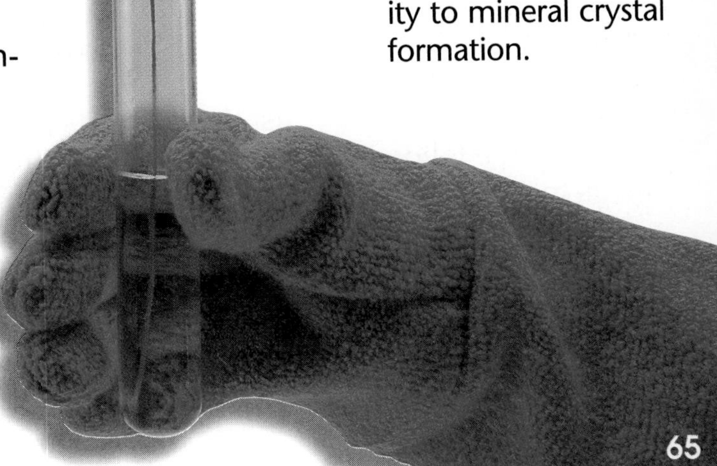

Mineral Compositions and Groups

Some 90 elements occur naturally in Earth's crust. Ninety-eight percent of the crust is made up of only eight of these elements, as shown in **Figure 3-4**. Of the 4000 known minerals, only a few dozen are common, and these are composed of the eight common elements of Earth's crust. Most of these common rock-forming minerals are silicates, but other groups are also included. Feldspar and quartz, both silicates, and calcite, a carbonate, are examples of rock-forming minerals.

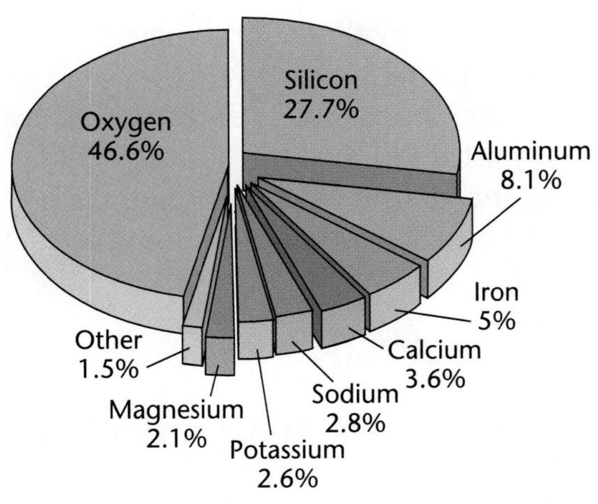

Figure 3-4

Most of Earth's crust is composed of these elements.

Silicates

Silicates are minerals that contain silicon and oxygen and usually one or more other elements. Silicon and oxygen are the two most abundant elements in Earth's crust. Other minerals are classified according to their composition. Major groups of minerals include the carbonates, oxides, sulfides, sulfates, halides, hydroxides, phosphates, and native elements.

You now know a lot more about common rock salt than you did before reading this section. You know that its real name is halite and that it is a mineral. You also know that halite has a cubic crystal shape that is formed when salt water evaporates.

Section Wrap-up

Review

1. What five conditions must a substance meet to be a mineral?
2. Describe two ways that minerals can form.
3. **Think Critically:** What two elements are found within all silicate minerals? In your Science Journal, explain why silicates are so common in Earth's crust.

Skill Builder

Classifying

Using Appendix F, list the elements that make up the following minerals. Organize the elements into metals, nonmetals, and metalloids. If you need help, refer to Classifying in the **Skill Handbook.**

siderite ($FeCO_3$)
talc ($Mg_3Si_4O_{10}(OH)_2$)
silver (Ag)
barite ($BaSO_4$)
enstatite ($MgSiO_3$)
sylvite (KCl)

Using Computers

Spreadsheet Refer to **Figure 3-4** above. Use your computer to make a graph of your own design that shows the relative percentages of the eight most common elements in Earth's crust.

Mineral Identification

3•2

Physical Properties

How can you tell the difference between one of your classmates and another? You can tell the difference between them without even thinking about it because you observe things about them that make them different. The color of a classmate's hair or the shape of his or her face helps you tell him or her from the rest of your class. Hair color and face shape are two properties unique to individuals.

Appearance

Individual minerals also have unique properties. These properties help us tell the difference between minerals. Look at **Figure 3-5.** Color and appearance are just two of the clues that are used to identify minerals.

But these clues alone aren't enough to tell most minerals apart. The minerals pyrite and gold are both gold in color and can appear to be the same. But gold is worth a lot of money, whereas pyrite has little value. You need to look at other properties of minerals to tell them apart.

Science Words

hardness
luster
streak
cleavage
fracture

Objectives

- List the physical properties used to identify minerals.
- Describe how physical properties such as hardness and streak are used to identify minerals.

Figure 3-5

Like people, minerals have many different characteristics.

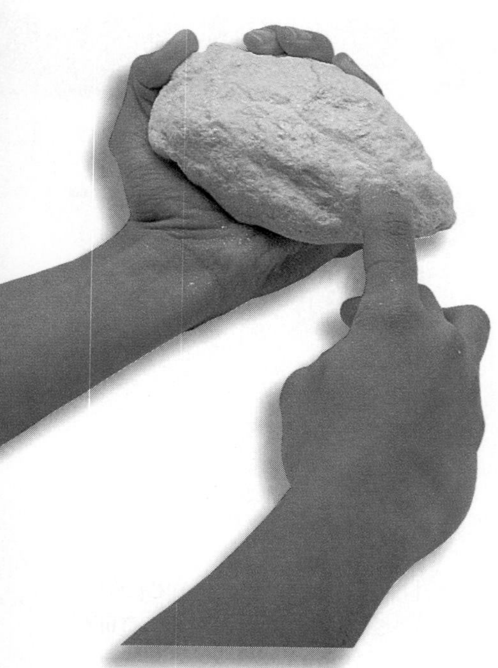

Figure 3-6

Talc can be scratched with your fingernail. *Why is this true?*

Hardness

A measure of how easily a mineral can be scratched is its **hardness.** The mineral talc, shown in **Figure 3-6,** is so soft you can scratch it loose with your fingernail. You might be familiar with talcum powder made from this mineral. Diamonds, on the other hand, are the hardest mineral. Some diamonds are used as cutting tools. A diamond can be scratched only by another diamond.

In order to compare the hardnesses of minerals, a list of common minerals and their hardnesses was developed. The German scientist Friedrich Mohs developed the Mohs scale of hardness, as seen in **Table 3-2.** The scale lists the hardnesses of ten minerals, with 1 being the softest and 10 the hardest.

Here's how the scale works. Let's say you have a clear or whitish-colored mineral that you know is either calcite or quartz. You scratch it on your fingernail and then on a bright, shiny piece of copper. You find that the mineral scratches your fingernail but doesn't scratch the copper. Because the hardness of your fingernail is 2.5 and that of a piece of copper is 3.5, you can determine the unknown mineral's hardness to be about 3. Because quartz has a hardness of 7, your mystery mineral must be calcite.

Table 3-2

Mineral Hardness

Mohs Hardness Scale

	softest	**Hardness of Common Objects**
Talc	1	
Gypsum	2	fingernail (2.5)
Calcite	3	piece of copper (3.5)
Fluorite	4	iron nail (4.5)
Apatite	5	glass (5.5)
Feldspar	6	steel file (6.5)
Quartz	7	streak plate (7)
Topaz	8	
Corundum	9	
Diamond	10	
	hardest	

Luster

Luster describes how light is reflected from a mineral's surface. Luster is defined as either metallic or nonmetallic. Minerals with a metallic luster, like the galena shown in **Figure 3-7,** always shine like metal. Metallic luster can be compared to the shine of a fancy metal belt buckle or the shiny chrome trim on some cars.

When a mineral does not shine like metal, its luster is nonmetallic. Examples of terms for nonmetallic luster include dull, pearly, and silky.

Figure 3-7

Galena has a metallic luster, which means that it shines like metal.

Color

The color of a mineral can also be a clue to its identity. A mineral whose color helps in its identification is sulfur. Sulfur has a distinctive yellow color. However, just remember that, as you learned with gold and pyrite, color alone usually isn't enough to identify a mineral.

Streak

Streak is the color of the mineral when it is broken up and powdered. When a mineral is rubbed across a piece of unglazed porcelain tile, as in **Figure 3-8,** a streak is left behind. This streak is the powdered mineral. Gold and pyrite can be identified with the streak test. Gold has a yellow streak and pyrite has a greenish black or brown-black streak.

The streak test works only for minerals that are softer than the streak plate. Very soft minerals will leave a streak even on paper. The last time you used a pencil to write on paper, you used the streak of the mineral graphite. Graphite is used in pencil lead because it is soft enough to leave a streak on paper.

MiniLAB

How can special properties be used to identify a mineral?

Observe the differences in print viewed through clear minerals.

Procedure

1. Obtain samples of the following clear minerals: gypsum, muscovite mica, halite, and calcite.
2. Place each over the print on this page and observe the letters.

Analysis

1. What happens to light as it passes through these minerals?
2. What mineral can be identified by the print's double image? What special property is used to identify this mineral?

Figure 3-8

Color is not always useful for mineral identification. Hematite, for example, can be dark red, gray, or silvery in color. Its streak, however, is always dark red-brown.

Problem Solving

Sometimes It's Gold, Sometimes It's Not

On a recent family vacation, Maria visited several areas of California. Because she had worked in a jewelry store that summer, she decided to explore Sutter's Mill, the place where gold was discovered in 1849. Maria saw some bright yellow metallic objects glistening in the clear water of a fast-moving stream. She reached into the water and withdrew what appeared to be four or five nuggets of gold.

Relying on her experience in the jewelry store, Maria excitedly tested the nuggets. She found that the gold nuggets left a greenish black powder when rubbed across a piece of white porcelain. The nuggets scratched a piece of copper she found in her room.

Solve the Problem:

1. **What color streak does gold have?**
2. **Is gold hard enough to scratch copper?**
3. **Did Maria strike it rich?**

Think Critically:

Streak and hardness are just two of the clues that we use to identify minerals.

1. Name two other clues that Maria used when she spied the minerals in the stream.
2. What other tests could she have performed to determine whether her sample was gold?

Figure 3-9

Weak bonds within the structure of mica allow it to be broken along smooth, flat cleavage planes. *If you broke quartz, would it look the same? Explain.*

Cleavage and Fracture

The way a mineral breaks is another clue to its identity. Minerals that break along smooth, flat surfaces have **cleavage.** Cleavage, like hardness, is determined by the arrangement of the mineral's atoms. Mica is a mineral that has perfect cleavage. You can see in **Figure 3-9** how it breaks along smooth, flat surfaces. If you were to take a layer cake and separate its layers, you would show that the cake has cleavage. But not all minerals have cleavage. Minerals that break with rough or jagged edges have **fracture.** Quartz is a mineral with fracture. If you were to grab a chunk out of the side of that cake, it would be like breaking a mineral with fracture.

Other Properties

Some minerals have unique properties. Magnetite, as you can guess by its name, is attracted to magnets. Lodestone, a form of magnetite, will pick up iron filings like a magnet. Light bends in two directions when it passes through some calcite specimens, causing you to see a double image, as in **Figure 3-10.** Calcite can also be identified because it fizzes when hydrochloric acid is put on it.

You can see that you sometimes need more information than just color and appearance to identify most minerals. You might also need to test its streak, its hardness, its luster, and its cleavage or fracture. You can be just as good at identifying minerals as you are at recognizing your friends in class.

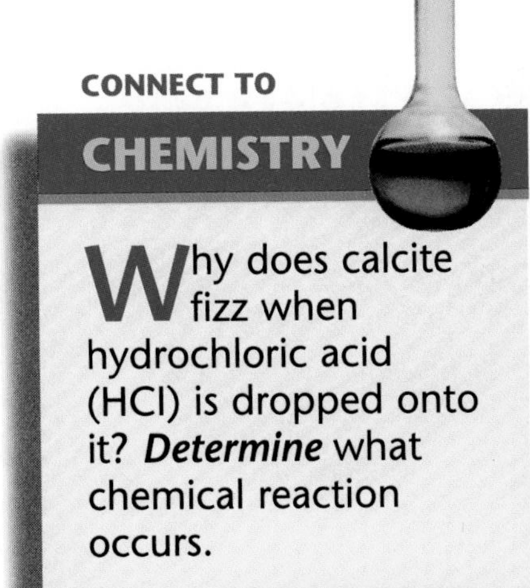

CONNECT TO CHEMISTRY

Why does calcite fizz when hydrochloric acid (HCl) is dropped onto it? ***Determine*** what chemical reaction occurs.

Figure 3-10

Iceland spar, a clear specimen of calcite, can be identified by its unique ability to bend light in two directions, causing a double image.

Section Wrap-up

Review

1. What's the difference between a mineral that has cleavage and one that has fracture?
2. How can an unglazed porcelain tile be used to identify a mineral?
3. **Think Critically:** What hardness does a mineral have if it is scratched by glass but scratches an iron nail?

Skill Builder

Making and Using Graphs

Make a bar graph of the hardnesses of the common objects used for comparison with the minerals in the Mohs scale of hardness. If you need help, refer to Making and Using Graphs in the **Skill Handbook.**

Science Journal

In your Science Journal, make a list of five minerals. Next to the name of each mineral, list at least two physical properties that can be used to identify the mineral.

Activity 3-2

Design Your Own Experiment

Mineral Identification

Some properties of minerals make them easy to identify, while others are not much help at all. Some minerals can be identified by one property. Others require the testing of several properties for identification.

PREPARATION

Problem

Which property would you try first to identify an unknown mineral?

Form a Hypothesis

State a hypothesis about which property you think will be most useful in identifying minerals.

Objectives

- Determine which properties of a mineral you will use for identification purposes.
- Design an experiment that uses these properties to help in identifying minerals.

Possible Materials

- mineral samples
- hand lens
- pan balance
- graduated cylinder
- water
- piece of copper
- glass slide
- small iron nail
- steel file
- streak plate
- 5 percent hydrochloric acid (HCl) with dropper
- goggles
- Mohs scale of hardness
- Appendices M and N

Safety Precautions

Review the safe use of acids. Wear goggles and an apron. Be careful when handling materials with sharp edges.

CAUTION: *HCl may cause burns. If spillage occurs, rinse with water.*

PLAN THE EXPERIMENT

1. As a group, agree upon and write your hypothesis statement.
2. As a group, list the steps that you will need to test your hypothesis. Be specific, describing exactly what you will do at each step.
3. Make a list of the materials that you will need to complete your experiment.
4. Make a list of the various properties of minerals that you will test.
5. Make a list of any special properties you expect to observe or test.

Check the Plan

1. Read over your entire experiment to make sure that all steps are in a logical order.
2. Should you test any of the properties more than once for any of the minerals?
3. Will you summarize data in a graph, table, or chart?
4. How will you determine whether certain properties indicate a specific mineral?
5. *Make sure your teacher approves your plan and that you have included any changes suggested in the plan.*

DO THE EXPERIMENT

1. Carry out the experiment as planned.
2. While conducting the experiment, write down any observations that you or other members of your group make and summarize your data in your Science Journal.

Analyze and Apply

1. Which property was most useful in identifying your samples? Which property was least useful?
2. **Compare** the properties that worked best for you with those that worked best for other students.
3. **Discuss** reasons why one property is useful and others are not.

Go Further

If all you had were a piece of paper, a steel knife, and a glass bottle, could you distinguish between calcite and quartz? What other test would help you identify calcite?

3•3 Uses of Minerals

Science Words

gem
ore

Objectives

- Discuss characteristics gems have that make them different from and more valuable than other minerals.
- List the conditions necessary for a mineral to be classified as an ore.

Gems

What makes one mineral more useful or valuable to us than another? Why are diamonds and rubies considered to be valuable? The next time you go shopping, look in the window of a jewelry store. Chances are you will see rings, bracelets, and maybe even a watch with diamonds or other gems on them. What properties do gems possess that make them valuable?

Properties of Gems

Gems or gemstones like the ones in **Figure 3-11** are highly prized minerals because they are rare and beautiful. Many gems are cut and polished and used for jewelry. They are brighter and more colorful than common samples of the same mineral.

The difference between a gem and the common form of the same mineral can be slight. Amethyst, a gem form of quartz, contains just traces of manganese in its structure. This manganese gives amethyst a desirable purple color. And sometimes, a gem has a crystal structure that allows it to be cut and polished to a higher quality than that of a non-gem mineral.

Figure 3-11

It is easy to see why gems are prized—for their beauty and rarity. Shown here are watermelon tourmaline, turquoise, polished amethyst, and unpolished amethyst. Shown above are a ruby and an opal.

Figure 3-12

Ore must be processed and refined into more useful materials. *When is a mineral considered an ore?*

Ores

If you look around your room at home, you will find many things made from mineral resources. See how many you can name. Is there anything in your room with iron in it? If so, the iron may have come from the mineral hematite. The aluminum in soft-drink cans comes from the mineral bauxite.

Bauxite and hematite are minerals that can also be called ores. A mineral is an **ore** if it contains a useful substance that can be mined at a profit. Aluminum can be refined from bauxite, as shown in Figure **3-12,** and made into the useful products shown in **Figure 3-13.** These products are worth more money than the cost of the mining, so bauxite is an ore.

Figure 3-13

Bauxite ore is used to make aluminum products.

USING TECHNOLOGY

On the Cutting Edge

You probably knew that diamonds were used in jewelry, but did you know that diamonds are used on drill bits and other instruments? Diamond-tipped drill bits can cut through hard substances such as steel and rock. Diamonds are used on the bits because they are the hardest mineral.

Scientists have developed a way to put a thin coating of diamonds on objects such as the record needle shown in this photo. They use methane gas and microwaves to make the diamonds. A methane molecule consists of a carbon atom surrounded by four hydrogen atoms. Microwaves are used to strip the hydrogen atoms away from the molecule. Then the carbon atoms link together on the surface of the object being coated, forming tiny rows of diamonds. This process can be used to make diamond-edged surgical scalpels, razor blades, dental drills, diamond-tipped record needles, and diamond-coated computer parts.

Think Critically: The diamonds scientists make using methane gas and microwaves aren't true minerals. Why not?

Diamond-Tipped Record Needle

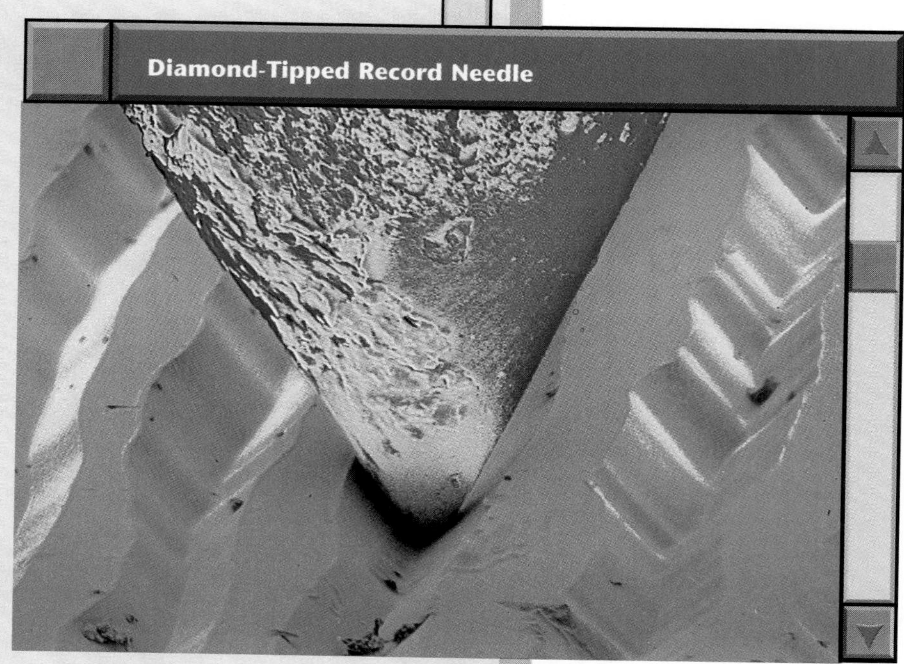

Waste Rock Removal

Many ores are obtained from large open-pit mines like the one shown in **Figure 3-14.** When the ore is mined, it contains unwanted material along with the valuable material. The waste rock or material must be removed before the ore can be used. Removing the waste rock can be expensive, and in some cases, harmful to the environment. If the cost of removing the waste rock gets higher than the value of the desired material, the mineral will no longer be classified as an ore.

Think back to the last soft drink you had in an aluminum can. What do you think would happen if people stopped using aluminum cans? This would cause the demand for aluminum to go down. Some bauxite mines would close down because they would no longer be able to make money. The value of a mineral can change if the supply of or the demand for that mineral changes.

Figure 3-14

Many ores are obtained from large open-pit mines. *What would happen to this mine if the demand for its ore decreased? What would happen if the demand increased?*

You'd be surprised to find out all of the things that come from ores. Iron, used in everything from frying pans to ships, is obtained from its ore, hematite. Zinc is obtained from sphalerite. Copper, used in coins and electrical wires, among other things, is obtained from chalcopyrite. Some other ores of copper, malachite, and azurite are gemstones. As you can see, gems and ores are important mineral resources.

Decide how the physical property of density could be used to determine whether a possible ore contains gold or pyrite ("fool's gold").

Section Wrap-up

Review

1. Compare and contrast gem-quality amethyst with regular quartz.
2. **Think Critically:** Why couldn't a company stay in business if the mineral it was mining were no longer an ore?

Skill Builder
Concept Mapping

Make an events chain showing why bauxite can be an ore. Use the following terms and phrases: *gives aluminum, ore, mined at profit, bauxite.* If you need help, refer to Concept Mapping in the **Skill Handbook.**

Investigate and show how early prospectors used density to determine whether they had found gold or pyrite.

TECHNOLOGY: 3•4 Uses of Titanium

Science Words

titanium

Objectives

- List the properties of titanium that make it so useful in biomedicine, sporting equipment, and other applications.
- Identify the minerals that are mined for titanium.

Titanium

You might know someone who uses golf clubs with titanium shafts or a racing bicycle containing titanium. Perhaps you know someone who has needed a hip replacement made from titanium. **Titanium** is a durable, lightweight metal derived from minerals such as ilmenite or rutile.

Properties of Titanium

Titanium is used in automobile body parts, such as connecting rods, valves, and suspension springs. It is also used in the production of aircraft. The properties of titanium that make it so valuable to the automobile and aircraft industries are its durability and light weight. These same properties make titanium valuable in the production of sports equipment, such as tennis rackets. In fact, wheelchairs used by people who want to race or play basketball are often made from titanium. The light weight and durable nature of titanium, as well as the fact that it is a nontoxic metal, also make it useful in the production of hip replacements and other artificial body parts. **Figure 3-15** illustrates several uses for titanium.

Figure 3-15

Titanium's durability and light weight make it useful in a variety of applications.

A Wheelchairs used for racing and playing basketball often have parts made from titanium.

B Titanium is used in the pivots on airplane wings, allowing the wings to move forward and back for maximum lift and speed.

Ilmenite and Rutile

Two minerals that are classified as ores of the element titanium are ilmenite and rutile. There are two separate methods presently used by manufacturers to obtain the element titanium from each of its ores. One method is to dissolve ilmenite in sulfuric acid, thereby producing titanium dioxide and iron sulfate. In the other method, rutile, which contains a higher concentration of titanium, is combined with chlorine at high temperatures. This process produces titanium tetrachloride ($TiCl_4$), often called "tickle." The second method is growing in use because the iron sulfate produced by the first method is harmful to the environment.

C Titanium is strong enough to be used for total hip replacements. *What other quality of titanium makes it useful in the production of artificial body parts?*

Section Wrap-up

Review

1. What properties of titanium make it so useful in sports equipment and biomedical research?
2. Which two minerals are ores of titanium?

Explore the Technology

Manufacturers are more likely to produce titanium tetrachloride by combining the mineral rutile with chlorine at high temperatures. Why is this true?

SCIENCE & SOCIETY

Science & ART

This yak, painted by A. Kahn, is outlined in gold paint. The yak's tongue, nose, and eyes are tinted with cinnabar.

Minerals as Paint Pigments

Ultramarine, burnt sienna, vermilion, crimson, azure, ochre, cerulean blue—these exotic-sounding names describe colors artists use in their paintings and drawings. The substances that give color to paints, crayons, and pastel chalks are called pigments. Today, most artists' pigments are made artificially, using synthetic dyes from petroleum or using chemical processes that create desired combinations of minerals. But many pigments were originally made by digging minerals from the ground and grinding them into fine powders.

Some traditional artists prefer the more subtle colors produced by ground minerals and still use ancient methods to make their own paints. Many of the minerals that are used as pigments contain poisonous substances, such as lead, arsenic, copper, and mercury, and must be handled with skill and care.

White pigments are made from compounds containing calcium, zinc, titanium, or lead.

Red colors can be made from cinnabar, a mineral containing mercury and sulfur. Depending on how the powder is ground, it can produce colors ranging from pale pink to bright vermilion or deep maroon. Tiny particles of mercury give cinnabar pigments a silvery sparkle.

Shades of green, yellow-green, and blue-green are made from copper-containing minerals such as malachite and azurite.

The rich, dark blue known as ultramarine is made from lapis lazuli, a mineral containing silicon, sulfur, and sodium. Lapis pigments sometimes include spangles of iron pyrite, also called fool's gold.

Yellow comes from many sources. Ochre is a yellow clay that contains iron. When heated in an oven, it turns brick red. Sienna is a dark yellow clay that contains iron and manganese. It turns brown—becoming burnt sienna—when roasted in an oven. Other yellow sources include arsenic combined with sulfur and lead combined with antimony.

Gold, silver, and a few other metals can be made into shimmering paints. Gold that has been painted onto a canvas can be rubbed or burnished with stone to make it gleam.

Science Journal

Research how artists make their own paints. In your Science Journal, write a paragraph describing the process.

Chapter 3 Review

Summary

3-1: Minerals

1. All minerals are formed by natural processes and are inorganic solids with unique compositions and distinct internal structures.
2. Minerals have crystal structures in one of six major crystal systems.
3. Minerals can form when magma cools, when liquids containing dissolved minerals evaporate, or when particles precipitate from solution.

3-2: Mineral Identification

1. Hardness is a measure of how easily a mineral can be scratched. Luster describes how light is reflected from a mineral's surface. Color is a property that sometimes can be used to identify a mineral.
2. Streak is the color of the powder left by a mineral on an unglazed porcelain tile. Minerals that break along smooth, flat surfaces have cleavage. Minerals that break with rough or jagged surfaces have fracture.

3-3: Uses of Minerals

1. Gems are minerals that are more rare and beautiful than common minerals.
2. An ore is a mineral or group of minerals that can be mined at a profit.

3-4: Science and Society: Uses of Titanium

1. Titanium is a durable, lightweight metal derived from minerals such as ilmenite and rutile.
2. The properties of titanium make it ideal for use in sports equipment and as hip replacements, among other things.
3. Two methods exist for manufacturers to obtain titanium from its ores ilmenite and rutile.

Key Science Words

a. cleavage
b. crystal
c. fracture
d. gem
e. hardness
f. luster
g. magma
h. mineral
i. ore
j. silicate
k. streak
l. titanium

Reviewing Vocabulary

Match each phrase with the correct term from the list of Key Science Words.

1. naturally occurring, inorganic, crystalline solid with a definite composition
2. hot, melted rock material
3. mineral containing silicon, oxygen, and usually one or more other elements
4. how light is reflected from a mineral
5. an element obtained from ilmenite or rutile
6. valuable, rare mineral
7. a mineral mined at a profit
8. breakage along smooth, flat planes
9. solid with a repeating arrangement of atoms
10. the color of a mineral's powder

Chapter 3 Review

Checking Concepts

Choose the word or phrase that completes the sentence.

1. Minerals _______ .
   a. can be liquids c. are organic
   b. have no crystal structure d. are inorganic
2. All silicates contain _______ .
   a. magnesium
   b. silicon and aluminum
   c. silicon and oxygen
   d. oxygen and carbon
3. _______ is hot, melted rock material.
   a. Magma c. Salt water
   b. Quartz d. A mineral
4. Quartz is a(n) _______ .
   a. oxide c. silicate
   b. carbonate d. sulfide
5. _______ is a measure of how easily a mineral can be scratched.
   a. Luster c. Cleavage
   b. Hardness d. Fracture
6. The color of a powdered mineral on an unglazed porcelain tile is its _______ .
   a. luster c. hardness
   b. density d. streak
7. Quartz breaks with _______ .
   a. cleavage c. luster
   b. fracture d. smooth surfaces
8. _______ can be used in hip replacements.
   a. Iron c. Titanium
   b. Carbon d. Steel
9. The largest group of minerals is the ______ .
   a. oxides c. silicates
   b. halides d. sulfides
10. Halite forms _______ crystals.
    a. triclinic c. monoclinic
    b. hexagonal d. cubic

Understanding Concepts

Answer the following questions in your Science Journal using complete sentences.

11. Describe the environmental concerns involved with obtaining titanium.
12. Explain why air is not a mineral.
13. Compare and contrast the properties of cleavage and fracture. Give an example of a mineral that cleaves and one that fractures.
14. Why is titanium useful in sports equipment?
15. If you found a nearly perfect hexagonal crystal of quartz, what could you conclude about its formation?

Thinking Critically

16. Water is a nonliving substance formed by natural processes on Earth. It has a unique composition. Sometimes water is a mineral and other times it is not. Explain.
17. How many sides are there to a perfect salt crystal, such as the one shown below?
18. Suppose you let a sugar solution evaporate, leaving sugar crystals behind. Are these crystals minerals? Explain.
19. Will diamond leave a streak on a streak plate? Explain.
20. Explain how you would use **Table 3-2** on p. 68 to determine the hardness of any mineral.

Developing Skills

If you need help, refer to the **Skill Handbook.**

21. **Observing and Inferring:** Suppose you found a white, nonmetallic mineral that was harder than calcite. You identify the sample as quartz. What are your observations? What is your inference?
22. **Interpreting Data:** Suppose you were given these properties of a mineral: pink color, nonmetallic, softer than topaz and quartz, scratches apatite, harder than fluorite, has cleavage, and is scratched by a steel file. What is it?
23. **Outlining:** Make an outline of how at least seven physical properties can be used to identify unknown materials.
24. **Measuring in SI:** The volume of water in a graduated cylinder is 107.5 mL. A specimen of quartz, tied to a piece of string, is immersed into the water. The new water level reads 186 mL. What is the volume of the piece of quartz?
25. **Concept Mapping:** Make a network tree concept map showing six crystal systems and examples from each group. Use the following words and phrases: *orthorhombic, hexagonal, cubic, tetragonal, monoclinic, triclinic, topaz, corundum, halite, wulfenite, gypsum, and albite.*

Performance Assessment

1. **Carrying Out a Strategy and Collecting Data:** Collect samples of objects that you think are minerals. Following procedures learned in Activity 3-2, identify each mineral. Use a mineral identification book for help in listing some uses of the minerals you identify.
2. **Model:** Go to the library to research the six major mineral crystal systems. Use cardboard to make models of each major crystal shape for each system.
3. **Lab Report:** In Activity 3-1, you observed two ways in which crystals can form. Perform the experiment again, but this time add only one-half as much salt and sugar to the separate solutions you are making. Write a lab report that describes your results and compares the results with those you obtained in the original experiment.

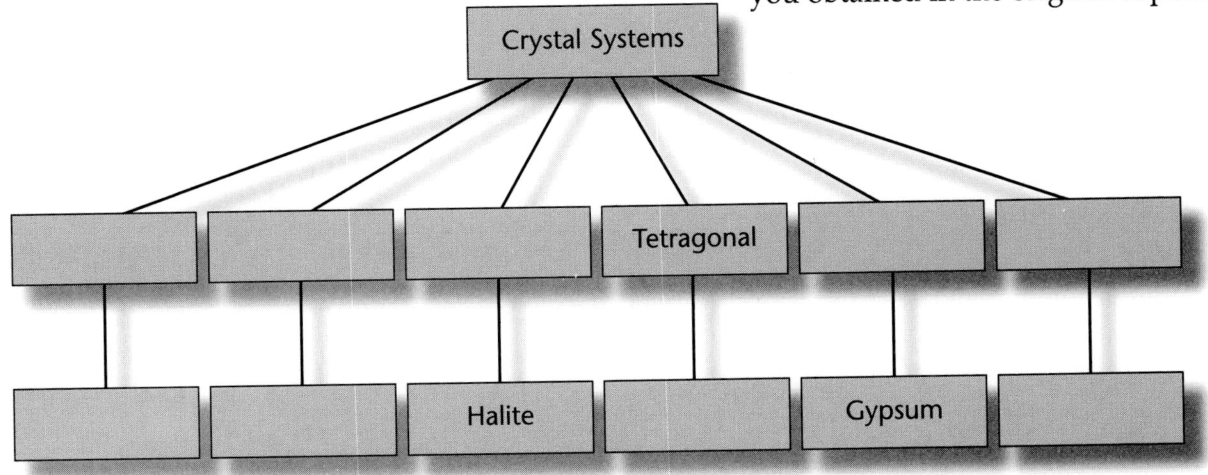

Previewing the Chapter

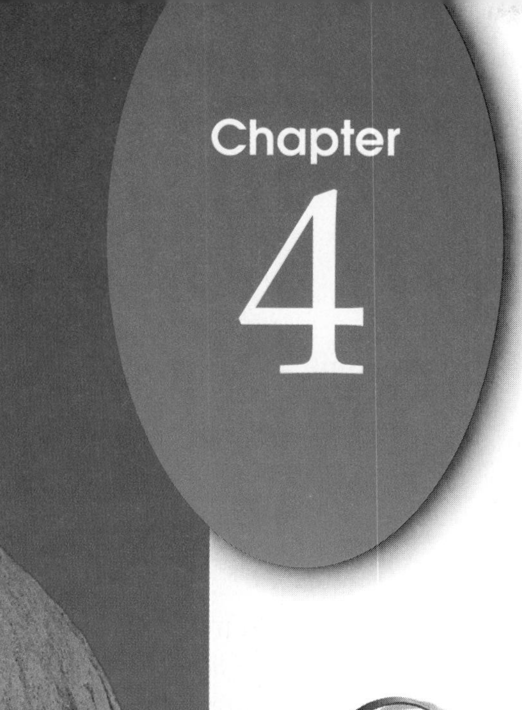

Rocks

This boulder stands like a sentinel in the desert sun of Arizona. Though it looks large and imposing, it is made of tiny particles of sand and other materials tightly pressed together. Other rocks are formed from molten lava ejected from volcanoes or from the remains of once-living plants and animals. If you examine rocks close up, you can sometimes tell what they are made of.

EXPLORE ACTIVITY

Determine what rocks are made from.

Rocks are all around you. You may have collected samples of rock or tried skipping a stone across a lake. You probably have noticed that not all of them look the same. Can you describe what rocks are made of?

1. Collect three or four different fragments of rocks from around your home.
2. Draw a picture of the details in each rock.
3. Look for different types of material within the same rock; review the term mixture from Chapter 2.

Observe: What do you notice about each rock? Are your rocks mixtures? If so, what are they mixtures of? Write your observations in your Science Journal.

Previewing Science Skills

- In the **Skill Builders**, you will **map concepts, interpret scientific illustrations, sequence events,** and **interpret data.**
- In the **Activities**, you will **classify, analyze,** and **draw conclusions.**
- In the **MiniLABs**, you will **classify, observe,** and **infer.**

4•1 The Rock Cycle

Science Words

rock
rock cycle

Objectives

- Differentiate between a rock and a mineral.
- Describe the rock cycle and the changes that a rock may undergo.

What is a rock?

Imagine that you're on your way home from a friend's house when you notice an unusual rock in a driveway. You pick it up, wondering why it looks different from most of the other rocks there. The other rocks are flat and dull, but this one is rounded and has shiny crystals in it. You decide to stick the interesting rock in your pocket and ask your Earth science teacher about it tomorrow.

What exactly should you ask your teacher? You might begin by asking, "What is a rock?" and "Why are rocks so different from one another?"

Figure 4-1

Granite is a mixture of feldspar (A), quartz (B), mica (C), hornblende (D), and other minerals.

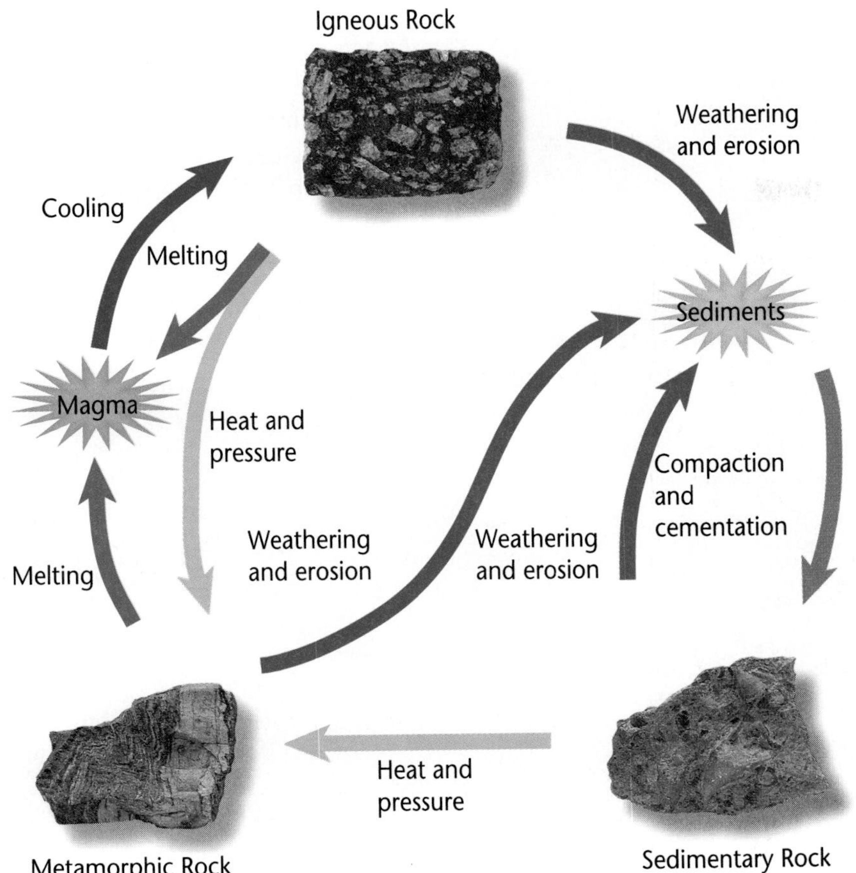

Figure 4-2

This model of the rock cycle shows how rocks are constantly changed from one form to another. *How are rocks different from minerals?*

Forming Rocks

A **rock** is a mixture of minerals, mineraloids, glass, or organic matter. You learned about the mineral quartz in the last chapter. You know that it's a common mineral found in rocks. Other common rock-forming minerals include feldspar, hornblende, and mica. **Figure 4-1** shows all these minerals mixed together to form the rock granite.

But how do these minerals mix together? And once they've formed a rock, do they stay in that same rock forever? Let's find out.

The Rock Cycle

Figure 4-2 is a model of the **rock cycle,** showing the three types of rock and how they are formed. You'll learn more about sedimentary, igneous, and metamorphic rocks later in this chapter. For now, look at the rock cycle and notice that rocks are changed by processes such as weathering, erosion, compaction, cementation, melting, and cooling. For example, a sedimentary rock can be changed by heat and pressure to form a metamorphic rock. The metamorphic rock can then melt and later cool to form an igneous rock. The igneous rock may then weather and erode, and the fragments from it might form another sedimentary rock. Heat and pressure can also change igneous rocks into metamorphic rocks.

MiniLAB

How do rocks change during the rock cycle?

Procedure

1. Have your teacher crush several different samples of rock into a fine powder.
2. Place the powder into a flat pan.
3. Mix a solution of water and white glue.
4. Pour the solution into the pan and mix thoroughly with the powder.
5. Place a layer of waxed paper over the mixture, then place a heavy weight on top of the waxed paper.

Analysis

1. After several days, remove the weight and examine the mixture.
2. How is the process used to make the mixture similar to one portion of the rock cycle?
3. Describe other processes that might be used to model other parts of the rock cycle.

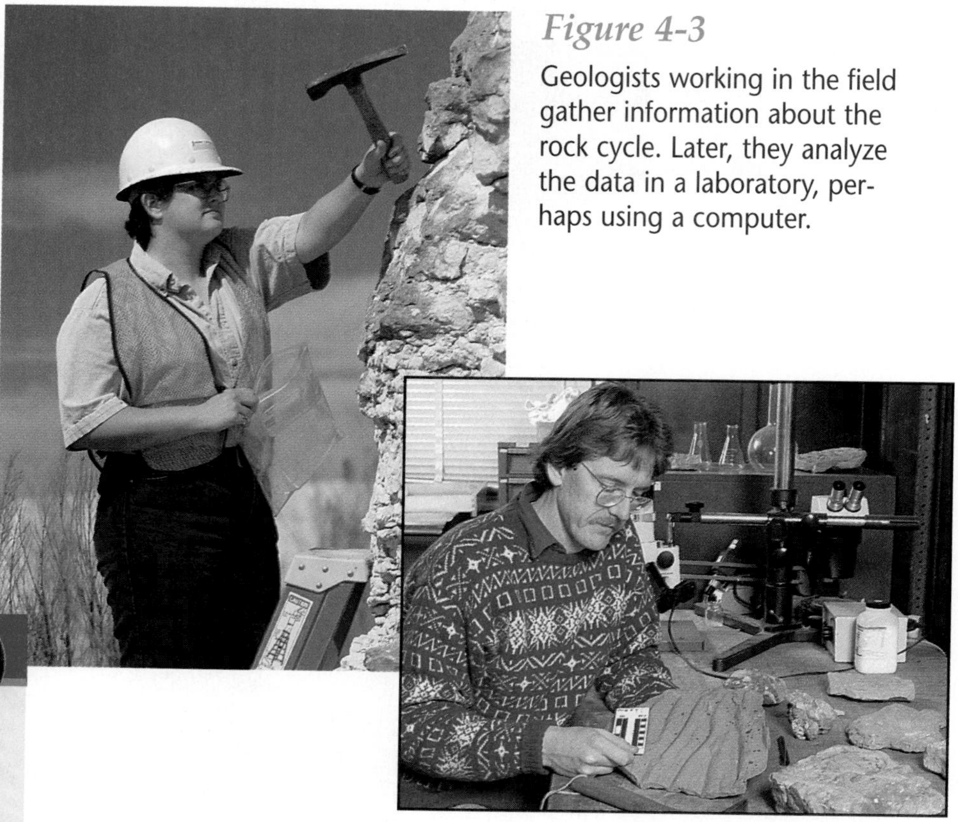

Figure 4-3

Geologists working in the field gather information about the rock cycle. Later, they analyze the data in a laboratory, perhaps using a computer.

CONNECT TO

CHEMISTRY

As heat increases inside Earth, materials contained in rocks begin to melt. ***Decide*** what is happening to the atoms and molecules to cause this change of state.

In Chapters 5 and 6, you will explore weathering, erosion, deposition, and other processes involved in the rock cycle. The rock cycle shows how all of these processes interact to form and change the rocks around you. Like the geologists in **Figure 4-3,** let's now investigate how igneous, metamorphic, and sedimentary rocks fit into the rock cycle.

Section Wrap-up

Review

1. What materials mix together to form a rock?
2. What is the rock cycle?
3. **Think Critically:** Look at the model of the rock cycle. How would you define *magma* based on **Figure 4-2?** How would you define *sediments* and *sedimentary rock?*

Skill Builder

Concept Mapping

Make a concept map that explains how igneous rocks can become sedimentary, then metamorphic, and finally, other igneous rocks. If you need help, refer to Concept Mapping in the **Skill Handbook.**

Science Journal

As magma cools, it solidifies. In your Science Journal, write a description that explains what happens to the atoms and molecules to cause this change of state.

Science & ART

Set in Stone

Since humans first appeared on Earth, they have relied on rocks in many ways. Stone Age people used rocks as tools and weapons. Medieval sculptors carved intricate designs into the limestone and sandstone façades of numerous buildings. Today, masons engrave names, dates, and spiritual sayings onto granite cemetery markers. Read on to learn about a unique way in which glacial rocks were once used by the Inuit of the northern Hudson Bay area.

Cairns

If you have ever been hiking in the wilderness, you may have stumbled upon small piles of rocks left by previous hikers. These rock piles are known as cairns. Cairns are often erected to serve as landmarks. They can aid even the most experienced hikers in finding their way, especially in landscapes with few natural landmarks.

Inukshuit

Now, imagine yourself hiking through glaciated lands of the Arctic—without a compass. Surrounded by lakes, snow, and unchanging horizons, how could you navigate? You might rely on cairns erected thousands of years ago by various Inuit groups. These cairns are called inukshuit (singular, inukshuk), which means "something acting in the capacity of man."

An inukshuk is a humanlike stone figure ranging from about one to four meters in height. The figures were used to aid travelers crossing the frozen tundra. For instance, the inukshuk shown on this page alerted travelers that there was only one path leading into and out of the valley in which the cairn stands.

Stone Scarecrows

Probably one of the most important uses of these stone figures was in the hunt for caribou, which provided food and fur for the Inuit. Hunters would mimic the sounds of wolves to frighten the caribou, setting the herd in motion between the long lines of inukshuit. Thinking the stone sentinels were actually people, the caribou would stay between the cairns. Hunters would hide among the inukshuit to ensure that the herd continued to move forward. Once the animals reached the top of a hill or the end of a valley, other hunters used bows and arrows to kill their prey.

Science Journal

Sketch an original inukshuk that might be used by a modern traveler navigating between two known points in your area—for example, between your school and a nearby building. Have one or two other students figure out the symbolism in your drawing.

4•2 Igneous Rocks

Science Words

igneous rock
lava
intrusive
extrusive
basaltic
granitic

Objectives

- Recognize magma and lava as the materials that cool to form igneous rocks.
- Contrast the formation of intrusive and extrusive igneous rocks.
- Contrast granitic and basaltic igneous rocks.

Origin of Igneous Rocks

In June of 1991, Mount Pinatubo in the Philippines erupted. Perhaps you've heard of other recent volcanic eruptions. When most volcanoes erupt, they eject a thick, gooey flow of molten material. This material is similar to fudge candy before it has cooled. You know that if you allow the fudge to cool, it

Figure 4-4

Magma trapped below Earth's surface is insulated by the rocks surrounding it. This holds in the heat and causes the magma to cool slowly. The atoms have time to arrange into large crystals called mineral grains.

A Intrusive rocks such as gabbro and diorite have large crystal grains. *In your Science Journal, explain how these rocks were formed.*

becomes hard and you have to cut it with a knife. When molten material from a volcano or from deep inside Earth cools, it forms **igneous rocks** such as the ones in **Figure 4-4.** But why do volcanoes erupt, and where does the molten material come from?

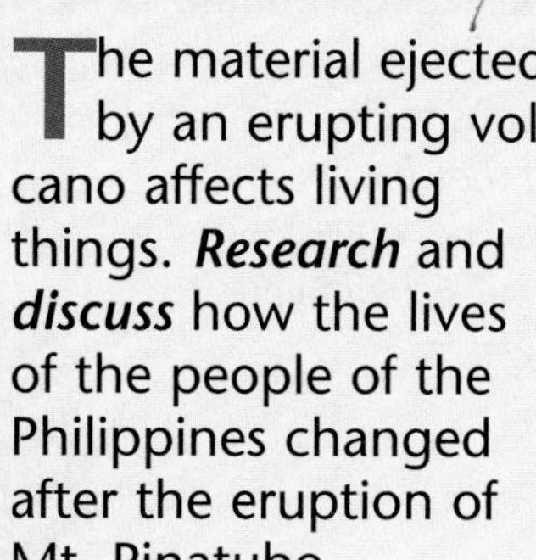

CONNECT TO

LIFE SCIENCE

The material ejected by an erupting volcano affects living things. ***Research*** and ***discuss*** how the lives of the people of the Philippines changed after the eruption of Mt. Pinatubo.

Magma

Temperatures reach about 1400°C at depths between 60 to 200 km below Earth's surface. The rocks at this depth are under great pressure from overlying rocks. Radioactive elements in the rocks generate thermal energy, heating the rocks. In certain locations on Earth, the temperature and pressure are just right to melt the rocks and form magma.

The magma is less dense than the surrounding solid rock, so it is forced upward toward Earth's surface. Magma that eventually reaches Earth's surface flows from volcanoes as **lava.**

obsidian

rhyolite

B Extrusive rocks such as rhyolite and obsidian form from fast-cooling lava. Magma is forced upward toward Earth's surface, where it cools to form extrusive igneous rocks.

Figure 4-5

Pumice (A), obsidian (B), and scoria (C) are actually glass, but are classified as extrusive igneous rocks. *Did the lava that formed these rocks cool quickly or slowly? Explain your answer.*

Intrusive Rocks

Magma is made up of atoms of melted minerals. When it cools, the atoms rearrange themselves into new crystals called mineral grains. Rocks form as these mineral grains grow together. Rocks that form below Earth's surface are **intrusive** igneous rocks. Generally, intrusive igneous rocks have large mineral grains. Intrusive rocks are found at Earth's surface when the kilometers of rock and soil that once covered them have been removed, or when forces in Earth have pushed them up.

Extrusive Rocks

Extrusive igneous rocks are formed when lava cools on or near Earth's surface. When lava flows on Earth's surface, it is exposed to air and moisture. Lava cools quickly under these conditions. The quick rate of cooling keeps large mineral grains from growing. The atoms don't have time to arrange into large crystals. Extrusive igneous rocks have a fine-grained texture.

Figure 4-5 shows pumice, obsidian, and scoria. These objects cooled so quickly that no visible mineral grains formed. Most of the atoms in these objects are not arranged into neat crystal patterns. Obsidian, scoria, and pumice are actually glass but are classified as extrusive igneous rocks.

In the case of pumice and scoria, air and other gases become trapped in the gooey molten material as it cools. Many of these gases eventually escape, but holes are left behind where the rock formed around the pockets of gas.

Science Journal

In your Science Journal, make a list of minerals you would expect to find in granitic, andesitic, and basaltic rocks.

Classification of Igneous Rocks

You've learned that igneous rocks are called intrusive or extrusive depending on where they formed. A way to further classify these rocks is by magma type. As shown in **Table 4-1,** an igneous rock can form from basaltic, granitic, or andesitic magma.

Table 4-1

Common Igneous Rocks

Type of Magma or Lava	Intrusive	Extrusive
Basaltic	Gabbro	Basalt
		Scoria
Andesitic	Diorite	Andesite
Granitic	Granite	Rhyolite
		Pumice
		Obsidian

Basaltic Rocks

Basaltic igneous rocks are dense, heavy, dark-colored rocks that form from basaltic magma. Basaltic magma is rich in iron and magnesium. Basaltic lava flows from the volcanoes in Hawaii as shown in **Figure 4-6.** How does this explain the black beach sand common in Hawaii?

Granitic and Andesitic Rocks

Granitic igneous rocks are light-colored rocks of a lower density than basaltic rocks. Granitic magma is thick and stiff and contains a lot of silicon and oxygen. Granitic magma can build up a great deal of pressure, which is released during violent volcanic eruptions.

Andesitic rocks have mineral compositions between those of granitic and basaltic rocks. Many volcanoes in the Pacific Ocean are andesitic.

Figure 4-6

Basalt is the most common extrusive rock. Sediments from weathered and eroded basalt form the black-sand beaches of the Hawaiian Islands.

Activity 4-1

Design Your Own Experiment
Igneous Rocks

One way that rocks can form is from melted rock material or magma. Some rocks formed in this way cool quickly from lava at or near Earth's surface. Others cool slowly from magma deep inside Earth. Can igneous rocks be classified and studied based on how they cool? Is the color of the rocks associated with mineral content, and can this property also be used to help classify igneous rocks?

PREPARATION

Problem

How can we classify igneous rocks?

Form a Hypothesis

Based on observations made of rocks supplied with this exercise, state a hypothesis about how their characteristics can be used to classify them.

Objectives

- Observe differences in rocks that have formed quickly compared with those that have formed slowly. Also, observe any differences in the rocks' color.
- Design an experiment that tests whether a variable such as crystal size is observable and whether it indicates how the rock may have formed.

Possible Materials

- igneous rock samples (9)
- hand lens
- Table 4-1

PLAN THE EXPERIMENT

1. As a group, agree upon and write your hypothesis statement.
2. As a group, list the steps that you need to take to test your hypothesis. Be specific, describing exactly what you will do at each step.
3. Make a list of the materials that you will need to complete your experiment.
4. Design a data table with two columns in your Science Journal. The first column should be labeled *Color,* the second column *Texture.*

Check the Plan

1. Read over your entire experiment to make sure that all steps are in a logical order.
2. Identify any constants and the variables of the experiment.
3. Have you allowed for a control? How will the control be treated?
4. Do you have to run any tests more than once?
5. Will the data be summarized in a graph?
6. *Make sure your teacher approves your plan and that you have included any changes suggested in the plan.*

DO THE EXPERIMENT

1. Carry out the experiment as planned.
2. While the experiment is going on, write down any observations that you make and complete the data table in your Science Journal.

Analyze and Apply

1. Dark-colored igneous rocks are classified as basaltic, light-colored ones as granitic, and intermediate-colored ones as andesitic. Based on this, how would your rocks be classified?
2. Would you **classify** obsidian as basaltic or granitic? Look closely at one edge before you make up your mind.
3. **Decide** what minerals might cause the varying colors found in your rocks.
4. Igneous rocks that form slowly have large crystals. Those that form quickly have small crystals. Did yours form quickly or slowly?

Go Further

What process could form a rock that has large crystals surrounded by small crystals? Place some pumice and some scoria in two separate containers of water. Explain the cause of what you observe.

Figure 4-7

In the Philippines, the eruption of Mt. Pinatubo caused tons of volcanic ash to settle over the land.

The classification of an igneous rock tells you a lot about its origin, formation, and composition. Basalt is an extrusive, basaltic igneous rock. This means that it formed on Earth's surface, where cooling was fast, giving the rock a fine-grained texture. Basalt has a high concentration of iron and magnesium because it formed from basaltic magma.

Igneous rocks are the most abundant type of rock on Earth. In this chapter, you have learned how igneous rocks are formed from the magma of volcanoes. In Chapter 10, you will learn more about volcanoes, and their effects on people. **Figure 4-7** shows one way erupting volcanoes impact people's lives and property.

Section Wrap-up

Review

1. Why do some types of magma and lava form igneous rocks that are dark colored and dense?
2. How do intrusive and extrusive igneous rocks differ?
3. **Think Critically:** How are granite and rhyolite similar? How are they different?

Skill Builder

Interpreting Scientific Illustrations

Suppose you are given a photograph of two igneous rocks. You are told one is an intrusive rock and one is extrusive. By looking only at the photographs, how could you know which is which? If you need help, refer to Interpreting Scientific Illustrations in the **Skill Handbook.**

USING MATH

Four elements make up more than 85% of all the rocks in the world.

Oxygen—46.6%
Silicon—27.7%
Aluminum—8.1%
Iron—5.0%

Make a circle graph of these data.

Metamorphic Rocks 4•3

Origin of Metamorphic Rocks

You wake up, go into the kitchen, and pack a lunch for school. You place a sandwich and a cream-filled cake in the bag. As you leave for school, you decide to throw in an apple. At lunchtime, you open your lunch bag and notice things have changed. Your cream-filled cake doesn't look good anymore. The apple was resting on the cake all morning. The heat in your locker and the pressure from the apple have changed the form of your lunch. Rocks like those in **Figure 4-8** can also be affected by temperature changes and pressure.

Science Words

metamorphic rock
foliated
nonfoliated

Objectives

- Describe conditions that cause metamorphic rocks to form.
- Classify metamorphic rocks as foliated or nonfoliated.

Metamorphic Rocks

Rocks that have changed due to temperature and pressure increases or that undergo changes in composition are **metamorphic rocks.** Metamorphic rocks can be formed from changes in igneous, sedimentary, or other metamorphic rocks. What occurs in Earth to change these rocks?

Figure 4-8

The mineral grains in granite (A) are flattened and aligned when pressure is applied to them. Gneiss (B) is formed. *What other conditions can cause metamorphic rocks to form?*

USING TECHNOLOGY

Solid as a Rock

Rocks are used as building materials because of their durability and appearance. Large slabs of rock, however, are expensive and difficult to work with. Their weight alone requires special handling and strong support.

What if a way were found to reduce the weight, yet keep the strength of rock? A company in Italy has made a material that looks like granite, wears like granite, and in fact, is about 95 percent granite. Granirex is a human-made material made of crushed granite that's glued together. This material weighs much less than granite and ranges in thickness from less than 0.5 cm to almost 1 cm. Granirex can be used on countertops, as floors, to make furniture, and to cover walls. It comes in a variety of colors. In addition to this, Granirex costs about 50 percent less than granite.

Think Critically:

Granite is made of interlocking mineral crystals that have grown together. How does Granirex differ from this?

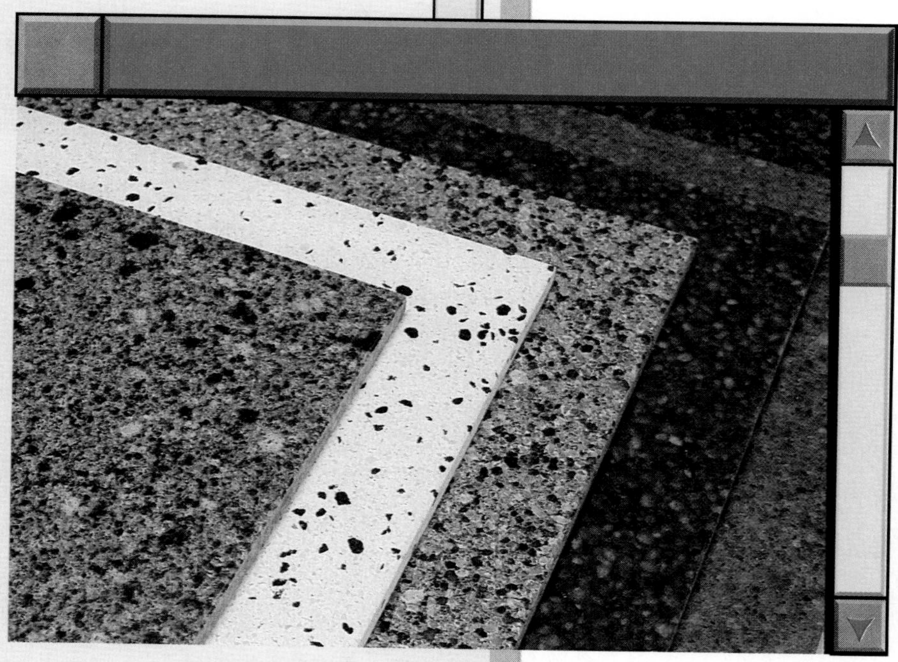

Heat and Pressure

Rocks beneath Earth's surface are under great pressure from overlying rock layers. Once the heat and pressure reach a certain point, the rocks melt and magma forms. But what happens if the heat isn't high enough or the pressure isn't the amount needed to cause melting?

In areas where melting doesn't occur, some mineral grains are flattened like the cake in your lunch bag. Sometimes, minerals exchange atoms with surrounding minerals and new or bigger minerals form.

In Section 4-1, you saw that an igneous rock can be transformed into a metamorphic rock. For example, the igneous rock granite can be changed into the metamorphic rock gneiss (NISE).

Depending upon the amount of pressure applied, one type of rock can change into several different metamorphic rocks. Shale, for example, will change into slate. As more and more pressure

Figure 4-9

The different minerals in shale result in the many different colors of slate. *In your Science Journal, explain why slate is useful as patio and stepping stones, and roofing shingles.*

is applied, the slate can change into phyllite, then schist, and eventually gneiss. Schist can also form when basalt is metamorphosed.

Classification of Metamorphic Rocks

You've learned that metamorphic rocks can be formed from changes in igneous, sedimentary, or other metamorphic rocks. Heat and pressure trigger the changes. The resulting rocks can be classified according to their texture.

Foliated Rocks

When mineral grains flatten and line up in parallel bands, the metamorphic rock has a **foliated** texture. Two examples of foliated rocks are slate and gneiss. Slate forms from the sedimentary rock shale. The minerals in shale are arranged into layers when they're exposed to heat and pressure. As **Figure 4-9** shows, slate is easily separated along these foliation layers. The minerals in slate are so tightly compacted that water can't pass between them.

Gneiss, another foliated rock, forms when granite and other rocks are changed. Quartz, feldspar, mica, and other minerals in granite aren't changed much, but they are rearranged into alternating bands.

Nonfoliated Rocks

In some metamorphic rocks, no banding occurs. The mineral grains change, grow, and rearrange, but they don't form bands. This process produces a **nonfoliated** texture.

MiniLAB

How do metamorphic rocks form?

Procedure

1. From your teacher, obtain samples for four metamorphic rocks and four nonmetamorphic rocks; be sure each of the metamorphic rocks formed from one of the nonmetamorphic rocks.
2. Observe each of the rocks.
3. In your Science Journal, make a list of the characteristics of each.

Analysis

1. Compare your observations of the characteristics of the metamorphic rocks with your observations of the nonmetamorphic rocks.
2. Which of the nonmetamorphic rocks was the "parent" of each of the metamorphic rocks?

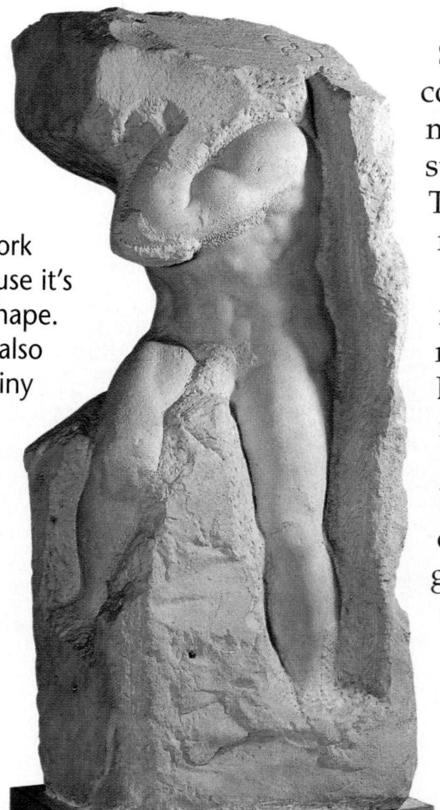

Figure 4-10

Sculptors often work with marble because it's soft and easy to shape. Its calcite crystals also give it a glassy, shiny luster.

Sandstone is a sedimentary rock that's often composed mostly of quartz minerals. When its mineral grains are changed by heat and pressure, the nonfoliated rock quartzite is formed. The only change that occurs is in the size of the mineral grains.

Another nonfoliated metamorphic rock is marble. Marble forms from the sedimentary rock limestone, which is composed of calcite. Look at **Figure 4-10.** The calcite crystals give marble the glassy, shiny luster that makes it a popular material for sculpturing. Usually, marble contains several other minerals besides calcite. For example, hornblende and serpentine give marble a greenish tone, whereas hematite makes it red.

So far, we've traveled through only a portion of the rock cycle. We still haven't observed how sedimentary rocks are formed and how igneous and metamorphic rocks evolve from them. The next section will complete our investigation of the rock cycle.

Section Wrap-up

Review

1. How is the formation of igneous rock different from that of metamorphic rock?
2. How are metamorphic rocks classified? What are the characteristics of rocks in each of these classifications?
3. **Think Critically:** Marble is used to make sculptures. What properties of marble make it useful for this purpose?

Skill Builder

Sequencing

Put the following events in a sequence that could explain how a metamorphic rock might form from an igneous rock. (HINT: Start with *igneous rock forms.*) Use each event just once. If you need help, refer to Sequencing in the **Skill Handbook.**

Events: *sedimentary rock forms, weathering occurs, heat and pressure are applied, igneous rock forms, metamorphic rock forms, erosion occurs, sediments are formed, deposition occurs*

Using Computers

Graphing With **Figure 4-8** as a model, use the graphic mode on a computer to illustrate how applied pressure causes alignment of mineral grains. Be sure to indicate from which direction pressure was applied.

Sedimentary Rocks 4•4

Origin of Sedimentary Rocks

Most of the rocks below Earth's surface are igneous rocks. Igneous rocks are the most common rocks on Earth. But chances are, you've seen more sedimentary rocks than igneous rocks. Seventy-five percent of the rocks at Earth's surface are sedimentary rocks.

Science Words

sedimentary rock
sediment
compaction
cementation

Objectives

- Explain how sedimentary rocks form from sediments.
- Classify sedimentary rocks as detrital, chemical, or organic in origin.

Sedimentary Rocks

Sedimentary rocks form when sediments become pressed or cemented together or when sediments precipitate out of solution. **Sediments** are loose materials such as rock fragments, mineral grains, and bits of plant and animal remains that have been moved by wind, water, ice, or gravity. Minerals that are dissolved in water are also sediments. But where do sediments come from? If you look at the model of the rock cycle, you will see that they come from already-existing rocks that are weathered and eroded.

Weathering is the process that breaks rocks into smaller pieces. **Table 4-2** shows how these pieces are classified by size. The movement of weathered material is called erosion.

Table 4-2

Sediment Sizes

Sediment	Clay	Silt	Sand	Gravel
Size Range	<0.004 mm	0.004-0.06 mm	0.06-2 mm	>2 mm
Examples of rock formed from	Shale	Siltstone	Sandstone	Conglomerate

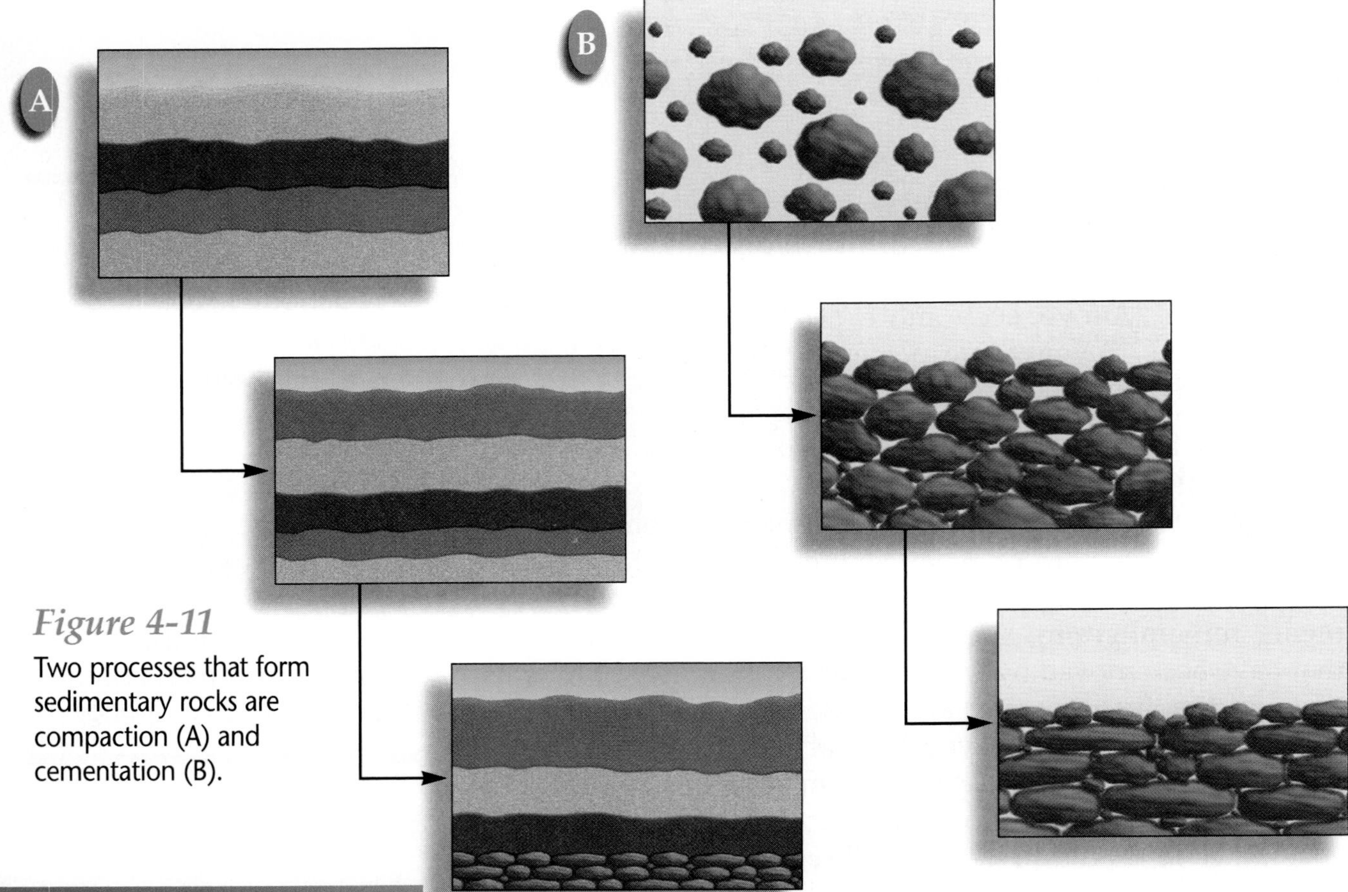

Figure 4-11
Two processes that form sedimentary rocks are compaction (A) and cementation (B).

MiniLAB

What are rocks made of?

Procedure

1. Spread different samples of sediment on a sheet of paper.
2. Use tweezers or a dissecting probe to separate the sediments.
3. Classify the sediments based on size—large, medium, or small.
4. Separate each of these three piles into two more piles based on shape—rounded or angular.

Analyze

1. Compare the different piles of sediments.
2. Describe each pile.
3. What rock is probably made from each type of sediment that you have?

Compaction

Erosion moves sediments to a new location, where they are then deposited. Here, layer upon layer of sediment builds up. Pressure from the upper layers pushes down on the lower layers. If the sediments are small, they can stick together and form solid rock. This process, shown in **Figure 4-11,** is called **compaction.**

You've compacted sediments if you've ever made mud pies. Mud is made of small, clay-sized sediments. They easily stick together under the pressure applied by your hands.

Cementation

If sediments are large, like sand and pebbles, pressure alone can't make them stick together. Large sediments have to be cemented together. **Cementation** occurs in the following way. Water soaks through soil and rock. As it moves, it dissolves minerals in the rock such as calcite, hematite, and limonite. These minerals are natural cements. The solution of water and dissolved minerals moves through open spaces between sediments. The natural cements are deposited around the pieces of sediment, and they stick together. A group of sediments cemented together in this way forms a sedimentary rock.

Sedimentary Rock Layers

Sedimentary rocks often form as layers, like those shown in **Figure 4-12.** The older layers are on the bottom because they were deposited first. Then, more sediments pile up and they, too, become compacted and cemented together to form another layer of rock.

Sedimentary rock layers are a lot like the papers in your locker. The oldest papers are on the bottom, and the ones you get back today will be deposited on top of them. However, if you disturb the papers by searching through them for a pencil at the bottom of the pile, the older ones may come to the top. Sometimes, layers of rock are disturbed by forces within Earth. The layers are overturned, and the oldest are no longer on the bottom.

Classification of Sedimentary Rocks

Sedimentary rocks can be composed of just about anything. Sediments come from weathered and eroded igneous, metamorphic, and sedimentary rocks. Sediments also come from the remains of plants and animals. The composition of a sedimentary rock depends upon the composition of the rocks and living things its sediments came from.

Like igneous and metamorphic rocks, sedimentary rocks are classified by their composition and by the way they formed. Sedimentary rocks are usually classified as detrital, chemical, or organic.

Detrital Sedimentary Rocks

The word *detrital* comes from the Latin word *detritus*, which means "to wear away." Detrital sedimentary rocks are made from the broken fragments of other rocks. These sediments are compacted and cemented together.

Figure 4-12

These sedimentary rock layers formed from the compaction and cementation of many layers of sediments. *In your Science Journal, compare and contrast compaction and cementation.*

Clastic Texture

Detrital sedimentary rocks are often referred to as clastic rocks because of their texture. The word *clastic* comes from the Greek word *klastos,* meaning "broken." Detrital sedimentary rocks have a clastic texture, but as you will see, so do some organic rocks.

CONNECT TO

CHEMISTRY

You may know of large caves or caverns that have formed underground in limestone deposits. ***Decide*** what causes these caverns to form.

Shape and Size of Sediments

Detrital rocks are named according to the shape and size of the sediments. For example, conglomerate and breccia both form from large sediments. If the sediments have been well rounded, the rock is called conglomerate. If the sediments are not rounded and have sharp angles, the rock is called breccia.

The gravel-sized sediments in both conglomerate and breccia may consist of any type of rock or mineral. Often, they are chunks of the minerals quartz or feldspar. They can also be pieces of rocks such as gneiss, granite, or limestone. The cement holding them all together is usually quartz or calcite.

Have you ever looked at the concrete in sidewalks and driveways? Look at **Figure 4-13.** Concrete is made of gravel and sand grains that have been cemented together. The structure is similar to that of naturally occurring conglomerate, but is it considered a rock?

Figure 4-14 shows the sedimentary rock sandstone. Sandstone is formed from smaller particles than conglomerates and breccias. Its sand-sized sediments are usually grains of the minerals quartz and feldspar, but can be just about any mineral. These sand grains can be compacted together if clay particles are also present, or they can be cemented.

Figure 4-13

The concrete making up the sidewalk is similar to naturally occurring conglomerate (above).

Shale is a detrital sedimentary rock that requires no cementation to hold its particles together. Its sediments are clay-sized particles. Clay-sized sediments are compacted together by pressure from overlying layers.

Chemical Sedimentary Rocks

Chemical sedimentary rocks form when minerals are precipitated from a solution or are left behind when a solution evaporates. Think back to our discussion of minerals in Chapter 3. We found that salt is deposited in the bottom of a glass when saltwater solution evaporates. In a similar way, minerals collect when seas or lakes evaporate. The deposits of minerals that precipitate out of solution, or remain after evaporation, form rocks.

Limestone

The mineral calcite is carried in solution in ocean water. When calcite comes out of solution and its many crystals grow together, limestone is formed. Limestone may also contain other minerals and sediments, but it's at least 50 percent calcite. Limestone is usually deposited on sea or ocean floors. Large areas of the United States are underlain by limestone because oceans once covered much of the country for millions of years.

Rock Salt

When lakes and seas evaporate, they often deposit the mineral halite. Halite, mixed with a few other minerals, forms rock salt. Rock-salt deposits range in thickness from a few meters to more than 400 m. People mine these deposits because rock salt is an important resource. It's used in the manufacturing of glass, paper, soap, and dairy products. The halite in rock salt is used as table salt.

USING MATH

It took 300 million years for a layer of plant matter about 0.9 to 2.1 m thick to produce a bed of bituminous coal 0.3 m thick. At this rate, estimate the depth of plant matter that produced a bed of coal 0.6 m thick.

Figure 4-14

This rock formed from sand deposited in layers by desert winds. *Why does this sandstone look so much like desert sand dunes?*

The Geology of Buildings

While on a field trip in the city, Peter and his classmates observed rocks used as building materials. In the city square, Peter noticed flowers arranged in a rock terrace. The rocks were light colored and contained many small fossils.

Next, the class entered an historical district. Peter saw a building made from light-pink rocks with small crystals of quartz. Continuing down the street, he observed another building with columns. This one was constructed of a light-colored rock containing large mineral grains. Further down, he saw a wooden building with a roof made of dark, layered tiles.

Solve the Problem

1. **Why were so many different rocks used in the buildings?**
2. **What property of rocks was shown by the dark, layered tiles?**

Think Critically:

You have learned much in this chapter about the different properties of minerals and rocks.

1. Based on this knowledge, name the four different types of rock that Peter observed.
2. What rocklike materials do people make and use for buildings and other structures?

Organic Sedimentary Rocks

When rocks form from the remains of once-living things, they are organic sedimentary rocks. One of the most common organic sedimentary rocks is fossil-rich limestone. Like chemical limestone, fossil-rich limestone is made of the mineral calcite. But fossil-rich limestone also contains remains of once-living ocean animals instead of just calcite that has precipitated from ocean water.

Useful Sedimentary Rocks

Animals such as mussels, corals, and snails make their shells from the mineral calcite. When they die, their shells accumulate on the ocean floor. These calcite shells are compacted and cemented together, and fossil-rich limestone is formed. If the shell fragments are relatively large, the rock is called coquina (koh KEE nuh). If the shells are microscopic, the rock is called chalk. Look at **Figure 4-15.** When your teachers use naturally occurring chalk to write with, they're actually crushing and smearing the calcite shells of once-living ocean animals.

Another useful sedimentary rock is coal. Coal forms when pieces of dead plants are buried under other sediments in swamps. These plant materials are chemically changed by microorganisms. The resulting sediments are compacted over millions of years to form coal. In Section 4-5, you will learn that coal is used as a source of energy.

Figure 4-15

The White Cliffs of Dover, England, are composed mostly of chalk. *What is chalk made of?*

Another Look at the Rock Cycle

You have seen that the rock cycle has no beginning and no end. Rocks are continually changing from one form to another. Sediments come from rocks and other objects that have been broken apart. Even the magma that forms igneous rocks comes from the melting of rocks that already exist.

All of the rocks that you've learned about in this chapter formed through the processes of the rock cycle. And all of the rocks around you, including those used to build houses and monuments, are part of the rock cycle. They are all changing. The rock cycle is a continuous, dynamic process.

Section Wrap-up

Review

1. Where do sediments come from?
2. List chemical sedimentary rocks that are essential to your health or used to make life more convenient. How is each used?
3. **Think Critically:** Use the rock cycle to explain how pieces of granite and slate could be found in the same piece of conglomerate.

Skill Builder

Interpreting Data

You are told that a detrital sedimentary rock is composed of sediments of the following sizes: gravel, sand, and silt. The larger sediments are surrounded by the smaller sediments, which are cemented together by quartz. What is the name of this rock? If you need help, refer to Interpreting Data in the **Skill Handbook.**

USING MATH

Sediment sizes are presented in Table 4-2. How many times larger than clay are the largest grains of silt and sand?

SCIENCE & SOCIETY

TECHNOLOGY: 4•5 Burning Waste Coal

Science Words

waste coal
cogeneration

Objectives

- Realize that new technologies are enabling companies of today to solve problems caused by mining operations of the past.
- Describe the process of cogeneration, and show how it is beneficial.

Cogeneration from Waste Coal

Recall from Section 4-4 that coal is an organic sedimentary rock. It has been and still is economically important to the technological advancement of the United States. It supplies fuel for electricity. Although mines of today are technologically advanced and the companies involved work within regulations to minimize the effects of mining on the environment, this has not always been so. The effects of old mining can be seen in many places in the United States. In some areas of the country, large piles of **waste coal,** like that shown in **Figure 4-16,** lie near coal mines that have been abandoned for years.

Burning Waste Coal

Waste coal is a by-product of coal mining. It is poor-quality coal that could not be used at the time it was mined and was usually piled up on one side of the mine. The piles of waste coal not only are unsightly, but because of high sulfur content, they also can generate acid runoff when rainwater flows through them.

With processes available today, this coal can be burned without creating air pollution. The coal is trucked to a burning facility where it is pulverized and mixed with pulverized limestone. The limestone removes more than 90 percent of sulfur dioxide emissions. This surpasses current government-mandated levels. Compliance with pollution-control laws is regulated by monitoring air-quality levels.

Figure 4-16

Waste coal piles can still be found in many areas of the country. *Explain the process by which this coal can be burned without polluting the air.*

Once the coal-limestone mixture has been burned, the residue ash is trucked back to its original pile. Mixing with limestone has changed the acidic waste coal to alkaline ash. It creates no acid mine drainage and works to alleviate the acidic problems caused by the pile of waste coal. The ash has a high affinity for water, and when it is hydrated, it changes to a low-grade cement. The resulting concrete mound is buried and covered by layers of topsoil. The topsoil is seeded to establish vegetation. The lay of the land has been returned to what it was, and the mound can be used as a location for a park or playground.

Figure 4-17

Burning waste coal generates both thermal and electrical energy.

Cogeneration

Some companies use waste coal to produce and export electricity. In addition to the electricity produced by burning waste coal, low-pressure steam from the process can be exported to meet community heating needs.

When power is generated in a power plant such as the one shown in **Figure 4-17,** both electrical and thermal energy are produced. The electrical power is used, but often the remaining power is lost as wasted thermal energy. In the process of **cogeneration,** both the electrical and thermal energy are used by the plant producing it.

Section Wrap-up

Review

1. Even though the ash produced by the burning of waste coal is returned to the waste site, why is this process considered good for the environment?
2. How does the process of cogeneration benefit not only the company using it, but others in the community as well?

Explore the Technology

Although the technology involved in burning waste coal and disposing of the ash is relatively new, there already are several other possible uses for the idea. Coal seams in surface mines vary in quality. Some have a greater potential for acid mine drainage. In order to work these surface mines, the ash from burning waste coal could be used in the mine to alleviate the acidity.

Activity 4-2

Sedimentary Rocks

Sedimentary rocks are formed by the compaction and cementation of sediment. Because sediment is found in all shapes and sizes, do you think these characteristics could be used to classify detrital sedimentary rocks? Sedimentary rocks can also be classified as chemical or organic.

Problem

How are rock characteristics used to classify rocks as detrital, chemical, or organic?

Materials 

- unknown sedimentary rock samples
- marking pen
- 5 percent hydrochloric acid (HCl)
- dropper
- goggles
- hand lens
- paper towels
- water

Procedure

1. In your Science Journal, make a Data and Observations chart similar to the one shown below.
2. Determine the types of sediments in each sample. Using **Table 4-2**, classify the sediments in the detrital rocks as gravel, sand, silt, or clay.
3. Put a few drops of HCl on each rock sample. Bubbling on a rock indicates the presence of carbonate minerals.

   CAUTION: *HCl is an acid and can cause burns. Wear goggles. Rinse spills with water. Wash hands afterwards.*
4. Look for fossils and describe them if any are present.
5. Determine whether each sample has a clastic or nonclastic texture.
6. Classify your samples as detrital, chemical, or organic. Identify each rock sample.

Analyze

1. Why did you test the rocks with hydrochloric acid? What minerals react with hydrochloric acid?
2. What is needed in order for sedimentary rocks to form from fragments?
3. The mineral halite forms by evaporation. Would you classify halite as a detrital, a chemical, or an organic rock?

Conclude and Apply

4. **Determine** how sedimentary rocks with a clastic texture differ from rocks with a nonclastic texture.
5. **Explain** how you can classify sedimentary rocks.

Data and Observations

Sample	Observations	Minerals or Fossils Present	Sediment Size	Detrital, Chemical, or Organic	Rock Name
A					
B					
C					
D					
E					

Chapter 4 Review

Summary

4-1: The Rock Cycle

1. A rock is a mixture of one or more minerals, mineraloids, glass, or organic matter.
2. The rock cycle includes all processes by which rocks form.

4-2: Igneous Rocks

1. Magma is molten material that hardens to form igneous rocks.
2. Intrusive igneous rocks form when magma cools below Earth's surface. Extrusive igneous rocks form when lava cools at or near Earth's surface.
3. Basaltic rocks are dense, heavy, dark-colored rocks. Granitic rocks are light colored and less dense than basalts. Andesitic rocks are intermediate between basaltics and granitics.

4-3: Metamorphic Rocks

1. Increases in heat and pressure can cause metamorphic rocks to form.
2. Slate and gneiss are classified as foliated, or banded, metamorphic rocks. When banding is not visible, as in quartzite, metamorphic rocks are classified as nonfoliated.

4-4: Sedimentary Rocks

1. Sedimentary rocks form when fragments of rocks, minerals, and/or organic materials are compacted and cemented together.
2. Detrital sedimentary rocks form when fragments of other rocks are compacted and/or cemented together. Chemical sedimentary rocks precipitate out of solution or are left behind by evaporation. Organic sedimentary rocks are made mostly of once-living organisms.

4-5: Science and Society: Burning Waste Coal

1. Pulverized waste coal is mixed with pulverized limestone and burned. The ash produced is alkaline and can be used to alleviate mine acid drainage.
2. Cogeneration occurs when both electrical and thermal energy produced by a power plant are used.

Key Science Words

a. basaltic
b. cementation
c. cogeneration
d. compaction
e. extrusive
f. foliated
g. granitic
h. igneous rock
i. intrusive
j. lava
k. metamorphic rock
l. nonfoliated
m. rock
n. rock cycle
o. sediment
p. sedimentary rock
q. waste coal

Reviewing Vocabulary

Match each phrase with the correct term from the list of Key Science Words.

1. mixture of one or more minerals, mineraloids, glass, or organic matter
2. processes that form and change rocks
3. molten material at Earth's surface
4. igneous rocks that form when lava cools
5. a rock formed by heat and pressure
6. quartzite has this kind of texture
7. fragments of rocks, minerals, plants, and animals
8. sediments are pressed together
9. sediments become glued together
10. poor-quality coal piled up near old mines

Chapter 4 Review

Checking Concepts

Choose the word or phrase that completes the sentence or answers the question.

1. Magma reaches Earth's surface because it is _______ than the surrounding rocks.
   a. more dense c. cooler
   b. more massive d. less dense
2. Igneous rocks form from _______ .
   a. sediments c. gravel
   b. mud d. magma
3. _______ rocks have large mineral grains.
   a. Intrusive c. Obsidian
   b. Extrusive d. Basaltic
4. During metamorphism of shale into slate, minerals _______.
   a. partly melt c. grow larger
   b. become new minerals d. align into layers
5. Gneiss is a(n) _______ rock.
   a. foliated c. intrusive
   b. nonfoliated d. extrusive
6. _________ is a rock made of large, angular pieces of sediments.
   a. Conglomerate c. Limestone
   b. Breccia d. Chalk
7. Conglomerates form in much the same way as shales, sandstones, and _______.
   a. limestones c. breccias
   b. evaporites d. precipitates
8. _______ occurs when a power plant uses both the electrical and thermal power produced by the plant.
   a. Wasted thermal energy
   b. Inefficiency
   c. Cogeneration
   d. Loss of power
9. _______ forms when water carries sulfur from coal in solution.
   a. An open pit c. A surface mine
   b. Soil d. Acidic water
10. Which of these is not an organic rock?
   a. shale c. chalk
   b. coal d. coquina

Understanding Concepts

Answer the following questions in your Science Journal using complete sentences.

11. Explain why the rock cycle has no beginning and no end.
12. Compare magma and lava.
13. Compare and contrast detrital rocks with organic and chemical rocks.
14. Explain why waste coal can be harmful to the environment.
15. Explain the process of cogeneration.

Thinking Critically

16. Granite, pumice, and scoria are igneous rocks. Why doesn't granite have airholes like the other two?
17. Why are only a few fossils found in marble?
18. Explain why coquina is classified as an organic rock with a clastic texture.
19. What is true about the texture of all detrital sedimentary rocks?
20. Why are granitic igneous rocks light in color?

Chapter 4 Review

Developing Skills

If you need help, refer to the **Skill Handbook.**

21. **Concept Mapping:** Copy and complete the concept map shown below. Add rectangles and connecting lines so you can include examples of each classification of rock.

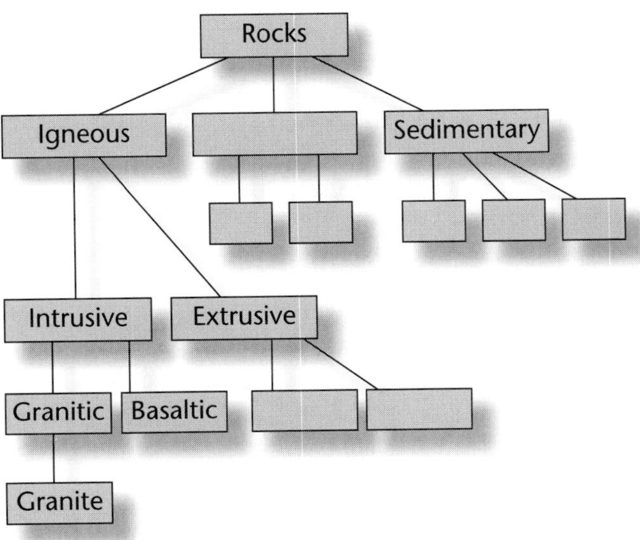

22. **Comparing and Contrasting:** Compare and contrast basaltic and granitic magmas.
23. **Hypothesizing:** A geologist found a sequence of rocks in which 200-million-year-old shales were lying on top of 100-million-year-old sandstones. Hypothesize how this could happen.
24. **Measuring in SI:** The rock shown on the opposite page is a limestone that contains fossils. Find the average length of the fossils.
25. **Recognizing Cause and Effect:** Explain the cause and effects of pressure and temperature on shale.

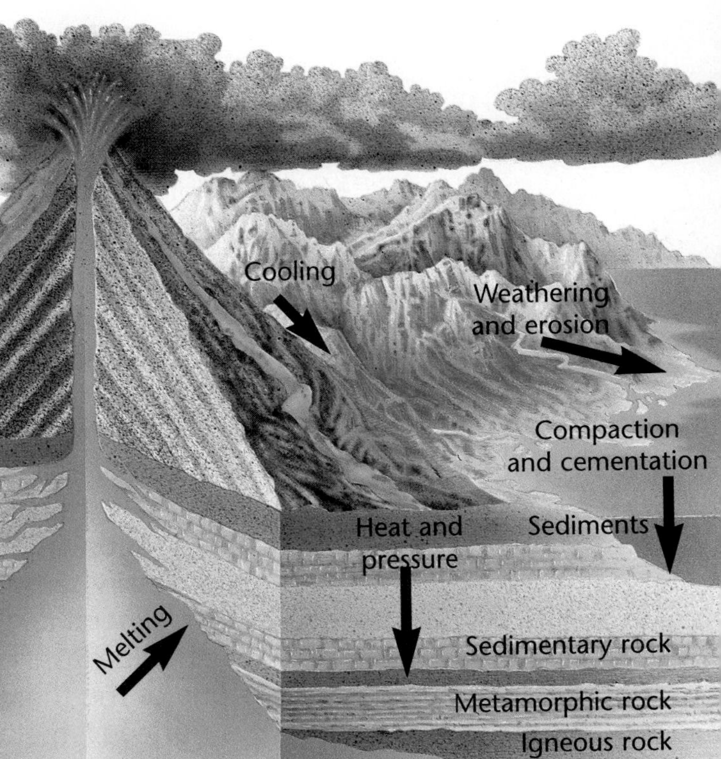

Performance Assessment

1. **Formulating a Hypothesis:** Based on what you learned in Activity 4-1, what would you expect to see if an intrusive igneous rock that contained even more iron and magnesium than basaltic rocks were placed on the table in front of you? Describe what you would expect to see as far as crystal size and color of the rock.
2. **Making and Using a Classification System:** In Activities 4-1 and 4-2, you learned how to classify igneous rocks and sedimentary rocks. Devise and carry out an activity to classify metamorphic rocks.
3. **Model:** Using skills from the MiniLAB on page 87, use sand, gravel, mud, clay, and a salt solution to make at least three metamorphic and three sedimentary "rocks." Label your rocks and explain how each forms in nature.

UNIT PROJECT 1

Rock and Roll

You take a walk along a well-traveled path and find a shiny, black rock in the path. You reach down and pick it up. It's interesting, but what will you do with it? You might wonder if it's worth anything.

Earth's surface is covered with treasures. Some are valuable, but the value of others only becomes apparent when they're viewed as part of a collection of Earth's rocks and minerals.

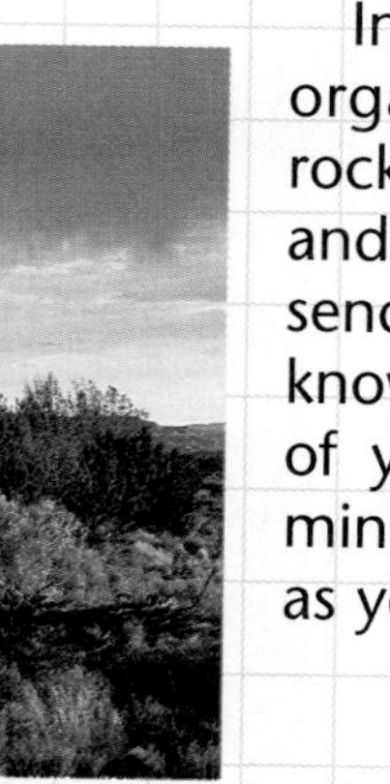

In this project you'll collect and organize samples of minerals and rocks from around your school, home, and neighborhood. You could also send requests to people you might know around the country. Make it part of your project to collect rocks and minerals from as many different places as you can.

Procedures

1. As you collect samples, think of a way to separate them into groups. You might separate your samples based on where they originated.
2. Determine if you're dealing with single minerals or rocks containing a mixture of minerals. Your collection of rocks and minerals should have two basic subdivisions. One should classify samples based on their mineral content. The other should classify rocks into one of the three major types (igneous, metamorphic, or sedimentary). Further subdivide your rock samples according to the classification system you learned in Chapter 4. If you find that most of your samples are of one type, think of ways to increase the number of samples of other types. Your goal is to build a well-rounded, complete rock and mineral collection.
3. Make a wood or cardboard box in which you can display each of

your samples. Think of a way to protect your samples so they can be displayed for other students to enjoy. When you make/display your box, allow room near each sample to display information obtained during research in the library.

4. Create labels for each of your samples that include answers to the following questions. If necessary, go to the library and research the answers.
   - Where was each sample found?
   - What type of geologic feature is associated with the location of each sample? For example, the rocks may be igneous rocks from a lava flow.
   - What is the composition of each sample? Identify the primary mineral contained in each mineral sample and the more prominent minerals found in each of your rock samples.
   - Once you determine the mineral content of each sample, find out how people use them.
   - To which crystal systems do your mineral samples and the minerals in your rocks belong?
5. Based on the information you've collected, write a short description of each sample in your collection. Include information about each of your samples that describes its origin, composition, and use. Describe everyday uses that people have for the rocks and minerals you've collected.

Going Further

Each year new and rare minerals are discovered. Many of these minerals are found only in one or two locations. Even so, they are classified by their crystal structure. Based on the research you did on the crystal systems involved in your samples, classify each of your samples by its crystal structure. Near where you've mounted each sample, draw a diagram of the crystal system to which each one belongs.

UNIT 2

Unit Contents

The Changing Surface of Earth

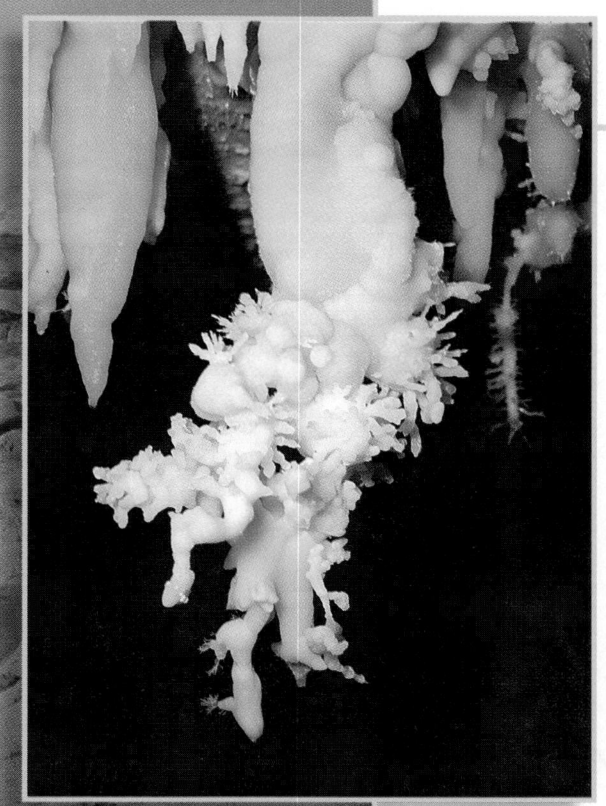

What's Happening Here?

Earth's surface is constantly changing. Some processes of change alter entire landscapes. Wind, for example, can blow sediments that slowly wear away at Earth's surface, leaving behind beautiful pillars of sculpted rock spread across a vast landscape. Huge boulders of hard rock may balance delicately upon thin shafts of worn, softer rock. Some processes may alter Earth's surface at a microscopic level. Gypsum flowers may form beneath Earth's surface in caves as minerals build up molecules at a time. As you study this unit, you'll learn about the many processes that change the shape of our world.

Science Journal

Investigate the shape of Earth's surface where you live. How do you think the surface may have changed over time? In your Science Journal, write about what it may be like to go far back in time to the exact same spot. What would Earth's surface look like? What might it look like far in the future?

Previewing the Chapter

Views of Earth

Did you know that we are able to use cameras attached to satellites to take pictures of Earth from space? This method of studying our planet is a form of a technology called remote sensing. Photographs taken from these cameras show us many details of Earth's surface. Look at the picture again. Can you pick out the rivers and flat areas in this photograph? How can you recognize the higher elevations?

EXPLORE ACTIVITY

Use a globe or world map.

1. Find the Andes Mountains on the globe or map.
2. Locate the Amazon, the Ganges, and the Mississippi Rivers.
3. Locate the Indian Ocean, the Sea of Japan, and the Baltic Sea.
4. Now, find the continents of Australia, South America, and North America.
5. Locate your own country.

Observe: Choose one country on the globe or map and describe its major physical features in your Science Journal.

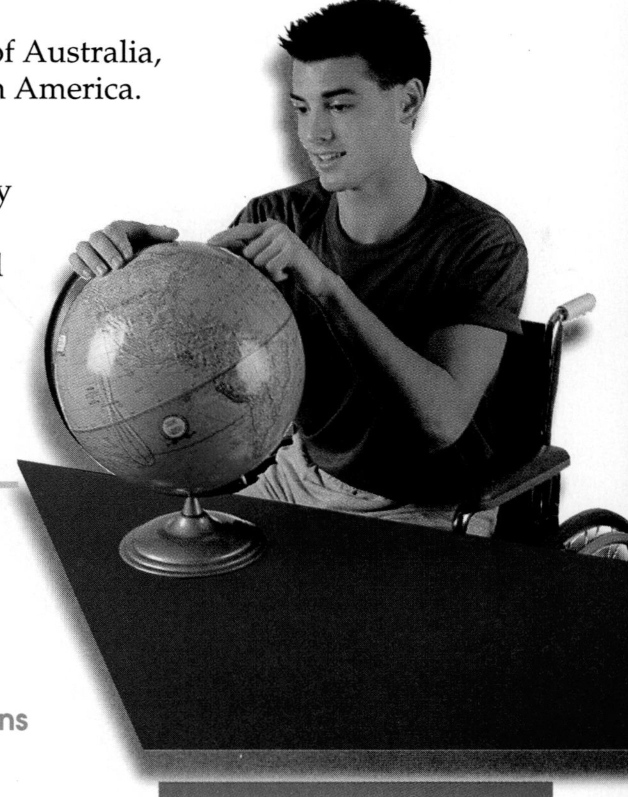

Previewing Science Skills

- In the **Skill Builders,** you will map concepts, interpret scientific illustrations, and measure in SI.
- In the **Activities,** you will observe, measure, use numbers, and interpret data.
- In the **MiniLABs,** you will interpret scientific illustrations and make a model.

5•1 Landforms

Science Words

plain
plateau
folded mountain
upwarped mountain
fault-block mountain
volcanic mountain

Objectives

- Differentiate between plains and plateaus.
- Compare and contrast folded, upwarped, fault-block, and volcanic mountains.

Plains

There are a lot of interesting landforms around the world. A landform is a feature that makes up the shape of the land at Earth's surface. **Figure 5-1** shows the three basic types of landforms: plains, plateaus, and mountains.

We all know what mountains are. In our minds, we can see tall peaks reaching towards the sky. But what do you think of when you hear the word *plains?* You might think of endless flat fields of wheat or grass. That would be accurate, because many plains are used to grow crops. **Plains** are large, relatively flat areas. Plains found near the ocean are called coastal plains. Flat, grassy areas used to grow crops or for grazing are also plains. Together, these two types of plains make up one-half of all the land in the United States.

Coastal Plains

Coastal plains are broad areas along the ocean's shore. They are often called lowlands because of their low elevations. Elevation refers to distance above or below sea level. As you might guess, sea level has zero elevation. The Atlantic Coastal Plain is a good example of this type of landform. It stretches along the East Coast of the United States. This area is characterized by low rolling hills, swamps, and marshes. A marsh is grassy wetland, usually flooded with water.

Figure 5-1

Three basic types of landforms are plains, plateaus, and mountains.

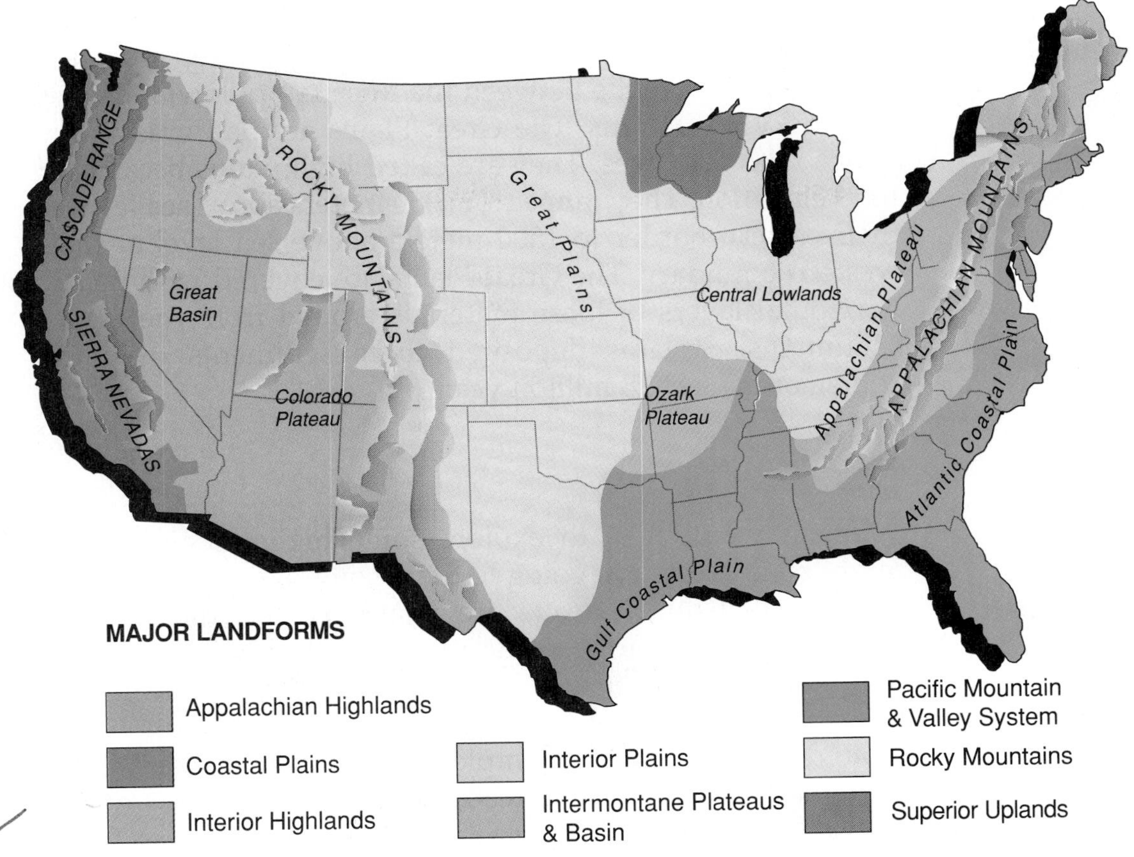

Figure 5-2

The United States is made up of eight major landforms. *In your Science Journal, describe the region that you live in.*

If you hiked along the Atlantic Coastal Plain, you would realize it isn't perfectly flat. Many low hills and valleys have been carved by rivers. What do you suppose caused the Atlantic Coastal Plain to form? It actually began forming under water about 70 million years ago from sediments composed of marine organisms that fell to the ancient ocean floor. When sea level dropped, the plain was exposed.

Another example of this landform is the Gulf Coastal Plain. It includes the lowlands in the southern United States that surround the Gulf of Mexico. Much of this plain was formed from sediments deposited by the Mississippi River as it entered the Gulf of Mexico.

The elevation of Denver, Colorado, is about 1624.5 meters above sea level. The elevation of New Orleans, Louisiana, is 1626 meters lower than Denver's. Find the elevation of New Orleans.

Interior Plains

A large portion of the center of the United States is called the interior plains. The interior plains of the United States are shown in **Figure 5-2.** They extend from the Appalachian Mountains in the east to the Rocky Mountains in the west, to the Gulf Coastal Plain in the south. They include the rolling hills of the Great Lakes area and the Central Lowlands around the Missouri and Mississippi Rivers.

MiniLAB

What would a profile of the major landforms of this country look like?

Make a profile, or side view, of the United States.

Procedure

1. Place the bottom edge of a piece of paper across the middle of **Figure 5-2**, extending from the west coast to the east coast.
2. Mark where different landforms are located along this edge.
3. Use Appendix H and the descriptions of the landforms in Section 5-1 to help you draw the profile.

Analysis

1. Describe how your profile changed shape as you moved from west to east.
2. Describe how the shape of your profile would be different if you moved from north to south.

A large portion of the interior plains is referred to as the Great Plains. They lie between the Mississippi lowlands and the Rocky Mountains. The Great Plains are flat, grassy, dry plains with few trees. They are called high plains because of their elevation. They range from 350 meters above sea level at their eastern border to 1500 meters above sea level at their western boundary. The Great Plains are covered with nearly horizontal layers of loose materials eroded from the Rocky Mountains. Streams deposited these sediments over the course of the last 28 million years.

Plateaus

If you would like to explore some higher regions, you might be interested in going to the second basic type of landform—a plateau. **Plateaus** are relatively flat, raised areas of land. They are areas made up of nearly horizontal rocks that have been uplifted by forces within Earth. Plateaus are different from plains in that they rise steeply from the land around them. A good example of a plateau in the United States is the Colorado Plateau, which lies just west of the Rocky Mountains. As **Figure 5-3** shows, the Colorado River has cut deeply into the rock layers of the plateau, forming the Grand Canyon. Because the Colorado Plateau is located in a very dry region, only a few permanent rivers have developed on its surface. If you hiked around on this plateau, you would see a desert landscape.

Figure 5-3 Rivers cut deep into the Colorado Plateau. *How are plateaus different from plains?*

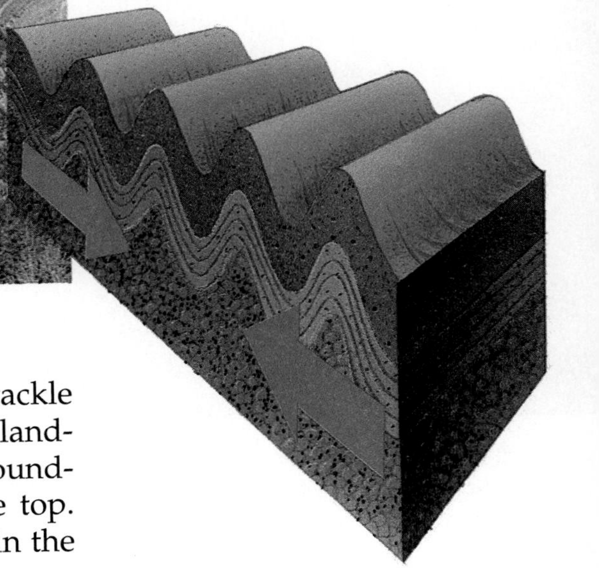

Figure 5-4

Folded mountains form when rock layers are squeezed from opposite sides.

Mountains

Plains and plateaus are relatively flat. If you want to tackle a steep rock face, you must go to the third basic type of landform—a mountain. Mountains rise high above the surrounding land, often providing a spectacular view from the top. The world's highest mountain peak is Mount Everest, in the Himalayan Mountains. It is more than 8800 meters above sea level. By contrast, mountain peaks in the United States reach just over 6000 meters. Mountains vary greatly in size and in how they are formed. The four main types of mountains are folded, upwarped, fault-block, and volcanic.

Folded Mountains

The first mountains we will investigate are folded mountains. If you ever travel through a road cut in the Appalachian Mountains, you'll see rock layers that are folded like the ones in **Figure 5-4.** Folded rock layers look like a rug that has been pushed up against a wall. What do you think caused this to happen?

Tremendous forces inside Earth force rock layers together. When rock layers are squeezed from opposite sides, they buckle and fold into **folded mountains.** You'll learn more about the forces that create mountains in Chapters 9, 10, and 11.

The Appalachian Mountains are folded mountains that formed in this way 300 to 250 million years ago. They are the oldest mountains in North America and also one of the longest ranges, stretching from Quebec, Canada, south to Alabama. At one time, the Appalachians were higher than the Rocky Mountains, but weathering and erosion have worn them down to less than 2000 meters above sea level.

Science Journal

The Himalayan Mountains are still growing taller. In your Science Journal, describe why you think this is happening. Then, find out if you were right. Record your findings.

Figure 5-5

Upwarped mountains form when crust is pushed up by forces inside Earth.

Upwarped Mountains

The southern Rocky Mountains in Colorado and New Mexico, the Black Hills in South Dakota, and the Adirondak Mountains in New York are **upwarped mountains.** These mountains were formed when crust was pushed up by forces inside Earth. Over time, the sedimentary rock on top of the crust eroded, leaving behind the igneous and metamorphic rock underneath. These igneous and metamorphic rocks then eroded to form sharp peaks and ridges, as shown in **Figure 5-5.**

Fault-Block Mountains

The Grand Teton Mountains of Wyoming and the Sierra Nevada Mountains in California formed in yet another way. **Fault-block mountains** are made of huge tilted blocks of rocks that are separated from surrounding rock by faults. A fault is a large crack in rocks along which there is movement. As **Figure 5-6** shows, when these mountains formed, one block was tilted and pushed up. The other block was pushed down. If you decide to go to the Tetons or to the Sierra Nevadas, you'll see the sharp, jagged peaks that are characteristic of fault-block mountains.

Figure 5-6

Fault-block mountains are created when faults occur. Some rock blocks move up, while others move down. *What types of peaks are characteristic of these mountains?*

Figure 5-7

Volcanic mountains form when molten material oozes from Earth's crust and forms a cone-shaped structure.

Volcanic Mountains

Mount St. Helens in Washington and Mauna Loa in Hawaii are two of many volcanic mountains in the United States. **Volcanic mountains** like the one shown in **Figure 5-7** begin when molten material reaches the surface through a weak area of the crust. The materials pile up, one layer on top of another, until a cone-shaped structure forms. The Hawaiian Islands are just the peaks of huge volcanoes that stick out above the water.

Plains, plateaus, and mountains offer a wide variety of landforms to explore. They range from low coastal plains and high desert plateaus to mountain ranges thousands of meters high.

CONNECT TO

CHEMISTRY

The composition of erupting materials helps determine a volcano's shape. High amounts of water and silica create violently erupting volcanoes with steep sides. Find a picture of Mauna Loa and ***observe*** its shape. Do its eruptions usually contain high amounts of water and silica?

Section Wrap-up

Review

1. Describe the eight major landforms in the United States.
2. What causes some mountains to be folded and others to be upwarped?
3. **Think Critically:** If you wanted to know whether a particular mountain was formed by a fault, what would you look for?

Skill Builder

Concept Mapping

Make an events chain concept map to explain how upwarped mountains form. If you need help, refer to Concept Mapping in the **Skill Handbook.**

Using Computers

Spreadsheet Design a spreadsheet that compares the origin and features of folded, upwarped, fault-block, and volcanic mountains.

5•2 Viewpoints

Science Words

equator
latitude
prime meridian
longitude
International Date Line

Objectives

- Differentiate between latitude and longitude.
- Describe how latitude and longitude are used to identify locations.
- Calculate the time and date in different time zones.

Latitude and Longitude

If you are going to explore landforms, you might want to learn how to find locations on Earth. If you wanted to go to the Hawaiian Islands, how would you describe their location? You might say that they are located in the Pacific Ocean. That's correct, but there is a more precise way to locate places on Earth. You could use lines of latitude and longitude. These lines form an imaginary grid system that enables points on Earth to be located exactly.

Latitude

First, look at **Figure 5-8.** The **equator** is an imaginary line that circles Earth exactly halfway between the North and South Poles. The equator separates Earth into two equal halves, called the northern hemisphere and the southern

Figure 5-8

The degree value used for latitude is the measurement of the imaginary angle created between the equator, the center of Earth, and that location (A). Likewise, longitude is the measurement of the angle created between the prime meridian, the center of Earth, and that location (B).

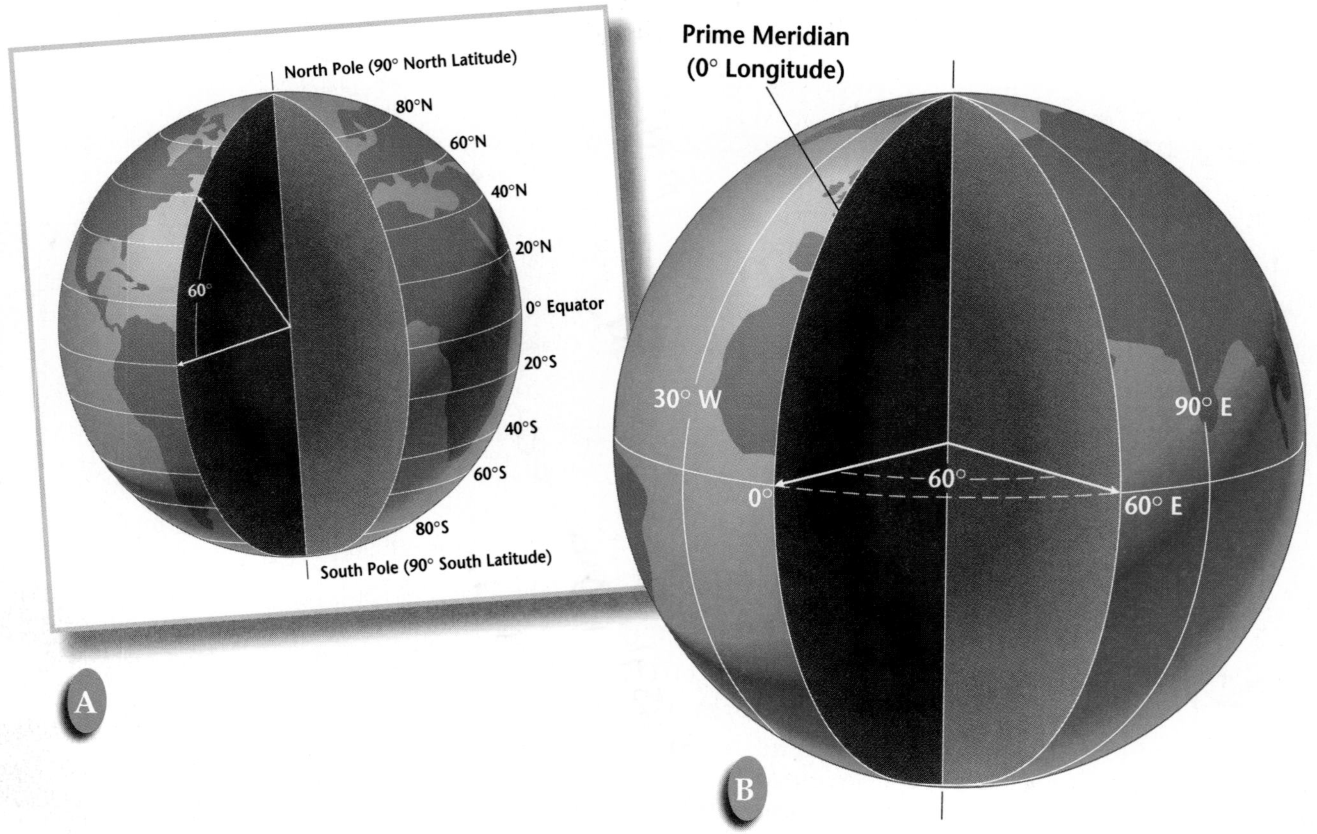

hemisphere. The lines running parallel to the equator are called lines of latitude, or parallels. **Latitude** refers to distance in degrees either north or south of the equator. Just as other parallel lines do not intersect, lines of latitude do not intersect.

The equator is numbered 0° latitude. The poles are each numbered 90°. Therefore, latitude is measured from 0° at the equator to 90° at the poles. Locations north of the equator are referred to by degrees north latitude. Locations south of the equator are referred to by degrees south latitude. For example, Minneapolis, Minnesota, is located at 45° north latitude.

Longitude

Latitude lines are used for locations north and south of the equator, but what about locations in east and west directions? These vertical lines, seen in **Figure 5-8,** have two names—meridians and lines of longitude. Just as the equator is used as a reference point for north/south grid lines, there's a reference point for east/west grid lines. This reference point is the **prime meridian.** This imaginary line represents 0° longitude. In 1884, astronomers decided the prime meridian should go through the Greenwich (GREN itch) Observatory near London. **Longitude** refers to distances in degrees east or west of the prime meridian. Points west of the prime meridian have west longitude measured from 0° to 180°, while points east of the prime meridian have east longitude, also measured from 0° to 180°.

The prime meridian does not circle Earth as the equator does. Rather, it runs from the North Pole through Greenwich, England, to the South Pole. The line of longitude on the opposite side of Earth from the prime meridian where east lines of longitude meet west lines of longitude is the 180° meridian. This line is also known as the International Date Line.

Using latitude and longitude, you can locate Hawaii more accurately, as illustrated in **Figure 5-9.** Hawaii is located at 20° north latitude and about 155° west longitude, or 20°N, 155°W. Note that latitude comes first when we discuss the coordinates (latitude and longitude) of a particular location. Read on to see how longitude lines are used to tell time on Earth.

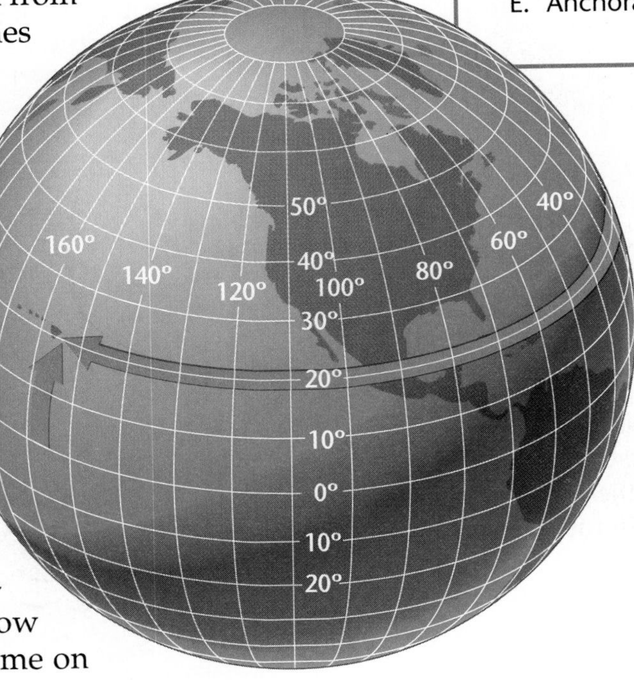

Figure 5-9

Hawaii is located at about 20°N, 155°W. *In your Science Journal, compare and contrast latitude and longitude.*

MiniLAB

How are latitude and longitude used to locate places on a map?

Procedure

1. Find the equator and prime meridian on a world map.
2. Move your finger to latitudes north of the equator, then south of the equator. Move your finger to longitudes west of the prime meridian, then east of the prime meridian.

Analysis

1. Identify the cities that have the following coordinates:
   A. 56°N, 38°E
   B. 34°S, 18°E
   C. 23°N, 82°W
   D. 13°N, 101°E
   E. 38°N, 9°W
2. Determine the latitude and longitude of the following cities:
   A. London, England
   B. Melbourne, Australia
   C. Buenos Aires, Argentina
   D. Paris, France
   E. Anchorage, Alaska

CONNECT TO

LIFE SCIENCE

Many organisms use biological clocks to "tell time." The timing mechanisms are triggered by changes in the environment and changes within the organism. ***Infer*** what environmental change causes trees to become dormant in the fall.

Earth Time

What time is it right now? That depends on where you are on Earth. We keep track of time by measuring Earth's movement in relation to the sun. Earth rotates one full turn every 24 hours. When one half of Earth is facing the sun, the other half is facing away from it. For the half facing the sunlight, it is daytime. For the half in darkness, it is nighttime. And because Earth is constantly spinning, time is always changing.

Time Zones

How can you know what time it is at any particular location on Earth? Earth is divided into time zones. Because Earth takes 24 hours to rotate once, it is divided into 24 time zones, each one hour different. Each time zone is 15 degrees wide. There are six different time zones in the United States. Because Earth is rotating, the sun rises earlier in Atlanta, Georgia, than it does in Los Angeles, California. In Atlanta at 7:00 A.M. you would be on your way to school, like the students in **Figure 5-10A.** But in Los Angeles, it's 4:00 A.M. and you would be asleep, like the student in **Figure 5-10B.** If you lived in Los Angeles, and were in your first or second period class, a student in Atlanta would be at lunch.

As you can see in **Figure 5-11,** the time zones do not strictly follow lines of longitude. Time zone boundaries have been adjusted in local areas. For example, if a city were split by a time

Figure 5-10

Earth is divided into 24 time zones, each one hour different. There are six time zones in the United States.

A Atlanta, Georgia, lies in the Eastern time zone. Students there would be on their way to school at 7:00 A.M. in the morning.

B But a student in Los Angeles, California, which lies in the Pacific time zone, would still be fast asleep. *What time would it be in Los Angeles when the students in Atlanta returned home at 3:00 P.M.?*

zone boundary, great confusion would result. In such a situation, the time zone boundary is moved to outside of the city.

Calendar Dates

We all know that a day ends and the next day begins at 12 midnight. If it is 11:59 P.M. Tuesday, two minutes later it is 12:01 A.M. Wednesday. Every time zone experiences this transition from one day to the next. The calendar advances to the next day in each time zone at midnight.

You gain or lose time each time you travel through a time zone until at some point you gain or lose a whole day. The **International Date Line** is the 180 degree meridian that is the transition line for calendar days. If you were traveling west across the International Date Line, you would advance your calendar one day. If you were traveling east you would move your calendar back one day.

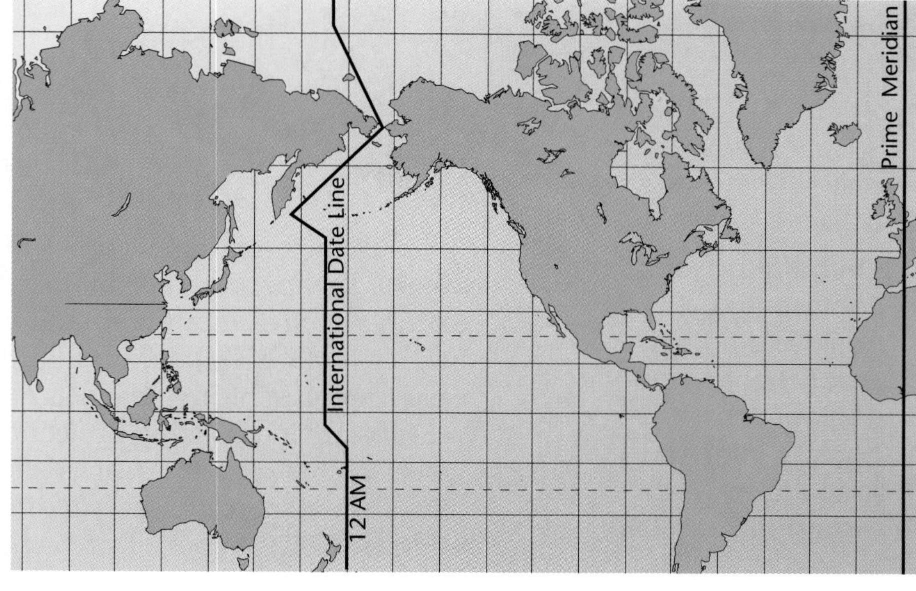

Figure 5-11

Lines of longitude roughly determine the locations of time zone boundaries. These boundaries are adjusted locally to avoid splitting cities and other political subdivisions (for example, counties) into different time zones.

Section Wrap-up

Review

1. How do lines of latitude and longitude help us find locations on Earth?
2. What are the longitude and latitude of New Orleans?
3. **Think Critically:** How could you leave home on Monday to go sailing, sail for an hour on Sunday, and return home on Monday?

Skill Builder

Interpreting Scientific Illustrations

Use Appendix G to find the approximate longitude and latitude of the following locations: Sri Lanka; Tokyo, Japan; and the Falkland Islands. If you need help, refer to Interpreting Scientific Illustrations in the **Skill Handbook.**

USING MATH

If you left London on the Concord jet airplane at 8 A.M., London time, you would arrive in New York at 6 A.M., New York time. You would have crossed five time zones during your flight. How long did the trip take?

5•3 Maps

Science Words

Mercator projection
Robinson projection
conic projection
topographic map
contour line
contour interval
map legend
map scale

Objectives

- Differentiate among Mercator, Robinson, and conic projections.
- Describe how contour lines and contour intervals are used to illustrate elevation on a topographic map.
- Explain why topographic maps have scales.

Map Projections

Think of the different types of maps you have seen. There are road maps, weather maps, and maps that show physical features such as mountains and valleys. They are all models of Earth's surface. But because Earth's surface is curved, it is not easy to represent on a flat piece of paper.

Maps are made using projections. A map projection is made when points and lines on a globe's surface are transferred onto paper. There are several different ways to make map projections. But all types of projections have some sort of distortion in either the shapes of landmasses or their areas.

Mercator Projection

One type of map is a Mercator projection. A **Mercator projection** has correct shapes of continents, but their areas are distorted. Lines of longitude are projected onto the map parallel to each other. As you learned earlier, only latitude lines are parallel. Longitude lines meet at the poles. When longitude lines are projected as parallel, areas near the poles are

Figure 5-12

Because Earth's surface is curved, all types of map projections distort either the shapes of landmasses or their areas.

A — A Mercator projection exaggerates the areas near the poles.

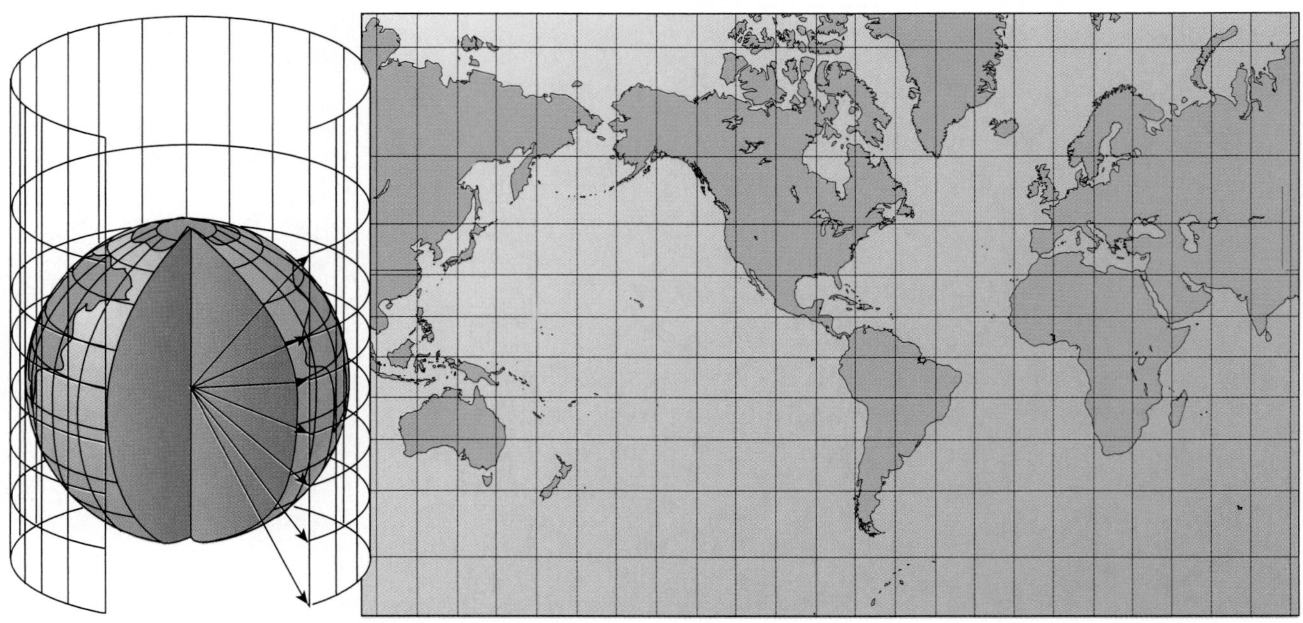

exaggerated. Look at Greenland in the Mercator projection in **Figure 5-12A.** It appears to be larger than South America. Greenland is actually much smaller than South America. Mercator projections are mainly used on ships and airplanes.

Robinson Projection

A map that has accurate continent shapes and shows accurate land areas is the **Robinson projection.** As in **Figure 5-12B,** lines of latitude remain parallel, and lines of longitude are curved as they would be on a globe. This results in less distortion near the poles.

Conic Projection

A third type of projection is a conic projection. You use this type of projection, shown in **Figure 5-12C,** whenever you look at a road map or a weather map. **Conic projections** are used to produce maps of small areas. They are made by projecting points and lines from a globe onto a cone.

C A conic projection is very accurate for small areas of Earth. *What could you use this type of map for?*

B A Robinson projection shows less distortion near the poles than a Mercator projection.

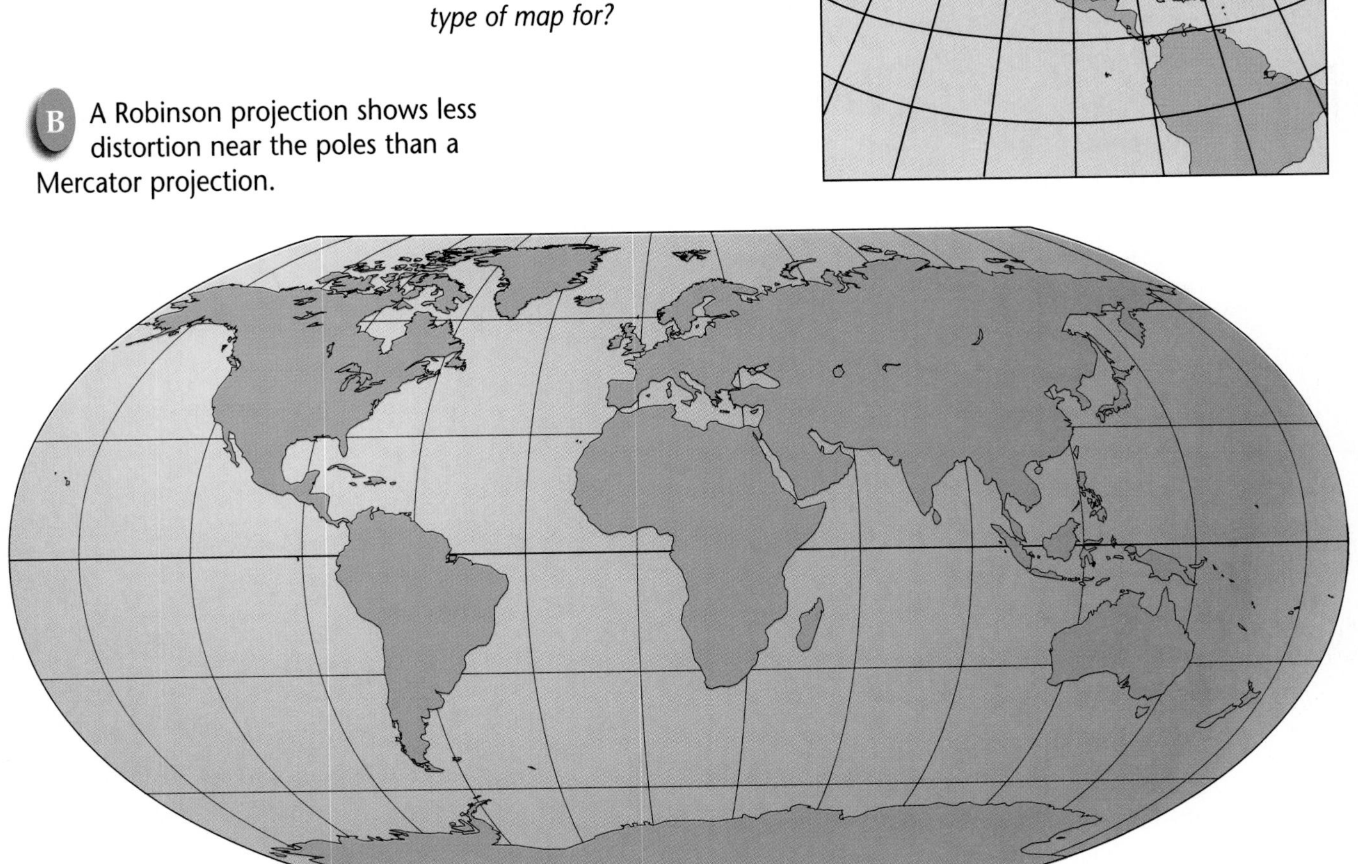

Topographic Maps

If you wanted to go hiking, a conic map projection would get you to the mountain. Next, you would need a detailed map showing the hills and valleys of that specific area. A **topographic map** shows the changes in elevation of Earth's surface. With this map, you could tell how steep the mountain trail is. It would also show natural features such as mountains, hills, plains, lakes, and rivers, and cultural features such as roads, cities, dams, and other structures built by people.

Contour Lines

Before starting your hike up the mountain, you would look at the contour lines on your topographic map to see the trail's changes in elevation. A **contour line** is a line on a map that connects points of equal elevation. Elevation refers to the distance of a location above or below sea level. The difference in elevation between two side-by-side contour lines is called the **contour interval.** If the contour interval was 10 meters, then when you walked between those two lines on the trail, you would have climbed or descended 10 meters.

Figure 5-13

This sequence shows how a topographic map of a landform is made.

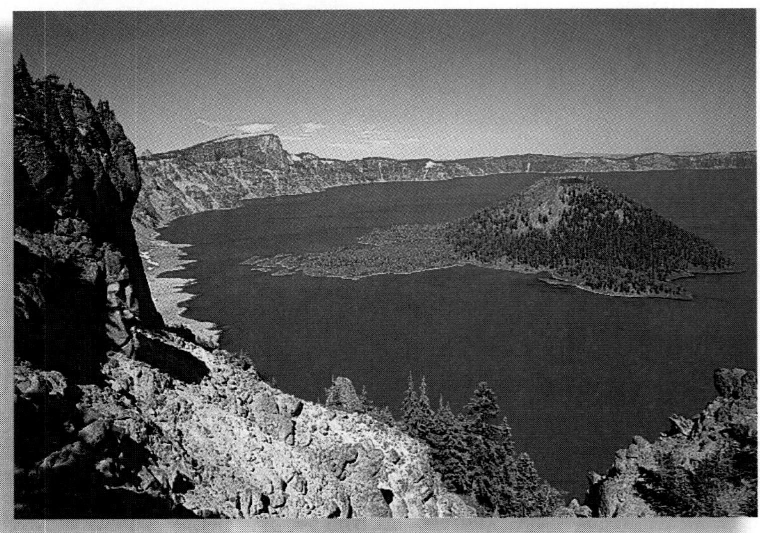

A Wizard Island is a volcanic cone that forms an island in Crater Lake, Oregon.

B Different points of elevation are projected onto paper.

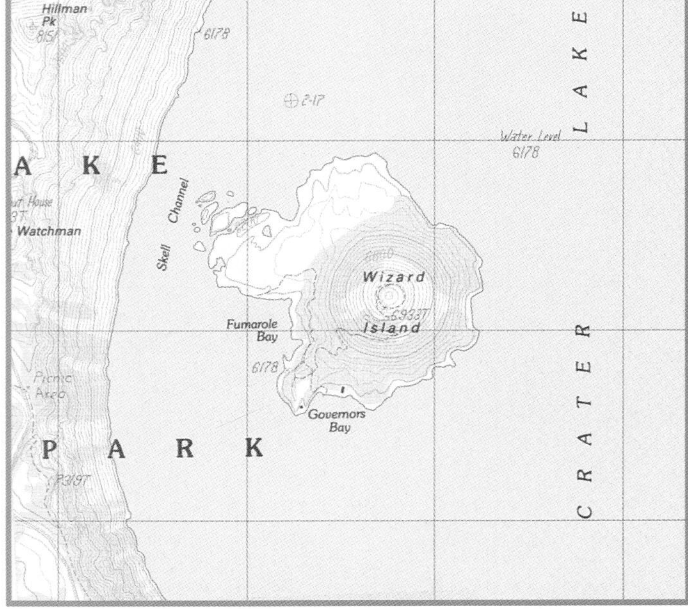

C The points of elevation are connected to form a topographic map of the island. *What do contour intervals tell us about elevation?*

As **Figure 5-13** shows, the elevation of the contour interval can vary. For mountains, the contour lines might be very close and the contour interval might be as great as 100 meters. This would tell you that the land is very steep because there is a large change in elevation between lines. However, if there isn't a great change in elevation and the contour lines are far apart, your map might have a contour interval of 5 meters.

Index Contours

Some contour lines, called index contours, are marked with their elevation. If the contour interval is 5 meters, you can tell the elevation of other lines around the index contour by adding or subtracting 5 meters from the elevation indicated on the index contour.

USING TECHNOLOGY

Rocks in 3-D!

You have learned that topographic maps are two-dimensional models of Earth. On such maps, contour lines connect points of equal elevation. Topographic maps are used to study features on Earth's surface.

To unravel Earth's complex structure, however, geologists also need to know what the rock beds "look like" in three dimensions. With computers, topographic maps are digitized to get the "top layer" of data for the 3-D maps. Digitizing is a process by which points are located on a coordinate grid.

Digging Deeper

Geologists then use data from wells drilled into the crust to obtain information about other layers of rocks beneath the surface. The computer analyzes all the data and generates a three-dimensional map that shows what the rocks of an area look like. The map shown in the photograph illustrates how the drainage basin of an ancient river system was changed when forces at Earth's surface and forces within Earth acted on the river in Montana.

River System in 3-D

Think Critically:

List one advantage a 3-D map has over a 2-D map.

Figure 5-14

Contour lines form Vs whenever they cross streams. *In your Science Journal, explain why this is the case.*

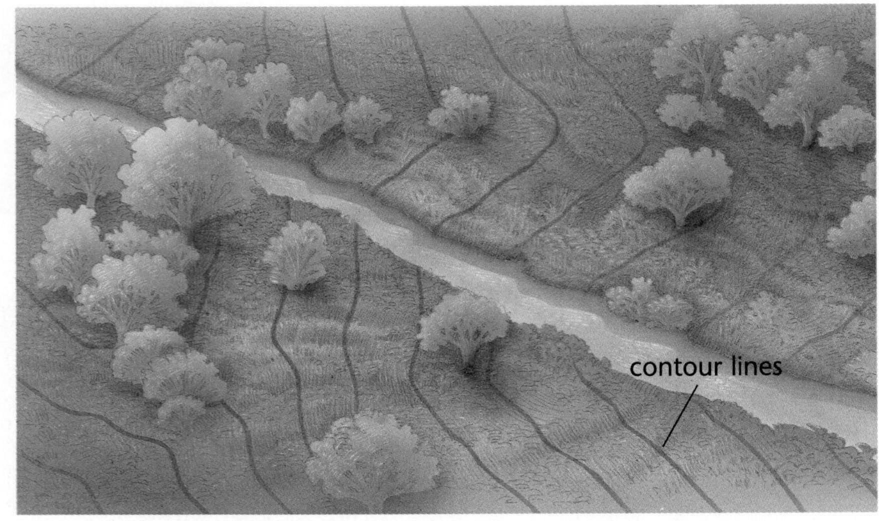

Here are some rules to remember when examining contour lines.

1. **Contour lines close around hills and basins or depressions.** To decide whether you're looking at a hill or basin, you can read the elevation numbers or look for hachures. Hachures are short lines at right angles to the contour line that are used to show depressions. These lines point toward lower elevations.
2. **Contour lines never cross.** If they did, it would mean that the spot where they cross would have two different elevations.
3. **Contour lines form Vs that point upstream whenever they cross streams.** This is because streams flow in depressions that are beneath the elevation of the surrounding land surface. Look at **Figure 5-14.** When the contour lines follow the depression, they appear as Vs pointing upstream on the map.

Figure 5-15

Here are some typical symbols used on topographic maps.

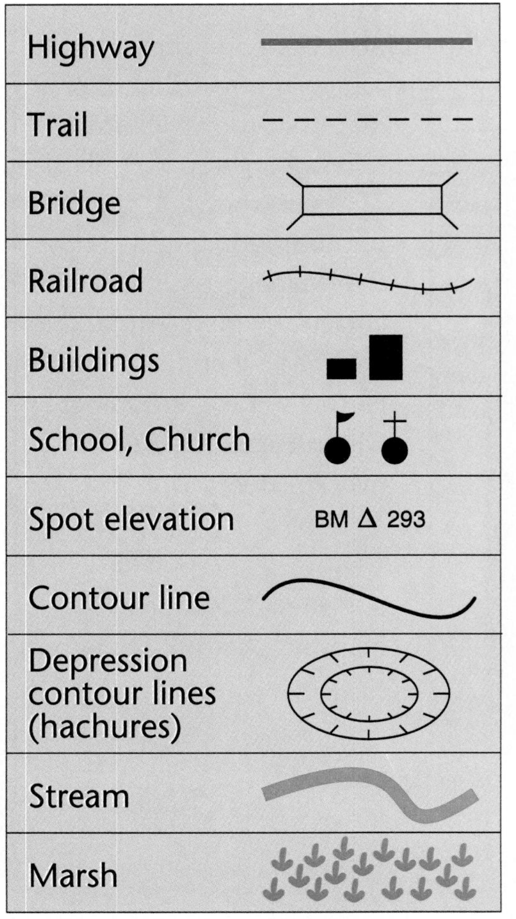

Map Legend and Scale

Another thing you would want to know before you set out on your hike is, "How far is it to the top of the mountain?" Because maps are small models of Earth's surface, distances and sizes of things on a map are proportional to the real thing on Earth. This is accomplished using "scale" distances.

Map Legend

Topographic maps and most other maps have a legend. A **map legend** explains what the symbols used on the map mean. Some frequently used symbols for topographic maps are shown in **Figure 5-15** and Appendix E.

Map Scale

The **map scale** is the relationship between the distances on the map and actual distances on Earth's surface. Scale is often represented as a ratio. For example, a topographic map of the Grand Canyon may have a scale that reads "1:80 000." This means that one unit on the map represents 80 000 units on land. If the unit you wanted to use was a centimeter, then one cm on the map would equal 80 000 cm on land. The unit of distance may be in feet or millimeters or any other measure of distance. However, the units of measure on each side of the ratio must always be the same. A map scale may also be in the form of a small bar graph that is divided into a number of units. The units are the scaled-down equivalent distances to real distances on Earth.

Using a Topographic Map

The map below is a topographic map of an area in California. One sunny day, three hikers started from the point marked with the + in the center of the map. One hiker tackled the peak of Cedar Mountain, another the peak of Orr Mountain, while the third headed for Cedar Well.

All three traveled at the same rate on flat or gentle slopes. Their climb slowed as the ground grew steeper.

Solve the Problem:

1. **Which peak is the highest?**
2. **Which hiker had the steepest climb? Explain using contour lines.**
3. **Name three items found in a map legend that the hiker heading for Orr Mountain crossed before reaching his or her goal.**

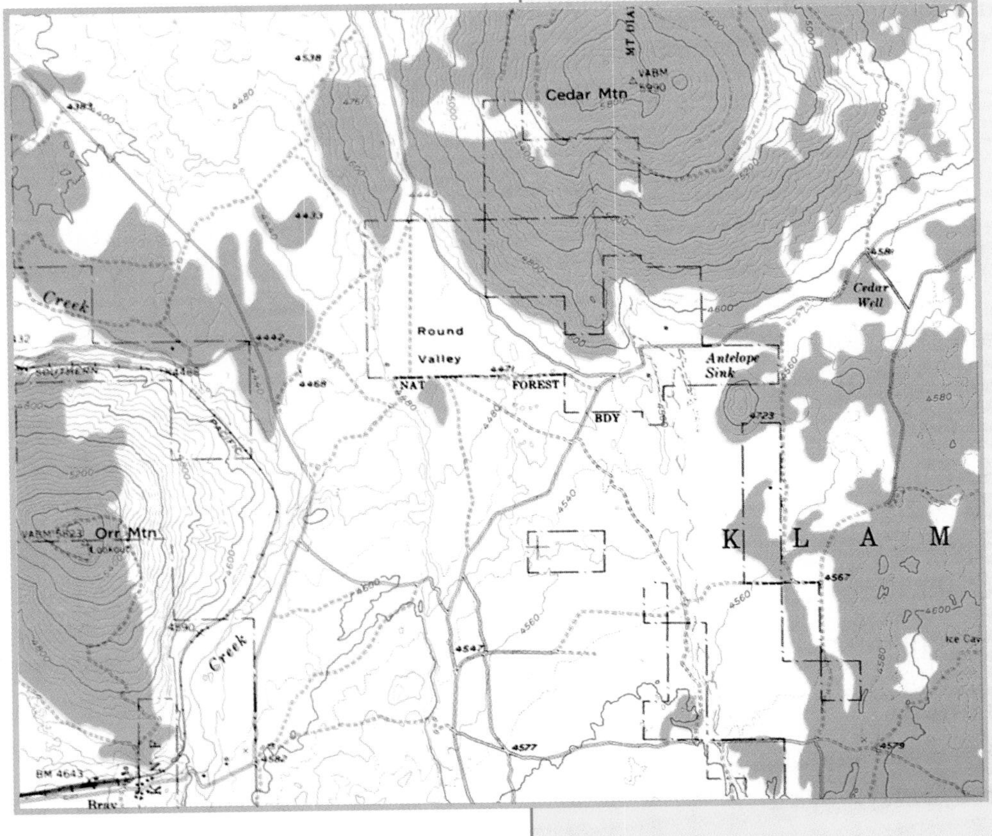

Think Critically:

1. If each hiker could choose any route to his or her destination, which one do you think reached his or her goal first? Explain.
2. Once at the top, could the hiker on Cedar Mountain see the hiker on Orr Mountain? Why or why not?

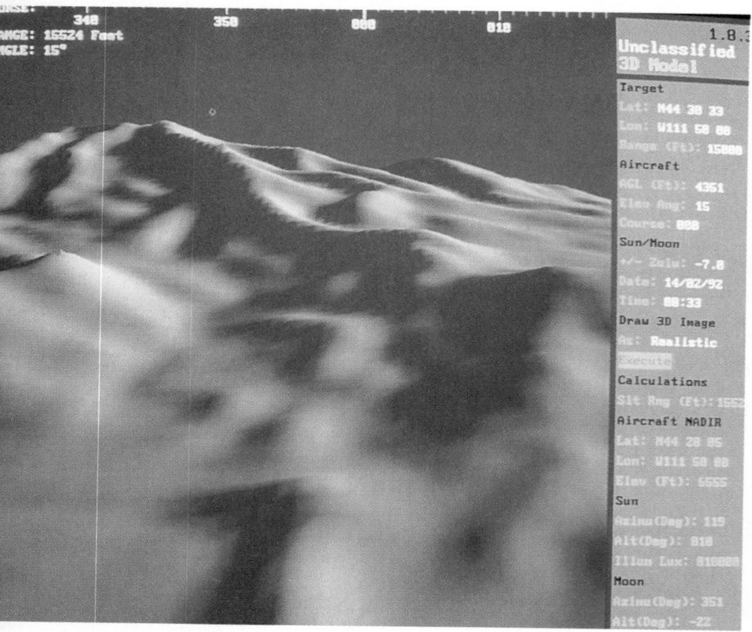

Figure 5-16

Computers have revolutionized the way cartographers make maps.

Uses of Maps

As you have learned, there are many different ways to view Earth. The map you choose to use will depend upon your need. For instance, if you wanted to determine New Zealand's location relative to Canada, you would probably examine a Mercator projection. In your search, you would use lines of latitude and longitude, and map scale. If you wanted to travel across the country, you would rely on a conic projection. And if you wanted to scale the highest peak in your county, you would take along a topographic map.

As **Figure 5-16** shows, mapmaking or cartography has experienced a technological revolution in the past few decades. Remote sensing and computers have changed the way maps are made. Read more about the history of cartography in this chapter's Science & History feature.

Section Wrap-up

Review

1. Why does Greenland appear to be larger on a Mercator projection than it does on a Robinson projection?
2. Why can't contour lines ever cross?
3. **Think Critically:** Suppose you have a topographic map with a contour interval of 50 m. According to the map scale, 1 cm on the map equals 1 km. The distance between points A and B on the map is 8 cm. Four contour lines lie between them. How far apart are the points and what is the change in elevation?

Skill Builder

Measuring in SI

Use the topographic map in Problem Solving to practice measuring in SI.

1. How far is it from Cedar Well to Orr Mountain?
2. How big is Antelope Sink at its widest place?

If you need help, refer to Measuring in SI in the **Skill Handbook.**

Science Journal

Draw a map in your Science Journal that your friends could use to get from school to your home. Include symbols and a map scale.

Activity 5-1

Making a Topographic Map

Have you ever wondered how topographic maps are made? Today, radar and remote sensing devices aboard satellites collect data, and computers and graphic systems make the maps. In the past, surveyors and aerial photographers collected data. Then maps were hand drawn by cartographers, or mapmakers. In this activity, you can try your hand at cartography.

Problem

How is a topographic map made?

Materials  

- plastic model landform
- water tinted with food coloring
- transparency
- clear plastic storage box with lid
- beaker
- metric ruler
- tape
- transparency marker

Procedure

1. Using the ruler and the transparency marker, make marks up the side of the storage box 2 cm apart.
2. Secure the transparency to the outside of the box lid with tape.
3. Place the plastic model in the box. The bottom of the box will be zero elevation.
4. Using the beaker, pour water into the box to a height of 2 cm. Place the lid on the box.
5. Use the transparency marker to trace the top of the water line on the transparency.
6. Using the scale 2 cm = 10 m, mark the elevation on the line.
7. Remove the lid and add water until a depth of 4 cm is reached.
8. Map this level on the storage box lid and record the elevation.
9. Repeat the process of adding water and tracing until you have the hill mapped.
10. Transfer the tracing of the hill onto a white sheet of paper.

Analyze

1. What is the contour interval of this topographic map?
2. How does the distance between contour lines on the map show the steepness of slope on the landform model?
3. **Determine** the total elevation of the hill.
4. How was elevation represented on your map?

Conclude and Apply

5. How are elevations shown on topographic maps?
6. Must all topographic maps have a 0-m elevation contour line? **Explain.**
7. **Compare** the contour interval of an area of high relief with one of low relief on a topographic map.

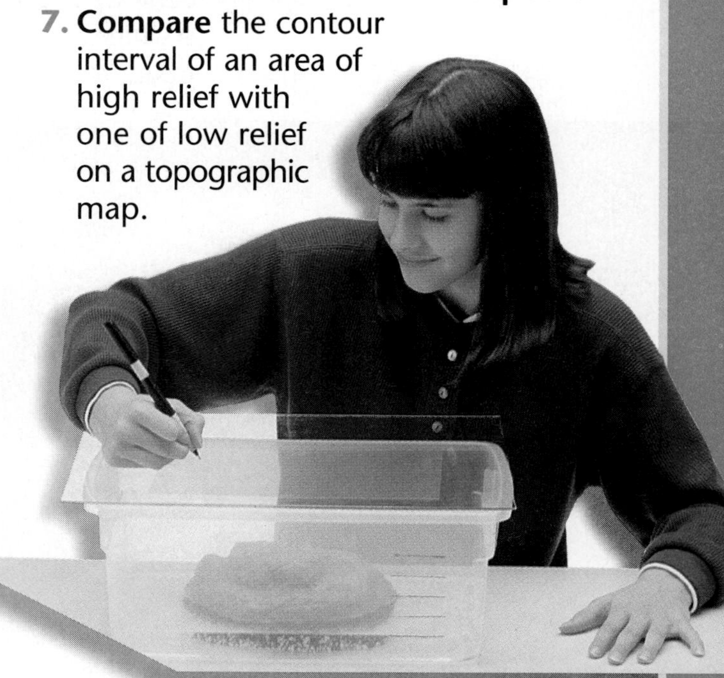

TECHNOLOGY: 5•4 Mapping Our Planet

Science Words

Landsat Satellite
Topex-Poseidon Satellite
sonar

Objectives

- Describe how Landsat Satellites collect information about Earth's surface.
- Compare the data-collecting technique of the Topex-Poseidon Satellite with sonar.

INTEGRATION
Physics

Remote Sensing from Space

As you learned in Activity 5-1, today we rely on remote sensing and radar to collect much of the data used for making maps.

Landsat Satellites

Landsat Satellites detect different wavelengths of energy reflected or emitted from Earth's surface. Each Landsat has a mirror that moves to scan Earth's surface. On this mirror are rows of detectors that measure the intensity of the energy they receive from Earth. This information is transmitted to Earth as a series of 0s and 1s. The result is a digital image. Digital images are computer images that show each of the different wavelengths of energy as different colors. These images show landforms in great detail.

Topex-Poseidon Satellite

The **Topex-Poseidon Satellite** (Topex stands for topographic experiment) uses radar to compute the distance to the ocean's surface. Radar waves are high-frequency radio signals that are sent from the satellite to the ocean. As **Figure 5-17** illustrates, a receiving device then picks up the returning echo as it bounces off the water. The distance to the water's surface is calculated using the speed of radar and the time it takes for the radar to be reflected back. Since there is more gravitational force between ocean water and massive structures on the ocean floor, water forms bulges over mountains and depressions over valleys. With this information, computers draw ocean-bottom maps that show features as small as 13 km.

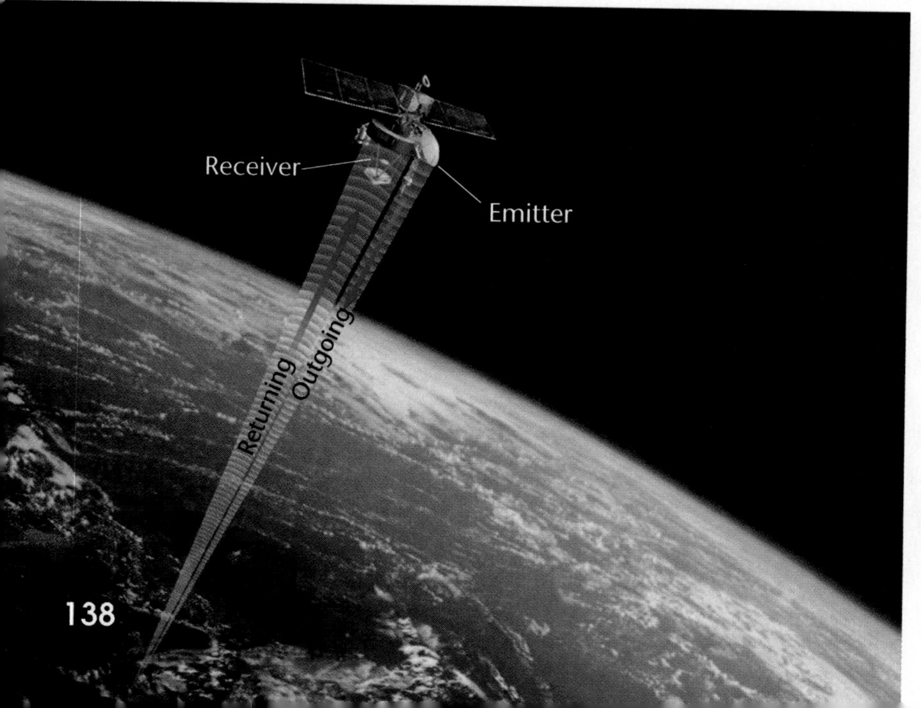

Figure 5-17

Using high-frequency radio waves, the Topex-Poseidon Satellite can map small features on the ocean floor.

Remote Sensing Under Water

Sonar refers to the use of sound waves to detect ocean-bottom structures. First, a sound wave is sent from a ship towards the ocean floor. A receiving device then picks up the returning echo when it bounces off the bottom. The distance to the bottom is calculated by a computer on the ship using the speed of sound in water and the time it takes for the sound to be reflected.

Sea Beam

Using a new technology called Sea Beam, scientists are making very detailed, accurate maps of the ocean floor. A ship with Sea Beam, like the one shown in **Figure 5-18,** has more than a dozen sonar devices, each aimed at different parts of the seafloor. As the ship goes back and forth across the seafloor, readings of depth overlap. Computers take this information and make a detailed, continuous map of the ocean floor.

These new Sea Beam maps are very useful. For example, fishing companies can use them to locate underwater canyons where certain fish might be found. Oceanographers use them to study the ocean.

Originally, all Sea Beam maps were classified as secret. Because they were so accurate, government officials didn't want other countries to have detailed maps of the ocean floor around the United States. They feared that hostile submarines could use the maps to navigate our coastal areas. Now, most of the Sea Beam maps have been released to the public.

Figure 5-18

Sea Beam can also make detailed maps of the ocean floor. *What are some uses for these maps?*

Section Wrap-up

Review

1. How do Landsat Satellites collect information about Earth's surface?
2. In your Science Journal, explain the similarities between the data-collecting methods of Topex-Poseidon and sonar.

Explore the Technology

Landsat satellite information is used for more than just making maps. The digital images are used to locate soil erosion problems, monitor desert and forest growth, explore for minerals, study ocean currents, measure pollution, determine population density, and even track migrating animals. Explain how Landsat information can be used to plan the growth of cities.

USING MATH

Sound travels through water at 1454 meters per second. If a ship sends a sonar signal to the ocean floor and it reflects back to the ship in 2.25 seconds, how deep is the ocean at this point?

SCIENCE & SOCIETY

Activity 5-2

Design Your Own Experiment

Modeling Earth

Have you ever built a model plane, train, or car? Modeling is more than just fun. Models are used to help engineers and designers build actual planes, trains, and cars. A topographic map is a two-dimensional model—on flat paper. How can you build a three-dimensional model of a landform?

PREPARATION

Problem

How can a 3-D model be made of an area shown on a topographic map?

Form a Hypothesis

Based on the drawing below, state a hypothesis about how you can make a large model of Blackberry Hill, such that its base is as long as a piece of notebook paper.

Objectives

- Design and make a 3-D model that shows the relationship between topographic maps and landforms.
- Interpret data from your model.

Possible Materials

- transparency marker (fine point)
- blank transparency
- overhead projector
- sheet of white paper
- pencil
- tape
- corrugated cardboard sheets or foam-board sheets
- scissors or razor knife
- glue
- metric ruler

Safety Precautions

Be careful while working near the overhead projector light. It can get very hot. While using scissors or a knife, be careful not to cut yourself or the desk top.

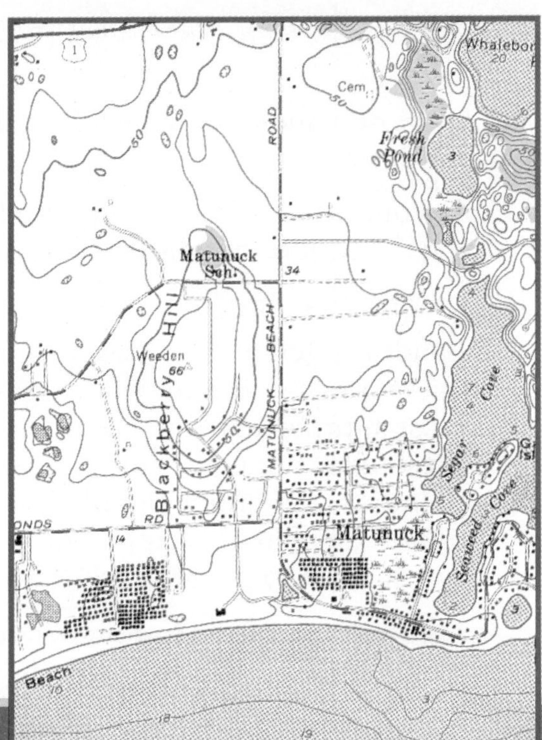

PLAN THE EXPERIMENT

1. With your partner, design a way that you can make an enlarged copy of the topographic features of Blackberry Hill using a transparency marker, overhead projector, pencil, sheet of white paper, and tape. Write down the steps you will take.
2. Explain how you can use the contour lines on your white paper as patterns for making the different layers of your model.
3. Describe a way to make your 3-D model using stacked sheets of cardboard or foam board.

Check the Plan

1. Read over your entire plan to make sure that all steps are in a logical order.
2. *Make sure your teacher approves your plan before you proceed.*

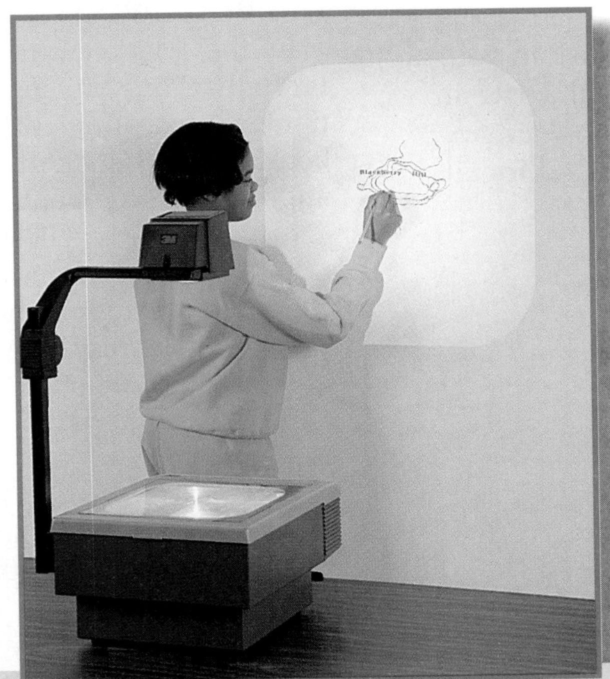

DO THE EXPERIMENT

1. Build your model as planned.
2. While the activity is going on, write down any observations that you make in your Science Journal.

Analyze and Apply

1. **Compare** your model with other students'. How are they similar? How are they different?
2. **Determine** the horizontal scale of your model.
3. **Infer** what the height of each sheet in your model represents.
4. **Describe** the most difficult part of making your model.
5. **Describe** the easiest part.

Go Further

Design a way to make your scale model have gentle slopes instead of the step effect you have with your sheet model.

Science & History

The Art of Mapmaking

As early as 2300 B.C., the Babylonians drew maps of individual plots of land. Their maps served as the basis for mapmaking in Europe in the Middle Ages. Later, the Babylonians constructed simple maps of the world. One Babylonian map carved in stone about 500 B.C. showed seven circles containing the names of countries and cities. A pair of lines in the center probably represented the Euphrates River.

Science Journal

How have you used a map lately? In your Science Journal, write an essay about how a map helped you in a difficult situation.

The Beginning of Modern Mapmaking

In the 13th century, mapmaking changed dramatically. From that time on, maps began to be based more on observation and measurement. The best mapmakers charted the seas, because mariners had compasses and other devices to make angular measurements.

The earliest mapmakers were usually artists. They often decorated their maps with artistic details. Notice those on the map of the western hemisphere shown here. This map was made in the 1600s. How does it differ from present-day maps? How is it similar?

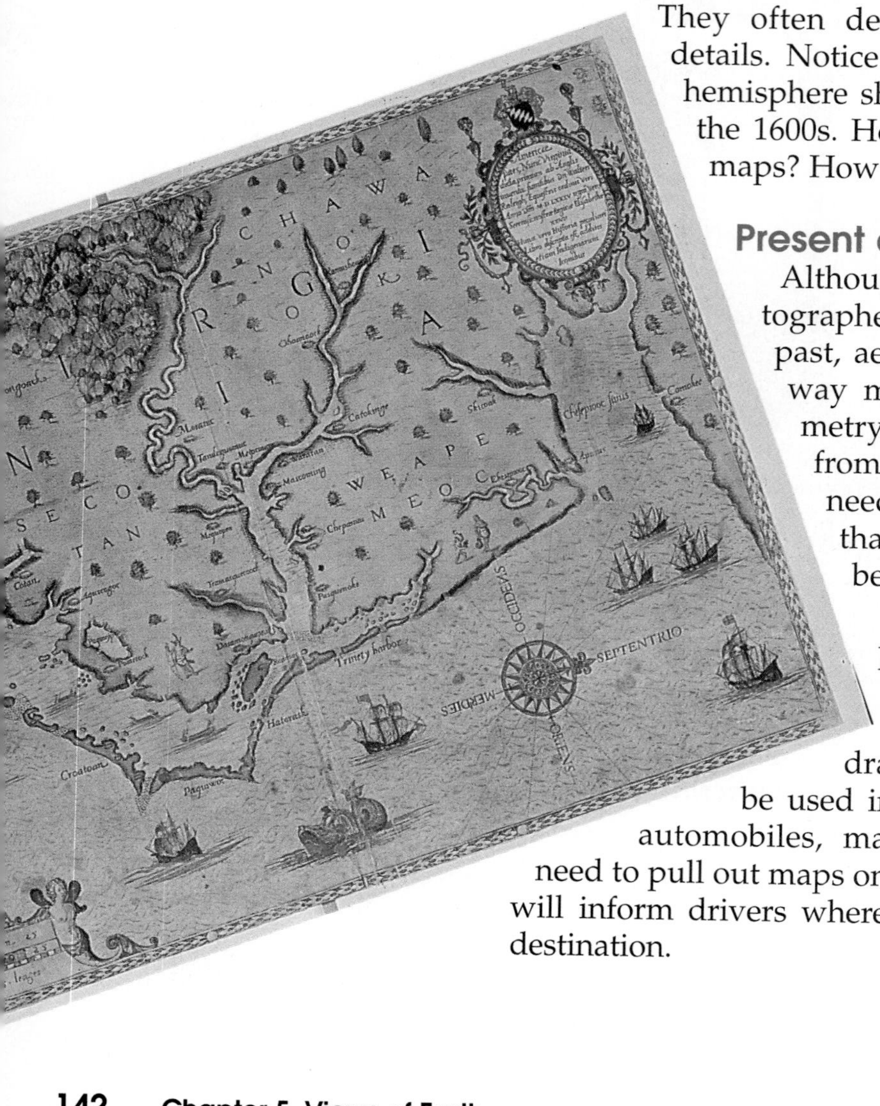

Present and Future Maps

Although present-day mapmakers or cartographers owe a great deal to those of the past, aerial photography has changed the way maps are made today. Photogrammetry, the ability to take measurements from photographs, has eliminated the need for a cartographer to visit a site that is being mapped. Maps can now be made from aerial photographs.

Today's cartographers use computer technology to make and update maps. When they enter data on a computer, the computer draws the map. Digital map data can be used in a number of ways. In the latest automobiles, map data will soon eliminate the need to pull out maps on a motor trip. Computer programs will inform drivers where they are and how to reach their destination.

Chapter 5 Review

Summary

5-1: Landforms

1. Plains are large, flat areas that cover much of the United States. Plateaus are high, relatively flat areas next to mountains.
2. There are four types of mountains: folded, upwarped, fault-block, and volcanic.

5-2: Viewpoints

1. Latitude refers to the distance in degrees north or south of the equator. Longitude refers to distance in degrees east or west of the prime meridian.
2. Earth is divided into 24 time zones, each one hour ahead or behind the adjacent zone.

5-3: Maps

1. Because Earth is curved, map projections are distorted. The main types of projections are Mercator, Robinson, and conic.
2. A contour line on a topographic map connects points of equal elevation. The difference in elevation between contour lines is the contour interval.
3. A scale is used to show the relationship between map distances and the actual distances on Earth's surface.

5-4: Science and Society: Mapping Our Planet

1. Landsat Satellites detect energy reflected from Earth's surface, while the Topex-Poseidon Satellite uses radar to gather information for maps.
2. Sonar uses sound waves bounced off the ocean floor to detect ocean-bottom features.

Key Science Words

a. conic projection
b. contour interval
c. contour line
d. equator
e. fault-block mountain
f. folded mountain
g. International Date Line
h. Landsat Satellite
i. latitude
j. longitude
k. map legend
l. map scale
m. Mercator projection
n. plain
o. plateau
p. prime meridian
q. Robinson projection
r. sonar
s. Topex-Poseidon Satellite
t. topographic map
u. upwarped mountain
v. volcanic mountain

Reviewing Vocabulary

Match each phrase with the correct term from the list of Key Science Words.

1. high, flat area next to mountains
2. mountain formed when Earth's crust is squeezed from opposite sides
3. imaginary line parallel to the equator
4. 0° longitude
5. projection used mainly for navigation
6. map that shows changes in elevation
7. line connecting points of equal elevation
8. change in elevation between adjacent contour lines
9. instrument that uses sound waves to detect features
10. instrument that uses radar to study ocean-bottom structures

Chapter 5 Review

Checking Concepts

Choose the word or phrase that completes the sentence.

1. _______ make up about 50 percent of all land areas in the United States.
   a. Plateaus c. Mountains
   b. Plains d. Volcanoes
2. The North Pole is located at _______ degrees north latitude.
   a. 0 c. 50
   b. 180 d. 90
3. The Hawaiian Islands are________ mountains.
   a. fault-block c. upwarped
   b. volcanic d. folded
4. Lines parallel to the equator are _______.
   a. lines of latitude c. lines of longitude
   b. prime meridians d. contour lines
5. Earth can be divided into 24 time zones that are _______ degrees apart.
   a. 10 c. 15
   b. 34 d. 25
6. _______ projections are very distorted at the poles.
   a. Conic c. Robinson
   b. Topographic d. Mercator
7. A _______ map shows changes in elevation at Earth's surface.
   a. conic c. Robinson
   b. topographic d. Mercator
8. _______ is measured with respect to sea level.
   a. Contour interval c. Conic projection
   b. Elevation d. Sonar
9. _______ are used to show depressions.
   a. Vs c. Hachures
   b. Scales d. Legends
10. The Grand Canyon is part of the_______.
    a. Great Plains
    b. Colorado Plateau
    c. Rocky Mountains
    d. Appalachian Mountains

Understanding Concepts°

Answer the following questions in your Science Journal using complete sentences.

11. Compare and contrast the Atlantic Coastal Plain and the Great Plains. Be sure to discuss how they were formed.
12. Compare and contrast volcanic and folded mountains. Draw a diagram of each type of mountain.
13. Maps are made by projecting points and lines on a globe's surface onto paper. Why would Antarctica appear to be larger on a Mercator map than it actually is?
14. Would a topographic map of the Great Plains have a large or small contour interval? Explain.
15. How does the information gathered by ships using Sea Beam sonar differ from that collected with standard methods that use one sonar device?

Thinking Critically

16. How would a topographic map of the Atlantic Coastal Plain differ from a topographic map of the Rocky Mountains?

17. If you left Korea early Wednesday morning and flew to Hawaii, on what day of the week would you arrive?

18. If you were flying directly south from the North Pole and reached 70° north latitude, how many more degrees of latitude would be left to pass over before reaching the South Pole?

19. Why can't two contour lines overlap?

20. What does a map scale of 1:50 000 mean?

Developing Skills

If you need help, refer to the **Skill Handbook.**

21. Measuring in SI: What is the area in square kilometers of the topographic map in the Problem Solving feature in Section 5-3?

22. Comparing and Contrasting: Compare and contrast Mercator, Robinson, and conic map projections.

23. Concept Mapping: Make a network tree concept map that explains how topographic maps are used. Use the following terms: *topographic maps, mountains, rivers, natural features, contour lines, changes in elevation, equal elevation, hills, plains.*

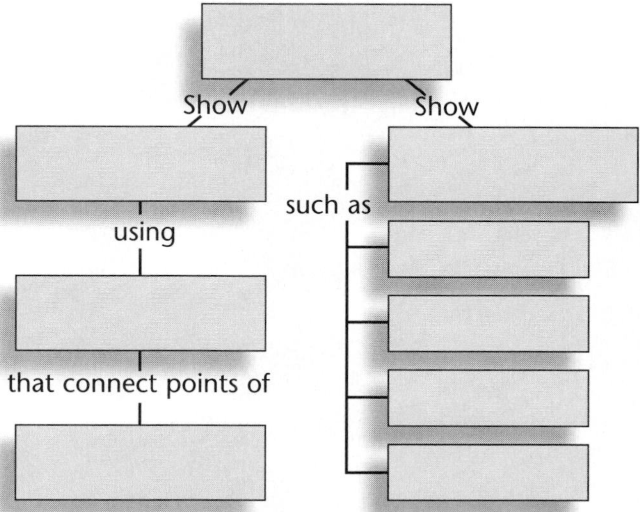

24. Sequencing: Arrange these cities in order from the city with the earliest time to that with the latest time on a given day: Anchorage, AK; San Francisco, CA; Bangor, ME; Columbus, OH; Houston, TX.

25. Interpreting Data: If a map has a scale of 1 cm = 80 000 cm, how far apart on the map would two cities be if they were 3 km apart on Earth's surface?

Performance Assessment

1. Checking Lab Skills: In the MiniLAB on page 122, you learned how to make a profile. Make a profile of the model you made of Blackberry Hill in Activity 5-2.

2. Model: Make a 3-D model of a canyon using the same materials as those listed in Activity 5-2. To begin, draw a topographic map of a canyon on a sheet of note paper. Have at least four contour lines on your map.

3. Problem Solving: Use modeling clay to construct a physiographic map to scale of the United States. Be sure to include all of the landforms shown in **Figure 5-2.**

Previewing the Chapter

Chapter 6

Weathering and Soil

The mysterious statues shown here are on Easter Island in the South Pacific Ocean. Probably long before anyone can unravel their meaning, nature will wear them away by the process of weathering. Weathering pits and roughens the stone of monuments and rocks in nature.

All stone and rock weathers over time. Grains of sediment fall out. Cracks develop on the surface and deep inside. Weathering causes the stone of monuments to crumble away gradually. Weathering also causes rocks in nature to fracture, buckle, and crumble into soil and sediment.

EXPLORE ACTIVITY

Compare different rates of weathering.

1. Obtain several kinds of rocks.
2. Over white paper, rub one rock against a metal file.
3. Repeat this process with each kind of rock.
4. Compare the amount of broken-off particles from each rock.

Observe: Which rock breaks apart most easily? In your Science Journal, describe each rock based upon your observations.

Previewing Science Skills

- In the **Skill Builders,** you will recognize cause and effect and map concepts.
- In the **Activities,** you will hypothesize, observe, infer, interpret data, and measure.
- In the **MiniLABs,** you will observe and compare and contrast.

6•1 Weathering

Science Words

weathering
mechanical weathering
ice wedging
chemical weathering
oxidation
climate

Objectives

- Contrast mechanical weathering and chemical weathering.
- Explain the effects of climate on weathering.

Evidence of Weathering

The next time you take a walk or a drive, notice the sand and grit along the sidewalk and curb and the layer of gritty dirt lying at the bottom of road cuts. Much of the gritty sediment you see comes from small particles that break loose from concrete curbs and from rocks exposed to the natural elements. These sediments are evidence of weathering.

Conditions and processes in the environment cause the weathering of rock and concrete. In Chapter 4, the rock cycle showed how sedimentary rocks form. But did you know that sediments can also form soil? Both soil formation and the development of sedimentary rocks are dependent on the process of weathering.

Weathering is the process that breaks down rocks into smaller and smaller fragments. Over millions of years, the process of weathering helped change Earth's surface. **Figure 6-1** illustrates how weathering wears down mountains to hills. Weathering blurs the writing on tombstones and slowly breaks down statues. Weathering can also cause potholes in streets. The two types of weathering are mechanical and chemical. Although mechanical and chemical weathering are discussed separately, the two processes work together to break down rock.

Figure 6-1

Over long periods of time, weathering helps change sharp, jagged mountains into smooth, rolling mountains and hills.

Figure 6-2

Tree roots can break up sidewalks.

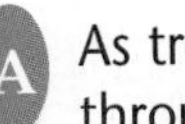

As trees grow, their roots spread throughout the soil.

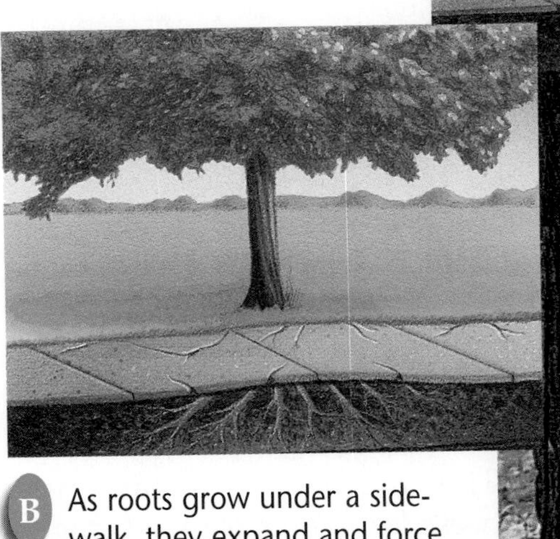

As roots grow under a sidewalk, they expand and force the concrete to crack.

Over time, the sidewalk buckles and breaks apart.

Mechanical Weathering

Mechanical weathering breaks apart rocks without changing their chemical composition. Each fragment and particle weathered away by a mechanical process retains the same characteristics as the original rock. Mechanical weathering can be caused by growing plants, expanding ice, growing crystals, lightning, and expansion and contraction with heating and cooling. These physical processes create enough force to break rocks into smaller pieces.

Plants

Plant roots grow into cracks of rocks where they find water and nutrients. As roots grow, they wedge rocks apart. If you've skated on a sidewalk and tripped over a crack near a tree, you have experienced the results of mechanical weathering. The sidewalk near the tree in **Figure 6-2** shows signs of weathering. How could cracks in rocks occur in the same way?

Animals burrowing in soil contribute to weathering. They carry materials to the surface, where more weathering takes place.

Gabriella Saves the Freezer

On a hot day, Gabriella baby-sat six-year-old Tanya and her baby brother. While Gabriella tended the baby, Tanya made lemonade. Tanya filled a glass jar with lemonade concentrate and water, put a lid on it, and stuck the glass jar in the freezer to cool quickly. Then she forgot about it. When Gabriella brought the baby into the kitchen, she discovered the glass bottle of lemonade and removed it from the freezer before a messy accident could occur.

Solve the Problem:

1. **What could have happened to the glass bottle of lemonade in the freezer?**
2. **Why?**
3. **How does the volume of water change when it freezes?**

Think Critically:

1. How does this same process weather rocks?
2. In what parts of the world would this kind of weathering occur? In what parts of the world would this mechanical weathering process not occur?

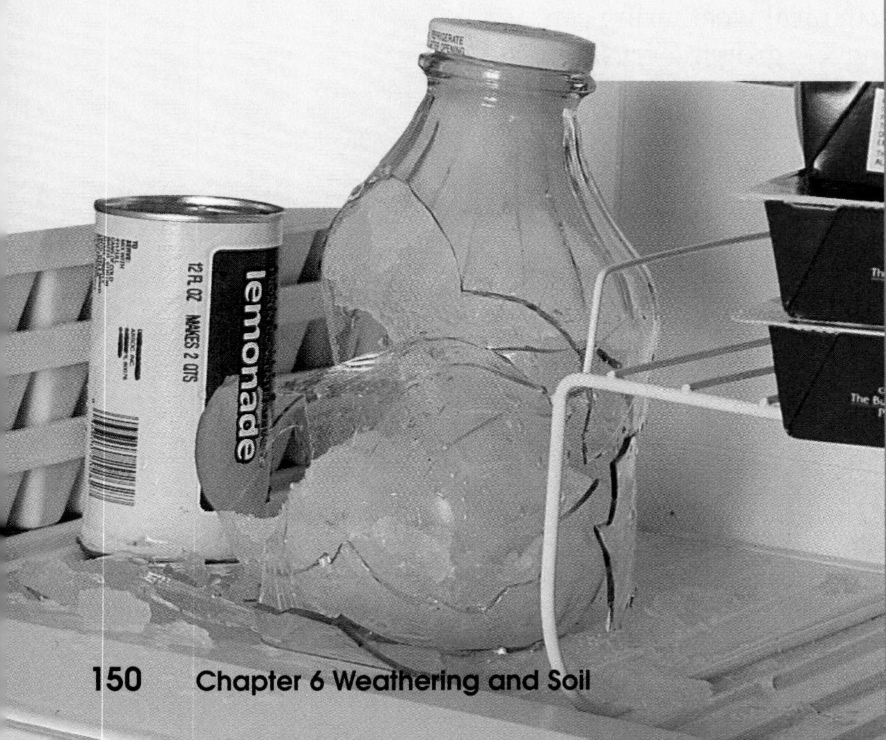

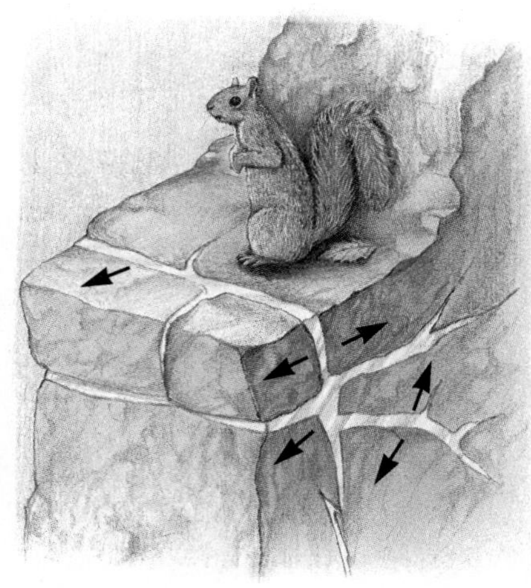

Figure 6-3

When water freezes in cracks of rocks, it expands. Pressure builds and breaks the rock apart. When the ice thaws and the water refreezes, this process reoccurs.

Ice Wedging

The mechanical weathering process known as **ice wedging** is illustrated in **Figure 6-3.** Ice wedging can wear down rocks rapidly. At night, in cold areas, low temperatures freeze water. During the day, warmer temperatures thaw the ice. Ice wedging is particularly noticeable in the mountains, where it contributes to the wearing down of sharp mountain peaks to rounded hills, a process pictured in **Figure 6-1.** This cycle of freezing and thawing not only breaks up rocks but also breaks up roads and highways. When water gets into cracks in road pavement and freezes, it forces the pavement apart. This can lead to the formation of potholes in roads. Weathering by both roots and ice wedging can rapidly reduce rocks to rubble. Breaking up rocks through mechanical weathering exposes a greater surface area to chemical weathering. This illustrates how mechanical and chemical weathering work together to break down rock.

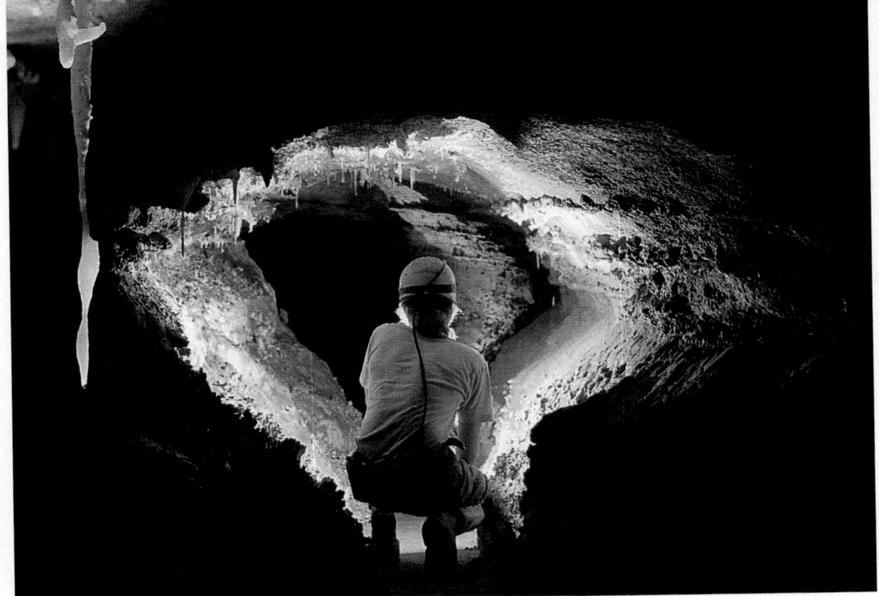

Figure 6-4

This limestone cave is a product of chemical weathering. *How did water help form this cave?*

Chemical Weathering

INTEGRATION Chemistry

The second type of weathering occurs when water, air, and other substances react with the minerals in rocks. This type of weathering is called **chemical weathering** because the chemical composition of the rock changes. Let's see how chemical weathering happens.

Water

Water is an important agent of chemical weathering. When the hydrogen and oxygen atoms in water react with the chemicals in some rocks, new compounds form. The resulting composition is much different from that of the original rock.

Acids

Naturally formed acids can weather rocks chemically. When water mixes with carbon dioxide from the air, a weak acid, carbonic acid, forms. Carbonic acid is the same weak acid formed when soft drinks are carbonated. Carbonic acid reacts with minerals such as calcite, the main mineral in limestone. The product of this reaction then dissolves and can be carried away. Over thousands of years, carbonic acid has weathered enough limestone to create caves, such as the one shown in **Figure 6-4.**

When carbonic acid comes in contact with granite rock, chemical weathering occurs. Over a long time, the mineral feldspar in the granite is broken down into silica, a potassium salt, and the clay mineral kaolinite. Kaolinite clay makes up a high percentage of the material in some soils. Clay is an end-product of weathering.

Some roots and decaying plants give off acids that can dissolve minerals in rock. Removing these minerals weakens the rock. Eventually the rock will break into smaller pieces. The next time you find a moss-covered rock, peel back the moss and look at the small pits underneath. Acids from plant roots caused the pits.

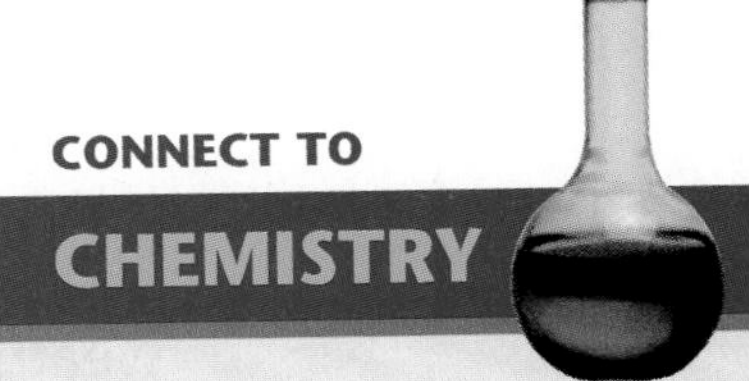

CONNECT TO

CHEMISTRY

The chemical formula for calcite, a common rock-forming mineral, is $CaCO_3$. Use the Periodic Table, Appendix F on page 712, to ***identify*** the elements in this formula.

Activity 6-1

Design Your Own Experiment

Weathering Chalk

Chalk is a type of limestone made of the shells of tiny animals and plants. When you write your name on the chalkboard or draw a picture on the driveway with a piece of chalk, what happens to the chalk? It is mechanically weathered. This experiment will help you understand how chalk can be chemically weathered.

PREPARATION

Problem

How can chalk be chemically weathered? What variables affect the rate of chemical weathering?

Form a Hypothesis

In this experiment, you will test three variables that can affect the rate of chemical weathering. Write hypotheses about how surface area, acidity, and heat affect the rate of chemical weathering of chalk.

For example, if you place a piece of chalk in water and another in acid (vinegar), what do you think will happen? If you break a piece of chalk and drop the pieces in acid, how will the size of the pieces of chalk affect the rate of weathering? If you heat the acid, how will that affect the rate of weathering?

Objectives

- To design experiments to compare the effect of acidity, surface area, and heat on the rate of chemical weathering of chalk.
- To understand some of the factors that affect chemical weathering.

Possible Materials

- 6 equal pieces of chalk
- 2 small beakers
- metric ruler
- water
- white vinegar
- hot plate
- graduated cylinder
- safety goggles

Safety Precautions   

Wear safety goggles when pouring acids. Be careful when using a hot plate and heated solutions.

PLAN THE EXPERIMENT

1. Develop hypotheses about the effects of acidity, surface area, and heat on chemical weathering.
2. Devise a way to test your first hypothesis. List the steps needed to test the hypothesis.
3. Repeat this for your other two hypotheses.
4. Design data tables in your Science Journal. Use one for acidity, one for surface area, and one for heat.

Check the Plan

1. Read over your plan to make certain that the steps are in a logical order.
2. Identify what remains constant in your experiment and what varies. Each test should have only one variable.
3. Have you allowed for a control in each experiment?
4. Summarize your data in a graph. Decide from reading the Skill Handbook which type of graph to use.
5. *Make sure your teacher approves your plan before you start the experiment.*

DO THE EXPERIMENT

1. Carry out the three experiments as planned.
2. While the experiments are going on, write your observations and complete the data tables in your Science Journal.

Analyze and Apply

1. **Analyze** your graph to find out which substance—water or acid—weathered the chalk more quickly. Was your hypothesis supported?
2. **Infer** from your data obtained in the surface-area experiment whether the amount of surface area makes a difference in the rate of chemical weathering. Explain why this occurs.
3. **Conclude** from the last experiment how heat affects the rate of chemical weathering. What does this imply about weathering in the tropics?

Go Further

Based on your results, predict where chemical weathering will be greater, in a cold climate or a warm climate.

Oxygen

Oxygen helps cause chemical weathering. You've seen rusty cars and swing sets. Rust is caused by oxidation. **Oxidation** occurs when a material such as iron is exposed to oxygen and water. When rocks containing iron are exposed to water and the oxygen in the air, the iron in the rock "rusts" and turns reddish, as seen in **Figure 6-5.**

MiniLAB

How does rust form?

Find out what happens when you leave steel wool in water.

Procedure

1. Place some steel wool in a glass dish with 1 cm of water.
2. Observe for several days.

Analysis

1. What changes occurred?
2. What caused the changes?
3. How are these changes related to weathering?

Effects of Climate on Weathering

Mechanical and chemical weathering occur everywhere. However, climate affects the rate and type of weathering. **Climate** is the pattern of weather that occurs in a particular area over many years. In cold climates where freezing and thawing are frequent, mechanical weathering breaks down rocks rapidly through the process of ice wedging.

Chemical weathering is more rapid in warm, wet climates. Thus, chemical weathering occurs rapidly in tropical areas such as the Amazon River region of South America. Lack of moisture in deserts and low temperatures in polar regions keep chemical weathering at a minimum in those climates. How weathering affects rock depends on the type of rock, as illustrated in **Figure 6-6.**

Mechanical and chemical weathering work together. For example, when rocks break apart because of mechanical weathering, exposing more surface area, this increases the rate of chemical weathering. This was shown in Activity 6-1.

Figure 6-5

These rocks have been chemically weathered. *What caused them to be a reddish color?*

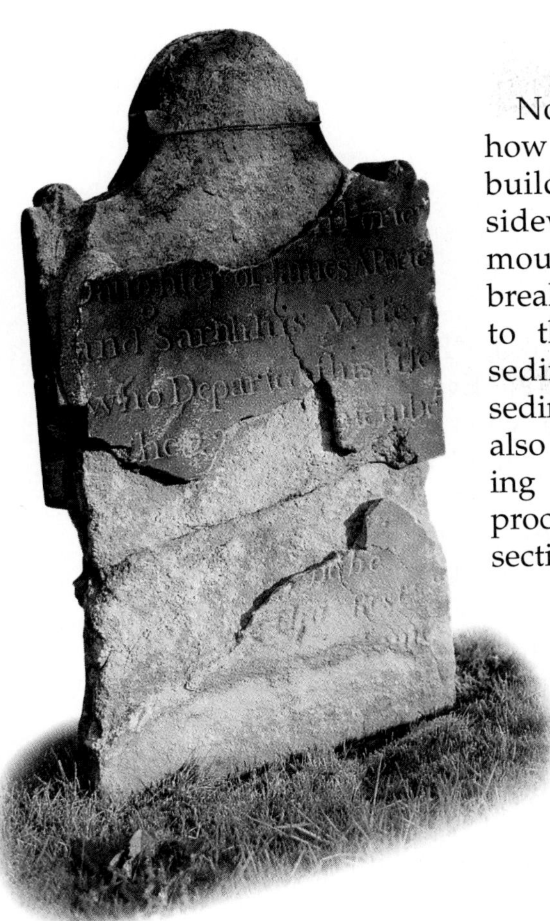

Now you can understand how weathering affects roads, buildings, streets, cemeteries, sidewalks, rocks, caves, and mountains. When weathering breaks down rocks, it contributes to the rock cycle by making sediment that can form into sedimentary rocks. Weathering also begins the process of breaking down rock into soil. This process is discussed in the next section.

Figure 6-6

These old tombstones are about the same age, but they have weathered differently. The type of rock also influences how fast a surface weathers. *Why?*

Section Wrap-up

Review

1. What is the difference between mechanical and chemical weathering?
2. How is mechanical weathering affected by climate?
3. **Think Critically:** How can water be a factor in both mechanical and chemical weathering?

Skill Builder

Recognizing Cause and Effect

Identify the cause and effect in each of these examples of weathering. If you need help, refer to Recognizing Cause and Effect in the **Skill Handbook.**

1. Acid rain has turned a bronze statue green.
2. Tree roots are exposed in cracks in your sidewalk.
3. A piece of limestone has a honeycomb appearance.

Using Computers

Spreadsheet Make a spreadsheet that (1) identifies examples of weathering that you see around your neighborhood and school and (2) classifies each example as the result of mechanical weathering, chemical weathering, or both.

6•2 Soil

Science Words

soil
humus
soil profile
horizon
leaching

Objectives

- Explain how soil evolves from rock.
- Describe soil by comparing the A, B, and C soil horizons.
- Discuss how environmental conditions affect the evolution of soils.

Formation of Soil

How often have you been told, "Take off those dirty shoes before you come into this house"? Ever since you were a child, you've had experience with what many people call dirt, which is actually soil. Soil is found in lots of places: empty lots, farm fields, gardens, and forests.

What is soil and where does it come from? The surface of Earth is covered by a layer of rock and mineral fragments produced by weathering. As you learned in Section 6-1, weathering gradually breaks rocks into smaller and smaller fragments. But these fragments are not soil until plants and animals live in them. Plants and animals add organic matter, such as leaves, twigs, and dead worms and insects to the rock fragments. Soil begins to evolve. **Soil** is a mixture of weathered rock, organic matter, mineral fragments, water, and air. **Figure 6-7** illustrates the process of soil-making. Soil is a material capable of supporting vegetation. Climate, types of rock, slope, amount of moisture, and length of time rock has

Figure 6-7

Soil is constantly evolving from rock.

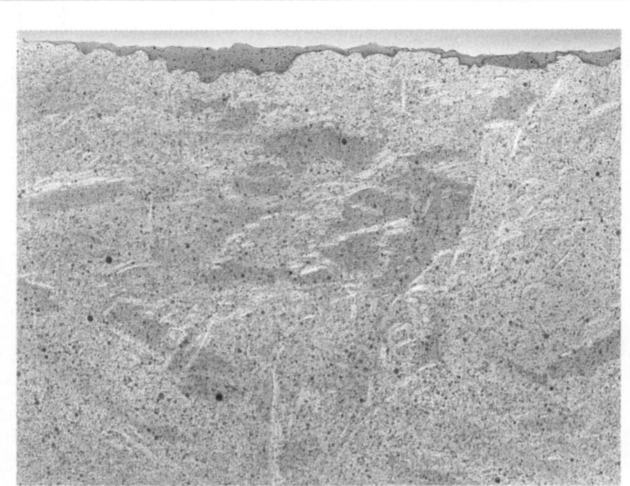

A Rock begins to fracture and break down.

B As rock weathers into smaller fragments, plants begin to grow in the weathered rock.

been weathering influence the formation of soil.

Soil can be considered a complex ecosystem. Soil may contain small rodents, insects, worms, algae, fungi, bacteria, and decaying organic matter. When soil is developing, organic material, such as plants, decays. This material decays until the original form of the matter has disappeared. The material turns into dark-colored matter called **humus.** Humus serves as a source of nutrients for plants, providing nitrogen, phosphorus, potassium, and sulfur. Humus also promotes good soil structure and helps soil retain water. As worms, insects, and rodents burrow throughout the soil, they mix the humus with the fragments of rock. In good-quality surface soil, about half of the volume is humus and half is broken-down rock.

Soil can take thousands of years to form and can range in thickness from 60 m in some areas to just a few centimeters in others. A fertile soil is one that supplies nutrients for plant growth. Soils that develop near rivers often are fertile. Other soils, such as those that develop on steep slopes, may be nutrient-poor and have low fertility.

Soils have small spaces in them. These spaces fill with air or water. In swampy areas, water may fill these spaces year-round. In other areas, soil may fill up with water after rains or during floods.

CONNECT TO

LIFE SCIENCE

In tropical areas with heavy rainfall, rain washes all the elements from the soil. High temperatures and bacteria destroy the humus. The soil becomes so empty of nutrients that plants cannot grow there. ***Infer*** from this whether it would be easy to farm in these areas.

C Worms, insects, bacteria, and fungi living among the plant roots add organic matter to the soil.

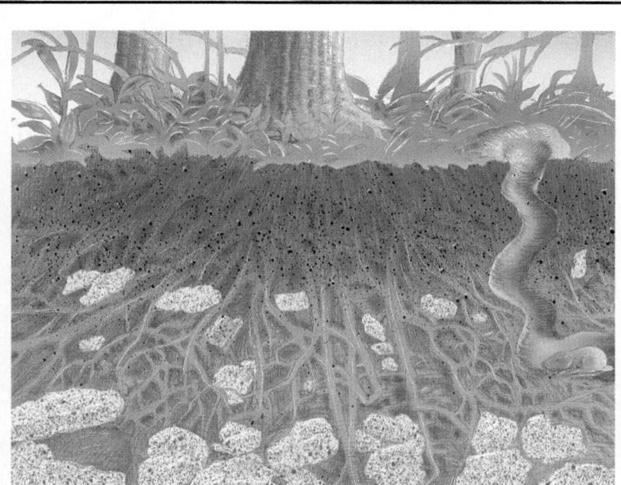

D When plants and animals in the soil die, they break down, or decay, and form dark humus.

Soil Profile

You may have seen layers of soil if you've ever dug a deep hole or driven by a steep slope such as a road cut where the soil and rock are exposed. You might have noticed that plants grow in the top layer of soil. The top layer of soil is darker than the soil layers below it. These different layers of soil make up what is called a **soil profile.** Each layer in the soil profile is called a **horizon.** There are generally three horizons. They are labeled A, B, and C, as in the diagram in **Figure 6-8.** The photo in **Figure 6-9** shows soil horizons in an eroded hillside. Look for examples of soil horizons in road cuts and along steep slopes.

MiniLAB

What is soil made of?

Examine a sample of soil closely to compare it with other students' samples.

Procedure

1. Collect a sample of soil.
2. Observe it closely with a magnifying glass or a microscope.

Analysis

1. Describe the different particles found in your sample. Did you find any remains of once-living organisms?
2. Compare and contrast your sample with those other students have collected. How are the samples the same? How are they different?

Figure 6-8

Under this meadow is soil with a specific profile.

A The soil profile of this meadow has at least three horizons.

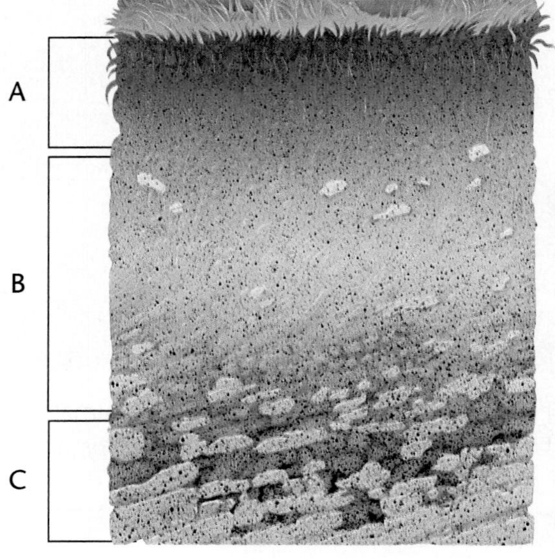

B The horizons in the soil profile of this meadow show different degrees of soil evolution. The A horizon contains humus and small grains of rock. It has changed the most from the underlying rock layer. The B horizon contains minerals leached from the A horizon and little or no humus. The C horizon contains minerals leached from the B horizon and partly weathered rock. *How do you know that this meadow has a well-developed A horizon?*

Horizon A

The A horizon is the top layer of soil. It's also known as topsoil. Topsoil has more humus and smaller rock and mineral particles than the other, less evolved layers in a soil profile. If you dig up a scoop of topsoil and look at it closely, you will see soil, dark in color and containing grains of rock and minerals, decayed leaves, roots of plants, and insects and worms. The A horizon is the most fully evolved soil layer in a soil profile. This means that the A horizon has changed the most from weathered rock.

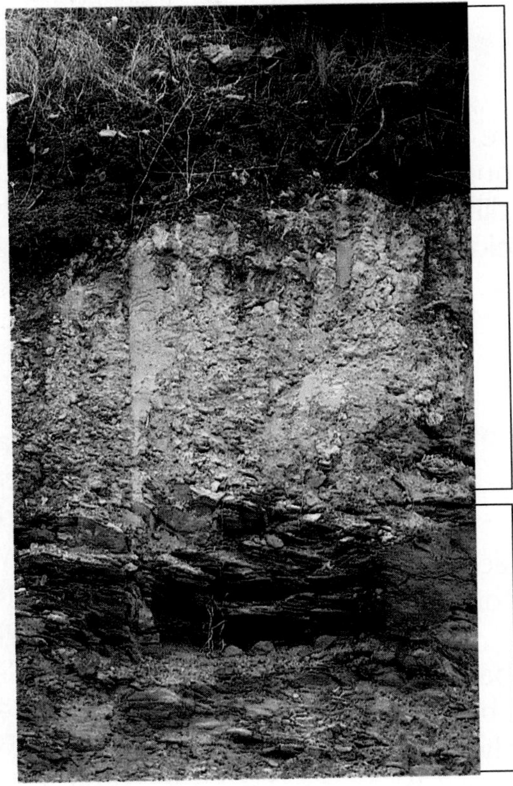

Figure 6-9

Each soil horizon represents a different stage of evolution of this soil.

Horizon B

The layer below the A horizon is the B horizon. This layer is less evolved. It is lighter in color than the A horizon because it has less humus. Some plant roots reach into this horizon, which usually contains elements washed down from the A horizon by the process of leaching. **Leaching** is the removal of soil materials dissolved in water. Water runs down through the soil horizons, leaching out soil materials and depositing them in the layers below. The process of leaching resembles making coffee in a drip coffeemaker. In a coffeemaker, water drips into ground coffee. In the soil, water seeps into the A horizon. In a coffeemaker, the water absorbs the flavor and color from the coffee and flows down into a coffeepot. In the soil, the water reacts with humus to form an acid. This acid dissolves some of the elements from the minerals in the A horizon and carries them into the B horizon. Some leaching also occurs in the B horizon and moves into the C horizon.

CONNECT TO

LIFE SCIENCE

Fungi and bacteria are decomposers, found mainly in horizon A. Decomposers break down the complex organic material in animals and plants into smaller molecules, creating humus. ***Infer*** from this what our world would be like without decomposers.

Horizon C

The C horizon is below the B horizon. It is the bottom layer in a soil profile. Some materials in this layer were leached from the B horizon. The C horizon also contains partly weathered rock that is beginning the long, slow process of evolving into soil. What would you find if you dug all the way to the bottom of the C horizon? As you might have guessed, there would be solid rock. This is the parent rock that gave rise to the soil horizons above it.

Activity 6-2

Soil Characteristics

There are thousands of different kinds of soils around the world. Around your area, you've probably noticed that there are a number of different soils. Collect samples of soil to compare from around your neighborhood and from designated areas of your school grounds.

Problem

What are the characteristics of soils?

Materials

- soil sample
- sand
- gravel
- clay
- water
- watch
- paper
- scissors
- thumbtack
- cheesecloth squares
- graduated cylinder
- plastic coffee can lids (3)
- rubber bands
- 250-mL beakers (3)
- large polystyrene or plastic cups (3)
- hand lens

Procedure

1. Spread your soil sample on a sheet of paper.
2. Describe the color of the soil, and examine the soil with a hand lens. Describe the different particles.
3. Rub a small amount of soil between your fingers. Describe how it feels. Also press the soil sample together. Does it stick together? Wet the sample and try this again. Record all your observations.
4. Test the soil for permeability (the ability of water to move through a substance). Label the three cups A, B, and C. Using a thumbtack, punch an equal number of holes in and around the bottom of each cup.
5. Cover the area of holes with a square of cheesecloth. Secure the cloth on each cup with a rubber band.
6. To hold the cups over the beakers, cut the three coffee-can lids so that the cups will just fit inside the hole (see photo). Place a cup and lid over each beaker.
7. Half-fill cup A with dry sand and cup B with clay. Half-fill cup C with a mixture of equal parts of sand, gravel, and clay.
8. Use the graduated cylinder to pour 100 mL of water into each cup. Record the time when the water is first poured into each cup and when the water first drips from each cup.
9. Allow the water to drip for 25 minutes, then measure and record the amount of water in each beaker.

Analyze

1. **Describe** your soil sample in as much detail as possible.
2. Which substance tested in steps 4-9 is most permeable? Least permeable?

Conclude and Apply

3. How does the addition of gravel and sand affect the permeability of clay?
4. **Describe** three characteristics of soil. Which characteristics affect permeability?
5. Use your observations to **explain** which soil sample would be best for growing plants.

Types of Soil

The texture of soil depends on the proportion of sand, silt, and clay. In Activity 6-2, you discovered that the texture of soil affects how water runs through it. That is not all you'll discover from examining soil profiles.

If you examine a soil profile in one place, it will not look exactly like a soil profile from another location. The two profiles will look different in a number of ways, as you would see if you examined samples of soil from different locations, such as from a desert, a prairie, and a temperate area. Deserts are dry. Prairies are semidry. The temperate zone profile represents a soil from an area with a moderate amount of rain and moderate temperatures.

The thickness of the soil horizons and the soil composition of the profiles depend on a number of conditions. One condition that strongly affects soil profiles and composition is climate. Examples of three

USING TECHNOLOGY

Land Reclamation

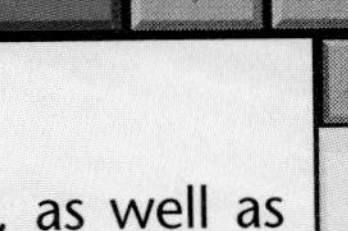

In the United States, sand and gravel, as well as over 60 percent of coal, are obtained by surface mining. In surface mining, huge machines remove the earth above a coal seam and dump it. In the past, the dumped material remained where it was; the mined area remained as desolate as a moon landscape.

Reclamation

The 1977 Federal Strip Mine Law requires that topsoil, the A horizon, now be replaced on surface-mined land. Today, when dozers remove topsoil, it is stockpiled. The material between the topsoil and the coal is blasted to fragments. Then the fragments are transported to a hollow or small valley, sorted, and regraded into land with a minimum slope.

Successful reclamation depends on getting a good topsoil on the land as quickly as possible. Machines carry the stockpiled topsoil to the valley and cover the fragmented rock. Replacing topsoil allows new vegetation to grow rapidly and helps prevent erosion, dirt running into streams, and desolate hillsides. Reclamation experts have used organic waste such as composted sewage and garbage, bark mulch, and other wastes to fertilize the topsoil of reclaimed areas. Some reclaimed surface-mined land is more productive than it was before mining.

Mine Reclamation

Think Critically:

Compare the soil profile of unreclaimed mined land with the soil profile of unmined land. How are they different?

soil profiles from different climates are shown in **Figure 6-10.**

Chemical weathering is slower in areas where there is little rainfall. Soils in dry climates contain little organic material. The soil horizons in dryer areas are thinner than soil horizons in wetter climates. The amount of precipitation affects how much leaching of minerals has occurred in the soil. Soils that have been leached are light in color. Some can be almost white.

Another condition that affects soil development is time. Time changes the characteristics of soil. If the weathering of the rock has been going on for a short time, the parent rock of the soil determines the soil characteristics. As the weathering continues for a longer time, the soil resembles the parent rock less and less.

Slope also affects soil profiles. On steep slopes, soil horizons are often poorly developed. In bottomlands, where there is a lot of water, soils are often very thick, dark, and full of organic material. A south-facing slope receives more solar radiation and consequently has a different soil development from a north-facing slope. The amount of humus in the soil also affects soil profiles. In the United States, there are nine groups of soils recognized as well as many subgroups. The map in **Figure 6-11** shows the nine main soil groups.

Figure 6-10

Prairie soils are brown and fertile with thick grass roots that fill the deep A horizon. Temperate soils are loose, brown soils with less-developed A horizons than prairie soils. Desert soils are coarse, light-colored, and contain a lot of minerals. Of the three soil types shown here, desert soils have the least-developed A horizon.

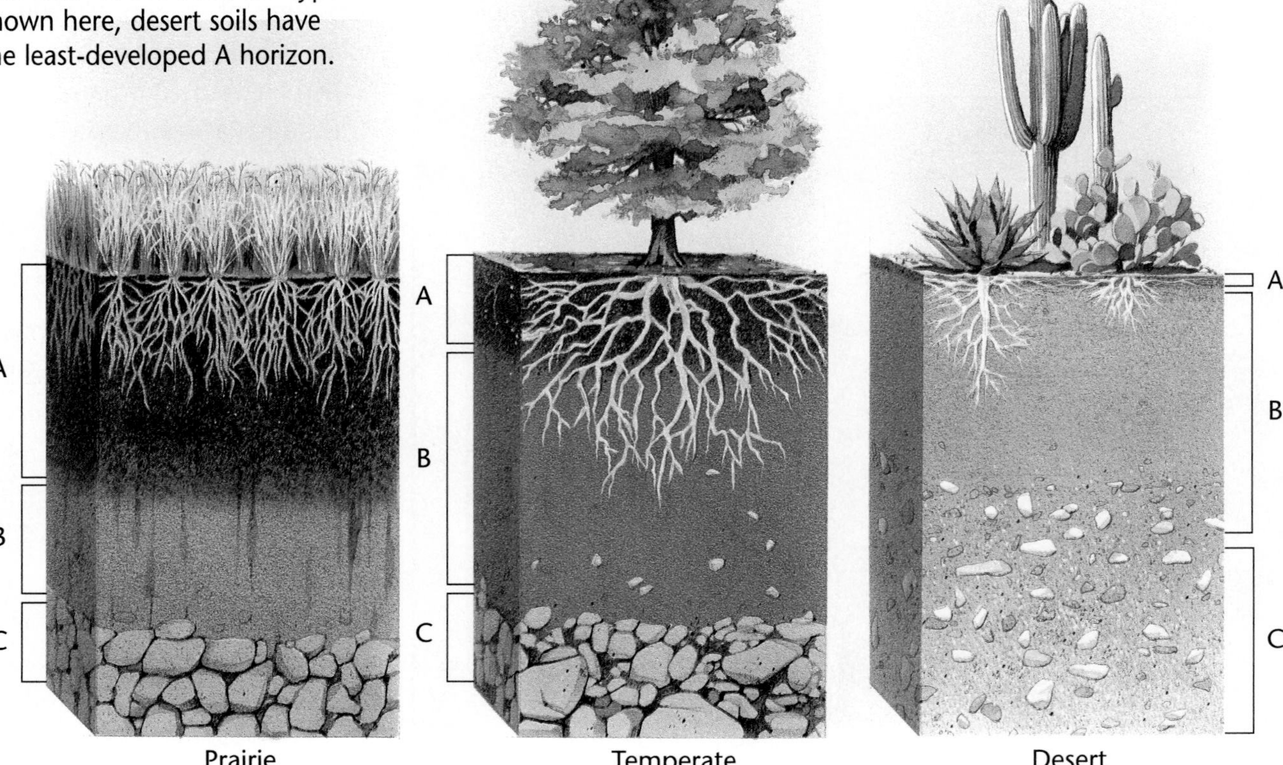

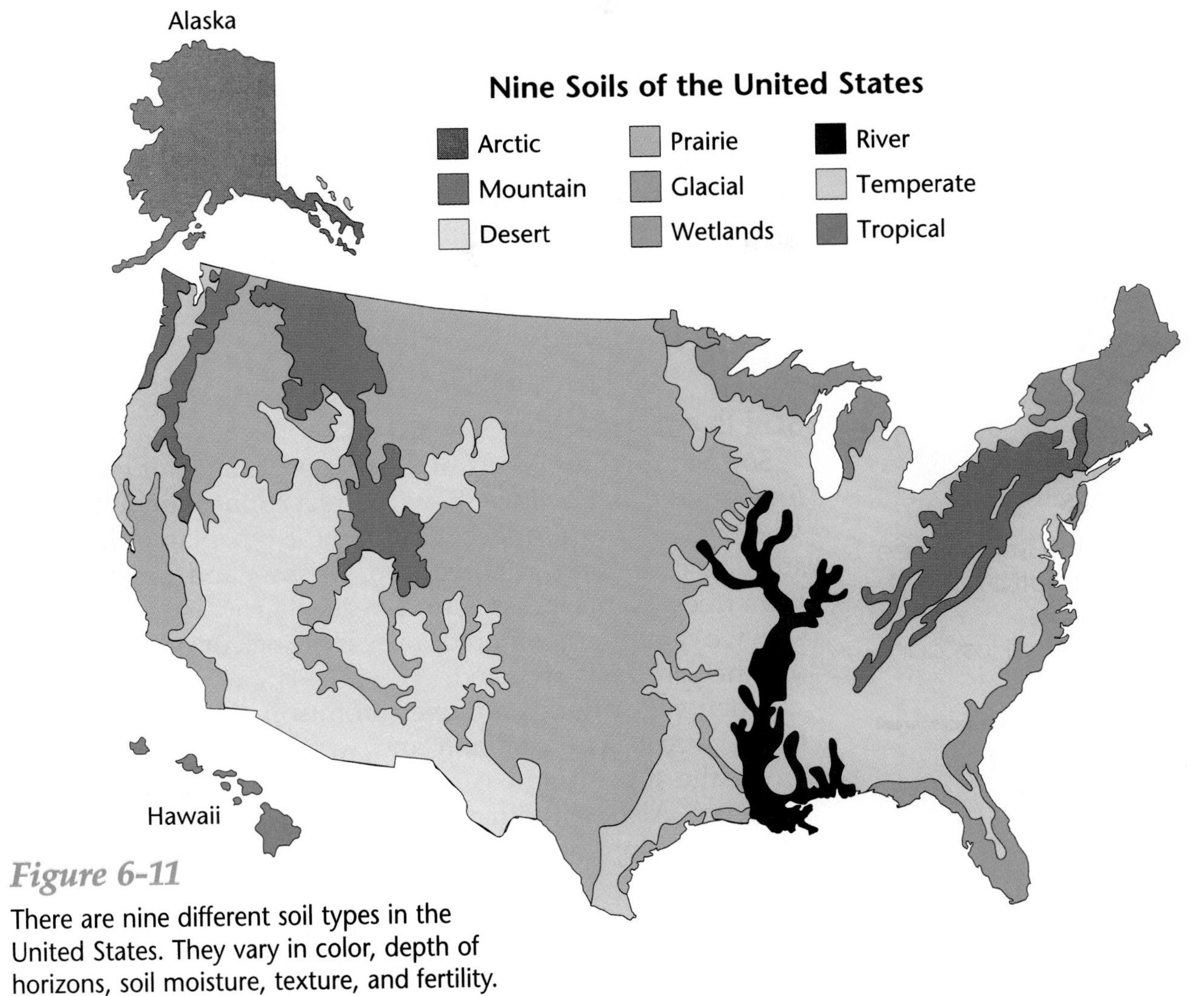

Figure 6-11

There are nine different soil types in the United States. They vary in color, depth of horizons, soil moisture, texture, and fertility.

Section Wrap-up

Review

1. How do organisms help soils develop?
2. Why do soil profiles contain layers or horizons?
3. Why does horizon B contain minerals from horizon A?
4. **Think Critically:** Why is the soil profile in a rain forest different from one in a desert?

Skill Builder

Concept Mapping

Make an events chain map that explains how soil evolves. Use the following terms and phrases: *soil is formed, humus develops, rock is weathered, plants grow, worms and insects move in, humus mixes with weathered rock.* If you need help, refer to Concept Mapping in the **Skill Handbook.**

Science Journal

Research how farmers can improve the quality of soil by planting legumes. Write a summary of what you learn in your Science Journal.

SCIENCE & SOCIETY

6•3 ISSUE: Land Use and Soil Loss

Science Words

desertification

Objectives

- Describe the importance of soil and activities that lead to its loss.
- Explain how soil loss can be minimized.

Agriculture

Soil Is an Important Resource

Soil is important. Many of the things we take for granted—food, paper, and cotton—have a direct connection to the soil. Vegetables, grains, and cotton come from plants, livestock such as cattle and pigs feed on grasses, and paper comes from trees. Plants, grasses, and trees all grow in soil. Without soil, we cannot grow food or raise livestock or produce paper or other products we need.

When vegetation is removed from soil, the soil is exposed to the direct action of rain and wind. Rain and wind can erode the topsoil and carry it away, destroying the soil's structure. Also, without plants, soil development slows and sometimes stops because humus is no longer being produced.

2 Points of View

Plowing Disrupts the Soil

Every year the population of Earth increases by 93 million people. More people need more food. Farmers plow more fields to raise more food for the increasing population. This increases pressure on our soil resources.

Soils have not always been managed effectively. Plowing soil is the mechanical turning and loosening of the soil to improve it for crops. Plowing soil removes the plant cover that holds soil particles in place, leaving soils open to wind and water erosion. Sometimes, as shown in Figure 6-12, the wind blows soil from a plowed field. Soil erosion in many places occurs at a much faster rate than the natural processes of weathering can replace it.

Figure 6-12

Some farming practices can increase soil loss.

Soil loss is particularly severe in the tropics. Tropical rains running down steep slopes quickly erode soil. Each year, thousands of square kilometers of tropical rain forest are cleared for farming and grazing. Soils in tropical rain forests appear rich in nutrients but are almost sterile below the first few centimeters. When the rain forest is removed, the soil is useful to farmers for only a few years before the nutrients are gone. The soil then becomes infertile. Farmers clear new land, repeating the process and increasing the damage to the soil.

Near the deserts of the world, sheep and cattle eat every bit of grass. When natural vegetation is removed from land that receives little rain, plants don't grow back. This leads to a loss of soil through wind erosion. Groundwater evaporates. The dry, unprotected surface can be blown away. The desert spreads. Desert formation, called **desertification,** is currently happening on every continent.

Figure 6-13

In no-till farming, the soil is not plowed before planting. *How does this conserve soil?*

Farmers Work to Minimize Soil Loss

All over the world, farmers take steps to minimize soil erosion. They plant shelter belts of trees to break the force of the wind and cover bare soils with decaying plants to hold soil particles in place. In dry areas, instead of plowing under the natural vegetation to plant crops, farmers graze animals on the vegetation. Proper grazing management can retain plants and reduce soil erosion.

Steep slopes, prone to erosion, can be taken out of cultivation or terraced. In the tropics, planting trees to block the force of rain falling on open ground reduces erosion. On gentle slopes, plowing along the natural contours of the land or planting crops in strips helps reduce water erosion. In strip cropping, a crop that covers the ground is alternated with a crop such as corn that leaves a considerable amount of land exposed. In recent years, many farmers have begun the practice of no-till farming. Normally, farmers till or plow their fields three or more times a year. In no-till farming, seen in **Figure 6-13,** plant stalks are left in the field. At the next planting, farmers seed crops without destroying these stalks and without plowing the soil. No-till farming provides cover for the soil all year round and reduces erosion.

USING MATH

About 18 square km of rain forests disappear each hour. At this rate, all rain forests in the world will be gone in about 175 years. How many square kilometers of rain forests exist on Earth currently?

Section Wrap-up

Review

1. Why is more and more land plowed?
2. What farming practices minimize soil loss?

Explore the Issue

Studies show that for every 3.5 cm of topsoil lost, crop yields drop by about 6 percent. How can we protect soil and still provide the necessities of life?

SCIENCE & SOCIETY

People and Science

Susan Colclazer, Naturalist
Bryce Canyon National Park

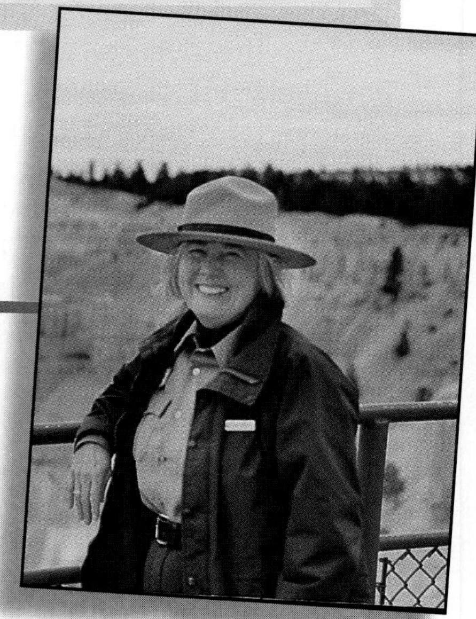

On the Job

Q **What would we see if we visited Bryce Canyon National Park?**

A As you come to the park, you make a tremendous transition, driving by grass and sagebrush, then entering the towering Ponderosa pine forest at the park boundary. As you reach an overlook parking area, suddenly everything seems to drop away in front of you. With just a short walk, you see canyons filled with incredibly-shaped, colored rock formations we call hoodoos. If you look across the valley and up, you see a set of pink cliffs that look just like the ones beneath your feet. Then you start thinking about what might have happened to create this place.

Q **Are the interesting colors of the canyon's rock formations due to oxidation?**

A Yes. Iron, manganese, and magnesium oxides give the shades of yellow, blue, purple, pink, and orange. The white is from limestone, which makes up most of the top rock layers in the park. Different weather and angles of light hitting the rocks make the colors almost seem to change right in front of you.

Personal Insights

Q **When we think of the effects of weathering on rock, we tend to think about very long-term changes. Have you witnessed any changes in the time you've worked in the park?**

A During storms, you might actually see a large chunk of the rim wash away. I have seen trails relocated because a rock slide completely buried them. One night, about half of the spire on a formation called the Sentinel broke off. People who've been around the area longer than I have can tell you about many formations that are now just piles of rubble.

Career Connection

Geologists aren't the only scientists who are interested in Bryce Canyon. Archaeologists, sociologists, and botanists have made studies there. Make a list of things the following scientists might study in the canyon, and tell how their findings would add to the overall picture of the canyon's history.

- **Archeologist**
- **Botanist**

Chapter 6

Review

Summary

6-1: Weathering

1. Mechanical weathering breaks apart rocks without changing their chemical composition. Ice and plant roots can cause mechanical weathering. Chemical weathering changes the composition of the rock. Chemical weathering can be caused by acidic water and oxygen.
2. Climate affects the rate and type of weathering.

6-2: Soil

1. Soil develops when rock is weathered and organic matter is added.
2. In a soil profile, the A horizon, or topsoil, is dark in color and contains weathered rock and organic matter. The B horizon has little or no humus and usually contains materials leached from the A horizon. The C horizon contains partly weathered rock and some materials leached from the B horizon.
3. Soil characteristics depend on the climate of the area, the type of parent rock, the slope of the land, the amount of humus in the soil, and the length of time the soil has been evolving.

6-3: Science and Society: Land Use and Soil Loss

1. Soil is important for food. Soil development stops when vegetation is removed from soil.
2. The increasing population on Earth leads to development of the land to provide food. More development can lead to soil erosion, but farmers try to stop this with soil conservation methods.

Key Science Words

a. chemical weathering
b. climate
c. desertification
d. horizon
e. humus
f. ice wedging
g. leaching
h. mechanical weathering
i. oxidation
j. soil
k. soil profile
l. weathering

Reviewing Vocabulary

Match each phrase with the correct term from the list of Key Science Words.

1. breaking down rocks without changing the chemical composition
2. weathering in which the composition of the rock is changed
3. mixture of weathered rock and organic matter
4. decayed organic matter
5. the A, B, and C layers of soil
6. each layer in a soil profile
7. a process that carries dissolved materials downward in soil
8. desert formation
9. rock is forced apart by ice
10. chemical weathering process due to exposure to water and oxygen

Chapter 6 Review

Checking Concepts

Choose the word or phrase that completes the sentence.

1. Acids that are produced by plants can cause _______.
   a. desertification
   b. overgrazing
   c. mechanical weathering
   d. chemical weathering
2. Freezing and thawing weathers rocks because water _______ as it freezes.
   a. contracts   c. expands
   b. gets more dense   d. percolates
3. _______ occurs when roots force rock apart.
   a. Mechanical weathering
   b. Leaching
   c. Ice wedging
   d. Chemical weathering
4. _______ causes rust to form.
   a. Oxygen   c. Feldspar
   b. Carbon dioxide   d. Paint
5. Poor farming practices in areas that receive little rain can result in _______.
   a. ice wedging   c. leaching
   b. oxidation   d. desertification
6. Chemical weathering is most rapid in _______regions.
   a. cold, dry   c. warm, moist
   b. cold, moist   d. warm, dry
7. _______ is a mixture of weathered rock and organic matter.
   a. Soil   c. Carbon dioxide
   b. Limestone   d. Clay
8. Decayed organic matter is called _______ .
   a. leaching   c. soil
   b. humus   d. sediment
9. Humus is found almost exclusively in the _______ .
   a. A horizon   c. C horizon
   b. B horizon   d. D horizon
10. No-till farming helps prevent _______ .
    a. Leaching   c. Overgrazing
    b. Crop rotation   d. Soil erosion

Understanding Concepts

Answer the following questions in your Science Journal using complete sentences.

11. How have farmers tried to stop soil loss from their land?
12. Explain how soil evolves from rock.
13. How can a worm help soil develop?
14. Explain why the B horizon is less weathered than the A horizon.
15. Explain how the clearing of forests, as in the photo below, contributes to soil loss.

Thinking Critically

16. Which type of weathering, mechanical or chemical, would you expect to be more effective in a desert region? Explain.
17. How does ice wedging damage roads and streets in cold climates?
18. Why is soil so important?
19. Why is it difficult to replace lost topsoil?
20. Explain how chemical weathering can form a cavern.

Developing Skills

For help, refer to the **Skill Handbook.**

21. **Sequencing:** Do a sequence chart of soil development.
22. **Concept Mapping:** Complete the events chain concept map that shows two ways in which acids can cause chemical weathering.

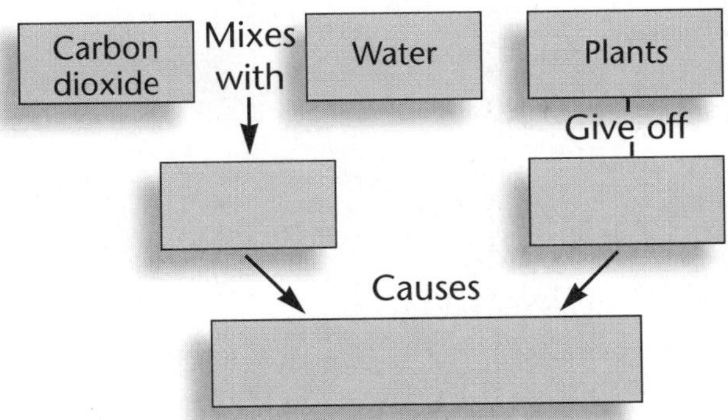

23. **Using Variables, Constants, and Controls:** Juan Carlos wanted to know if planting grass on a slope would prevent soil from being washed away. To find out, he put the same amount and kind of soil in two identical pans. In one of the pans, he planted grass. To create equal slopes for his test, he placed identical wooden wedges under one end of each pan. He was careful to pour the same amount of water at the same rate over the soil in the two pans. What is Juan's control? What factors in his activity are constants? What is the variable he is testing?
24. **Recognizing Cause and Effect:** In Juan Carlos's activity, he found that in the pan containing soil and grass, the least amount of soil washed away. What are the cause and the effect in his observation?
25. **Classifying:** Classify the following events as either chemical or mechanical weathering: rocks that contain iron rust; freezing and thawing of water cause cracks to form in the street; acids from mosses leave small pits in rocks; roots of trees break rocks apart; and water seeping through cracks in limestone dissolves away some of the rock.

Performance Assessment

1. **Design an Experiment:** In Activity 6-1, you observed how an acid (vinegar) causes chemical weathering of a rock (chalk). Now design an experiment to find out how the concentration (or strength) of an acid affects the rate of chemical weathering. State your hypothesis clearly and eliminate all variables except the one you want to test. Perform your experiment.
2. **Design a Model:** Demonstrate how leaching occurs in soil. Review the procedures of Activity 6-2 on page 160 before you begin. The following list of materials is suggested: a soil sample, red powdered tempera paint, plastic coffee-can lid, cheesecloth, rubber band, 250-ml beaker, thumbtack, large plastic cup, graduated cylinder, and water.
3. **Poster:** Plant three identical plants in three different types of soils. Keep everything constant in your experiment except the type of soil. Over a period of weeks, observe and measure plant growth. Then make a poster display to share your results with your classmates. The poster should show how the soil type affected plant growth.

Previewing the Chapter

Chapter 7

Erosional Forces

In Chapter Six, you learned about weathering of Earth's surface. Weathering breaks down rocks into smaller particles. Then forces of nature move these particles and deposit them elsewhere. One of these forces, wind, blew sand particles that eroded this natural rock sculpture in Vermillion Cliffs Wilderness in Utah. Erosional forces constantly change the surface of Earth, making changes in our landscape.

EXPLORE ACTIVITY

Find out how sediments move from one location to another.

1. Place a small pile of sand and gravel in one end of a large shoe box lid.
2. Move the pile to the other end of the lid without touching the particles with your hands.
3. Try to move the sand in a number of different ways.

Observe: How many methods did you use? How did your methods compare with forces of nature that move sediments? Explain your ideas in your Science Journal.

Previewing Science Skills

- In the **Skill Builders,** you will map concepts, compare and contrast, and sequence.
- In the **Activities,** you will design experiments, observe, infer, and analyze.
- In the **MiniLABs,** you will design experiments, observe, and infer.

7•1 Gravity

Science Words

erosion
deposition
mass movement
slump
creep

Objectives

- Define erosion and deposition.
- Compare and contrast slumps, creep, rockslides, and mudflows.

Erosion and Deposition

Have you ever been by a river just after a heavy rain and seen the water look as muddy as the water in **Figure 7-1?** Rivers look muddy when there is a lot of dirt and soil in them. Some of the dirt comes from along the riverbank, but the rest is carried to the river from much more distant sources.

Muddy water is a product of erosion. The process of **erosion** wears away surface materials and moves them from one location to another. The major agents of erosion are gravity, glaciers, wind, and water. The first three will be discussed in this chapter. Running water and wave erosion will be discussed in Chapter 8. Another kind of erosion is shown in **Figure 7-2.**

As you investigate the agents of erosion, you will notice that they have several things in common. Gravity, glaciers, wind, and water all wear away materials and carry them off. But these agents erode materials only when they have enough energy of motion to do work. For example, air can't erode sediments when the air is still. But once air begins moving and develops into wind, it can carry dust, soil, and even rock along with it.

Figure 7-1

The muddy look of some rivers comes from the load of sediment carried by the water.

Another thing that the agents of erosion have in common is that they all drop their load of sediments when their energy of motion decreases. This dropping of sediments is called **deposition.** Deposition is the final stage of an erosional process. Sediments and rocks are deposited. The surface of Earth is changed. But next year or a million years from now, those sediments may be eroded again.

If a rock falls from the edge of a cliff and touches nothing to slow it down, gravity causes it to accelerate (go faster) as it falls. The formula $v = gt$ can be used to find the speed, v, after t seconds. In the formula, g is the acceleration due to gravity, 9.8 m/s^2. Calculate the speed of a falling rock after two seconds of free-fall.

Erosion and Deposition by Gravity

Gravity is the force of attraction that exists between all objects. Because Earth has great mass, other objects are attracted to Earth. This makes gravity a force of erosion and deposition. Gravity causes loose materials to move down a slope. When gravity alone causes materials to move downslope, this type of erosion is called **mass movement.** Some mass movements are very slow; you hardly notice that they're happening. Others, however, happen very quickly. Let's examine some types of mass movements.

Figure 7-2

This hill has been eroded. *What was the agent of erosion?*

Figure 7-3

Slumps occur when material slips downslope as one large mass.

Slump

A **slump** is a mass movement that happens when loose materials or rock layers slip down a slope. In a slump, strong rock or sediment lies over weaker materials. The underlying material weakens and can no longer support the rock and sediment above. The soil and rock slip downslope in one large mass. Sometimes a slump happens when water penetrates the upper layer on a slope, but cannot flow through the lower layers. Water and mud build up. The upper layer of sediments slips along the mud and slides downslope. A curved scar is left where the slumped materials originally rested, as shown in **Figure 7-3.** Slumps happen most frequently after earthquakes or heavy, prolonged rains.

Figure 7-4

Perhaps you can find evidence of soil creep around your home or school. Look for tilted retaining walls and fences and even sod that has stretched apart.

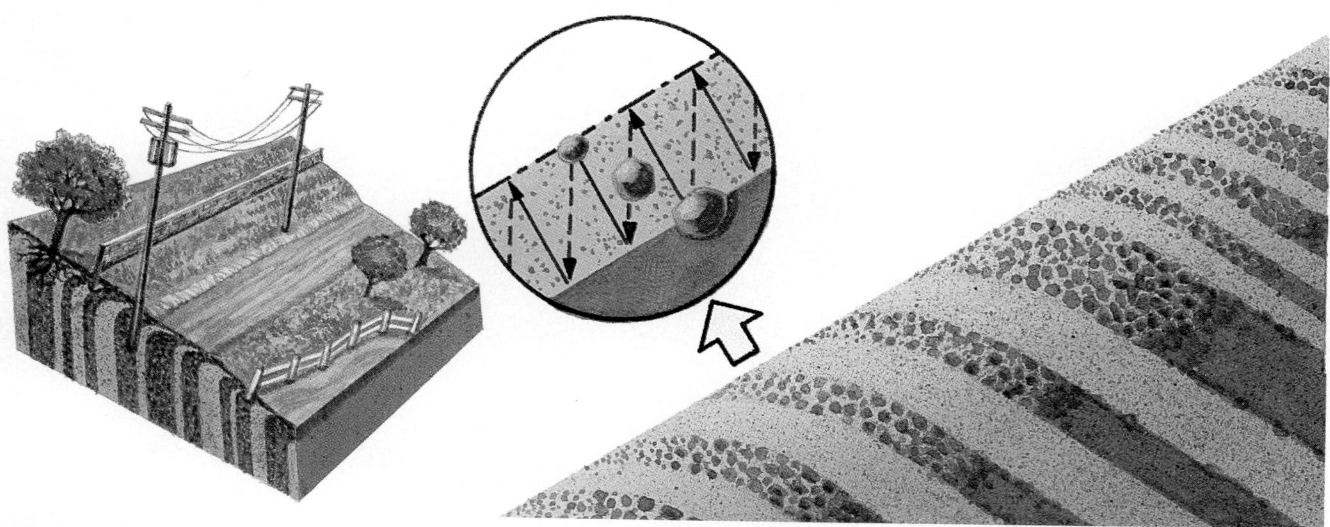

A Several years of creeping downslope can cause objects such as utility poles and fence posts to lean.

B Below the surface, as the ground freezes, expanding ice in the soil pushes up fine-grained sediment particles. Then, when the soil thaws, the sediment falls downslope, often less than a millimeter at a time.

Creep

On your next drive, look along the roadway for slopes where trees, utility poles, and fence posts lean downhill. Leaning poles indicate another mass movement called creep. **Creep** gets its name from the way sediments slowly inch their way down a hill. As **Figure 7-4** illustrates, creep is common in areas of freezing and thawing.

Rockslides

"Falling Rock" signs warn of another type of mass movement called a rockslide. Rockslides happen when large blocks of rock break loose from a steep slope and start tumbling. As they fall, these rocks crash into other rocks and knock them loose. More and more rocks break loose and tumble to the bottom. Rockslides are very fast and can be very destructive in populated mountain areas like the Alps. They commonly occur in mountainous areas or where there are steep cliffs, as portrayed in **Figure 7-5.** Rockslides happen most often after heavy rains or during earthquakes, but they can happen on any rocky slope at any time without warning.

Figure 7-5

Piles of broken rock at the bottom of a cliff tell you that rockslides have occurred and are likely to happen again.

MiniLAB

What causes mass movements?

Procedure

1. Put moist sand and gravel in a pan. Use these sediments to model landforms showing how mass movements of sediments occur.
2. Set up situations to model mudflows, slumps, and rockslides.

Analysis

1. What factors must be present for mass movements to occur?
2. In what ways are the three mass movements similar? Different?

The fall of a single, large rock down a steep slope can cause serious damage to structures at the bottom. Rockfalls like the one in **Figure 7-6** frequently occur in mountainous areas. During the winter, freezing and thawing of ice in the cracks of the rock cause pieces of the rock to fracture. In the spring, the pieces of rock eventually break loose and fall down the mountainside.

Mudflows

Imagine traveling along a mountain road during a rainstorm. Suddenly a wall of mud, the consistency of chocolate pudding, slides down a slope and threatens to cover your car. You've been caught in a mudflow, a thick mixture of sediments and water flowing down a slope. The mudflow in **Figure 7-7** caused a lot of destruction.

Mudflows usually occur in relatively dry areas where weathering accumulates thick layers of dry sediments. When heavy rains fall in these areas, water mixes with sediments and forms a thick, pasty substance. Gravity causes this mass to slide downhill. When a mudflow finally reaches the bottom of a slope, it loses its energy of motion and deposits all the sediments and debris it has been carrying. These deposits usually form a mass that spreads out in a cone shape.

Figure 7-6

In the Alps, a rock fell from the cliff above and struck this new apartment the day before the residents were supposed to move in.

Figure 7-7

A mudflow has enough energy to move almost anything in its path. *How do mudflows differ from slumps, creep, and rock-slides?*

Mudflows, rockslides, creep, and slump are similar in some ways. They're all more likely to happen on steep slopes. They all depend on gravity to make them happen. And, regardless of the type of mass movement, it will occur more often after a heavy rain because the water adds mass and lubricates the area where different layers of sediment meet.

Can you think of one more way these mass movements are alike? All mass movements erode sediments from the top of a slope and deposit them farther downslope. The result is that mass movements constantly change the shape of a slope so that it becomes less steep.

Section Wrap-up

Review

1. Define erosion and name the agents that cause it.
2. How does erosion change the surface of Earth?
3. What characteristics do all types of mass movements have in common?
4. **Think Critically:** When people build houses and roads, they often pile up dirt or cut into the sides of hills. Predict how these activities affect sediments on a slope.

Skill Builder

Concept Mapping

Make a concept map about mass movements using these terms: *gravity, slump, creep, rockslides, mudflows, curved scar, leaning trees and poles, rock piles, and cone-shaped masses.* If you need help, refer to Concept Mapping in the **Skill Handbook.**

Using Computers

Spreadsheet Using a spreadsheet, make a diagram comparing the four different types of mass movements. Compare at least three characteristics of these mass movements.

SCIENCE & SOCIETY

7•2 ISSUE: Developing Land Prone to Erosion

Objectives

- Describe ways that erosion can be reduced in some high-risk areas.
- Explain why problems develop when people live on land prone to excessive erosion.

Living on the Edge

Some people like to live in houses and apartments on the sides of hills and mountains. Realtors say that people want to live in these places for the good view. But when you consider gravity as an agent of erosion, do you think steep slopes are safe places to live?

2 Points of View

Steep Slopes Can Be Made Safe

People can have a beautiful view and reduce erosion on steep slopes. Planting vegetation is one of the best ways to reduce erosion. Plant roots hold soil in place. Plants also absorb large amounts of water. A person living on a steep slope might also build terraces or retaining walls to reduce erosion.

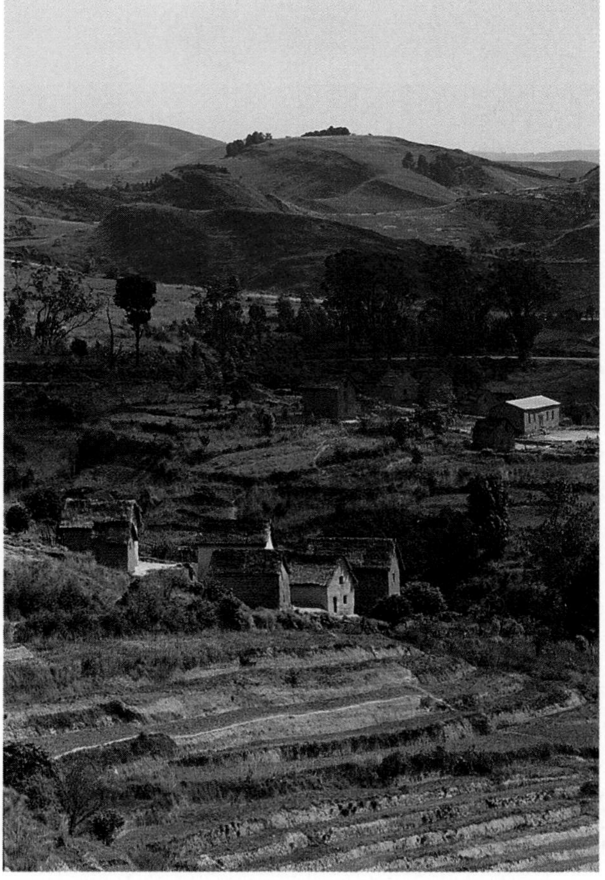

Figure 7-8
Terracing is a way to prevent erosion on slopes. In Madagascar, terraces have existed for a long time. *How do terraces prevent erosion?*

Terraces are broad, step-like cuts made into the side of a slope, as shown in **Figure 7-8.** When water flows into a terrace, it slows down and loses its energy. Terracing slows soil erosion. Retaining walls made of concrete or railroad ties can also reduce erosion by keeping soil and rocks from sliding downhill. However, preventing mass movements on a slope is difficult because rain or earthquakes can cause the upper layers of rock to slip over the lower layers.

Building on Steep Slopes Is Dangerous

When people build homes on steep slopes, they must constantly battle erosion that occurs naturally. Sometimes when they build, people make a slope steeper or remove vegetation. This speeds up the erosion process and creates additional problems. Some steep slopes are prone to slumps because of weak sediment layers underneath. Not even planting bushes and trees could save the houses in **Figure 7-9.**

Figure 7-9

Building on steep slopes can have severe consequences.

People who live in areas with erosion problems spend a lot of time and money trying to preserve their land. Sometimes they're successful in slowing down erosion, but they can never eliminate it and the danger of mass movement. Eventually gravity wins and cliffs cave in. Sediments constantly move from place to place, changing the shape of the land forever.

Section Wrap-up

Review

1. How can erosion be reduced on steep slopes?
2. How does building on slopes increase erosion?

Explore the Issue

Imagine that you live on a cliff overlooking a river. You love your house and the beautiful view. Unfortunately, the river below floods frequently. The river is undercutting the cliff where you live. Several times the city has evacuated your family. Finally, the mayor informs your family that you must move. She cites the danger of living on the cliff and the cost of evacuating you each time. Is this fair? Should communities be allowed to control where people live? How does your homesite affect taxes and insurance rates in the community? Construct a play in which students present both sides of this issue.

SCIENCE & SOCIETY

7•3 Glaciers

Science Words

glacier
plucking
till
moraine

Objectives

- Describe the movement of glaciers.
- Explain glacial erosion.
- Compare and contrast till and outwash.

Most glaciers move extremely slowly, sometimes as slowly as 25 centimeters per day. On average, how many meters will a slow glacier move in 1 year?

Continental and Valley Glaciers

Does it snow where you live? In some areas of the world, it is so cold that snow remains on the ground year-round. When snow doesn't melt, it begins piling up. As it accumulates, the weight of the snow becomes great enough to compress its bottom layers into ice. Eventually the snow can pile so high that the ice on the bottom partially melts and becomes putty-like. The whole mass begins to slide on this putty-like layer, and it moves downhill. This moving mass of ice and snow is a **glacier.**

Glaciers are agents of erosion. As glaciers pass over land, they erode it, changing its features. Glaciers then carry eroded material along and eventually deposit it somewhere else. Glacial erosion and deposition change large areas of Earth. There are two types of glaciers: continental glaciers and valley glaciers.

Continental Glaciers

Continental glaciers are huge masses of ice and snow. In the past, continental glaciers covered up to 28 percent of Earth. Scientists call the periods when glaciers covered much of the land *ice ages*. The most recent ice age began over a period of 2 to 3 million years ago. Then, about 20 000 years ago, the ice

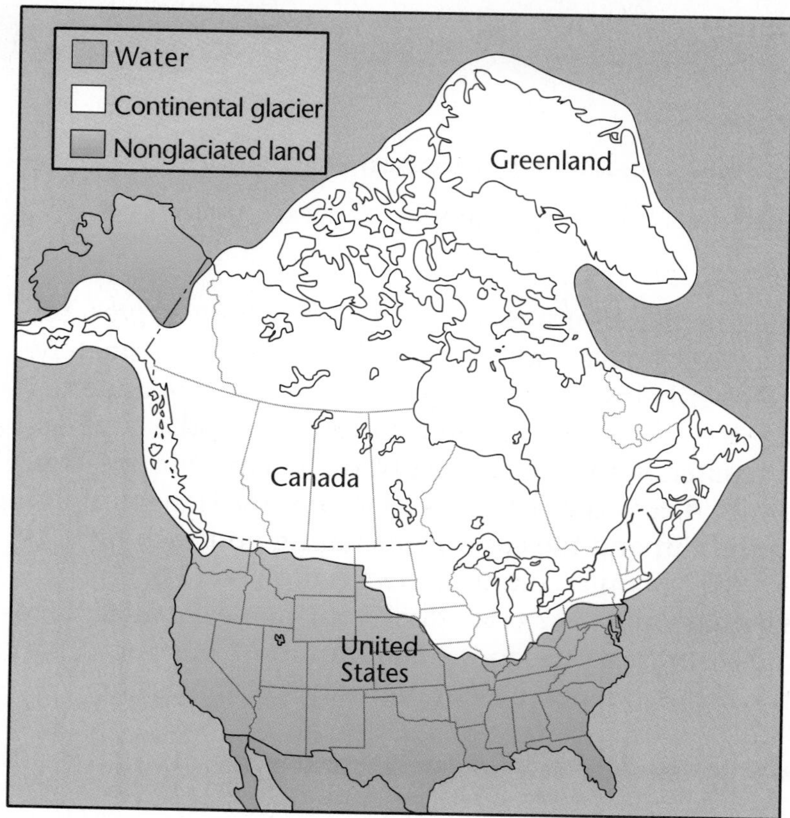

Figure 7-10

This map shows the extent of continental glaciation in North America about 20 000 years ago. *Was your area glaciated?*

Figure 7-11

Continental glaciers and valley glaciers are both agents of erosion.

A Today, a continental glacier covers Antarctica.

sheets began to melt. Today, glaciers like the one in **Figure 7-11A** cover only 10 percent of Earth, mostly near the poles in Antarctica and Greenland. Continental glaciers are such thick sheets of ice that they can almost bury mountain ranges on the land they cover. Glaciers make it impossible to see most of the land features in Antarctica. During the time when much of North America was covered by ice, the average air temperature on Earth was about 5°C lower than it is today.

Valley Glaciers

Valley glaciers occur even in today's warmer climate. In the high mountains where the average temperature is low enough to prevent snow from melting during the summer, valley glaciers grow and creep along. **Figure 7-11B** shows a valley glacier in Tibet.

How is it possible that something as fragile as snow or ice can become an agent of erosion that can push aside trees, drag along rocks, and erode the surface of Earth?

B Valley glaciers, like this one in Tibet, form in mountains whose peaks lie above the snowline, where snow lasts all year.

Figure 7-12

The diagram and photos show landforms characteristic of glacial erosion.

Glacial Erosion

As they move over land, glaciers are like bulldozers, pushing loose materials out of their path. **Figure 7-12A** shows this. Eroded sediments get pushed in front of a glacier, carried underneath it, and piled up along its sides. Glaciers also weather and erode rock and soil that isn't loose. When glacial ice melts, water flows into cracks in rocks. Later, the water refreezes in these cracks, expands, and fractures the rock into pieces. (**Figure 6-3** on page 150 illustrates this ice wedging process.) These rock pieces are then lifted out by the glacial ice sheet. This process, called **plucking,** results in boulders, gravel, and sand being added to the bottom and sides of a glacier.

As a glacier moves forward, plucked rock fragments and sand at its base scrape the soil and bedrock, eroding even more material than ice alone could. When bedrock is gouged deeply by dragged rock fragments, marks such as those in **Figure 7-12B** are left behind. These marks, called grooves, are deep, long, parallel scars on rocks. Less-deep marks are called striations (stri AY shuns). These marks indicate the direction the glacier moved.

Valley glaciers

Arête

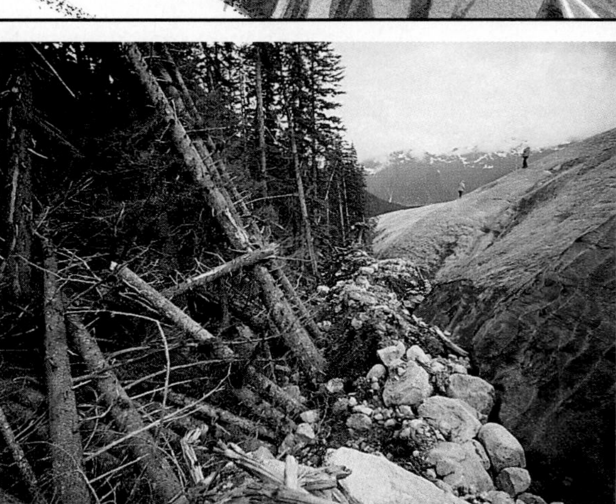

A No agent of erosion is more powerful than an advancing glacier.

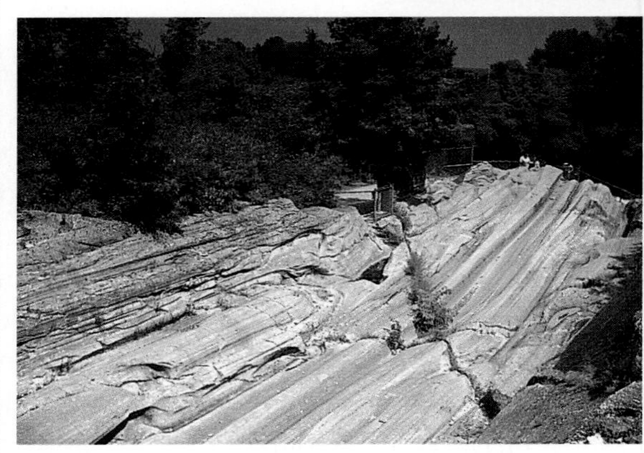

B When glaciers melt, striations or grooves may be found on the rocks beneath. These glacial grooves on Kelley's Island, Ohio, are 10 m wide and 5 m deep.

Evidence of Valley Glaciers

If you visit the mountains, you can see if valley glaciers ever existed there. You might look for striations, then search for evidence of plucking. Glacial plucking often occurs near the top of a mountain where a glacier is in contact with a wall of rock. Valley glaciers erode bowl-shaped basins, called cirques, in the sides of the mountains. A cirque (SURK) is shown in **Figure 7-12C.** If two or more glaciers erode a mountain summit from several directions, a ridge, called an arête (ah RET), or sharpened peak, called a horn, forms. These and other features are shown in **Figure 7-12.** The photo in **Figure 7-12D** shows a mountain horn.

Valley glaciers flow down mountain slopes and along valleys, eroding as they go. Valleys that have been eroded by glaciers have a different shape from those eroded by streams. Stream-eroded valleys are normally V-shaped. Glacially eroded valleys are U-shaped because a glacier plucks and scrapes soil and rock from the sides as well as from the bottom. **Figure 7-14,** on page 187, compares a U-shaped valley with a V-shaped valley.

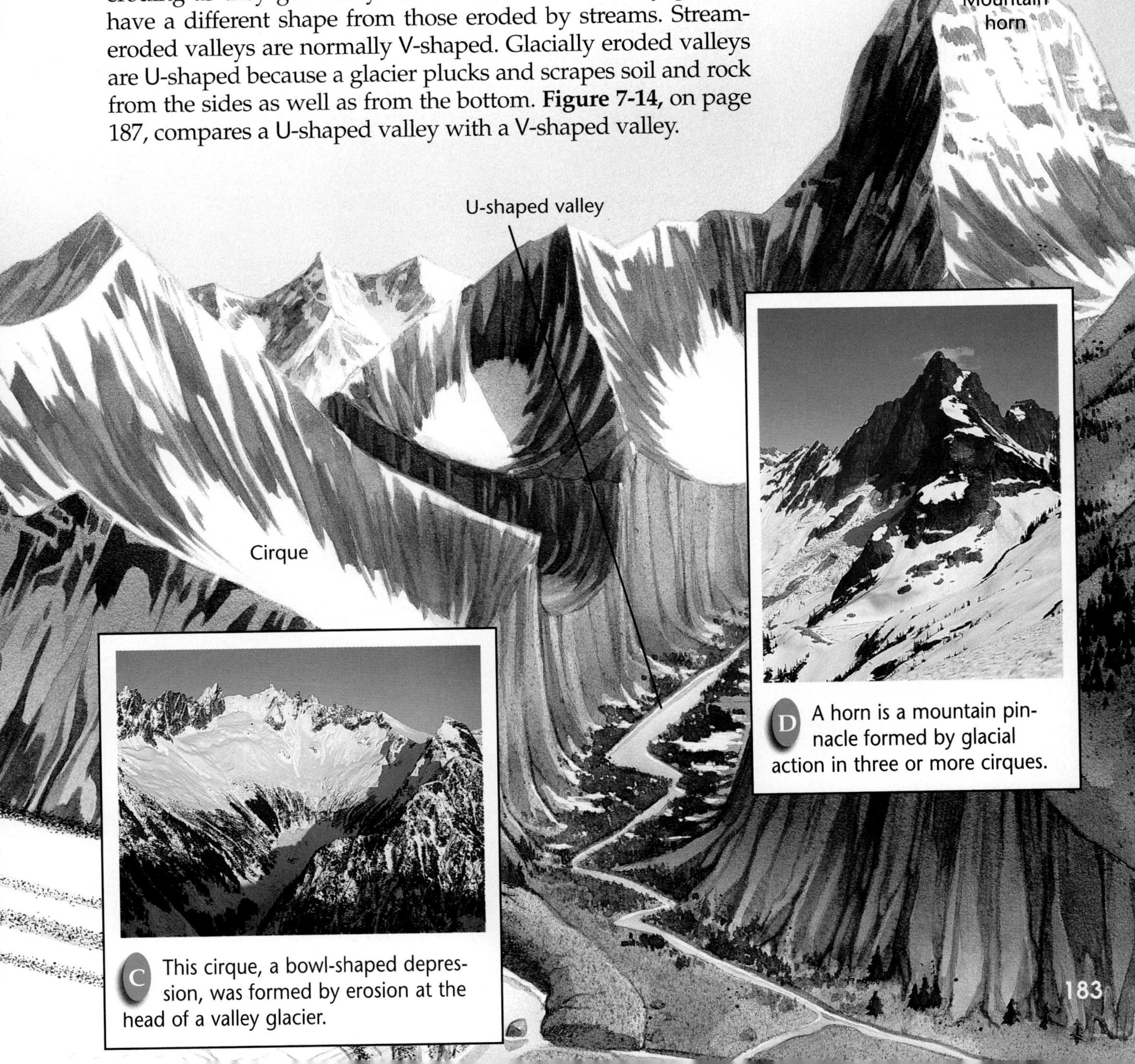

C This cirque, a bowl-shaped depression, was formed by erosion at the head of a valley glacier.

D A horn is a mountain pinnacle formed by glacial action in three or more cirques.

CONNECT TO

CHEMISTRY

Icebergs float because ice is less dense than water. Density relates mass to volume. The formula for finding density is $D = M/V$. When water freezes, the crystal arrangement changes and makes ice, a solid, less dense than water, a liquid. The density decreases. Using the formula, ***infer*** what happens to the volume.

Glacial Deposition

When glaciers begin to melt, they no longer have enough energy to carry much sediment. The sediment drops or is deposited on the land.

Till

When the glacier slows down, a jumble of boulders, sand, clay and silt drops from its base. This mixture of different-sized sediments is called **till.** As shown in **Figure 7-13**, till deposits can cover huge areas of land. During the last ice age, continental glaciers in the northern United States dropped enough till to completely fill valleys and make these areas appear flat. Till areas include the wide swath of wheat land running northwestward from Iowa to northern Montana, some farmland in parts of Ohio, Indiana and Illinois, and the rocky pastures of New England.

Till is also deposited in front of a glacier when it stops moving forward. Unlike the till that drops from a glacier's base, this second type of deposit doesn't cover a very wide area. Because it's made of the rocks and soil that the glacier has been pushing along, it looks like a big ridge of material left behind by a bulldozer. Such a ridge is called a **moraine.** Other moraines are deposited along the sides of glaciers and sometimes in the middle. A moraine is shown in **Figure 7-13.**

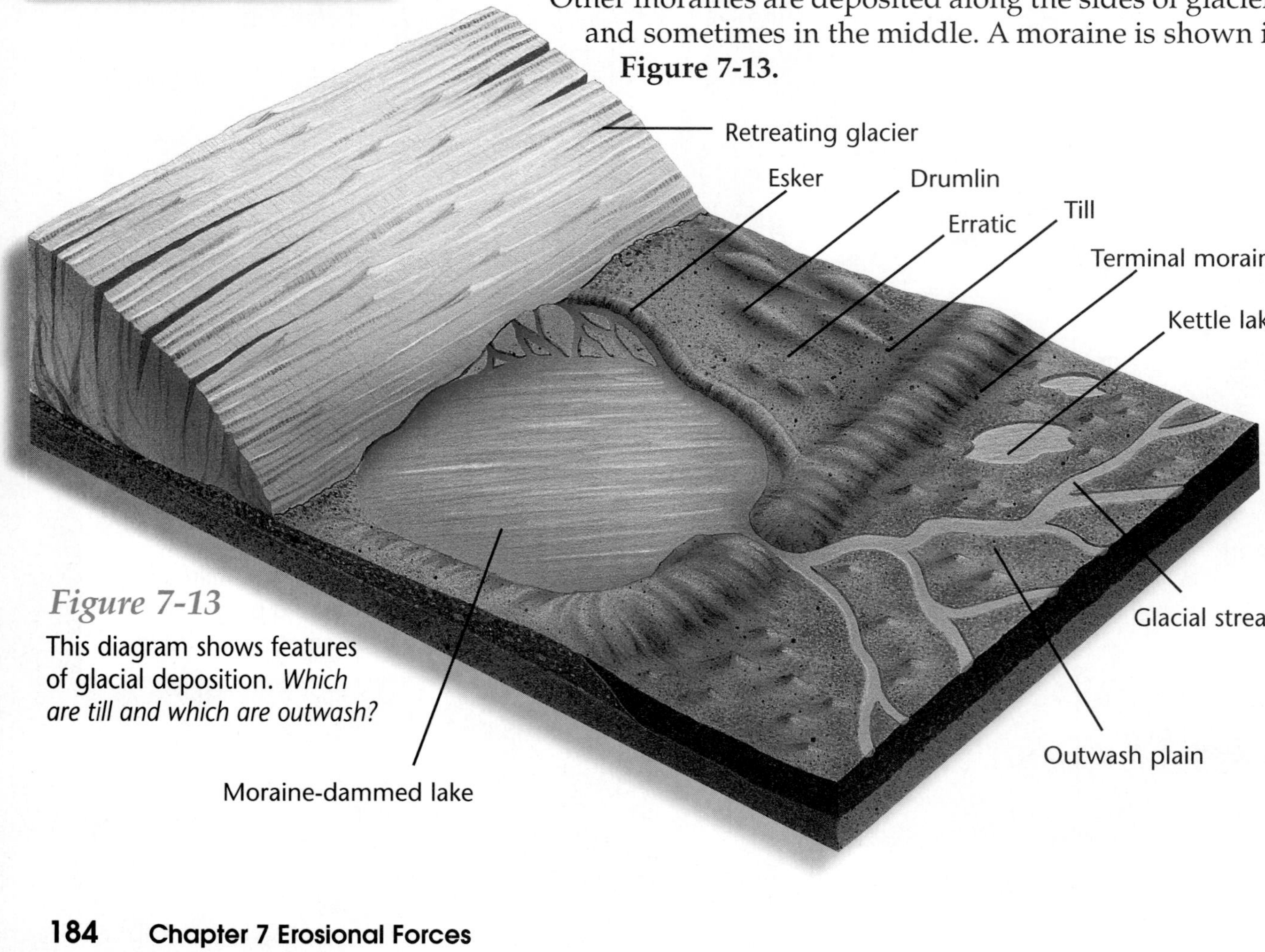

Figure 7-13
This diagram shows features of glacial deposition. *Which are till and which are outwash?*

Outwash

When melting exceeds snow accumulation, the glacier retreats or starts to melt. Material deposited by the meltwater from a glacier is called *outwash.* Outwash is shown in **Figure 7-13.** The meltwater carries sediments and deposits them in layers much as a river does. Heavier sediments drop first so the bigger pieces of rock are deposited closer to the glacier. The outwash from a glacier can also form into a fan-shaped deposit when the stream of meltwater drops sand and gravel in front of the glacier.

Another type of outwash deposit looks like a long winding ridge. This deposit forms beneath a melting glacier when meltwater forms a river within the ice. This river carries sand and gravel and deposits them within its channel. When the glacier melts, a winding ridge of sand and gravel, called an esker, is left behind. An esker is shown in **Figure 7-13.** Meltwater also forms *outwash plains* of deposited materials in front of a retreating glacier.

USING TECHNOLOGY

Bergy Bits?

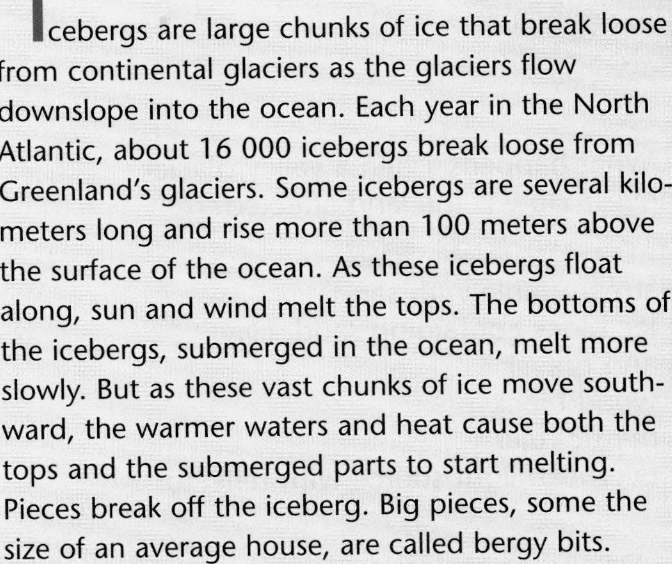

Icebergs are large chunks of ice that break loose from continental glaciers as the glaciers flow downslope into the ocean. Each year in the North Atlantic, about 16 000 icebergs break loose from Greenland's glaciers. Some icebergs are several kilometers long and rise more than 100 meters above the surface of the ocean. As these icebergs float along, sun and wind melt the tops. The bottoms of the icebergs, submerged in the ocean, melt more slowly. But as these vast chunks of ice move southward, the warmer waters and heat cause both the tops and the submerged parts to start melting. Pieces break off the iceberg. Big pieces, some the size of an average house, are called bergy bits.

More than 85 percent of Earth's fresh water is locked up in glacial ice. Scientists estimate that an iceberg about 270 square kilometers would provide over 7 billion cubic meters of fresh water! A tug could tow this iceberg at a rate of about 32 kilometers per day. This may be the only way to bring drinking water to areas of the world that need fresh water, but at this slow rate of towing, not much ice would be left.

A Bergy Bit

Think Critically:

What areas of the world would benefit most from using icebergs as a source of fresh water? How would towing icebergs affect the environment?

Activity 7-1

Glacial Grooving

Throughout the world's mountainous regions there are 200 000 valley glaciers moving in response to local freezing and thawing conditions, as well as gravity.

Problem

What happens when a valley glacier moves? How is the land affected?

Materials  

- stream table with sand
- ice block containing sand, clay, and gravel
- wood block
- metric ruler
- overhead light source with reflector

Procedure

1. Set up the stream table as shown. Place the wood block under one end of the table to give it a slope.
2. Cut a narrow channel, like a river, through the sand. Measure and record its width and depth. Draw a sketch that includes these measurements.
3. Position the overhead light source to shine on the channel as shown.

4. Force the ice block into the river channel at the upper end of the stream table.
5. Gently push the "glacier" along the river channel until it's halfway between the top and bottom of the stream table and is positioned directly under the light.
6. Turn on the light and allow the ice to melt. Observe and record what happens. Does the meltwater change the original channel?
7. Measure and record the width and depth of the glacial channel. Draw a sketch of the channel and include these measurements.

Analyze

1. **Explain** how you can determine the direction a glacier traveled from the location of deposits.

Conclude and Apply

2. Explain how you can **determine** the direction of glacial movement from sediments deposited by meltwater.
3. How do valley glaciers affect the surface over which they move?

Data and Observations

Sample Data	Width	Depth	Observations
Original channel	6 cm	3 cm	stream channel looked V-shaped
Glacier channel			
Meltwater channel			

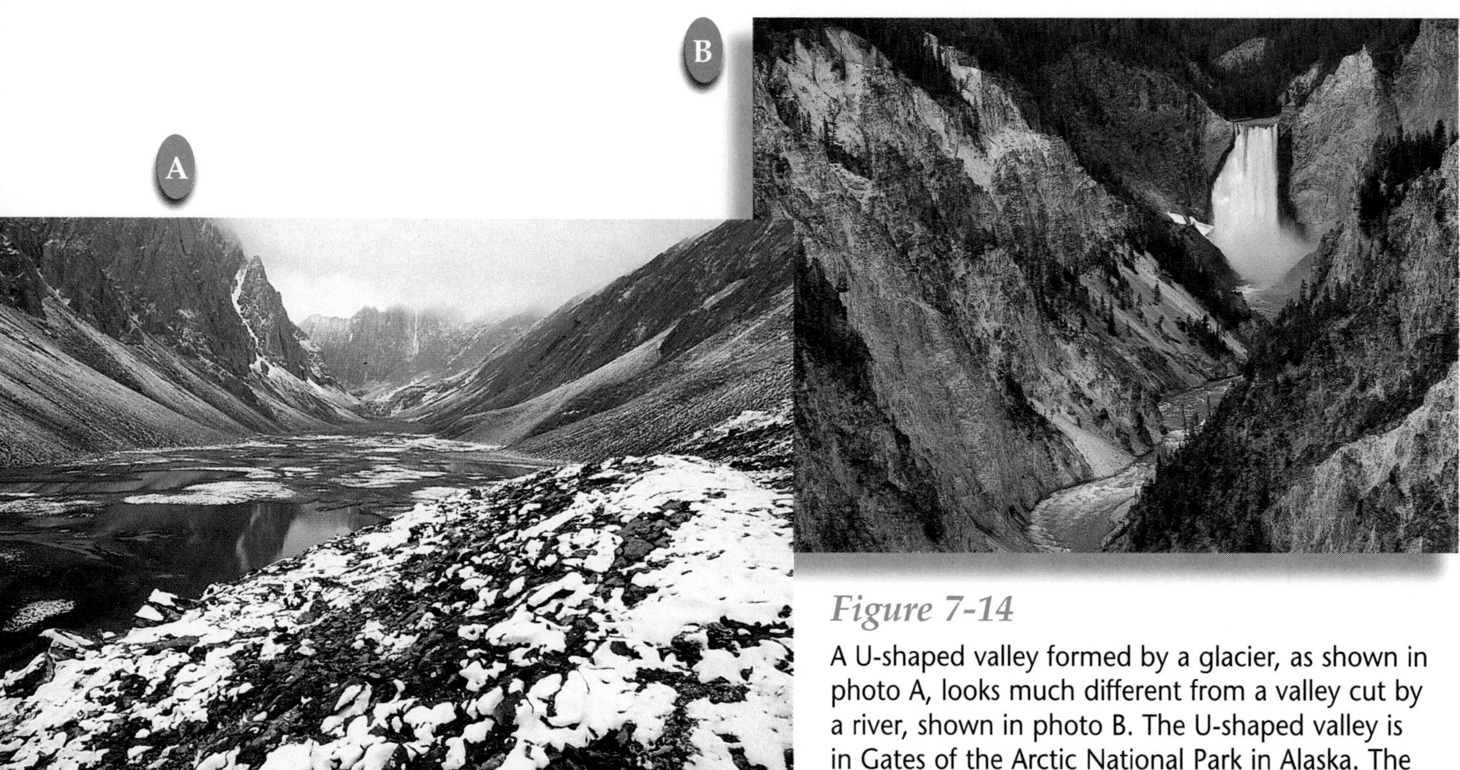

Figure 7-14
A U-shaped valley formed by a glacier, as shown in photo A, looks much different from a valley cut by a river, shown in photo B. The U-shaped valley is in Gates of the Arctic National Park in Alaska. The V-shaped valley was cut by the Yellowstone River.

Glaciers from the last ice age changed the surface of Earth. Glaciers eroded mountaintops and scoured out valleys like the one in **Figure 7-14A.** In Chapter 8 you'll learn about water-eroded valleys like the one in **Figure 7-14B**. Glaciers also deposited sediments over vast areas of North America and Europe. Today, glaciers in the polar regions and in mountains continue to change the surface features of Earth.

Section Wrap-up

Review

1. How does a glacier cause erosion? What processes come into play?
2. Explain how till and outwash are different.
3. **Think Critically:** Suppose that in a field in Indiana, you find a large rock that matches rocks normally found in Canada. Give one explanation of how it got there.

Skill Builder

Sequencing

Put the stages of glacial development and erosion into a correct sequence. If you need help, refer to Sequencing in the **Skill Handbook.**

Science Journal

An *erratic* is a rock fragment deposited by a glacier. Erratic comes from the Latin word *errare* "to wander." Research how glaciers erode and deposit erratics. In your Science Journal, write a poem about the "life" of an erratic.

7-4 Wind

Science Words

deflation
abrasion
loess

Objectives

- Explain how wind causes deflation and abrasion.
- Discuss how loess and dunes form.

Wind Erosion

When air moves, it can pick up loose material and transport it to other places. Air differs from other erosional forces in that it cannot pick up heavy sediments. But because wind is not confined to a river channel or a glacial valley, wind can transport and deposit sediments over large areas. Sometimes wind carries dust from fields or volcanoes high into the atmosphere and deposits the dust far away.

Deflation and Abrasion

Wind erodes Earth's surface by deflation and abrasion. When wind erodes by **deflation**, it blows across loose sediment, removing small particles such as clay, silt, and sand, leaving coarse material behind. When these windblown sediments strike rock, they erode by **abrasion.** Both deflation and abrasion happen to all land surfaces but occur mostly in deserts, beaches, and plowed fields. In these areas, few plants anchor the sediments. When winds blow over them, there is nothing to hold them down. **Figure 7-15** illustrates deflation.

Wind erosion by abrasion is similar to the action of a crew of restoration workers sandblasting a building. These workers use machines that spray a mixture of sand and water against a building. The blast of sand wears away dirt from stone, concrete, or brick walls. It also polishes the building walls by breaking away small pieces and leaving an even, smooth finish.

Figure 7-15

Deflation produces airborne sediments and leaves behind what is called desert pavement.

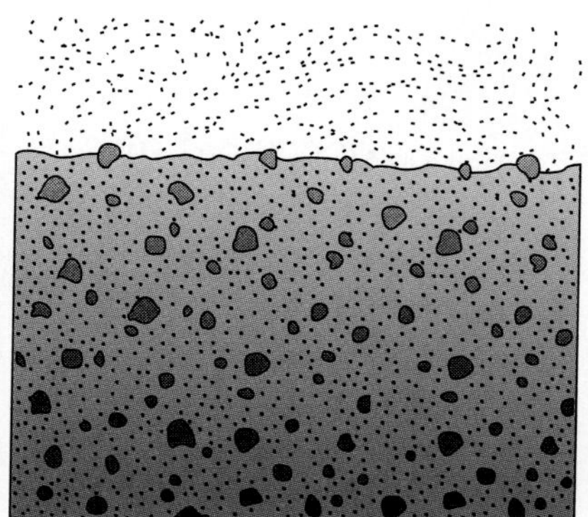

A As deflation begins, wind blows away silt and sand.

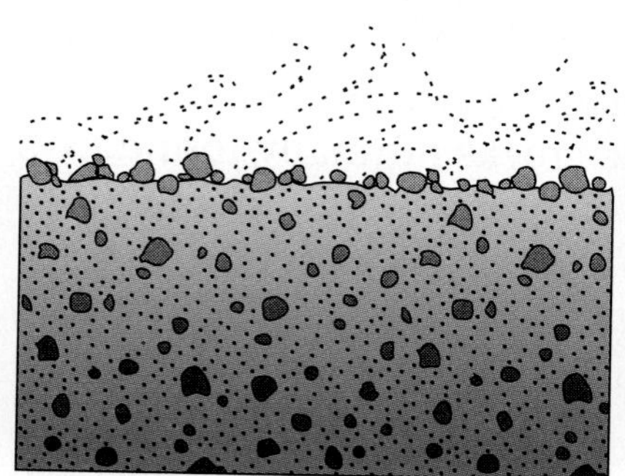

B Deflation continues to remove finer particles. Deflation lowers the surface.

C After the finer particles are blown away, larger pebbles and rocks form a pavement that prevents further deflation.

Figure 7-16

Photo A shows how Egypt's desert winds have abraded the Sphinx. Photo B shows a worker using abrasion to clean and smooth a limestone waterfall.

Wind acts like a sandblasting machine rolling and blowing sand grains along. These sand grains strike rocks and break off small fragments. The rocks become pitted or worn down. **Figure 7-16** shows a comparison of machine and wind abrasion.

Sand Storms

Even when the wind blows strongly, it seldom bounces sand grains higher than one-half meter from the ground. Sand grains are too heavy for wind to lift high in the air. However, sandstorms do occur. When the wind blows with great force in the sandy parts of deserts, sand grains bounce along and collide with other sand grains, causing more and more grains to rise into the air. These wind-blown sand grains form a low cloud, just above the ground. Most sandstorms occur in deserts and sometimes on beaches and in dry riverbeds.

Figure 7-17

During the 1930s, the southern part of the Great Plains of the United States was known as the Dust Bowl because dust storms swept the soil away. Dust storms still occur around the world in places such as Mongolia, in western India, in the Sahel region of Africa, and in the United States.

Dust Storms

When soil is moist, it stays packed on the ground. But when it has not rained for a long while, the soil dries out and can be eroded by wind. Because soil particles weigh less than sand, wind can pick them up and blow them high into the atmosphere. But because silt and clay particles are flat, the wind must blow faster to lift these fine particles of soil than to lift grains of sand. But once the wind does lift them, it can hold these particles and carry them long distances. In the 1930s, silt and dust picked up in Kansas fell in New England and in the North Atlantic Ocean.

Dust storms play an important part in soil erosion. Where the land is dry, dust storms can cover hundreds of miles. The storms blow topsoil from open fields, overgrazed areas, and places where vegetation has disappeared. A dust storm is shown in **Figure 7-17.**

INTEGRATION
Life Science

Reducing Wind Erosion

As you've learned, wind erosion is most common where plants do not exist to protect the soil. One of the best ways to slow or stop wind erosion is to plant vegetation.

Windbreaks

People in many countries plant vegetation to reduce wind erosion. For centuries, farmers have purposely planted trees along their fields to act as windbreaks to prevent soil erosion. As the wind hits the trees, its energy of motion is reduced. The wind no longer has the energy to lift particles.

In a study, a thin belt of cottonwood trees reduced the effect of a 25-kilometer-an-hour wind to about 66 percent of its normal speed. Tree belts reduce wind erosion and also trap snow and hold it on land, adding to the moisture of the soil.

Roots

Along many steep slopes, seacoasts, and deserts, people plant vegetation to hold the soil grains in place, thus reducing wind erosion. The best vegetation to plant to stop wind erosion has a fibrous root system as grasses do. Grass roots are shallow, slender, and have many fibers. They twist and turn between particles in the soil and hold the soil in place.

Planting vegetation is a good way to reduce the effects of deflation and abrasion. But if the wind is strong and the soil is dry, nothing can stop it completely. **Figure 7-18** shows a project to stop wind erosion.

MiniLAB

How do plant roots help hold soil in place?

Procedure

1. Obtain a piece of grass sod.
2. Carefully remove the soil from the sod roots by hand. Examine the roots with a magnifying glass or stereoscopic microscope.

Analysis

1. Draw several roots in your Science Journal.
2. What characteristics of grass roots help hold soil in place and thus help reduce erosion?

Figure 7-18

This marram grass was planted to stabilize sand dunes in Cornwall, England. *How does grass slow erosion?*

Figure 7-19
These loess cliffs formed along the Yellow River in China.

Deposition by Wind

Sediments blown away by wind are eventually deposited. These windblown deposits develop into several types of landforms.

Loess

Some large deposits of windblown sediments are found near the Mississippi River. These wind deposits of fine-grained sediments are known as **loess** (LES). Strong winds that deflated glacial outwash areas carried the sediments and deposited them. The sediments settled on hilltops and in valleys. Once there, the particles were packed together, creating a thick, buff-colored deposit lacking layers.

Loess is as fine as talcum powder. Many farmlands of the midwestern United States are on the fertile soils that have evolved from loess deposits. Loess is also found along the Yellow River in China, where the sand and silt blow in from the Gobi and Ordos deserts.

Problem Solving

Fighting Deflation

If you travel in a desert, you might see large rocks piled several feet high at the bases of utility poles, fence posts, and houses. In places where this hasn't been done, deflation can occur at the bottoms of structures. In fact, some poles might lean. Windblown sediments cause fine scratch marks on unprotected structures.

In some places, sand may have blown across the highway. In areas where there are fences lining the highway, there is very little sand on the road.

Solve the Problem

1. **Infer from the information above why people pile rocks against the bases of poles, posts, and houses in desert regions.**
2. **Explain how fences help control the natural deposition of sand on desert highways.**

Think Critically:

If you lived in the desert, what are some ways that you could reduce wind erosion around your home?

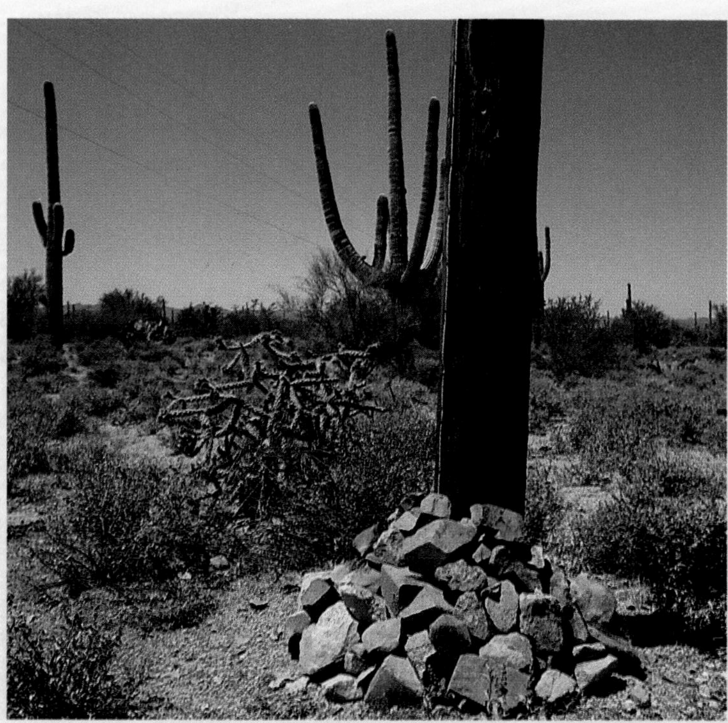

Activity 7-2

Design Your Own Experiment

Blowing in the Wind

Have you ever played a sport outside and suddenly had the wind blow dust into your eyes? What did you do? Turn your back? Cover your eyes? How does wind pick up sediment? Why does wind pick up some sediments and leave others on the ground?

PREPARATION

Problem

What factors affect wind erosion? Do both sediment moisture and speed of wind affect the rate of wind erosion?

Form a Hypothesis

How does the amount of moisture in the sediment affect the ability of wind to erode the sediments? Does the speed of the wind limit the size of sediments it can transport? Form a hypothesis about how sediment moisture affects wind erosion. Form another hypothesis about how wind speed affects the size of the sediment the wind can transport.

Objectives

- Observe the effects of soil moisture and wind speed on wind erosion.
- Design and carry out experiments that test the effects of soil moisture and wind speed on wind erosion.

Possible Materials

- goggles
- flat pans (4)
- fine sand (400 mL)
- powdered clay (400 mL)
- gravel (400 mL)
- hair dryer
- sprinkling can
- water
- cardboard sheets (4)
- tape
- mixing bowl
- metric ruler

Safety Precautions

Wear your safety goggles at all times when using the hair dryer on sediments.

PLAN THE EXPERIMENT

1. As a group, agree upon and write out your hypothesis statements and work on the experiment.
2. List the steps needed to test your first hypothesis. Plan specific steps and vary only one factor at a time. Then, list the steps needed to test your second hypothesis. Test only one factor at a time.
3. Mix the sediments in the pans. Plan how you will fold cardboard sheets and attach them to the pans to keep sediments from flying everywhere.
4. Design data tables in your Science Journal. Use them as your group collects data.

Check the Plan

1. Read over your plan to make certain that all steps are in logical order.
2. Identify the constants and the variables of the experiments.
3. Do you have to run any test more than one time?
4. Will the data be summarized in graphs?
5. *Make sure your teacher approves your plan before you proceed.*

DO THE EXPERIMENT

1. Carry out the experiments as planned.
2. During the experiments, write observations that you make and complete the data tables in your Science Journal.

Analyze and Apply

1. **Compare** your results with those of other groups.
2. What relationship is there between the sediment moisture and the amount of sediment eroded?
3. **Graph** the relationship that exists between the speed of the wind and the size of the sediments it transports.
4. How does the energy of motion explain the results of your experiment?

Go Further

Repeat your second experiment. This time, instead of mixing sediments together, test the sediments separately. How did this change your results?

CONNECT TO

LIFE SCIENCE

Red clover has a taproot system consisting of one main root that grows directly downward. Grasses have a fibrous root system. ***Infer*** which type of plant would be the best to plant along coastal sand dunes to prevent wind erosion.

Dunes

What happens when wind blows sediments against an obstacle such as a rock or a clump of vegetation? The sediments settle behind the obstacle. More and more sediments build up and eventually a dune is formed. A dune is a mound of sand drifted by the wind. **Figure 7-20** shows sand dunes.

Sand dunes constantly move as wind erodes them and deposits the sand elsewhere. On Cape Cod, Massachusetts, and along the Gulf of California, the coast of Oregon, and the eastern shore of Lake Michigan, you can see beach sand dunes. Sand dunes accumulate where there are sand and prevailing winds or sea breezes that blow daily.

To visualize how a dune forms, think of the sand at the back of a beach. That sand is dry because the ocean and lake waves do not reach these areas. The wind blows this dry sand farther inland until an obstruction such as a rock slows the wind. The wind sweeps around or over the rock. Like a river, air drops its load of sediment when its energy of motion slows. Sand starts to accumualte behind the rock. As the sand continues to accumulate, the mound of sand becomes an even greater obstacle and traps even more sand. If the wind blows long enough, the mound of sand will become a sand dune.

Sand will continue to accumulate and build a dune until the sand runs out or the obstruction is removed. Some sand dunes may grow to 50 to 180 meters high, but most are much lower.

Figure 7-20

Dunes are the most common wind deposits. Sand dunes form in deserts and beaches where sand is abundant. They are constantly changing and moving as wind erodes them.

Dune Migration

A sand dune has two sides. The side facing the wind has a gentler slope. The side away from the wind is steeper. Examining the shape of a dune tells you the direction the wind usually blows from.

Unless sand dunes are planted with grasses, most dunes don't stay still. They migrate or move away from the direction of the wind. This migration process is shown in **Figure 7-21.** Some dunes are known as traveling dunes because they move across desert areas as they lose sand on one side and accumulate it on the other.

When dunes and loess form, the landscape is changed. Wind, like gravity, running water, and glaciers, shapes the land as it erodes sediments. But the new landforms created by these agents of erosion are themselves being eroded. Erosion and deposition are part of a cycle of change that constantly shapes and reshapes the land.

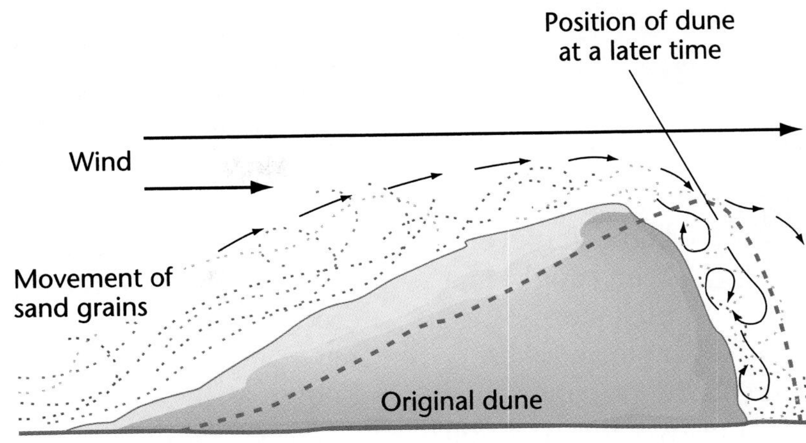

Figure 7-21

As wind blows the sand, the grains jump, roll, and slide up the gentler slope of the dune, accumulate at the top, then slide down the steeper slope on the side away from the wind.

Section Wrap-up

Review

1. Compare and contrast abrasion and deflation. How do they affect the surface of Earth?
2. Explain the differences between dust storms and sandstorms. How does this affect the deposition of sand and dust?
3. **Think Critically:** You notice that snow is piling up behind a fence outside your apartment building. Why?

Skill Builder

Sequencing

Sequence the following events that describe how a sand dune forms. If you need help, refer to Sequencing in the **Skill Handbook.**

a. Grains collect to form a mound.

b. Wind blows sand grains until they hit an obstacle.

c. Wind blows over an area and causes deflation.

d. Vegetation grows on the dune.

USING MATH

Between 1972 and 1992, the area of the Sahara Desert in Africa increased by nearly 700 square kilometers in Mali and the Sudan. Much of this desertification was due to drought and the removal of vegetation. Calculate the average number of square kilometers the desert increased each year between 1972 and 1992.

Science & ART

Science Journal

Have you ever been caught in a dust storm? Or in some other storm where you couldn't see more than a foot or two ahead of you? In your Science Journal, write a description of the experience. Tell how you found your way without being able to see well.

The Music of Erosion

In the 1930s, the soil in large parts of Texas, Oklahoma, Colorado, Kansas, and New Mexico was blowing away. Overgrazing and overfarming had stripped vast expanses of the land of protective grass. Drought had dried the soil. Wind eroded the land. It blew away soil one day, then deposited a new pile of dust a few days later. This dry region became known as the Dust Bowl.

Woody Guthrie wrote and sang folk songs about the hardships of the Dust Bowl. His song "The Great Dust Storm" describes a region where, according to Guthrie, "the dust flows and the farmer owes."

The storm took place at sundown
It lasted through the night.
When we looked out next morning
We saw a terrible sight.

We saw outside our window
Where wheatfields they had grown,
Was now a rippling ocean
Of dust the wind had blown.

It covered up our fences,
It covered up our barns,
It covered up our tractors
In this wild and dusty storm.©

Guthrie's songs tell of hardship and pain. But they also tell of courage and humor. "Dust Can't Kill Me" is a song about a man who refuses to be destroyed even after the dust storms take everything he has. In "Dust Pneumonia Blues," he sings about a man's trip to Texas. His girlfriend isn't used to water anymore, so she faints in the rain. The man has to throw a bucket of dirt in her face to revive her.

Chapter 7

Review

Summary

7-1: Gravity

1. Erosion is the process that wears down and transports sediments. Deposition occurs when an agent of erosion loses its energy of motion and can no longer carry its load.
2. Slump, creep, rockslides, and mudflows are all mass movements related to gravity. Slump and creep occur slowly; rockslides and mudflows occur rapidly.

7-2: Science and Society: Developing Land Prone to Erosion

1. Vegetation, terraces, and retaining walls can reduce erosion on slopes.
2. Building on steep slopes and removing vegetation from slopes increase erosion.

7-3: Glaciers

1. Glaciers are powerful agents of erosion.
2. Plucking adds rock and soil to a glacier's sides and bottom as water freezes and thaws and breaks off pieces of surrounding rocks.
3. Till is a jumble of sediments deposited directly from glacial ice and snow. Outwash is glacial debris deposited by meltwater of the glacier.

7-4: Wind

1. Deflation occurs when wind erodes only fine-grained sediments, while coarse sediments like pebbles and boulders are left behind. The pitting and polishing of rocks and sediments by windblown sediments is called abrasion.
2. Deposits of fine-grained particles that are tightly packed form loess. Dunes begin to form when windblown sediments pile up behind an obstacle.

Key Science Words

a. abrasion
b. creep
c. deflation
d. deposition
e. erosion
f. glacier
g. loess
h. mass movement
i. moraine
j. plucking
k. slump
l. till

Reviewing Vocabulary

Match each phrase with the correct term from the list of Key Science Words.

1. process of wearing down and transporting weathered sediments
2. slow movement of sediments downhill because of freezing and thawing
3. downward movement of rock and soil under influence of gravity
4. mass movement in which materials move as one large mass
5. erosion caused by freezing and thawing of glacial ice
6. wind erosion that leaves coarse sediments behind
7. mixture of rocks and sediments deposited in ridges by glacial snow and ice
8. a thick mass of ice and snow that flows over land
9. erosion caused by natural sandblasting
10. thick, densely packed deposits of dust

Chapter 7 Review

Checking Concepts

Choose the word or phrase that completes the sentence or answers the question.

1. Which of the following is the slowest type of mass movement?
   a. abrasion
   b. creep
   c. slump
   d. mudflow
2. The best vegetation to plant to reduce erosion has a _______.
   a. taproot system
   b. striated root system
   c. fibrous root system
   d. sheet root system
3. Where a valley glacier started, a(n) _______ is created.
   a. esker
   b. moraine
   c. till
   d. cirque
4. Glacial erosion _______.
   a. happens in warm places
   b. leaves behind landforms such as arêtes and grooves
   c. happened only in the past
   d. happens only in the present
5. A mass of snow and ice in motion is a(n) _______.
   a. loess deposit
   b. glacier
   c. outwash
   d. abrasion
6. Glacier-created valleys are _______.
   a. V-shaped
   b. L-shaped
   c. U-shaped
   d. S-shaped
7. An example of a structure created by deposition is a(n) _______.
   a. cirque
   b. abrasion
   c. striation
   d. dune
8. One characteristic that all agents of erosion have in common is _______.
   a. they carry sediments when they have enough energy of motion
   b. they are most likely to erode when sediments are moist
   c. they create deposits called dunes
   d. they erode large sediments before they erode small ones
9. Wind erosion in which pebbles and boulders are left behind is called _______.
   a. deflation
   b. loess
   c. abrasion
   d. sandblasting
10. A(n)_______ is a bowl-shaped erosional feature formed by glacial plucking.
    a. striation
    b. esker
    c. cirque
    d. moraine

Understanding Concepts

Answer the following questions in your Science Journal using complete sentences.

11. Discuss the causes and effects of slumping.
12. How does the deposition of till differ from the deposition of outwash?
13. Compare and contrast weathering and erosion.
14. How does the amount of soil moisture affect mass movement?
15. Discuss ways people can reduce erosion on slopes.

Thinking Critically

16. How can striations give information about the direction a glacier moved?
17. How effective would a retaining wall made of fine wire mesh be against erosion?
18. Sand dunes often migrate. What can be done to prevent the migration of beach dunes?
19. Scientists have found evidence of movement of ice within a glacier. Explain how this may occur. (HINT: Recall how putty-like ice forms at the base of a glacier.)
20. The front end of a valley glacier is at a lower elevation than the tail end. How does this explain melting at its front end while snow is still accumulating at its tail end?

Developing Skills

If you need help, refer to the **Skill Handbook.**

21. **Making Tables:** Make a table to contrast continental and valley glaciers.
22. **Designing an Experiment:** Explain how to test the effect of glacial thickness on a glacier's ability to erode.
23. **Sequencing:** Copy and complete the events chain to show how a sand dune forms. Use the terms *sand rolls, migrating, wind blows, sand accumulates, dune, dry sand, obstruction traps,* and *stabilized.*

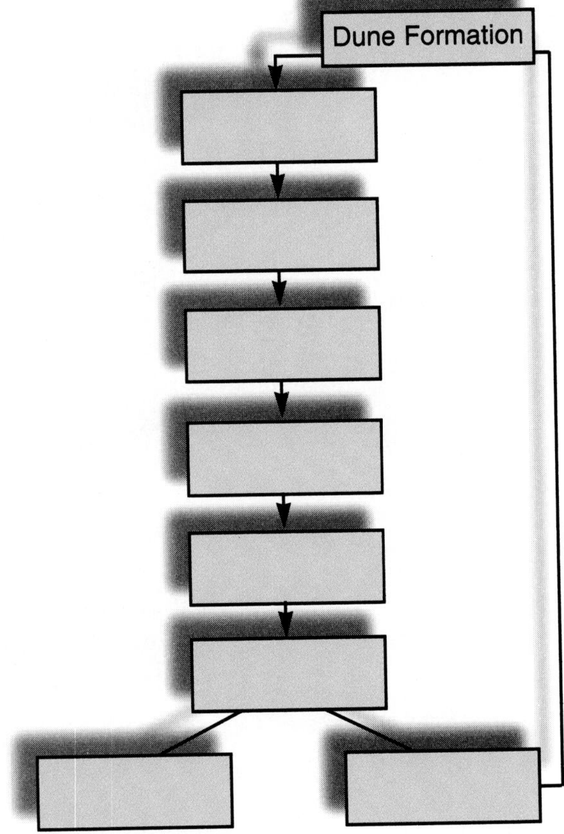

24. **Hypothesizing:** Hypothesize why the materials of loess deposits were transported farther than those of sand dune deposits.
25. **Interpreting Scientific Illustrations:** Look at the map of Ohio on the right. In which part of Ohio would you find erosion and deposition caused by glaciers of the last ice age? How would you expect the terrain to differ throughout the state?

Performance Assessment

1. **Design an Experiment:** In the MiniLAB on page 176, you made models of landforms of different mass movements. Based on what you learned, design an experiment to see how the amount of moisture added to sediments affects slump erosion. Design your experiment to keep all variables constant except the amount of moisture in the sediment. Try your experiment.
2. **Design a Study:** In Activity 7-2, on pages 192 and 193, you learned how sediment moisture and the speed of the wind affect erosion. Based on your observations, design and carry out an experiment to see how vegetation or small fences placed in the sediments affect wind erosion. Use the results to help your class reduce erosion around your school and home. If you live in a snowy area, use these ideas to help control blowing snow.
3. **Poster:** Make a poster of magazine photos showing glacial features in North America. Include photos and descriptions of glacial features not covered in this chapter. Add a map showing where each feature is located.

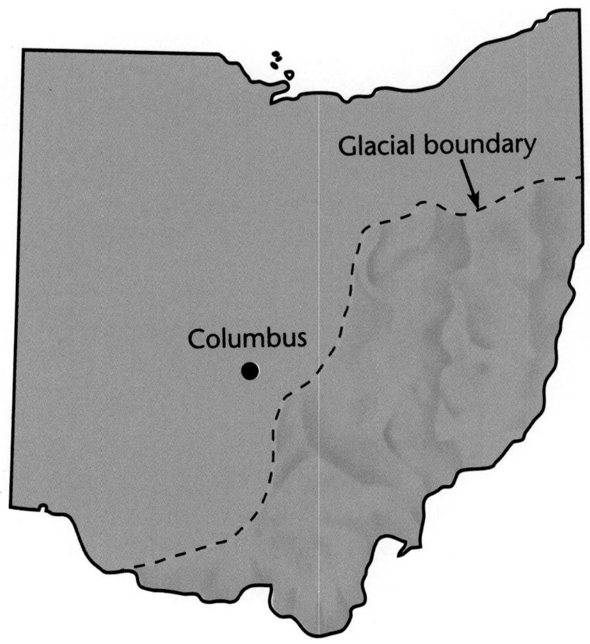

Previewing the Chapter

Chapter 8

Water Erosion and Deposition

Do you know what caused the patterns in this landscape? Water! It erodes more sediments than any other agent of erosion. It's easy to see why—moving water has great energy of motion. Sometimes rainwater just soaks into sediments, other times it moves down the slope because of gravity's pull. What factors do you think determine whether rain soaks into the ground or runs off and erodes the surface?

EXPLORE ACTIVITY

Things That Affect Water Erosion

1. Place a piece of waxed paper on your desktop.
2. Make a mound of dry soil on top of the paper.
3. Slowly drip water from an eyedropper onto your mound and observe what happens.
4. Drip the water more quickly and continue to observe what happens.
5. Repeat the 4 previous steps, but this time change the slope of your mound. Start again with dry soil.

Observe: In your Science Journal, record how the time over which rain falls and the slope of the land affect how much erosion takes place.

Previewing Science Skills

- In the **Skill Builders,** you will compare and contrast, and map concepts.
- In the **Activities,** you will hypothesize, design an experiment, observe, classify, and interpret data.
- In the **MiniLABs,** you will make a model, observe and infer, and record observations.

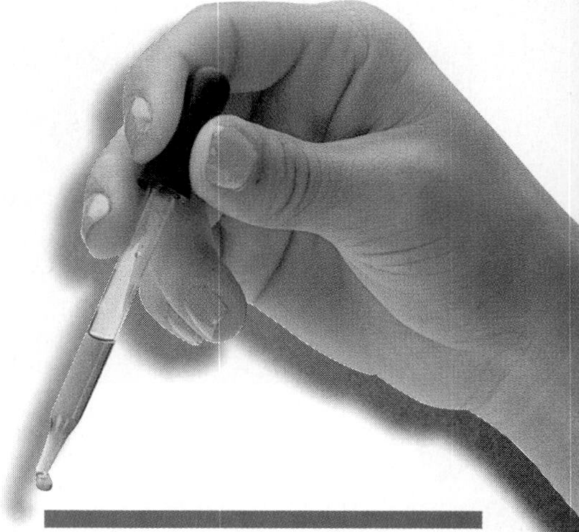

8•1 Surface Water

Science Words

runoff
rill erosion
gully erosion
sheet erosion
drainage basin
meander
floodplain
alluvial fan
delta

Objectives

- Explain what causes runoff.
- Compare rill, sheet, gully, and stream erosion.
- Discuss the three different stages of stream development.
- Describe how alluvial fans and deltas form.

Runoff

Water that doesn't soak into the ground or evaporate flows across Earth's surface and is called **runoff.** If you've ever spilled milk while pouring it, you've experienced something similar to runoff. You can picture it in your mind. You start pouring a glass of milk, but it overflows, spilling all over the table. Then, before you can grab a towel to clean up the mess, the milk runs off the table and onto the floor. This is similar to what happens to rainwater that doesn't soak into the ground or evaporate. It runs along the ground and eventually enters streams, lakes, or the ocean.

Factors Affecting Runoff

But what factors determine whether rain soaks into the ground or runs off? The amount of rain and the time span over which it falls are two factors that affect the amount of runoff. Light rain falling over several hours will probably have time to soak into the ground. Heavy rain falling in less than an hour or so will run off because it doesn't have time to soak in.

Another factor that affects the amount of runoff is the slope of the land. Gentle slopes and flat areas hold water in place, giving it a chance to evaporate or sink into the ground. On steep slopes, however, gravity causes the water to run off before either of these things can happen.

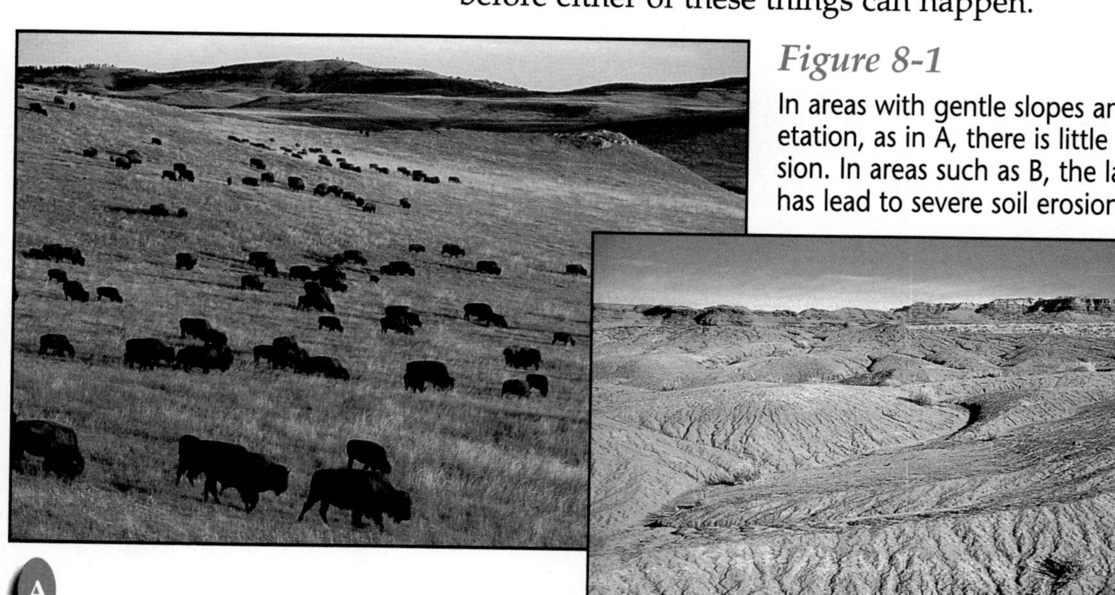

Figure 8-1

In areas with gentle slopes and abundant vegetation, as in A, there is little runoff and erosion. In areas such as B, the lack of vegetation has lead to severe soil erosion.

As you can see in **Figure 8-1,** the amount of vegetation, such as grass, also affects the amount of runoff. Just like milk running off the table, water will tend to run off smooth surfaces that have little or no vegetation. Plants and their roots act like sponges to soak up and hold water. By slowing down runoff, plants and roots help prevent the erosion of soil.

Figure 8-2

During floods, the high volume of fast-moving water erodes large amounts of soil.

The Effects of Gravity

Gravity is the attracting force all objects have for one another. The greater the mass of an object, the greater its force of gravity. Because Earth has a much greater mass than any of the objects on it, Earth's gravitational force pulls objects toward its center of mass. Water falling down a slope is evidence of gravity. As objects drop to Earth's surface, they pick up speed. This happens because gravity is a constant accelerating force. On Earth, for example, a falling object moves 9.8 meters per second faster than it did during the previous second. Thus, when water falls down a slope, it picks up speed. Its energy of motion is much greater, and, as shown in **Figure 8-2,** it erodes more quickly than slower-moving water.

CONNECT TO

PHYSICS

Energy is the ability to cause change or to do work. There are two types of energy, potential energy and kinetic energy. Potential energy is stored energy. Kinetic energy is energy in motion. ***Infer*** which type of energy the water eroding the stream channel below is demonstrating.

Water Erosion

Suppose you and several friends walk the same way to school each day through a field or empty lot, always walking in the same footsteps as you did the day before. By now, you've worn a path through the field. Water also wears a path as it travels down a slope.

Rill and Gully Erosion

You may have seen a scar or small channel on the side of a slope that was left behind by running water. This is evidence of rill erosion. **Rill erosion** begins when a small stream forms during a heavy rain. As this stream flows along, it has enough energy to carry away plants and soil. There's a scar left on the slope where the water eroded the plants and soil. If a stream frequently flows in the same path, rill erosion may evolve into gully erosion.

In **gully erosion,** a rill channel becomes broader and deeper. Large amounts of soil are removed to form a gully as shown in **Figure 8-3**.

Figure 8-3

Heavy rains can remove large amounts of soil and sediment, forming a deep gully in the side of a slope.

Sheet Erosion

Water often erodes without being in a stream channel. For example, when it rains over a fairly flat area, the rainwater accumulates until it eventually begins moving down a gentle slope. **Sheet erosion** happens when rainwater flows into lower elevations, carrying sediments with it, as shown in **Figure 8-4.** In these lower elevations, the water loses some of its energy of motion and it drains into the soil or slowly evaporates.

Figure 8-4

The runoff that causes sheet erosion eventually loses its energy of motion. The sediments left behind cover the soil like a sheet.

Stream Erosion

Sometimes water continues to flow along a depression it has created. It then becomes a stream like the one shown in **Figure 8-5.** As the water in a stream moves along, it constantly picks up sediments from the bottom and sides of its channel. Water picks up and carries some of the lightweight sediments, while large, heavy particles just roll along the bottom of the stream channel. All of these different-sized materials scrape against the bottom and sides of the channel, where they continue to knock loose more sediments. Because of this, a stream continually cuts a deeper and wider channel.

Figure 8-5

This is a cross section of a typical stream channel. *How will the shape of this channel change over time?*

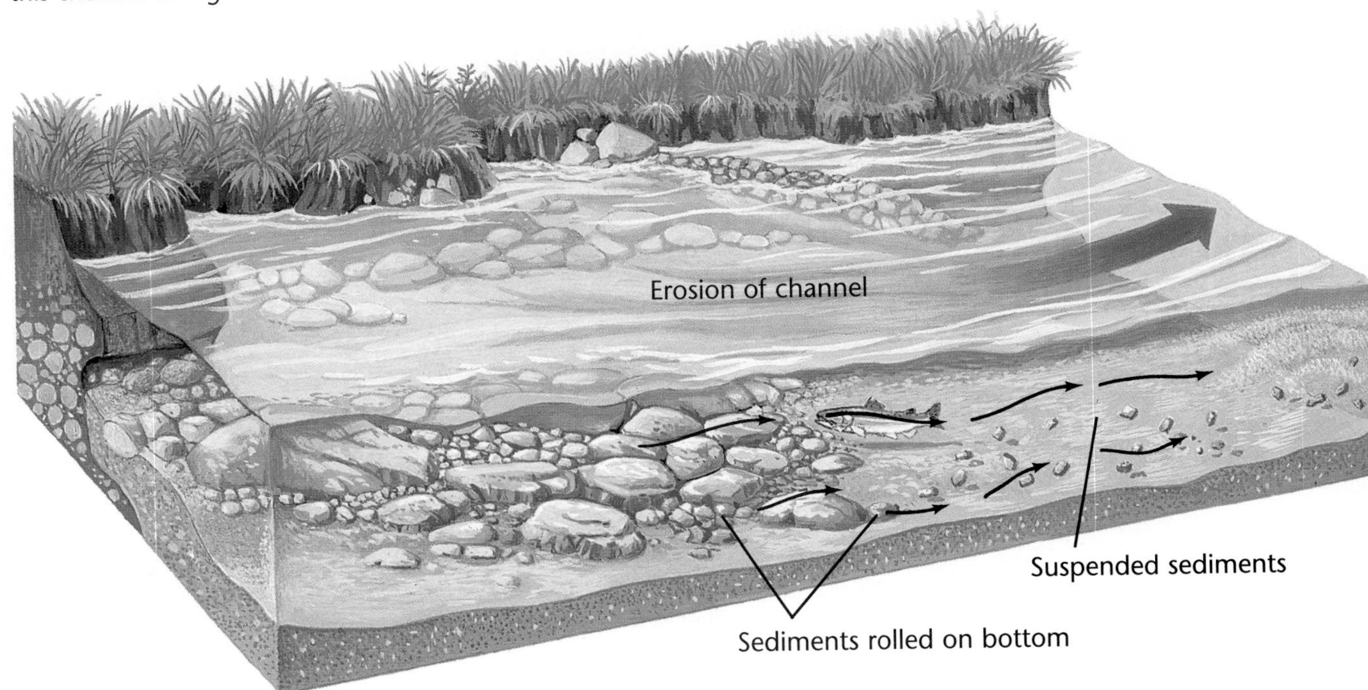

MiniLAB

Where does most erosion occur in a stream?

Procedure

1. Fill a large pan with soil.
2. Use the edge of a ruler or a pencil to level the soil, and use your hands to gently pack it down.
3. With your finger or a pencil, make a stream channel in the soil. Put several bends or turns in the channel so that the water will not be able to flow along a straight path.
4. Lay a pencil under one end of the pan so that you create a gradual slope.
5. Slowly pour water from a beaker into the channel from the higher end of the pan.

Analysis

1. Describe where the stream erodes the most sediments.
2. Describe how the path of a stream of running water affects the rate at which erosion takes place.

River System Development

Is there a stream in your neighborhood or town? Maybe you've been fishing or swimming in that stream. Each day, thousands of liters of water flow through your neighborhood or town in that stream. Where does all the water come from?

River Systems

The stream in your neighborhood is really a part of a river system. The water in the stream came from rills, gullies, and smaller streams located upstream. Just as the tree in **Figure 8-6** is a system containing twigs, branches, and a trunk, a river system also has many parts. Water runs off of the ground and enters small streams. Where small streams join, a larger stream forms. Finally, these larger streams merge, forming a larger body of water called a *river*.

Figure 8-6

River systems can be compared with the structure of a tree.

A A large portion of the streams and rivers in the United States are part of the Mississippi River drainage basin. *What river represents the "trunk" of the river system?*

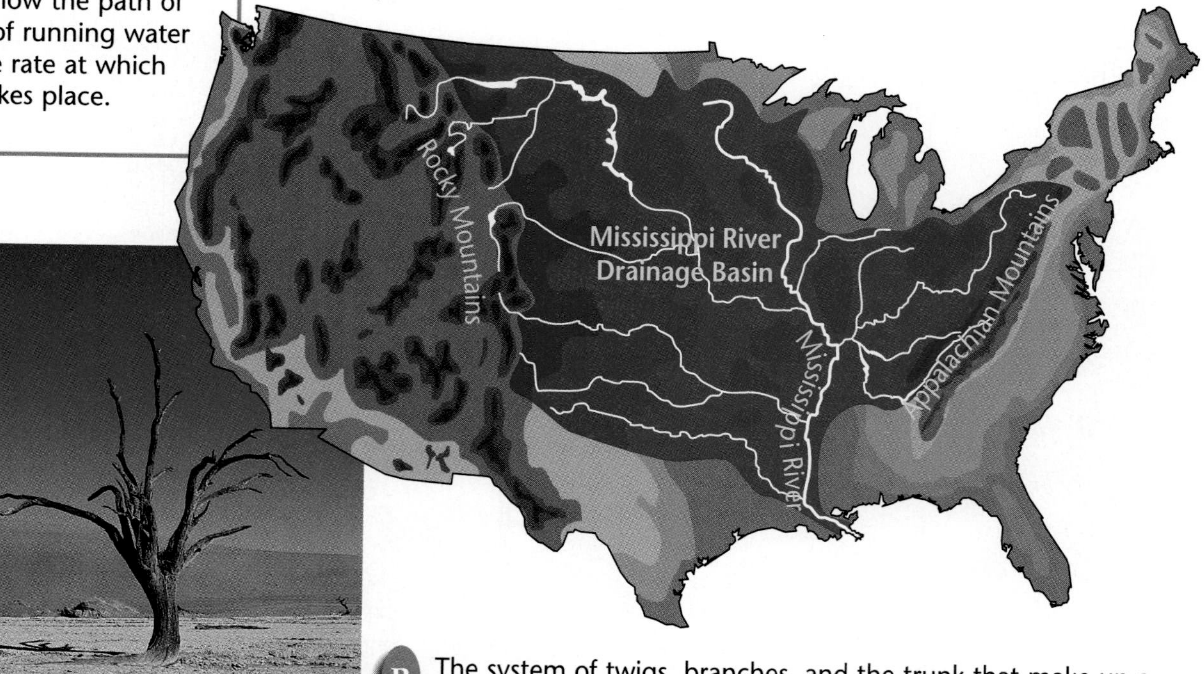

B The system of twigs, branches, and the trunk that make up a tree is similar to the system of streams and rivers that make up a river system.

Drainage Basins

The land area from which a stream gets its water is called a **drainage basin.** A drainage basin can be compared to a bathtub. Water that collects in a bathtub flows toward the drain. Likewise, all of the water in a river system eventually flows to one location—the main river. The largest drainage basin in the United States is the Mississippi River drainage basin.

Stages of Stream Development

There are many different types of streams. Some are narrow and swift–moving, and others are very wide and slow–moving. Streams differ because they are in different stages of development. These stages depend on the slope of the ground over which the stream flows. Streams are classified as young, mature, or old. **Figure 8-7,** on the next two pages, describes how each of the stages comes together to form a drainage basin.

The stages of development aren't always related to the age of a river. The New River in West Virginia, for example, is one of the oldest rivers in North America. However, it has a steep valley and cascades through raging rapids and, as a result, is classified as a *young stream.*

Problem Solving

Washed Out

Suppose you are really looking forward to Friday, when your class plans to have a picnic at the park next to the river. You live in an area of the state near where two smaller rivers meet to form the larger river next to the park. Weather reports tell you that northern areas of the state have had heavy rain, and more is expected over the next few days. It's been really cloudy where you live, but so far it hasn't rained.

Friday comes. There's a fine mist in the air, but it still hasn't rained in your town. It's early in the morning, but you decide to go to the park early to help set up. You pedal your bike to the park. Much to your surprise, the picnic area next to the river is underwater and the picnic must be canceled.

Think Critically:

How can the park be flooded if it didn't rain in your area?

Figure 8-7

Streams of all stages of development can be found in most major river systems.

A Young streams are found in mountainous or hilly regions and flow down steep slopes. *Would you expect to see the water moving rapidly or slowly in a young stream?*

Young Streams

A stream that flows swiftly through a steep valley and has steep sides is a young stream. A young stream may have whitewater rapids and waterfalls. Because the water is flowing rapidly downhill, it has a high level of energy and erodes the stream bottom more than its sides.

Mature Streams

The next stage in the development of a stream is the mature stage. A mature stream flows less swiftly through its valley. Most of the rocks in the streambed that cause waterfalls and rapids have been eroded away.

The erosional energy of a mature stream is no longer concentrated on its bottom. Now, the stream starts to erode more along its sides, developing curves. These curves form because the speed of the water varies throughout the width of the channel.

Water in shallow areas of a stream is slowed down by the friction created by the bottom of the river. In deep areas, less of the water comes in contact with the bottom. This means that deep water has less friction with the bottom of the stream and therefore flows faster. This faster-moving water erodes the side of the stream where the current is strongest, forming curves. A curve that forms in this way is called a **meander. Figure 8-7C** shows what a meandering stream looks like from the air.

B A curving stream that flows down a gradual slope is a mature stream. *Would you expect to see many large rapids or waterfalls along a mature stream? Explain.*

The broad, flat valley floor carved by a meandering stream is called a **floodplain.** When a stream floods, it will often cover a part of or the whole floodplain.

Old Streams

The last stage in the development of a stream is the old stage. An old stream flows slowly through a broad, flat floodplain that it has carved. The lower Mississippi River is in the old stage.

Major river systems usually contain streams in all stages of development. At the outer edges of a river system, you find whitewater streams moving swiftly down mountains and hills. At the bottom of mountains and hills, you find streams that are starting to meander and are in the mature stage of development. These streams meet at the trunk of the drainage basin to form a major river.

C Old streams flow slowly through flat, broad floodplains. *What happens to the area surrounding an old-stage stream during floods?*

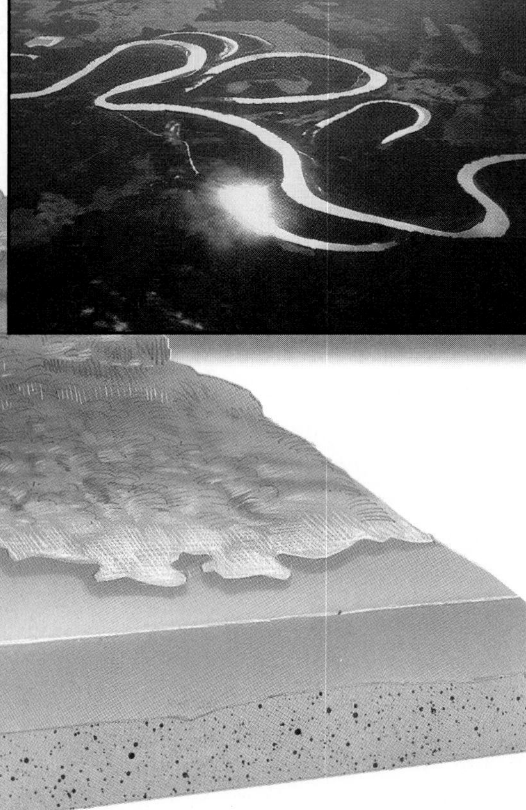

Floodplain

Activity 8-1

Design Your Own Experiment

Stream Speed

Have you ever wondered what it would be like to make a raft and use it to float on a river? Do you think that guiding a raft would be easy? Probably not. You'd be at the mercy of the current. The greater the speed of the river, the stronger the current.

PREPARATION

Problem

What factors affect a stream's speed? How does the speed of a stream affect the water's ability to erode?

Form a Hypothesis

Think about streams you have observed, then consider various factors that may affect how fast a stream flows. Do you think the slope of a stream affects its speed? Do you think the speed of a stream affects the amount of erosion that takes place? Form a hypothesis about how the slope of a stream affects the speed of a stream. Form another hypothesis about how the speed of a stream affects erosion.

Objectives

- Design and carry out an experiment to show the relationship between the slope of a streambed and a stream's speed.
- Design and carry out an experiment to show the relationship between the speed of a stream and a stream's ability to erode.
- Observe the effects of slope of a stream on speed of the stream.
- Observe the effects of speed of a stream on erosion.

Possible Materials

- a large pan or stream table
- sand to fill the pan or stream table
- plastic pails (2)
- rubber tubing
- meterstick
- small cork
- water
- stopwatch
- wooden blocks (2)

PLAN THE EXPERIMENT

1. As a group, agree upon and write out the hypotheses statements.
2. As a group, list the steps that you need to take to test your hypotheses. Include in your plan how you will (a) make your stream channel, (b) adjust the height of the slope, (c) measure the length of the stream channel, (d) use tubing and one of the pails to begin the water flowing down the stream, (e) catch the overflow water at the other end of the stream table or pan with your other pail, (f) determine the time it takes the cork to flow down the stream channel, and (g) observe the amount of erosion that takes place.
3. You will need a data table. Design one in your Science Journal so that it is ready to use as your group collects data.

Check the Plan

1. Read over your entire plan to make sure that all steps are in logical order.
2. Identify any constants and the variables of the experiment.
3. Have you allowed for a control in your experiment?
4. Do you have to run any tests more than one time?
5. Will the data be summarized in graphs? Decide which type of graph to use.
6. *Make sure your teacher approves your plan and that you have included any changes suggested in the plan.*

DO THE EXPERIMENT

1. Carry out the experiment as planned.
2. While the experiment is going on, write down any observations that you make and complete the data tables in your Science Journal.

Analyze and Apply

1. **Calculate** the speed of your streams using the formula: speed $= \frac{\text{distance}}{\text{time}}$.
2. Were sediments carried along with the water to the end of your stream channels?
3. Did the slope affect the speed of the streams in your experiment?
4. Did the streams with the greatest or least slopes erode the most sediments?

Go Further

Repeat your experiment, but this time see if you can devise a way to collect and measure the amount of erosion that takes place on different slopes.

USING TECHNOLOGY

Taming a River

A river that overflows its banks can bring disaster. Such was the case in the Mississippi River flood of 1993 when parts of nine Midwestern states were under water.

For many years, people living along the Mississippi River have relied on levees and dams to keep out floodwaters. Levees are earthen barriers built along rivers. Dams are structures that block the flow of water, causing it to build up behind the dam into a reservoir lake.

Both levees and dams can prevent damage to cities and farms when the flooding is not too extreme. But during the flood of 1993, the levees and dams couldn't keep out the water. In fact, the funneling effect caused by the levees caused the floodwaters to move faster and deeper than they would have if the levees had not been there.

Think Critically:

Describe in your Science Journal how levees that were built upstream contributed to increased erosion problems downstream.

Mississippi River Levee Breaking

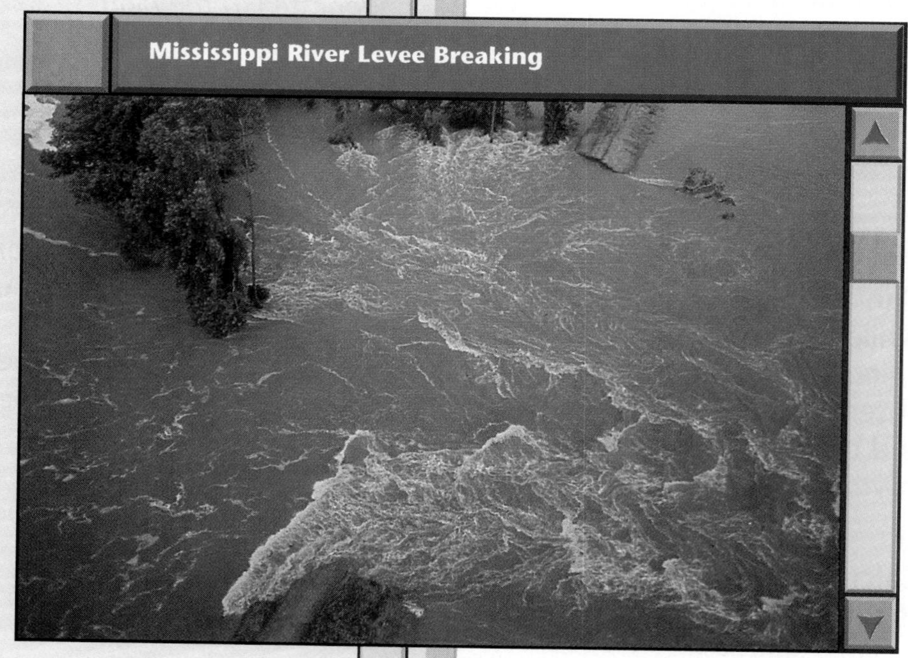

Deposition by Surface Water

As water moves throughout a river system, what do you suppose happens as it loses some of its energy of motion? The water can no longer carry some of its sediments, and they are deposited.

Some stream sediments aren't carried very far at all before they are deposited. In fact, many sediments are deposited within the stream channel itself. Other stream sediments travel great distances before they are deposited. Sediments picked up when rill and gully erosion occur are examples. Water usually has a lot of energy of motion as it moves down a steep slope. When the water begins flowing on a level surface, it slows down, loses energy of motion, and drops its sediments.

Figure 8-8

Alluvial fans commonly occur at the base of steep mountain slopes. *How does the amount and type of vegetation affect the alluvial fan?*

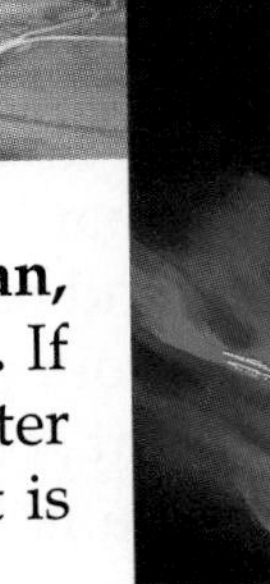

Figure 8-9

This satellite image of the Mississippi Delta shows how sediments accumulate when the Mississippi River empties into the Gulf of Mexico.

One type of deposit that results, an **alluvial fan,** is shaped like a triangle as shown in **Figure 8-8.** If the sediments are not deposited until the water empties into an ocean, gulf, or lake, the deposit is known as a **delta.**

Let's use the Mississippi River as an example to tie all of this section together. Runoff causes rill and gully erosion as it picks up sediments and carries them into the larger streams that flow into the Mississippi River. The Mississippi is quite large and has a lot of energy. It can erode many sediments. As it flows, it cuts into its banks and picks up more sediments. In other places, where the land is flat, the river deposits some of its sediments in its own channel.

Eventually, the Mississippi River reaches the Gulf of Mexico. There it flows into the gulf, loses most of its energy of motion, and dumps its sediments in a large deposit on the Louisiana coast. This deposit, shown in **Figure 8-9,** is the Mississippi Delta.

Section Wrap-up

Review

1. How does the slope of an area affect its runoff?
2. Describe the three stages of stream development.
3. **Think Critically:** How is a stream's rate of flow related to the amount of erosion it causes and the size of the sediments it deposits?

Skill Builder

Comparing and Contrasting

Compare and contrast alluvial fans and deltas. If you need help, refer to Comparing and Contrasting in the **Skill Handbook.**

Using Computers

Spreadsheet Design a table using spreadsheet software to compare and contrast sheet, rill, gully, and stream erosion.

8•2 Groundwater

Science Words

groundwater
permeable
impermeable
aquifer
zone of saturation
water table
artesian well
spring
hot spring
geyser
cave

Objectives

- Describe the groundwater system.
- Explain the effect that soil and rock permeability have on groundwater movement.
- Describe ways that groundwater erodes and deposits sediments.

Groundwater System Development

What would have happened if the spilled milk in Section 8-1 had run off the table onto a carpeted floor? It would have quickly soaked into the carpet. Water that falls on Earth can also soak into the ground.

But what happens to the water then? Water that soaks into the ground becomes a part of a system, just as water that stays above ground becomes a part of a river system. You already know that soil is made up of many small rock fragments and that there is weathered rock beneath the soil. Between these fragments and pieces of weathered rock are spaces called pores, as shown in **Figure 8-10.** Water that soaks into the ground collects in these pores and becomes part of what is called **groundwater.**

Figure 8-10

Some soils and rocks have connected pores through which water can move.

Permeability

A groundwater system is similar to a river system. However, instead of having channels that connect different parts of the drainage basin, the groundwater system has connecting pores. Soil and rock are **permeable** if water can pass through them. Sandstone is an example of a permeable rock.

Soil or rock that has many large, connected pores is highly permeable. Water can pass through it easily. Soil or rock that has few or small pores is less permeable because water can't easily pass through. Some material, such as clay, has very small pores or no pores at all. This material is **impermeable,** which means that water cannot pass through it.

USING MATH

In an experiment, you find that 100 mL of gravel (Total Volume) can hold 31 mL (Volume of Pore Spaces, VPS) of water. Calculate the percentage of pore space (porosity) using the following formula.

$$\frac{\text{VPS}}{\text{Total Volume}} \times 100 = \text{\% Porosity}$$

Groundwater Movement

How deep into Earth's crust do you suppose groundwater can go? Groundwater will keep going to lower elevations until it reaches a layer of impermeable rock. When this happens, the impermeable rock acts like a barrier and the water can't move down any deeper. As a result, water begins filling up the pores in the rocks above the impermeable layer. A layer of permeable rock that transmits water freely is an **aquifer.** The area where all of the pores in the rock are filled with water is the **zone of saturation.** The upper surface of this zone is the **water table,** as seen in **Figure 8-11.**

Figure 8-11

If you want to know where the water table is in a particular area, find a stream. A stream's surface is the water table. Below that is the zone of saturation.

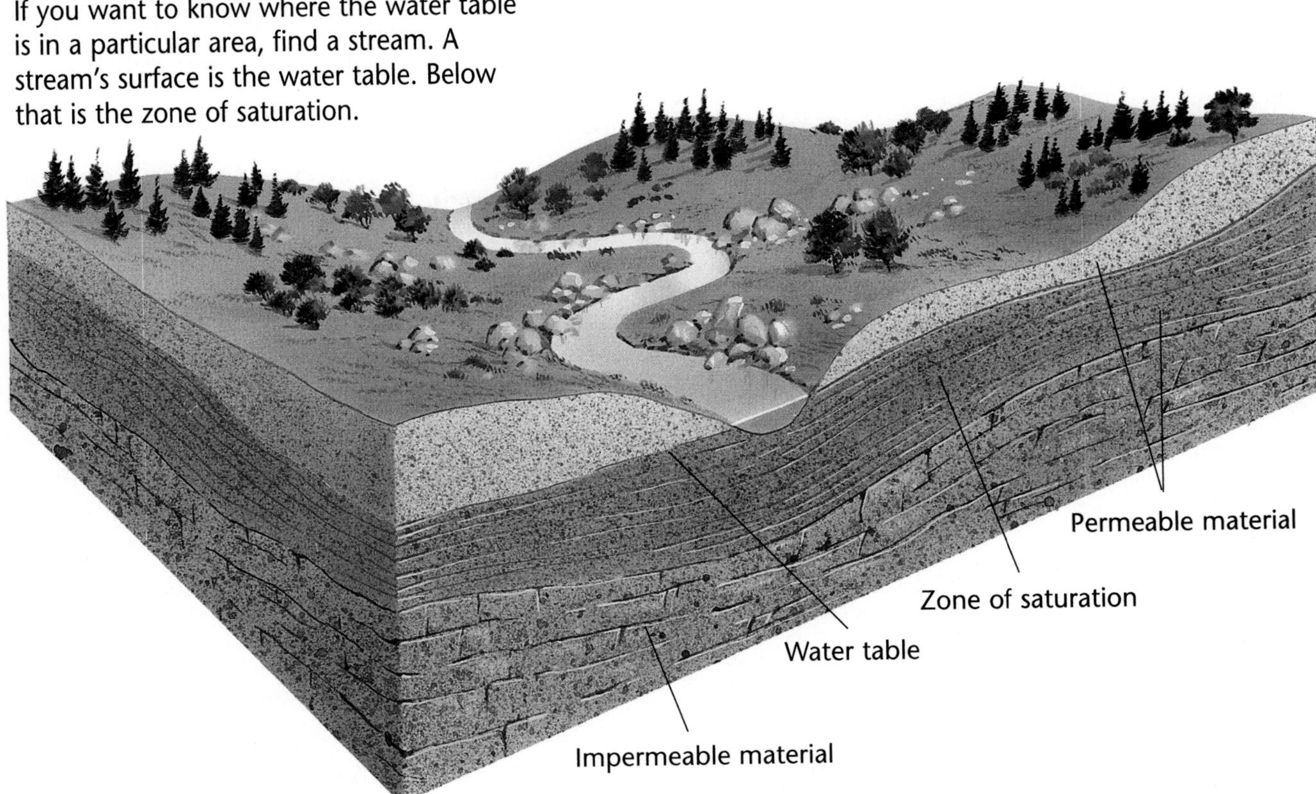

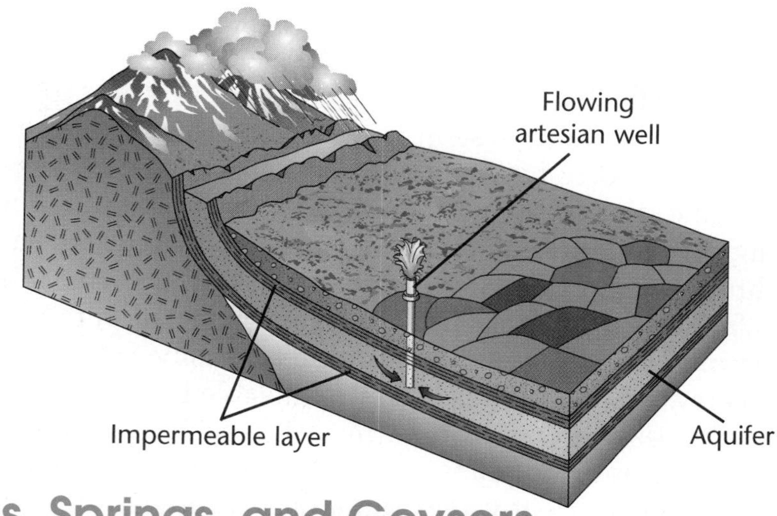

Figure 8-12

The pressure of water in a sloping aquifer keeps an artesian well flowing. *What limits how high an artesian well could be placed in an aquifer?*

MiniLAB

How can you measure pore space?

Procedure

1. Put 100 mL of sand in one beaker and 100 mL of gravel in another beaker.
2. Fill a graduated cylinder with 100 mL of water.
3. Pour the water slowly into the beaker with the sand and stop when the water just covers the top of the sand.
4. Record the volume of the water used.
5. Repeat the procedure with the gravel.

Analysis

1. Which substance has more pore space—sand or gravel? How do you know?
2. If you repeated your experiment, but first crushed your gravel into smaller pieces, how would the pore space change?

Wells, Springs, and Geysers

What's so important about the zone of saturation and the water table? Many people get drinking water from groundwater in the zone of saturation. Water wells are drilled down into the zone of saturation.

Wells

Water flows into a well, and a pump brings it to the surface. A well must go down at least past the top of the water table to reach water. Sometimes during dry seasons, a well can go dry because the water table drops. Having too many wells in one area can also cause the water table to drop. This happens because more water is taken out of the ground than can be replaced by rain.

There is a type of well that doesn't need a pump to bring water to the surface. An **artesian well** is a well in which water under pressure rises to the surface. Artesian wells are less common than other wells because of the unique conditions they require, as shown in **Figure 8-12.**

An artesian well requires a sloping aquifer located between two impermeable layers. Water will enter at the high part of the sloping aquifer. Water in the higher part of the aquifer puts pressure on the water in the lower part. If a well is drilled into the lower part of the aquifer, the pressurized water will flow to the surface. Sometimes there is enough pressure to force the water into the air, forming a fountain.

Springs

In some places, the water table meets Earth's surface. When this happens, water flows out and forms a **spring.** Springs are found on hillsides or in any other place where the water table is exposed at the surface. Springs can be used as a source of fresh water.

The water from most springs is cold. But in some places, groundwater is heated and comes to the surface as a **hot spring.** The groundwater is heated by rocks that come in contact with molten material under Earth's surface.

Geysers

One of the places where groundwater is heated is in Yellowstone National Park in Wyoming. In Yellowstone, there are hot springs and geysers. A **geyser,** like the one in **Figure 8-13A,** is a hot spring that erupts periodically, shooting water and steam into the air. Groundwater is heated to very high temperatures, causing it to expand underground. This expansion forces some of the water out of the ground, taking the pressure off of the remaining water. The remaining water boils quickly, with much of it turning to steam. The steam shoots out of the opening like steam out of a teakettle, forcing the remaining water out with it. Yellowstone's famous geyser, Old Faithful, in the photo in **Figure 8-13B,** shoots about 40 000 liters of water and steam into the air once each hour.

CONNECT TO

CHEMISTRY

Although limestone is commonly said to *dissolve* in groundwater, that doesn't accurately describe what happens. Weakly acidic groundwater reacts with limestone. This reaction produces compounds that dissolve more easily in water than limestone does. ***Infer*** what effect acid rain might have on a statue made of limestone.

Groundwater Erosion and Deposition

Just as water is the most powerful agent of erosion on Earth's surface, it can also have a great effect underground. As mentioned in Chapter 6, when water mixes with carbon dioxide in the air, it forms a weak acid. One type of rock that is easily dissolved by this acid is limestone. As acidic groundwater moves through natural cracks in limestone, it dissolves the rock. Gradually, the cracks in the limestone are enlarged until an underground opening called a **cave** is formed.

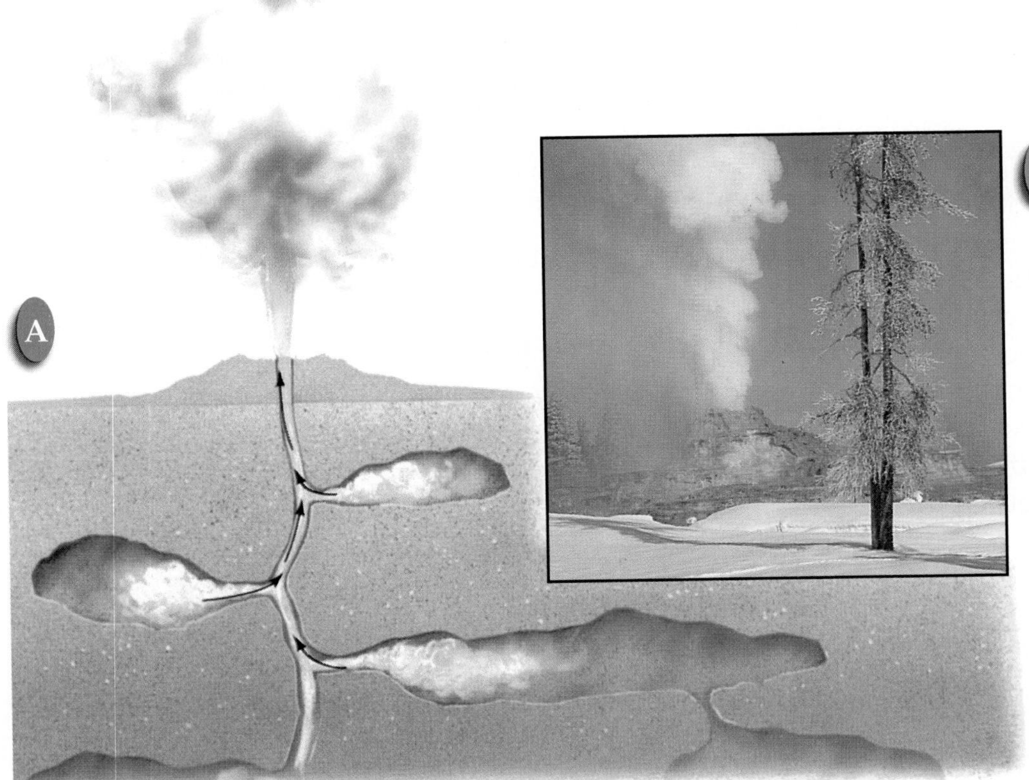

Figure 8-13

After a geyser erupts (A), water runs back into underground openings where it is heated and erupts again. Yellowstone's famous geyser, Old Faithful (B), erupts once each hour.

Figure 8-14

Water containing dissolved calcium ions forms interesting features in caves. Look at the features in the cave shown above and try to infer how they formed.

Cave Formation

You've probably seen a picture of the inside of a cave, or perhaps you've visited one. Groundwater not only dissolves limestone to make caves, but it also can make spectacular deposits on the insides of caves, as shown in **Figure 8-14.**

Water often drips slowly from cracks in the cave walls and ceilings. This water contains calcium ions dissolved from the limestone. If this water evaporates while hanging from the roof of a cave, a deposit of calcite is left behind. Stalactites form when this process happens over and over. Where drops of water fall to the floor of the cave, a stalagmite forms.

Sinkholes

If underground rock is dissolved near the surface, a sinkhole may form. A sinkhole is a depression that forms when the roof of a cave collapses.

In summary, when rain falls and becomes groundwater, it might dissolve a cave, erupt from a geyser, or be pumped from a well to be used at your house.

Section Wrap-up

Science Journal

Read an article about geothermal energy and draw a diagram in your Science Journal explaining how it works. Also list the limitations of this energy source.

Review

1. How does water enter the groundwater system?
2. How does the permeability of soil and rocks affect the flow of groundwater?
3. Explain how caves form.
4. **Think Critically:** Would you expect water in wells, geysers, and hot springs to contain eroded materials? Why or why not?

Skill Builder

Comparing and Contrasting

Compare and contrast wells, geysers, and hot springs. If you need help, refer to Comparing and Contrasting in the **Skill Handbook.**

Science & History

Huang Ho, China's Sorrow

Any study of erosion by rivers will be somehow incomplete if you don't learn about the Huang Ho, or Yellow River, of China. This river got its name from the soft yellow earth it carries. It empties 1 billion tons of this sediment into the sea every year. Its deposit creates an average of 25 km^2 of newly formed land at the river's mouth annually.

Science Journal

If you were placed in charge of flood control in your area, what measures would you propose? In your Science Journal, write your proposal.

History of Flooding

In the past 4000 years, the Huang Ho has flooded 1500 times. It is responsible for killing more people than any other river in the world with its floods. The first recorded flooding of this river occurred in 2297 B.C. By 602 B.C., the Chinese had begun to build dikes, which are walls or banks erected along a river for flood protection.

In 1887, heavy rains fell in Honan Province. The Huang Ho burst through 21-m-high dikes. About a million people died in this flood. Another million may have died in the famine that followed. The worst flood of the Huang Ho in terms of the people killed occurred more recently, in 1931. About 3.7 million died in that flood. A flood of a different sort occurred in 1938. That's when the Chinese General Chiang Kai-shek had his troops destroy the river's dikes in order to slow down the Japanese army. The general achieved his purpose, but the flood killed more than 500 000 people and left 13 million homeless.

For thousands of years, floods caused major changes of the river's course, keeping it from returning to its original position after a flood. The lower Huang Ho has changed the most, flowing northeastward to the Po Gulf after one flood and southeastward into the main body of the Yellow Sea after another.

Figure 8-15

A new flood control dam on the Huang Ho River.

Respite from Floods

Since 1949 there have been no breaks in the dikes that would cause serious floods. This is due in great part to a special program that enlists people to collect data comparing the level of the riverbed to the height of the dikes in every area along the river. If the riverbed has risen, the people set to work raising the level of the dikes to hold back the waters. They also build reservoirs on the floodplains to control the river's flow, as shown in **Figure 8-15.** Still, the riverbed continues to rise at the rate of 1.8 m per year in some areas.

SCIENCE & SOCIETY

8•3 ISSUE: Water Wars

Objectives

- Give examples of ways people use water.
- Explain why some communities must rely on water diversion for their water supply.
- Identify a problem caused by water diversion.

Water as a Resource

How much water do you use each day? An average person in the United States uses about 397 liters every day. That's enough to fill 1118 soft-drink cans!

Just imagine how much water an entire community uses each day for fire hydrants, street cleaning, and water fountains like the one shown in **Figure 8-16.** Where does your town's water come from? While some towns have easily accessible aquifers, large rivers, reservoirs, or lakes nearby, others must use pipelines to transport water from distant locations. The city of Los Angeles, California, gets its water from several sources. Among these sources are streams more than 400 km north of Los Angeles in the Sierra Nevada Mountains.

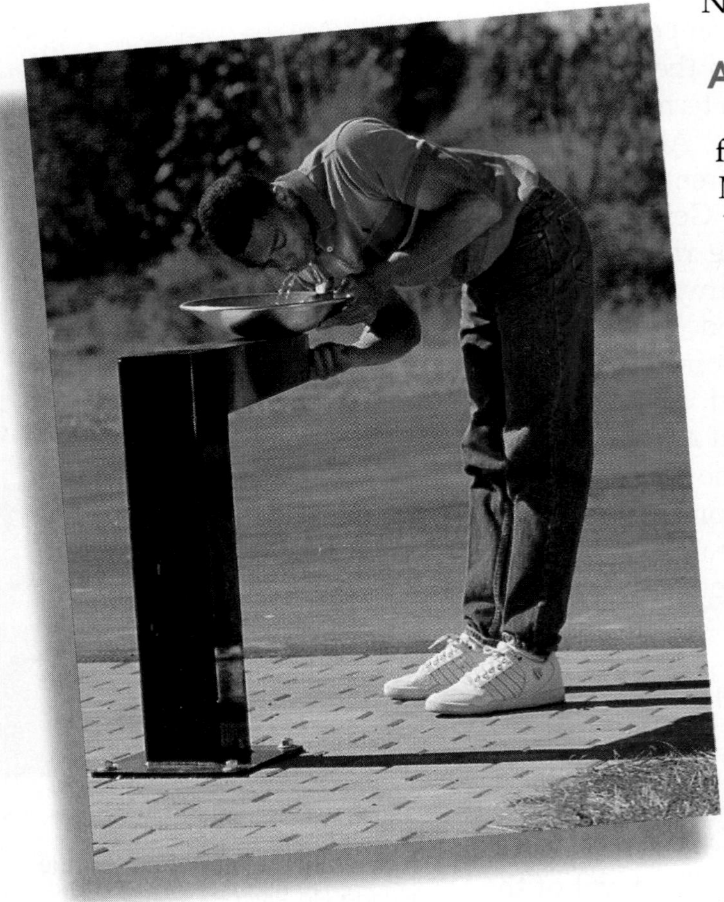

Figure 8-16

The availability of water is not something to be taken for granted.

A Bitter Battle

The streams that Los Angeles has diverted for its water needs feed into the unique Mono Lake. Since the diversion began in 1941, the lake's level has dropped 13.5 meters and its volume has been cut in half. People began to worry that Mono Lake might dry up. If this happened, not only would a beautiful lake be lost, but an entire ecosystem would be extinguished as well. Thus began a battle between the city of Los Angeles and a host of geologists and ecologists.

In 1994, after hearing arguments from both scientists and Los Angeles city officials, the California Water Resources Control Board decided that Los Angeles must allow the lake level to rise 5 meters above the 1994 water level. While this reduces the amount of water Los Angeles can divert from Mono Lake, it doesn't cut it off completely.

2 Points of View

Conserve the Natural Resource

One of the concerns about lowering Mono Lake's water levels is the increased mineral and salt concentrations, shown in **Figure 8-17.** With less fresh water flowing into the lake, it became nearly twice as salty. Many ecologists hypothesized that the increased salt concentrations threatened the populations of brine shrimp and alkali flies, which are an important food source for many migratory birds that move through the area. Aside from possibly losing their food, migratory waterfowl lost much of their shoreline habitat.

Figure 8-17
These structures, called tufa towers, are formed by mineral-rich springs. As Mono Lake's water level dropped, many of these structures were revealed.

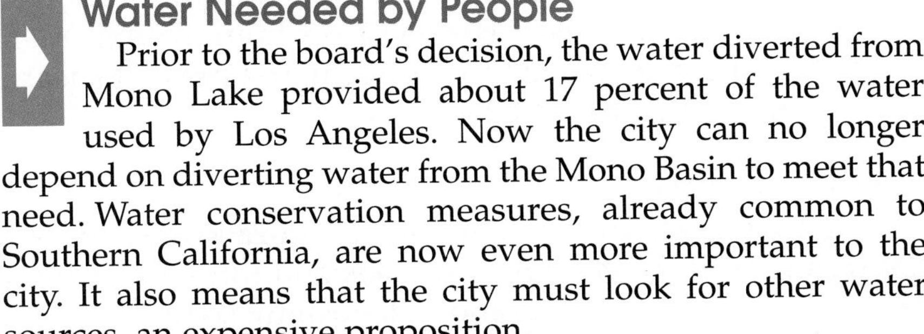

Water Needed by People

Prior to the board's decision, the water diverted from Mono Lake provided about 17 percent of the water used by Los Angeles. Now the city can no longer depend on diverting water from the Mono Basin to meet that need. Water conservation measures, already common to Southern California, are now even more important to the city. It also means that the city must look for other water sources, an expensive proposition.

As Earth's population continues to increase, greater demands are placed on Earth's freshwater supply. With the increased demand, there will likely be more conflicts between conserving natural resources and the need for water for agriculture, industry, and recreation. Where will we get the fresh water we will need?

Science Journal

You drink only a small part of the water you use. Much more goes into each shower or bath you take. In your Science Journal, write about other ways you use water directly or indirectly.

Section Wrap-up

Review

1. What are four ways that you use water each day?
2. What is a problem caused by water diversion?

Explore the Issue

The human population in the southwestern United States is increasing rapidly. As a result, there is a greater demand for water in a dry climate. To solve this problem, some have proposed building a giant canal to divert water from the Great Lakes. Infer what some of the problems might be if the canal is built.

SCIENCE & SOCIETY

8•4 Ocean Shoreline

Science Words

longshore current
beach
barrier island

Objectives

- Describe forces that cause shoreline erosion.
- Compare and contrast different types of shorelines.
- List some origins of sand.

The Shore

Picture yourself sitting on a beautiful white sand beach. Palm trees sway in the breeze above your head, and small children play in the quiet waves lapping at the water's edge. It's hard to imagine a place more peaceful than this shore. Now picture yourself sitting along another shore. You're on a high cliff, overlooking waves crashing onto huge boulders below. Both of these settings are shorelines. This is where land meets the ocean.

The two shorelines we just described are very different. Why are they so different? Both are subjected to the same forces. Both experience the same surface waves, tides, and currents. These cause both shorelines to constantly change. Sometimes you can see these changes from hour to hour. We'll look at why these shorelines are different, but first, let's learn about the forces that carve shores.

Shoreline Forces

Along all shorelines, like the one in **Figure 8-18A,** surface waves constantly move sediments back and forth. The waves shape shorelines by eroding and redepositing sediments. The tides also shape shorelines. Every day tides raise and lower the place on the shoreline where surface waves erode and deposit sediments.

Figure 8-18

A Waves, tides, and currents cause shorelines to constantly change.

B Waves approaching the shoreline at an angle create a longshore current. When the water built up by waves returns to sea, a rip current forms. *Why is it wise to avoid rip currents when swimming at a beach?*

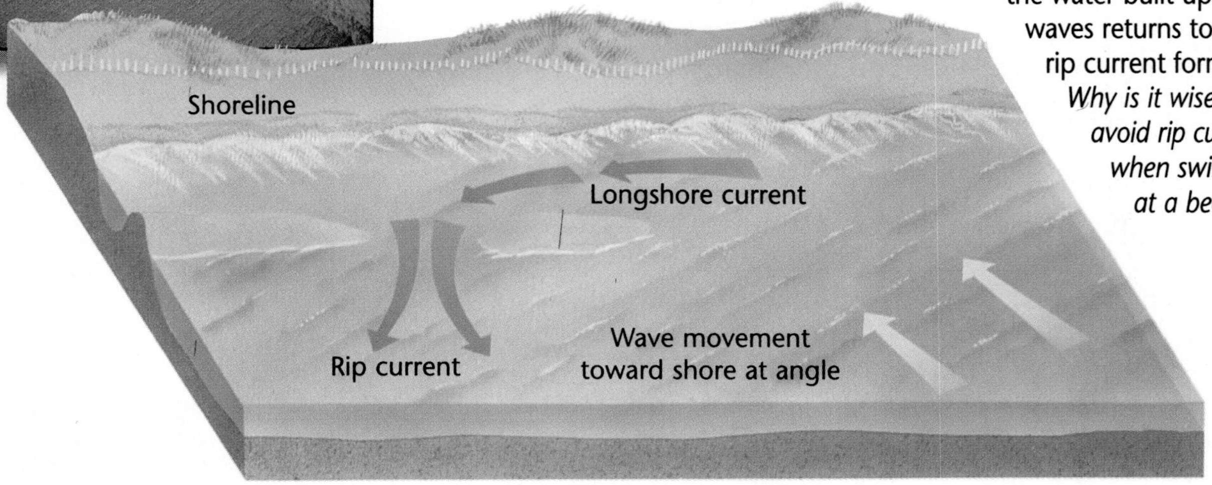

Shoreline

Longshore current

Rip current

Wave movement toward shore at angle

Figure 8-19

Along a rocky shoreline, the force of pounding waves breaks rock fragments loose and grinds them into smaller and smaller sediments.

Waves usually collide with a shore at slight angles. This creates a **longshore current** of water that runs parallel to the shore, as shown in **Figure 8-18B.** Longshore currents carry many metric tons of loose sediments and act like rivers of sand in the ocean. What do you suppose happens if a longshore current isn't carrying all of the sand it has the energy to carry? It will use this extra energy to erode more shoreline sediments.

You've seen the forces that affect all shorelines. Now we'll look at the differences that make one shore a flat, sandy beach and another shore a steep, rocky cliff.

Rocky Shorelines

Along rocky shorelines like the one in **Figure 8-19,** rocks and cliffs are the most common features. Waves scour the rocks to form hollows or notches. Over time, these enlarge and become caves. Rock fragments broken from the cliffs are ground up by the endless motion of waves. These fragments act like the sand on sandpaper.

Softer rocks are eroded away before harder rocks, leaving islands of harder rocks. This takes many years, but remember that the ocean never sleeps. In a single day, about 14 000 waves will crash onto any shore.

The rock fragments produced by eroding waves are sediments. When rocky shorelines are being eroded away, where do you think the sediments go? Waves carry them away and deposit them where water is quieter. If you want to relax on a nice wide, sandy beach, you probably won't go to a rocky shoreline.

Figure 8-20

Sand varies in size, color, and composition. This sand contains shell fragments and tiny grains of quartz.

Sandy Beaches

Smooth, gently sloping shorelines are quite different from steep, rocky shorelines. Beaches are the main feature here. **Beaches** are deposits of sediment that run parallel to the shore. They extend inland as far as the tides and waves are able to deposit sediment.

Beaches are made of different materials like the sand shown in **Figure 8-20.** Some are made of rock fragments from the shoreline, and others consist of seashell fragments. These fragments range from pebbles large enough to fill your hand to fine sand. Sand grains are 0.07 mm to 2 mm in diameter. Why do many beaches have sand-sized particles? This is because waves break rocks and seashells down to sand-sized particles. The constant wave motion bumps sand grains together and, in the process, rounds their corners.

What kinds of materials do you think make up most beach sands? Most are made of resistant minerals such as quartz. However, sand in some places is composed of other things. For example, Hawaii's black sands are made of basalt, and green sands are made of the mineral olivine. Jamaica's white sands are made of coral and shell fragments.

Sand Erosion and Deposition

Sand constantly is carried down beaches by longshore currents to form features such as those seen in **Figure 8-21.** Sand is also moved by storms and the wind. Thus, beaches are fragile, temporary features that are easily damaged by storms and human activities such as construction.

Figure 8-21

Longshore currents move sediment along shorelines, changing beaches and creating baymouth bars and spits, like those in this aerial photograph of Martha's Vineyard, Massachusetts.

What's so special about sand?

You know that sand is made of many different kinds of grains, but did you realize that the slope of a beach is actually related to the size of its grains? The coarser the grain size, the steeper the beach. Did you know that many sands are mined because they have economic value?

Problem

What characteristics can be used to classify beach sands?

Materials

- 3 samples of different beach sands
- stereomicroscope
- magnet

Procedure

1. Design a data table in which to record your data when you compare the three sand samples. You will need five columns in your table. One column will be for the samples, and the others for the characteristics you will be examining. If you need help in designing your table, refer to the Skill Handbook.
2. Use the diagram below to determine the average roundness of each sample.

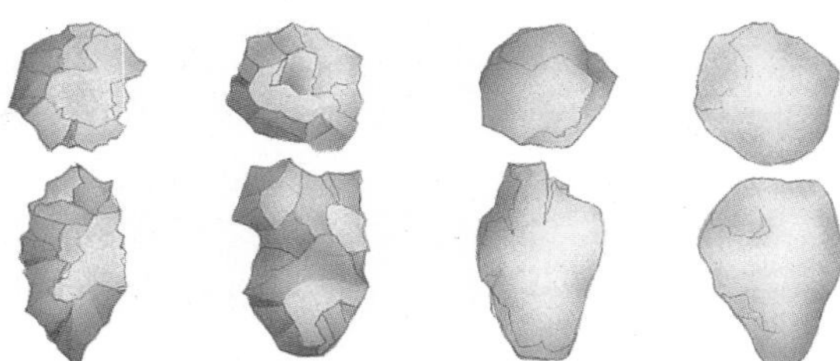

3. Identify the grain size of your samples by using the sand gauge on this page. To determine the grain size, place sand grains in the middle of the circle of the sand gauge. Use the upper half of the circle for dark-colored particles, and the bottom half of the circle for light-colored particles.
4. Decide on two other characteristics to examine that will help you classify your samples.

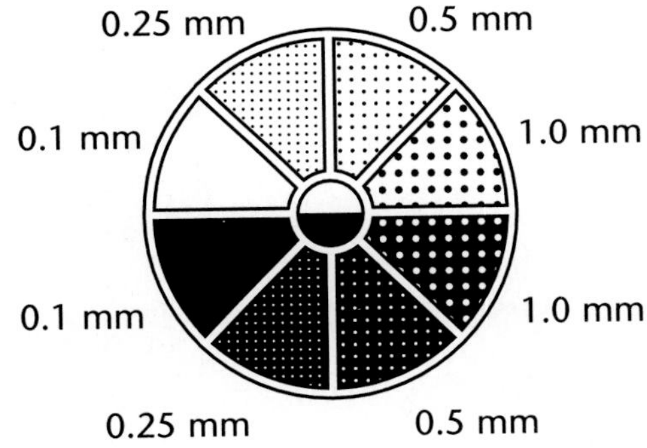

Sand Gauge

Analyze

1. Were the grains of a particular sample generally the same size? Explain.
2. Which of these sand samples do you think came from the steepest beach?
3. Were the grains in your three samples generally the same shape?

Conclude and Apply

4. What are some characteristics of beach sand?
5. Why are there differences in the characteristics of different sand samples?

Barrier Islands

Barrier islands are sand deposits that parallel the shore but are separated from the mainland, as shown in **Figure 8-22.** These islands start as underwater sand ridges formed by breaking waves. Hurricanes and storms add sediment to them, raising some to sea level. Once a barrier island is exposed, the wind blows the loose sand into dunes, keeping the island above sea level. As with all seashore features, barrier islands are temporary, lasting from a few years to a few centuries.

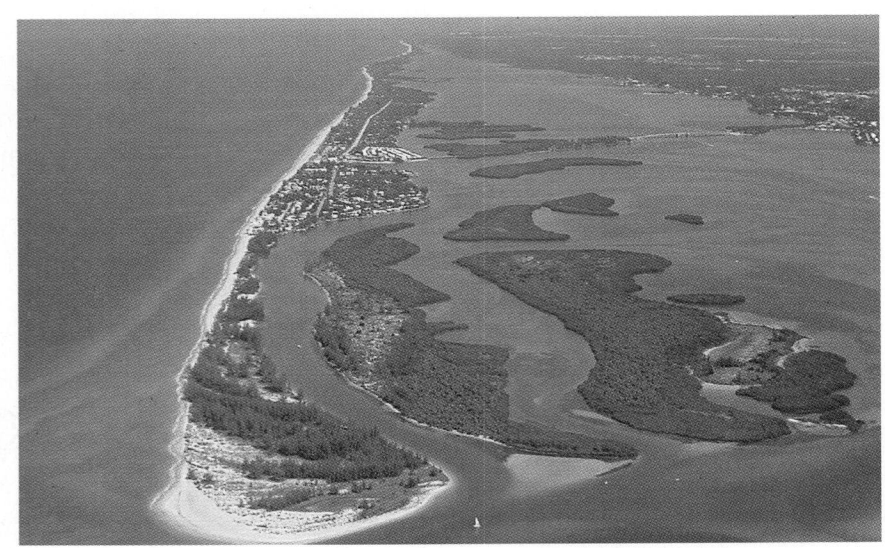

Figure 8-22

There are many barrier islands along North America's Atlantic Coast and the Gulf of Mexico. The size and shape of these islands constantly change due to wave action.

Section Wrap-up

Review

1. What are the forces that cause shoreline erosion?
2. Contrast the features you'd find along a steep, rocky shoreline with the features you'd find along a gently sloping shoreline.
3. **Think Critically:** Why is there no sand on many of the world's shorelines?

Skill Builder

Concept Mapping

Make a cycle concept map that discusses how the sand from a barrier island that is currently in place can become a new barrier island 100 years from now. Use these terms: *barrier island, breaking waves, wind, longshore currents,* and *new barrier island.* If you need help, refer to Concept Mapping in the **Skill Handbook.**

USING MATH

If in a single day, about 14 000 waves crash onto a shore, how many waves crash onto a shore in one year? Calculate how many have crashed onto a shore since you were born.

Chapter 8

Review

Summary

8-1: Surface Water

1. The amount of runoff in an area depends on the amount of rain that falls, the time period over which it falls, the slope of the land, and the amount of vegetation.
2. Rill, gully, and sheet erosion are all types of surface water erosion due to runoff.
3. Water in a stream picks up sediments from the bottom and sides of its channel.
4. Young streams flow swiftly through steep-sided valleys and erode the bottoms of their channels more than the sides. Mature streams erode the sides of their channels and form meanders. A stream in the old stage has less energy than a young stream because it moves very slowly.
5. Alluvial fans and deltas are triangular deposits that form when water loses energy of motion.

8-2: Groundwater

1. Water that soaks into the ground becomes part of the groundwater system.
2. Springs, artesian wells, hot springs, and geysers are all a part of the groundwater system.
3. Groundwater erosion results in the formation of caves and sinkholes. Groundwater deposition results in the formation of cave formations like stalactites and stalagmites.

8-3: Science and Society: Water Wars

1. People use water by drinking it, bathing, and in manufacturing and irrigation.
2. Some communities must divert water from other locations. With water diversion comes environmental problems.

8-4: Ocean Shoreline

1. Shorelines are subjected to waves, tides, and currents.
2. Rocks and cliffs are common along rocky shorelines.
3. Sandy beaches form along gently sloping shorelines. Sand forms when rocks, coral, and seashells are broken into small particles by wave action.

Key Science Words

a. alluvial fan
b. aquifer
c. artesian well
d. barrier island
e. beach
f. cave
g. delta
h. drainage basin
i. floodplain
j. geyser
k. groundwater
l. gully erosion
m. hot spring
n. impermeable
o. longshore current
p. meander
q. permeable
r. rill erosion
s. runoff
t. sheet erosion
u. spring
v. water table
w. zone of saturation

Chapter 8 Review

Reviewing Vocabulary

Match each phrase with the correct term from the list of Key Science Words.

1. water that flows over Earth's surface
2. water that flows parallel to the shore
3. a curve in a stream channel
4. a body of permeable rock through which water can flow freely
5. the area where all the pores in the rock are filled with water
6. hot springs that erupt
7. a triangular deposit that forms when a river enters a gulf or lake
8. any kind of sediment that is deposited parallel to a shore
9. water that soaks into the ground
10. a large underground opening formed when limestone dissolves

Checking Concepts

Choose the word or phrase that completes the sentence or answers the question.

1. An example of a structure created by deposition is a(n) _______.
   a. beach c. cave
   b. rill d. geyser
2. A deposit that forms when a mountain river runs out onto a plain is called _______.
   a. subsidence c. infiltration
   b. an alluvial fan d. water diversion
3. A layer of rock that water flows through is _______.
   a. an aquifer c. a water table
   b. a pore d. impermeable
4. The network formed by a river and all the smaller streams that contribute to it is a _______.
   a. groundwater system
   b. zone of saturation
   c. river system
   d. water table
5. Soils through which fluids can easily flow are _______.
   a. impermeable c. saturated
   b. meanders d. permeable
6. Streams in mountains tend to be in the _______ stage of development.
   a. young c. old
   b. mature d. meandering
7. A(n) _______ forms when the water table is exposed at the surface.
   a. meander c. aquifer
   b. spring d. stalactite
8. Heated groundwater that reaches Earth's surface is a(n) _______.
   a. water table c. aquifer
   b. cave d. hot spring
9. Beaches are most common along _______.
   a. rocky shorelines c. aquifers
   b. flat shorelines d. young streams
10. Water rises in an artesian well because of _______.
   a. a pump c. heat
   b. erosion d. pressure

Understanding Concepts

Answer the following questions in your Science Journal using complete sentences.

11. Would you expect more runoff from a parking lot or a grassy field with the same slope? Explain.
12. Compare and contrast the valleys of young, mature, and old streams.
13. Explain why shorelines are constantly changing.

Thinking Critically

14. Explain why the Mississippi River has meanders.
15. What determines whether a stream, like the one shown on the right, erodes its bottom or its sides?
16. Why would you be concerned if developers of a new housing project started drilling wells near your well?
17. A stack is an island of rock along certain shorelines. Along what kind of shoreline would you find stacks? Explain.
18. Explain why beach sands collected from different locations will differ in composition, color, and texture.

Developing Skills

If you need help, refer to the **Skill Handbook.**

19. **Concept Mapping:** Complete the concept map below using the following terms: developed meanders, gentle curves, gentle gradient, old, rapids, steep gradient, wide floodplain, young

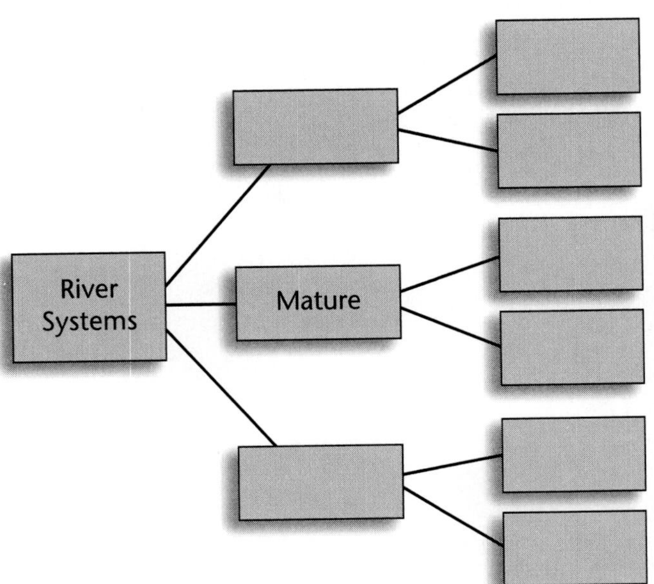

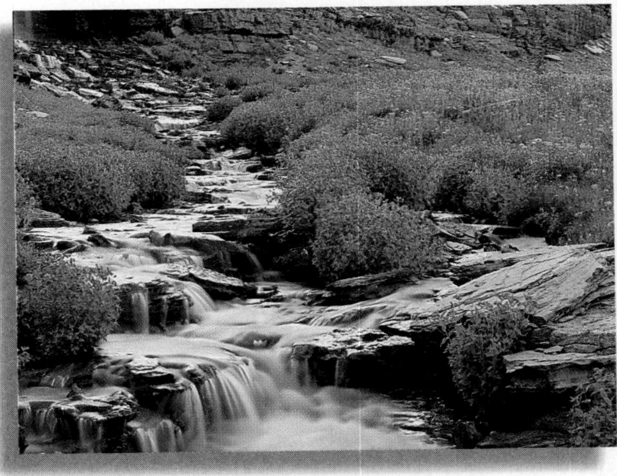

20. **Interpreting Data:** If the Brahmaputra River flows at a rate of 19 800 m^3/s and the La Plata River flows at 79 300 m^3/s, infer which river erodes more sediments.
21. **Hypothesizing:** Hypothesize why most of the silt in the Mississippi Delta is found farther out to sea than the sand-sized particles.
22. **Outlining:** Make an outline that explains the three stages of stream development.
23. **Using Variables, Constants, and Controls:** Explain how you could test the effect of slope on the amount of runoff produced.

Performance Assessment

1. **Designing an Experiment:** Design a way to compare the speeds of a mature, meandering stream and a young, straight stream.
2. **Making Observations and Inferences:** Experiment to find out the pore space of 100 mL of clay. Repeat your experiment to measure the pore space when you mix 50 mL of gravel with 50 mL of clay. Repeat once more with 50 mL of clay mixed with 50 mL of sand.
3. **Display:** Use a shoe box and clay to make a model of a limestone cave. Include stalactites, stalagmites, columns, and a sinkhole.

UNIT PROJECT 2

The Oregon Trail

Thousands of people traveled the Oregon Trail from 1841 to the 1870s. It was a long and dangerous journey from Independence, Missouri, where the trail began, to destinations throughout Oregon. The pioneers traveled in wagons pulled by oxen over high mountains, prairies, deserts, plateaus, and rivers. If everything went smoothly along the trail, it took five months to complete the journey. But for many, the trip took much longer, and some never completed the journey at all.

The Oregon Trail in the 1800s

Research the path of the Oregon Trail. What landforms did it cross? Where were settlements along the trail? Find out why people wanted to travel over 2000 miles of wilderness to get to the Pacific Northwest. How much would it have cost a person to make the trip? What types of diseases were common in those days? What other health problems existed? What materials would they have to take with them? How did they get water? What hardships would have existed while they were crossing prairies, mountains, deserts, rivers, and plateaus? What weather problems would they have had to face?

Procedures

1. Use the information in this unit and your research on the Oregon Trail to make a colorful map board game of the trail. Have spaces along the trail where players can move. Be sure to draw all important landforms and settlements along the trail.
2. Design game pieces and a way for players to move along the trail.
3. You may decide to use a spinner, dice, or another way to determine the spaces to move.
4. Label some spaces with problems people might have encountered, like "Quicksand in the river, lose your turn" or "Mudslide covers the trail, go back two spaces."
5. Design cards that players draw at various places along the trail. Cards might say, "You must hunt for food, lose one turn" or "You discover shortcut, jump ahead 3 spaces." In making your labels for the spaces and cards, consider how weather and landforms affected travel. Also, remember that oxen needed to feed on grass, sick people needed to rest, and the pioneers had to take time to hunt for food and water.
6. After you've designed your game, write directions for playing the game and describe the goal(s) of the game.

Going Further

After all members of your class have completed their board games, swap your game with another team, and play their game. After you've finished, describe similarities and differences in the games. Based on what you've learned, would you have enjoyed being in a wagon train on the Oregon Trail in the 1800s?

UNIT 3

Unit Contents

Earth's Internal Processes

What's Happening Here?

Deep under the surface of Earth's crust, constant changes are taking place. Continents are shifting at a very small rate. Intense heat and pressure are melting solid rock into magma. On the surface of the planet, we go about our lives unaware of the forces at work beneath the crust—until a spectacular event, such as the volcanic eruption shown here, reminds us of the magnitude of Earth's internal processes. Read on to learn more about volcanoes, earthquakes, and plate tectonics, and how these processes impact life on the surface of the planet.

Science Journal

A volcanic eruption is a major news event. Imagine that you are a reporter sent to cover a volcanic eruption in the Philippines. In your Science Journal, make a list of the questions you would use to interview people who witnessed the event.

Previewing the Chapter

Earthquakes

You may have seen pictures such as the one shown here of the 1995 Hanshin-Awaji (Kobe), Japan earthquake. Have you ever thought about what caused this earthquake, or wondered if it could happen where you live? As you read this chapter, you will explore how movement and forces inside Earth produce earthquakes.

EXPLORE ACTIVITY

Show how forces inside Earth cause rocks to deform.

1. Place three pieces of clay flat on a table.
2. Place your hands on opposite ends of one of the pieces of clay.
3. Begin pushing your hands together, compressing the clay.
4. Now hold another piece of clay in your hands.
5. Begin to apply tension by gradually pulling the clay apart.
6. Holding the third piece of clay flat on the table, begin pushing so that one hand slides past the other hand.

Observe: What happens to the clay layer when your hands compress it from directly opposite directions? What happens to the clay layer when you gradually pull it apart? What happens to the clay when you push on it so that your hands eventually slide past each other? Draw a picture of each of the three clay pieces in your Science Journal.

Previewing Science Skills

- In the **Skill Builders**, you will **compare and contrast, map concepts, make and use graphs,** and **make and use tables.**
- In the **Activities**, you will **measure, construct and interpret graphs, analyze data,** and **draw conclusions.**
- In the **MiniLABs**, you will **interpret graphs, analyze data,** and **evaluate hypotheses.**

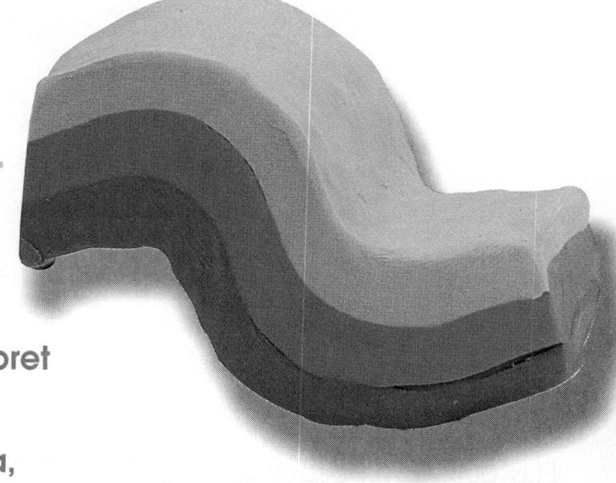

9•1 Forces Inside Earth

Science Words

fault
earthquake
normal fault
reverse fault
strike-slip fault

Objectives

- Explain how earthquakes result from the buildup of stress in Earth's crust.
- Contrast normal, reverse, and strike-slip faults.

Causes of Earthquakes

Think about the last time you used a rubber band to hold a roll of papers together. You knew you could stretch the rubber band only so far before it would break. Rubber bands bend and stretch when force is applied to them. Because they are elastic, they return to their original shape once the force is released. Plastic or play putty, as shown in **Figure 9-1,** acts in much the same way. With a certain amount of applied stress, plastic will undergo elastic deformation, but then it returns to normal. With additional stress, the plastic will undergo plastic deformation. Once it bends, it remains in the deformed shape.

Passing the Elastic Limit Causes Faulting

There is a limit to how far a rubber band will stretch or plastic will deform. Once this elastic limit is passed, the rubber band or plastic breaks. Rocks act in much the same way. Up to a point, applied stresses cause rocks to bend and stretch, undergoing elastic or plastic deformation. Once their elastic limit is passed, the rocks remain bent and may break. Rocks break and move along surfaces called **faults.** The rocks on either side of a fault move in different directions.

What produces the forces that cause faults to form? Obviously, something must be causing the rocks to move; otherwise, the rocks would just rest quietly without any stress

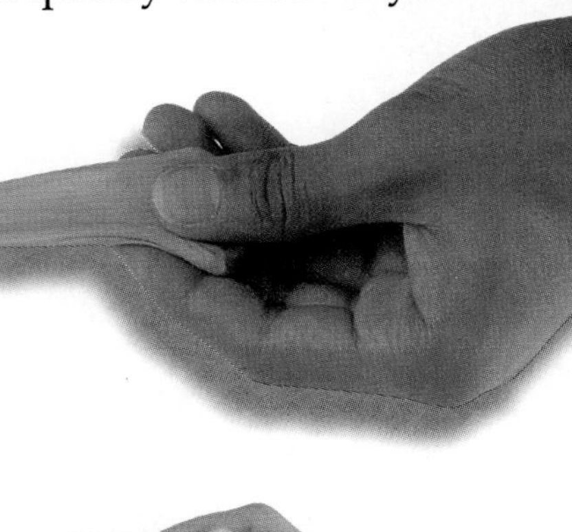

Figure 9-1

Play putty can be used to demonstrate applied stresses and the elastic limit.

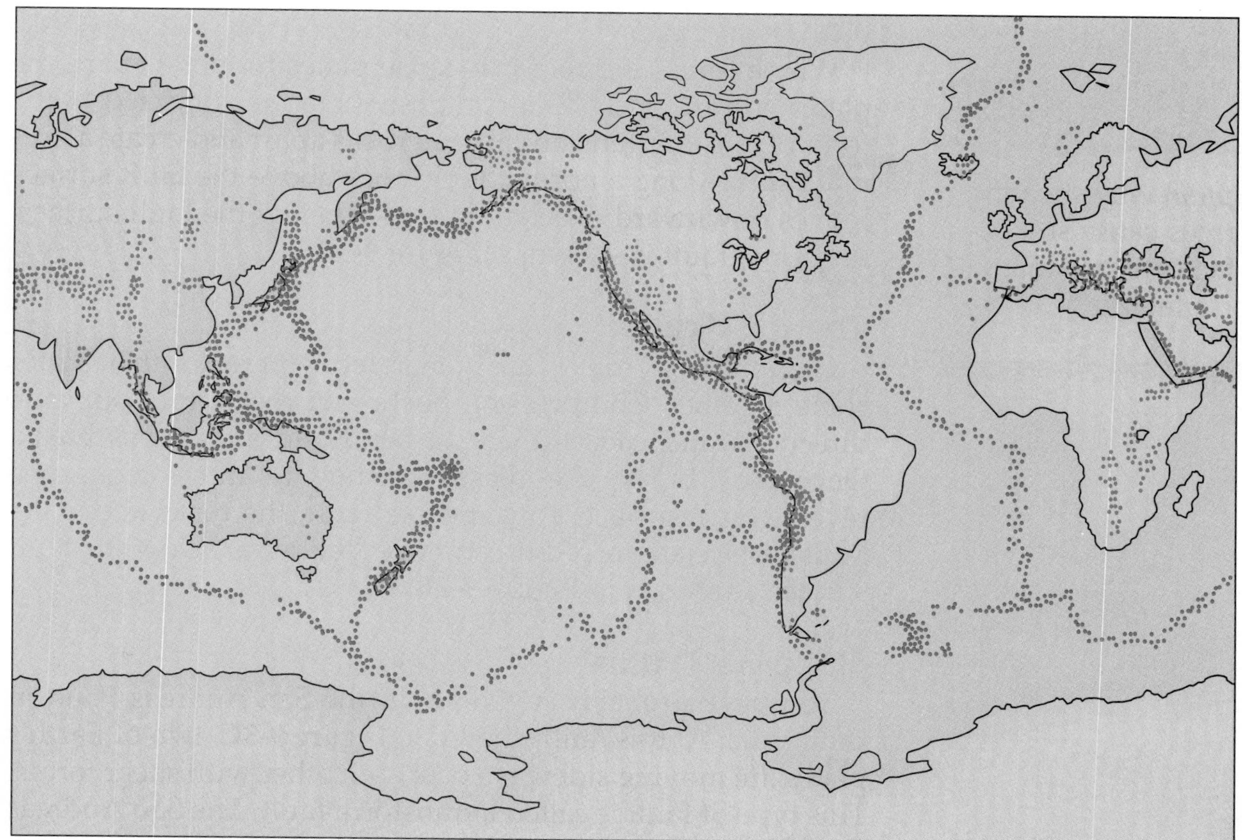

Figure 9-2

The dots represent the epicenters of the major quakes over a ten-year period. Eighty percent of earthquakes occur along the edges of the area known as the Pacific Plate. *Based on what you know of the area, why is this earthquake-prone area called the Pacific Ring of Fire?*

building up in them. However, Earth's crust is in constant motion because of forces inside Earth. These forces cause sections of Earth's crust to move, putting stress on rocks. To relieve this stress, the rocks tend to bend, compress, and stretch like plastic or rubber bands. But if the force is great enough, the rocks break. This breaking produces vibrations, called **earthquakes.**

Types of Faults

Rocks experience several types of forces where sections or plates of Earth's crust and upper mantle meet. In the Explore Activity at the beginning of this chapter, you experimented with three forces—compression, tension, and shear. Compression is a force or stress that squeezes and compresses, while tension is the stress that causes stretching and elongation. Shear is the force that causes slippage and the rocks on either side of the fault to move past each other. Let's take a look at these three forces and the types of faults they create.

E*xplain* why shearing forces cause strike-slip faults rather than reverse faults.

Normal Faults

Where forces inside Earth cause plates to move apart, the plates and the rocks that compose them are subjected to the force of tension. Tension can pull rocks apart and create a **normal fault.** Along a normal fault, rock above the fault surface moves downward in relation to rock below the fault surface. A normal fault is shown in **Figure 9-3A.**

Reverse Faults

Compression forces are generated where Earth's plates come together. Compression pushes on rocks from opposite directions and causes them to bend and sometimes break. Once they break, the rocks continue to move along the reverse fault surface. At a **reverse fault,** the rocks above the fault surface are forced up and over the rocks below the fault surface as shown in **Figure 9-3B.**

Strike-Slip Faults

You have probably heard about the San Andreas Fault in California. At this fault, shown in **Figure 9-3C,** two of Earth's plates are moving sideways past each other with shear forces. This type of fault is called a transform fault. The San Andreas Fault is a transform or strike-slip fault. At a **strike-slip fault,** rocks on either side of the fault surface are moving past each

Figure 9-3

As you learned in the Explore Activity, rock layers are affected differently by tension, compression, and shear forces. *With which type of force(s) would Earth's crust be stretched and thinned? With which type(s) would Earth's crust be folded and thickened?*

A When rock moves along a fracture caused by tension forces, the break is called a normal fault. Rock above the fault moves downward in relation to the rock below the fault surface. Normal faults can form mountains such as the Sierra Nevadas, which border California on the east.

other without much upward or downward movement. Compare the faults in **Figure 9-3.** How do they differ?

As the rocks move past each other at a strike-slip fault, their irregular surfaces snag each other, and the rocks are twisted and strained. Not only are they deformed, but also, the snagging of the irregular surface hinders the movement of the plates. As forces keep driving the plates to move, the stress builds up and the rocks reach their elastic limit. When the rocks are stressed past their elastic limit, they break and an earthquake results.

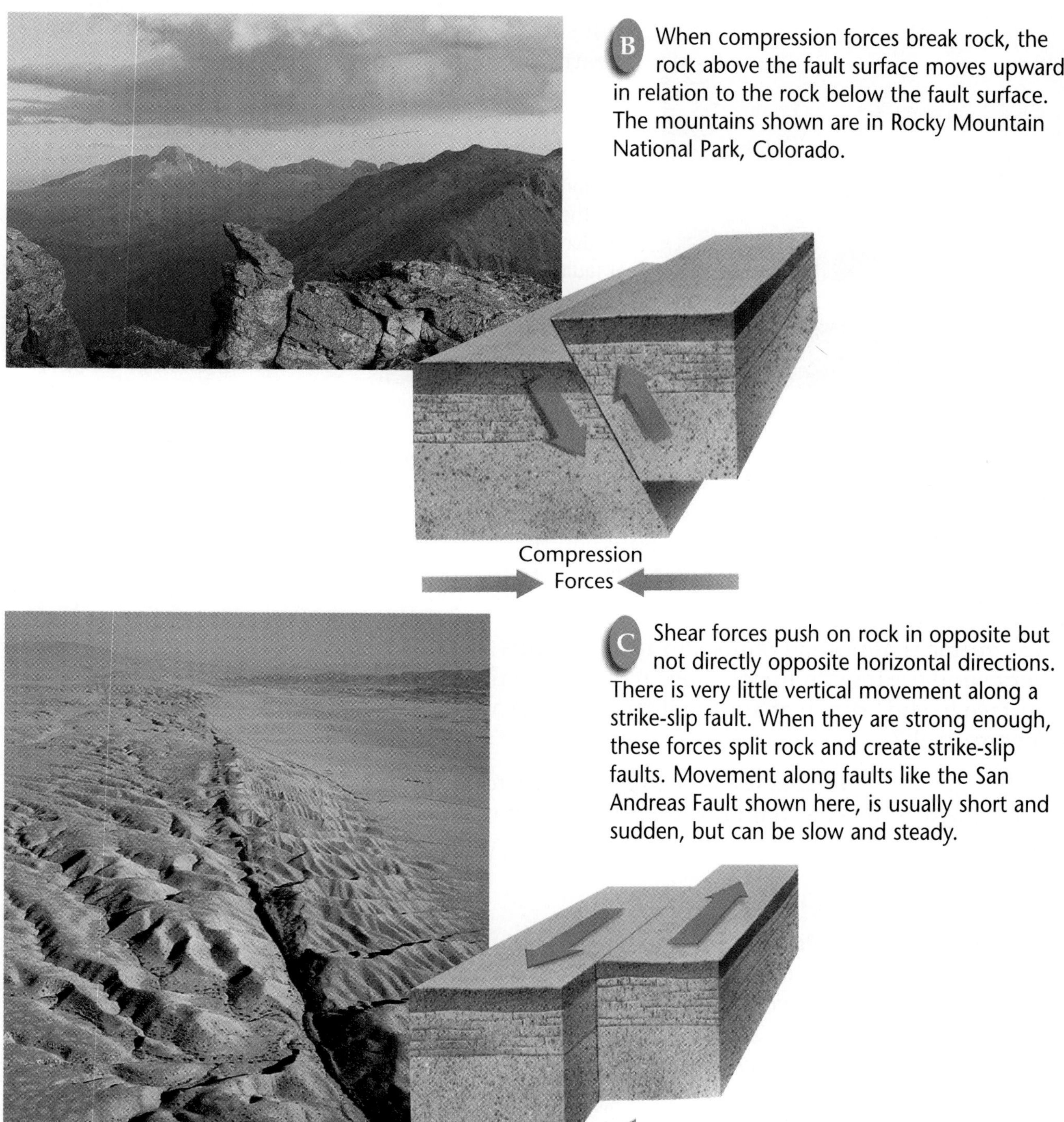

B When compression forces break rock, the rock above the fault surface moves upward in relation to the rock below the fault surface. The mountains shown are in Rocky Mountain National Park, Colorado.

C Shear forces push on rock in opposite but not directly opposite horizontal directions. There is very little vertical movement along a strike-slip fault. When they are strong enough, these forces split rock and create strike-slip faults. Movement along faults like the San Andreas Fault shown here, is usually short and sudden, but can be slow and steady.

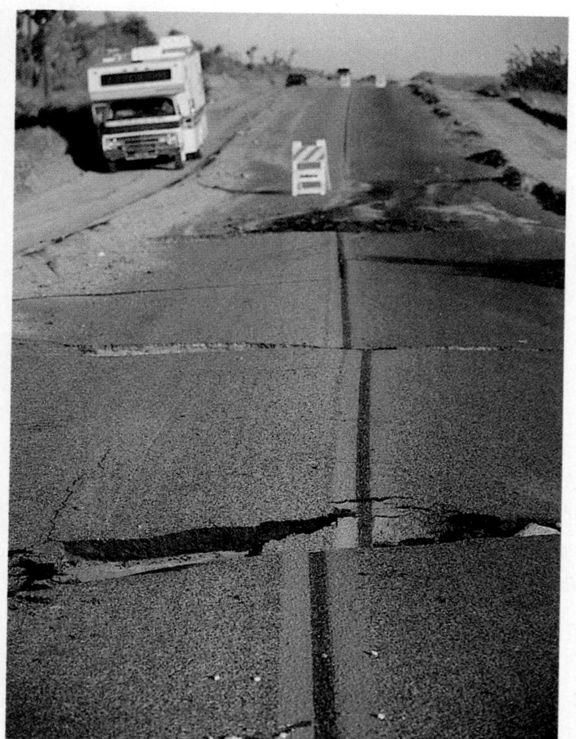

Earthquakes can be dramatic events. Some of them have devastating effects, while others go almost unnoticed. Regardless of their magnitudes, most earthquakes are the result of the plates moving over, under, and around each other. If these plates simply slid smoothly past each other, the tension, compression, and shear forces would not build up stress. But in actuality, rocks do experience these forces and stresses build up in them, causing small amounts of deformation, as seen in **Figure 9-4.** When rocks break because of the stress, energy is released along the fault surfaces and we observe the effects in the form of earthquakes.

Figure 9-4

Geologists use surface evidence like these cracks when searching for dangerous hidden fractures or faults.

Section Wrap-up

Review

1. Why is a shearing force usually generated at strike-slip fault boundaries?
2. The Appalachian Mountains formed when two of Earth's plates collided. Explain why reverse faults are the most common in these mountains.
3. The surfaces of normal faults and reverse faults look very similar. How do forces cause rocks above the fault surface to slide down at a normal fault and up at a reverse fault?
4. **Think Critically:** Why is it easier to predict *where* an earthquake will occur than it is to predict *when* it will occur?

Using Computers

Graphics Use the graphics capabilities of a computer to make simple working models of the three types of faults: normal, reverse, and strike-slip.

Skill Builder

Concept Mapping

Make a cycle concept map that shows why many earthquakes occur along the San Andreas Fault. Use the following terms and phrases: *rocks, stress, bend and stretch, elastic limit reached, earthquakes.* If you need help, refer to Concept Mapping in the **Skill Handbook.**

Earthquake Information

9•2

Science Words

seismic wave
focus
primary wave
secondary wave
epicenter
surface wave
inner core
outer core
mantle
crust
Moho discontinuity

Objectives

- Compare and contrast primary, secondary, and surface waves.
- Explain how an earthquake epicenter is located using seismic wave information.
- Describe how seismic wave studies indicate the structure of Earth's interior.

Types of Seismic Waves

Have you ever seen a coiled-spring toy? When children play with a coiled-spring toy, they send energy waves through it. **Seismic waves,** generated by an earthquake, are similar to the waves of the toy. Where are seismic waves formed? How do they move through Earth and how can we use the information that they carry? Let's investigate how scientists have answered these questions.

Earthquake Focus

As you have learned, when rocks move along a fault surface, energy is released and damage occurs, as seen in **Figure 9-5.** The point in Earth's interior where this energy release occurs is the **focus** of the earthquake. Seismic waves are produced and travel outward from the earthquake focus.

Figure 9-5

As shown in this photograph of buildings damaged during the 1989 Loma Prieta, California earthquake, surface waves cause most earthquake damage. This earthquake was caused by the Pacific Plate slipping past the North American Plate by only 2 m. *Why would surface waves generate so much damage?*

Activity 9-1

Design Your Own Experiment

Earthquake Depths

You learned earlier in this chapter that Earth's crust is broken into sections called plates. Movement of these plates generates stress within rocks that must be released. When this release of stress is sudden and rocks break, an earthquake occurs.

PREPARATION

Problem

Can a study of the foci of earthquakes tell us anything about how stress builds up in rocks and how it may be released?

Form a Hypothesis

State a hypothesis about how the movement of plates at Earth's crust could cause stress within rocks to build up and where rocks will break to relieve this stress.

Objectives

- Observe any connection between earthquake-focus depth and epicenter (the point on Earth's surface directly above the focus) location using the data provided on the next page.
- Describe any observed relationship between earthquake-focus depth and the movement of plates at Earth's surface.

Possible Materials

- graph paper
- pencil

PLAN THE EXPERIMENT

1. As a group, agree upon and write out your hypothesis statement.
2. As a group, list the steps that you need to take to test your hypothesis. Be very specific, describing exactly what you will do at each step.
3. Make a list of the materials that you will need to complete your experiment.
4. Determine how best to plot the data and observations on graph paper.

Check the Plan

1. Read over your entire experiment to make sure that all steps are in a logical order.
2. Will you summarize data in a table, or some other form of organization?
3. *Make sure your teacher approves your plan and that you have included any changes suggested in the plan.*

DO THE EXPERIMENT

1. Carry out the experiment as planned.
2. While conducting the experiment, write down any observations that you or other members of your group make and complete any data tables used in your Science Journal.

Analyze and Apply

1. Describe any observed relationship between the location of earthquake epicenters and the depth of earthquake foci.
2. Based on the graph you have completed, **hypothesize** what is happening to the plates at Earth's surface in the vicinity of the plotted earthquake foci.
3. In your opinion, what process is causing the earthquakes plotted on your graph paper?

Go Further

Based on the data you have plotted from the data table to the left, is the continent located east or west of the edge of the section of Earth's crust? Explain what you based your answer on. Hypothesize why none of the plotted earthquakes occurs below 700 km. Based on what you have observed and plotted, do all earthquakes occur at the same depth? Is there a relationship between earthquake depth and the movement of sections of Earth's crust?

Data and Observations

Quake	Focus Depth	Distance of Epicenter from Coast (km)
A	- 55 km	0
B	- 295 km	100 east
C	- 390 km	455 east
D	- 60 km	75 east
E	- 130 km	255 east
F	- 195 km	65 east
G	- 695 km	400 east
H	- 20 km	40 west
I	- 505 km	695 east
J	- 520 km	390 east
K	- 385 km	335 east
L	- 45 km	95 east
M	- 305 km	495 east
N	- 480 km	285 east
O	- 665 km	545 east
P	- 85 km	90 west
Q	- 525 km	205 east
R	- 85 km	25 west
S	- 445 km	595 east
T	- 635 km	665 east
U	- 55 km	95 west
V	- 70 km	100 west

Science Journal

In your Science Journal, write a brief explanation of how you might build a building so that it would be able to withstand surface waves caused by an earthquake.

Seismic Waves

Waves that move through Earth by causing particles in rocks to move back and forth in the same direction the wave is moving are called **primary waves.** If you squeeze one end of a coiled-spring toy and then release it, you cause it to compress and then stretch as the primary wave travels through it. Particles in rocks also compress together and stretch apart, transmitting primary waves through the rock.

Now, if you and a friend stretch the coiled-spring toy between you, and then move one end up and down, a different type of wave will pass through the toy. The spring will move up and down as the wave moves along it. **Secondary waves** move through Earth by causing particles in rocks to move at right angles to the direction of the wave.

The point on Earth's surface directly above an earthquake's focus is the **epicenter** (see **Figure 9-6**). Energy that reaches the surface of Earth generates waves that travel outward from the epicenter. These waves, called **surface waves,** move by

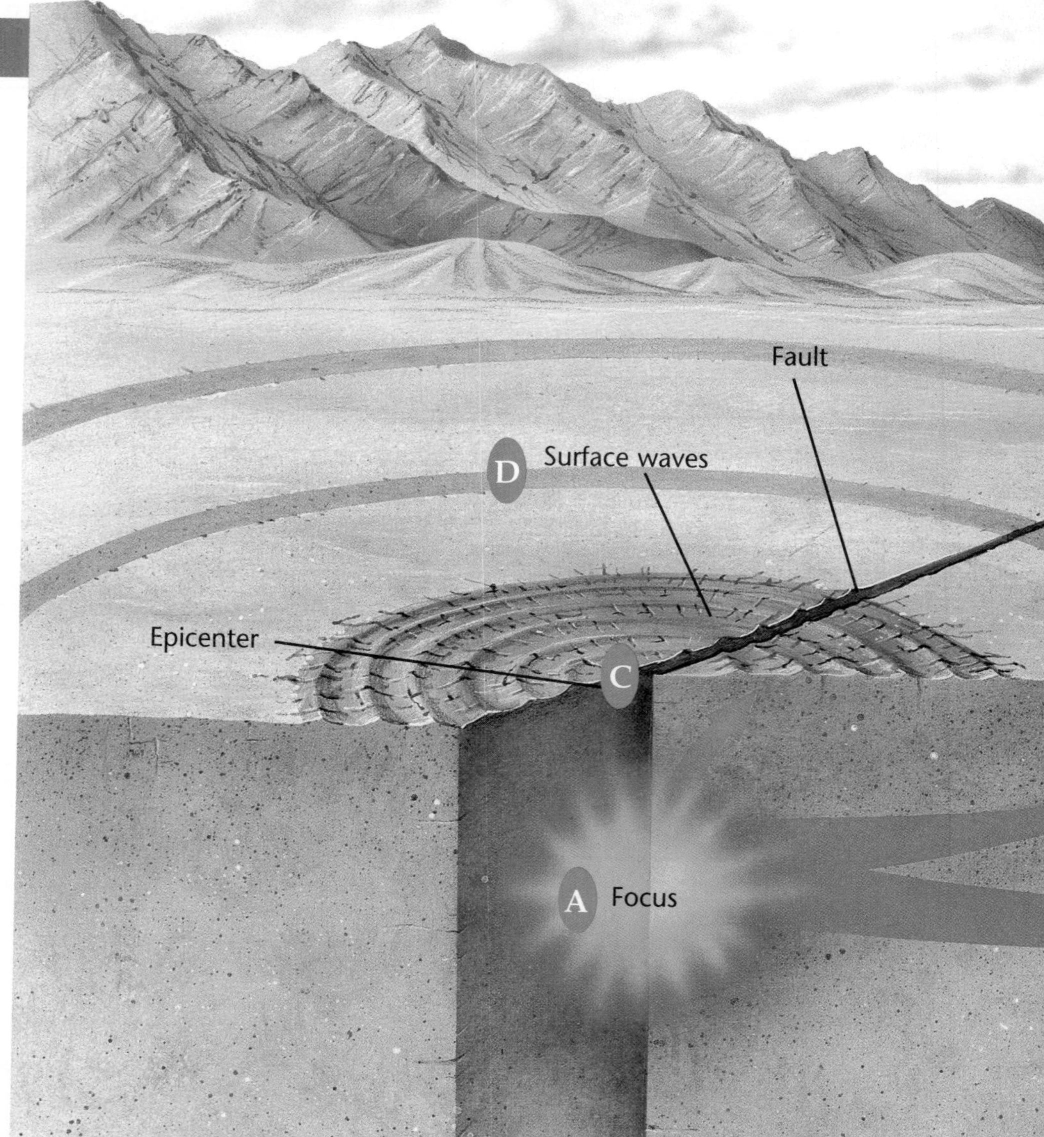

Figure 9-6

Primary and secondary waves travel outward from the focus. Surface waves move outward from the epicenter.

A Sudden movement along a fault releases energy that causes an earthquake. The point beneath Earth's surface where the movement occurs is the focus of the earthquake.

B Primary waves and secondary waves originate at the focus and travel outward in all directions. Primary waves travel about twice as fast as secondary waves.

giving particles an elliptical motion, as well as a back-and-forth swaying motion.

Surface waves cause most of the destruction during an earthquake. Because most buildings are very rigid, they begin to fall apart when surface waves pass. The waves cause one part of the building to move up while another part moves down.

USING MATH

Primary waves travel at about 6 km/s through granitic crust. The distance from Phoenix, Arizona to Los Angeles, California is about 600 km. How long would it take primary waves to travel between these two cities?

Locating an Epicenter

Primary, secondary, and surface waves don't travel through Earth at the same speed. Primary waves are the fastest; surface waves are the slowest. Can you think of a way this information could be used to determine how far away an earthquake epicenter is? Think of the last time you and two friends rode your bikes to the store. You were fastest so you arrived first. In fact, the longer you rode, the farther ahead of your friends you became. Scientists use the different speeds of seismic waves to determine the distance to the earthquake epicenter.

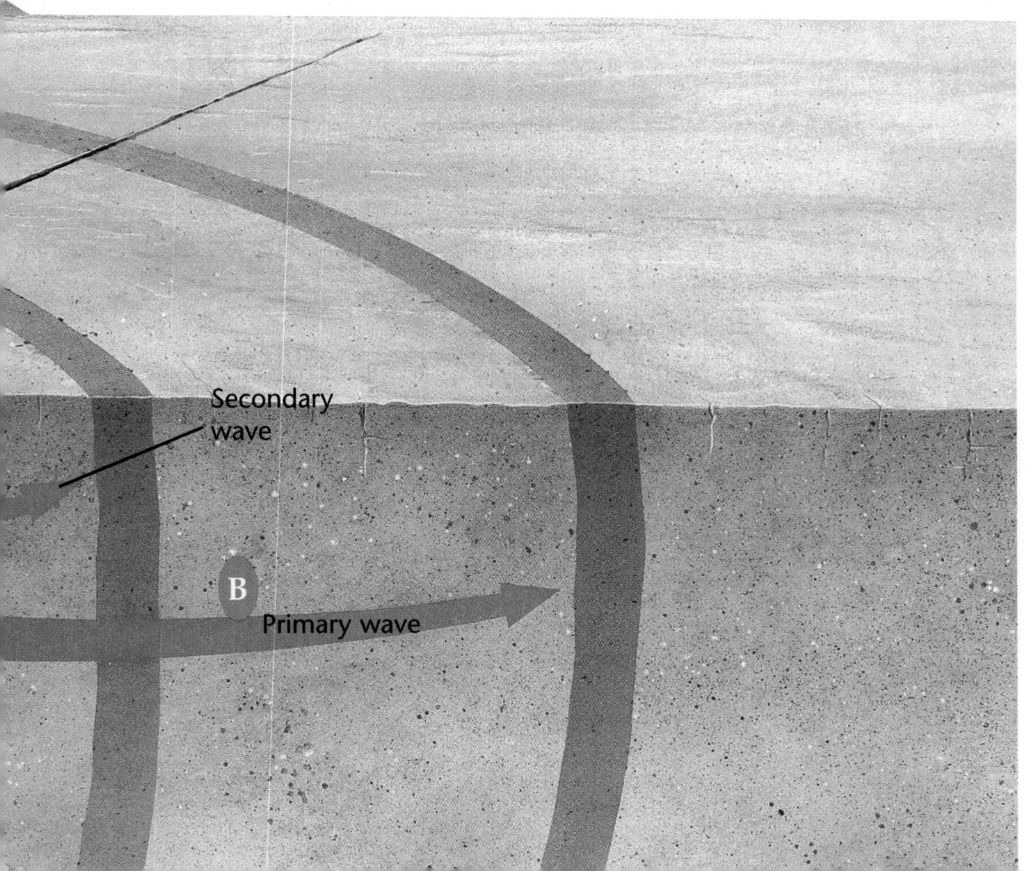

C The place on Earth's surface directly above the earthquake focus is called the epicenter. When primary and secondary waves reach the epicenter, they generate the slowest kind of seismic waves, surface waves. If you've ever floated on an inner tube in a wave pool, you have experienced waves similar to surface waves.

D Surface waves travel outward from the epicenter along Earth's surface in much the same way that ripples travel outward from a stone thrown into a pond.

MiniLAB

Can the travel time of seismic waves be used to find the distance to an earthquake?

Procedure

1. Use the graph in **Figure 9-7** to determine the difference in arrival times for primary and secondary waves at the distances listed in the data table below. Two examples are provided for you.
2. Repeat step 1 for at least two other distances of your choice.

Analysis

1. Interpret what happens to the difference in arrival times as the distance from the earthquake increases.
2. Do the distances you chose fit with what you interpreted in question 1?

Distance (km)	Difference in Arrival Time
1500	2 min; 45 s
2250	
2750	
3000	
4000	5 min; 35 s
7000	
9000	

Seismograph Stations

Based on the different speeds of seismic waves, primary waves arrive first at seismograph stations, secondary waves second as seen in **Figure 9-7,** and surface waves last. This enables scientists to determine the distance to an earthquake epicenter. The farther apart the waves, the farther away the epicenter is. When epicenters are far from the seismograph station, the primary wave has more time to put distance between it and the secondary and surface waves.

Epicenter Location

If seismic wave information is obtained at three seismograph stations, the location of the epicenter can be determined, as shown in **Figure 9-8.** To locate an epicenter,

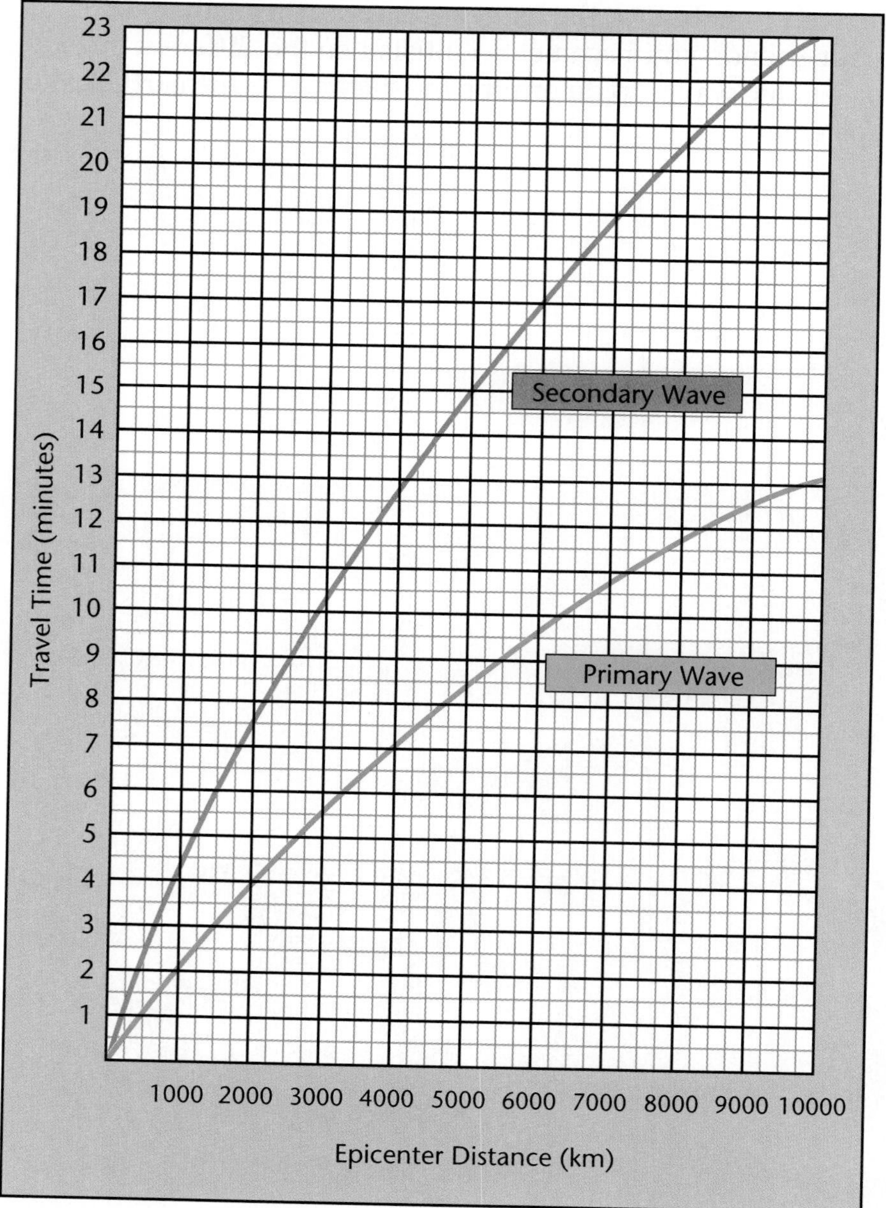

Figure 9-7

This graph shows the distance that primary and secondary waves travel over time. By measuring the difference in arrival times, a seismologist can determine the distance to the epicenter.

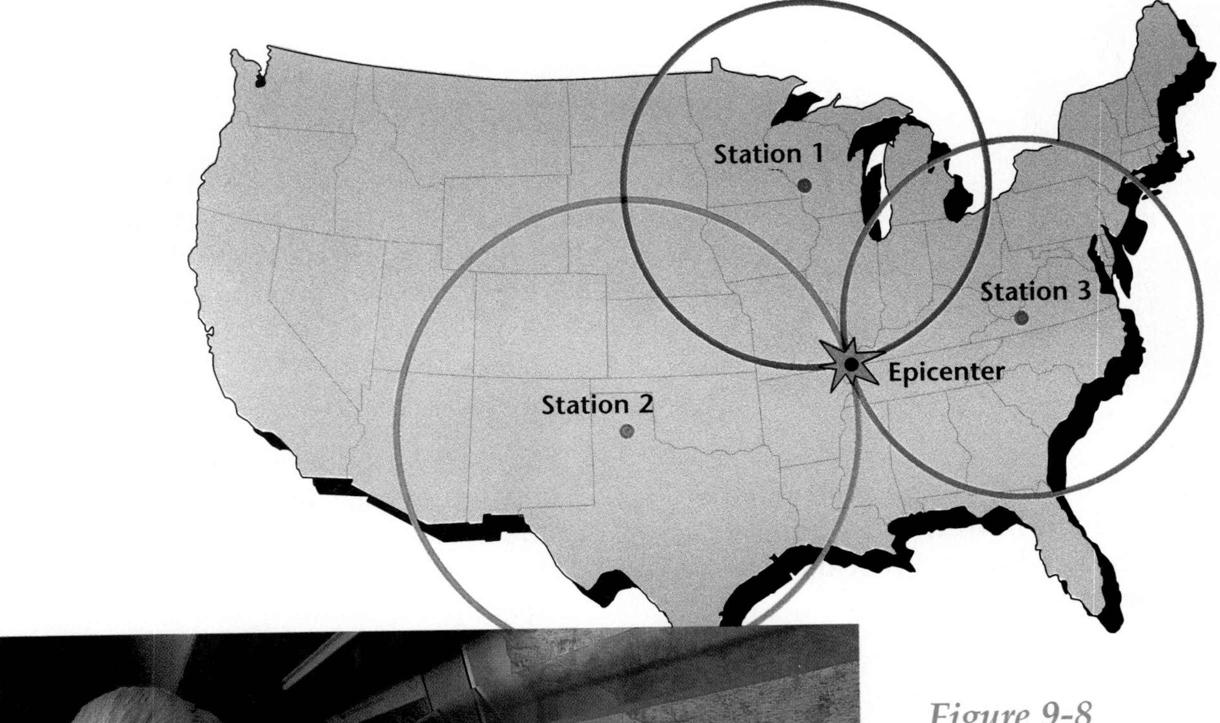

Figure 9-8

The radius of each circle is equal to the distance from the epicenter to each seismograph station. The intersection of the three circles is the location of the epicenter. *Why is one seismograph station not enough?*

scientists draw circles around each station on a map. The radius of each circle equals that station's distance from the earthquake epicenter. The point where all three circles intersect is the location of the earthquake epicenter.

Using Seismic Waves to Map Earth's Interior

Scientists have found that at certain depths within Earth, the speed and path of seismic waves change. These changes mark the boundaries of the layers in Earth with different densities, as shown in **Figure 9-10** on page 250. The Problem Solving exercise you will do later in this section demonstrates how scientists have learned about Earth's interior without ever having been there.

Activity 9-2

Epicenter Location

Try this activity to see how to plot the distance of several seismograph stations from the epicenters of two earthquakes and how to use that data to interpret where the earthquake epicenters were located. You will be using a globe and **Figure 9-7** to help you locate the epicenters.

Problem

Can plotting the distance of several seismograph stations from two earthquake epicenters be used to interpret the locations of the two epicenters?

Materials

- **Figure 9-7**
- metric ruler
- chalk
- string
- globe
- paper

Procedure

1. Determine the difference in arrival time between the primary and secondary waves at each station for each quake from the data table below.
2. Once you determine the arrival times of seismic waves for each seismograph station, use the graph in **Figure 9-7** to determine the distance in kilometers of each seismograph from the epicenter of each earthquake. Record these data in a data table provided by your teacher. The difference in arrival times in Paris for earthquake B is 10.0 minutes. On the graph, the primary and secondary waves are separated along the vertical axis by 10.0 minutes at 9750 km.
3. Using the string, measure the circumference of the globe. Determine a scale of centimeters of string to kilometers on Earth's surface. (Earth's circumference = 40 000 km.)
4. For each earthquake, A and B, place one end of the string at each seismic station location on the globe. Use the chalk to draw a circle with a radius equal to the distance to the earthquake's epicenter.
5. Identify the epicenter for each quake.

Analyze

1. How is the distance of a seismograph from the earthquake related to the arrival time of the waves?
2. What is the location of each earthquake epicenter?
3. How many stations were needed to accurately locate each epicenter?

Conclude and Apply

4. **Predict** why some seismographs didn't receive secondary waves from some quakes.
5. **Discuss** how epicenters are located.

Data and Observations

Location of Seismograph	Wave	Wave Arrival Times	
		Earthquake A	Earthquake B
(1) New York	P	2:24:05 P.M.	1:19:00 P.M.
	S	2:28:55 P.M.	1:24:40 P.M.
(2) Seattle	P	2:24:40 P.M.	1:14:37 P.M.
	S	2:30:00 P.M.	1:16:52 P.M.
(3) Rio de Janeiro	P	2:29:00 P.M.	—
	S	2:38:05 P.M.	—
(4) Paris	P	2:30:15 P.M.	1:24:05 P.M.
	S	2:40:15 P.M.	1:34:05 P.M.
(5) Tokyo	P	—	1:23:30 P.M.
	S	—	1:33:05 P.M.

Structure of Earth

Seismic wave studies have enabled scientists to construct a model of Earth's interior, as shown in **Figure 9-9.** At the very center of Earth is a solid, very dense **inner core** composed mostly of iron and nickel. Above the solid inner core lies the liquid **outer core,** also composed of iron and nickel. Earth's **mantle** is the largest layer, lying directly above the outer core. It is made mostly of silicon, oxygen, magnesium, and iron. Earth's outermost layer is the **crust.** It is separated from the mantle by the Moho discontinuity.

Moho Discontinuity

Seismic waves speed up when they reach the bottom of the crust. This boundary between the crust and the mantle is called the **Moho discontinuity.** The boundary was discovered by the Yugoslavian scientist Andrija Mohorovičić, who inferred that seismic waves speed up because they're passing into a denser layer of the lithosphere. The lithosphere is made up of the rigid crust and upper mantle.

What's in there?

Your teacher has placed five closed and sealed boxes in front of the class. The sizes of the boxes vary but none is larger than 30 cm × 18 cm × 12 cm. Your teacher has challenged you to complete problem-solving exercises in order to earn points to be awarded. In order to earn the full amount of points available, you must list at least three facts about the contents of each box. You are permitted to do anything you need to do except open the boxes and look directly at the enclosed objects. The other exception is that you cannot damage any of the boxes in any way that might in turn damage the contents of the boxes.

Solve the Problem:

1. **Determine what tests you can conduct on each box that will reveal facts about the box's contents.**
2. **You may wish to work with another student in case more than one person is required to perform any of your chosen tests.**
3. **In your Science Journal, list any data you collect as well as your inferences concerning the contents of each box.**

Think Critically:

How many facts did you and your partner correctly discover about the contents of each box? How is the challenge presented by your teacher related to earthquakes and mapping Earth's interior?

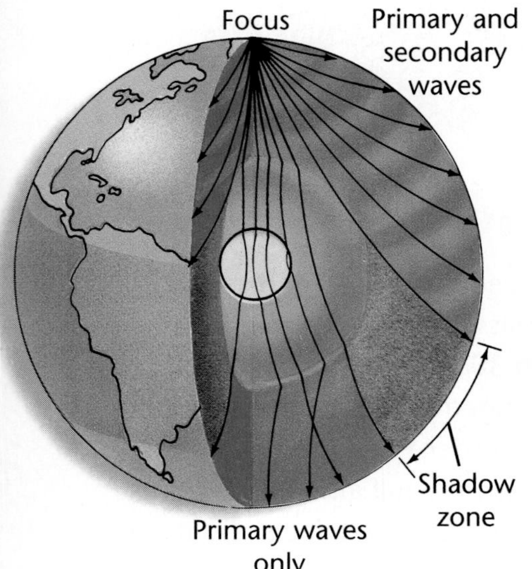

Figure 9-9

Primary waves bend when they contact the outer core, and secondary waves are stopped completely. Primary waves also bend and speed up when they enter the inner core. In fact, as shown, seismic waves gradually bend and change speed as the density of rock changes.

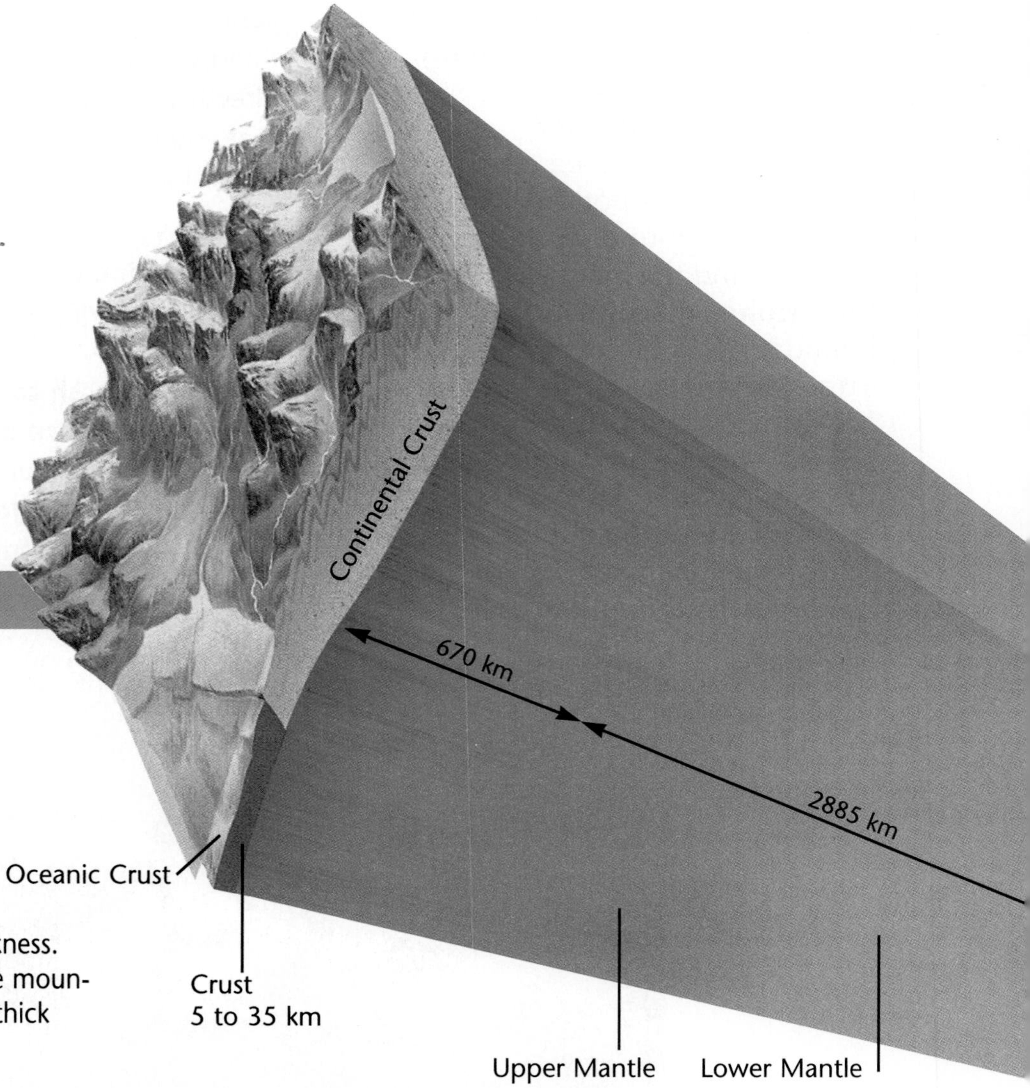

Primary and secondary waves slow down when they hit the plasticlike asthenosphere, which is part of the upper mantle, and then speed up again as they're transmitted through the solid lower mantle. In this way, the mantle is subdivided into an upper and lower layer.

There's an area on Earth, between 105° and 140° from the focus, where no waves are detected. This area is called the shadow zone. Secondary waves aren't transmitted through liquid, so they're stopped completely when they hit the liquid outer core. Primary waves are slowed and deflected but not stopped by the liquid outer core. The deflection of the primary waves and the stopping of the secondary waves create the shadow zone, as shown in **Figure 9-9.** These primary waves again speed up as they travel through the solid inner core.

Figure 9-10

This wedge shows the layers inside Earth from the inner core. The inner core, outer core, and mantle are shown at the correct scale, but the crust is shown much thicker than it actually is.

Continental Crust
670 km
2885 km
Oceanic Crust
Crust 5 to 35 km
Upper Mantle
Lower Mantle

A The crust of Earth varies in thickness. It is greater than 60 km in some mountainous regions, and less than 5 km thick under the oceans.

B Rock material in the upper mantle is described as plasticlike. It has characteristics of a solid, but also flows like a liquid when under pressure. Some kinds of taffy have this plasticlike characteristic. The taffy can be pulled apart slowly, but if you hit it on the edge of a table, it will break.

Section Wrap-up

Review

1. Which type of seismic wave does the most damage to property? Explain why.
2. Why is a seismic record from three locations needed to determine the position of an epicenter?
3. **Think Critically:** Suppose an earthquake occurs at the San Andreas Fault. What area on Earth would experience no secondary waves? Would China experience primary and secondary waves? Explain your answers.

Skill Builder

Making and Using Graphs

Use the data table below to make a graph of some travel times of earthquake waves. Which line represents primary waves? Which line represents secondary waves? If you need help, refer to Making and Using Graphs in the **Skill Handbook.**

Distance from Earthquake (km)	1500	2000	5000	5500	8600	10 000
Time (minutes)	5.0	2.5	14.0	7.0	11.0	23.5

Science Journal

When sound is produced, waves move through the air by pressing molecules together and then spreading them apart. In your Science Journal, research sound waves and describe seismic wave types to which they are similar.

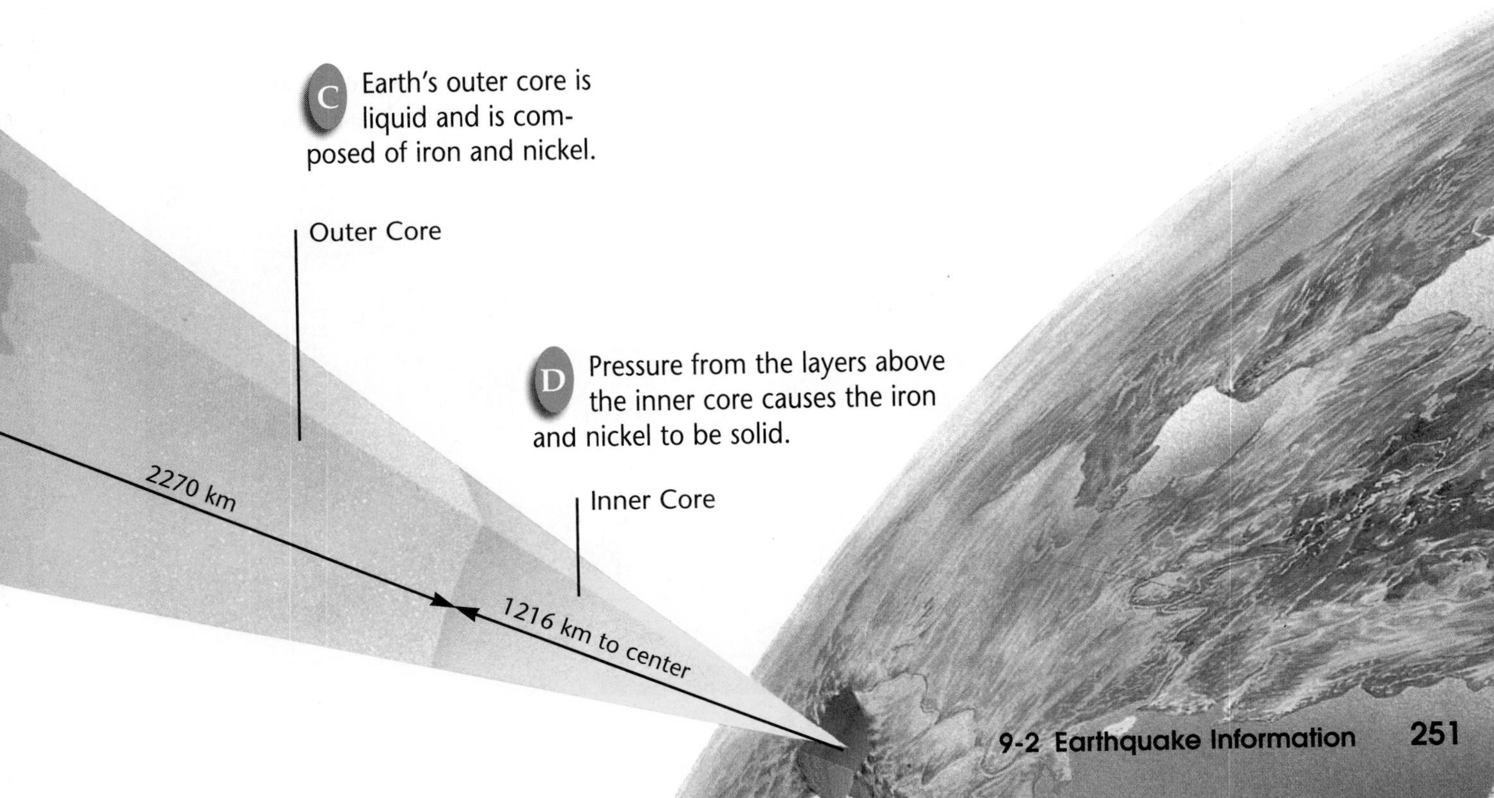

9•3 Destruction by Earthquakes

Science Words

seismologist
seismograph
magnitude
tsunamis

Objectives

- Define magnitude and the Richter Scale.
- List ways to make your classroom and home more earthquake-safe.

Measuring Earthquakes

On January 17, 1995, a major earthquake occurred in Kobe, Japan, causing about $100 billion of property damage and 5378 deaths. On January 17, 1994, a major earthquake occurred in Northridge, California, causing billions of dollars of property damage, as seen in **Figure 9-11,** and 61 deaths. At least 30 000 people died in the September 1993 earthquake that struck India. On October 17, 1989, an earthquake shocked the San Francisco Bay area. The quake killed 62 people and damaged billions of dollars of property. In 1988, more than 28 000 people died when an earthquake struck Armenia. What determines the amount of damage done by an earthquake, and what can you do to protect yourself from the effects? With so many lives lost and such potential destruction, as shown in **Table 9-1**, it is important for scientists to learn as much as possible about earthquakes to try and minimize their damage.

Seismology

Scientists who study earthquakes and seismic waves are **seismologists.** They use an instrument called a **seismograph** to record primary, secondary, and surface waves from earthquakes all over the world.

One type of seismograph has a drum holding a sheet of paper on a fixed frame. A pendulum with an attached pen is suspended from the frame. When seismic waves occur at the station, the drum vibrates but the pendulum remains at rest. The pen on the pendulum traces a record of the vibrations on a sheet

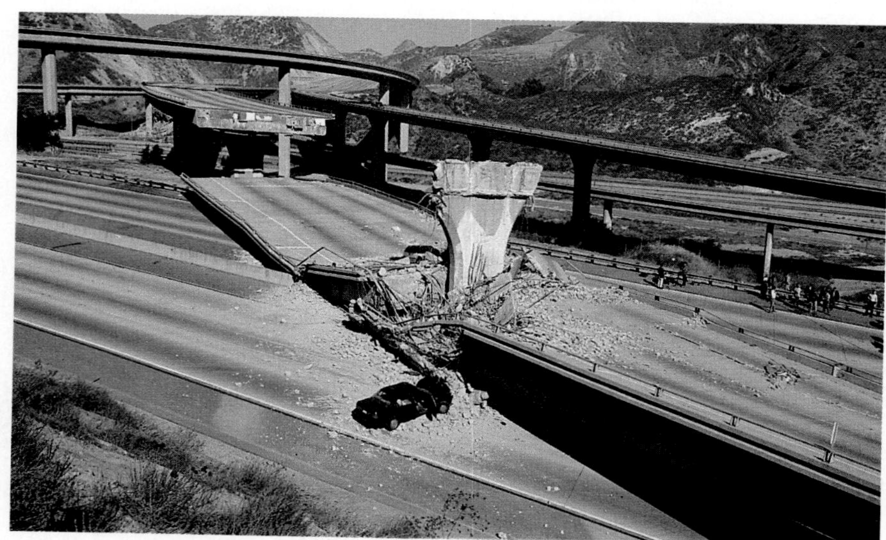

Figure 9-11

Several major highways were damaged in the January 17, 1994 earthquake in Northridge, California. *What happens during an earthquake that causes so much damage to highway overpasses?*

of paper. The height of the lines traced on the paper is a measure of the energy released, or **magnitude** of the earthquake.

Earthquake Magnitude

Not all seismographs measure vibrations in the same way. Because the Richter scale measures only local intensity on a specific type of seismograph, seismologists refer to magnitude in several other ways as well.

The Richter magnitude is based on seismic waves that travel through Earth. It deals mainly with the strength of the break, not with the length or breadth of the fault. The Richter scale describes how much energy is released by the earthquake. For

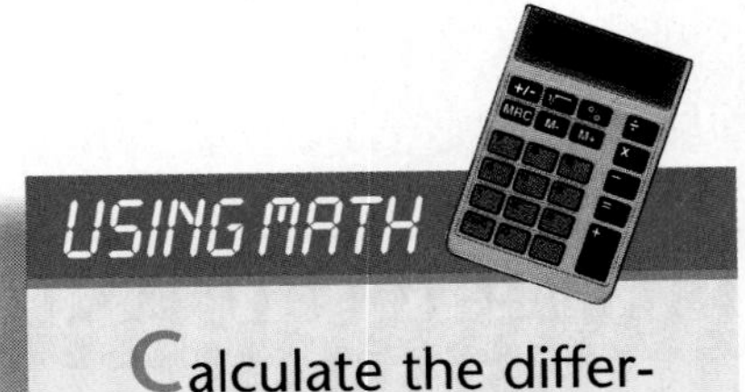

USING MATH

Calculate the difference in energy released between an earthquake of Richter magnitude 7.5 and one of magnitude 5.5.

Table 9-1

Strong Earthquakes

Year	Location	Richter Value	Deaths
1556	Shensi, China	?	830 000
1737	Calcutta, India	?	300 000
1755	Lisbon, Portugal	8.8	70 000
1811-12	New Madrid, MO	8.3	few
1886	Charleston, SC	?	60
1906	San Francisco, CA	8.3	1500
1920	Kansu Province, China	8.5	180 000
1923	Tokyo, Japan	8.3	143 000
1939	Concepción, Chile	8.3	30 000
1960	Southern Chile	8.6	5 700
1964	Prince William Sound, AK	8.5	131
1970	Peru	7.8	66 800
1975	Laoning Province, China	7.5	few
1976	Tangshan, China	7.6	240 000
1985	Mexico City, Mexico	8.1	9 500
1988	Armenia	6.9	28 000
1989	Loma Prieta, CA	7.1	62
1990	Iran	7.7	50 000
1990	Luzon, Philippines	7.8	1621
1993	Guam	8.1	none
1993	Marharashtra, India	6.4	30 000
1994	Northridge, CA	6.7	61
1995	Kobe, Japan	6.9	5378

USING TECHNOLOGY

Predicting Quakes

The year 1992 came and passed without the predicted earthquake of magnitude 6.0 or more occurring along the San Andreas Fault near Parkfield, California. Based on the fact that moderate earthquakes had occurred an average of once every 22 years since 1857, seismologists had predicted that another would occur sometime between 1988 and 1992. Instruments such as lasers, creepmeters, survey alignment instruments, and other instruments were installed in the area as part of the Parkfield Earthquake Prediction Experiment. In the creepmeters, a wire is stretched across the fault to record even minute movements.

Parkfield Quakes

Allan Lindh of the USGS believes that a magnitude-6 earthquake is bound to hit Parkfield "any old day now." One scientist, Robert Nadeau of the University of California, Berkeley, predicts that 13 earthquakes of about magnitude 1 should strike Parkfield by the end of 1996. Scientists are using the interval between jolts in a specific place as a means of determining a window of time in which earthquakes should occur. It is hoped that this research will help scientists predict larger earthquakes.

Think Critically:

Prediction of earthquakes deals with windows of time during which earthquakes should occur at a specific location. Why can't seismologists be more specific about the predicted date on which an earthquake should occur?

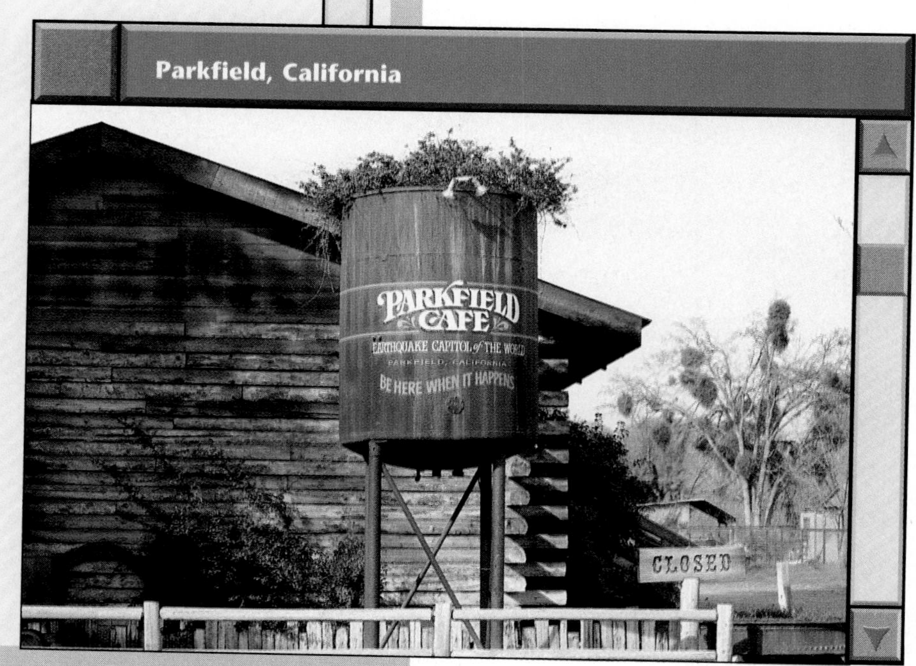

Parkfield, California

each increase of 1.0, the amplitude of the largest surface waves is 10 times greater. However, about 32 times as much energy is released for every increase of 1.0 on the scale. For example, a magnitude-8.5 earthquake releases about 32 times as much energy as a magnitude-7.5 earthquake. **Table 9-2** shows how often various magnitude earthquakes are expected to occur.

Another magnitude used by seismologists is based on Earth movement or surface waves. Seismologists also use a magnitude called the moment magnitude. It is derived by multiplying the length of the fault rupture by the amount of rock movement and then again by the rock stiffness. The moment magnitude is related to the strength and size of fault movement. The magnitude usually first reported is Richter scale magnitude modified for modern equipment. After further study, the moment magnitude can be determined.

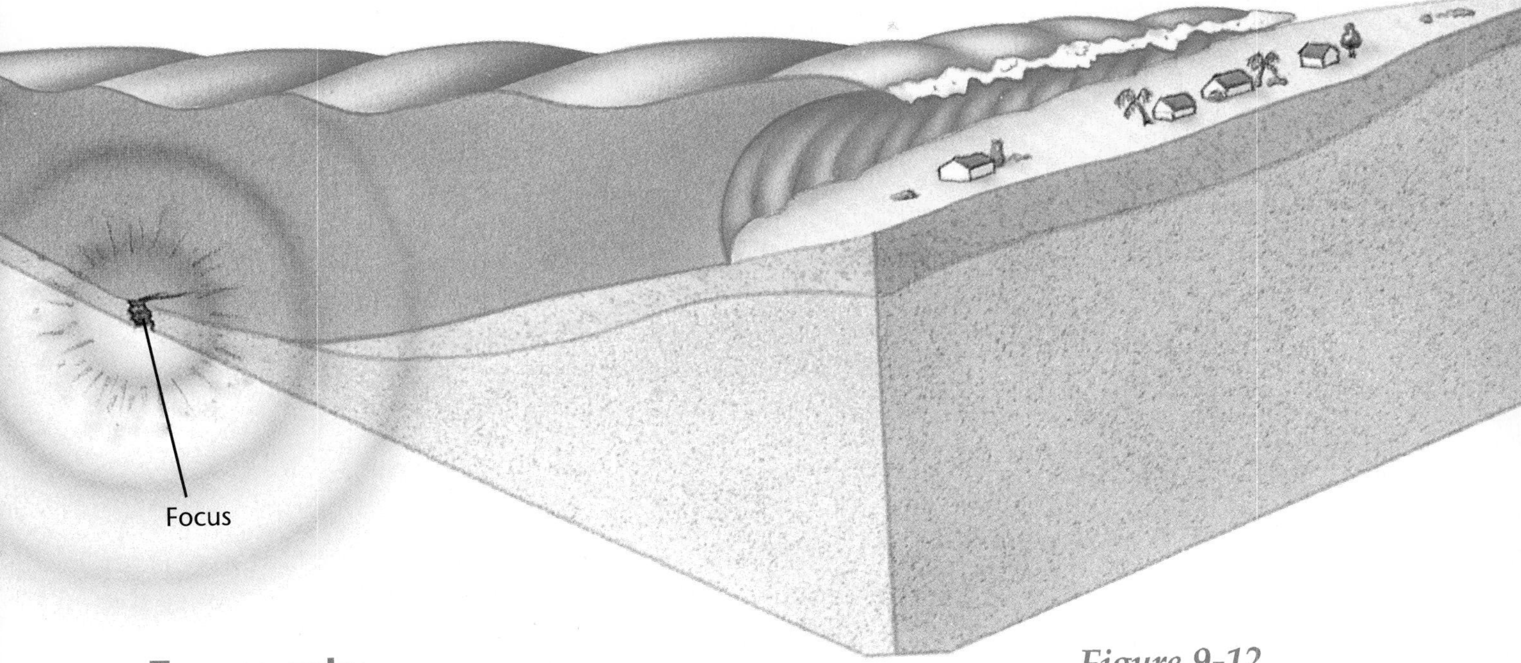

Figure 9-12

A tsunami begins over the earthquake focus. *What might happen to towns located near the shore?*

Tsunamis

Most earthquake damage happens when surface waves cause buildings, bridges, and roads to collapse. People living near the seashore, however, have another concern. An earthquake under the sea causes abrupt movement of the ocean floor. The movement pushes against the water, generating a powerful wave that travels to the surface, as shown in **Figure 9-12.** After reaching the surface, the wave can travel thousands of kilometers in all directions.

Far from shore, an earthquake-generated wave is so long that a large ship may ride over it without anyone noticing. But when one of these waves breaks on a shore, it forms a towering crest that can reach 30 m high. Ocean waves generated by earthquakes are called seismic sea waves, or **tsunamis** (soo NAHM eez).

Table 9-2

Earthquake Occurrences

Richter Magnitude	Number Expected per Year
1.0 to 3.9	> 949 000
4.0 to 4.9	6200
5.0 to 5.9	800
6.0 to 6.9	226
7.0 to 7.9	18
8.0 to 8.9	< 2

MiniLAB

How can structures be made earthquake-safe?

Procedure

1. Obtain a set of building blocks from your teacher.
2. On a tabletop, build one structure out of the blocks by simply placing one block on top of another.
3. Build a second structure by wrapping sections of three blocks together with rubber bands. Then wrap larger rubber bands around the entire completed structure.
4. Set the second structure on the tabletop and pound on the side of the table with a steady rhythm.

Analysis

1. Which of your two structures was better able to withstand the "earthquake" caused by pounding on the table?
2. How might the idea of wrapping the blocks with rubber bands be used in construction of supports for elevated highways?

Earthquake Safety

You've seen the destruction that earthquakes can cause. However, there are ways to minimize the damage and loss of life.

One of the first steps in earthquake safety is to study the earthquake history of a region, such as the one illustrated in **Figure 9-14.** If you live in an area that's had earthquakes in the past, you can expect them to occur there in the future. As you know, most earthquakes happen along plate boundaries. **Figure 9-2** shows where severe earthquakes have happened. Being prepared is an important step in earthquake safety.

Quake-Proofing Your Home

Make your home as earthquake-safe as possible, like the one shown in **Figure 9-13.** Take heavy objects down from high shelves and place them on lower shelves. To reduce the chance of fire from broken gas lines, see that hot-water heaters and gas appliances are held securely in place. During an earthquake, keep away from windows and avoid anything that could fall on you. Watch for fallen power lines and possible fire hazards. Stay clear of rubble that could contain sharp edges.

Figure 9-13

You can make your home more earthquake-safe by doing some very simple things, such as moving heavy objects from high shelves.

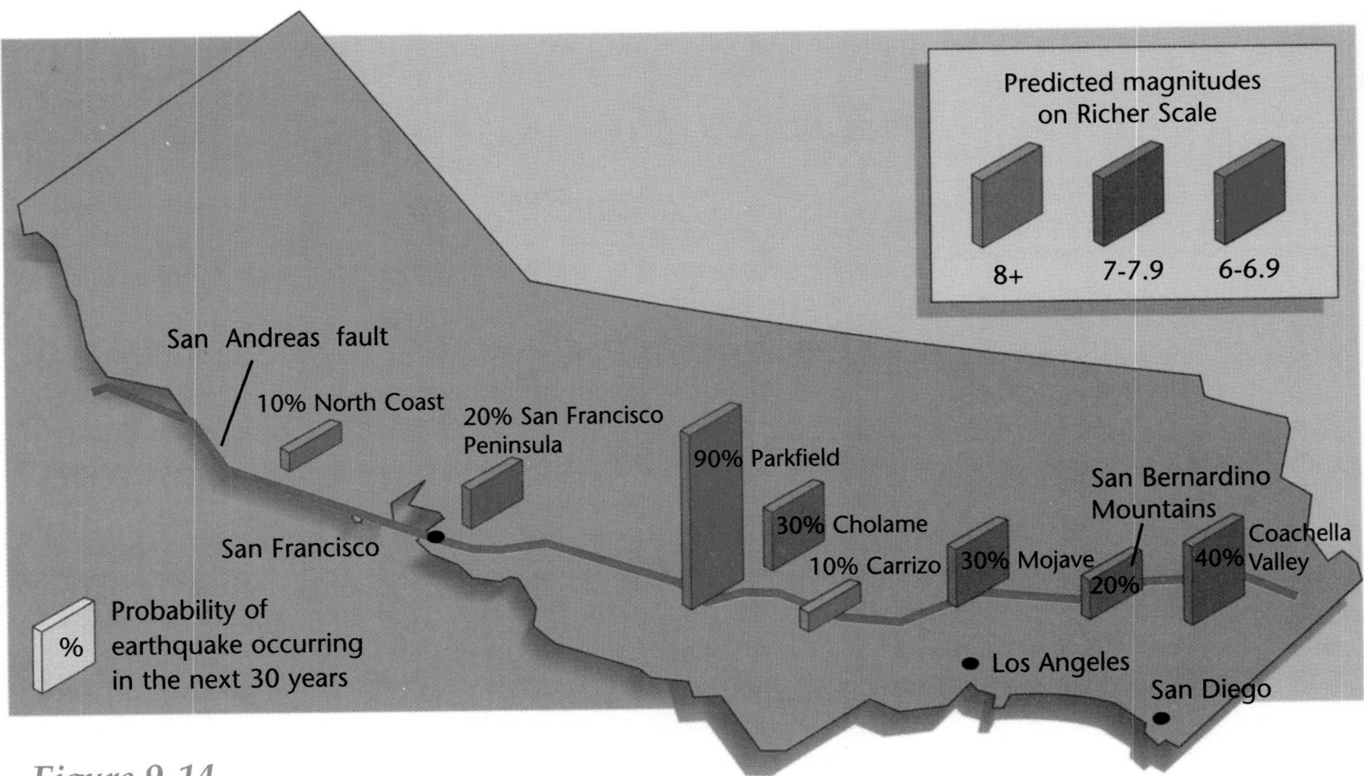

Figure 9-14

The bars on the graphic show the probabilities that an earthquake of the specified magnitude will strike these areas within the next 30 years. California residents are preparing for the major earthquakes predicted there. *What other areas besides California should be concerned about the probability of an earthquake?*

Section Wrap-up

Review

1. What can you do to prepare for an earthquake?
2. Research how animal behavior has been studied to predict earthquakes. What changes may be occurring in the environment to cause animals to act differently?
3. **Think Critically:** Explain why a seismograph wouldn't work if the pen vibrated along with the rest of the machine.

Skill Builder

Making and Using Tables

Use **Table 9-1** to determine which quake listed had the highest magnitude. Hypothesize why the 1975 China quake resulted in fewer deaths than the 1976 quake. If you need help, refer to Making and Using Tables in the **Skill Handbook.**

USING MATH

How much more energy would be released by an earthquake with a Richter magnitude of 8.9 than the earthquake that hit Kobe, Japan? (Refer to **Table 9-1**.)

SCIENCE & SOCIETY

9•4 TECHNOLOGY: Living on a Fault

Objectives

- Recognize that most loss of life in an earthquake is caused by the destruction of human-made structures.
- Consider who should pay for making structures seismic-safe.

Who should pay for earthquake preparation?

Throughout this chapter, you have seen pictures of the aftermath of earthquakes. What kind of damage did you see? Buildings, bridges, and highways were cracked and broken. Some were totally destroyed.

Most loss of life in an earthquake occurs when people are trapped in and on these crumbling structures. What can be done to reduce loss of life? Who should be responsible for the cost of making structures seismic-safe?

Seismic-Safe Structures

Seismic-safe structures are resistant to vibrations that occur during an earthquake. **Figures 9-15** and **9-16** show how buildings and highways can be built to resist earthquake damage.

Will making structures seismic-safe reduce the loss of life in an earthquake? Look again at **Table 9-1.** Notice that earthquakes in Armenia (December 1988), in Loma Prieta (October 1989), and in Iran (June 1990) were all close in magnitude. However, the loss of life in each of these earthquakes was quite different. Why were so many more lives lost in Armenia and Iran than in Loma Prieta?

Figure 9-15

The rubber portions of this building's moorings absorb most of the wave motion of an earthquake. The building itself only sways gently. *What purpose does the rubber serve?*

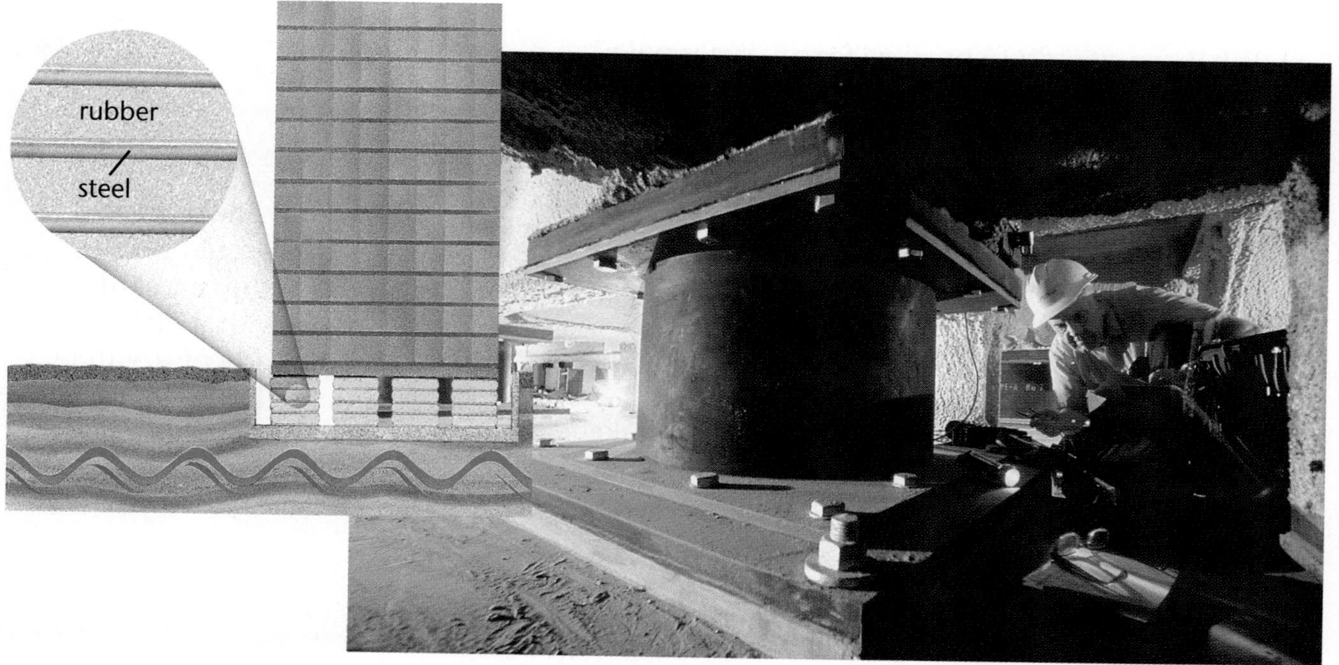

People in Earthquake-Prone Areas Should Prepare

Loma Prieta and countries like Japan are very susceptible to earthquakes. People living in California and Japan have been getting ready for big earthquakes for many years. Since 1971, stricter building codes have been enforced and older buildings have been reinforced. In other parts of the world, such seismic-safe structures are rare or don't exist at all.

Today in California, some new buildings are anchored to flexible, circular moorings made of steel plates filled with alternating layers of rubber and steel. **Figure 9-15** shows how they work. Tests have shown that buildings supported with these moorings should be able to withstand an earthquake measuring up to 8.3 on the Richter scale without major damage.

Sharing Seismic-Safe Technology

Highways and buildings in earthquake-prone areas can be made seismic safe. Lives and property can be saved by replacing underground water and gas pipes with ones that will bend, but not break, during an earthquake. However, seismic-safe structures are expensive and many communities in earthquake-prone areas simply can't afford them. In these areas, seismic-safe structures can be constructed only if people outside of the region are willing to help.

Figure 9-16

Seismic-safe highways are supported by vertical steel rods wrapped with reinforcing rods encased in concrete. The highway that collapsed in the Loma Prieta earthquake was not built this way.

Section Wrap-up

Review

1. What conditions can cause greater loss of life from one earthquake than from another of the same magnitude?
2. Why did supports for the upper deck of Highway 880 break during the 1989 Loma Prieta earthquake?

Explore the Technology

Should governments worldwide accept the responsibility of sharing the expense of providing seismic-safe structures to all earthquake-prone areas? If yes, where would funding come from, especially in countries that do not have earthquake-prone areas? If no, how can the safety of the people in earthquake-prone areas be assured?

CONNECT TO PHYSICS

How are shock absorbers on a car similar to the circular moorings described on this page? How do they absorb shock?

Science & History

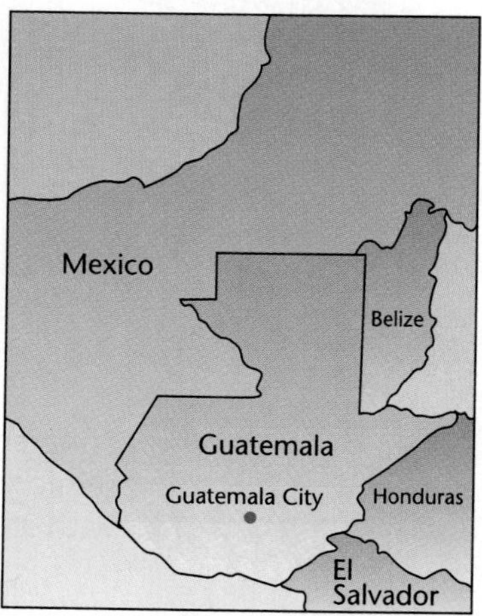

Figure 9-17

Map showing location of Guatemala City.

Guatemala City

Every so often you may read of a city being destroyed by an earthquake. It has happened more than once to Guatemala City. But in a way, Guatemala City was also created by earthquakes.

It was an earthquake in 1773 that caused Guatemala City to be founded. That quake leveled Santiago, the capital city of what was then a Spanish colony. Santiago barely survived seven major earthquakes from its founding in 1573 through 1773. The 1773 quake finally convinced the Spanish governor that the city had to be moved to a safer location. Guatemala City was built in 1776 as Santiago's replacement.

Even as Guatemala City grew, the old city of Santiago refused to die. Many residents would not leave, despite the earthquake dangers. They stuck to their homes even when the Spanish government forced all stores to close. Today it remains as the city of Antigua Guatemala, a major tourist site.

Although Guatemala City was the official capital of the area, another city became the unofficial capital. That was Quezaltenango, where a large number of the country's leading families lived. Once again an earthquake changed things, destroying Quezaltenango in 1902. Guatemala City became the new home of many of the county's most influential families and finally, the undisputed capital.

For much of its history, Guatemala City has had to contend with the constant threat of quakes without modern building technology. The builders relied on the construction techniques of the early settlers. Low buildings with massive walls were common—as were frequent repairs. Taller steel and concrete buildings didn't appear until after a series of earthquakes in late 1917 and early 1918. Even then, the new type of construction was confined to the newest sections of the city.

But low buildings and massive walls can do only so much to prevent tragedy. When the 1976 earthquake hit, 58 000 buildings were destroyed in Guatemala City. About 23 000 people died and another 80 000 people were injured, and more than a million people were left homeless. It was a terrible way to mark the city's 200th anniversary.

Science Journal

Research one of the historical earthquakes shown on **Table 9-1,** and the changes that resulted from it, in your Science Journal.

Chapter 9

Review

Summary

9-1: Forces Inside Earth

1. Plate movements put stress on rocks. To a point, the rocks bend and stretch. But if the force is great enough, rocks will break and produce earthquakes.
2. Normal faults form when rocks undergo tension. Compressional forces produce reverse faults. Strike-slip faults result from shearing forces.

9-2: Earthquake Information

1. Primary waves compress and stretch rock particles as the waves move. Secondary waves move by, causing particles in rocks to move at right angles to the direction of the waves. Surface waves move by, giving rock particles an elliptical and side-to-side motion.
2. Scientists can locate earthquake epicenters by measuring seismic wave speeds.
3. By observing the speeds and paths of seismic waves, scientists are able to determine the boundaries among Earth's layers.

9-3: Destruction by Earthquakes

1. The magnitude of an earthquake is a measure of the energy released by the quake. The Richter scale describes how much energy is released by an earthquake.
2. Removing objects from high shelves and securing hot-water heaters and gas appliances help prevent earthquake damage.

9-4: Science and Society: Living on a Fault

1. Most lives lost during a quake are due to the destruction of human-made structures.
2. Money for seismic-safe structures might come from people who live in the earthquake-prone area or from people in other parts of the country or the world.

Key Science Words

a. crust
b. earthquake
c. epicenter
d. fault
e. focus
f. inner core
g. magnitude
h. mantle
i. Moho discontinuity
j. normal fault
k. outer core
l. primary wave
m. reverse fault
n. secondary wave
o. seismic wave
p. seismograph
q. seismologist
r. strike-slip fault
s. surface wave
t. tsunamis

Reviewing Vocabulary

Match each phrase with the correct term from the list of Key Science Words.

1. a fault formed due to tension on rocks
2. fault due to shearing forces
3. point where earthquake energy is released
4. point on Earth's surface directly above the origin of an earthquake
5. wave that produces the most surface damage
6. layer of Earth directly below the crust
7. instrument that records seismic waves
8. measure of energy released by an earthquake
9. seismic sea wave
10. scientist who studies earthquakes

Chapter 9 Review

Checking Concepts

Choose the word or phrase that completes the sentence.

1. Earthquakes can occur when the _______ of rocks is passed.
   a. tension c. elastic limit
   b. compression d. strength
2. A _______ fault forms when the rock above the fault surface moves down relative to the rock below the fault surface.
   a. normal c. reverse
   b. strike-slip d. shearing
3. Seismic waves move outward from the _______.
   a. epicenter c. Moho discontinuity
   b. focus d. tsunami
4. _______ waves stretch and compress rocks.
   a. Surface c. Secondary
   b. Primary d. Shear
5. _______ waves are the slowest.
   a. Surface c. Secondary
   b. Primary d. Pressure
6. At least _______ seismograph stations are needed to locate the epicenter of an earthquake.
   a. two c. four
   b. three d. five
7. Primary waves _______ when they go through solids.
   a. slow down c. stay the same
   b. speed up d. stop
8. The _______ of a seismograph remains still.
   a. sheet of paper c. drum
   b. fixed frame d. pendulum
9. An earthquake of magnitude 7.5 has _______ energy than a quake of 6.5.
   a. 32 times more c. twice as much
   b. 32 times less d. about half as much
10. Most lives lost during an earthquake are due to _______.
    a. tsunamis c. collapse of buildings
    b. primary waves d. broken gas lines

Understanding Concepts

Answer the following questions in your Science Journal using complete sentences.

11. Compare and contrast normal faults with reverse faults.
12. How are primary and secondary waves alike? How are they different?
13. Explain how seismic records were used to determine that Earth's outer core is liquid.
14. What is the relationship between earthquakes with magnitudes on the Richter scale of 1.0 and 3.0?
15. In 1906, an earthquake with a magnitude of 8.6 struck San Francisco, as seen below. Most of the damage done was due to fire. Hypothesize why this occurred.

Thinking Critically

16. What kind of faults would you expect to be most common along the Mid-Atlantic Ridge? Explain.
17. Where is earthquake damage greater—nearer the focus or nearer the epicenter? Explain.
18. Explain why the pendulum of a seismograph remains at rest.
19. Tsunamis are often called tidal waves. Explain why this is incorrect.
20. Which would probably be more stable during an earthquake—a single-story wood-frame house or a brick building? Explain.

Developing Skills

If you need help, refer to the **Skill Handbook.**

21. **Concept Mapping:** Complete the concept map on the right showing what faults result from the three forces. Use the following terms: *tension, compression, shear, normal faults, reverse faults,* and *strike-slip faults.*
22. **Using Variables, Constants, and Controls:** Leah investigated how waves are reflected from curved and flat surfaces using water, a dropper, and flexible cardboard. She filled two flat pans half-full of water and produced ripples with water from the dropper. One pan held the flat cardboard, the other the curved piece. What are her variables? What should she keep constant?
23. **Communicating:** You are a science reporter assigned to interview the mayor about the earthquake safety of city buildings. Make a list of questions about earthquake safety that you will ask the mayor.
24. **Measuring in SI:** Use an atlas and metric ruler to answer the following question. Primary waves travel at about 6 km/s in continental crust. How long would it take a primary wave to travel from San Francisco, CA to Reno, NV?
25. **Interpreting Scientific Illustrations:** The illustration below is a typical record of earthquake waves made on a seismograph. How many minutes passed between the arrival of the first primary wave and the first secondary wave?

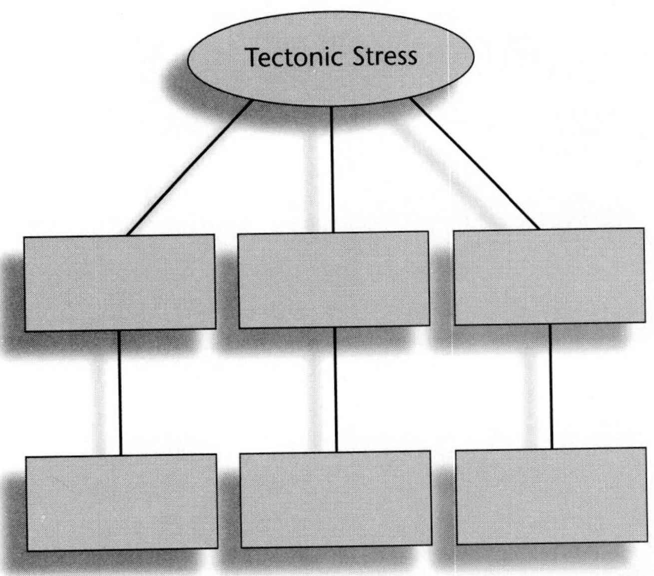

Performance Assessment

1. **Formulating a Hypothesis:** In Activity 9-1, you learned how earthquake depths can be related to the theory of plate tectonics. You learned that subduction of one plate under another at a convergent boundary will produce earthquakes with shallow, intermediate, and deep foci. Use the concepts you have learned to hypothesize whether earthquake foci at divergent plate boundaries would be deep, shallow, or intermediate.
2. **Science Fair Display:** Make a display showing why data from two seismograph stations are not enough to determine the location of an earthquake epicenter.
3. **Model:** Construct a working model of a tiltmeter. Demonstrate how it might predict that an earthquake is about to occur.

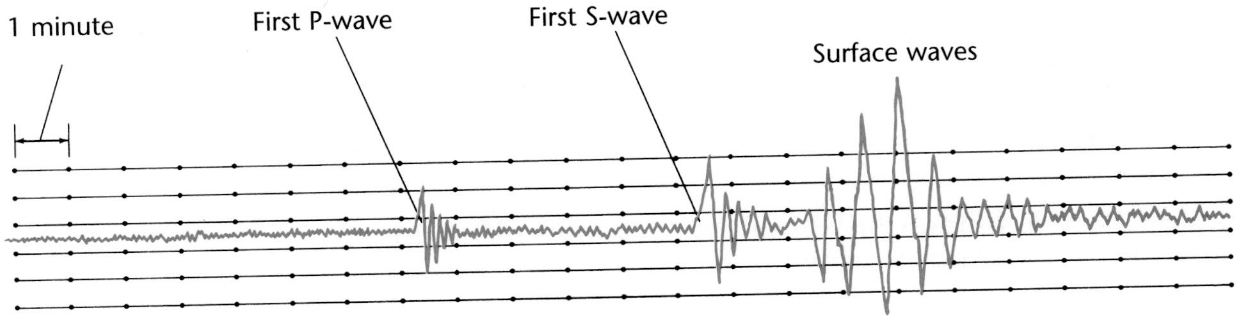

Previewing the Chapter

Volcanoes

The explosive eruption of Mount Pinatubo in the Philippines in 1991 ejected tons of volcanic ash into the atmosphere. Volcanoes can be spectacular and dangerous. Massive ejections of volcanic ash into Earth's atmosphere can cause drastic changes in the environment. On a smaller scale, volcanic eruptions affect humans in many ways. Can you list harmful and also beneficial effects that volcanoes have?

EXPLORE ACTIVITY

Make a model volcano. 

Volcanoes erupt in many different ways. What happens inside a volcano to cause an eruption?

1. Use clay to make a small model volcano with a crater at the top.
2. Place a small amount of baking soda (less than ¼ teaspoon) and a drop of red food coloring in the crater.
3. After putting on safety goggles to protect your eyes, add approximately 20 mL of vinegar to the baking soda in the crater.

Observe: What happens to the baking soda and food coloring when the vinegar is added? Infer how your model eruption is similar to an actual eruption, and how it is different.

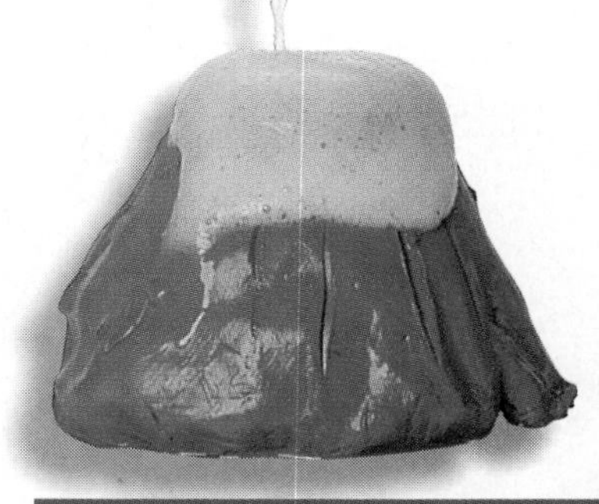

Previewing Science Skills

- In the **Skill Builders**, you will map concepts, sequence events, and compare and contrast.
- In the **Activities**, you will hypothesize, measure, observe, predict, classify, infer, and interpret data.
- In the **MiniLABs**, you will formulate models, observe, and compare and contrast.

10•1 Volcanoes and Earth's Moving Plates

Science Words

volcano
vent
crater
Pacific Ring of Fire
hot spot

Objectives

- Describe how volcanoes can affect people.
- Describe conditions that cause volcanoes.
- Describe the relationship between volcanoes and Earth's moving plates.

Volcanoes and You

A **volcano** is an opening in Earth's surface that often forms a mountain when layers of lava and volcanic ash erupt and build up. Most of Earth's volcanoes are dormant, which means that they are not currently active, but more than 600 are active. Active volcanoes sometimes spew smoke, steam, ash, cinders, and flows of lava.

In 1980, Mount Saint Helens in Washington state erupted. It was one of the largest recent volcanic eruptions in North America. Geologists warned people to leave the area surrounding the mountain. Most people left, but a few stayed. A total of 63 people were killed as a result of the eruption. Heat from the eruption melted snow, which also caused flooding in the area.

Eruption of the Century

In June, 1991, Mount Pinatubo erupted in the Philippines, killing nearly 900 people. The eruption is considered the largest of any volcano this century. As much as 27 million metric tons

Figure 10-1

Kilauea in Hawaii has been continually erupting since January 3, 1983, becoming the most active volcano on Earth. Living with volcanoes as active as Kilauea can create serious problems for homeowners. Losses have reached $61 million as 181 homes have been destroyed.

of sulfur dioxide and ash were thrown into Earth's upper atmosphere. It's possible that this material was the cause of the lowered global temperatures and record ozone losses that were observed as recently as 1993.

Just prior to the eruption of Mount Pinatubo, Mount Unzen in Japan erupted. Forty-one people lost their lives, including several volcanologists who were studying the erupting volcano and producing an educational program.

Figure 10-2

Volcanic ash covered several buildings in Iceland during an eruption in 1973. *With the obvious danger, why do people insist on living near volcanoes?*

Most Active Volcano

For centuries, the Kilauea volcano in Hawaii has been erupting, but not explosively. Most of the town of Kalapana Gardens was destroyed in May 1990. No one was hurt because the lava moved slowly. The most recent series of eruptions from Kilauea, as seen in **Figure 10-1**, began in January 1993. Kilauea is the world's most active volcano.

Approximately 7000 flows of quickly moving, hot gas and volcanic debris called pyroclastic flows have occurred on Mount Unzen from 1991 through 1994. On average, how many pyroclastic flows can be expected to occur on Mount Unzen each month?

What causes volcanoes?

What happens inside Earth to create volcanoes? Why are some areas of Earth more likely to have volcanoes than others?

You learned in Chapter 4 that magma forms deep inside Earth. Heat and pressure cause rock to melt and form magma. Some deep rocks already are molten. Others are hot enough that a small rise in temperature or drop in pressure can melt them to form magma.

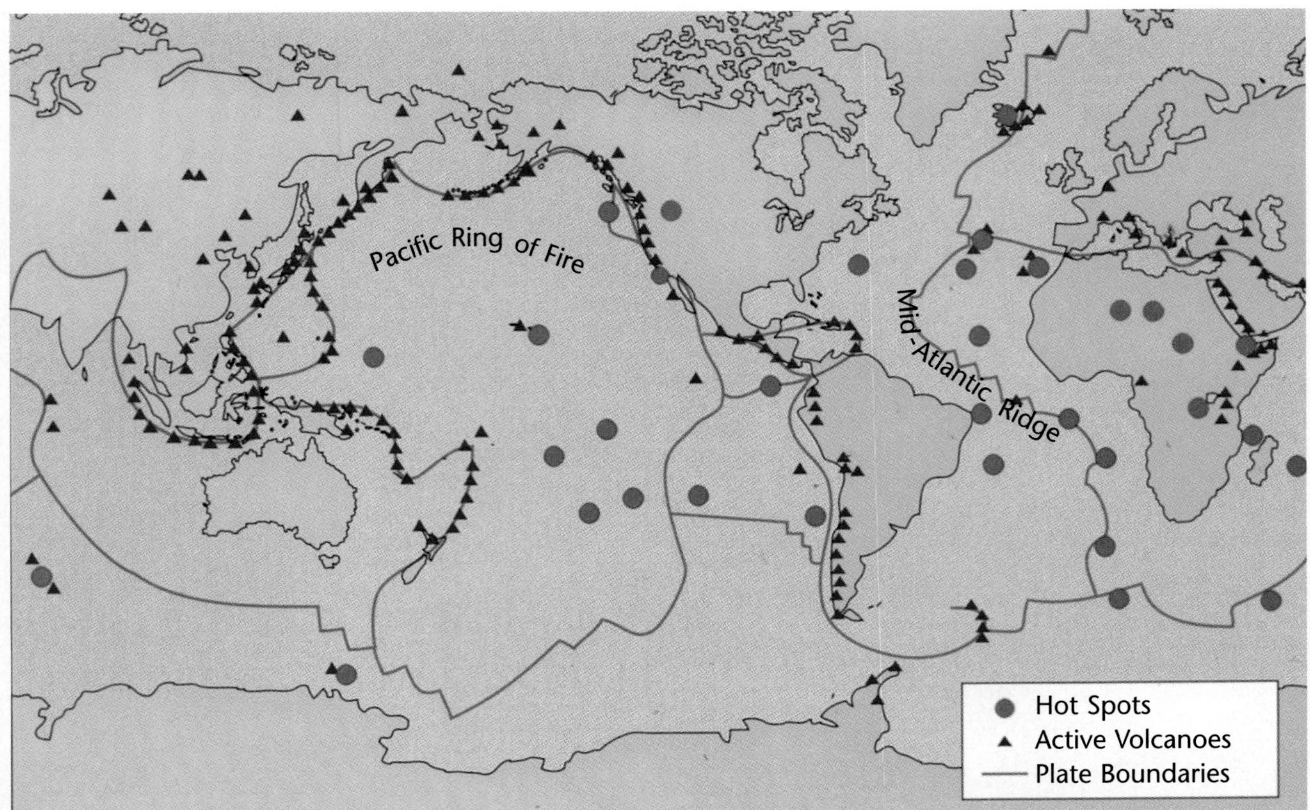

Figure 10-3

The diagram above shows active volcanoes and hot spots around the world. All of the earthquakes and volcanoes in the Pacific Ring of Fire can be attributed to tectonic movement where the Pacific Plate meets other plates. *What process might be occurring at these boundaries that causes rock material to melt and be forced upward to form volcanoes?*

Magma Forced Upward

Magma is less dense than the rock around it, so it is very slowly forced upward toward Earth's surface. You can see this process if you turn a bottle of cold syrup upside down. Watch the dense syrup force the less-dense air bubbles slowly upward to the top.

After many thousands or even millions of years, magma reaches Earth's surface and flows out through an opening called a **vent.** As lava flows out, it cools quickly and becomes solid, forming layers of igneous rock around the vent. The steep walled depression around a volcano's vent is the **crater.**

Where do volcanoes occur?

Volcanoes form in three kinds of places that are directly related to the movement of Earth's plates. Volcanoes occur where plates are moving apart, where plates are moving together, and at locations called hot spots. There are many examples of volcanoes around the world at these three different types of locations. Let's explore Iceland, Mount Saint Helens, and Hawaii.

Divergent Boundaries

Iceland is a large island in the North Atlantic Ocean. It is near the Arctic Circle and has some glaciers. But, as seen in

Figure 10-2, on page 267, it also has volcanoes. Iceland has volcanic activity because it sits on top of the Mid-Atlantic Ridge.

The Mid-Atlantic Ridge is a divergent plate boundary, an area where Earth's plates are moving apart. Where plates separate, they form long, deep cracks called rifts. Magma flows from rifts as lava and is quickly cooled by the seawater. As more lava flows, it builds up from the seafloor. Sometimes the volcanoes and rift eruptions rise above sea level, forming islands such as Iceland.

Convergent Boundaries

Mount Saint Helens is one of several volcanoes that make up the Cascade Mountain Range of Oregon and Washington in the northwestern United States. Mount Saint Helens and the other volcanic peaks in the Cascade Range formed because of a convergent plate boundary. Earth's plates move together at convergent plate boundaries.

USING TECHNOLOGY

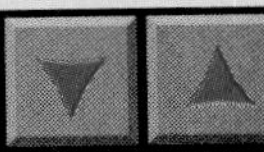

Mudflow Protection

On November 13, 1985, Colombia's Nevado del Ruiz volcano erupted. On that day, nearly 23 000 people lost their lives, not from the exploding mountain or from lava flows, but from mudflows. Mudflows are another hazard of volcanoes. Many volcanic peaks have a high enough elevation that they are covered with snow fields and glaciers. The heat from volcanic activity melts this snow and ice during an eruption. This can mean disaster for people living below such a volcano. The melted snow mixes with ash from the eruption and soil from the mountain and flows rapidly downhill, as it did in Colombia.

The Japanese have developed mudflow control technology to protect populated areas below active volcanoes. Their technology is designed to slow the mud as it flows down valleys. They have installed concrete and steel damlike structures in valleys where mudflows have occurred before.

Think Critically: In addition to these structures, the Japanese have developed other methods designed to provide protection from volcanic eruptions. For example, television cameras are used as earthquake sensors to help them detect eruptions. What will mudflow control enable villagers to do in case of eruptions? Why would earthquake sensors be used to detect volcanic eruptions?

Mudflow Dams, Japan

Activity 10-1

Design Your Own Experiment

Locating Active Volcanoes

Volcanoes form when hot, melted rock material is forced upward to Earth's surface. As the melted rock moves inside Earth, vibrations occur, which are felt as earthquakes. In this activity, you will design an experiment to see whether the locations of active volcanoes are correlated to the locations of recent earthquakes.

PREPARATION

Problem

Is there a correlation between the locations of active volcanoes and the locations of earthquake epicenters?

Form a Hypothesis

State a hypothesis about whether you expect to see a correlation between the locations of active volcanoes and the locations of earthquake epicenters.

Objectives

- Plot the locations of several active volcanoes.
- Describe any correlation you see between locations of volcanoes and locations of earthquake epicenters.

Possible Materials

- Appendix G
- tracing paper
- **Figure 9-2**, page 237
- **Figure 10-3**, page 268

PLAN THE EXPERIMENT

1. As a group, agree upon and write out your hypothesis statement.
2. As a group, list the steps that are needed to test your hypothesis. Be specific, describing exactly what you will do at each step.
3. Determine how best to plot the provided volcano latitude and longitude data on a tracing of Earth's surface.
4. Make a list of any special conditions or facts you expect to observe or test.

Check the Plan

1. Read over your experiment to make sure that all steps are in a logical order.
2. Determine the manner in which you will summarize data and state conclusions about any correlations you discover.
3. How will you determine whether certain facts or conditions will indicate a correlation of volcano locations with locations of earthquake epicenters?
4. *Make sure your teacher approves your plan and that you have included any changes suggested in the plan.*

DO THE EXPERIMENT

Data and Observations

Volcano	Latitude	Longitude
#1	64° N	19° W
#2	28° N	34° E
#3	43° S	172° E
#4	35° N	136° E
#5	18° S	68° W
#6	25° S	114° W
#7	20° N	155° W
#8	54° N	167° W
#9	16° N	122° E
#10	28° N	17° W
#11	15° N	43° E
#12	6° N	75° W
#13	64° S	158° E
#14	38° S	78° E
#15	21° S	56° E
#16	38° N	26° E
#17	7° S	13° W
#18	2° S	102° E
#19	38° N	30° W
#20	54° N	159° E

1. Carry out the experiment as planned.
2. While conducting the experiment, write down any observations that you make in your Science Journal.
3. Compare the plot of volcanoes you have constructed with **Figures 9-2** and **10-3** on pages 237 and 268.

Analyze and Apply

1. **Describe** any patterns of distribution that active volcanoes form on Earth.
2. **Describe** any patterns of distribution of earthquake epicenters shown in **Figure 9-2** on page 237.
3. **Compare and contrast** any patterns that you observe with the locations of Earth's plates and hot spots.

Go Further

Write a hypothesis to explain any patterns you observed for locations of volcanoes, earthquake epicenters, Earth's moving plate boundaries, and hot spots. Propose ways in which geologists might test your hypothesis.

Figure 10-4

The Hawaiian Islands have formed as the Pacific Plate moves over a hot spot. Continued movement of the Pacific Plate formed Kauai, Oahu, Molokai, Maui, and Hawaii over a period of about five million years. Compare parts A and B of this figure. *Based on the position and latitude of the Hawaiian Emperor Seamounts, has the Pacific Plate always moved in the same direction? Explain.*

Here, the Juan de Fuca Plate is converging with and sliding under the North American Plate. Magma is created in the subduction zone when the plate being pushed down under the North American Plate gets deep enough and hot enough to partially melt. The magma is then forced upward to the surface, forming the volcanoes of the Cascades.

Such volcanoes have formed all around the Pacific Plate where it collides with other plates. This area around the Pacific Plate where earthquakes and volcanoes are common is called the **Pacific Ring of Fire,** as seen in **Figure 10-3** on page 268. Mount Saint Helens is just one volcano in the Pacific Ring of Fire. What are the names of some others?

Hot Spots

Like Iceland, the Hawaiian Islands are volcanic islands. But unlike Iceland, they haven't formed at a plate boundary. The Hawaiian Islands are in the middle of the Pacific Plate, far from its edges. What process could be forming them?

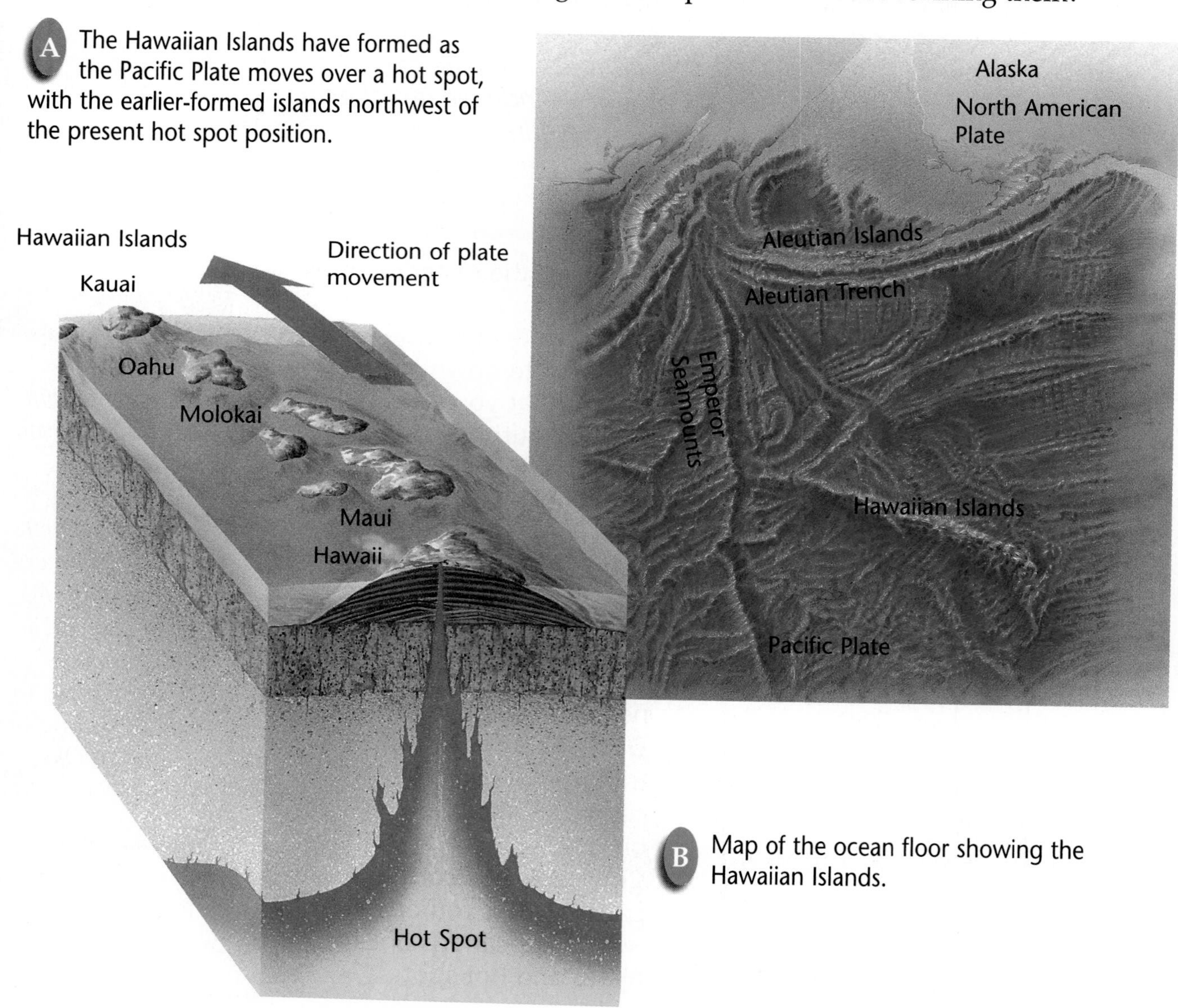

A The Hawaiian Islands have formed as the Pacific Plate moves over a hot spot, with the earlier-formed islands northwest of the present hot spot position.

B Map of the ocean floor showing the Hawaiian Islands.

Geologists believe that some areas in the mantle are hotter than other areas. Some geologists believe that hot spot magma originates at the mantle-outer core boundary. These **hot spots** melt rock, which is then forced upward toward the crust as magma. The Hawaiian Islands sit on top of a hot spot under the Pacific Plate. Magma from deep in Earth's mantle has melted through the crust to form several volcanoes. Those that rise above the water form the Hawaiian Islands, as shown in **Figure 10-5.**

Hawaii

Loihi
(Submarine volcano)

Figure 10-5

Computer image showing the island of Hawaii and Loihi, a submarine volcano. If Loihi reaches the surface, it will form a new island.

As you can see in **Figure 10-4,** the Hawaiian Islands are all in a line. This is because the Pacific Plate is moving over the stationary hot spot. The island of Kauai is the oldest Hawaiian island and was once located where the big island of Hawaii is today. As the plate moved, Kauai moved away from the hot spot and became dormant. Continued movement of the Pacific Plate formed Oahu, Molokai, Maui, and Hawaii over a period of about five million years.

Section Wrap-up

Review

1. How are volcanoes related to Earth's moving plates?
2. As rock material melts, it becomes less dense. Explain what's happening to the atoms and molecules to cause this.
3. **Think Critically:** If the Pacific Plate stopped moving, what would happen to the island of Hawaii?

Skill Builder

Concept Mapping

Make a concept map that shows how the Hawaiian Islands formed over a hot spot. Use the following terms and phrases: *volcano forms, plate moves, volcano becomes dormant, new volcano forms.* If you need help, refer to Concept Mapping in the **Skill Handbook.**

Science Journal

Scientists were able to predict approximately when Mount Pinatubo in the Philippines would erupt. They based their predictions on changes in Earth's crust, shallow earthquakes, and changes in gaseous output. Research this and other recent eruptions and, in your Science Journal, write a report on what equipment volcanologists use to predict volcanic eruptions.

SCIENCE & SOCIETY

TECHNOLOGY:

10•2 Energy from Earth

Science Words

geothermal energy
hot dry rock (HDR)

Objectives

- List the benefits of using geothermal energy to produce electricity.
- Describe the process of using hot dry rock to produce electricity.
- Compare and contrast geothermal energy from magma bodies with geothermal energy from hot dry rock.

Electricity from Geothermal Energy

Because of explosive eruptions like that at Mount Saint Helens, we usually think of igneous activity as being destructive. However, generating electricity with heat from magma bodies inside Earth has been successful in California, Nevada, Utah, Hawaii, and in more than 20 foreign countries. In this process, heat is extracted from water or steam that naturally circulates through rock located near magma bodies. The problem with this type of geothermal energy is that it works only where magma bodies exist close enough to Earth's surface. This occurs only at certain locations around the planet. A new technology that generates electricity from hot dry rock and does not rely on pre-existing natural reservoirs of hot water may eventually decrease human dependence on fossil fuels.

Geothermal Energy from Magma

Magma bodies inside Earth hold tremendous amounts of thermal energy. This **geothermal energy** can be used to generate electricity. The heat from magma can be used to heat water and produce steam in a power plant. The steam is pressurized and then spins turbines that run generators to make electricity.

The United States currently leads the world in the use of geothermal energy. In the four states mentioned above, 70 hydrothermal plants have a capacity to generate 2800 megawatts of electricity. Unless a way is found to enable geothermal energy to be used worldwide, it cannot come close to matching the convenience of energy from fossil fuels. If geothermal energy can be used in place of resources such as fossil fuels, several benefits could be realized. These benefits include reductions in the threat of oil spills on coastlines, mine waste, radioactive waste, and pollutants from burning fossil fuels. Also, geothermal plants presently being used are very reliable.

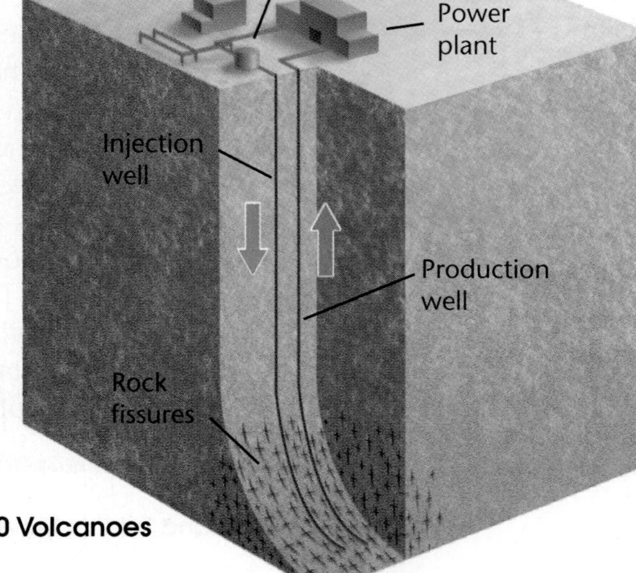

Figure 10-6

A new technology, called hot dry rock (HDR), could greatly increase the use of geothermal energy. Heat from hot rock material is used to heat water and make steam, which is used to generate electricity.

Geothermal Energy from Hot Dry Rock (HDR)

Geothermal energy can also be derived from Earth without relying on the presence of magma. This new technology, called **hot dry rock (HDR),** involves pumping water into hot dry rock that has been fractured, as seen in **Figure 10-6.** The water is pumped down one well, circulated through fractures made in the rock, then forced up a second well. The water returning to Earth's surface is hot water. As with geothermal energy from magma bodies, the hot water is used to produce steam. The steam is pressurized and used to spin turbines, which in turn run generators to make electricity.

HDR is more available as a resource than heat from magma. This is because temperature increases with depth into Earth, and hot dry rock occurs almost everywhere. However, if the hot dry rock is very deep, the cost of drilling forces the overall cost of this energy source upwards. In contrast, harm to the environment caused by geothermal energy is much lower than for other resources such as fossil fuels, as shown in **Figure 10-7.** Does the lower amount of harm to the environment outweigh the much higher cost of HDR? Presently, HDR is too expensive even though it is a much cleaner energy source. But this could change in the future.

In addition to the problem of drilling, HDR requires that the rock have fissures or cracks in it in order for the water to pass through and be heated. If the cracks don't already exist, they must be created. If pollution problems become so severe that a cleaner energy source must be used no matter the cost, or if improved drilling methods bring the cost of HDR down, it could prove to be an extremely valuable resource.

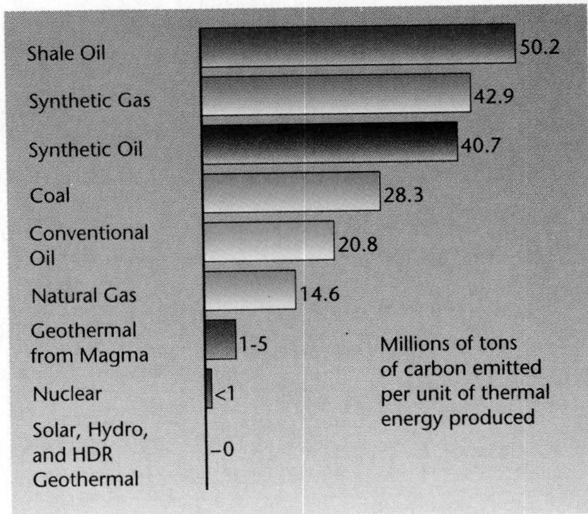

Figure 10-7

Compared with other energy sources, geothermal energy produces very little environmental pollution. *Which type of geothermal energy, geothermal from magma or HDR, produces less carbon-dioxide emissions?*

Section Wrap-up

Review

1. List two things that make geothermal energy beneficial.
2. Describe how hot dry rock is used to generate electricity.

Explore the Technology

HDR sounds like a great way to provide electricity without harming the environment. But there are concerns. HDR systems require water, and the water that is heated and brought to the surface may contain toxic chemicals. Still, if the problems can be handled, HDR may be a useful alternative energy source.

SCIENCE & SOCIETY

10•3 Eruptions and Forms of Volcanoes

Science Words

shield volcano
tephra
cinder cone
composite volcano

Objectives

- Relate the explosiveness of a volcanic eruption to the silica and water vapor content of its lava.
- Describe three forms of volcanoes.

Types of Eruptions

Some volcanic eruptions are explosive and violent, like those from Mount Pinatubo and Mount Saint Helens. But in others, the lava quietly flows from a vent, as in the Kilauea volcano eruptions. What causes this difference?

There are two important factors that determine whether an eruption will be explosive or quiet. One is the amount of water vapor and other gases that are trapped in the magma. The other factor is whether the magma is basaltic or granitic, as you learned in Chapter 4. Let's look first at the gas content of the magma.

Trapped Gases

Have you ever shaken a soft-drink container and then quickly opened it? The pressure from the gas in the drink builds up and is released suddenly when you open the can, spraying the drink. In the same way, gases such as water vapor and carbon dioxide are trapped in magma by the pressure of the surrounding magma. As the magma nears the surface, the pressure is reduced. This allows the gas to escape from the magma. Gas escapes easily from some magma during quiet eruptions. Gas that gets trapped under high pressure eventually escapes, causing explosive eruptions such as that seen in the photographs of Mount Saint Helens in **Figure 10-8.**

Approximate time: 8:32 A.M.

Approximate time: 38 seconds

Figure 10-8

A calm day in Washington was suddenly interrupted when Mount Saint Helens erupted at 8:32 A.M. on May 18, 1980, as shown in this sequence of photographs. *Why was the eruption so violent compared with eruptions of volcanoes like Kilauea?*

Magma Composition

The second major factor that affects the type of eruption is the composition of the magma. Basaltic magma contains less silica, is very fluid, and produces quiet, nonexplosive eruptions such as those at Kilauea. This type of lava pours from volcanic vents and runs down the sides of the volcano. These quiet eruptions form volcanoes over hot spots such as Hawaii. They also flow from rift zones such as those in Iceland. Because the magma is very fluid when it is forced upward in a vent, trapped gases can escape easily in a nonexplosive manner. Sometimes gas causes lava fountains during quiet eruptions, as illustrated in **Figure 10-10** on page 279.

Granitic magma, on the other hand, produces explosive, violent eruptions such as those at Mount Saint Helens. It often forms in the subduction zones where Earth's plates are converging. Granitic magma, as you learned in Chapter 4, is very thick and contains a lot of silica. Because it is thick, it gets trapped in vents, causing pressure to build up beneath it. When an explosive eruption occurs, the gases expand rapidly, often carrying pieces of lava in the explosion.

CONNECT TO

LIFE SCIENCE

Opposition to geothermal energy in Hawaii concerns environmental damage. Find out how severe the damage is expected to be. ***Describe*** how the changes in the rain forest could affect animal habitats.

Approximate time: 42 seconds

Approximate time: 53 seconds

Figure 10-9

The form of volcano that is produced is determined by the nature of its eruption.

A When hot, thin lava flows from one or more vents without violent eruptions, it builds into a gentle slope when it cools. This creates a shield volcano such as Mauna Ulu in Hawaii.

Magma Water Content

Another factor that causes granitic magma to erupt explosively is its high water content. The magma at subduction zones contains a great deal of water vapor. This is because of the wet oceanic crust that is carried into the subduction zone. The trapped water vapor in the thick magma causes explosive eruptions. Some magmas have an andesitic composition (between granitic and basaltic). Because of their higher silica content, they also produce eruptions more violent than those from basaltic magmas.

Forms of Volcanoes

A volcano's form depends on whether it is the result of a quiet or an explosive eruption and the type of lava it is made of—basaltic, granitic, or an intermediate composition. Volcanoes are of three basic forms—shield volcanoes, cinder cone volcanoes, or composite volcanoes, as seen in **Figure 10-9.**

B Explosive eruptions throw lava high into the air. The lava cools and hardens into tephra. When tephra falls to the ground, it forms a steep-sided, loosely consolidated cinder cone volcano. Pictured here is a cinder cone in Arizona.

Shield Volcano

Quiet eruptions spread out basaltic lava in flat layers. The buildup of these layers forms a broad volcano with gently sloping sides called a **shield volcano.** Examples of shield volcanoes are the Hawaiian Islands.

Cinder Cone Volcano

Explosive eruptions throw lava high into the air. The lava cools and hardens into different sizes of volcanic material called **tephra.** Tephra varies in size from volcanic ash—the smallest—to cinders, to larger rocks called bombs. When tephra falls to the ground, it forms a steep-sided, loosely consolidated **cinder cone** volcano.

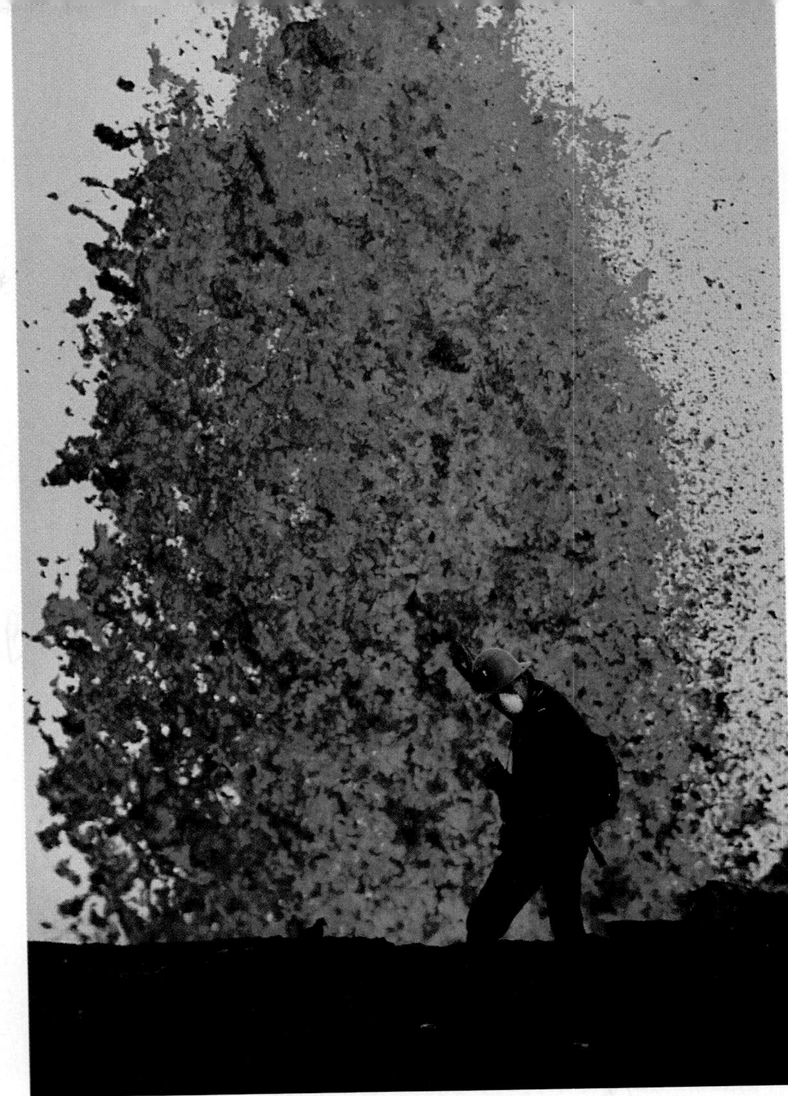

Figure 10-10

Usually the hot, thin lava flows of Kilauea are nonviolent eruptions. *What could be causing the lava fountain shown above?*

Layers of tephra and lava

Magma

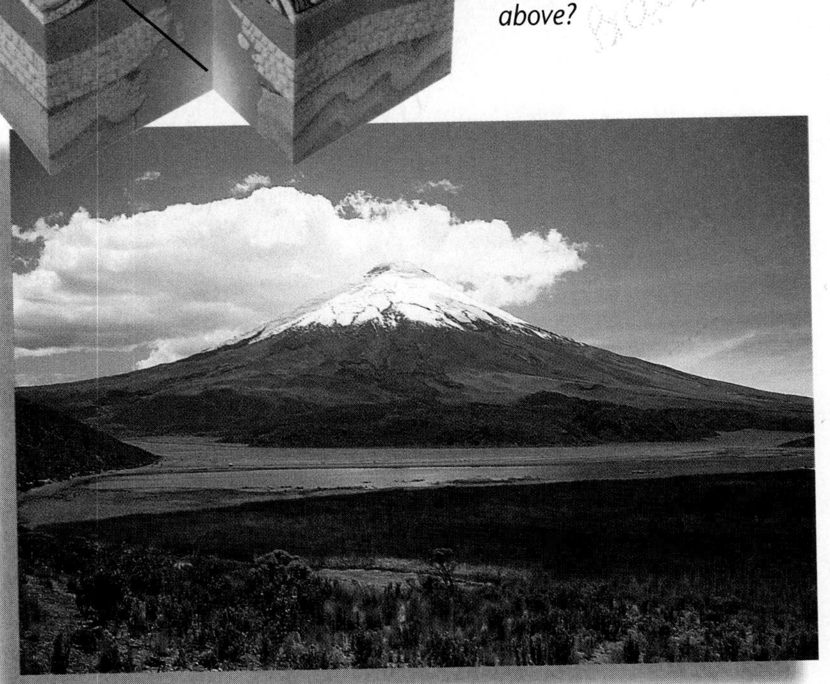

C Whenever volcanic eruptions vary between violent and quieter times, tephra layers alternate with lava layers. A volcano built by this layering of tephra and lava has a composite form, such as Cotopaxi in Ecuador shown here or Mount Saint Helens.

Comparing Volcanic Rocks

During your study of volcanoes and the material that is ejected from volcanoes, you are given two different igneous rocks. One is dark-colored, the other is light-colored. Your task is to determine how the rocks formed and what elements they likely contain.

The two rocks are very fine-grained and full of holes. The holes were caused by escaping gases during the cooling of these rocks. You learned in Chapter 4 that the color of volcanic rocks can indicate what minerals each rock contains. Dark-colored rocks tend to contain minerals high in iron and magnesium, whereas light-colored rocks tend to have a higher concentration of silica-rich minerals.

Solve the Problem:

1. **Based on the size of grains in the rocks, do you think the igneous rocks are intrusive or extrusive?**
2. **Based on the overall color of the rocks, what elements do you think they are likely to contain?**

Think Critically:

Because both rocks are full of holes formed by gases as the rock cooled, it is possible that gases are also trapped inside both rocks. If you are correct in your assumption about the element content of each rock, is there a method to further test element content and possible presence of trapped gas inside the rocks? Explain.

A Mexican farmer learned about cinder cones one morning when he went to his cornfield. He noticed that a hole in his cornfield that had been there for as long as he could remember was emitting smoke that smelled like sulfur. Throughout the night, hot glowing cinders were thrown high into the air. In just a few days, a cinder cone several hundred meters high covered his cornfield. This is the volcano named Parícutin.

Composite Volcano

Some volcanic eruptions can vary between quiet and violent. An explosive period can release gas and ash, forming a tephra layer. Then, the eruption can switch over to a quiet period, erupting lava over the top of the

Figure 10-11
Mount Pinatubo in the Philippines erupted violently in 1991.

tephra layer. When this cycle of lava and tephra is repeated over and over in alternating layers, a **composite volcano** is formed. Composite volcanoes are found mostly where Earth's plates converge and form subduction zones. Mount Saint Helens is an example.

As you can see, there are many factors that affect volcanic eruptions and the form of a volcano, as seen in **Table 10-1.** Mount Saint Helens was formed as the Juan de Fuca Plate was subducted under the North American Plate. The basaltic ocean floor was melted. As it was forced upward, the basaltic magma mixed with surrounding rock material and became more silica-rich. Successive eruptions of lava and tephra produced the composite volcano that towers above the surrounding landscape. Magma inside the volcano solidified, blocking the opening to the surface. Pressure continued to build until in May of 1980, Mount Saint Helens released the pressure in a series of explosive eruptions, as seen in **Figure 10-8** on pages 276 and 277.

The same forces that caused the Mount Saint Helens volcanic activity also caused the 1991 eruptions of Mount Unzen and Mount Pinatubo, as seen in **Figure 10-11.** These volcanoes violently erupted after lying dormant for 200 and over 600 years, respectively. The islands of Japan and the Philippines are volcanic island arcs, formed as the Pacific and the Philippine Plates converged with the Eurasian Plate.

MiniLAB

How can you model different types of volcanic cones?

Procedure

1. Pour a granulated substance such as sand onto one spot on a paper plate, forming a model of a cinder cone volcano.
2. Mix a batch of plaster of paris and pour it onto one spot on another paper plate, forming a model of a shield volcano.
3. Use a protractor to measure the slope angles of the sides of the volcanoes. Allow the model of the shield volcano to dry before measuring its slope angle.

Analysis

1. Which of your volcano models has steeper sides?
2. What form of volcano is represented by the model with steeper sides?
3. Infer why this is so.

Table 10-1

Twelve Selected Eruptions in History							
Volcano and Location	**Year**	**Type**	**Eruptive Force**	**Magma Content**		**Ability of Magma to Flow**	**Products of Eruption**
				Silica	**H_2O**		
Etna, Sicily	1669	composite	moderate	high	low	medium	lava, ash
Tambora, Indonesia	1815	cinder	high	high	high	low	cinders, gas
Krakatoa, Indonesia	1883	cinder	high	high	high	low	cinders, gas
Pelée, Martinique	1902	cinder	high	high	high	low	gas, ash
Vesuvius, Italy	1906	composite	moderate	high	low	medium	lava, ash
Mauna Loa, Hawaii	1933	shield	low	low	low	high	lava
Parícutin, Mexico	1943	cinder	moderate	high	low	medium	ash, cinders
Surtsey, Iceland	1963	shield	moderate	low	low	high	lava, ash
Saint Helens, WA	1980	composite	high	high	high	low	gas, ash
Kilauea Iki, Hawaii	1989	shield	low	low	low	high	lava
Pinatubo, Philippines	1991	composite	high	high	high	low	gas, ash
Galeras, Colombia	1993	cinder	high	high	high	low	gas, ash

Section Wrap-up

Review

1. Some volcanic eruptions are quiet and others are violent. What causes this difference?
2. Why are granitic magmas thicker than basaltic magmas?
3. **Think Critically:** In 1883, Krakatoa in Indonesia erupted. Which kind of lava did Krakatoa erupt—basaltic, granitic, or intermediate? How do you know?

Skill Builder

Sequencing

Arrange the following events of the history of Mount Saint Helens in correct order in your Science Journal: erupts in 1980, composite volcano formed, subduction zone formed, silica-rich magma forced upward, pressure builds, magma in volcano solidifies. If you need help, refer to Sequencing in the **Skill Handbook.**

Science Journal

In your Science Journal, write a brief description of conditions in the Philippines just after Mount Pinatubo erupted. Include an account of how people have coped with the devastation following the eruption.

Igneous Rock Features

Science Words

batholith
dike
sill
volcanic neck
caldera

Objectives

- Give examples of intrusive igneous features and how they form.
- Explain how a volcanic neck and a caldera form.

Intrusive Features

We can observe volcanic eruptions because they are examples of igneous activity on the surface of Earth. But there is far more igneous activity underground because most magma never reaches the surface to form volcanoes. As you learned in Chapter 4, magma that cools underground forms intrusive igneous rock. What forms do intrusive igneous rocks take on? You can look at some of these features in **Figure 10-12.**

Batholiths

The largest intrusive igneous rock bodies are **batholiths.** They can be many hundreds of kilometers wide and long and several kilometers thick. Batholiths form when magma cools underground before reaching the surface. However, not all of them are hidden in Earth. Some batholiths have been exposed at Earth's surface by erosion. The granite domes of Yosemite National Park, as seen in **Figure 10-13A,** are exposed parts of a huge batholith that extends across much of the length of California.

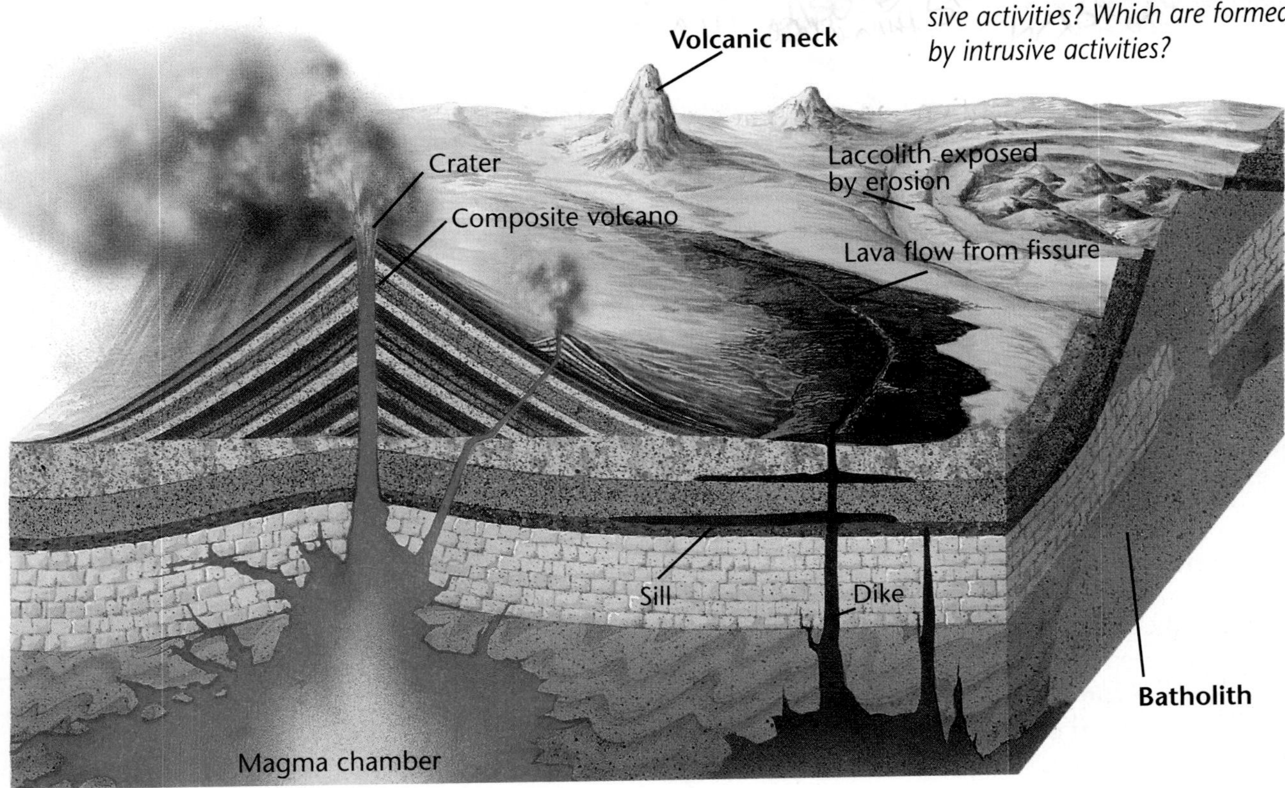

Figure 10-12

This diagram shows intrusive and other features associated with volcanic activity. *Which features shown are formed by extrusive activities? Which are formed by intrusive activities?*

MiniLAB

What types of materials are ejected by volcanoes?

Procedure

1. Obtain some samples of products of volcanic eruptions.
2. Use a pin or dissecting probe to scratch away particles of volcanic rock.
3. Use a magnifying lens and tweezers to separate the products into different piles.
4. Rub the different-sized particles, volcanic ash, and beach sand from volcanic areas slowly between your fingers.

Analysis

Place the volcanic material under an illuminated microscope and take note of the physical properties of each type of material. 1. Describe what you see and feel. 2. What physical properties of these materials might be used to identify them as having a volcanic origin?

Dikes and Sills

Magma sometimes squeezes into cracks in rock below the surface. This is like squeezing toothpaste into the spaces between your teeth. Magma that is squeezed into a generally vertical crack that cuts across rock layers and hardens is called a **dike.** Magma that is squeezed into a horizontal crack between rock layers and hardens is called a **sill.** Both features are shown in **Figures 10-13C** and **10-13D.** Dikes and sills run from a few meters to hundreds of meters long. Some magma that forms a sill may continue to push the rock layers upward. This forms a dome of rock called a laccolith.

Other Features

When a volcano stops erupting, the magma hardens inside the vent. Erosion begins to wear away the volcano. The cone is much softer than the solid igneous rock in the vent. Thus, the cone erodes away first, leaving behind the solid igneous core as a **volcanic neck.** Ship Rock, New Mexico, is a volcanic neck, as seen in **Figure 10-13B.** It is just one of many volcanic necks in the southwestern United States.

Figure 10-13

Intrusive igneous bodies can form in many different sizes and shapes. Some of the most common are batholiths, dikes, sills, and volcanic necks. *What characteristics of dikes and sills probably caused them to be named as they are?*

A Most of the bare rock visible in Yosemite National Park is a batholith that has been exposed by erosion.

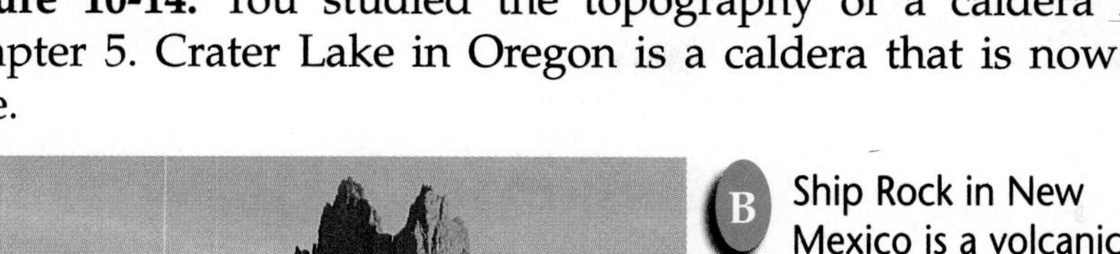

Sometimes after an eruption, the top of a volcano may collapse down into the partially emptied magma chamber. This produces a very large opening called a **caldera,** as shown in **Figure 10-14.** You studied the topography of a caldera in Chapter 5. Crater Lake in Oregon is a caldera that is now a lake.

CONNECT TO

PHYSICS

You have learned that large bodies of magma underground are gradually forced upward toward Earth's surface. ***Determine*** what forces inside Earth are responsible for forcing the magma upward.

B Ship Rock in New Mexico is a volcanic neck.

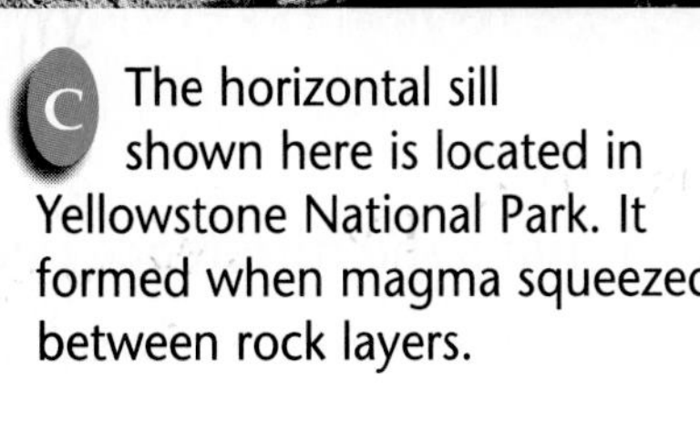

C The horizontal sill shown here is located in Yellowstone National Park. It formed when magma squeezed between rock layers.

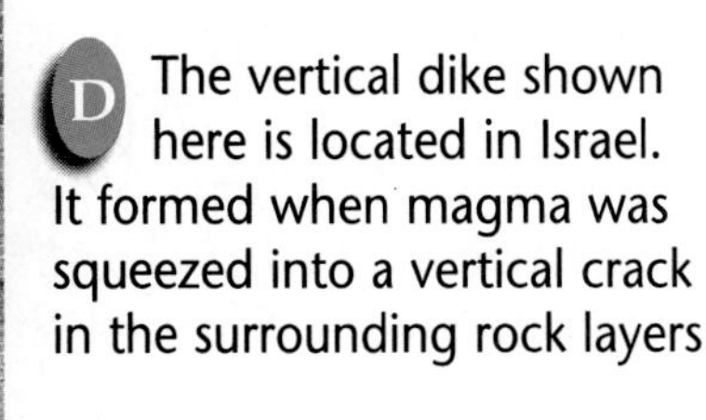

D The vertical dike shown here is located in Israel. It formed when magma was squeezed into a vertical crack in the surrounding rock layers.

Figure 10-14
Crater Lake formed when the top of the volcano collapsed, forming a caldera, as shown in the sequence above.

You have learned in this chapter about one way that Earth's surface is continually built up and how it is worn down. The surface is built up by volcanoes. Also, igneous rock is formed when magma hardens below ground. Eventually, the processes of erosion wear down the rock, exposing batholiths and forming volcanic necks.

Section Wrap-up

Review

1. What's the difference between a caldera and a crater?
2. What is a volcanic neck and how does it form?
3. Explain how calderas form.
4. **Think Critically:** Why are the dome features of Yosemite National Park actually intrusive volcanic features when they are exposed at the surface in the park?

Skill Builder

Comparing and Contrasting

Compare and contrast dikes, sills, batholiths, and laccoliths. If you need help, refer to Comparing and Contrasting in the **Skill Handbook.**

Using Computers

Graphics Use the graphics capabilities of your computer to produce an illustration of igneous rock features based on **Figure 10-12.**

Identifying Types of Volcanoes

You have learned that certain properties of magma are related to the type of eruption and the form of the volcano that will develop. Try this activity to see how to make and use a table that relates the properties of magma to the form of volcano that develops.

Problem

Are the silica and water content of a volcano related to the form of volcano that develops?

Materials

- **Table 10-1** on page 282
- paper
- pencil

Procedure

1. Copy the graph shown at right.
2. Using the information from **Table 10-1**, plot the magma content data for each of the volcanoes listed by writing the name of the basic type of volcano in the appropriate spot on the graph. The data for the 1669 eruption of Mount Etna has already been plotted for you on the diagram.
3. When the plotting of all 12 volcanoes has been completed, analyze the patterns of volcanic types on the diagram to answer the questions.

Analyze

1. What relationship appears to exist between the ability of the magma to flow and the eruptive force of the volcano?
2. Which would be more liquid in its properties, a magma that flows easily or one that flows with difficulty?
3. What relationship appears to exist between the silica or water content of the magma and the nature of the material ejected from the volcano during the eruptions?

Conclude and Apply

4. How is the ability of the magma to flow related to its silica and water content?
5. **Infer** which of the two variables (silica or water content) appears to have the greater effect on the eruptive force of the volcano.
6. **Describe** the relationship that appears to exist between the silica and water content of the magma and the type of volcano that is produced.

Data and Observations

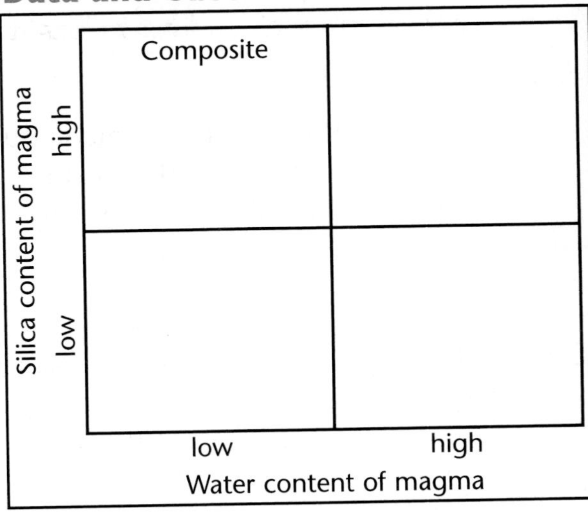

Science & ART

One-Hundred Views of Mount Fuji

Why would an artist draw one hundred pictures of an inactive volcano? Japanese artist Katsushika Hokusai never seemed to tire of drawing Mount Fuji, the highest mountain in Japan. Sometimes he made the mountain fill the picture. One time, it served as the backdrop for a slender crane perched amid low, snow-frosted bushes. One drawing showed Mount Fuji through a thicket of bamboo. In still another, the mountain was a dim outline, barely visible through a summer shower.

Many of Hokusai's ink drawings of Mount Fuji were made into woodcuts by gluing the drawing to a block of hardwood. Then a carver whittled away the wood between the lines, leaving the design Hokusai had drawn. After ink was applied to the raised design, the block was used to make prints.

Born in 1760 in what is now Tokyo, Hokusai signed his drawings and paintings with many different names, choosing a new name at least every ten years. He also moved often, living in more than 90 places during his 89 years of life. This restlessness extended to his art. Hokusai experimented with a wide range of subjects, from drawings of Japanese actors, to book illustrations, to his exquisite landscapes. Only Mount Fuji seemed to hold his attention.

The artist worked from early morning until after dark, creating thousands of prints during his lifetime. His most famous work is *Thirty-Six Views of Mount Fuji,* actually a collection of 46 color prints published from 1826 to 1833. Hokusai followed this in 1834 and 1835 with *One-Hundred Views of Mount Fuji,* drawings in black, white, and gray.

Fuji last erupted in 1707, and its symmetrical, snow-capped volcanic crater is recognizable in every drawing in both of Hokusai's collections. Each time he drew the mountain, though, he changed his perspective and the composition of the picture. Hokusai did not feel confined to what he could see and sometimes inserted people into his drawings. Several scenes show Mount Fuji as viewed from merchants' shops.

Mount Fuji is considered sacred and a symbol of Japan. Thousands climb it every year. Perhaps Hokusai's many woodcuts gave Japanese who lived far from the mountain an opportunity to feel closer to this sacred spot. The drawings definitely earned Hokusai a reputation as a master artist and a major influence on both Japanese and Western art.

Science Journal

Why did Hokusai, otherwise so restless, draw one mountain over and over again? Write in your Science Journal why you think Mount Fuji had so much power over this artist.

Chapter 10

Review

Summary

10-1: Volcanoes and Earth's Moving Plates

1. Volcanoes can be dangerous to people, causing deaths and destroying property.
2. Rocks in the mantle melt to form magma, which is forced upward toward Earth's surface. When the magma flows through vents, it's called lava and forms volcanoes.
3. Volcanoes along rift zones form when magma flows onto the seafloor. The lava builds up from the seafloor to form an island. Volcanoes over hot spots form when the magma breaks through the crust. Volcanoes also form when an ocean plate is subducted under another plate. Here, the subducted plate melts to form magma.

10-2: Science and Society: Energy from Earth

1. Geothermal energy can reduce our dependence on oil and reduce air pollution.
2. A new technology, hot dry rock (HDR), is being developed to use Earth's internal heat to generate energy. Heat from hot, dry rock is used to heat water and make steam, which is used to generate electricity.

10-3: Eruptions and Forms of Volcanoes

1. Basaltic lavas are thin and flow easily, producing quiet eruptions. Silica-rich lavas are thick and stiff, and thus produce very violent eruptions. Water vapor in magma adds to its explosiveness.
2. Shield volcanoes are mountains made of basaltic lava and have gently sloping sides. Cinder cones are steep-sided and are made of tephra. Composite volcanoes, made of silica-rich lava and tephra, are steep-sided.

10-4: Igneous Rock Features

1. Batholiths, dikes, sills, and laccoliths form when magma solidifies underground.
2. A caldera forms when the top of a volcano collapses, forming a very large depression.

Key Science Words

a. batholith
b. caldera
c. cinder cone
d. composite volcano
e. crater
f. dike
g. geothermal energy
h. hot dry rock (HDR)
i. hot spot
j. Pacific Ring of Fire
k. shield volcano
l. sill
m. tephra
n. vent
o. volcanic neck
p. volcano

Reviewing Vocabulary

Match each phrase with the correct term from the list of Key Science Words.

1. mountain made of lava and/or volcanic ash
2. large depression formed by the collapse of a volcano
3. solid magma core of a volcano
4. volcano with gently sloping sides
5. ash and cinders thrown from a volcano
6. steep-sided volcano of lava and tephra
7. largest igneous intrusion
8. an opening through which lava flows
9. an igneous intrusion formed between rock layers
10. energy that comes from magma or HDR

Chapter 10 Review

Checking Concepts

Choose the word or phrase that completes the sentence.

1. Composite volcanoes form near locations where Earth's plates are _______.
   a. diverging
   b. sticking and slipping
   c. converging
   d. sliding past each other
2. Hawaii is made of volcanoes due to _______.
   a. plates moving apart
   b. a hot spot
   c. plates moving together
   d. rift zones
3. Lavas ______ produce violent volcanic eruptions.
   a. rich in silica
   b. that are fluid
   c. made of basalt
   d. rich in iron
4. Magma that is rich in iron produces ______ eruptions.
   a. thick
   b. caldera
   c. quiet
   d. explosive
5. A _______ is made of tephra.
   a. shield volcano
   b. caldera
   c. cinder cone
   d. composite volcano
6. Kilauea is a _______.
   a. shield volcano
   b. composite volcano
   c. cinder cone
   d. rift zone
7. Magma that squeezes into a vertical crack in rock layers and then hardens is a _______.
   a. sill
   b. dike
   c. volcanic neck
   d. batholith
8. A _______ is a dome-shaped igneous intrusive body.
   a. dike
   b. laccolith
   c. sill
   d. batholith
9. HDR geothermal energy comes from _______.
   a. fossil fuels
   b. electricity
   c. the sun
   d. hot dry rock
10. Geothermal energy _______.
    a. relies on heat from the sun
    b. reduces need for fossil fuels
    c. increases air pollution
    d. isn't possible in the United States

Understanding Concepts

Answer the following questions in your Science Journal using complete sentences.

11. Why do volcanoes differ in size and shape?
12. Contrast volcanoes that form where Earth's plates diverge with those that form where one of Earth's plates slides under another.
13. Explain how these terms are related: vent, caldera, and crater.
14. What conditions about the 1985 eruption of Colombia's Nevado del Ruiz caused most of the nearly 23 000 deaths?
15. Why does the State of Hawaii have many volcanic islands instead of just one large island?

Thinking Critically

16. Explain how glaciers and volcanoes can exist on Iceland.
17. What kind of eruption is produced when basaltic lava flows from a volcano? Explain.
18. How are volcanoes related to earthquakes?
19. A mountain called Misti is a volcano in Peru. Peru is on the western border of South America. How might this volcano have formed?
20. In addition to Iceland and Hawaii, where else on Earth do you think people could use geothermal energy? What if HDR became economical?

Developing Skills

If you need help, refer to the **Skill Handbook.**

21. Concept Mapping: Make a network tree concept map that compares quiet eruptions with explosive eruptions. Use the following words and phrases: *high silica, flows easily, granitic, explosive, cinder cone, Parícutin, shield, low silica,* and *basaltic.*

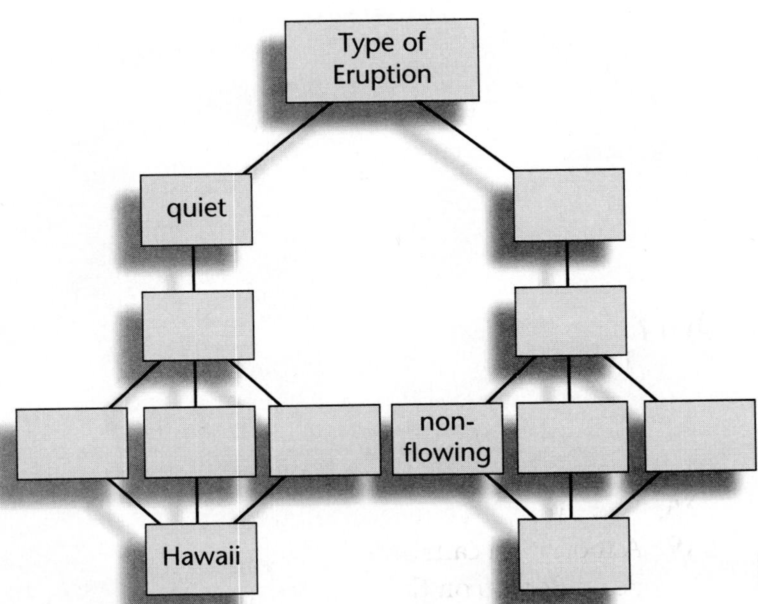

22. Observing and Inferring: A volcano violently erupted in Indonesia in 1883. What can you infer about the magma's composition? If people saw the eruption, what were they able to observe about the flow of the lava?

23. Comparing and Contrasting: Compare and contrast batholiths and laccoliths.

24. Classifying: Mount Fuji's steep sides are made of layers of silica-rich lava and ash. Classify Mount Fuji.

25. Measuring in SI: The base of the volcano Mauna Loa is about 5000 m below sea level. The total height of the volcano is 9560 m. What percentage of the volcano is above sea level? Below sea level?

Performance Assessment

1. Model: Build a cut-away model of a geothermal-energy power plant that shows how water might be pumped into hot dry rock, changed to steam, and then returned to the surface to run a turbine.

2. Asking Questions: In Activity 10-1, you learned that the locations of active volcanoes are related to earthquakes, boundaries of Earth's moving plates, and hot spots. Based on what you have learned, what could be said about an area in the midst of a continent that shows evidence of a large number of ancient, extinct volcanoes?

3. Data Table: In Activity 10-2, you identified types of volcanoes and plotted them on a table based on the composition of the magma that was forced out of them. Based on what you have learned, research the 1991 eruption of Mount Unzen in Japan and the 1982 eruption of El Chichón in Mexico and plot them on your table.

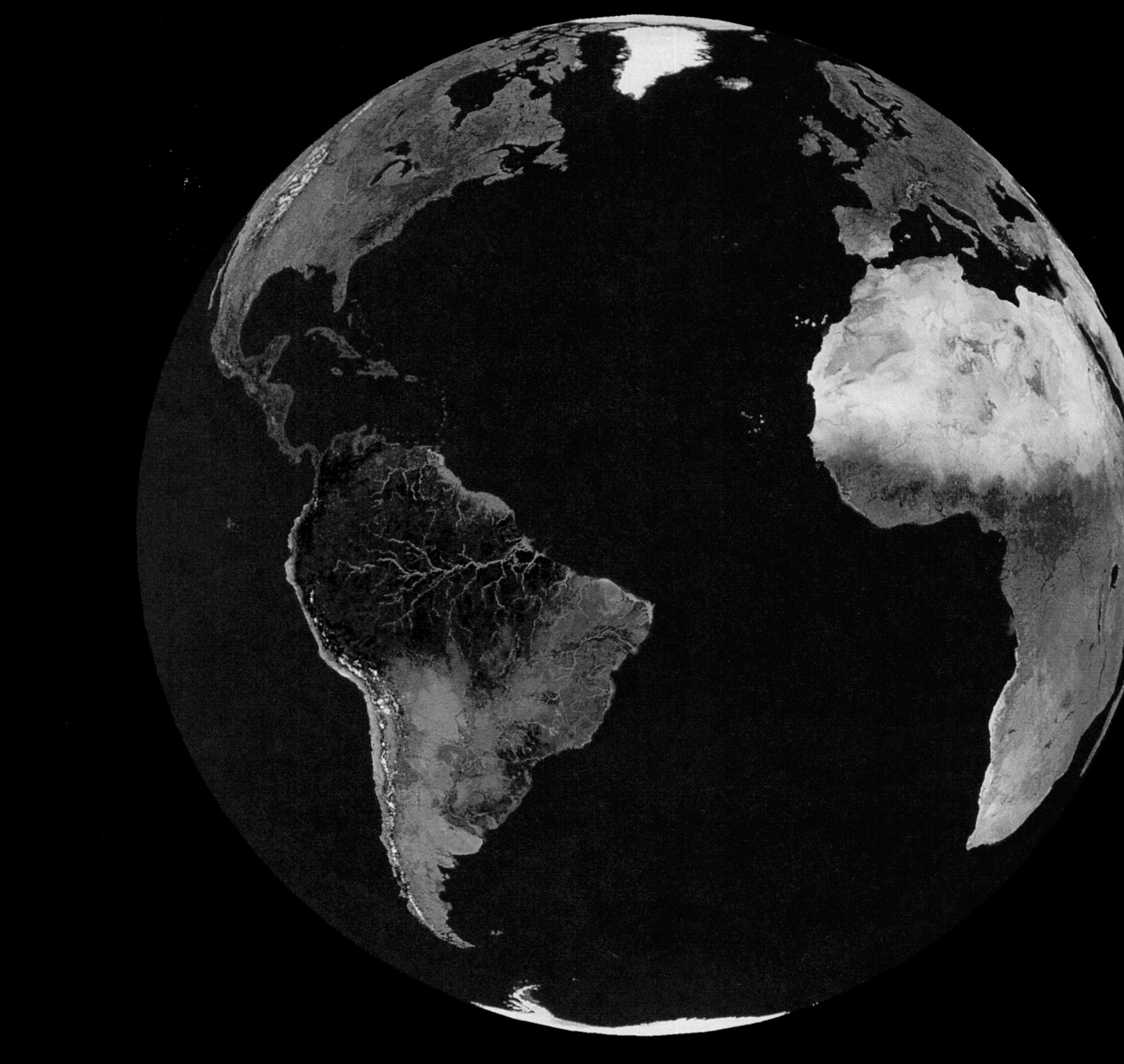

Previewing the Chapter

Chapter 11 Plate Tectonics

This photograph of Earth is unique because the clouds have been removed using a computer. You can see the shapes of the continents just like on a map. Look very closely at the general shapes of the continents. Do you see any relationship between continents? If this photograph of Earth were cut into pieces, could you fit the pieces back together? What clues might you use?

EXPLORE ACTIVITY

Do the following activity to see how clues on adjoining pieces of a cut-up photograph can be used to re-form the image.

1. Working with a partner, obtain photographs of interest to you from an old magazine. Do not look at each other's photographs.
2. You and your partner are each to cut one picture into small pieces.
3. Exchange pictures with your partner.
4. Using clues on surrounding pieces, re-form the image of the photograph your partner has cut into small pieces.

Observe: What characteristics of the cut-up photograph did you use to re-form the image? Can you think of other examples in which characteristics of objects are used to match them up with other objects?

Previewing Science Skills

- In the **Skill Builders**, you will **outline, map concepts,** and **recognize cause and effect.**
- In the **Activities,** you will **analyze and interpret data, make a model, observe,** and **infer.**
- In the **MiniLABs,** you will **experiment, compare and contrast, observe and infer,** and **hypothesize.**

11•1 Continental Drift

Science Words

continental drift
Pangaea

Objectives

- Explain the theory of continental drift.
- Discuss four pieces of evidence for the theory of continental drift.

Evidence for Continental Drift

When you look at a map of Earth's surface, one thing is very obvious. In **Figures 11-1** and **11-2** you see that the edges of some continents look as if they would fit together like a puzzle. In the early 1800s, as accurate maps of Earth's surface were first being developed, other people also noticed this fact.

Pangaea

Alfred Wegener thought that the fit of the continents wasn't just a coincidence. He believed that all the continents were joined together at some point in the past, and in a 1912 lecture, he proposed the idea of continental drift. The theory of **continental drift** states that continents have moved horizontally to their current locations. Wegener believed that all continents were once connected as one large landmass that broke apart about 200 million years ago. When the continents broke apart, they drifted to their present positions. He called this large landmass **Pangaea** (pan JEE uh), which means "all land."

Figure 11-1

Glossopteris, Mesosaurus, and other organisms shown in this illustration lived in many areas of Pangaea. Their fossils are now found on separate continents.

A Fossils and fossil remains of the plants and animals shown here have been found on more than one continent. *Mesosaurus* fossils, for example, have been discovered in South America and in Africa.

Since Wegener's death in 1930, his basic theory, that the continents moved, has been accepted. The evidence Wegener had to support his theory wasn't enough to convince many people during his lifetime. However, Wegener's early evidence has since been joined by other important proofs. Let's explore both Wegener's clues and some newer ones.

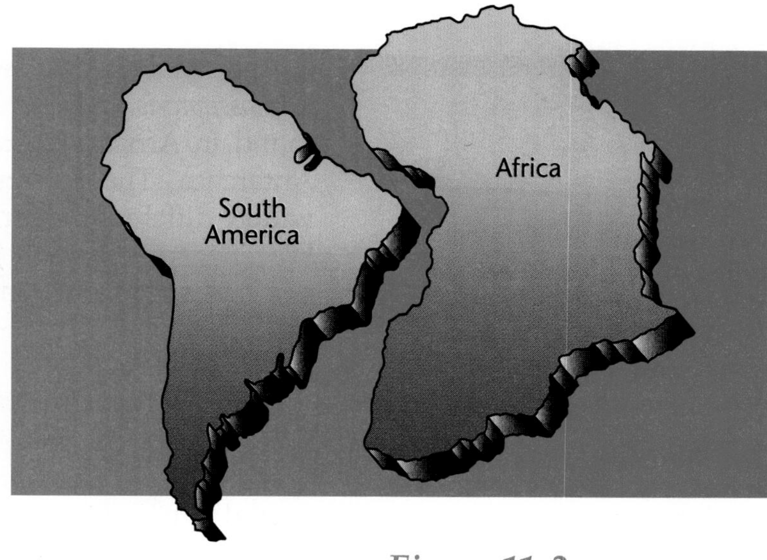

Figure 11-2

The eastern coastline of South America and the western coastline of Africa look like they would fit together as puzzle pieces.

Fossil Clues

Besides the puzzlelike fit of the continents, other clues were found from fossils. Fossils of the reptile *Mesosaurus* have been found in South America and Africa, as shown in **Figure 11-1.** This swimming reptile lived in fresh water and on land. How could fossils of the *Mesosaurus* be found so far apart? It's very unlikely that it could have swum between the continents. Wegener thought this reptile lived on both continents when the continents were connected.

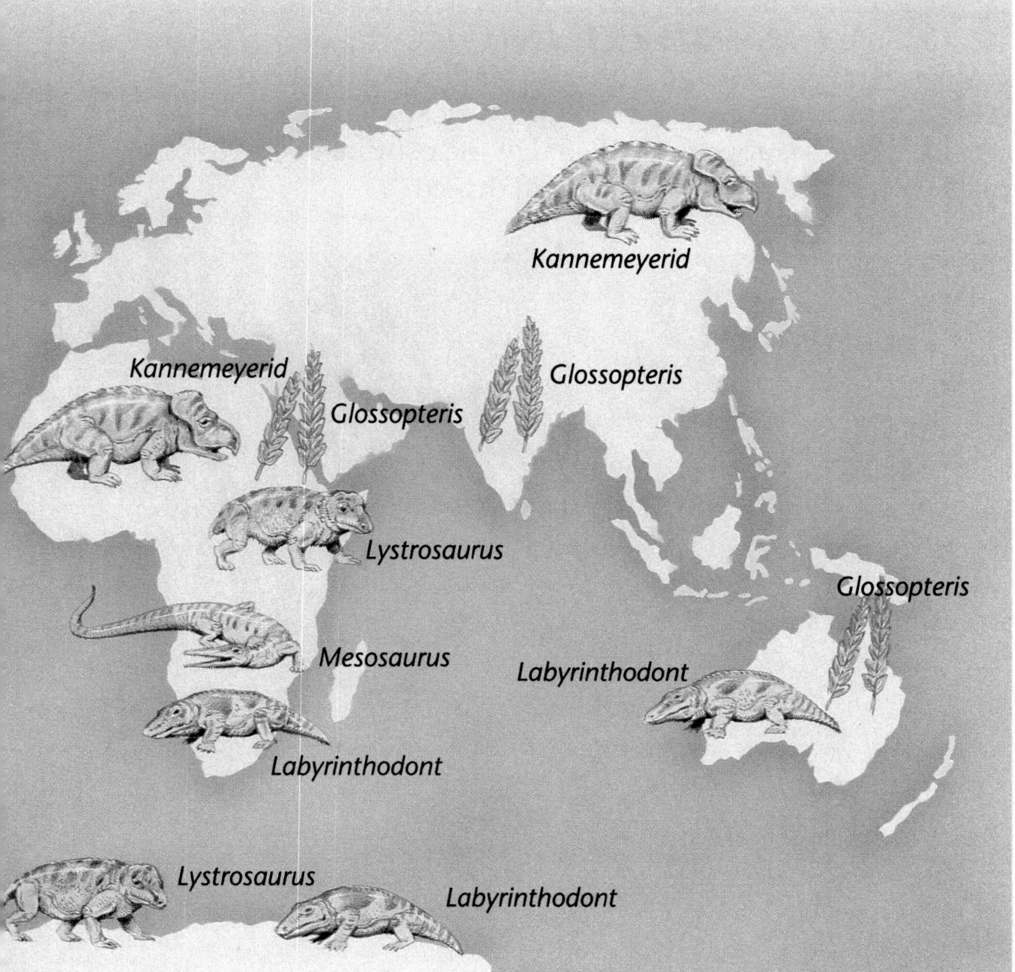

B *How does the study of* Glossopteris, Mesosaurus, Kannemeyerid, Labyrinthodont, *and other fossils support Wegener's hypothesis of continental drift?*

MiniLAB

Can clues such as fossils, types of life, and rock structures be used to determine whether continents now separated were once joined?

Procedure

1. Build a three-layer landmass using clay.
2. Mold the clay into mountain ranges.
3. Place similar "fossils" into the clay at various locations around the landmass.
4. Form five continents from the one landmass. Also, form two smaller landmasses out of different clay with different mountain ranges and fossils.
5. Place the five continents and two smaller landmasses around the room.
6. Students who did not make or place the landmasses will locate the drifted continents and construct a drawing that shows how they were once positioned.

Analysis

1. Were you able to reconstruct all or part of the original clay landmass?
2. What clues, if any, were useful in reconstructing the original landmass?
3. How did you deal with continents that initially didn't seem to fit?

Another fossil that helps support the theory of continental drift is *Glossopteris*. In **Figure 11-3** you see this fossil fern, which was found in Africa, Australia, India, South America, and later in Antarctica. The presence of this fern in so many areas with widely different modern climates, as you see in **Figure 11-1,** led Wegener to believe that all of these areas were once connected and had a similar climate.

Climate Clues

Fossils of warm-weather plants were found on the island of Spitzbergen in the Arctic Ocean. Wegener believed that Spitzbergen drifted from the tropic regions. He also used glacial clues to support his theory. Glacial deposits and grooved bedrock found in southern areas of South America, Africa, India, and Australia indicated that these continents were once covered with glaciers. How could you explain why glacial deposits were found in these areas where no glaciers exist today? Wegener thought that these continents were all connected and covered with ice near Earth's South Pole at one time.

Rock Clues

If the continents were connected at one time, then rocks that make up the continents should be the same. Similar rock structures *are* found on different continents. Parts of the Appalachian Mountains of the eastern United States are similar to those found in Greenland and western Europe. If you were to travel to South America and western Africa, you would find rock structures that are very similar. These clues, found in rocks, support the idea that the continents were connected when these rock structures formed.

Figure 11-3

This fossil fern, *Glossopteris*, grew in a warm tropical climate.

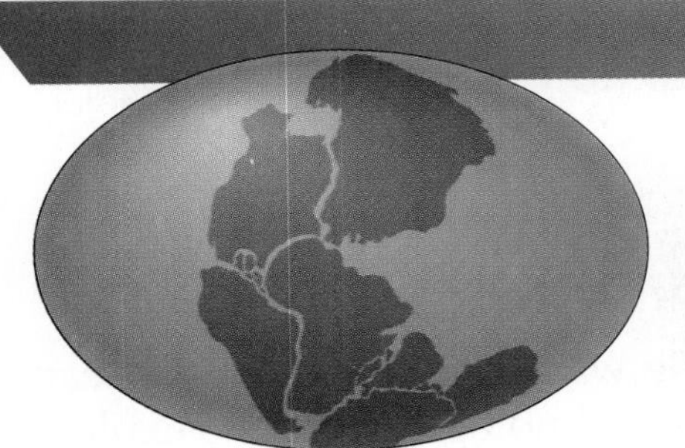

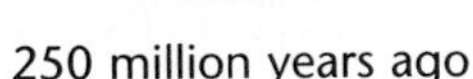

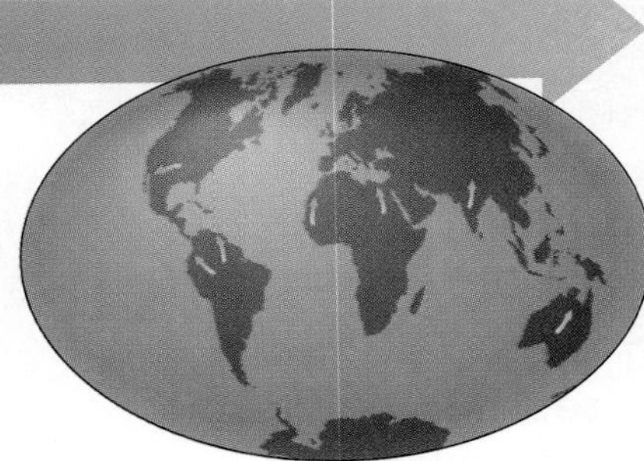

How could continents drift?

Although Wegener provided evidence to support his theory of continental drift, he couldn't explain how, when, or why these changes in the position of the continents, shown in **Figure 11-4,** had taken place. Because other scientists at that time could not provide these explanations either, Wegener's idea of continental drift was rejected. The idea was so different that most people closed their minds to it.

Rock, fossil, and climate clues were the main points of evidence for the continental drift theory. Later, after Wegener's death, more clues were found and new ideas related to continental drift were discovered. One of these new ideas, seafloor spreading, helped provide an explanation of how the continents could move.

Figure 11-4
These computer models show the probable course that continents have taken. On the far left is their position 250 million years ago, in the middle is their position 180 million years ago, and at right is their current position. *Based on what is presented in the diagrams and assuming that the rate of movement will stay the same, what will happen to the Atlantic Ocean during the next 100 million years?*

Section Wrap-up

Review

1. State one reason why Wegener's ideas about continental drift were not accepted.
2. How did Wegener use climate clues to support his hypothesis about continental drift?
3. **Think Critically:** Why would you expect to see similar rocks and rock structures on two landmasses that were connected at one time?

Skill Builder
Comparing and Contrasting

Compare and contrast the location of fossils of the tropical plant *Glossopteris,* as shown in **Figure 11-1,** with the climate that exists at each location today. If you need help, refer to Comparing and Contrasting in the **Skill Handbook.**

Science Journal

Pretend you are Alfred Wegener in the year 1912. In your Science Journal, write a letter to another scientist explaining your idea about continental drift. Try to convince this scientist that your theory is correct.

11 • 2 Seafloor Spreading

Science Words

seafloor spreading
magnetometer

Objectives

- Describe seafloor spreading.
- Relate how age and magnetic clues confirm seafloor spreading.

Clues on the Ocean Floor

Up until the early 1950s, little was known about the ocean floors. Scientists didn't have the technology needed to explore the deep oceans. But the invention of echo-sounding devices allowed the construction of accurate maps of the ocean floor. Soon, scientists discovered a complex ocean floor that had mountains and valleys just like continents have above water. They also found a system of ridges and valleys extending through the center of the Atlantic and in other oceans around the world. The mid-ocean ridges form an underwater mountain range that extends through the center of much of Earth's oceans. This discovery raised the curiosity of many scientists. What formed these mid-ocean ridges?

Figure 11-5

As the seafloor spreads apart at a mid-ocean ridge, new seafloor is created. The older seafloor moves away from the ridge in both directions. *If seafloor spreading is happening, what evidence should you expect to find by studying rocks taken from the seafloor?*

Age of ocean floor in millions of years

150-200	100-150	50-100	0-50	50-100	100-150	150-200

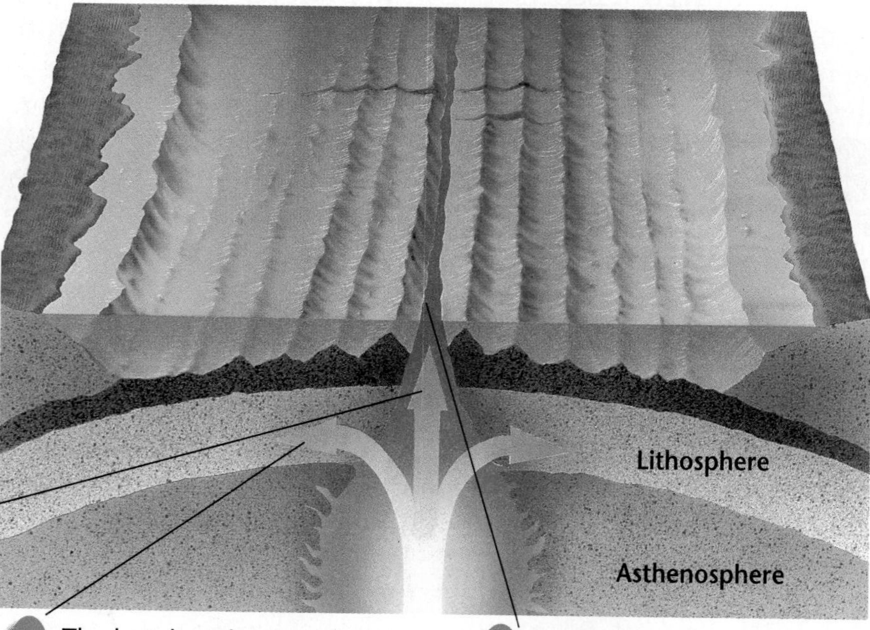

A Hot, less-dense, partially molten rock material from deep inside Earth is forced upward by the cooler, denser surrounding rock.

B The hot, less-dense rock material is forced upward into the plasticlike portion of the upper mantle. As it approaches the more rigid upper mantle and crust, it is deflected in opposite directions.

C Plates of Earth's crust and rigid upper mantle are forced apart and carried along with the moving hot rock material in the asthenosphere. A rift forms into which molten rock from the upper mantle is forced until it finally flows out onto Earth's surface as lava.

Figure 11-6
The research ship *Glomar Challenger* helped in the exploration of the world's oceans and the seafloor.

The Seafloor Moves

In the early 1960s, Princeton University scientist Harry Hess suggested an explanation. His now-famous and accepted theory is known as **seafloor spreading.** Hess proposed that hot, less-dense material in the mantle is forced upward to the surface at a mid-ocean ridge. Then it turns and flows sideways, carrying the seafloor away from the ridge in both directions, as seen in **Figure 11-5.** As the seafloor spreads apart, magma located a few kilometers below the ridge moves upward and flows from the cracks. It solidifies and forms new seafloor. The seafloor that is carried away from the ridge cools, contracts, and becomes more dense than the plasticlike asthenosphere below it. The seafloor begins to sink downward, forming trenches. The theory of seafloor spreading was later shown to be correct by the two following pieces of evidence.

Age Evidence

In 1968, scientists aboard the research ship *Glomar Challenger* began gathering information about the rocks in the seafloor. The *Challenger,* as shown in **Figure 11-6,** was equipped with a drilling rig that allowed scientists to drill into the seafloor to obtain rock samples. The scientists began drilling to study the age of rocks in the seafloor and made a remarkable discovery. They found no rocks older than 180 million years. In contrast, some continental rocks are almost four billion years old. Why were these seafloor rocks so young?

Scientists also found that the youngest rocks were located at the mid-ocean ridges. The age of the rocks became increasingly older farther from the ridges on both sides. The evidence for seafloor spreading was getting stronger.

USING TECHNOLOGY

New Picture of Earth

What does the interior of Earth look like? Since the early 1900s, scientists have been using earthquakes to create models of the structure of Earth's interior. When an earthquake occurs, it generates seismic waves that travel through Earth.

Seismic Tomography

Using a new technique called seismic tomography, scientists have made a new model of Earth's interior. In seismic tomography, a computer combines data from earthquakes all around the world. The models from the computer show hot, rising masses of rock (red) and cool, sinking masses of rock (blue). Cool rock material beneath Manchuria seems to be sliding under a warm mass beneath Japan. A large spike of cold rock is located under the crust of much of North America. A large collection of hot rock material is located directly under Iceland.

Think Critically:

Consider the spike of cold rock located under North America and the hot rock material under Iceland. Explain how these features could be related to earthquakes and/or volcanoes. How might the data obtained with seismic tomography help support the theories concerning plate movement at Earth's surface?

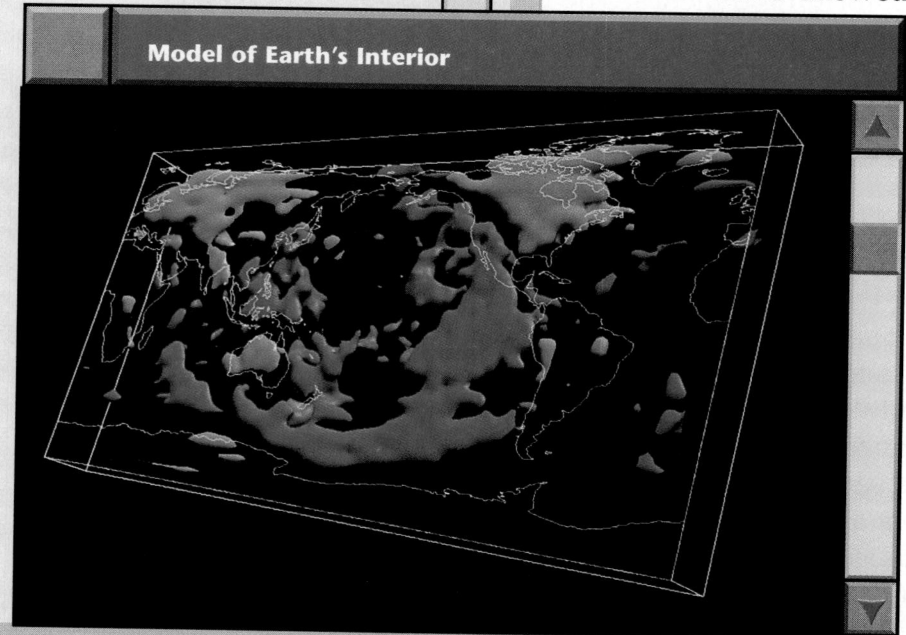

INTEGRATION
Physics

Magnetic Clues

The final bit of evidence in support of the theory of seafloor spreading came from magnetic clues found in the iron-bearing basalt rock from the ocean floor.

Scientists know that Earth's magnetic field has reversed itself several times in the past, as illustrated in **Figure 11-7.** Iron minerals in rocks such as basalt align themselves according to the magnetic field orientation at the time that they form. If Earth's magnetic field is reversed, new iron minerals being formed would reflect that magnetic reversal.

Scientists found that rocks on the ocean floor showed many magnetic reversals. They used a **magnetometer,** a sensitive instrument that records magnetic data. The magnetic alignment of the rocks reverses back and forth in strips parallel to the mid-ocean ridge.

This discovery provided strong support that seafloor spreading was indeed happening. The magnetic reversals showed that new rock was being formed at the mid-ocean ridges.

The ideas of Alfred Wegener and Harry Hess changed the way people think about Earth's crust. Fossil, rock, and climate evidence supporting the theory of continental drift is too strong to be discounted. Seafloor spreading proves that ocean floors change too. You'll see in Section 11-3 how these two ideas are closely related.

CONNECT TO

CHEMISTRY

Find out what the Curie point is and ***describe*** in your Science Journal what happens to iron minerals when they are heated to the Curie point.

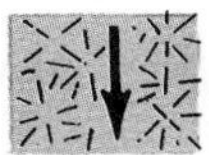

Figure 11-7

Changes in magnetic polarity of the rock on both sides of mid-ocean ridges reflect the past reversals of Earth's magnetic poles. *Why is this considered evidence for seafloor spreading?*

Section Wrap-up

Review

1. How does the magnetic alignment of iron minerals help support the theory of seafloor spreading?
2. What eventually happens to seafloor that is carried away from a mid-ocean ridge?
3. **Think Critically:** How is seafloor spreading different from continental drift?

Skill Builder

Concept Mapping

Make a concept map that discusses the evidence for continental drift using the following terms and phrases: *continental edges, same fossils, climate, rock structures, on different continents, mountains with similar features, continental ice sheets,* and *puzzle pieces.* If you need help, refer to Concept Mapping in the **Skill Handbook.**

On average, North America is moving 1.25 cm per year away from the Mid-Atlantic Ridge. Using this rate, determine how far apart the continents of North America and Africa will be after 200 million years.

Activity 11-1

Design Your Own Experiment

Seafloor Spreading

Magnetic data support the theory of seafloor spreading. When magnetic iron minerals, such as magnetite, are heated to the Curie point (about 580°C) in a magma body, they lose their magnetism. When these iron minerals once again cool off, they become magnetized in the direction of Earth's magnetic field at that time. In this activity, develop a model showing how magnetic clues indicate that the seafloor is spreading apart.

PREPARATION

Problem

Develop a model showing how magnetic clues indicate that the seafloor is spreading apart.

Form a Hypothesis

State a hypothesis about how a model can show how magnetic clues are used to indicate that the seafloor is spreading apart.

Objectives

- Make a model of how magnetic clues are used to indicate seafloor spreading.
- Describe how the Curie point is involved in a study of magnetic clues on the seafloor.

Possible Materials

- metric ruler
- paper
- small magnetic compasses (2)
- tape
- bar magnets
- pen or marker

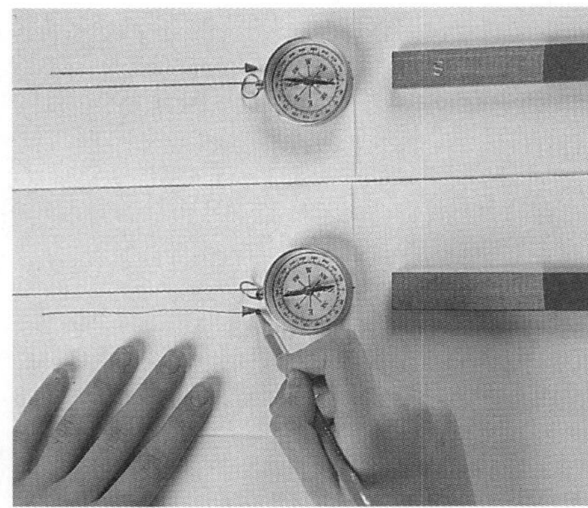

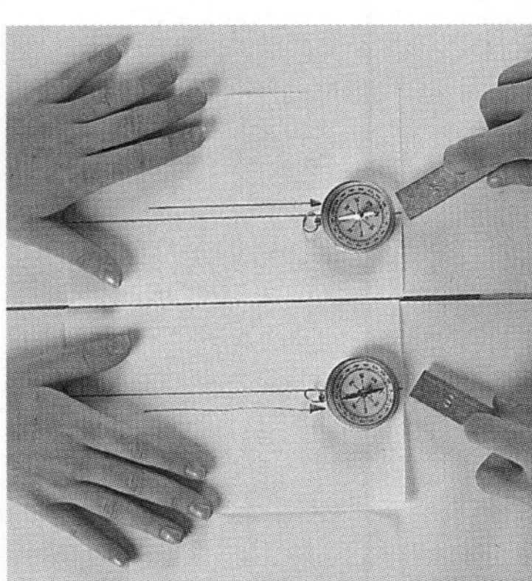

PLAN THE EXPERIMENT

1. As a group, agree upon and write out your hypothesis statement.
2. As a group, list the steps needed to test your hypothesis. Be very specific, describing exactly what your group will do at each step.
3. Make a list of the materials that your group will need to make your model and complete your experiment.
4. Determine how to make your model of seafloor spreading.
5. Determine how your model will demonstrate magnetic clues from the ocean floor.

Check the Plan

1. Read over your entire experiment to make sure that all steps are in a logical order.
2. Using the illustration provided as an example, determine whether the plan for your model will be able to demonstrate seafloor spreading.
3. Will you use the small magnetic compasses to demonstrate how magnetic clues, including alternating directions of magnetic north, indicate seafloor spreading?
4. How will you use the paper to demonstrate seafloor spreading?
5. *Make sure your teacher approves your plan before you proceed.*

DO THE EXPERIMENT

1. Carry out the experiment as planned and construct your model.
2. While constructing your model, write down in your Science Journal any observations that you or other members of your group make.
3. Demonstrate how your model uses magnetic clues, such as those caused by alternating directions of magnetic north, to support the theory of seafloor spreading.

Analyze and Apply

1. When you have developed your model, **compare** your ideas with those of other groups of students in your class.
2. You may wish to **modify** your model to incorporate ideas that other teams are successfully using.
3. Where are the "oldest" marks on the strip of paper?
4. On the seafloor, is the youngest rock near or far from the mid-ocean ridges?

Go Further

Does your model support the concept of seafloor spreading? Explain the role of the bar magnets in your model. What do they represent? Explain how flipping the model's bar magnets is a model of the behavior of Earth's magnetic field.

11•3 Theory of Plate Tectonics

Science Words

plate tectonics
plate
lithosphere
asthenosphere
divergent boundary
convergent boundary
subduction zone
transform fault
convection current

Objectives

- Compare and contrast divergent, convergent, and transform plate boundaries.
- Describe how convection currents might be the cause of plate tectonics.
- Describe the effects of plate tectonics found at each type of boundary.

Plate Tectonics

With the discovery of seafloor spreading, scientists began to understand what was happening to Earth's crust and upper mantle. The idea of seafloor spreading showed that more than just continents were moving, as Wegener had thought. It was now evident to scientists that sections of the seafloor and continents move around in relation to one another.

Plate Movements

By 1968, scientists had developed a new theory that combined the main ideas of continental drift and seafloor spreading. The theory of **plate tectonics** states that Earth's crust and upper mantle are broken into sections. These sections, called **plates,** move around on the mantle. The plates can be thought of as rafts that float and move around on the mantle.

Composition of Earth's Plates

Plates are composed of the crust and a part of the upper mantle, as seen in **Figure 11-8.** These two parts together are called the **lithosphere** (LITH uh sfihr). This rigid layer is about 100 km thick and is less dense than material underneath. The plasticlike layer below the lithosphere is called the **asthenosphere** (as THEN uh sfihr). The less-dense plates of the lithosphere "float" and move around on the denser asthenosphere. The interaction of the plates on the asthenosphere is similar to

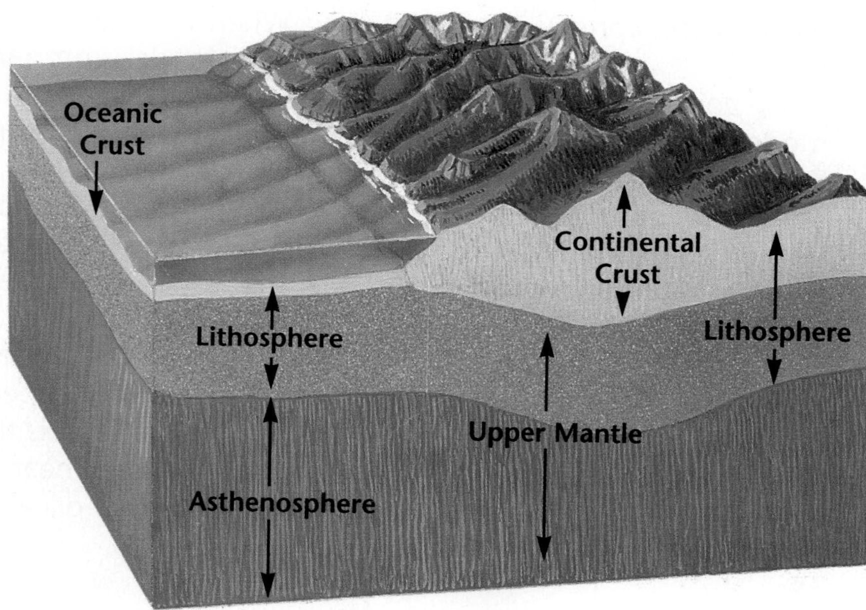

Figure 11-8

Plates of the lithosphere are carried along by convection currents within Earth's asthenosphere.

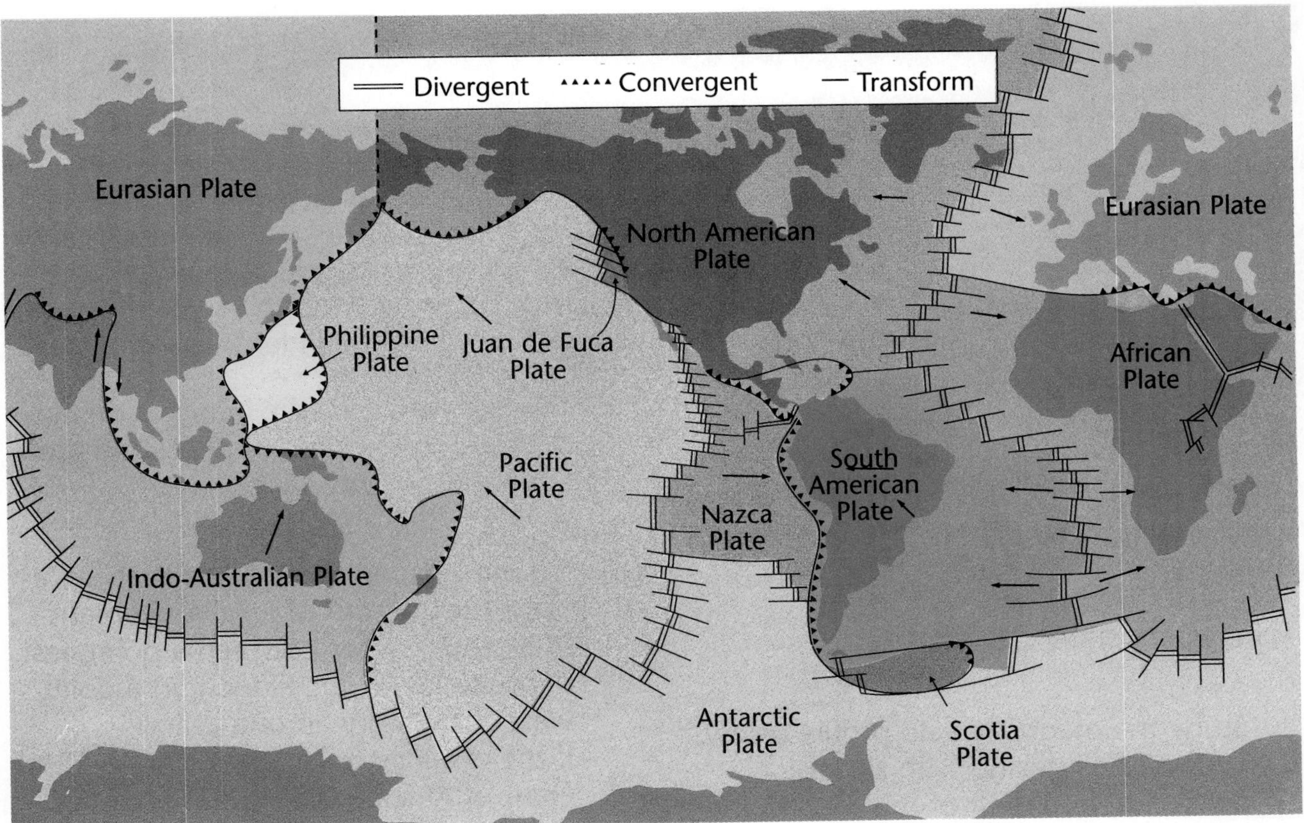

Figure 11-9

This diagram shows the major plates of the lithosphere, their direction of movement, and the type of boundary between them. *Based on what is shown in this figure, what is happening to the Nazca Plate where it meets the South American Plate?*

that of large, flat stones when they are placed in plastic. By applying force, you can move the stones around in putty.

Plate Boundaries

What happens when plates move? They can interact in three ways. They can move toward each other and collide, they can pull apart, or they can simply move past one another. When the plates interact, the result of their movement is seen at the plate boundaries, as in **Figure 11-9.** Movement along any plate boundary requires that adjustments be made at the other boundaries. What is happening to the Atlantic Ocean between the North American and African plates? Compare this with what is happening along the western margin of South America.

Divergent Boundaries

The boundary between two plates that are moving apart from one another is called a **divergent boundary.** You learned about divergent boundaries when you read about seafloor spreading. In the Atlantic Ocean, the North American Plate is moving away from the Eurasian and the African plates, as seen in **Figure 11-9.** That divergent boundary is called the Mid-Atlantic Ridge. The Great Rift Valley in eastern Africa is another example of a divergent plate boundary. Here, a valley has formed where two continental plates are separating.

Activity 11-2

Seafloor Spreading Rates

So far, you've learned a lot about seafloor spreading, magnetic reversals, and plate tectonics. How can you use your knowledge to reconstruct Pangaea? Try this activity to see how you can determine where a continent may have been located in the past.

Problem

Can magnetic clues, such as magnetic field reversals on Earth, be used to help reconstruct Pangaea?

Materials

- metric ruler
- pencil

Procedure

1. Study the magnetic field profile below. You will be working with six major peaks east and west of the Mid-Atlantic Ridge for both normal and reversed polarity.
2. Place the ruler through the first peak west of the main rift. Determine and record the distance in kilometers to the Mid-Atlantic Ridge.
3. Repeat step 2 for each of the six major peaks east and west of the main rift, for both normal and reversed polarity.
4. Find the average distance from peak to ridge for each pair of corresponding peaks on either side of the ridge. Record these values.
5. Use the normal polarity readings to find the age of the rocks at each average distance.
6. Using normal polarity readings, calculate the rate of movement in cm/year. Use the formula *rate = distance/time* to calculate the rate. You must convert kilometers to centimeters.

Analyze

1. Compare the age of the igneous rock found near the mid-ocean ridge with that of the igneous rock found farther away from the ridge.
2. In what way does the information shown on the graph relate to the procedure in Activity 11-1?

Conclude and Apply

3. On your paper, **draw** a line that represents the amount of total movement that would occur between a point east of the Mid-Atlantic Ridge and a point west of the ridge in one year.
4. If the distance from a point on the coast of Africa to the Mid-Atlantic Ridge is approximately 2400 km, **calculate** how long ago that point in Africa was at or near the Mid-Atlantic Ridge.
5. Using the data you have recorded, **determine** the position of Africa in relation to other continents when the point in Africa mentioned in question 4 was at or near the Mid-Atlantic Ridge.

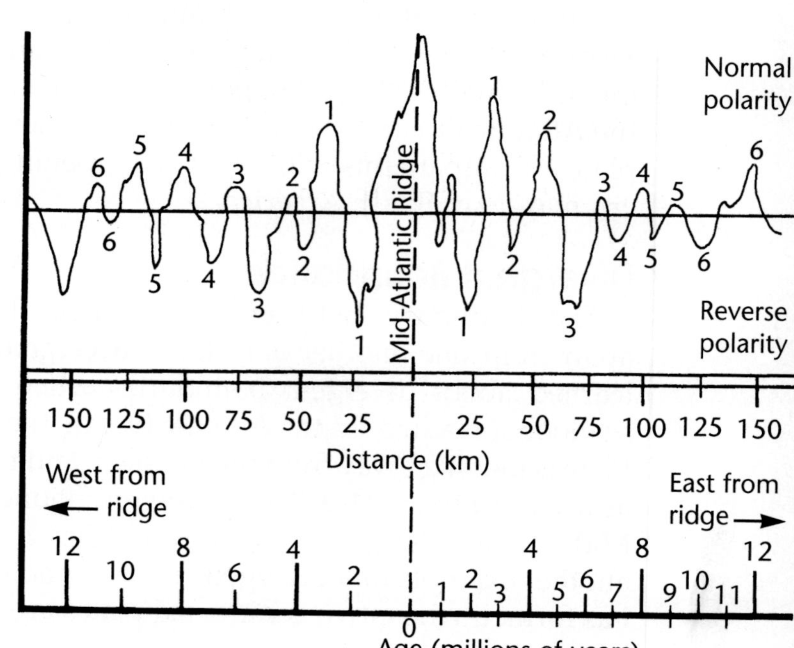

Convergent Boundaries

If new crust is being added at one location, why doesn't Earth's crust keep getting thicker? As new crust is added in one place, it disappears at another. The disappearance of crust can occur when seafloor cools, becomes denser, and sinks, or where two plates collide at a **convergent boundary.**

There are three types of convergent boundaries. When an ocean-floor plate collides with a less-dense continental plate, the denser ocean plate sinks under the continental plate. The area where an oceanic plate descends into the upper mantle is called a **subduction zone.** Volcanoes occur above subduction zones.

Figure 11-10 on pages 308-309 shows how this type of convergent boundary creates a deep-sea trench where one plate is subducting under the other. High temperatures and pressures cause the subducted plate to melt as it descends under the other plate. The newly formed magma is forced upward along these plate boundaries, forming volcanic mountains. The Andes Mountains of South America contain many volcanoes. They were formed at the convergent boundary of the Nazca and the South American plates.

The Fit Isn't Perfect

Recall the Explore Activity you performed at the beginning of this chapter. While you were trying to fit pieces of a cut-up photograph together, what clues did you use?

Take an old map of the world and cut out each of the continents. Lay them out on a tabletop and try to fit them together, using techniques you used in the Explore Activity. You will find that the pieces of your Earth puzzle, the continents, do not fit together very well. Yet several of the areas on some continents fit together extremely well.

Solve the Problem:

If you have ever worked on a puzzle, you have probably used scientific methods to solve the puzzle. With each piece, you look at its general shape and specific hints on its surface that help you find where the piece fits into the puzzle. In this activity, you will find out how continents can be thought of as pieces of a much larger puzzle, and how their shapes can be used to show how they are related.

1. **Take out another old map—one that shows the continental shelves as well as the continents.**
2. **Cut the continents, including the continental shelves, out of the second map.**

Think Critically:

When the continents are pieced together with the continental shelves attached, almost all of them fit together well. Why did this slight change of including the continental shelves solve the problem dealing with how the continents fit together? What is true about the continental shelves that indicates they should be included with maps of the continents?

Ocean-Ocean Collisions

The second type of convergent boundary occurs when two ocean plates collide, or when seafloor that has become denser due to cooling begins to sink. Whether caused by two ocean plates colliding or by the sinking of denser seafloor, one plate bends and slides under the other, forming a subduction zone as shown in **Figure 11-10C.** A deep-sea trench is formed, and the new magma that is produced rises to form an island arc of volcanoes. The islands of Japan are volcanic island arcs formed when two oceanic plates collided.

The third type of convergent boundary occurs when two continental plates collide. Because both of these plates are less dense than the material in the asthenosphere, usually no subduction occurs. The two plates just collide and crumple up,

Figure 11-10

As Earth's plates pull apart at some boundaries, they collide at others, forming mountains and volcanoes.

A As one continental plate converges with another, subduction stops and mountains are formed.

B Where continental lithospheric plates are pulled apart at divergent boundaries, a rift valley forms. If the rift valley separates further, it may flood and become an ocean.

C As an oceanic plate converges with a less-dense continental plate, the continental plate is forced upward and the oceanic plate slides under it at a subduction zone. As the oceanic plate slides into the subduction zone, it starts to melt. The melted rock is less dense than surrounding rock and is forced upward, forming volcanoes.

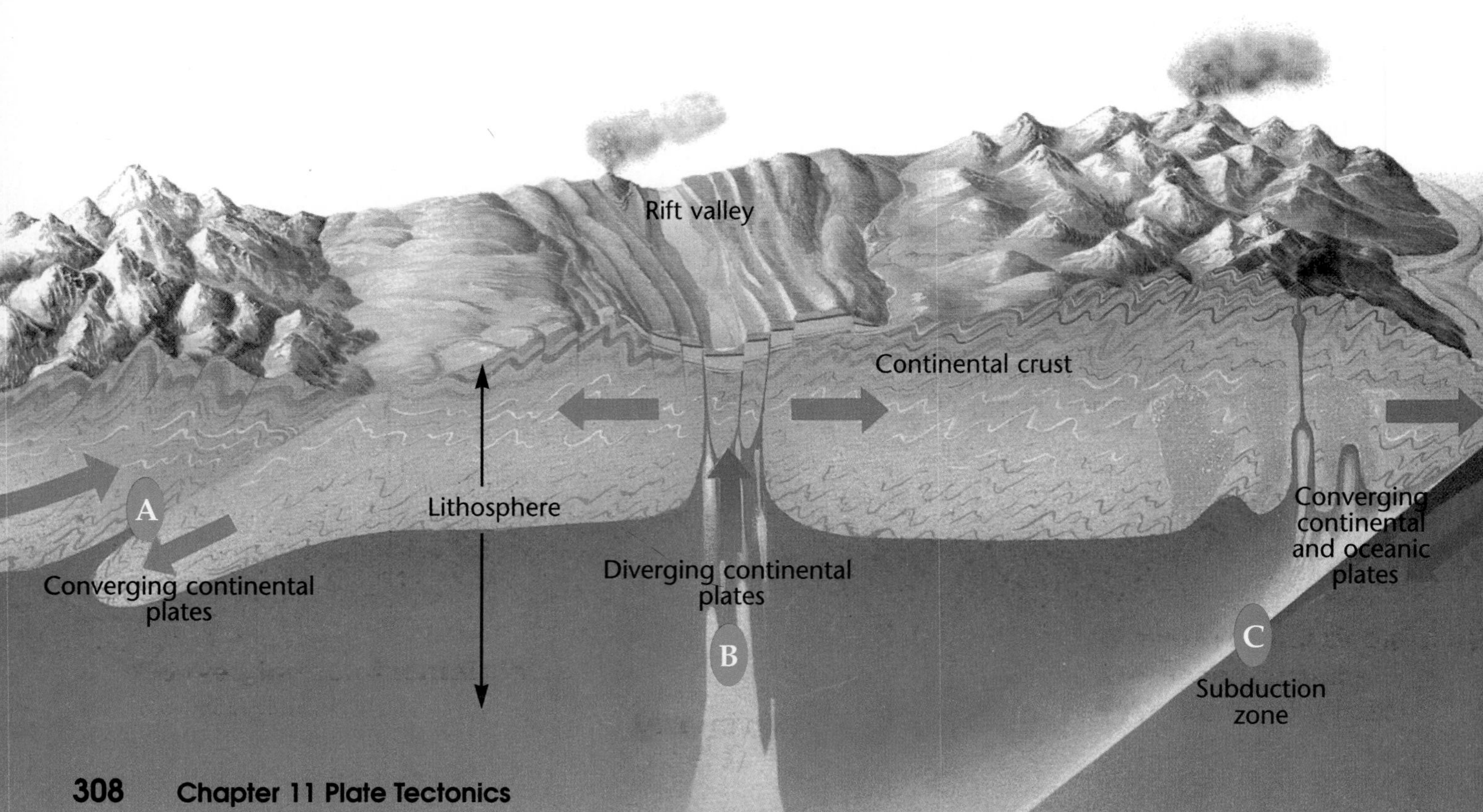

forming mountain ranges. Earthquakes are common at these convergent boundaries. The Himalayan Mountains in Asia were formed when the Indo-Australian Plate crashed into the Eurasian Plate.

Transform Fault Boundaries

The third type of plate boundary occurs at a **transform fault.** Transform faults occur when two plates slide past one another and are moving either in opposite directions or in the same direction at different rates. As one plate slides past another, earthquakes occur. The Pacific Plate is sliding past the North American Plate, forming the famous San Andreas Fault, as seen in **Figure 11-11** on the next page. The San Andreas Fault is a transform-fault plate boundary. It has been the site of many earthquakes.

D A mid-ocean ridge forms whenever diverging plates continue to separate after being flooded. As the rising magma cools, it forms new ocean crust.

E A subduction zone forms whenever two oceanic plates collide or one oceanic plate that has become denser due to cooling begins to sink under another. Volcanic island arcs form above the subduction zone, as the descending plate reaches depths where melting occurs.

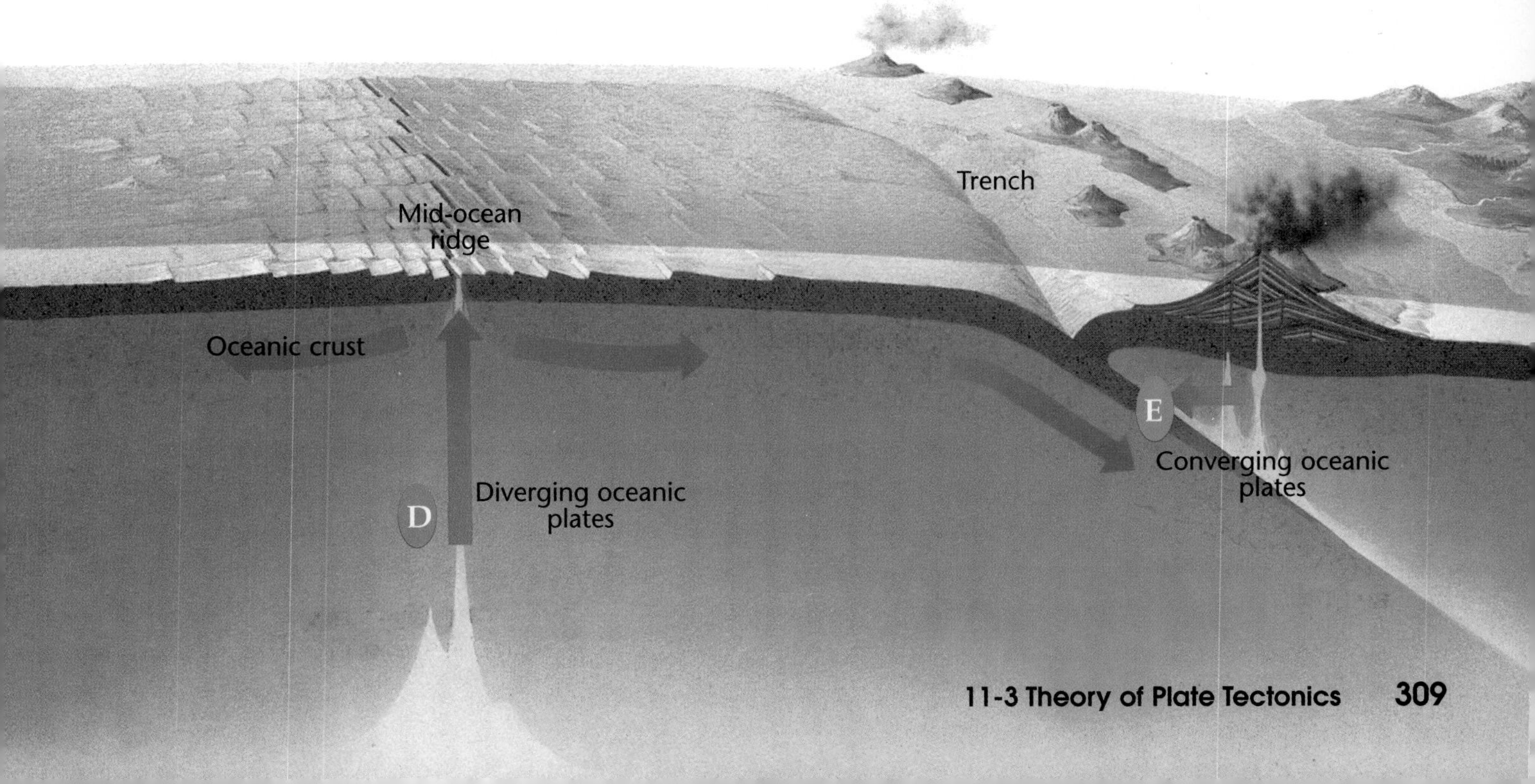

Trench
Mid-ocean ridge
Oceanic crust
E
Converging oceanic plates
D
Diverging oceanic plates

MiniLAB

How do convection currents form?

Procedure 

1. Fill a clear, colorless casserole dish with water to 5 cm from the top.
2. Center the dish on a hot plate and heat. **CAUTION:** *Wear thermal mitts to protect your hands.*
3. Add a few drops of food coloring to the water directly above the hot plate.
4. Observe what occurs in the water.
5. In your Science Journal, describe what you observe. If possible, make an illustration of what you observe.

Analysis

1. Determine whether any currents form in the water.
2. If so, infer what causes the currents to form.
3. If not, determine how to change the experiment in order to cause currents to form.
4. After receiving permission from your teacher, try to redesign the experiment and check to see if currents form.

Causes of Plate Tectonics

There have been many new discoveries about Earth's crust since Wegener's day. But one question still remains. What causes the plates to move and the seafloor to spread? Scientists now think they have a pretty good idea. They think that plates are moved by the same basic process that is used to heat some buildings.

Convection Currents

In a forced air heating system, air is warmed in the furnace and a blower forces it into each room of the building. The air is forced upward from the register and releases its heat to surrounding air. The cooler air, which is more dense, sinks to the floor of the room. It returns to the furnace, through the cold air return, to be reheated. This entire cycle of heating, rising, cooling, and sinking is called a **convection current.** This same process occurring in the mantle is thought to be the force behind plate tectonics.

Scientists believe that differences in density cause hot, plasticlike rock in the asthenosphere to be forced upward toward the surface. When this plasticlike rock reaches Earth's lithosphere, it moves horizontally and carries plates of the lithosphere with it, as described earlier. As it cools, the plasticlike rock becomes denser. It then sinks into the mantle, taking overlying crust with it.

These huge convection cells provide the energy to move plates in the lithosphere, as shown in **Figure 11-12.** They are also the cause of many of Earth's surface features.

Figure 11-11

The San Andreas Fault forms a transform-fault boundary where the Pacific Plate is sliding past the North American Plate.

A This photograph shows an aerial view of the San Andreas Fault.

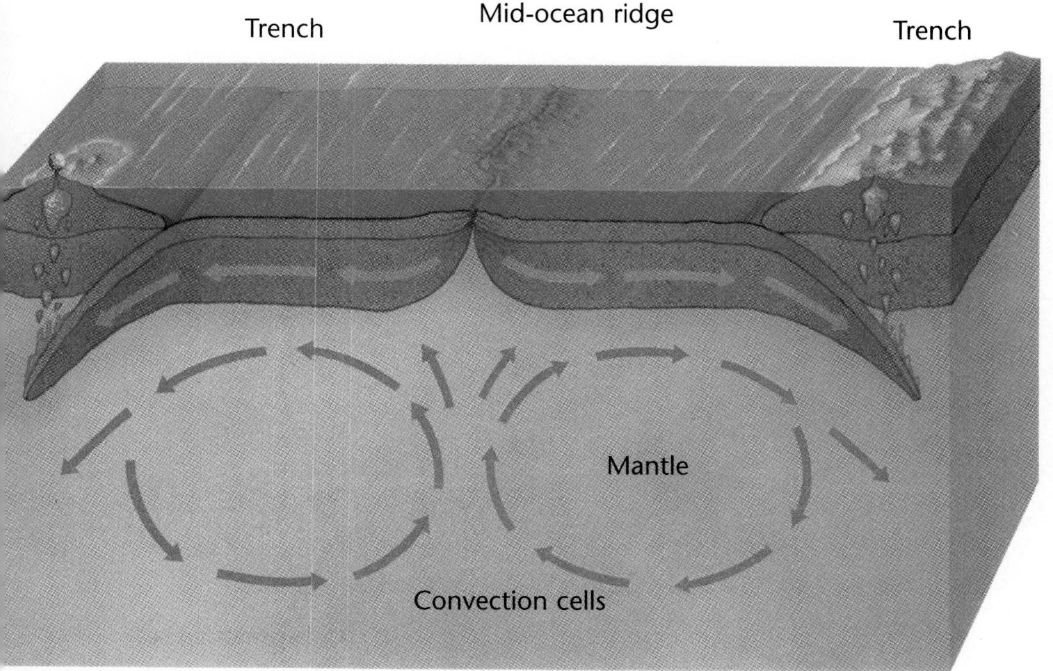

Figure 11-12

Pictured is one theory of how convection currents (see arrows) are the driving force of plate tectonics. In this theory, convection is limited to the upper mantle only. In another theory, convection currents occur throughout the mantle. In yet another, convection is confined to hot plumes being forced upward from near the core.

Effects of Plate Tectonics

Many of the landforms caused by plate tectonics are shown and described in **Figure 11-10** on pages 308-309. As you can see, many different landforms are caused by plate tectonics. However, two of the main effects of plate tectonics are volcanoes and earthquakes. Earth is a dynamic planet with convection currents inside that power the movement of plates. As the plates move, they interact. The interaction of plates produces forces that build mountains, rift ocean basins, and cause volcanoes and earthquakes.

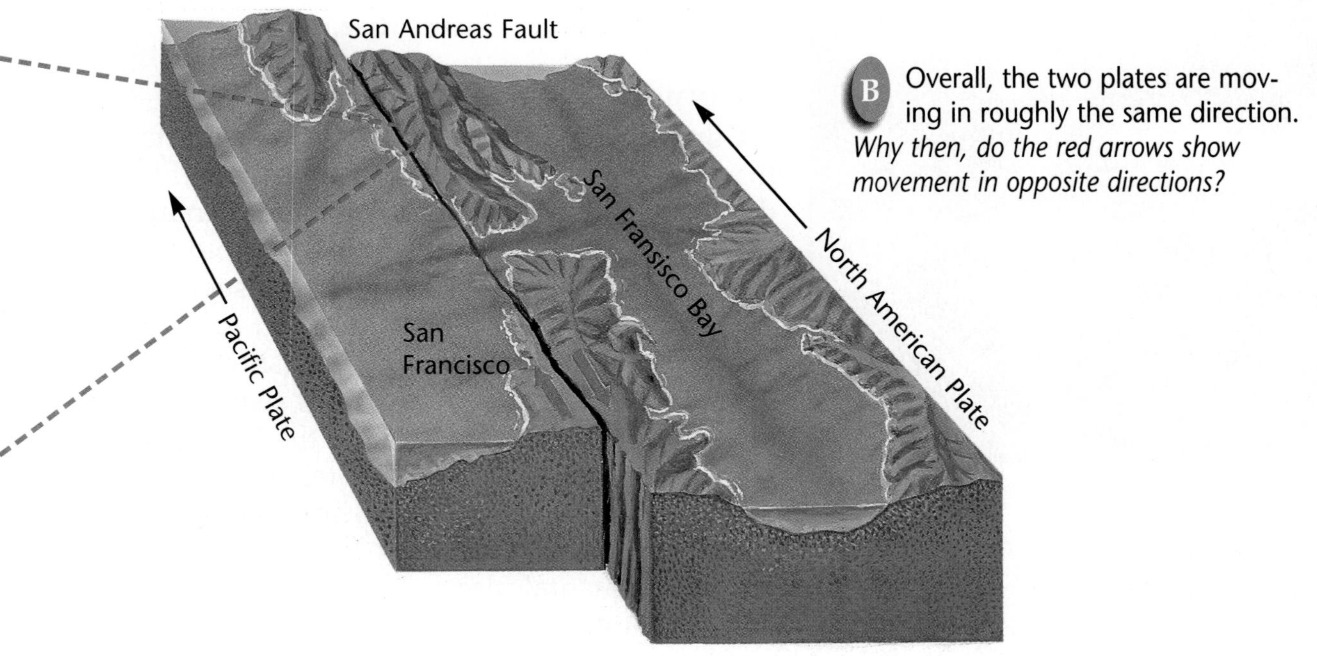

B Overall, the two plates are moving in roughly the same direction. *Why then, do the red arrows show movement in opposite directions?*

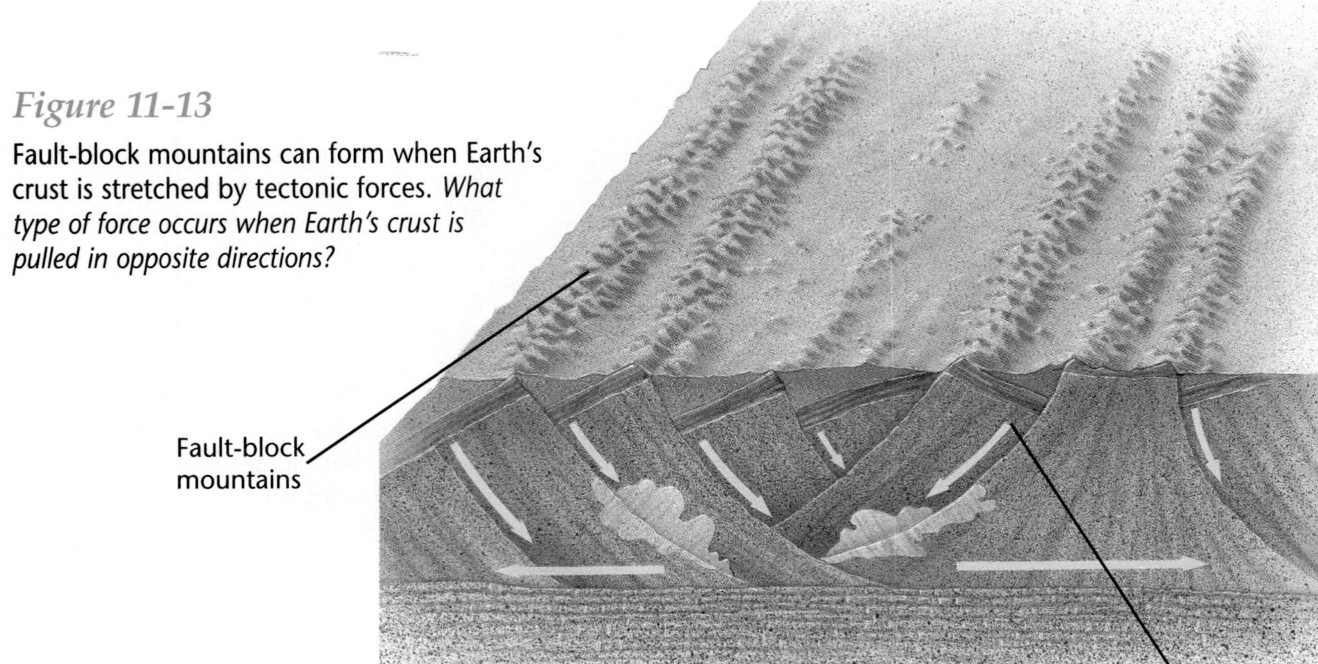

Figure 11-13

Fault-block mountains can form when Earth's crust is stretched by tectonic forces. *What type of force occurs when Earth's crust is pulled in opposite directions?*

Faults and Rift Valleys

Tension forces associated with diverging plates tend to stretch Earth's crust, causing fault blocks to tilt or slide down and form fault-block mountains, as shown in **Figure 11-13.** Generally, the faults that form from tension are normal faults. Once the divergence causes a separation, rift valleys can form. Examples of rift valleys are the Great Rift Valley in Africa, shown in **Figure 11-14,** and the valleys that occur in the middle of mid-ocean ridges, which are mountains that form on either side of rift valleys. Examples include the Mid-Atlantic Ridge and the East Pacific Rise.

Figure 11-14

The Great Rift Valley of Africa may become an ocean if the rifting continues.

Mountains, Arcs, and Volcanoes

Compression forces produce several effects that can take place when plates converge. If the converging plates are both continental, the forces generated cause massive folding of rock layers into mountain ranges such as the Himalayan Mountains or the greatly eroded Appalachian Mountains, shown in **Figure 11-15.** Reverse faults may also occur if the forces are great enough. If the two converging plates are oceanic plates, one plate slides under the other and island arcs and volcanoes form. If an oceanic plate converges on a continental plate, the oceanic plate slides under the continental plate, melting occurs, and volcanoes form. Entire mountain ranges can form at this type of convergent boundary.

Strike-Slip Faults

If one plate is sliding past another, the forces are not directly opposite. The plates stick and then slide along massive strike-slip faults. One such example is the San Andreas Fault. When plates move suddenly, vibrations are generated inside Earth that are felt as an earthquake. Plate tectonics explains how activity inside Earth can affect Earth's crust differently in different locations. We have seen how plates have moved since Pangaea separated. What was Earth like before that?

CONNECT TO

PHYSICS

In what directions are forces applied at convergent, divergent, and transform fault boundaries? ***Demonstrate*** these forces using wooden blocks or your hands.

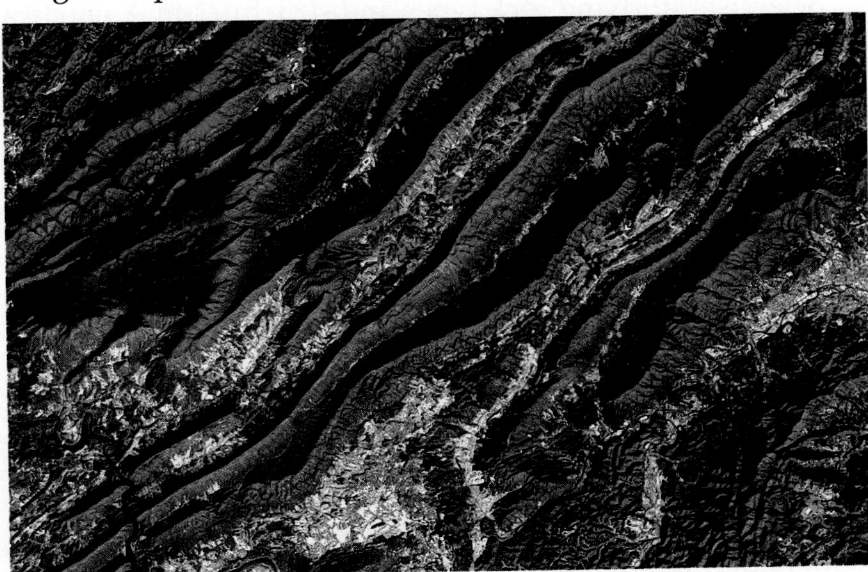

Figure 11-15

Satellite photo of Valley and Ridge Province in the Appalachian Mountains.

Section Wrap-up

Review

1. What happens to plates at a transform fault boundary?
2. What occurs at plate boundaries associated with seafloor spreading?
3. Think Critically: Using **Figure 11-9** and Appendix G, determine what natural disasters might be likely to occur in Iceland.

Skill Builder

Recognizing Cause and Effect

What causes a divergent boundary to form? What is the effect of collision on the edges of continental plates? Answer these questions, then write a cause-and-effect statement explaining the relationship between convection currents and plate movement. If you need help, refer to Recognizing Cause and Effect in the **Skill Handbook.**

Science Journal

Research the 1992 Landers, California earthquake. In your Science Journal write a brief summary explaining why this earthquake is so important to the study of plate tectonics and plate boundaries.

SCIENCE & SOCIETY

11•4 TECHNOLOGY: Before Pangaea, Rodinia

Objectives

- Describe evidence in support of the separation of the North American Plate from the Antarctic Plate.
- Track the North American Plate after separation from the Antarctic Plate and before the formation of Pangaea.

Antarctic Collisions

Scientists studying rock types and structures in Antarctica use computer models to show how this new evidence supports formation of a supercontinent long before Pangaea. Evidence presented in the computer model indicates that this landmass, named Rodinia, formed more than 750 million years ago.

Fossil Clues

As Rodinia broke apart, a rift valley formed in an area now occupied by the Pensacola Mountains in Antarctica. The rift deepened and rivers poured in water containing sediment. Shallow seas formed over the area of the rift. The presence of sandstone and fossils of the warm-water animal *Archaeocyatha* found in rocks dating back to the Cambrian Period suggest a separation of East Antarctica from another continental landmass, probably the North American Plate, sometime before 540 million years ago.

Quartz sandstones found in the area are full of worm burrows known as *Skolithus*. These trace fossils of ancient filter feeders are very similar to fossils found in western North America. This supports the theory that the North American Plate was joined to the Antarctic Plate sometime before 540 million years ago.

Figure 11-16

The Transantarctic Mountains, shown here, mark a very old plate boundary between East Antarctica and another continental plate.

Magnetic, Mountain, and Margin Clues

Whenever paleomagnetic evidence from North American and African rocks dated before the Mesozoic Era is entered into the computer model, movement of these two continents is different from that measured since Pangaea broke up. When data from volcanic rocks and early Paleozoic fossils located in the Appalachian Mountains and in mountains of Argentina and Africa are entered into the computer model, the model illustrates that the Atlantic Ocean has opened more than once and that movement of these landmasses has changed.

The computer model now indicates that Rodinia split apart about 750 million years ago. The continental plates then moved around and came back together as Pangaea about 260 million years ago. Pangaea then split apart and the continental plates began moving to their present positions.

A 750 million years ago

North America on the Move

Simulations on the computer model indicate that the North American Plate broke away from East Antarctica and then moved up the western side of the South American Plate, as shown in **Figure 11-17.** During this journey the eastern side of the North American Plate evidently collided with the western side of the South American Plate several times before finally colliding with Africa in the formation of Pangaea. Each time a collision occurred, some rock containing fossils of one continent would be left on the other continent. Also, mountain ranges may have formed in this way, providing the evidence used in present-day computer simulations.

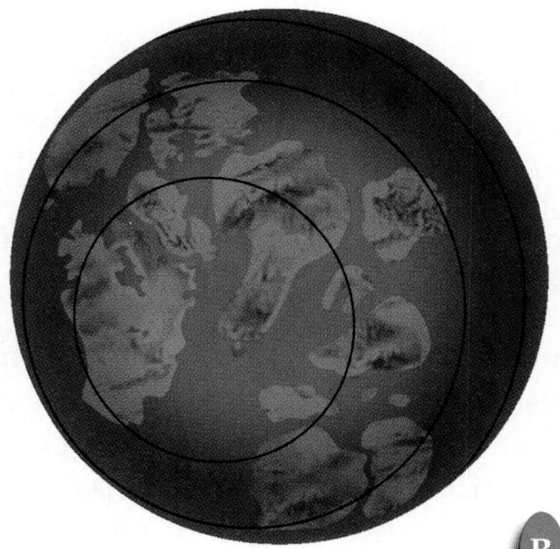

B 550 million years ago

Figure 11-17

Evidence indicates that a supercontinent, called Rodinia, existed long before Pangaea. *Why is it likely that rocks and rock formations in eastern North America are similar to rocks and rock formations found in western South America?*

Section Wrap-up

Review

1. What two continental plates were probably joined, forming the Transantarctic Mountains?
2. List two pieces of evidence supporting the theory that a supercontinent formed long before Pangaea.

Explore the Technology

A great deal of data were needed before these new ideas about an earlier supercontinent could be presented. In fact, research is still going on concerning the interpretation of the collected data. The use of computer simulations of continental movement helps make this study easier to understand. Based on what you have learned and what you know about computer simulations, how do you think this process is accomplished?

People and Science

VICKI HANSEN, *Associate Professor of Geological Science*

On the Job

Q What are you trying to learn about plate tectonics on Earth?

A I'm very interested in what happens at convergent boundaries, at depths deeper than where earthquakes occur (10 to 15 km). I want to know how things move in these environments deep inside Earth.

Q How can you study something that takes place so deep in Earth?

A You look at old mountain belts that preserve the deep parts of convergent boundaries. When you look at the rocks in these belts, they have beautiful patterns, because they flowed like putty when they were deep inside Earth. The patterns in those rocks are used to figure out how that part of the crust was flowing.

Q Explain how your studies of Venus led you to reject one theory and come up with another one.

A Some people proposed that deep troughs on Venus are subduction zones, as on Earth. However, the deep troughs are circular, and their shapes and fracture patterns are more easily explained by a theory in which these features are blisters in the crust, formed by hot plumes from deep inside, bubbling up. As we look at more data, this theory holds up better than a subduction theory.

Personal Insights

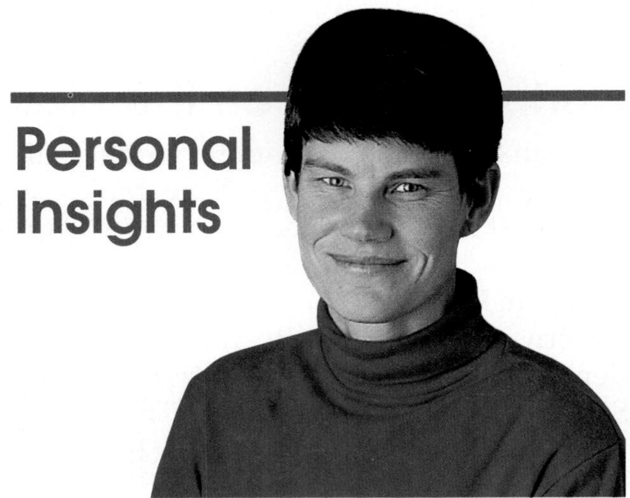

Q Your work has taken some unexpected turns. What has that taught you?

A I never expected to be studying Venus. But that allowed me to come to it with a fresh, childlike view. I think that's a very valuable thing to bring to science. A scientist is interested in puzzles, and you can't predict what shape those puzzles are going to come in. I hope in the next 10 years, I'll be working on something else I couldn't have predicted.

Career Connection

The U.S. Geological Survey has maps of Venus based on data from NASA's *Pioneer* and *Magellan* Venus orbiters. Look at one of these maps and compare the features of Venus and Earth.

Write questions about Earth you'd like the following scientists to answer:

- **Structural Geologist**
- **Seismologist**

Chapter 11

Review

Summary

11-1: Continental Drift

1. The theory of continental drift states that continents have moved to their present positions on Earth.
2. The puzzlelike fit of the continents, fossils, climatic evidence, and similar rock structures support Wegener's idea of continental drift.

11-2: Seafloor Spreading

1. Seafloor spreading is the spreading apart of the seafloor at the mid-ocean ridges.
2. Seafloor spreading is supported by magnetic evidence in rocks and in the age of rocks on the ocean floor.

11-3: Theory of Plate Tectonics

1. Plates move away from each other at divergent boundaries. Plates collide at convergent boundaries. At a transform fault, two plates move horizontally past each other.
2. Hot, plasticlike material from the mantle is forced upward to the lithosphere, moves horizontally, cools, and then sinks back into the mantle. The movement of this material sets up convection currents, the driving force of plate tectonics.
3. Convergent boundaries are the location of most mountain belts, volcanoes, and earthquakes. Mid-ocean ridges and rift zones occur at divergent boundaries. Major earthquakes occur at transform fault boundaries.

11-4: Science and Society: Before Pangaea, Rodinia

1. Fossil evidence in the Transantarctic Mountains and paleomagnetic data support this interpretation.
2. It is thought that the North American Plate broke away from an ancient supercontinent called Rodinia, moved around the other continental plates, and collided with them again to form Pangaea.

Key Science Words

a. asthenosphere
b. continental drift
c. convection current
d. convergent boundary
e. divergent boundary
f. lithosphere
g. magnetometer
h. Pangaea
i. plate tectonics
j. plates
k. seafloor spreading
l. subduction zone
m. transform fault

Reviewing Vocabulary

Match each phrase with the correct term from the list of Key Science Words.

1. instrument for studying Earth's magnetic field
2. plasticlike layer below the lithosphere
3. idea that continents moved to their current positions on Earth by plowing through the ocean floor
4. large landmass made of all continents
5. force behind plate tectonics
6. process that forms new seafloor
7. large sections of Earth's crust and upper mantle
8. boundary where plates move apart from each other
9. boundary at which plates collide
10. place where plates slide past one another

Chapter 11 Review

Checking Concepts

Choose the word or phrase that completes the sentence.

1. Earth's asthenosphere is found in the _______.
   a. crust c. outer core
   b. mantle d. inner core
2. The San Andreas Fault is a _______.
   a. divergent boundary
   b. subduction zone
   c. convergent boundary
   d. transform fault boundary
3. The theory of _______ states that continents moved to their present positions.
   a. subduction
   b. seafloor spreading
   c. continental drift
   d. erosion
4. During the break-up of Rodinia, East Antarctica is thought to have separated from the _______ Plate.
   a. South American c. Indo-Australian
   b. African d. North American
5. Evidence from _______ indicates that many continents were near Earth's South Pole.
   a. glaciers
   b. mid-ocean ridges
   c. volcanoes
   d. plates
6. _______ of iron in rocks supports the theory of seafloor spreading.
   a. Plate movement
   b. Subduction
   c. Magnetic alignment
   d. Weathering
7. The Great Rift Valley is a _______ boundary.
   a. convergent c. transform fault
   b. divergent d. lithosphere
8. The theory of _______ states that plates move around on the asthenosphere.
   a. continental drift
   b. seafloor spreading
   c. subduction
   d. plate tectonics
9. A _______ forms when a plate slides under another plate.
   a. transform fault
   b. divergent boundary
   c. subduction zone
   d. mid-ocean ridge
10. When two oceanic plates collide, _______ form.
    a. folded mountains
    b. island arcs
    c. transform faults
    d. mid-ocean ridges

Understanding Concepts

Answer the following questions in your Science Journal using complete sentences.

11. Compare and contrast continental drift and plate tectonics.
12. Why was the theory of continental drift initially not accepted by many scientists?
13. Explain the relationship between a mid-ocean ridge and seafloor spreading.
14. Compare and contrast divergent, convergent, and transform fault boundaries.
15. Explain how island arcs form.

Thinking Critically

16. Why are there few volcanoes in the Himalayas, but many earthquakes?
17. Glacial deposits often form at high latitudes near the poles. Explain why glacial deposits have been found in Africa.
18. How is magnetism used to support the theory of seafloor spreading?
19. Explain why volcanoes do not form along the San Andreas Fault.
20. Why wouldn't the fossil of an ocean fish found on two different continents be good evidence of continental drift?

Developing Skills

If you need help, refer to the **Skill Handbook.**

21. **Hypothesizing:** Mount St. Helens in the Cascade Mountain Range is a volcano. Use **Figure 11-9** and Appendix H to hypothesize how it may have formed.
22. **Outlining:** Outline the major points in Section 11-1.
23. **Measuring in SI:** Movement along the African Rift Valley is about 2.1 cm per year. If plates continue to move apart at this rate, how large will the rift be (in meters) in 1000 years? In 15 500 years?
24. **Comparing and Contrasting:** Compare and contrast the formation of the Andes Mountains and the Himalayas, as shown below.
25. **Concept Mapping:** Make an events chain concept map that describes seafloor spreading.

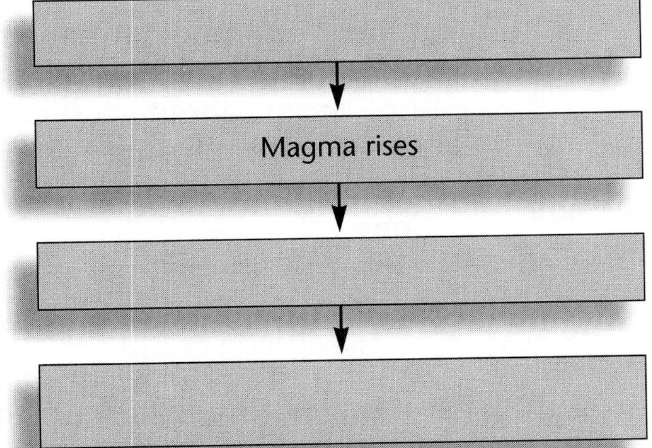

Performance Assessment

1. **Asking Questions:** In Activity 11-1, you constructed a model that supported the theory of seafloor spreading. Based on the model you constructed, what additional information about the seafloor would be helpful to add support to the ideas presented in your model?
2. **Making Observations and Inferences:** In the MiniLAB on page 310, you observed convection currents produced in water as it was heated. Try the experiment again placing pieces of wood, small pieces of rubber bands, or sequins into the water. Do their movements support your observations and inferences from the MiniLAB?
3. **Model:** Use plywood and a map of the world to construct a working model of plate tectonics. Use a sheet of plywood as a base and attach movable wood continents. When the model is complete, the continents should move from Pangaea to their present positions.

UNIT PROJECT 3

Researching Via the Internet

How do you use resources to find out more about a topic in which you are interested? In the past, you might have gone to the library and used a card catalog or periodical listing to find books, magazine and newspaper articles that were related to your topic. Now, at most libraries, you can find these same book and article listings on a computer. Many libraries have put their holdings in one large database to make research less time consuming and more successful.

Within the last few years, a new and more powerful source for information retrieval has become widely available. This source of information is called the Internet. If you have a computer or have access to a school computer, you can do research on the Internet without ever leaving your chair.

What is the Internet?

The internet is made up of millions of computers all around the world that are linked to each other. The information available in these computers is varied and vast. You can find information on almost any imaginable topic. Using electronic mail, or e-mail, you can speak directly to an expert in a particular field or to a person in another country. Some computers store articles or entire books about particular subjects. Others contain thousands of photographs and illustrations.

Getting Started

To use the Internet you must have a way to access it. Many schools have direct access to the Internet or have a connection through a local college or university. A phone line can also give you access by way of a commercial server. Each server differs slightly in the way it works. Time, practice, and perhaps some cooperative work with a few friends will help you locate the information you need.

Reference books on using the Internet will be helpful as will consulting with persons already proficient at research using the Internet.

Beginning Your Search

For this project you will use the Internet to find out more about earthquakes or volcanoes. Decide on a topic you wish to research and report on to the rest of your class. You may choose to investigate a particular earthquake or volcano, or a more general topic such as seismographs or volcano prediction. Make sure your teacher approves your research topic before beginning. Pick a topic that isn't too broad or too narrow.

Internet Yellow Pages

An Internet yellow pages contains the Internet addresses for locating information on any imaginable topic. However, because no one is in charge of the Internet, an Internet address may be available for a time, and then the people managing the address may decide to remove it.

Internet Search Tips

One important method for finding information is a search tool. A search tool allows you to type in a key word or words. The tool then searches thousands of sites on the Internet for those that contain your key words. After a few seconds, the search tool returns many addresses for other links related to your topic. The gopher is one type of search tool. There are thousands of gopher servers available on the Internet.

Making Your Presentation

After you've completed your search for information and gathered all of your data, you should be ready to prepare a presentation for your class. Your presentation should be put together using a multimedia computer program. Your presentation should consist of at least five different screens with text, graphics, and perhaps sound and photo. Design each screen so that it is appealing to the audience. Your report should include information from at least five sites on the Internet. Identify your sites and explain to the class how you got to them.

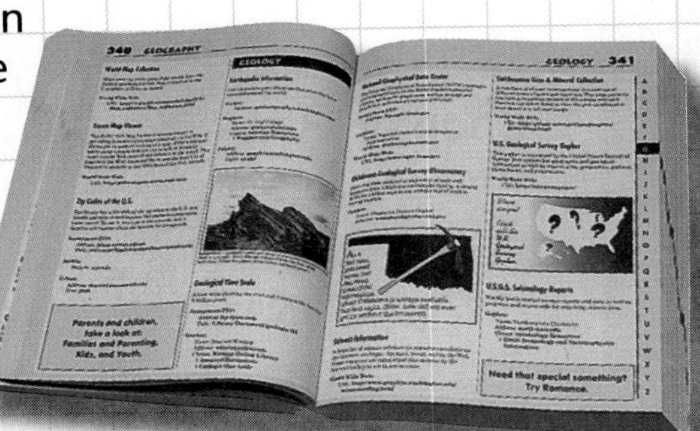

A word of caution. Not everything on the Internet is valid. Much of the information has not been edited, reviewed, or even verified as true. If you have questions about anything you find, ask a parent, teacher, or other adult to help you determine if it is useful or not.

UNIT 4

Unit Contents

Change and Earth's History

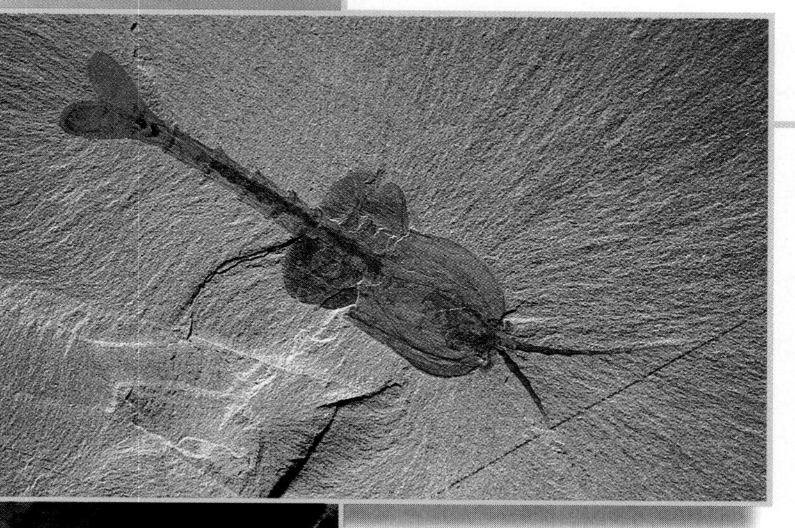

What's Happening Here?

The spiral shown here is a petroglyph or rock drawing. Thousands of years ago, humans etched these figures into stones, thus taking the first steps toward creating a written language. Though this may seem like a long time ago, those years represent less than one percent of Earth's history.

What was happening on Earth during the other 99 percent of time? Oceans covered much of the planet, glaciers formed then receded, and the continents broke apart. Some 600 million years ago, fossils of the first marine animals, such as the arthropod shown in the smaller photo, made an abrupt appearance in the fossil record. In this unit, you will explore this mysterious occurrence and other changes that took place during Earth's vast geological history.

Science Journal

In the library, research the geologic eras that make up Earth's history. In your Science Journal, describe some of the major events that occurred during each period.

Previewing the Chapter

Clues to Earth's Past

Pictured here is a fossil ammonite. This marine invertebrate animal lived during the Cretaceous Period. Finding particular fossils indicates the age of the rock in which they are found. What else do you think we could learn from fossils? How do they form? What evidence of their past life do we have?

EXPLORE ACTIVITY

Make a model of a fossil to see one way they might form. 

1. Cut the top off of a small milk carton and add enough plaster of paris to fill it halfway.
2. Mix enough water with the plaster so that it's smooth and thick.
3. Coat a leaf, shell, or bone with a thin layer of petroleum jelly.
4. Press it into the plaster.
5. Allow the plaster to dry at least 24 hours and then remove the leaf, shell, or bone.

Observe: In your Science Journal, discuss how the imprints compare with the original object. Can you determine, from the imprints alone, what object made them? How do you think imprints of once-living organisms are made?

Previewing Science Skills

- In the **Skill Builders**, you will **map concepts**, **interpret data**, and **make and use tables**.
- In the **Activities**, you will **interpret data**, **analyze**, and **conclude**.
- In the **MiniLABs**, you will **observe and infer**.

12•1 Fossils

Science Words

fossil
petrified remains
carbonaceous film
mold
cast
index fossil

Objectives

- List the conditions necessary for fossils to form.
- Describe processes of fossil formation.
- Explain how fossil correlation is used to determine rock ages.

USING MATH

Scientists often study the step angle made by animal footprints.

Step angle

Design a way to find your step angle.

Traces from Our Past

The dense forest reverberates as an *Allosaurus* charges forward in pursuit of an evening meal. On the other side of the swamp, a herd of apatosaurs moves slowly and cautiously onward. The adults surround the young to protect them from predators. Soon, night will fall on this prehistoric day, 160 million years ago.

Does this scene sound familiar to you? It's likely that you've read about dinosaurs and other past inhabitants of Earth before. But how do you know they really existed? What evidence do we have of past life on Earth? Scientists reconstruct what an animal looked like from its fossil remains, as in **Figure 12-1.**

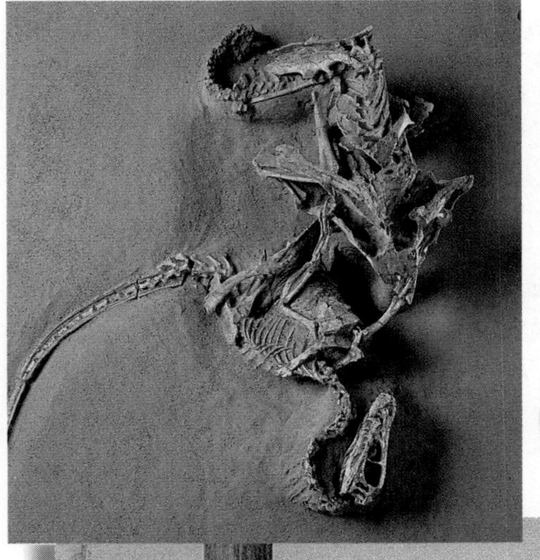

Figure 12-1

Scientists and artists can reconstruct what dinosaurs looked like using fossil remains (A). The fossils being reconstructed are a *Velociraptor* attacking a *Protoceratops* (B). The fossils were actually found in this position by paleontologists.

Figure 12-2
The hard parts of this fossil leaf made its preservation possible.

Fossil Formation

In the Explore Activity on page 325, you made imprints of parts of organisms. The imprints are records, or evidence, of life. Evidence such as the remains, imprints, or traces of once-living organisms preserved in rocks are **fossils.** By studying fossils, geologists help solve mysteries of Earth's past. Fossils have helped geologists and biologists determine approximately when life began, when plants and animals first lived on land, and when certain types of organisms, such as the dinosaurs, disappeared. Fossils tell us not only *when* and *where* organisms once lived, but also *how* they lived.

Usually the remains of dead plants and animals are quickly destroyed. Scavengers eat the dead organisms, or fungi and microorganisms cause them to decay. If you've ever left a banana on the shelf too long, you've seen this process begin. Compounds in the banana cause it to become soft and moist, and microorganisms move in and cause it to decay quickly. What keeps some plants and animals from decaying so they become fossils?

Necessary Conditions

First of all, to become a fossil, the body of a dead organism must be protected from scavengers and microorganisms. One way this can occur is when the body is buried quickly by sediments. If a fish dies and sinks to the bottom of a pond, sediments carried into the pond by a stream will rapidly cover the fish. As a result, no animals or microorganisms can get to it. However, quick burial alone isn't enough to make a fossil.

Organisms have a better chance of being preserved if they have hard parts such as bones, shells, or teeth. As you may know, these hard parts are less likely to be eaten by other organisms, they decay more slowly, and they are less likely to weather away. Most fossils, such as the fossil leaf in **Figure 12-2,** are composed of the hard parts of organisms. Fossils are usually found in sedimentary rocks. The heat and pressure involved in forming igneous and metamorphic rocks often destroy fossil material.

MiniLAB

What type of fossils might be preserved from our time?

Procedure

1. Take a brief walk outside and observe the area near your school.
2. Look around and notice what type of litter has been discarded on the school grounds. Note whether there is a paved road near your school. Note anything else that is human-made.

Analysis

1. Predict what human-made or natural objects from our time might be preserved far into the future.
2. Explain what conditions would need to exist for these objects to be preserved as fossils.

Figure 12-3

Much of the original matter in these petrified plant remains has been replaced by quartz and other minerals. *Why have the fossils retained the shape of the original plant?*

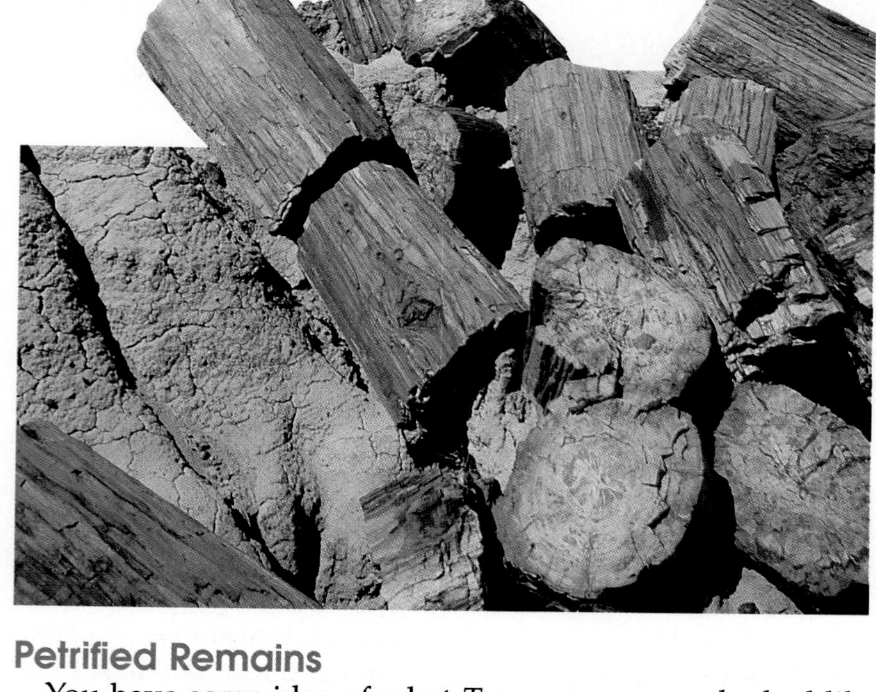

Petrified Remains

You have some idea of what *Tyrannosaurus rex* looked like because you've seen illustrations of this dinosaur. Perhaps you've seen skeletal remains of other dinosaurs towering above you in museums. Artists who draw *Tyrannosaurus rex* and other dinosaurs base their illustrations on fossil bones. These bones are usually petrified.

Petrified remains are hard and rocklike. Some or all of the original materials in the remains have been replaced by minerals. For example, a solution of water and dissolved quartz may flow through the bones of a dead organism. The water dissolves the calcium in the bone and deposits quartz in its place. Quartz is harder than calcium, so the petrified bone is rocklike.

We learn about past life-forms from bones, wood, and other remains that become petrified, like those in **Figure 12-3.** But there are many other types of fossils to consider.

Figure 12-4

This is a fossil that has been preserved as a carbonaceous film. This organism, called *eurypterid,* lived hundreds of millions of years ago and was a major predator in the seas.

Carbonaceous Films

The tissues of most organisms are made of compounds that contain carbon. Sometimes, the only fossil remains of a dead plant or animal is this carbon. As you know, fossils usually form when a dead organism is buried in sediments. As more and more sediments pile up, the organism is subjected to pressure and heat. These conditions force gases and liquids from the body. A thin film of carbon residue is left, forming an outline of the original organism, resulting in a type of fossil called a **carbonaceous film.** The process of chemically changing organic material is called carbonization, and is shown in **Figure 12-4.**

In swamps and deltas, large volumes of plant matter accumulate. Over millions of years, these deposits become completely carbonized, forming the sedimentary rock coal. Coal is more important as a source of fuel than as a fossil because the structure of the original plant is often lost when the coal forms.

Molds and Casts

Think again about the impressions in plaster of paris you made earlier. In nature, such impressions are made when seashells or other hard parts of organisms fall into soft sediments such as mud. The object and sediments are then buried by more sediments. Compaction and cementation turn the sediments into rock. Pores in the rock let water and air reach the shell or hard part and it then decays, leaving behind a cavity in the rock called a **mold.** Later, other sediments may fill in the cavity, harden into rock, and produce a **cast** of the original object, as shown in **Figure 12-5.**

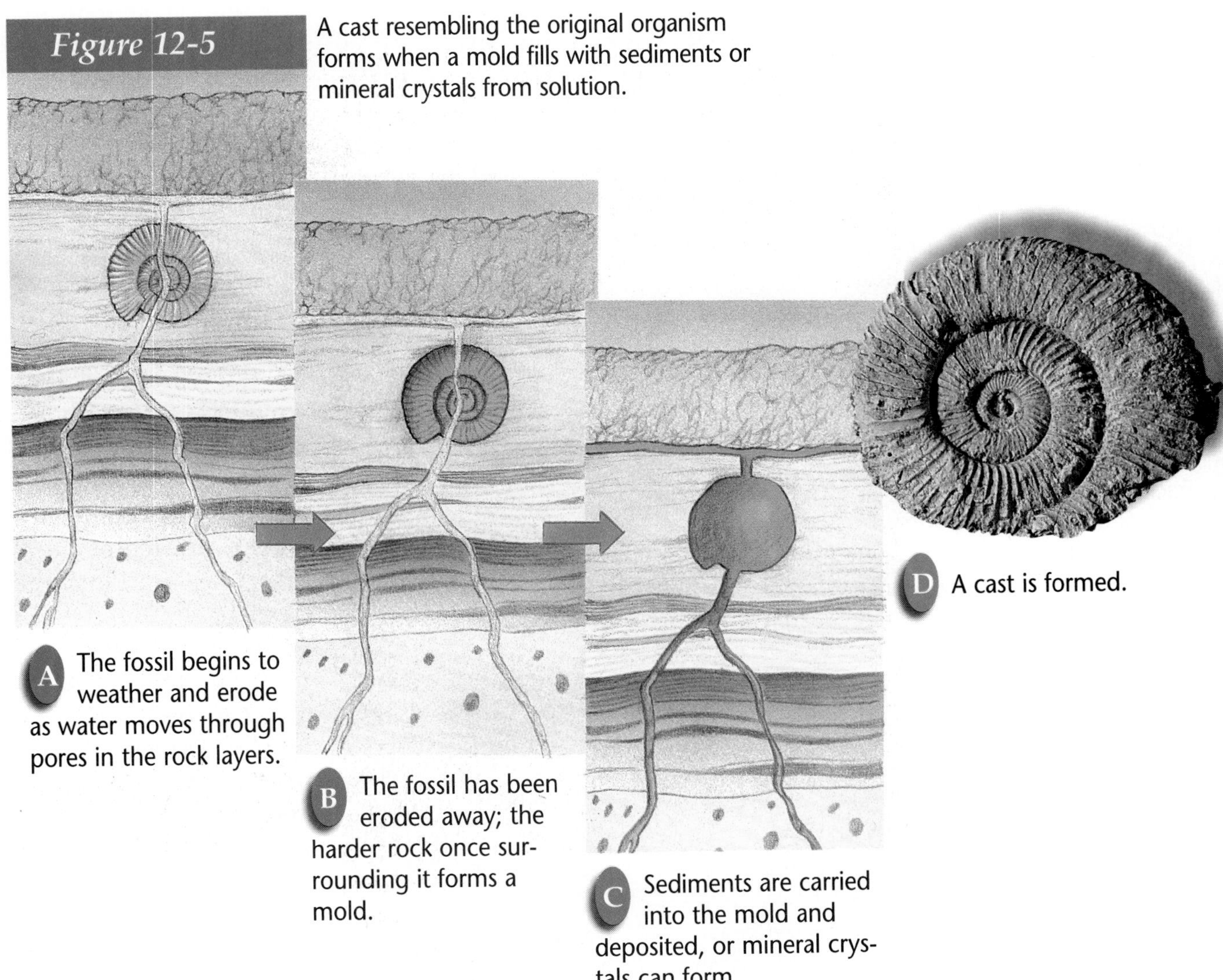

Figure 12-5 A cast resembling the original organism forms when a mold fills with sediments or mineral crystals from solution.

Figure 12-6

This 40-million-year-old grasshopper was trapped in the sticky resin produced by a plant. Over time, the resin crystallized into amber, preserving the insect inside.

Original Remains

Sometimes the actual organism or parts of the organism are found. **Figure 12-6** shows an insect trapped in amber, a crystallized form of the sticky resin produced by some trees. The amber protects the insect's body from decay and petrification. Other organisms, such as woolly mammoths, have been found preserved in frozen ground. In 1991, the completely intact body of a man who lived 5300 years ago was found frozen in glacial ice in the southern Alps. It is the oldest intact human body ever discovered. Original remains have also been found in tar seeps such as the La Brea tar pit in California.

Trace Fossils

Fossilized tracks and other evidence of animal activity are called *trace fossils.* Perhaps your parents made your handprint or footprint in plaster of paris when you were born. If so, it's a record that tells something about you. From it, we can guess your approximate size and maybe your weight at that age. Animals walking on Earth long ago have left similar tracks, such as those in **Figure 12-7.** In some cases, tracks can tell us more about how an organism lived than any other type of fossil.

For example, from a set of tracks at Davenport Ranch, Texas, we have learned something about the social life of *Apatosaurus,* one of the largest known dinosaurs. The largest tracks of the herd are on the outer edges and the smallest are on the inside. This suggests that the adult apatosaurs surrounded the young as they traveled—probably to protect them from enemies. In fact, a nearby set of allosaur tracks indicates that one was stalking the herd.

Other trace fossils include worm holes and burrows made by marine animals. These, too, tell us something about the lifestyle of these animals. As you can see, a combination of fossils can tell us a great deal about the individuals that inhabited Earth before us.

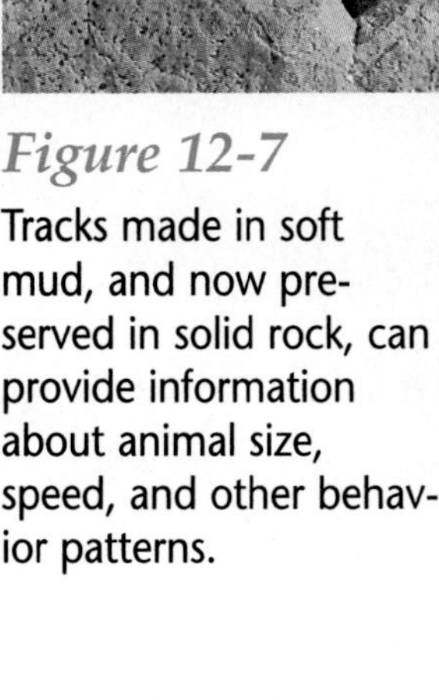

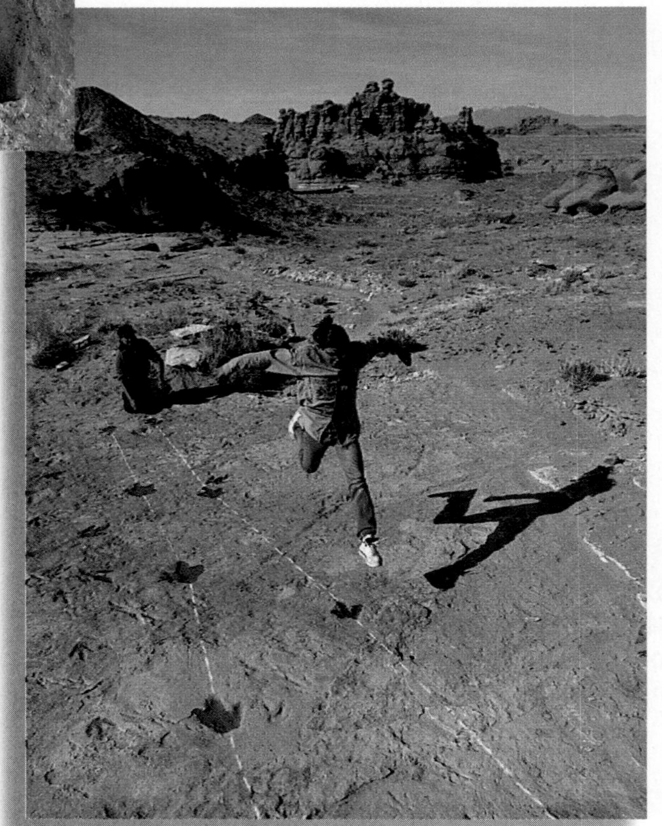

Figure 12-7

Tracks made in soft mud, and now preserved in solid rock, can provide information about animal size, speed, and other behavior patterns.

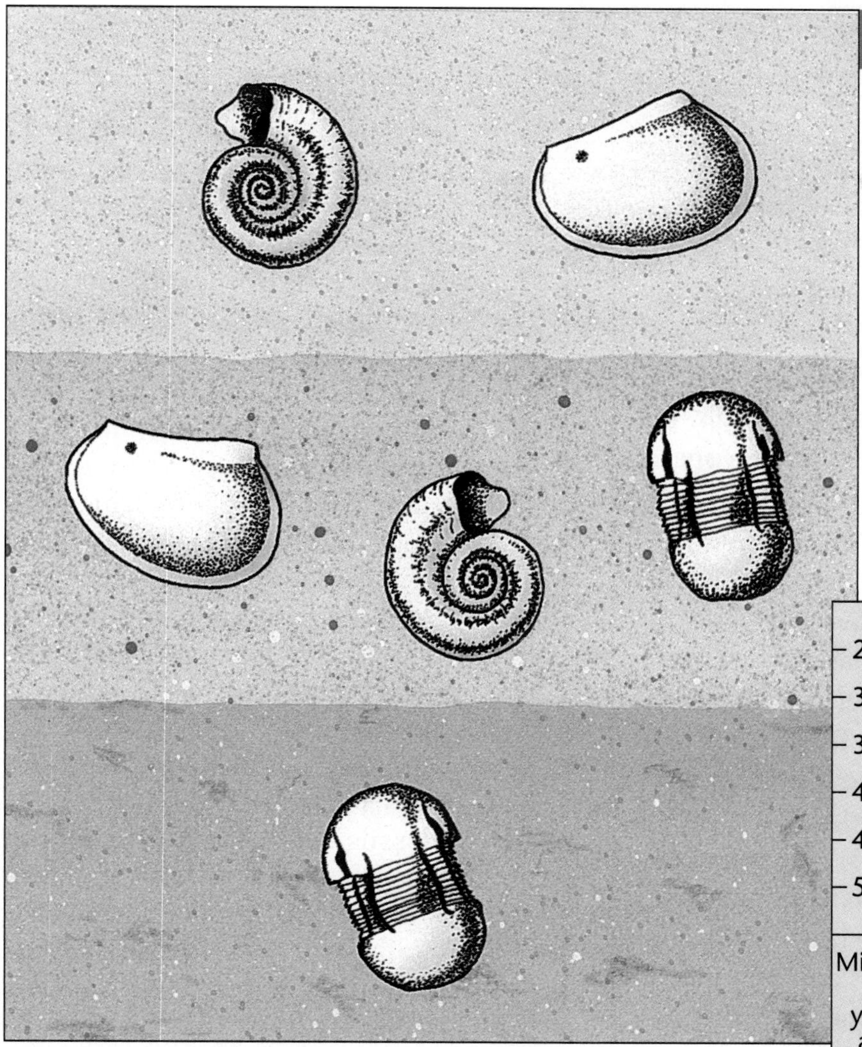

Figure 12-8

The sequence of sedimentary rock (A) and the fossils each contains can be used to date the rocks. The chart (B) shows when each organism inhabited Earth. *Why is it possible to say that the middle layer of rock had to be deposited between 438 and 408 million years ago?*

A Illustration of a sequence of rocks and the fossils they contain.

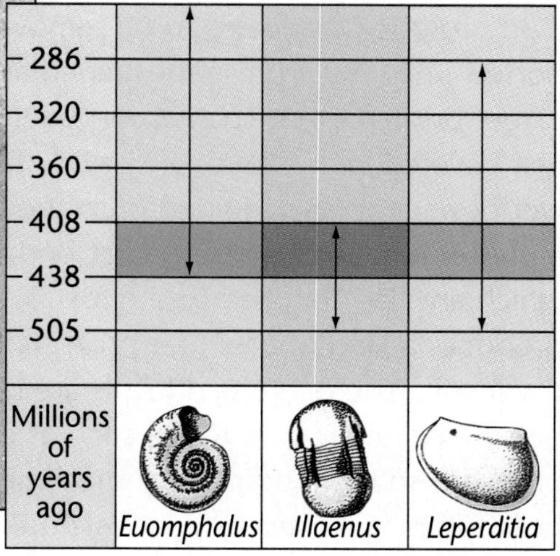

286
320
360
408
438
505
Millions of years ago
Euomphalus
Illaenus
Leperditia

B Fossil Range Chart

Index Fossils

One thing we've learned by studying fossils is that organisms are constantly changing, or evolving. Evidence indicates that species inhabit Earth for a certain period of time before they evolve into new species or they die out completely. Some species of organisms inhabit Earth for very long periods of time without changing much. Other species remain unchanged for only a short time. It is these organisms that produce index fossils.

Index fossils are from species that existed on Earth for relatively short periods of time, were abundant, and were widespread geographically. Scientists use index fossils to determine the age of rock layers. Because few fossils meet all the requirements to be an index fossil, groups of fossils are generally used to date rocks. This is how the rock layer in **Figure 12-8** was dated.

CONNECT TO

LIFE SCIENCE

You have learned that original remains of animals can be found in tar seeps. ***Hypothesize*** why so many animals became trapped in tar seeps.

USING TECHNOLOGY

Recovering Fossils

When scientists locate an area thought to have fossils, large equipment or explosives are used to remove overlying rocks and soil. Smaller tools are then used as the excavation nears the fossils. In the final phases of recovery, tiny picks and brushes are used to remove soil from the fossils.

Preparing Fossils

In order for the fossils to be removed or transported, they must be strengthened and protected. Large, brittle bones, such as dinosaur remains, are first covered with a layer of shellac. Then, strips of wet newspaper are molded onto the fossils. Finally, a plaster mixture is used to coat burlap strips, which are then applied to the fossils. This final step produces a kind of cast that protects and supports the fossils. The reinforced fossils are now ready to be transported. Some fossils such as dinosaur bones often are so large that heavy machinery, and even helicopters, are used to transport the fossils. They then go to museums or universities, where they will undergo final preparations before being exhibited.

Think Critically:

Why are remains such as dinosaur bones so fragile?

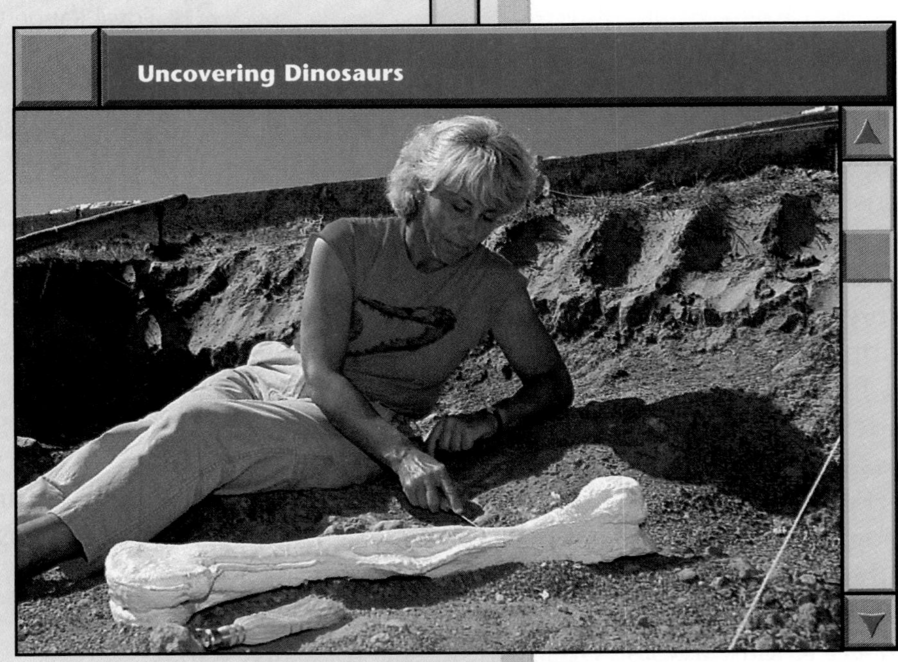

Uncovering Dinosaurs

Fossils and Ancient Environments

Fossils can also be used to determine what the environment of an area was like long ago. For example, rocks in Antarctica contain fossils of tropical plants. The environment of Antarctica today certainly isn't tropical, but we know that it was at the time these fossilized plants were living.

How would you explain the presence of fossilized brachiopods, animals that lived in shallow seas, in the rocks of the midwestern United States? A brachiopod fossil is shown in **Figure 12-9D.** The central portion of North America was covered by a shallow sea when the brachiopods were living, as shown in **Figure 12-9.**

Fossils tell us not only about past life on Earth, but also about the history of the rock layers that contain them. Fossils can provide information about environment, climate, and animal behavior, as well as dating the rocks.

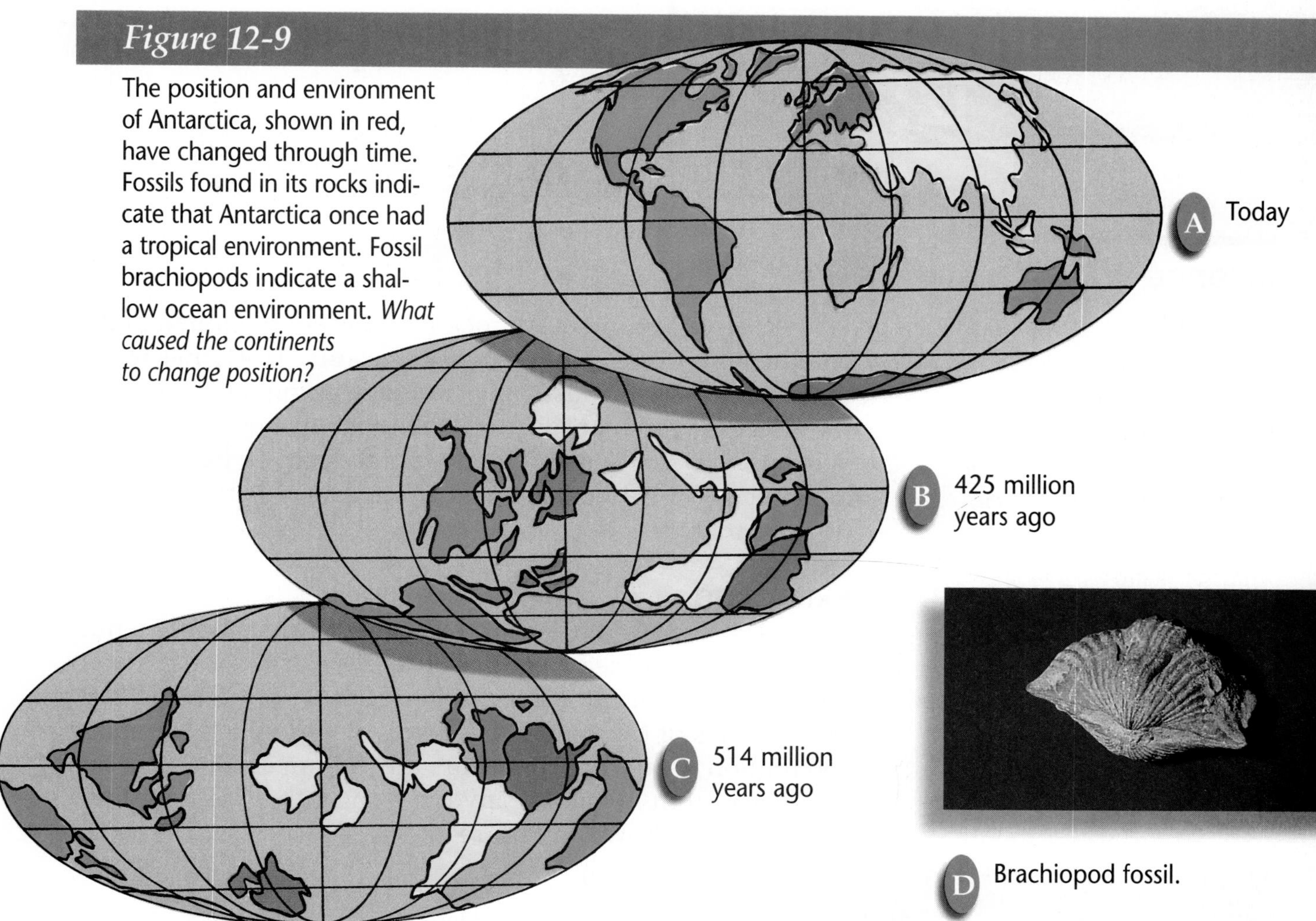

Figure 12-9

The position and environment of Antarctica, shown in red, have changed through time. Fossils found in its rocks indicate that Antarctica once had a tropical environment. Fossil brachiopods indicate a shallow ocean environment. *What caused the continents to change position?*

Section Wrap-up

Review

1. What conditions are needed for most fossils to form?
2. Describe how a mold and cast fossil might form.
3. **Think Critically:** What can be said about the ages of two widely separated layers of rock that contain the same type of fossil?

Skill Builder

Concept Mapping

Make a concept map that compares and contrasts petrified remains and original remains. Use the following terms and phrases: *types of fossils, original remains, evidence of former life, petrified remains, materials replaced by minerals,* and *actual parts of organisms.* If you need help, refer to Concept Mapping in the **Skill Handbook.**

Science Journal

Collect samples of fossils or visit a museum that has fossils on display. In your Science Journal, make an illustration of each fossil. Write a brief description, noting key facts about each. Also, write about how each fossil might have been formed.

SCIENCE & SOCIETY

12•2 ISSUE: Extinction of Dinosaurs

Science Words

extinct

Objectives

- Discuss the meteorite-impact hypothesis of dinosaur extinction.
- Describe several hypotheses about why dinosaurs became extinct.

What killed the dinosaurs?

In layers of sedimentary rock in the western United States rest the remains of thousands of dinosaurs. The bones tell us that dinosaurs were fast, agile, intelligent animals who ruled the land longer than any other organism before or since. Why are they no longer a dominant life-form on Earth? What happened to the dinosaurs, and what can we learn from their disappearance?

Extinction

The last species of dinosaurs became extinct about 66 million years ago. When a species becomes **extinct,** there are no longer any living members of its kind. Before their extinction, species of dinosaurs had dominated the land for about 160 million years. The dinosaurs are no longer on Earth, but we learn from them by investigating what caused their extinction, as shown in **Figure 12-10.**

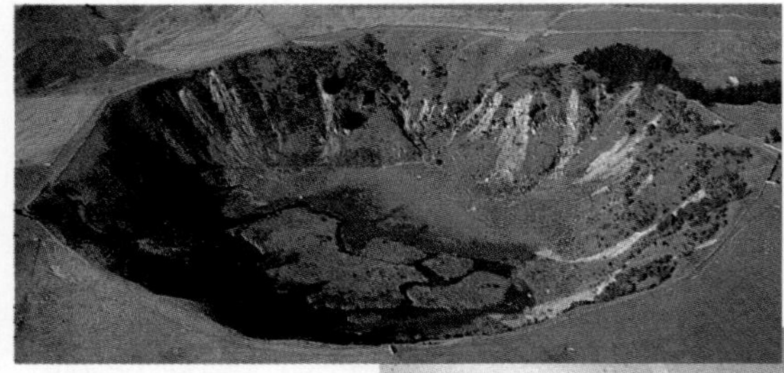

Figure 12-10

Evidence in the rock record indicates that a large meteorite may have struck Earth at about the same time the dinosaurs became extinct. The crater to the left is in New Zealand, and shows the force with which meteors can hit Earth.

2 Points of View

Meteorite Hypothesis

One of the many hypotheses about dinosaur extinction is that a large meteorite collided with Earth. The collision threw dust and debris into Earth's upper atmosphere. The collision also may have caused large forest fires that would have added smoke to the atmosphere.

If enough dust and smoke were released into Earth's upper atmosphere, the sun could have been completely blocked out. If the sun's energy remained blocked off for a long time, plants couldn't have carried on photosynthesis and would have died. With no food, plant-eating dinosaurs also would have died. With no plant-eating dinosaurs to prey on, meat-eating dinosaurs would have starved as well.

Figure 12-11

Volcanic eruptions such as this eject large amounts of ash, gas, and debris into the atmosphere.

Scientists have found evidence of an impact in a layer of clay in a column of sedimentary rock that was deposited at about the same time the dinosaurs became extinct. The layer contains small deformed grains of quartz very much like those found near meteorite craters elsewhere on Earth. But more importantly, the clay layer is rich in the element iridium. Iridium is rare on Earth's surface, but it is found in greater amounts in meteorites. The iridium in the clay layer may have come from the meteorite when it broke apart on impact.

Alternative Hypothesis

Some scientists disagree with the meteorite hypothesis and are looking for other explanations for dinosaur extinction. Perhaps the iridium-rich clay layer can be explained by large amounts of volcanic activity. The volcanic activity could have caused large amounts of dust to enter Earth's atmosphere, as shown in **Figure 12-11.** The rock record indicates that global temperatures started to decrease about 66 million years ago. Perhaps with colder temperatures, the dinosaurs could not adapt and eventually died.

Figure 12-12

Dinosaurs dominated Earth for 160 million years. But like us, they were dependent on their environment, and when it changed too drastically, they became extinct.

CONNECT TO

CHEMISTRY

Radiometric dating of tektites found near a buried crater on Mexico's Yucatan Peninsula indicates that the crater formed about 66 million years ago. If this age is correct, the meteorite impact may have been related to dinosaur extinction. ***Analyze*** what radiometric dating is and how it can be used to determine the tektites' age.

Another hypothesis suggests that there was nothing unusual about the extinction of dinosaurs and that they disappeared gradually because of slow changes in their environment. Perhaps these changes were brought about by shifts in the position of the continents caused by plate tectonics.

It's difficult to know for sure what caused Earth's environment to change 66 million years ago. But one thing that we have learned from the dinosaurs is that all organisms are dependent on their environments, as explained in **Figure 12-12.**

Section Wrap-up

Review

1. Discuss two theories that explain how iridium could have gotten into the clay layer deposited about 66 million years ago.
2. Explain how the inability of plants to photosynthesize would be harmful to the dinosaurs.

Explore the Issue

Often, the simplest explanation of an event is the most accurate. Based on this, how would you look at the issue of what caused dinosaur extinction? Which theory seems most likely to be correct? Explain your choice.

SCIENCE & SOCIETY

Relative Ages of Rocks

12 • 3

Science Words

principle of superposition
relative dating
unconformitiy

Objectives

- Describe several methods used to date rock layers relative to other rock layers.
- Interpret gaps in the rock record.
- Give an example of how rock layers may be correlated with other rock layers.

The Principle of Superposition

It's a hot summer day in July and you're getting ready to meet your friends at the local park. You put on your helmet and pads and grab your skateboard. But the bearings in one of the wheels are worn, and the wheel isn't spinning freely. You remember reading an article in a skateboarding magazine about how to replace wheels, and you decide to look it up. In your room is a stack of magazines from the past year, as seen in **Figure 12-13.** You know that the article came out in the January edition, so it must be near the bottom of the pile. As you dig downward, you find magazines from March, then February. January must be next.

How did you know that the January issue of the magazine would be on the bottom? To find the older edition under newer ones, you applied the principle of superposition.

Youngest Rocks on Top

The **principle of superposition** states that in an undisturbed layer of rock, the oldest rocks are on the bottom and the rocks become progressively younger toward the top. Why is this the case, and is it always true?

As you know, sediments are often deposited in horizontal beds, forming layers of sedimentary rock. The first layer to form is usually on the bottom. Each additional layer forms on top of the previous one. Unless forces, such as those generated by tectonic activity, overturn the layers, the oldest rocks are found at the bottom. When layers have been overturned, geologists use other clues in the rock layers to determine their original positions.

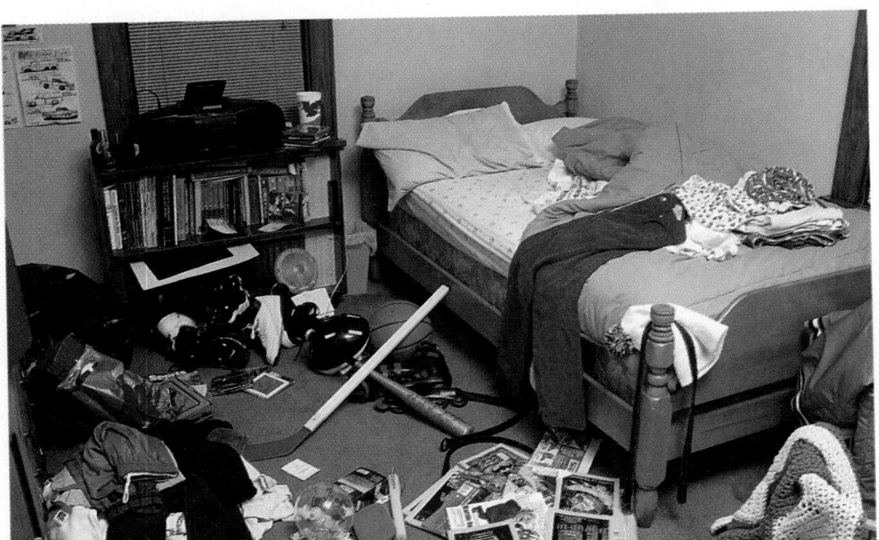

Figure 12-13

The pile of magazines illustrates the principle of superposition, which states that the oldest rock layer (or magazine) is on the bottom.

Relative Dating

Suppose you now want to look for another issue of a magazine. You're not sure exactly how old it is; all you know is that it arrived after the January issue. You can find it in the stack by using relative dating.

Relative dating is used in geology to determine the order of events and the relative age of rocks by examining the position of rocks in a sequence. For example, if layers of sedimentary rock are offset by a fault, you know that the layers had to be there first before a fault could cut through them. The relative age of the rocks is older than the relative age of the fault.

Relative dating doesn't tell you anything about the exact age of rock layers. You don't know if a layer is 100 million or 10 000 years old, only that it's younger than the layers below it and older than the fault cutting through it.

Figure 12-14

Illustration and photograph of a large-scale dome in sedimentary rocks showing the exposed rock layers. The oldest layers are folded up and exposed in the center.

Other Clues Help

Relative dating works well if rocks haven't been folded or overturned by tectonic processes. For example, look at **Figure 12-14.** Which layer is the oldest? In cases where rock layers

Figure 12-15

An angular unconformity results when horizontal layers overlie tilted layers.

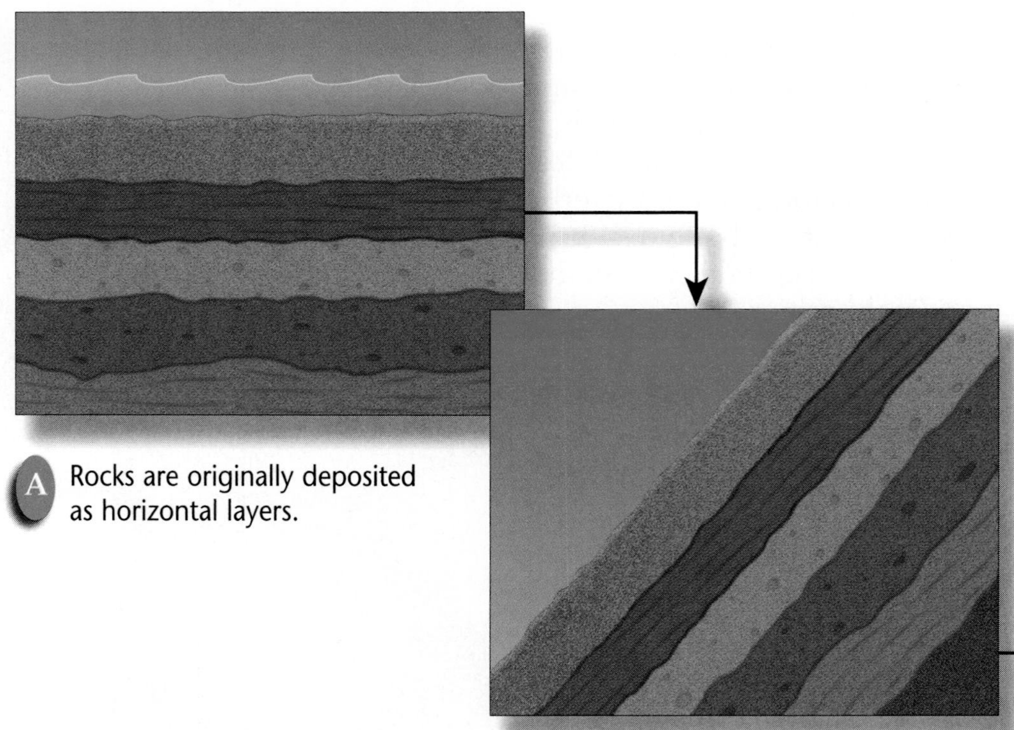

A Rocks are originally deposited as horizontal layers.

B The horizontal rock layers are tilted as they are deformed by tectonic forces.

have been disturbed, you may have to look for fossils and other clues to date the rocks. If you find a fossil in the top layer that's older than a fossil in a lower layer, you can hypothesize that layers have been overturned or faulted.

Unconformities

As you have seen, a layer of rock is a record of past events. But most rock records are incomplete—there are layers missing. These gaps in rock layers are called **unconformities.**

Unconformities develop when agents of erosion remove existing rock layers. They also form when a period of time passes without any new deposition occurring to form new layers of rock.

Angular Unconformities

Figure 12-15 illustrates one way an unconformity can form. Horizontal layers of sedimentary rock are tilted and uplifted, so that agents of erosion and weathering wear them down. Eventually, younger sediment layers are deposited horizontally on top of the eroded and tilted layers. Such an unconformity is called an angular unconformity.

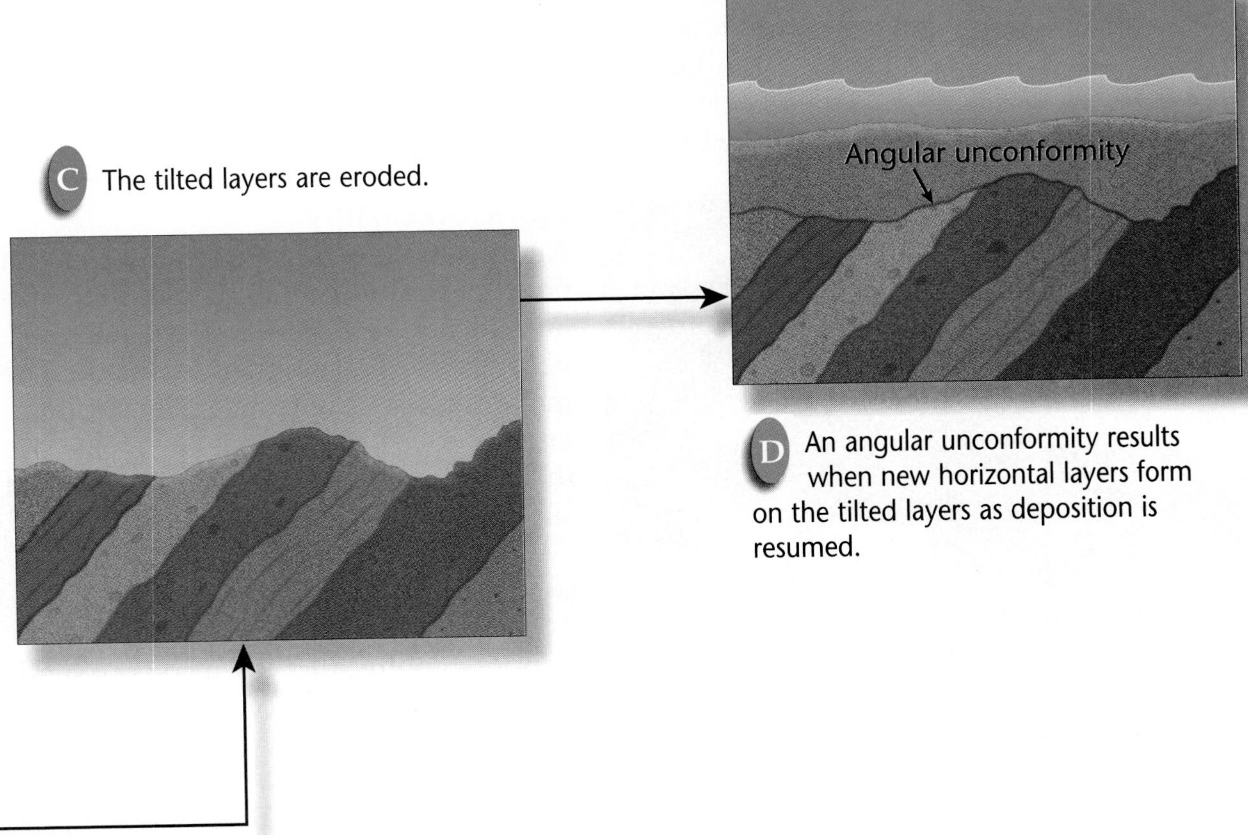

C The tilted layers are eroded.

D An angular unconformity results when new horizontal layers form on the tilted layers as deposition is resumed.

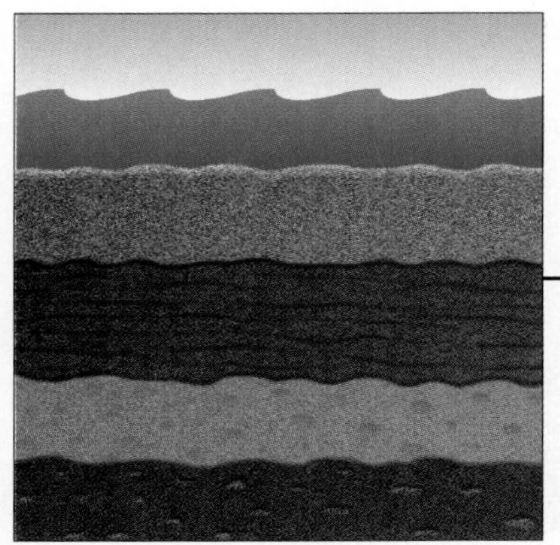

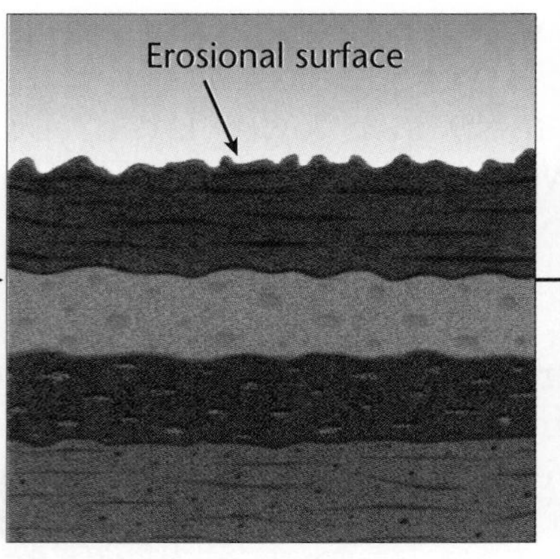

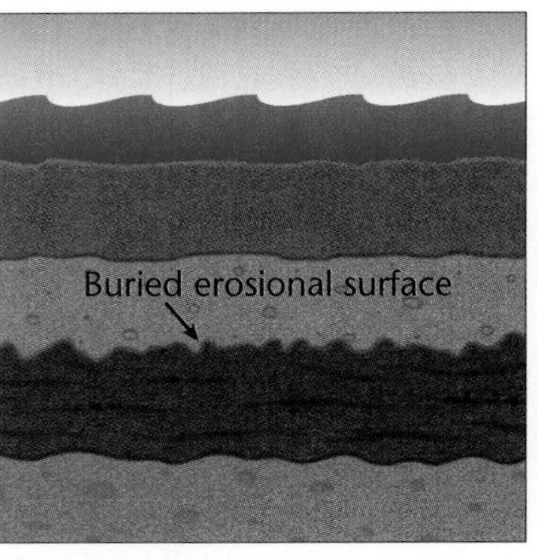

Sedimentary rock layers are deposited horizontally.

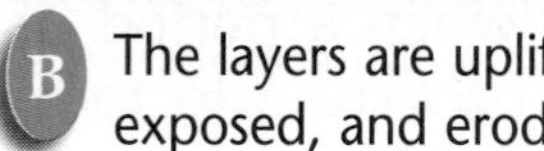
The layers are uplifted, exposed, and eroded.

C When deposition resumes, younger horizontal sediments are deposited on the buried erosional surface.

Figure 12-16

The buried erosional surface in the far right illustration is a disconformity. *How could you determine how much time and/or rock is missing?*

Disconformity

Suppose you're looking at a sequence of sedimentary rocks. They look complete, but actually, there are layers missing. If you look closely you may find an old erosional surface. This records a time when the rocks were exposed and eroded. Later, younger rocks formed above the erosional surface when sediment deposition began again. Even though all the layers are horizontal, there's still a gap in the record. This type of unconformity, called a disconformity, is illustrated in **Figure 12-16.**

Figure 12-17

These rock layers, exposed in the Grand Canyon, can be correlated across large areas of the western U.S.

Nonconformity

Another type of unconformity, called a nonconformity, occurs when sedimentary rock layers form above metamorphic or intrusive igneous rocks. The metamorphic or igneous rock is uplifted and eroded. Sedimentary rocks are then deposited on top of this erosional surface.

Correlating Rock Layers

Suppose you're studying a layer of sandstone in Bryce Canyon in Utah. Later, when you visit Canyonlands National Park, you notice that a layer of sandstone there looks just like the sandstone in Bryce Canyon, 250 kilometers away. Above the sandstone in the Canyonlands is a layer of limestone and then another sandstone layer. You return to Bryce Canyon and find the same sequence—sandstone, limestone, and sandstone. What do you conclude?

It's likely that you're actually looking at the same rocks at the two locations. These rocks are parts of huge deposits that covered this whole area of the western U.S. as seen in **Figure 12-17.** The sandstone and limestone you found at the two parks are the exposed surfaces of the same rock layers.

Evidence Used for Correlation

Geologists match up, or correlate layers of rocks over great distances, as seen in **Figure 12-18.** It's not always easy to say that a rock layer exposed in one area is the same as a rock layer exposed in another area. Sometimes it's possible to simply walk along the layer for kilometers and prove that it's a continuous unit. In other cases, such as at the Canyonlands area and Bryce Canyon, the rock layers are exposed only where rivers have cut down through overlying layers of rock and sediment. How can you prove that the limestone sandwiched between the two layers of sandstone in Canyonlands is the same limestone as at Bryce Canyon? One way is to use fossil evidence. If the same types of fossils are found in both outcrops of limestone, it's a good indication that the limestone at each location is the same age, and therefore, one continuous deposit.

Rock Correlation

Keiko and Masao spent part of their summer vacation on a field trip through Colorado and Utah. They observed many rock outcrops and recorded what they saw in journals. The geologic column on the left was drawn by Keiko from observations made in Green River, Utah. The column on the right was made by Masao from the data he collected in Westwater, Colorado. Help them reconstruct the geologic history of the area by answering the following questions.

Solve the Problem:

1. **How many unconformities, and what types, can you recognize in each column?**
2. **How are the rock types similar in the two locations?**

Think Critically:

Explain the geologic history of the Green River area in terms of erosion and deposition. Why are some formations missing from the Westwater column?

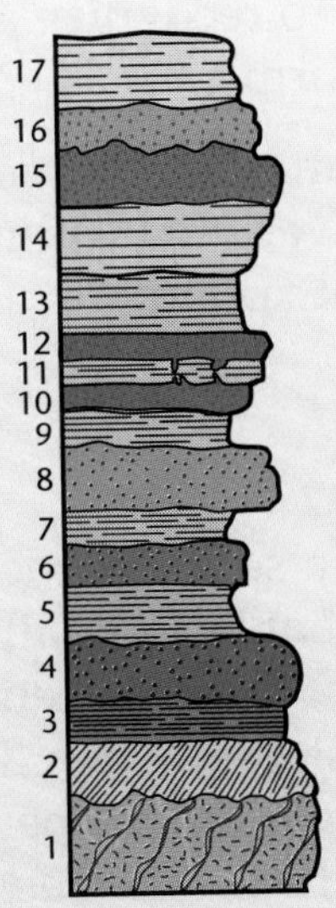

Green River, Utah

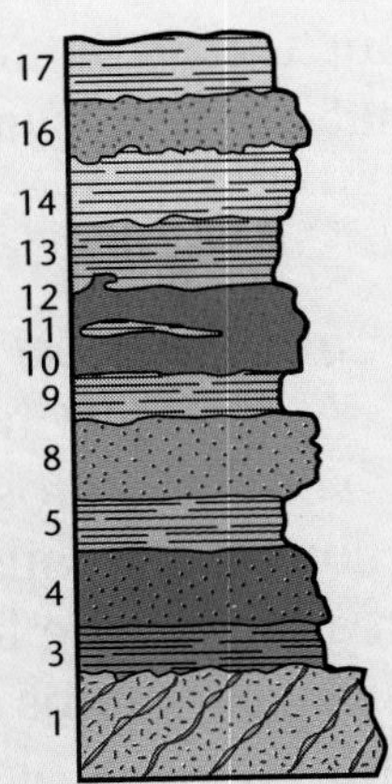

Westwater, Colorado

Figure 12-18

The many rock layers, or formations, in Canyonlands and Bryce Canyon have been dated and named. Some formations have been correlated between the two canyons. *Which layers are present at both canyons?* (NOTE: Fm = formation, Ss = sandstone, Gp = group.)

Are there other ways to correlate layers of rock? Sometimes relative dating isn't enough, and other dating methods must be used. In Section 12-4, you'll see how the actual age of rocks can be determined and how geologists have used this information to determine the age of Earth.

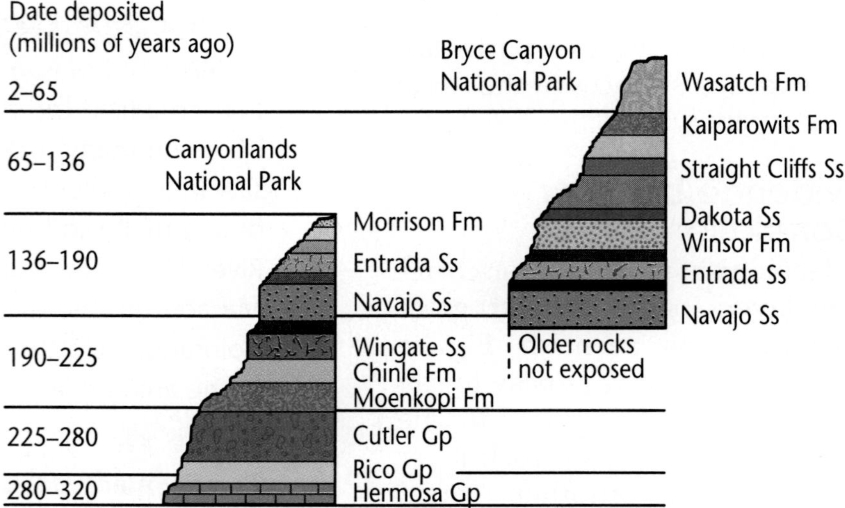

Section Wrap-up

Review

1. Suppose you haven't cleaned out your locker all year. Where would you expect to find papers from the beginning of the year? What principle in geology would you use to find these old papers?
2. Why is it more difficult to recognize a disconformity than an angular unconformity?
3. **Think Critically:** What are the relative ages of an igneous intrusion and overlying sedimentary rock layers that dome upward? Explain.

Using Computers

Spreadsheet Use the information about unconformities on pages 339 and 340 to prepare a spreadsheet listing the types of unconformities, their differences, and their similarities.

Skill Builder

Interpreting Data

A geologist finds a series of rocks. The sandstone contains a fossil that is 400 million years old. The shale contains fossils that are between 500 and 550 million years old. The limestone, which lies under the sandstone, contains fossils that are between 400 and 500 million years old. Which rock bed is oldest? Explain. If you need help, refer to Interpreting Data in the **Skill Handbook.**

Relative Age Dating of Geologic Features

Which of your two friends is older? To answer this question, you'd need to know the relative ages of the two. You wouldn't need to know the exact age of either of your friends, just who was born first. The same is sometimes true for rock layers. Geologists can also learn a lot about rock layers without knowing their exact ages.

Problem

Can the relative ages of rocks be determined by studying the rock layers and structures?

Procedure

1. Study Figures A and B. The legend will help you interpret the figures.
2. Determine the relative ages of the rock layers, unconformities, igneous dikes, and fault in each figure.

Analyze

Figure A

1. Were any layers of rock deposited after the igneous dike formed? Explain.
2. What type of unconformity is shown? Is it possible that there were originally more layers of rock than are shown here? Explain.
3. What type of fault is shown?
4. Is it possible to determine whether the igneous dike formed before or after the fault occurred? Explain.

Figure B

5. What type of fault is shown?
6. Is the igneous dike on the left older or younger than the unconformity nearest the surface? Explain.
7. Are the two igneous dikes shown the same age? How do you know?
8. Which two layers of rock may have been much thicker at one time than they are now?

Conclude and Apply

9. Make a sketch of Figure A. On it, **identify** the relative age of each rock layer, igneous dike, fault, and unconformity. For example, the shale layer is the oldest, so mark it with a *1*. Mark the next-oldest feature with a *2*, and so on.
10. Repeat the procedure in question 9 for Figure B.

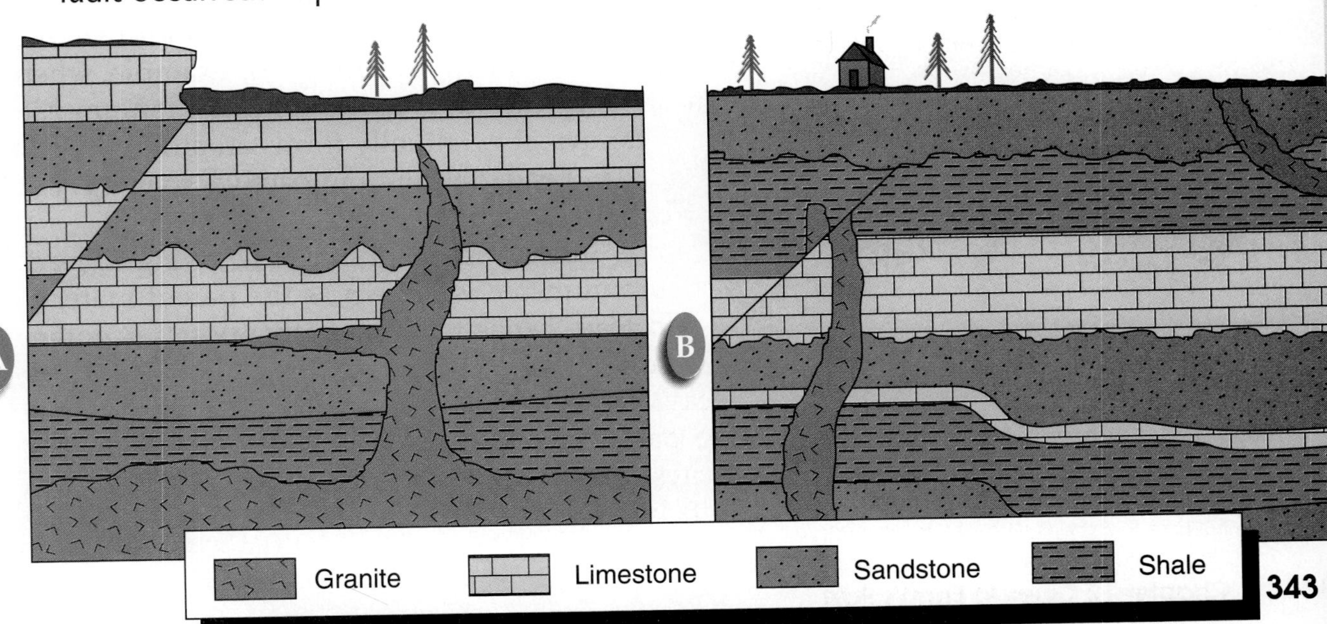

12•4 Absolute Ages of Rocks

Science Words

absolute dating
radioactive decay
half-life
radiometric dating
uniformitarianism

Objectives

- Identify how absolute dating differs from relative dating.
- Describe how the half-lives of isotopes are used to determine a rock's age.

INTEGRATION
Physics

Absolute Dating

As you continue to shuffle through your stack of magazines looking for articles about wheels and bearings, you decide you need to restack them into a neat pile. By now, they're a jumble and no longer in order of their relative ages, as shown in **Figure 12-19.** How can you stack them so the oldest are on the bottom and the newest on top? Fortunately, magazines have their dates printed on their covers. Thus, stacking magazines in order is a simple process. Unfortunately for geologists, rocks don't have their ages stamped on them. Or do they?

Absolute dating is a method used by geologists to determine the age, in years, of a rock or other object. Absolute dating is a process that uses the properties of atoms in rocks and other objects to determine their ages.

Radioactive Decay

In Chapter 2, you learned that an element can have atoms with different numbers of neutrons in their nuclei. Some of these isotopes undergo a process called radioactive decay. When an atom of some isotopes decays, one of its neutrons breaks down into a proton and an electron. The electron leaves the atom as a beta particle. The nucleus loses a neutron, but gains a proton. Other isotopes give off two protons and two neutrons in the form of an alpha particle as seen in **Figure 12-20.** As you know, when the number of protons in an atom is changed, as it is in **radioactive decay,** a new element is formed. For example, when an atom of the radioactive isotope uranium-238 decays, it eventually forms an atom of lead-206. Lead-206 isn't radioactive, so it will not decay any further.

In the case of uranium decaying to lead, uranium-238 is known as the parent material and lead-206 as the daughter product. Another example of a parent material is carbon-14, which decays to its daughter, nitrogen-14. Each radioactive parent material has a certain rate at which it decays to its daughter product. This rate is known as its half-life.

Figure 12-19

The magazines that have been shuffled through no longer illustrate the principle of superposition.

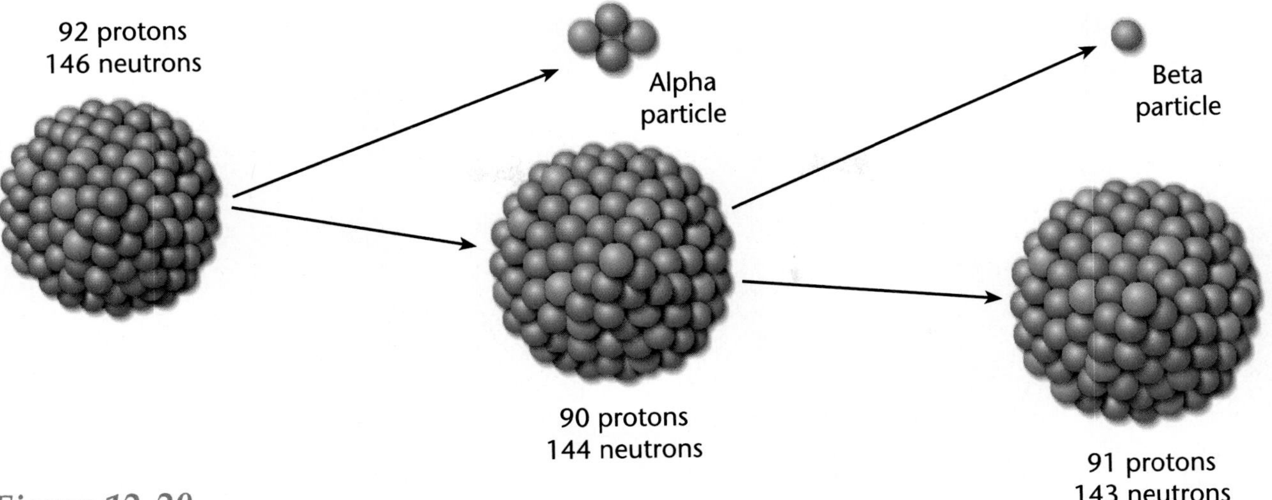

Figure 12-20

Uranium-238 decays by emitting alpha particles (two protons and two neutrons) and beta particles (one electron). A beta particle is produced when a neutron decays and becomes a proton. *Is any energy released during this process? If so, what?*

Half-Life

The **half-life** of an isotope is the time it takes for half of the atoms in the isotope to decay. For example, the half-life of carbon-14 is 5730 years. So it will take 5730 years for half of the carbon-14 atoms in an object to decay to nitrogen-14. You might guess that in another 5730 years, all of the remaining carbon-14 atoms will have decayed to nitrogen-14. However, this is not the case. Only half of the atoms of carbon-14 remaining after the first 5730 years will decay during the second 5730 years. So, after two half-lives, one-fourth of the original carbon-14 atoms still remain. Half of them will decay during another 5730 years. After three half-lives, one-eighth of the original carbon-14 atoms still remain. After many half-lives, such a small amount of the parent material remains that it may not be measurable.

Radiometric Dating

To a geologist, the decay of radioactive isotopes is like a clock ticking away, keeping track of time that's passed since rocks have formed. As time passes, the concentration of parent material in a rock decreases as the concentration of daughter product increases, as seen in **Figure 12-21** on the next page. By measuring the amounts of parent and daughter materials in a rock and by knowing the half-life of the parent, a geologist can calculate the absolute age of the rock. This process is called **radiometric dating.**

MiniLAB

What are the dates of some events in Earth's history?

Procedure

1. Sequence these events in Earth's history in relative order: Earth forms, first many-celled organisms, first land plants, first mammals, dinosaurs become extinct, first amphibians, first human ancestors, oldest known fossils.
2. Make a time line using these dates: 4.6 billion years, 3.5 billion years, 1.0 billion years, 630 million years, 408 million years, 225 million years, 66 million years, and 5 million years ago.
3. Match each event with the absolute date on your time line.

Analysis

1. Check your time line with your teacher.
2. Did you correctly list the events in relative order?
3. How does the age of Earth compare with the presence of humans on the time line?

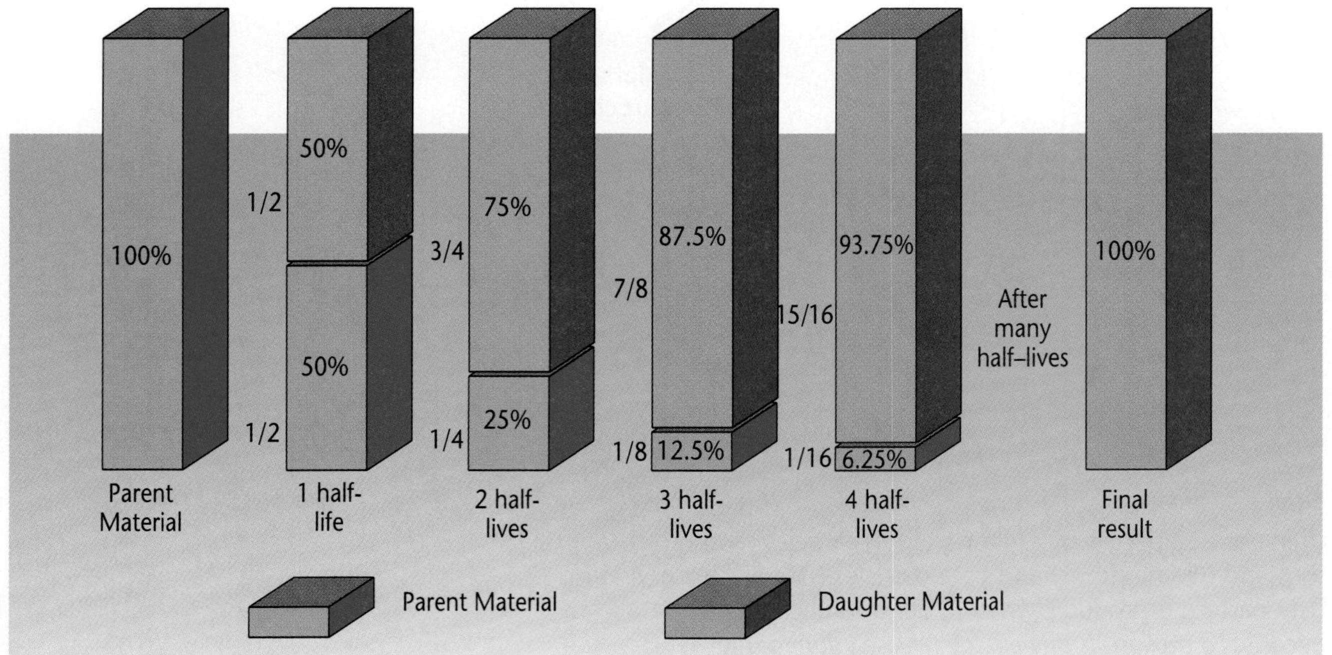

Figure 12-21

After each half-life, one-half the amount of parent material remains. Eventually, such a small amount of the parent material is left that it may not be measurable.

A scientist must decide which parent and daughter materials to measure when dating a rock or fossil. If the object to be dated is very old, then an isotope with a long half-life must be used. For example, if a fossil is 1 billion years old, there would be no carbon-14 left to measure. However, the half-life of uranium-238 is 4.5 billion years. Enough of the parent and daughter material would still be present to measure.

Radiocarbon Dating

Carbon-14 is useful for dating fossils, bones, and wood up to 50 000 years old. Organisms take in carbon from the environment to build tissues in their bodies. When the organism dies, some of the carbon-14 decays and escapes as nitrogen-14 gas. The amount of carbon-14 remaining can be measured to determine the age of the fossil or when humans used a fire site, as in **Figure 12-22**.

Other than for carbon-14 dating, rocks that can be radiometrically dated are mostly igneous and some recrystallized metamorphic rocks. Sedimentary rocks cannot be dated by this method

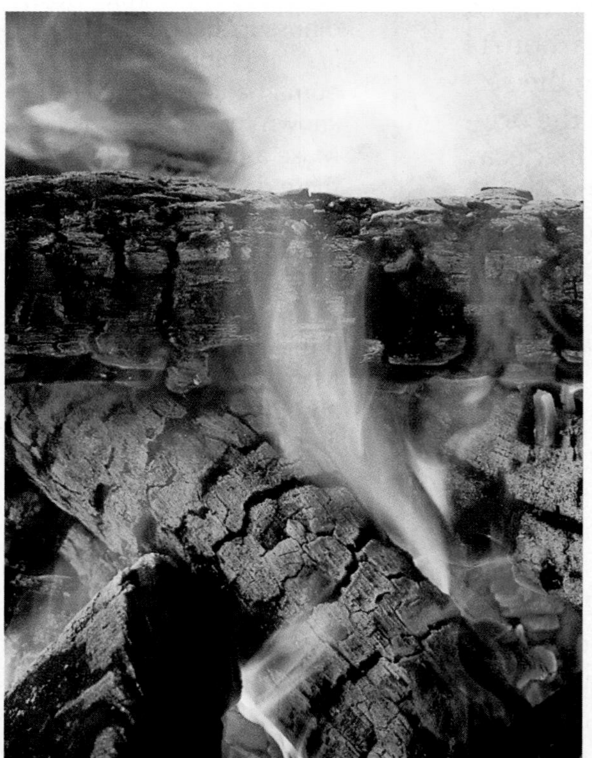

Figure 12-22

Human activity can also be dated with carbon-14. Things like this campfire or other types of charcoal can be dated. *What other events could leave charcoal behind and provide radiocarbon dates?*

because the absolute age of only the sediment grains in the rock can be determined, not the rock itself. Radiometric dating has been used to date the oldest rocks found on Earth. These rocks are 3.96 billion years old. Scientists have estimated the age of Earth at 4.6 billion years.

Uniformitarianism

Before radiometric dating was available, many people had estimated the age of Earth to be only a few thousand years old. But in the 1700s, Scottish scientist James Hutton estimated that Earth was much older. He used the principle of **uniformitarianism.** This principle states that Earth processes occurring today are similar to those that occurred in the past. He observed that the processes that changed the rocks and land around him were very slow, and he inferred that they had been just as slow throughout Earth's history. Hutton hypothesized that it took much longer than a few thousand years to form the layers of rock around him and to erode mountains that once towered kilometers high. John Playfair advanced Hutton's theories, but an English geologist, Sir Charles Lyell, is given the most credit for advancing uniformitarianism.

The half-life of radium-226 is 1600 years. How old is an object in which 1/32 of the original radium-226 is present?

Section Wrap-up

Review

1. You discover three rock layers that have not been overturned. The absolute age of the middle layer is 120 million years. What can you say about the ages of the layers above and below it?
2. How old would a fossil be if it had only one-eighth of its original carbon-14 content remaining?
3. **Think Critically:** Suppose you radiometrically date an igneous dike running through only the bottom two layers in question 1. The dike is cut off by the upper rock layer. The dike is 70 million years old. What can you say about the absolute age of the upper layer?

Skill Builder

Making and Using Tables

Make a table that shows the amounts of parent and daughter materials left of a radioactive element after four half-lives if the original parent material had a mass of 100 g. If you need help, refer to Making and Using Tables in the **Skill Handbook**.

Research John Playfair and Sir Charles Lyell in a geology book. In your Science Journal, write about their contributions to uniformitarianism.

Activity 12-2

Design Your Own Experiment
Radioactive Decay

Radioactive isotopes decay into their daughter elements in a certain amount of time. The rate of decay varies for each individual isotope. This rate can be used to determine the age of rocks that contain the isotopes under study. In this activity, you will develop a model that demonstrates how the half-life of certain radioactive isotopes can be used to determine absolute ages.

PREPARATION

Problem

What materials can be used to model age determination using radioactive half-lives?

Form a Hypothesis

State a hypothesis about what materials can be used in a model of radioactive decay and how they will be used to model radioactive half-lives.

Objectives

- Determine what materials can best be used to model radioactive half-lives.
- Design an experiment that models absolute age determination using the half-lives of radioactive isotopes.

Possible Materials

- shoe box with lid
- brass fasteners (100)
- paper clips (100)
- graph paper
- pennies (100)
- colored pencils (2)
- pipe cleaners (100)

Safety Precautions

Hold the lid of the box on tight to avoid having objects fly out of the box.

PLAN THE EXPERIMENT

1. As a group, agree upon and write out the hypothesis statement.
2. As a group, list the steps that you need to take to test your hypothesis. Be very specific, describing exactly what you will do at each step.
3. Make a list of the materials that you will need to complete your experiment.
4. Make a list of the various materials you will use to model radioactive decay.
5. Describe how the different materials you choose will be used in your model to demonstrate radioactive half-lives.

Check the Plan

1. Read over your entire experiment to make sure that all steps are in a logical order.
2. Should you use more than one type of material in your model?
3. Will you summarize data in a graph, table, or some other form of organization?
4. Will you use more than one graph or table or a combination of several?
5. How will you determine whether your model demonstrates absolute age determination from radioactive isotope half-lives?
6. *Make sure your teacher approves your plan and that you have included any changes suggested in the plan.*

DO THE EXPERIMENT

1. Carry out the experiment as planned.
2. While conducting the experiment, write down any observations that you or other members of your group make and complete any data tables or graphs in your Science Journal.

Analyze and Apply

1. What do the different materials you decided to use represent in your model of radioactive decay?
2. **Describe** how each of the processes or objects used in your model fits into the process of radioactive decay.
3. **Determine** what were the half-lives of the various parent materials used in your model of radioactive decay.

Go Further

Suppose you could perform the process from your model only once every 100 years. How might this affect the half-lives of the parent materials you selected?

People and Science

ENRIQUETA BARRERA, *Geochemist*

On the Job

Q How does your work shed light on Earth's past?

A I'm studying a period about 70 to 72 million years ago. Most people say the change in the biota and oceans at this time was one of the most dramatic changes in Earth's history. These changes may have influenced how organisms have evolved. I have found evidence for a tremendous change in the circulation of the oceans.

Q Some of your work has challenged the theory that a meteorite collision 66 million years ago caused a sudden, drastic change in the environment that led to widespread extinction of plants and animals. Please tell us what you found.

A I worked with another scientist to analyze layers of sediment deposited millions of years ago under a shallow sea in Texas, and in the deep ocean near Antarctica. My job was to analyze carbon and oxygen isotope ratios in fossils. We found evidence that some extinctions may have been caused by a gradual environmental change over a period of thousands of years, rather than by sudden, dramatic change.

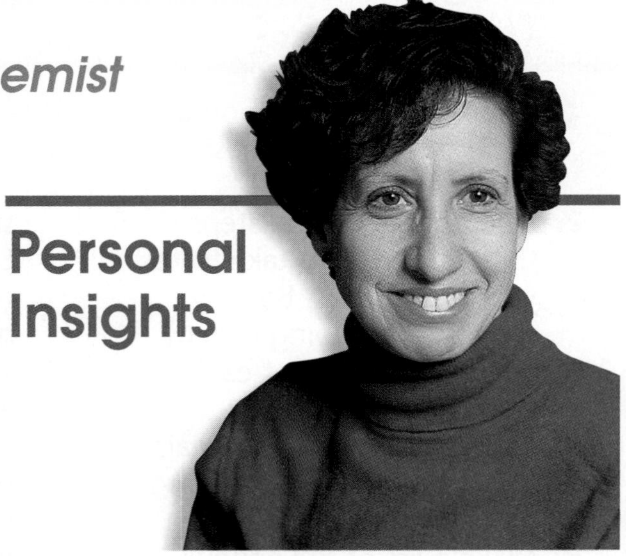

Personal Insights

Q What has been your biggest challenge as a scientist?

A There are many challenges. It takes years of hard work to become established in a field of research. Scientific research has a rigorous methodology that involves careful observations and analysis to formulate hypotheses. As we practice science, we are constantly revising our ideas and learning about our field. This process is very exciting. As an American woman of Latin heritage, it has been a little difficult to become established in my profession, partly because there are significantly fewer women than men in this field. It takes perseverance, but I love what I do—I am really excited about this profession!

Career Connection

Visit a natural history museum or university geology department. Talk to curators or professors and find out what kinds of work they do.

Write a job description for what kind of work you would like to do in this field and where you could work as a:

- **Paleontologist**
- **Geochemist**

Chapter 12

Review

Summary

12-1: Fossils

1. Fossils are more likely to form if hard parts of the dead organisms are covered quickly.
2. Some fossils form when original materials that made up the organisms are replaced with minerals. Other fossils form when remains are subjected to heat and pressure, leaving only a carbonaceous film behind. Some fossils are merely the tracks or traces left by former organisms.
3. Generally, a rock layer can be no older than the age of the fossils embedded in it.

12-2: Science and Society: Extinction of Dinosaurs

1. The meteorite-impact theory of dinosaur extinction states that a large object from space collided with Earth and caused vast climate changes. Dinosaurs weren't able to adapt and eventually became extinct.
2. Another theory suggests that volcanic activity led to the extinction of the dinosaurs.

12-3: Relative Ages of Rocks

1. The principle of superposition states that older rocks lie underneath younger rocks in areas where the rocks haven't been disturbed. Faults are always younger than the rocks they cross-cut.
2. Unconformities, or gaps in the rock record, are due to erosion, nondeposition, or both.
3. Fossils and rock types are often helpful when correlating similar rock bodies.

12-4: Absolute Ages of Rocks

1. Relative dating of rocks, unlike absolute dating, doesn't provide an exact age for the rocks.
2. The half-life of a radioactive isotope is the time it takes for half of the atoms in the isotope to decay. Because half-lives are constant, absolute ages of rocks containing radioactive elements can be determined.

Key Science Words

a. absolute dating
b. carbonaceous film
c. cast
d. extinct
e. fossil
f. half-life
g. index fossil
h. principle of superposition
i. mold
j. petrified remains
k. radioactive decay
l. radiometric dating
m. relative dating
n. unconformity
o. uniformitarianism

Reviewing Vocabulary

Match each phrase with the correct term from the list of Key Science Words.

1. thin film of carbon preserved as a fossil
2. rocklike fossils made of minerals
3. fossil of species that existed for a short time and was abundant and widespread
4. states that older rocks lie under younger rocks
5. states that natural processes occur today as they did in the past
6. gap in the rock record
7. evidence of once-living organisms
8. neutrons break down during this process
9. the time it takes for half of the atoms of a radioactive isotope to decay
10. this process measures the amounts of parent and daughter materials to determine age

Chapter 12 Review

Checking Concepts

Choose the word or phrase that completes the sentence.

1. Remains of organisms in rocks are _______.
   a. half-lives c. unconformities
   b. fossils d. extinctions
2. Fossils may form when dead organisms are _______.
   a. buried slowly
   b. exposed to microorganisms
   c. made of hard parts
   d. composed of soft parts
3. _______ are cavities left in rocks when a shell or bone decays.
   a. Casts c. Molds
   b. Petrified remains d. Carbon films
4. Dinosaurs lived _______ years ago.
   a. 1000 c. 5 million
   b. 10 000 d. more than 66 million
5. Another way to state the principle of _______ is to say "the present is the key to the past."
   a. superposition c. radioactivity
   b. succession d. uniformitarianism
6. A fault can be used to find the _______ age of a group of rocks.
   a. absolute c. index
   b. radiometric d. relative
7. An unconformity between horizontal rock layers is a(n) _______.
   a. angular unconformity c. disconformity
   b. fault d. nonconformity
8. During _______, new elements are formed.
   a. superposition c. evolution
   b. uniformitarianism d. radioactive decay
9. In one type of radioactive decay, a(n) _______ breaks down, releasing an electron.
   a. alpha particle c. beta particle
   b. proton d. neutron
10. Radiometric dating indicates that Earth is _______ years old.
    a. 2000 c. 3.5 billion
    b. 5000 d. 4.6 billion

Understanding Concepts

Answer the following questions in your Science Journal using complete sentences.

11. How do relative and absolute ages differ?
12. Why did James Hutton and others infer that Earth had to be much older than a few thousand years?
13. Explain why a clay layer rich in iridium might explain why the dinosaurs became extinct.
14. How many half-lives have passed in a rock containing 1/8 of the original radioactive material and 7/8 of the daughter product?
15. If large tracks of an animal are found surrounding smaller tracks of the same animal, what might this indicate?

Thinking Critically

16. We don't have a complete fossil record of life on Earth. Give some reasons why.
17. Suppose a lava flow were found between two sedimentary rock layers. How could the lava flow be used to date the rocks? (HINT: Most lava contains radioactive isotopes.)
18. Mammals began to evolve on Earth before the dinosaurs became extinct. Suggest a hypothesis explaining how the mammals may have caused the dinosaurs to become extinct.
19. Suppose you're correlating rock layers in the western United States. You find a layer of shale that contains volcanic dust deposits. How can this layer help you in your correlation over a large area?
20. Why is carbon-14 not suitable for dating fossils formed about 2 million years ago?

Developing Skills

If you need help, refer to the **Skill Handbook.**

21. **Concept Mapping:** Make a concept map listing the following possible steps in the process of making a cast of a fossil: organism dies, burial, protection from scavengers and bacteria, replacement by minerals, fossil erodes away, mineral crystals form from solution.
22. **Observing and Inferring:** Suppose you found a rock containing brachiopods. What can you infer about the environment in which the rock formed?
23. **Recognizing Cause and Effect:** Explain why some woolly mammoths have been found intact in frozen ground.
24. **Classifying:** Suppose you were given a set of ten fossils to classify. Make a table to classify each specimen according to type.
25. **Outlining:** Make an outline of Section 12-1 that discusses the ways in which fossils form.

Performance Assessment

1. **Making Observations and Inferences:** In Activity 12-1, you learned how to determine relative ages. Based on what you learned concerning relative ages of rocks, determine the probable relative ages of the rock layers shown below.
2. **Using a Classification System:** Start your own fossil collection. Label each find as to type, approximate age, and the place where it was found. Most state geological surveys can provide you with reference material on local fossils.
3. **Model:** Use various types of material to make molds and casts of clamshells and other animal or plant hard parts. Determine which type of material would preserve fossils best. Infer how this would relate to fossil formation in nature. Materials you might use include clay, sand, fine dust, and gravel.

Previewing the Chapter

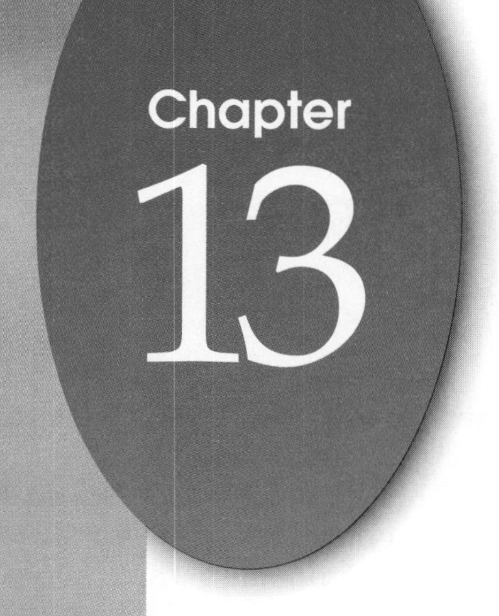

Chapter 13 Geologic Time

As new fossils are discovered, our ideas about how dinosaurs lived, like these *utahraptors,* change. Dinosaurs appear to have been much more intelligent and active than we previously thought. In the following activity, use a model to find out how traits might determine whether individuals can compete and survive in an environment; then learn how traits of actual organisms have affected their success or failure.

EXPLORE ACTIVITY

Make a model environment. 

1. Cut green, orange, and blue yarn into 3-cm lengths. You should have 15 pieces of each color.
2. Put them on a sheet of green construction paper.
3. Have your partner use a pair of tweezers to pick up as many pieces as he or she can in 15 seconds.

Observe: In your Science Journal, discuss whether one color was chosen more easily over another. Now suppose the construction paper represents grass, each piece of yarn represents an insect, and the tweezers represent a bird preying on the insects. Can you recognize the cause and effect as to which color of yarn your partner picks up or which color of insect would have a better chance to survive?

Previewing Science Skills

- In the **Skill Builders,** you will recognize cause and effect, make and use tables, and sequence events.
- In the **Activities,** you will measure in SI, hypothesize, formulate models, use numbers, and make inferences.
- In the **MiniLABs,** you will measure in SI, use numbers.

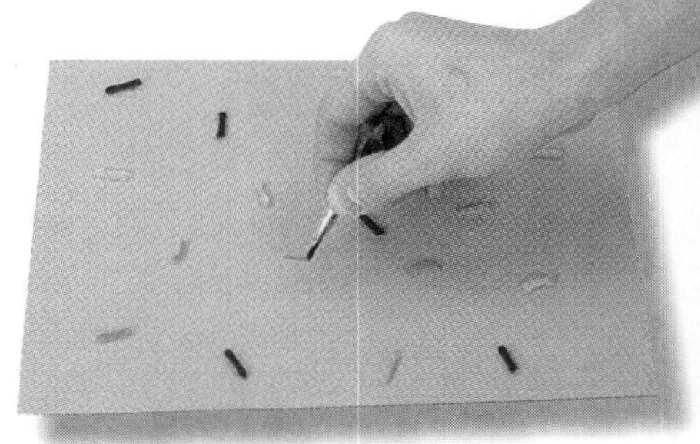

13•1 Evolution and Geologic Time

Science Words

geologic time scale
era
period
epoch
organic evolution
species
natural selection

Objectives

- Explain how geologic time is divided into units.
- Relate organic evolution to divisions on the geologic time scale.
- Describe how plate tectonics affects organic evolution.

Geologic Time

It's a rainy day in the prairie lands of the central United States. A herd of horses is moving toward a stream where they can find fresh drinking water. Their large, powerful muscles easily carry them across several kilometers of open grassland.

The horses are suited for this environment. Their hoofed feet allow them to run at great speeds to protect themselves from predators. The males use their speed and power to compete with other males for territory and mates. The horses' teeth allow them to grind up grass.

These characteristics allow horses to survive in the environment they live in. These same characteristics are what you would use to describe the horse—a large, powerful, hoofed animal with teeth made for grinding up grasses and grains.

Early Horses

Suppose you found a fossil animal that was the size of a dog. The animal had four toes on each of its front feet and three toes on each of its hind feet. Its teeth were sharp—suited for eating shrubs and bushes, but not for grinding grass. Would you classify this as a fossil of a horse? It may be just that.

At one time in Earth's history, horses were small, as seen in **Figure 13-1.** They had several toes and no hoofs, and they had teeth much different from the teeth of today's horses. Before that time, there were no animals we would classify as horses.

Figure 13-1

As the horse species evolved, horses increased in size. Another trait that horses evolved was the single-toed, hoofed foot. *What does this trait enable horses to do?*

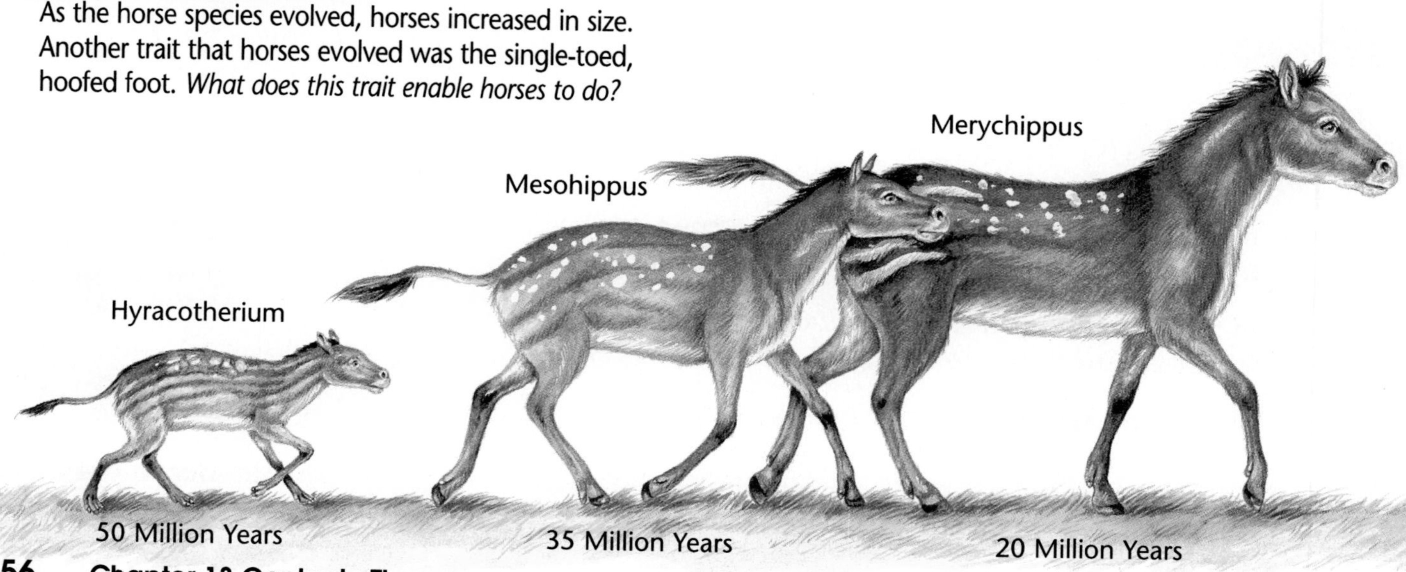

Before that, there weren't even mammals. If you look back far enough into time, there were no animals and no plants. At one point, there were only the molecules that combine to make life, but not life itself.

The Geologic Time Scale

The appearance and disappearance of types of organisms throughout Earth's history give us markers in time. We can divide Earth's history into smaller units based on the types of life-forms living during certain periods. The division of Earth's history into smaller units makes up the **geologic time scale.** Some of the divisions in the geologic time scale are also based on geologic changes occurring at the time.

The geologic time scale is a record of Earth's history, starting with Earth's formation about 4.6 billion years ago. Each period of time is named. When fossils and rock layers are dated, scientists can assign them to a specific place on the geologic time scale.

Subdivisions of Geologic Time

There are three types of subdivisions of geologic time—eras, periods, and epochs. **Eras** are major subdivisions of the geologic time scale based on differences in life-forms. As you can see in **Figure 13-2** on the next page, the Mesozoic Era began about 245 million years ago. Its end is marked by the extinction of the dinosaurs and many other organisms about 66 million years ago.

Eras are subdivided into **periods.** Periods are based on the types of life existing at the time and on geologic events, such as mountain building and plate movements.

CONNECT TO

LIFE SCIENCE

Research evolutionary changes that have occurred in horses throughout geologic time. In your Science Journal, ***write*** a report in which you ***describe*** how these changes are thought to be connected to changes in the environments in which horses have lived.

Figure 13-2

The geologic time scale is divided into subunits based on geologic events and the appearance and disappearance of types of organisms.

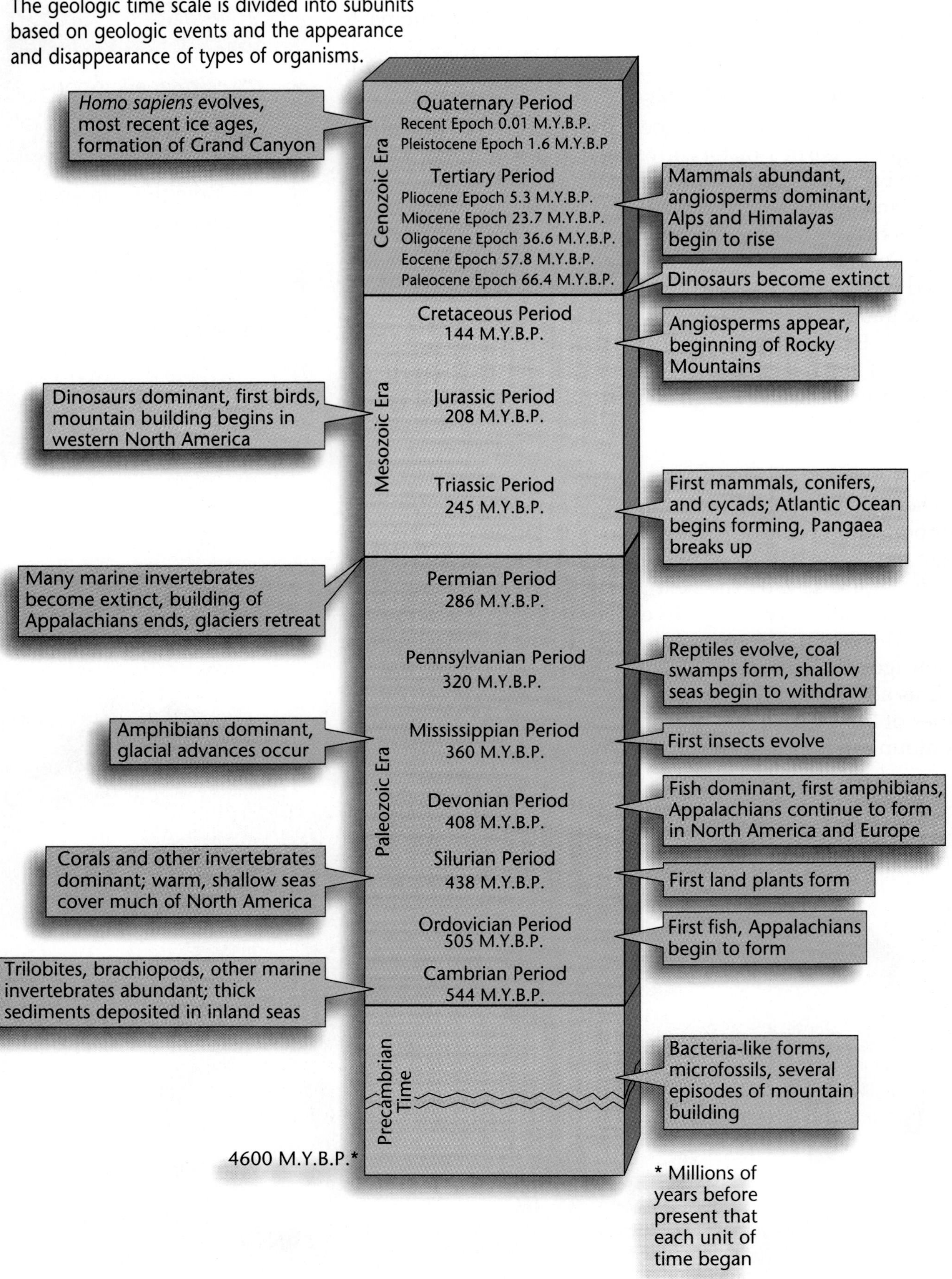

Periods may be divided into smaller units of time called **epochs. Figure 13-2** shows that only the Cenozoic Era is subdivided further into epochs. Why is this so? The fossil record and the record of geologic events is more complete in these recent rock layers. As a result, geologists have more markers with which to divide the time scale.

Organic Evolution

Based on the fossil record, organisms appear to have followed an ordered series of changes. This change in life-forms through time is known as **organic evolution.** Most theories describing the processes of organic evolution state that changes in the environment dictate who survives, and this in turn can result in changes in species of organisms.

Species

A **species** has often been defined as a group of organisms that normally reproduce only among themselves. For example, dogs are a species of animal because they mate and reproduce only with other dogs. As shown in **Figure 13-3,** there are some cases where organisms of different species can breed and produce offspring. These offspring, however, are unable to reproduce.

Natural Selection Within a Species

Suppose a bird species exists on an island. A few of the individuals have very hard beaks, but most have soft beaks. There are two food sources for the birds; one is soft and the

Figure 13-3

In some cases, animals of different species can breed and produce offspring. A lion (A) and a tiger (B) can produce a tigon (C). *However, because two different species are interbred, what is true of the offspring?*

A Lions have been known to mate with tigers in captivity.

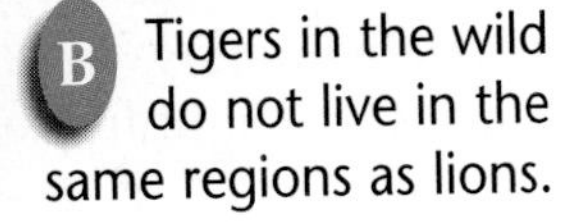

B Tigers in the wild do not live in the same regions as lions.

C The resulting offspring is called a tigon, liger, or tiglon.

other has a hard shell around it. Now suppose the climate changes, causing the soft food to become scarce. Which of the birds will be better suited, or more fit, to survive? The birds with the harder beaks will be better able to break the hard shell on the more common food. Some of the soft-beaked birds may die from lack of food. The hard-beaked birds have a better chance of surviving and reproducing. Their offspring will inherit the trait of having a hard beak. Gradually, the number of hard-beaked birds will become greater, and the number of soft-beaked birds will decrease. The species may evolve so that nearly all of its members have hard beaks.

Natural Selection

Because the selection of the hard-beaked birds was a natural process, this process is called natural selection. Charles Darwin, a naturalist who sailed around the world from 1831 to 1836 to study wildlife, published the theory of evolution by natural selection. He proposed that **natural selection** is the process by which organisms with traits that are suited to a certain environment have a better chance of surviving and reproducing than organisms whose traits are not suited to it.

Notice in the example of the birds that individual soft-beaked birds didn't change into hard-beaked birds. A new trait becomes dominant in a species only if some members already possess that trait. If no bird in the species had possessed a hard beak, evolution into a hard-beaked species could not have occurred. The birds may have been able to survive if they could find other food to eat or adapt to other food sources, or they may have died out completely. **Figure 13-4** shows another example of natural selection.

Figure 13-4

Before pollution darkened tree bark, light-colored members of the peppered moth species were abundant. Their light color hid them from birds in search of a meal. When pollution covered the trees, the darker members of the species began to survive and reproduce more often than the lighter members. *How might this natural selection affect future evolution of the moths?*

A The light-colored peppered moths were well camouflaged against the light tree bark.

B Against the pollution-darkened tree bark the light-colored moths stand out. They are then easier for birds to capture and eat.

The Evolution of New Species

Natural selection explains not only how characteristics develop within a species, but also how new species arise. For example, if the soft-beaked birds in our example had moved to a different part of the island where soft food was more available, they might have survived. The soft-beaked birds would continue to reproduce apart from the hard-beaked birds on a different part of the island. Over time, the soft-beaked birds would develop characteristics that were different from those of the hard-beaked birds, as seen in **Figure 13-5.** At some point, the birds might not be able to breed with each other. They would have evolved into two different species.

Changing Environments

Think again of the horses discussed at the beginning of this section. Why did they change over time? You have learned that fossil evidence shows that early

USING TECHNOLOGY

Embryology

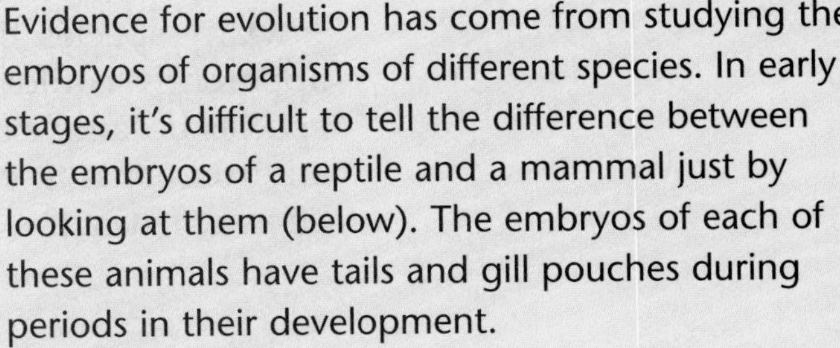

Evidence for evolution has come from studying the embryos of organisms of different species. In early stages, it's difficult to tell the difference between the embryos of a reptile and a mammal just by looking at them (below). The embryos of each of these animals have tails and gill pouches during periods in their development.

Development

Embryos of these widely different organisms display similar features because similar genes are at work during their early development. As the separate organisms continue to develop, genes that have changed during the process of evolution are at work. These genes cause the wide diversity seen in the mature organisms.

By studying the similarities among the embryos of different species, we have discovered clues about which species may be closely related. Those species with similar embryo development may have shared a common ancestor in geologic history.

Early Embryos

Mammalian embryo

Reptilian embryo

Think Critically: Mammals didn't evolve from dinosaurs, yet a mammalian embryo may have several features and developmental patterns in common with a dinosaur embryo. Explain why this might be the case.

horses were small, multi-toed animals adapted to grazing on shrubs and bushes. As environments on Earth changed from brushy fields to open grasslands, the horse species became bigger and developed hooves and complex molars. Over millions of years, horses became adapted to open grasslands by the process of natural selection. As the environment changed, the horse species adapted and survived.

Many species that lived on Earth during its long history apparently couldn't adapt to changing environments. Such species became extinct. What processes on Earth could cause environments to change so much that species must adapt or become extinct?

The Effect of Plate Tectonics on Earth History

Plate tectonics is one process that causes changing environments on Earth. As plates on Earth's surface moved over time, continents collided with and separated from each other many times. Collisions caused mountain building and the draining of seas. Separations caused deeper seas to develop between continents. This rearranging of land and sea still causes changes in climates. How might these changes affect organisms?

If species adapt to the changes, or evolve, they survive. If a species doesn't have individuals with characteristics needed for survival in the changing environment, the species becomes extinct, as seen in **Figure 13-6.**

Figure 13-5

Ten of the many species of finches on Isle Santa Cruz in the Galápagos archipelago are shown. All of these evolved from one ancestral species. Small groups of the ancestral species became isolated when they began to specialize in the types of food they ate and the areas in which they lived. These groups eventually evolved into different species.

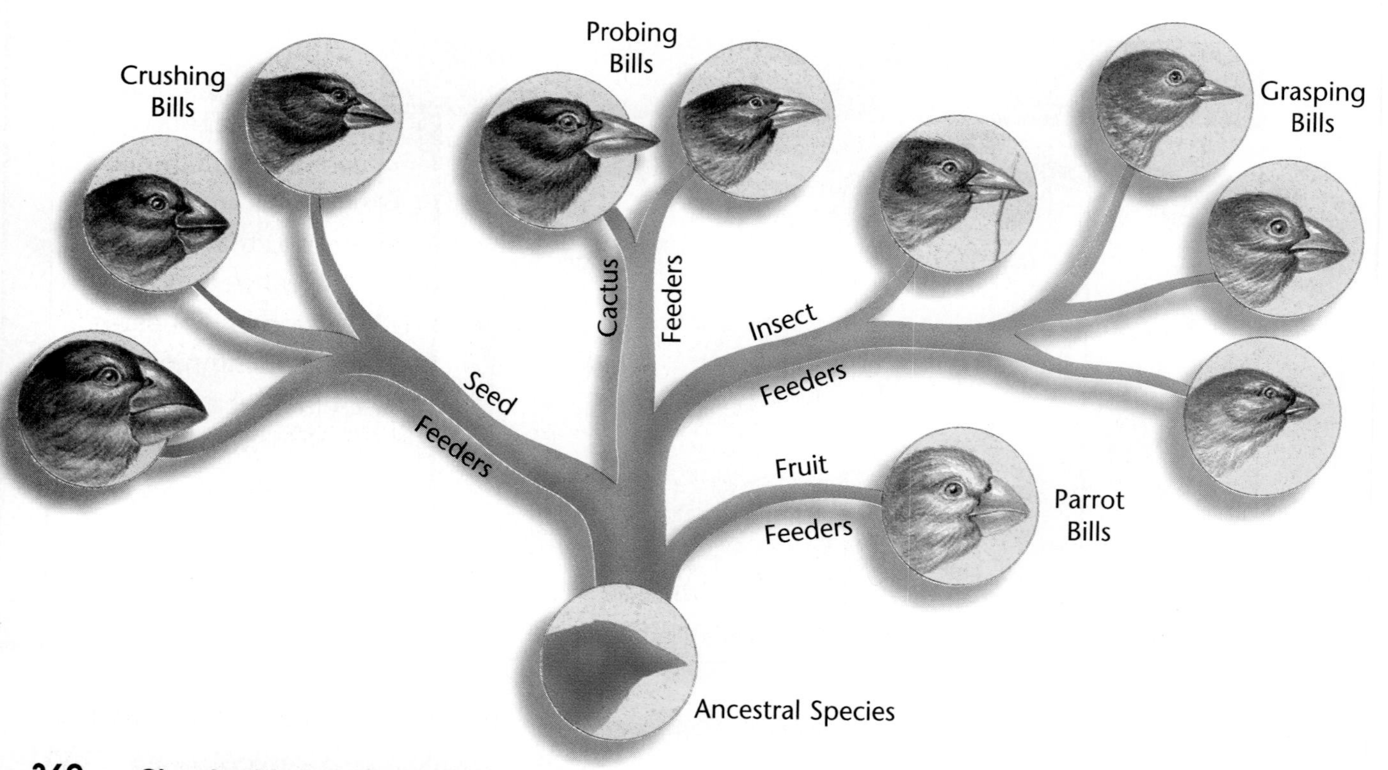

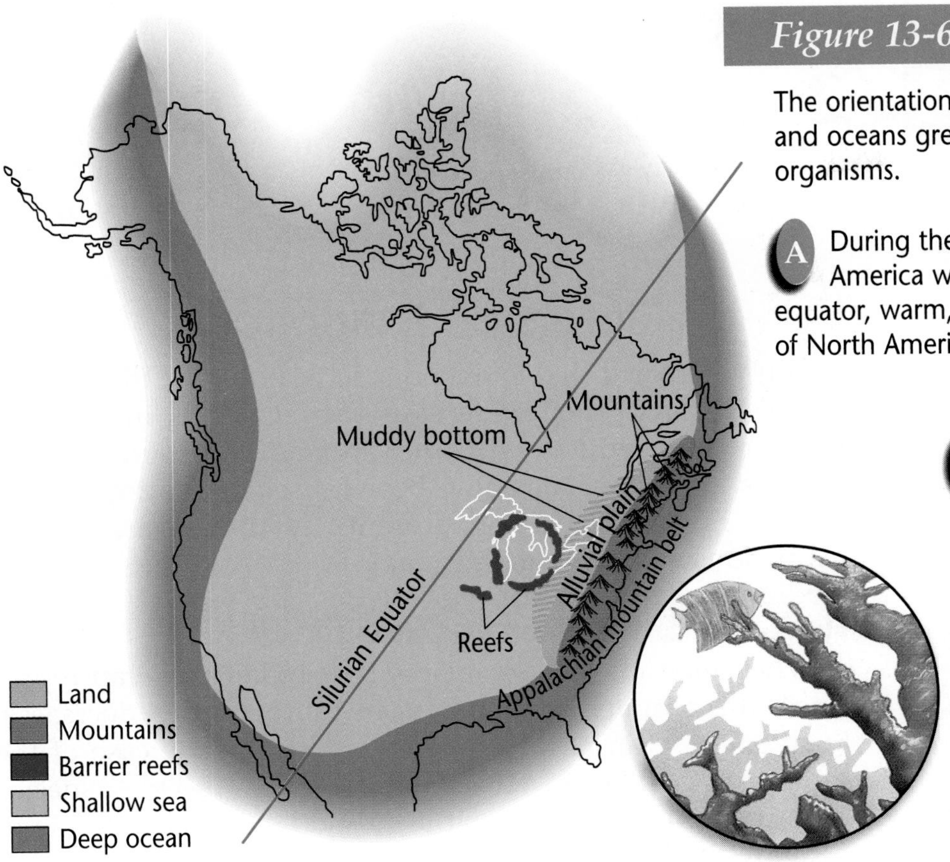

Figure 13-6

The orientation and position of continents and oceans greatly affect climate and organisms.

A During the Silurian Period when North America was positioned along the equator, warm, shallow seas covered much of North America.

B Corals, which are sea animals, were a dominant life-form in the Silurian Period. But as plate tectonic processes changed Earth's surface, climates also changed. Many species of coral did not adapt to the cooler climate. Thus, corals became less abundant, and some species became extinct. This example shows modern corals.

Section Wrap-up

Review

1. Relate organic evolution to the geologic time scale.
2. How might plate tectonics affect organic evolution?
3. **Think Critically:** Today, the moth species shown in **Figure 13-4** has many light-colored individuals. How can this be attributed to recent antipollution laws?

Skill Builder

Recognizing Cause and Effect

Answer the questions below. If you need help, refer to Recognizing Cause and Effect in the **Skill Handbook.**

1. Is natural selection a cause or effect of organic evolution?
2. How could the evolution of a trait within one species affect the evolution of a trait within another species? Give an example.

Science Journal

Write a paragraph in your Science Journal that explains Darwin's interpretation of the various beak types on the different species of Galápagos finches. Include illustrations to show the differences.

SCIENCE & SOCIETY

13•2 ISSUE: Present-Day Rapid Extinctions

Science Words

endangered
habitat

Objectives

- Recognize how humans have contributed to extinctions.
- Predict what might happen to the diversity of life on Earth if land is developed without protection of natural habitats.
- Hypothesize what can be done to stop or slow down the rate of species extinction.

Human Effects on Extinction Rates

You've learned that extinctions have occurred throughout Earth's history. They were caused by changes in environments or competition with other species for resources. Some of these extinctions may have been caused by the appearance of early humans. However, present-day humans are causing extinctions at a much greater rate.

How do humans promote extinctions?

When humans kill organisms faster than they can reproduce, the number of members in those species decreases. Such species can become endangered, as in the case of the mountain gorillas in Africa. (See **Figure 13-7**). A species becomes **endangered** when only a small number of its members are living. If the number of members of a species continues to dwindle, the species can become extinct. A species becomes extinct when no more of its members are living.

Humans contribute to extinctions by overhunting, through carelessness, and by making changes in the environment. Often, we take over the natural habitats of other species, leaving them with no food or space in which to live. A **habitat** is the place where organisms live, grow, and interact with each other and with the environment. Many species on Earth live in only one particular type of habitat. If that habitat is suddenly destroyed, the species does not have time to adapt, so all members of the species die.

Figure 13-7

These African mountain gorillas of Rwanda have become endangered because of habitat lost to human settlements.

2 Points of View

Development Without Habitat Protection

You may have heard about problems caused by the cutting or burning of tropical rain forests, shown in **Figure 13-8.** During the past decade, people have cleared much of these forests for farming, logging, and other industries. In doing so, they have destroyed many habitats.

The tropical rain forest habitat covers only about seven percent of Earth's surface, but it contains 50 to 80 percent of Earth's species. Think of what would happen to these species if the tropical rain forests were destroyed.

Figure 13-8

When forests are cleared, animal and plant habitat is lost. Many species may become extinct if too much forest is destroyed.

Development with Habitat Protection

Trees and other plants in a tropical rain forest derive many of their nutrients from a continuous supply of rapidly rotting organic matter. Whenever large areas of the tropical rain forests are destroyed, plants tend not to grow back because the soil in these areas is very poorly developed. Once the plants have been destroyed, there is nothing left to supply nutrients.

Governments could restrict construction to allow both development and preservation. Projects for various industries could be planned in ways that preserve habitats or disturb them as little as possible. When land is cleared, some could be left in its natural state.

Section Wrap-up

Review

1. List three ways humans contribute to extinctions.
2. How does developing land for construction and farming reduce the diversity of life on Earth?
3. **Think Critically:** How might humans work to slow the rate of present-day extinctions?

Explore the Issue

Should governments of developing countries require their people to preserve portions of their land in its natural state, thereby reducing the amount of land that can be developed for farming and logging? If so, how would you answer landowners who claim that they can grow enough food for their families and livestock to live on only if they develop all of their land? If not, then how can the rate of extinction of animals and plants that are so vital to our survival be reduced?

SCIENCE & SOCIETY

13•3 Early Earth History

Science Words

Precambrian time
cyanobacteria
Paleozoic Era
amphibian
reptile

Objectives

- Identify dominant life-forms in Precambrian time and the Paleozoic Era.
- Draw conclusions about how organisms adapted to changing environments in Precambrian time and the Paleozoic Era.
- Describe changes in Earth and its life-forms at the end of the Paleozoic Era.

Precambrian Time

Look again at **Figure 13-2. Precambrian** (pree KAM bree un) **time** represents the longest geologic time unit of Earth's history. This time lasted from 4.6 billion to about 544 million years ago. Although the Precambrian was the longest unit of geologic time, relatively little is known about Earth and the organisms that lived during this time. Why is the fossil record from Precambrian time so sparse?

Precambrian rocks have been deeply buried and changed by heat and pressure. They have also been eroded more than younger rocks. These changes affect not only the rocks, but the fossil record as well. Most fossils can't withstand the metamorphic and erosional processes that most Precambrian rocks have experienced.

Figure 13-9

Scientists look for evidence of what Earth looked like during Precambrian time and what life-forms were present.

A Lightning or the sun may have provided the energy necessary to build amino acids out of the simple compounds in Earth's Precambrian atmosphere.

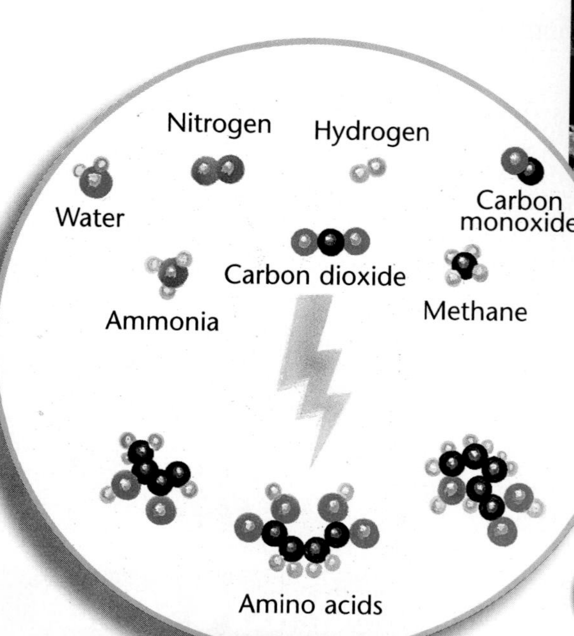

B Amino acids are the "building blocks of life." These amino acids reacted with each other and combined to form the compounds from which life may have evolved.

Figure 13-10

In some conditions, cyanobacteria produce mound-shaped layers of calcium carbonate called stromatolites. Stromatolites were common about 2.8 billion years ago and are still being formed today. *What does this imply about the life-form cyanobacteria?*

A Stromatolites have changed very little throughout geologic time. These modern ones in Australia look very much like ancient stromatolites.

B Cross section of an individual stromatolite head.

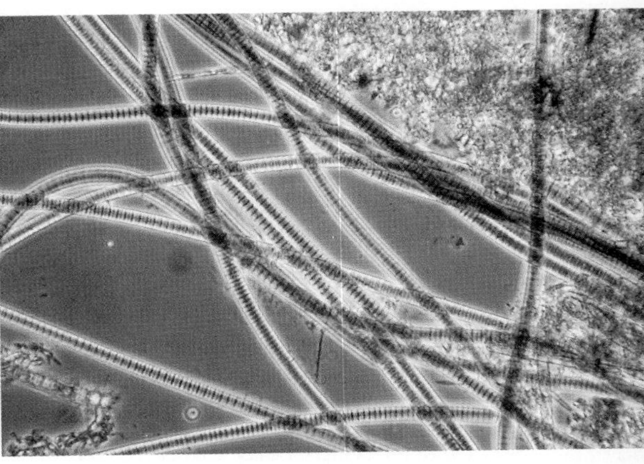

C Microscopic view of a cyanobacterium.

Early Life

It wasn't until fossilized cyanobacteria forming layered mats, called stromatolites, were found that scientists could begin to unravel Earth's complex history. **Cyanobacteria,** shown in **Figure 13-10,** are bacteria thought to be one of the earliest forms of life on Earth. Cyanobacteria first appeared on Earth about 3.5 billion years ago. As these organisms evolved, they contributed to changes in Earth's atmosphere. During the few billion years following the appearance of cyanobacteria, oxygen became a major gas in Earth's atmosphere. The ozone layer in the stratosphere also began to develop, shielding Earth from ultraviolet rays. These major changes in the air allowed species of single-celled organisms to evolve into more complex organisms.

Animals without backbones, called invertebrates, developed near the end of Precambrian time. Imprints of jellyfish and marine worms have been found in late Precambrian rocks. However, because these early invertebrates were soft-bodied, they weren't easily preserved as fossils. This is another reason the Precambrian fossil record is so sparse.

The Paleozoic Era

In Chapter 12, you discovered that fossils are more likely to form if organisms have hard parts. When organisms developed hard parts, the **Paleozoic** (pay lee uh ZOH ihk) **Era** began. Fossils were then more easily preserved.

CONNECT TO

CHEMISTRY

Cyanobacteria are thought to have been one of the mechanisms by which Earth's early atmosphere became richer in oxygen. ***Research*** the composition of Earth's early atmosphere and where these gases originated.

Activity 13-1

Evolution Within a Species

Use a model in this activity to observe how adaptation within a species might cause the evolution of a particular trait, leading to the development of a new species.

Problem

How might adaptation within a species cause the evolution of a particular trait?

Materials

- Deck of playing cards

Procedure

1. Remove all of the kings, queens, jacks, and aces from a deck of playing cards.
2. Each remaining card represents an individual in a population of animals called "varimals." The number on each card represents the height of the individual. For example, the 5 of diamonds is a varimal that's 5 units tall.
3. Calculate the average height of the population of varimals represented by your cards.
4. Suppose varimals eat grass, shrubs, and leaves from trees. A drought causes many of these plants to die. All that's left are a few tall trees. Only varimals at least 6 units tall can reach the leaves on these trees.
5. All the varimals under 6 units leave the area to seek food elsewhere or die from starvation. Discard all of the cards with a number value less than 6. Calculate the new average height of the population of varimals.
6. Shuffle the deck of remaining cards.
7. Draw two cards at a time. Each pair represents a pair of varimals that will mate and produce offspring.
8. The offspring of each pair reaches a height equal to the average height of his or her parents. Calculate and record the height of each offspring.
9. Repeat by discarding all parents and offspring under 8 units tall. Now calculate the new average height of varimals. Include both the parents and offspring in your calculation.

Analyze

1. How did the average height of the population change over time?
2. If you hadn't discarded the shortest varimals, would the average height of the population have changed as much? **Explain.**
3. What trait was selected for?

Conclude and Apply

4. Why didn't every member of the original population reproduce?
5. If there had been no varimals over 6 units tall in step 5, what would have happened to the population?
6. If there had been no variation in height in the population before the droughts occurred, would the species have been able to evolve into a taller species?
7. How does this activity **demonstrate** that traits evolve in species?

Era of Ancient Life

The Paleozoic Era, or era of ancient life, began about 544 million years ago. Warm, shallow seas covered much of Earth's surface during early Paleozoic time. Because of this, most of the life-forms were marine, meaning they lived in the ocean. Trilobites (TRI luh bites) were very common. Brachiopods (BRAY kee uh pahdz) and crinoids (KRI noyds), which still exist today, were also very common. Although these animals may not be familiar to you, one type of animal you are familiar with—the fish—also evolved during this era.

The Paleozoic Era is broken into seven periods. The start of the Ordovician Period is marked by the beginning of the Appalachian Mountain building process. This was probably caused by the collision of the Eurasian or African continental plate with the North American Plate.

The first vertebrates, animals with backbones, developed during the Ordovician Period. Plant life moved from the oceans onto land during the Silurian Period. Fish became dominant in the Devonian Period, as seen in **Figure 13-11.** By this time plant life had developed on land, and animals began to move onto land as well.

Figure 13-11

This giant fish, *Dunkleosteus,* lived in Ohio during the Devonian Period. It grew to over 9 m long.

MiniLAB

Fossil Correlation

In the activity, you will demonstrate how fossils are used to date rock layers. Draw a sequence of sedimentary rocks containing fossils.

Procedure

1. Number the layers 1 through 3, bottom to top.
2. Identify the fossils in each layer. Suppose the bottom layer, layer 1, contains fossils B and A. Layer 2 contains fossils A, B, and C. Layer 3 contains only fossil C.
3. Determine the geologic periods each of the fossils lived in. For example, fossil A lived from the Cambrian through the Devonian Periods. Fossil C lived from the Devonian through the Permian Periods, and so on.
4. Construct and interpret scientific illustrations to help you determine the ages of each rock layer.

Analysis

1. Which layer or layers were you able to date to a specific period?
2. Why isn't it possible to determine during which specific period the other layers formed?
3. What is the age or possible age of each layer?

Paleozoic Puzzle

Suppose that there are many different and interesting fossils displayed around your classroom. Many of the fossil samples have been collected by other students and by your teacher. Your teacher tells you that she has studied the outcrops of rocks in the city where she attended college. It was at this time that she began her excellent collection of fossils from that area, which is displayed in your room. Her favorite fossil was a particular species of brachiopod known as *Mucrospirifer.* She identified this fossil from pictures and descriptions in books about Paleozoic fossils she used while attending college.

When your teacher went on a fossil collecting trip last summer, she found what seemed to be the same type of *Mucrospirifer* fossil in a rock formation hundreds of kilometers from where she attended college.

Solve the Problem:

1. **Is it possible that both of the fossils found by your teacher are *Mucrospirifer*?**
2. **What one fact should your teacher know about *Mucrospirifer* before using her fossils to attempt dating the rock layers in which they were found?**

Think Critically: Once your teacher has determined that both samples she found were *Mucrospirifer*, what could she conclude about the rock layers near her college and those she studied on her more recent fossil collection trip?

Mucrospirifer
50 mm

Life on Land

One type of fish evolved a lung that enabled it to survive out of water. This fish had fins that allowed it to move across land. The fact that lungfish could move across land and breathe air has led scientists to theorize that lungfish and amphibians (am FIHB ee unz) evolved from the same ancestor. **Amphibians** live on land and breathe air, but they must return to water to reproduce. Their eggs must be kept moist in water. They first appeared during the Devonian Period and became the dominant form of vertebrate life on land by the Mississippian Period.

Over time, one species of amphibian evolved an egg with a membrane that protected it from drying out. Because of this, the species no longer needed to return to water to reproduce. By the Pennsylvanian Period, reptiles had evolved, probably from the same ancestor as amphibians. **Reptiles** do not need to return to water to reproduce as shown in **Figure 13-13** on page 372. Reptiles have skin with hard scales that prevent loss of body fluids. This adaptation enables them to survive farther from water. They can survive in relatively dry climates, whereas amphibians cannot.

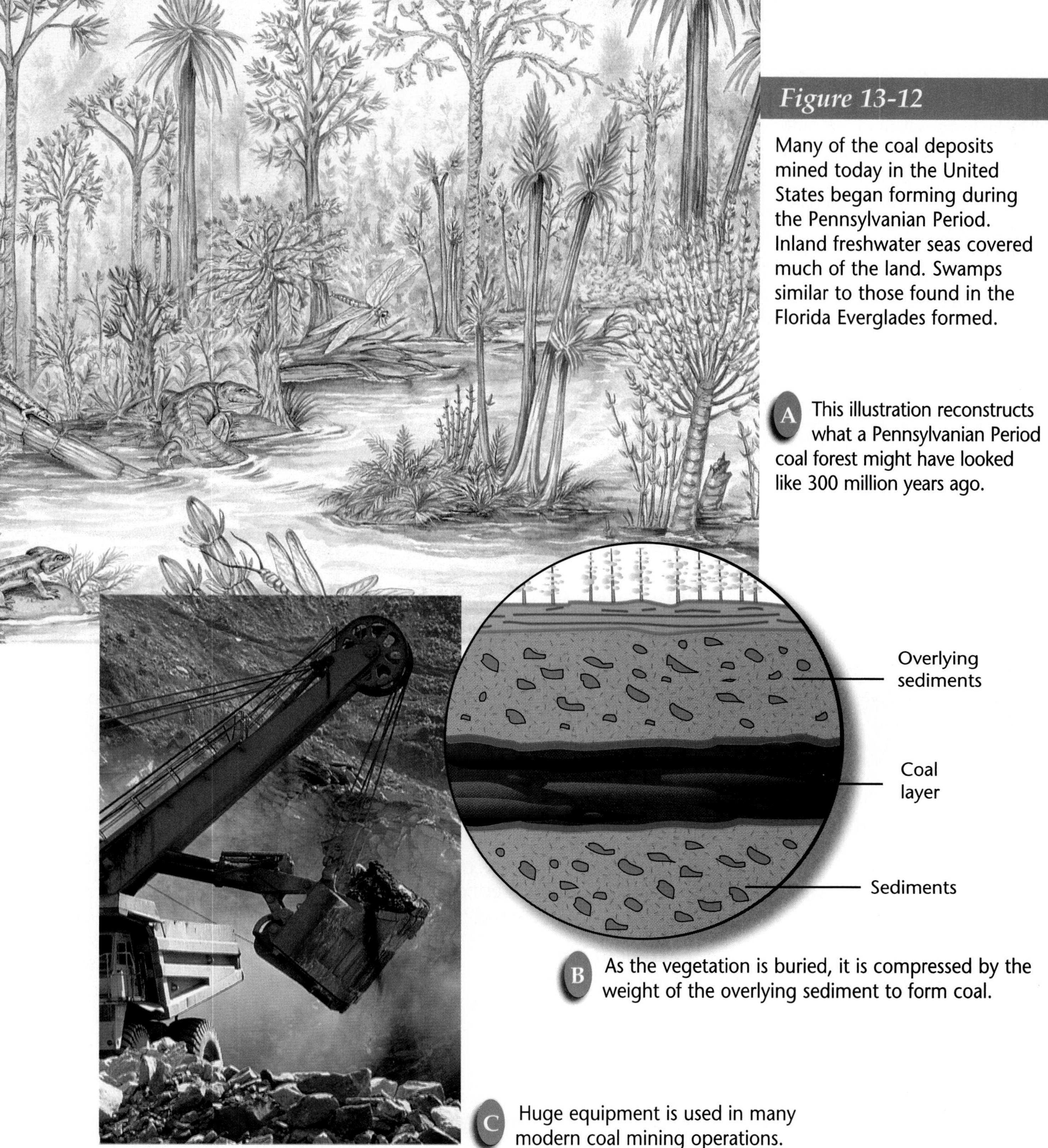

Figure 13-12

Many of the coal deposits mined today in the United States began forming during the Pennsylvanian Period. Inland freshwater seas covered much of the land. Swamps similar to those found in the Florida Everglades formed.

A This illustration reconstructs what a Pennsylvanian Period coal forest might have looked like 300 million years ago.

B As the vegetation is buried, it is compressed by the weight of the overlying sediment to form coal.

C Huge equipment is used in many modern coal mining operations.

End of an Era

Mass extinctions of many land and sea animals occurred, signalling the end of the Paleozoic Era. The cause of these extinctions may have been changes in the environment caused by plate tectonics. Near the end of the Permian Period, all continental plates came together to form the single landmass Pangaea and major glaciers formed.

Figure 13-13

Unlike frogs, salamanders, and other amphibians, reptiles can lay their eggs on land. This allows them to survive in relatively dry environments. *What characteristic of reptile eggs prevents the developing embryos from drying out?*

The slow, gradual collision of continental plates caused mountain building. Mountain building processes caused seas to drain away, and interior deserts spread over much of the United States and parts of Europe. Climates changed from mild and warm to cold and dry. Many species, especially marine organisms, weren't able to adapt to these and other changes and became extinct.

Section Wrap-up

Review

1. What geologic events occurred at the end of the Paleozoic Era?
2. How might geologic events at the end of the Paleozoic Era have caused the mass extinctions that occurred?
3. **Think Critically:** What major change in life-forms occurred to separate Precambrian time from the Paleozoic Era?

Skill Builder

Making and Using Tables

Use **Figure 13-2** to answer these questions about the Paleozoic Era. If you need help, refer to Making and Using Tables in the **Skill Handbook.**

1. When did the Paleozoic Era begin? When did it end?
2. How long did the Silurian Period last?
3. When did the Appalachian Mountains start to form?
4. When did the first insects appear on Earth?

Using Computers

Database Research trilobites, brachiopods, and crinoids in a historical geology book or computer database. Write a paragraph in your Science Journal describing each of these organisms and its habitat. Include hand-drawn illustrations and compare them with the illustrations in the computer database on historical geology.

Science & Literature

Dreamtime Down Under

In the traditional oral stories of the Aborigines of Australia, the people have their own explanation of how our world was formed. They say the world as we know it began when Dreamtime began, long ago, before anyone can remember. Then everything was one. At first, Earth was cold and dark, and the spirit Ancestors slept underground.

When the Ancestors awoke, they moved to Earth's surface and created the sun for warmth and light. Some Ancestors took the shapes of people; others became animals, plants, rain, clouds, or stars. As the Ancestors moved over Earth, they sang. Their movements and their singing shaped the land, creating hills, rivers, and other land features.

Leaving a Path

As they walked, the Ancestors left Dreaming Tracks that the Aborigines treasure and protect today. When the Ancestors tired, they returned underground. The bodies of some Ancestors remain on Earth's surface as rock outcroppings, trees, islands, and other features.

Ancient Aborigines drew maps in the sand to show where the Ancestors came out, walked, and returned underground. These sand drawings with their traditional dot patterns form the basis of Aboriginal art, like the bark painting shown above.

Science Journal

In your Science Journal, write a poem that expresses your own view of our relationship to nature and to the land.

Dreaming the Big Bang

Some compare the Dreamtime forces that shaped Earth to the big bang theory: huge fields of energy interacting and forming planets. Much later, more energy—more Dreaming—caused the movement of Earth's crust and eventually created today's continents, including Australia.

Aborigines believe that the Ancestors still live in the land. Dreamtime continues with no foreseeable end. Today, Aborigines are struggling to maintain their ancient traditions while living in modern Australia.

13•4 Middle and Recent Earth History

Science Words

Mesozoic Era
Cenozoic Era

Objectives

- Compare and contrast dominant life-forms in the Mesozoic and Cenozoic Eras.
- Explain how changes caused by plate tectonics affected the evolution of life during the Mesozoic Era.
- Identify when humans first appeared on Earth.

The Mesozoic Era

Some of the most fascinating life-forms ever to live on Earth evolved during the Mesozoic Era. One group of organisms you're familiar with—the dinosaurs—appeared during this geologic era.

The Breakup of Pangaea

The **Mesozoic** (mez uh ZOH ihk) **Era,** or era of middle life, began about 245 million years ago. At the beginning of the Mesozoic Era, all continents were joined as a single landmass. Recall from Chapter 11 that this landmass was called Pangaea, as shown in **Figure 13-14.** Pangaea separated into two large landmasses during the Triassic Period. The northern mass was *Laurasia,* and *Gondwanaland* was in the south. As the Mesozoic Era continued, *Laurasia* and *Gondwanaland* broke up and formed the present-day continents.

Species that survived the mass extinctions of the Paleozoic Era adapted to new environments. Recall that the hard scales of a reptile's skin help to retain body fluids. This trait, along with the hard shell of their eggs, enabled them to readily adapt to the drier climate of the Mesozoic Era. They became the dominant animal life-form in the Jurassic Period. Some of the reptiles evolved into archosaurs, the common ancestor of crocodiles, dinosaurs, and birds.

300 million years ago

Figure 13-14

The supercontinent Pangaea formed at the end of the Paleozoic Era. It began to break up in the Jurassic Period.

Dinosaurs

What were the dinosaurs like? Dinosaurs ranged in height from less than one meter to enormous creatures like *Apatosaurus* and *Tyrannosaurus*. One species of tyrannosaur stood as tall as a two-story building. Some dinosaurs ate meat, whereas others ate only plants.

The first small dinosaurs appeared during the Triassic Period. Larger species appeared during the Jurassic and Cretaceous Periods. Throughout the Mesozoic Era, new species of dinosaurs evolved as other species became extinct.

Recent studies indicate that dinosaurs may not have been cold-blooded, as are present-day reptiles. Tracks left in the mud by reptiles are usually close together. This indicates that reptiles generally move very slowly. Some dinosaur tracks that have been found indicate that they were much faster than most of the cold-blooded reptiles. This faster speed would be expected of warm-blooded animals, which need speed to be successful in hunting. *Gallimimus* was four meters long and could reach speeds of 80 km/h, as fast as a modern race horse.

Figure 13-15

Fossil evidence suggests that some dinosaurs, such as *Maiasaura,* may have nurtured their young. Fossil nests contain newly hatched and juvenile young. *What type of evidence might support this idea?*

Figure 13-16

Fossils of *Archaeopteryx* that are about 150 million years old show both birdlike features and dinosaurlike features. *What birdlike and dinosaurlike features can you recognize?*

A A detailed imprint of a fossil feather that is believed to be from *Archaeopteryx*.

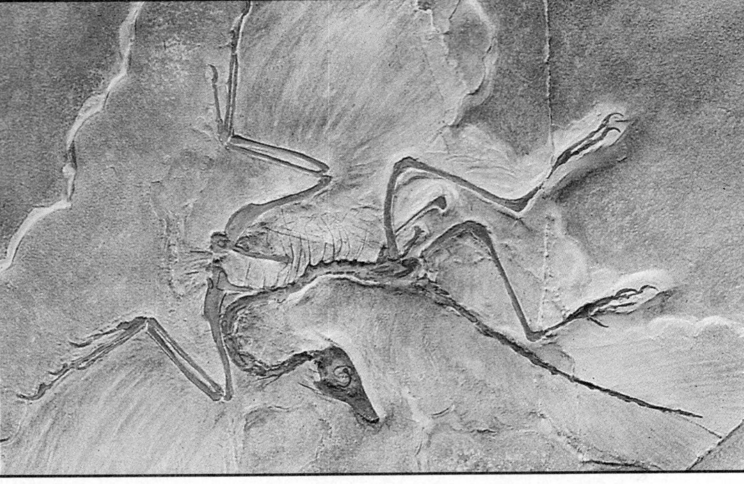

B Considered the world's most priceless fossil, *Archaeopteryx* was found in a limestone quarry in Germany in 1861.

C A reconstruction of what *Archaeopteryx* may have looked like.

Good Mother Dinosaurs

The fossil record indicates that some dinosaurs nurtured their young and traveled in herds with adults surrounding their young. One such dinosaur is *Maiasaura,* shown in **Figure 13-15** on page 375. This dinosaur built nests in which it laid its eggs and raised its offspring. Nests have been found in clusters, indicating that more than one family of dinosaurs built in the same area. Some fossils of hatchlings have been found very close to the adult animal. This has led some scientists to hypothesize that some dinosaurs nurtured their young. In fact, *Maiasaura* hatchlings may have stayed in the nest while they grew in length from about 35 cm to more than one meter.

Other evidence that leads scientists to think that dinosaurs may have been warm-blooded has to do with their bone structure. Cross sections of the bones of cold-blooded animals exhibit rings similar to growth rings in trees. The bones of some dinosaurs don't show this ring structure. Instead, they are similar to bones found in birds and mammals. These observations indicate that dinosaurs may have been warm-blooded, fast-moving, nurturing animals somewhat like present-day mammals and birds. They might have been quite different from present-day reptiles.

Birds

The first birds appeared during the Jurassic Period of the Mesozoic Era, as seen in **Figure 13-16.** The animal *Archaeopteryx* had wings and feathers like a bird but teeth and claws like a meat-eating dinosaur. *Archaeopteryx* may not have been a direct ancestor of today's birds. But modern birds and *Archaeopteryx* probably share a common ancestor. Scientists think that birds evolved from dinosaurs. Some scientists even think that birds are dinosaurs, evolved from the advanced theropod called *Troodon,* shown in **Figure 13-17.** Theropods form a group of meat-eating dinosaurs that walked mainly on their hind legs.

Mammals

Mammals first appeared in the Triassic Period. Mammals are warm-blooded vertebrates that have hair or fur covering their bodies. The females produce milk to feed their young. These traits have enabled mammals to survive in many changing environments.

Figure 13-17

A highly evolved dinosaur called *Troodon* had a birdlike stance, much like a modern ostrich.

MiniLAB

Seafloor Spreading

Geologists have measured the rate of seafloor spreading at the Mid-Atlantic Ridge at approximately 3.5 to 4.0 cm per year. This rate can be used to determine the age of the Atlantic Ocean.

Procedure

1. On a globe or world map, measure the distance in kilometers between a point near the east coast of South America and a corresponding point on the west coast of Africa.
2. Assuming that the rate of motion listed above has been relatively constant, calculate how many years it took to create the present Atlantic Ocean if the continents were once joined.
3. Measure in SI several times and take the average of your results.
4. Check your predictions with the information provided in **Figure 13-2.**

Analysis

1. Did the values used to obtain your average value vary much?
2. How close did your average value come to the accepted estimate for the beginning of the breakup of Pangaea?

Figure 13-18

Angiosperms and pollinating insects coevolved. The sweet nectar produced by many flowers attracts insects in search of food. The pollen of the flower sticks to the insect, which carries it to another flower. Some angiosperms wouldn't be able to reproduce without a particular species of insect.

Gymnosperms

During the Cretaceous Period, seas expanded inland and species of plants, animals, and other organisms continued to adapt to new environments. Gymnosperms (JIHM nuh spurmz), which first appeared in the Paleozoic Era, continued to adapt to their changing environment. Gymnosperms are called naked-seed plants because they have no fruit covering their seeds. Pines, sequoias, and firs are gymnosperms.

Angiosperms

A new type of plant, called angiosperms (AN jee uh spurmz), evolved from existing plants. Angiosperms, or flowering plants, produce seeds with hard, outer coverings. Common angiosperms are magnolias and willows.

Many angiosperms survived while other organisms did not because their seeds had hard coatings that protected them and allowed them to develop in varied environments. Angiosperms are so adaptive that they remain the dominant land plant today, as shown in **Figure 13-18.** Present-day angiosperms that evolved during the Mesozoic Era include maple and oak trees.

The end of the Mesozoic Era was a time when landmasses were breaking up and seas were draining from the land. There was also increased volcanic activity. Many life-forms, including the dinosaurs, became extinct. These extinctions were caused by changing environments. What caused the environments to change is still being actively investigated by scientists.

The Cenozoic Era

The **Cenozoic** (sen uh ZOH ihk) **Era**, or era of recent life, began about 66 million years ago when dinosaurs and many other life-forms became extinct. Many of the mountain ranges throughout North and South America began to form at this time.

During the Cenozoic Era, the climate became cooler and ice ages occurred. The Cenozoic Era is subdivided into two periods. The present-day period is the Quaternary. It began after the last ice age. Many changes in Earth, its climate, and its life-forms, shown in **Figure 13-19,** occurred in the Cenozoic Era.

Pangaea broke up during the Mesozoic Era, and continents continued to move toward their present positions.

Times of Mountain Building

The Alps formed as the African Plate collided with the Eurasian Plate. The Himalaya Mountains started to form when the Indian Plate collided with the Eurasian Plate.

As the number of flowering plants increased, their pollen and fruit provided food for the many insects and small, plant-eating mammals. The plant-eating mammals provided food for meat-eating mammals.

Figure 13-19

Many now-extinct animals lived in North America before and during the ice ages in the Pleistocene Epoch.

Activity 13-2

Design Your Own Experiment

Geologic Time Line

In Chapter 12, you learned how geologists date rocks, fossils, and geologic events. Using these dates, geologists have determined that Earth is 4.6 billion years old. Geologists have also used these dates to subdivide Earth's history into different units of time. Just as you can subdivide your past into shorter periods of time by dating major events, such as your first day of school, you can subdivide Earth's history by dating major events in its past.

PREPARATION

Problem

What kind of model can be used to show the subdivisions of Earth's past?

Form a Hypothesis

Based on information provided in **Figure 13-2** and the information presented in Sections 13-3 and 13-4, state a hypothesis about what type of model you can develop to show the subdivisions of Earth's history.

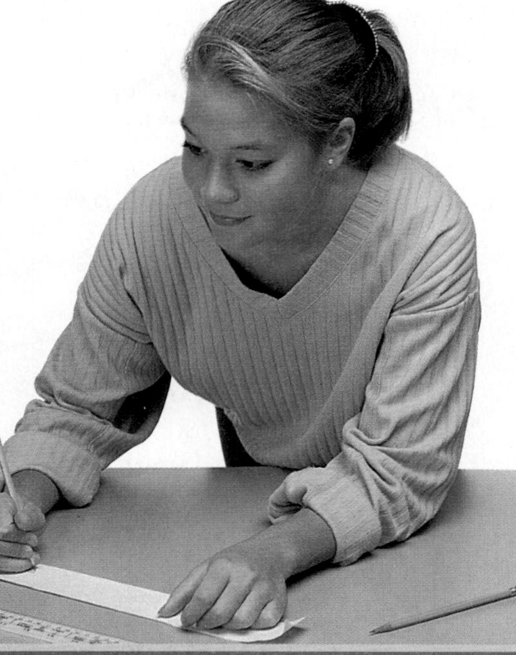

Objectives

- Observe how the ages of rocks, fossils, and geologic events are used to subdivide geologic time on the geologic time scale.
- Design an experiment in which a model is constructed that represents the subdivisions of geologic time.

Possible Materials

- adding machine tape
- meterstick or metric ruler
- set of colored pencils
- pencil
- scissors

Safety Precautions 

Take care not to cut yourself while using the scissors.

PLAN THE EXPERIMENT

1. As a group, agree upon and write out the hypothesis statement.
2. As a group, list the steps that you need to take to test your hypothesis. Be very specific, describing exactly what you will need to do at each step.
3. Make a list of the materials that you will need to complete your experiment.
4. Determine how you will construct your model of geologic time using the materials listed on the opposite page.
5. Determine a scale to use for your model of the subdivisions of geologic time.

Check the Plan

1. Read over your entire experiment to make sure that all steps are in a logical order.
2. Draw a small version of the model you plan to construct.
3. Will you summarize data in a graph, table, or some other form of organization?
4. *Make sure your teacher approves your plan and that you have included any changes suggested in the plan.*

DO THE EXPERIMENT

1. Carry out the experiment as planned.
2. Construct your model of the subdivisions of geologic time.
3. While conducting the experiment and constructing your model, write down any observations that you or other members of your group make in your Science Journal.
4. When collecting the data, be sure to include dates that are important to you as well as dates of geologic events.
5. For additional information to use on your model, use references such as geology books, history books, and encyclopedias.

Analyze and Apply

1. Which events were the most difficult to plot?
2. **Compare** events from your personal past or the existence of humans on Earth with the duration of geologic time.
3. Use a geologic time scale to determine the dates when each major division of geologic time started and ended. **Sequence** these ages on your model of geologic time.
4. **Estimate** what percent of geologic time occurred during Precambrian time?
5. During which era were you able to find the most in-depth information about geologic events for your model?

Go Further

Form a hypothesis as to why more is known about geologic events during the era for which you were able to find the most information.

Figure 13-20

Kangaroos are marsupials that live in Australia and carry their young (joey) in a pouch.

Further Evolution of Mammals

Mammals evolved into larger life-forms. Recall how horses have evolved from small, multi-toed animals into the much larger, hoofed animals of today. Not all mammals remained on land. Ancestors of the present-day whales and dolphins evolved to make their lives in the sea.

As Australia and South America separated from Antarctica in the continuing breakup of Pangaea, many life-forms became isolated. They evolved separately from life-forms in other parts of the world. Evidence of this can be seen today with the dominance of marsupials in Australia. Marsupials are mammals that carry their young in a pouch, as seen in **Figure 13-20.**

Our species, *Homo sapiens*, probably appeared about 500 000 years ago, but became a dominant animal only about 10 000 years ago. As the climate remained cool and dry, many larger mammals became extinct. Some scientists think the appearance of humans may have led to the extinction of other mammals. As their numbers grew, humans competed for food that other animals relied upon. Also, fossil records indicate that early humans were hunters.

Section Wrap-up

Review

1. In which era, period, and epoch did *Homo sapiens* first appear?
2. Did mammals become more or less abundant after the extinction of the dinosaurs? Explain why.
3. How did the development of hard seeds enable angiosperms to survive in a wide variety of climates?
4. **Think Critically:** Why are the periods of only the Cenozoic Era divided into epochs?

Skill Builder

Sequencing

Arrange these organisms in sequence according to when they first appeared on Earth: *mammals, reptiles, dinosaurs, fish, angiosperms, birds, insects, amphibians, first land plants,* and *bacteria.* If you need help, refer to Sequencing in the **Skill Handbook.**

USING MATH

Make a graph comparing the periods of time that make up the Mesozoic and Cenozoic Eras. Express how long dinosaurs were dominant compared with the time humans have been dominant.

Chapter 13

Review

Summary

13-1: Evolution and Geologic Time

1. Geologic time is divided into eras, periods, and epochs.
2. Divisions within the geologic time scale are based on major evolutionary changes in organisms and on geologic events such as mountain building and plate movements.
3. Plate movements cause changes in Earth's climate that affect organic evolution.

13-2: Science and Society: Present-Day Rapid Extinctions

1. Humans contribute to extinctions by eliminating natural habitats of organisms.
2. As land is developed by humans, the diversity of life on Earth is reduced.
3. Careful planning, concern for all organisms, and strict laws can help prevent extinctions.

13-3: Early Earth History

1. Cyanobacteria were an early form of life that evolved during Precambrian time. Trilobites, brachiopods, fish, and corals were abundant during the Paleozoic Era.
2. By the process of natural selection, bacteria evolved into higher life-forms, which evolved into many marine invertebrates during the early Paleozoic Era. Plants and animals began to move onto land once a protective ozone layer had been established.
3. During the Paleozoic Era, glaciers advanced, and seas withdrew from the continents. Many marine invertebrates became extinct.

13-4: Middle and Recent Earth History

1. Reptiles and gymnosperms were dominant land life-forms in the Mesozoic Era. Mammals and angiosperms began to dominate the land in the Cenozoic Era.
2. Plate tectonic changes in the Mesozoic Era caused climates to become drier and seas to expand. These changes affected life on land and in the sea.
3. *Homo sapiens* evolved during the Pleistocene Epoch.

Key Science Words

a. amphibian
b. Cenozoic Era
c. cyanobacteria
d. endangered
e. epoch
f. era
g. geologic time scale
h. habitat
i. Mesozoic Era
j. natural selection
k. organic evolution
l. Paleozoic Era
m. period
n. Precambrian time
o. reptile
p. species

Reviewing Vocabulary

Match each phrase with the correct term from the list of Key Science Words.

1. change in the hereditary features of a species over a long period of time
2. record of events in Earth history
3. largest subdivision of geologic time
4. geologic time with poorest fossil record
5. process by which the best-suited individuals survive in their environment
6. evolved from the same ancestors as amphibians
7. group of individuals that normally breed only among themselves
8. the geologic era in which we live
9. a species in which only a relatively small number of members exists
10. a place where organisms live and grow

Chapter 13 Review

Checking Concepts

Choose the word or phrase that completes the sentence.

1. The era in which you live began about _______ million years ago.
   a. 650 c. 1.6
   b. 245 d. 66
2. The process by which better suited organisms survive and reproduce is _______.
   a. endangerment c. gymnosperm
   b. extinction d. natural selection
3. The next smaller division of geologic time after the era is a(n) _______.
   a. period c. epoch
   b. stage d. eon
4. The beginning of the _______ Period was marked by the most recent ice age.
   a. Pennsylvanian c. Tertiary
   b. Triassic d. Quaternary
5. One of the earliest forms of life on Earth was the _______.
   a. gymnosperm c. angiosperm
   b. cyanobacterium d. dinosaur
6. Amphibians evolved from the same ancestors as _______.
   a. reptiles c. angiosperms
   b. lungfish d. gymnosperms
7. Dinosaurs lived during the _______ Era.
   a. Mesozoic c. Miocene
   b. Paleozoic d. Cenozoic
8. ________ have seeds without protective coverings.
   a. Angiosperms c. Gymnosperms
   b. Flowering plants d. Magnolias
9. _______ evolved to become the dominant land plant during the Cenozoic Era.
   a. Gymnosperms c. Ginkgos
   b. Angiosperms d. Algae
10. A key factor in preserving many species is _______.
    a. poaching c. construction
    b. law enforcement d. changing habitats

Understanding Concepts

Answer the following questions in your Science Journal using complete sentences.

11. How is natural selection related to evolution?
12. Briefly describe the major geologic and biological changes that took place during the Paleozoic Era.
13. Describe several causes for extinctions throughout geologic time.
14. Contrast the animal life of the Paleozoic Era with that of the early Cenozoic Era.
15. Compare and contrast reptile eggs and amphibian eggs.

Thinking Critically

16. Why couldn't plants move onto land prior to the establishment of an ozone layer?
17. Why are trilobites classified as index fossils?
18. What is the most significant difference between Precambrian and Paleozoic lifeforms?
19. How might the extinction of an edible species of plant from a tropical rain forest affect animals that live in the forest?
20. In the early 1800s, a naturalist proposed that the giraffe species has a long neck as a result of years of stretching their necks to reach leaves in tall trees. Explain why this isn't true.

Chapter 13 Review

Developing Skills

If you need help, refer to the **Skill Handbook.**

21. Observing and Inferring: Use the outlines of the present-day continents to make a sketch of the Mesozoic supercontinent Pangaea.

22. Hypothesizing: Why did trilobites become extinct at the end of the Paleozoic Era?

23. Interpreting Data: Fernando found what he thought was a piece of coral in a chunk of coal. Was he right? Explain.

24. Interpreting Scientific Illustrations: The circle graph below represents geologic time. Determine which era of geologic time is represented by each portion of the graph.

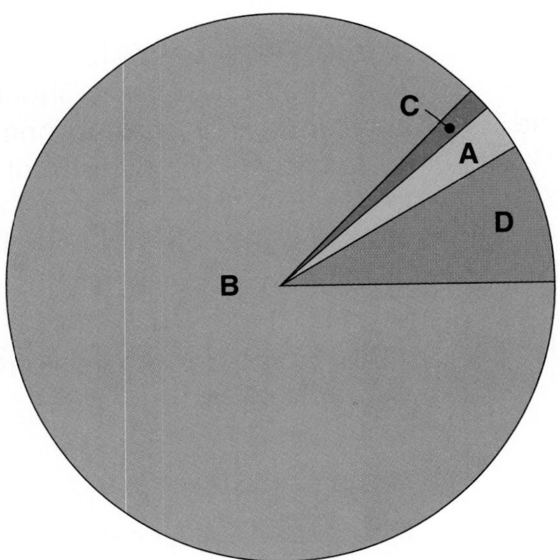

25. Interpreting Scientific Illustrations: The Cenozoic Era has lasted 66 million years. What percentage of Earth's 4.6 billion year history is that? How many degrees on the pie graph represent the Cenozoic Era?

Performance Assessment

1. **Formulate a Hypothesis:** Research the most recent theories on mass extinctions that occurred at the end of both the Paleozoic and Mesozoic Eras. Compare and contrast the most accepted theories. Form your own hypothesis as to why the extinctions occurred. Support your position with facts and models.
2. **Make a Model:** In Activity 13-1, you learned how a particular trait might evolve within a species. Modify the experimental model by using color instead of height as a trait. Design your activity with the understanding that varimals live in a dark-colored forest environment.
3. **Conduct a Survey:** Survey your classmates and have them share the important events from their time lines constructed in Activity 13-2. Generate one time line for the entire class. Using this newly constructed time line, repeat the Going Further section of the activity.

UNIT PROJECT 4

An Age-Old Question

What picture comes to mind when you think of a fossil? Many people think of dinosaurs. You've seen them in pictures, on television, and at the museum. But what other types of fossils exist? Do all fossils form in the same way? In this project, you will make "fossils" that represent each type described in Chapter 12. You'll use processes similar to those that occur in nature to form the following types of fossils: petrified remains, carbonaceous film, molds and casts, and original remains.

Procedures

1. With two or three classmates, go to the library and research different types of fossils. Find out where different fossil types are found and how each type forms. Compile your data in a table that shows the types of fossils your group has researched, how each type is formed, and where examples of each type have been found. Obtain photographs or make drawings in your Science Journal of the various types you have studied.
2. Work with other groups in the class to come up with possible ways of making models of the various types of fossils. Each group of students should concentrate on making one type of fossil. Can you think of a way of showing a carbon imprint of a tree leaf? To show how original remains are preserved, the group must devise a way that original parts of an organism would be preserved.

The group working with petrified remains must invent a method whereby original material in the organism is replaced by rock material. The group working on molds and casts must figure out some way to imitate the natural formation of molds and casts.

3. When your group decides how to make a model of your type of fossil, ask your teacher to review the method to be used. Using the method your group decided on, make your fossils. Once all groups have made fossils, the class should decide how to display its "fossil" collection.
4. If, during your library research, you discovered that fossils are found in areas near your home or school, ask your teacher or another adult to help you look for them. Remember, fossils found in rocks may not look anything like what you expect. If possible, have members of your class collect samples of each type of fossil all of the groups have made.

5. Each group should give a presentation that describes how they made their fossil. The presentation should include an explanation of each step and a description of the amount of time required to make the fossil model.

Going Further

Plan a trip to a local museum that has a fossil display. Ask the museum curator to show you examples of each type of fossil for which you made models. If possible, ask a local fossil collector to bring in his or her collection and share it with the class.

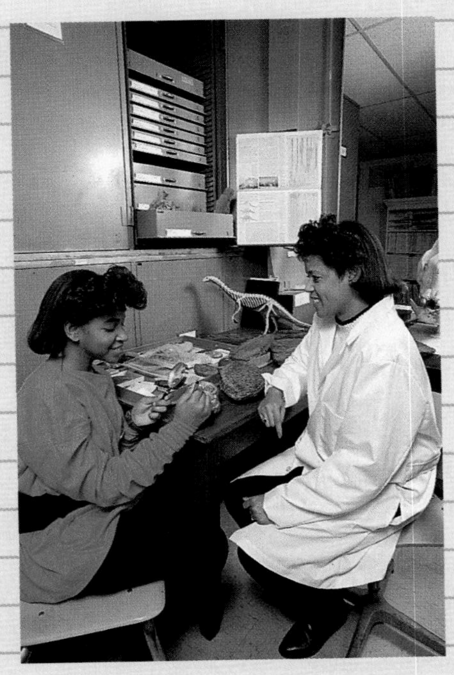

UNIT 5

Unit Contents

Earth's Air and Water

What's Happening Here?

When we think of tropical oceans, we may picture calm, clear waters under an azure sky. But tropical oceans give birth to hurricanes, the most powerful of storms. The swirling mass of energy in the smaller photograph is a hurricane gathering strength from the warm waters of the Atlantic. Far below, beneath the ocean's surface, tropical fish swim in lazy circles around a coral reef. Tropical organisms such as this sea star and anemone depend on the warm ocean water for survival. In this unit, you will learn about oceans and how they interact with Earth's atmosphere to affect life.

Science Journal

When hurricanes reach land, they can cause tremendous damage. In your Science Journal, write a report describing how scientists track the progress of these storms. Include a discussion on actions taken to minimize damage once a hurricane hits land.

Previewing the Chapter

Chapter 14

Atmosphere

Do you know how windsurfing works? A surfer uses the power of the wind to skim a surfboard across water. Wind comes from the uneven heating of the atmosphere, the air around Earth. When air is heated, its molecules absorb energy, move faster, and take up more space. The heated air becomes less dense. More-dense air slips underneath, forcing less-dense, heated air to rise. That creates wind.

EXPLORE ACTIVITY

Temperature affects the density of air.  

1. Your teacher will pour a small quantity of water into a soda can.
2. The can will be heated until the water boils.
3. Then the heat will be turned off, and the can submerged upside down in cold water.

Observe: What happens as the can cools? In your Science Journal, hypothesize why this happens. How is this related to windsurfing?

Previewing Science Skills

- In the **Skill Builders**, you will **make and use graphs, map concepts,** and **compare and contrast.**
- In the **Activities,** you will **record and analyze data, observe, draw conclusions,** and **design an experiment.**
- In the **MiniLABs,** you will **measure in SI, observe, infer,** and **compare and contrast.**

14•1 Earth's Atmosphere

Science Words

troposphere
ionosphere

Objectives

- Name the gases in Earth's atmosphere.
- Describe the structure of Earth's atmosphere.
- Explain what causes air pressure.

Composition of the Atmosphere

It's early morning in the future. You're getting dressed for work. As you eat breakfast, the weather report comes over the computer screen: "Smog levels higher than normal; temperatures near 38°C; the ozone layer in the stratosphere thinner than yesterday." You'll need your filter mask to protect your lungs from the smog. Pollution in the atmosphere has raised the temperature. You'll have to wear cool clothing. The thinner ozone layer requires you to use a strong sunblock lotion to protect yourself from skin cancer.

This scenario may not sound pleasant, but it's a future you may face. Because your life depends on the air you breathe, you need to know about the atmosphere, its composition, its structure, how it affects you, and how you affect it.

Atmospheric Gases

The atmosphere surrounding Earth extends from Earth's surface to outer space. The atmosphere is a mixture of gases with some suspended solids and liquids. **Figure 14-1** is a graph of the gases in Earth's atmosphere. Nitrogen is the most common gas. Oxygen makes up 21 percent of our atmosphere. We need oxygen to breathe. Water vapor makes up from zero to four percent of the atmosphere. When the percentage of water vapor is higher, the percentages of other gases are slightly lower.

The atmosphere also contains other gases and smog, a type of pollution. The kind of smog affecting an area depends on the pollutants. Car exhaust expels nitrogen oxides and unburned hydrocarbons into the air. These pollutants mixed

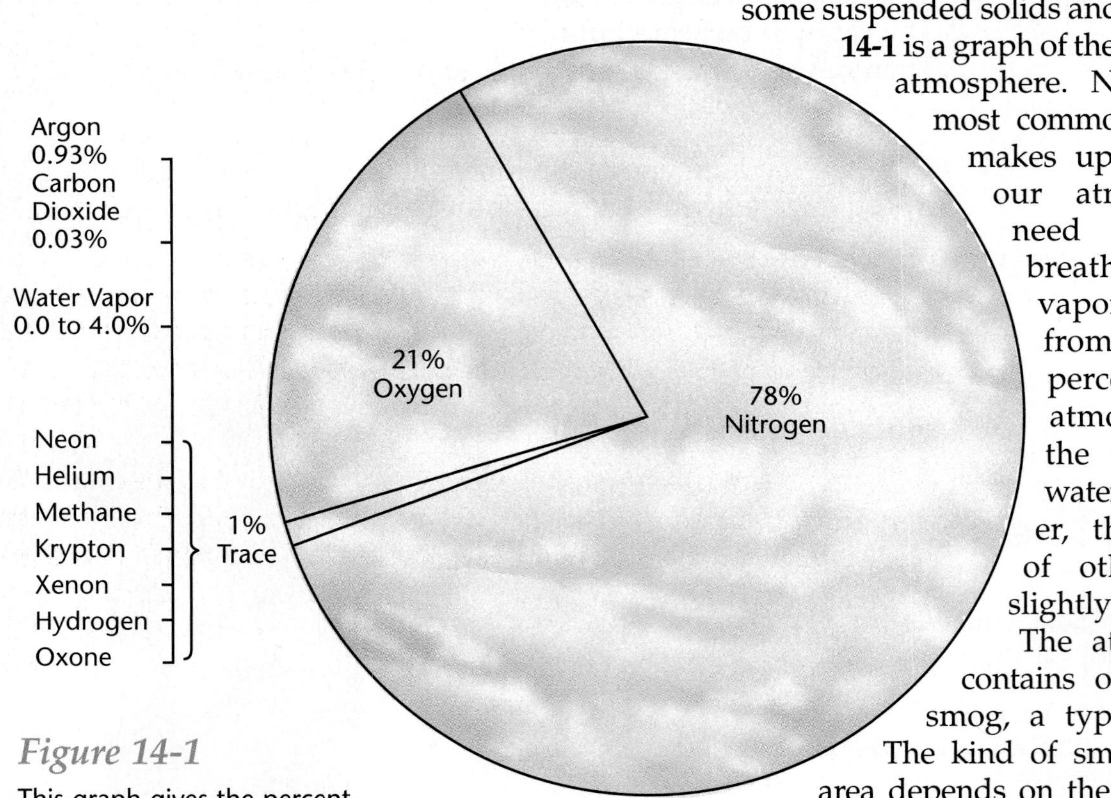

Figure 14-1

This graph gives the percentages of the gases that make up our atmosphere.

Figure 14-2

Smog lies over the Los Angeles area. Learning about the composition and structure of the atmosphere will help you understand pollution problems and help you make decisions about the atmosphere in the future.

with oxygen and other chemicals in the presence of sunlight cause an L.A.-type, brown smog, as shown in **Figure 14-2.** Sulfur oxides from burning coal or oil cause a gray smog.

Another component of smog is ozone. Ozone is a gas that occurs naturally in the stratosphere, but ozone is not normally found in the lower part of the atmosphere. When it is formed in the air above cities, ozone is considered a pollutant. In the lower atmosphere, ozone can harm plants and damage our lungs.

Atmospheric Solids and Liquids

Gases aren't the only thing making up Earth's atmosphere. Dust, salt, and ice are three common solids found in the atmospheric mixture. Dust gets into the atmosphere when wind picks it up off the ground and carries it along. Ice is common in the form of hailstones and snowflakes. Salt is picked up from ocean spray.

The atmosphere also contains liquids. The most common liquid in the atmosphere is water found in droplets of clouds. Water is the only substance that exists as a solid, liquid, and gas in Earth's atmosphere.

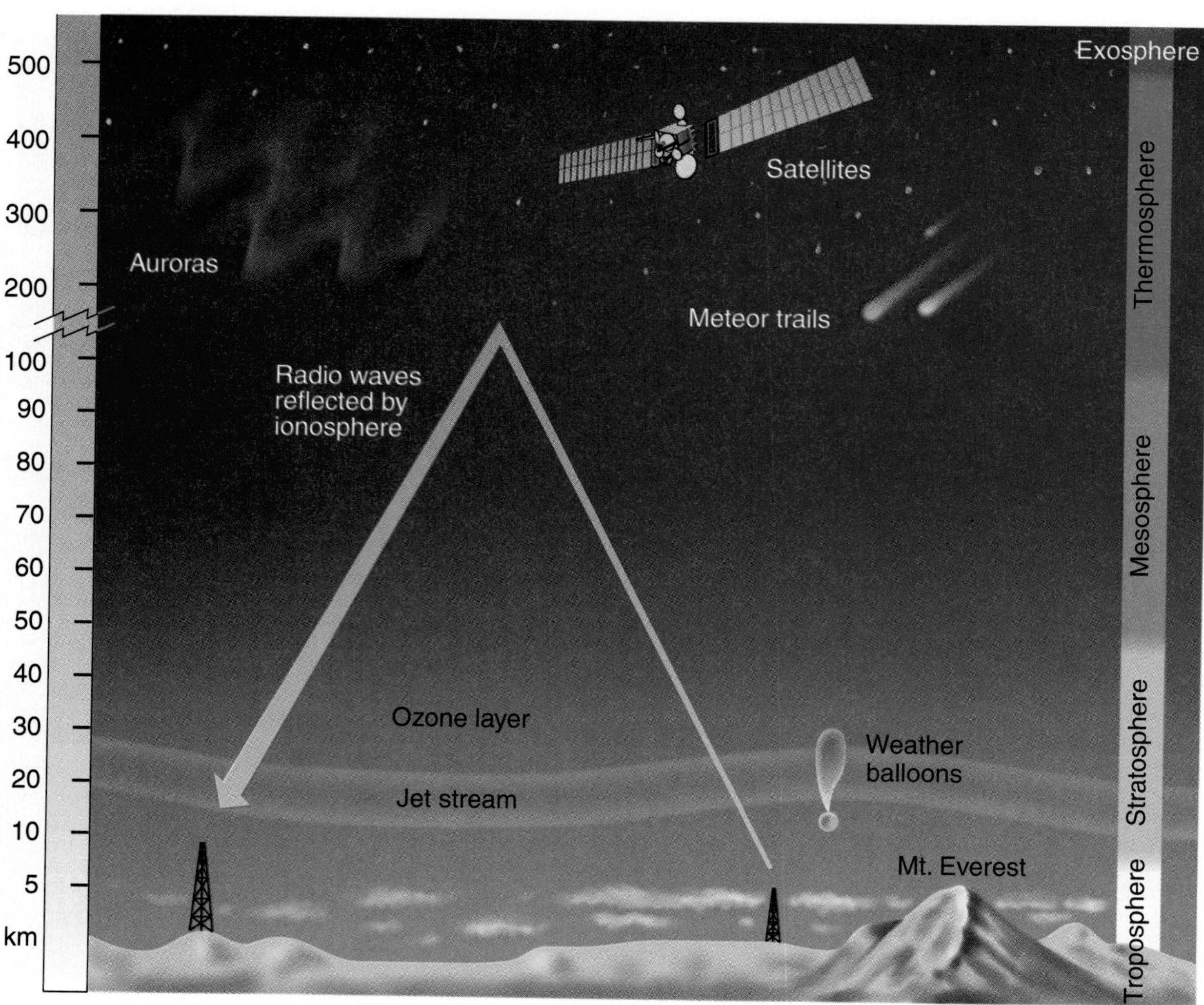

Figure 14-3

Although Earth's atmosphere extends hundreds of kilometers upward, 75 percent of all atmospheric gases are in the 15 km closest to Earth.

Structure of the Atmosphere

The weather forecast of the future predicted a high smog level and a thin ozone layer in the stratosphere. Both conditions affect your health, but in different ways. Where in the atmosphere does smog occur? Where can you find not just ozone but the ozone layer?

Figure 14-3 illustrates Earth's five main atmospheric layers: the troposphere, stratosphere, mesosphere, thermosphere, and exosphere. Each layer has unique characteristics.

Lower Layers of the Atmosphere

We live in the troposphere, the layer closest to the ground. The **troposphere** contains 75 percent of the atmospheric gases, as well as dust, ice, and liquid water. Weather, clouds, and smog occur in the troposphere.

Above the troposphere lies the stratosphere. As **Figures 14-3** and **14-6** show, a layer of ozone exists within the stratosphere. This ozone layer was mentioned in the future forecast because the ozone layer directly affects your health. You'll learn more about this layer in Section 14-2.

Upper Layers of the Atmosphere

Beyond the stratosphere are the mesosphere, thermosphere, and exosphere. One important layer of the thermosphere is the **ionosphere,** a layer of electrically charged particles. When solar energy bombards these particles, it creates ions and free electrons. These ions and free electrons can interfere with certain kinds of radio waves sent from Earth. During the daytime, energy from the sun interacts with the particles in the ionosphere, creating so many ions and free electrons that they absorb AM radio waves. At night, with fewer ions and free electrons, AM radio transmissions bounce off the ionosphere. This bouncing allows radio transmissions from one side of the globe to be received on the other side of the globe. **Figure 14-4** illustrates this.

The exosphere is the uppermost part of Earth's atmosphere. Beyond it lies space. If you were an astronaut traveling upward through the exosphere, you would encounter fewer and fewer molecules and ions. Eventually, you would find so few molecules and so few ions that, for all practical purposes, you would be out of Earth's atmosphere and in space. There's no clear boundary between the atmosphere and space.

The space shuttle orbits Earth in the exosphere, 280 km above Earth's surface. There are so few molecules and ions that the wings of the shuttle, used in the lower atmosphere, are useless. The spacecraft must rely on bursts from small rocket thrusters to maneuver.

Figure 14-4

Radio waves can be received by antennas around the globe (A). At night, AM radio waves that strike the ionosphere at sharp angles pass through to space, but other waves strike at lower angles and are reflected back toward Earth (B).

USING TECHNOLOGY

Weather Satellites

Weather forecasters rely on information collected by a variety of instruments, including weather satellites. Weather satellites measure visible and invisible radiation in the atmosphere. As satellites orbit, thrusters turn them to point cameras and take photos of Earth's surface. At night, satellites use heat radiation to produce images. Photos are stored on tape and transmitted by FM radio frequencies to a receiving station. High-speed computers assemble the information and make maps showing cloud cover. Weather forecasters use satellite pictures of clouds in forecasting. The photo below shows one kind of satellite.

There are two types of weather satellites. One type, the *geostationary satellite,* orbits Earth above the equator at speeds to position the satellite over the same spot on Earth. The satellite gives images of about half of Earth. The *polar-orbiting satellite* orbits from pole to pole and views the entire Earth twice each 24 hours. Polar orbiting satellites orbit closer to Earth and take more-detailed photos.

Think Critically: Explain why we need polar-orbiting satellites to take photos of Earth.

A Weather Satellite

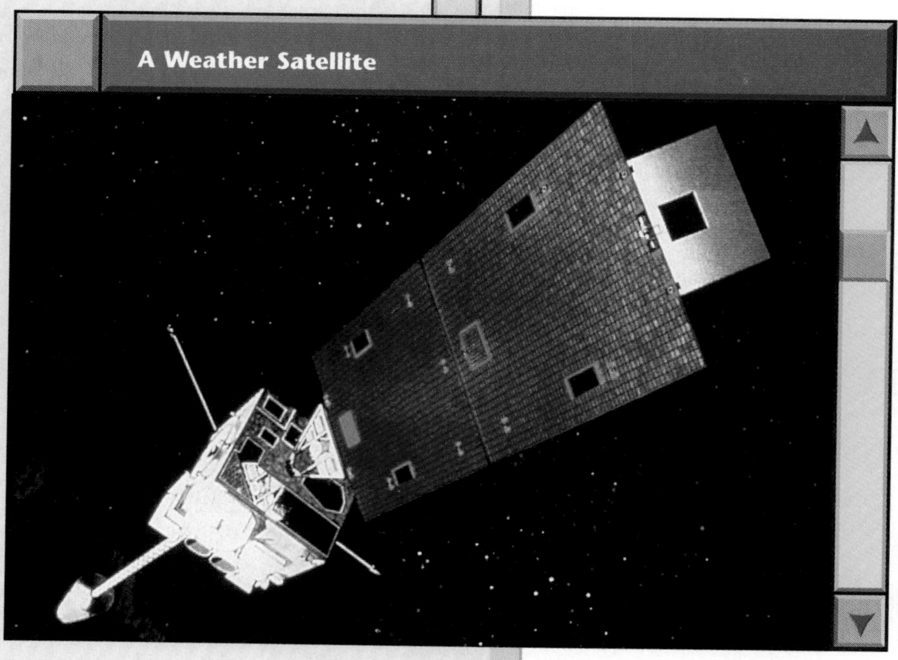

Atmospheric Pressure

Gases in the atmosphere, like all matter, have mass and a gravitational attraction to other matter. The gravitational attraction between Earth and molecules of gas causes atmospheric gases to be pulled toward Earth. Yet atmospheric gases extend upward hundreds of kilometers. The weight of the gases at the top of the atmosphere presses down on the air below, compressing the molecules and increasing the density of the air. Close to Earth, air is more dense. This dense air exerts more force than the less-dense air at the top of the atmosphere. Force exerted on an area is known as pressure.

Where do you think air pressure is greater—in the exosphere at the top of the atmosphere or in the troposphere near Earth's surface? Air pressure is greater nearer Earth. At sea level on Earth's surface, there are more molecules pushing down from above, as shown in **Figure 14-5.** In general, atmospheric pressure is greatest near Earth's surface and decreases as you move upward away from sea level. That means air pressure decreases as you go up in the mountains. That is why some people find it harder to breathe in high mountains. There are fewer molecules and consequently less air pressure.

Temperatures in the Atmosphere

You don't have to climb a mountain to find lower air pressure. Anywhere on Earth where the atmosphere is heated, air molecules move with greater energy. In heated air, fewer molecules occupy a cubic centimeter of space. The heated air becomes less dense and exerts a lower pressure. Colder air has a higher density of molecules and exerts a higher pressure. These areas of high and low pressure are often marked on weather maps because they affect our weather. Tracking the movement of these high and low pressure areas helps meteorologists forecast the weather.

Temperatures play another part in Earth's atmosphere. The atmosphere is divided into layers based on temperature differences. Earth's atmospheric gases are heated by absorbing energy from the sun. In the troposphere near Earth's surface, temperatures decrease with an increase in altitude. Just above

MiniLAB

Does air have mass?

Procedure

1. On a pan balance, determine the mass of an inflatable ball that is completely deflated.
2. Hypothesize about the change in the mass of the ball when it goes from being deflated to inflated.
3. Inflate the ball to its maximum recommended inflation pressure.
4. Determine the mass of the fully inflated ball.

Analysis

1. What change occurs in the mass of the ball from deflation to inflation?
2. Infer from your data whether air has mass.

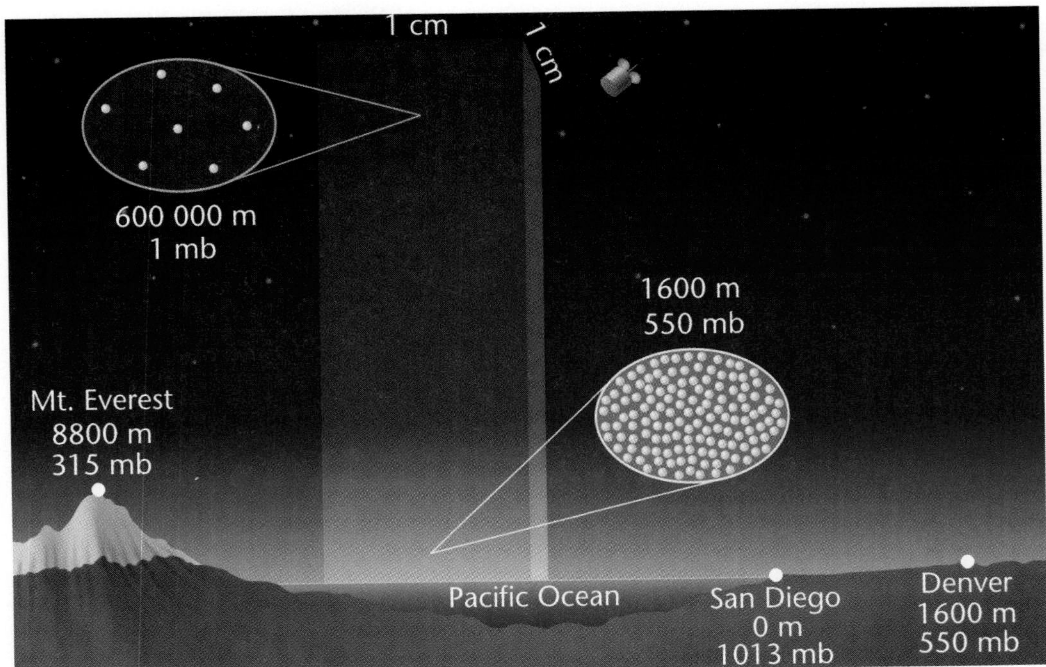

Figure 14-5

Air exerts pressure. The air at the bottom of the atmosphere is more dense and exerts more pressure than that at the top. Pressure is shown in millibars, altitude in meters. Note that Denver, at a higher altitude, experiences less air pressure than San Diego.

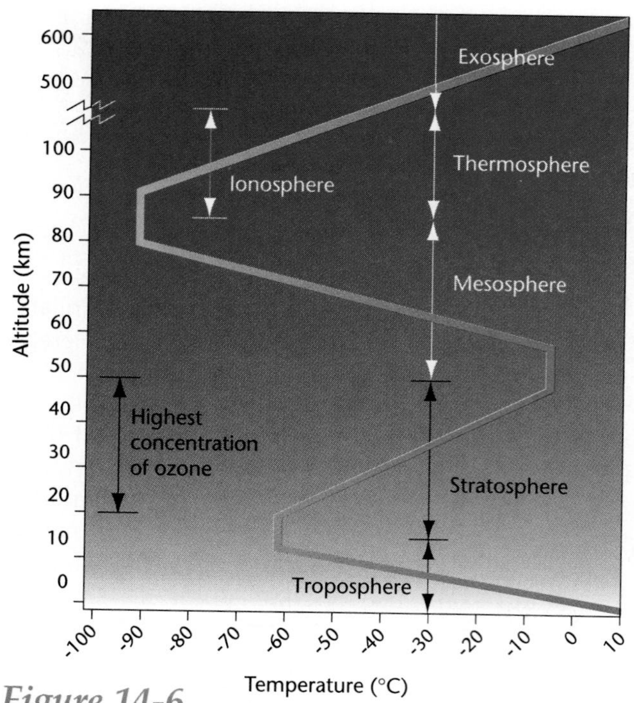

Figure 14-6

The division of the atmosphere into layers is based primarily on temperature variations.

it in the stratosphere, molecules of ozone absorb the sun's ultraviolet radiation, heating that layer. While some layers contain gases that easily absorb the sun's energy, other layers do not. As a consequence, the various layers contain different amounts of thermal energy. Notice in **Figure 14-6** how the temperature of the atmosphere fluctuates in different layers. Which layer is the coldest? Which layer is the warmest?

Understanding Our Atmosphere

Weather occurs in Earth's atmosphere. The sun heating Earth's atmosphere sets up uneven heating and activates a system of winds that moves energy from one area of Earth to another. Understanding the composition and structure of Earth's atmosphere helps us to forecast weather and to comprehend changes we are making to the atmosphere. In the next section, you will read how human activities may be affecting Earth's atmosphere.

Section Wrap-up

Review

1. Explain why the temperature of the atmosphere does not increase or decrease steadily as you move from Earth's surface toward space.
2. What causes air pressure?
3. **Think Critically:** Imagine you're a football player running with the ball. Six players tackle you and pile on top—one on top of the other. Relate the pressure that you and each player above you feels to the pressure in the layers of the atmosphere.

Skill Builder

Making and Using Graphs

Use Figure 14-1 to answer the following questions. If you need help, refer to Making and Using Graphs in the **Skill Handbook.**

1. What is the most abundant gas in Earth's atmosphere?
2. What percentage of the total volume do nitrogen and oxygen together represent? Argon and carbon dioxide?

Science Journal

The names of the atmospheric layers end with the suffix *sphere.* In these names, sphere means layer. Use a dictionary to find out what *meso-*, *thermo-*, and *exo-* mean. In your Science Journal, write the meaning of these prefixes and explain why the layers are appropriately or inappropriately named.

Activity 14-1 Making a Barometer

If you have flown in an airplane or zoomed to the top of a building in an elevator, you have experienced pressure inside your ears. When you rapidly increased your altitude, air pressure outside your eardrums became lower than the pressure inside your eardrums. The air inside your ears pushed against your eardrums. In this activity, you'll see that a barometer reacts as your eardrums did when exposed to differences in pressure.

Problem

How does a barometer react to a change in air pressure?

Materials 

- small coffee can
- drinking straw
- large rubber balloon
- construction paper
- transparent tape
- scissors
- rubber band

Procedure

1. Cut the balloon, and stretch it tightly over the can. Secure it in place with the rubber band.
2. Using tape, attach a piece of construction paper to the side of the coffee can as shown in the photo.
3. Trim one end of the straw to a point. Tape the other end of the straw to the balloon and point the trimmed end toward the paper.
4. Make a horizontal mark on the paper where the pointed end of the straw touches. Write *high* above this mark and *low* below it.
5. Design a data table to record your observations. Record the movement of the straw for a period of a week. Also record the weather conditions each day. Plot the movement of the straw on a graph.

Analyze

1. **Explain** how your barometer works. **Describe** any problems with accuracy.
2. What type of barometric readings would you expect if you took your barometer to the top of a mountain? What if you took it to the mesosphere?

Conclude and Apply

3. **Analyze** your data to see what type of weather was associated with the pressures you recorded.
4. **Conclude** from your activity how a weather forecaster can use barometric pressure to help predict weather.

TECHNOLOGY: 14•2 The Ozone Layer

Science Words

ozone layer
ultraviolet radiation
chlorofluorocarbons

Objectives

- Explain why exposure to ultraviolet radiation can be a problem for plants and animals.
- Describe how chlorofluorocarbons destroy ozone molecules.

Is the ozone layer in danger?

About 20 kilometers above your head lies the ozone layer. The **ozone layer** is an atmospheric layer with a high concentration of ozone. This layer, located in the stratosphere, is out of reach and unobservable, yet your life depends on it. The ozone layer shields you from harmful energy from the sun.

Ozone is a form of oxygen. The oxygen we breathe has two atoms per molecule. An ozone molecule, however, binds three oxygen atoms together. The layer of ozone molecules absorbs most of the ultraviolet radiation that enters the atmosphere. **Ultraviolet radiation** is one of the many types of energy that comes to Earth from the sun. Too much exposure to ultraviolet radiation can damage the skin. Ultraviolet radiation can cause cancer and other health problems in many types of plants and animals.

Each year, more than 800 000 Americans develop skin cancer, and more than 9000 die from it. If the ozone layer disappeared, cancer rates would shoot upward. **Figure 14-7** shows that the ozone layer is thinning and developing holes. In 1986, scientists found areas in the stratosphere where almost

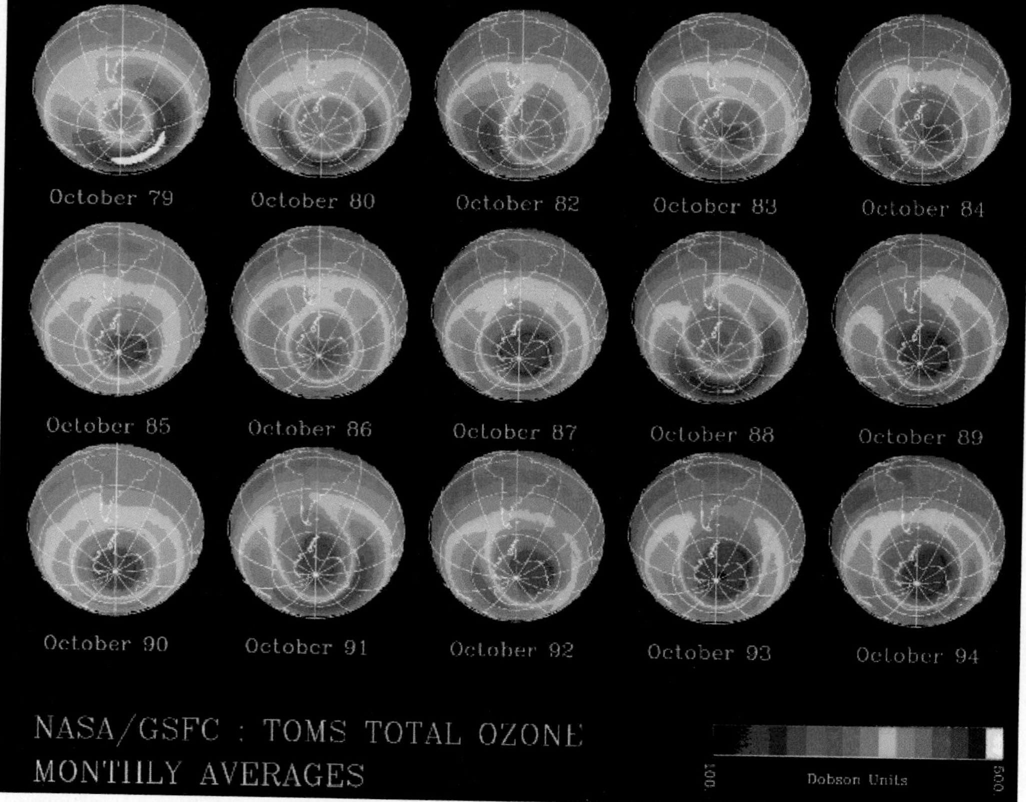

Figure 14-7

Every September and October, the ozone layer over Antarctica thins out by almost 50 percent. Not only does this hole appear, but the entire ozone layer has also become thinner around the world.

no ozone existed. One very large hole opened over Antarctica. A smaller hole was discovered over the North Pole. Since that time, every year, these holes appear during certain seasons and disappear during others. Not only is the ozone layer missing over the poles, but the entire ozone layer has also become thinner around the world.

Theory About Ozone Disappearance

Some scientists hypothesize that pollutants in the environment are destroying the ozone layer. Blame has fallen on **chlorofluorocarbons** (CFCs), a group of chemical compounds used in refrigerators, aerosol sprays, and foam packaging. When these products are manufactured and used, CFCs enter the atmosphere. Recently, governments have restricted the production and use of CFCs.

Figure 14-8 illustrates how chlorofluorocarbon molecules destroy ozone. Recall that an ozone molecule is composed of three oxygen atoms bonded together. When a chlorine atom from a chlorofluorocarbon molecule comes near a molecule of ozone, the ozone molecule breaks apart. It forms a regular two-atom molecule (O_2). This oxygen can't absorb ultraviolet radiation. The result is that more ultraviolet radiation reaches Earth's surface.

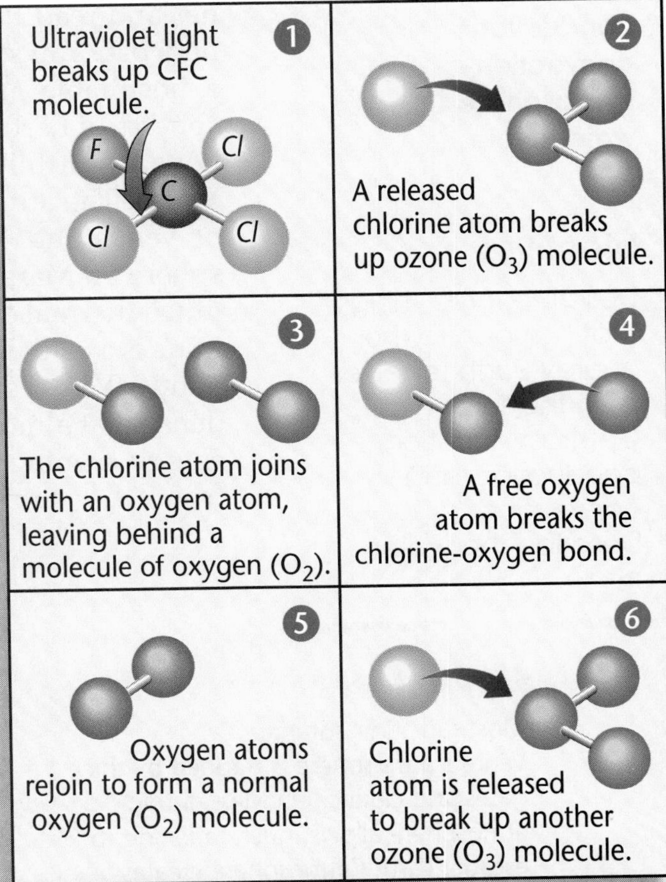

Figure 14-8

One atom of chlorine can destroy 100 000 ozone molecules. Each chlorofluorocarbon molecule has three chlorine atoms.

Section Wrap-up

1. Why do people other than those living near the poles need to be concerned about the ozone layer?
2. Describe how CFCs destroy ozone molecules.

Explore the Technology

In 1990, nations from around the world met in Montreal, Canada. They agreed to phase out the majority of chemicals that cause ozone depletion by the year 2000. Explain why it was important for all the nations to come to agreement.

SCIENCE & SOCIETY

14•3 Energy from the Sun

Science Words

radiation
conduction
convection
hydrosphere
water cycle

Objectives

- Describe three things that happen to the energy Earth receives from the sun.
- Contrast radiation, conduction, and convection.
- Describe the water cycle.

Energy Transfer in the Atmosphere

In the future scenario, you return from work. You eat dinner and read the evening news transmitted on the computer network. You see that the Space Agency is still trying to create a hospitable atmosphere on Mars. It's studying the atmospheres of Earth and Venus to understand how they work and how an Earthlike atmosphere might be produced on Mars or Venus. The atmospheres of Earth and its neighboring planets of Venus and Mars are depicted in **Figure 14-9.** The atmosphere on Mars is currently too thin to support life or to hold much thermal energy from the sun. As a result, Mars is a cold, lifeless world. On the other hand, Venus's atmosphere is so dense that almost no thermal energy coming in from the sun

Figure 14-9
Most radiation entering Venus's atmosphere is trapped by thick gases and clouds. On Mars, a thin atmosphere allows much radiation to escape. Earth's atmosphere creates a delicate balance between energy received and energy lost.

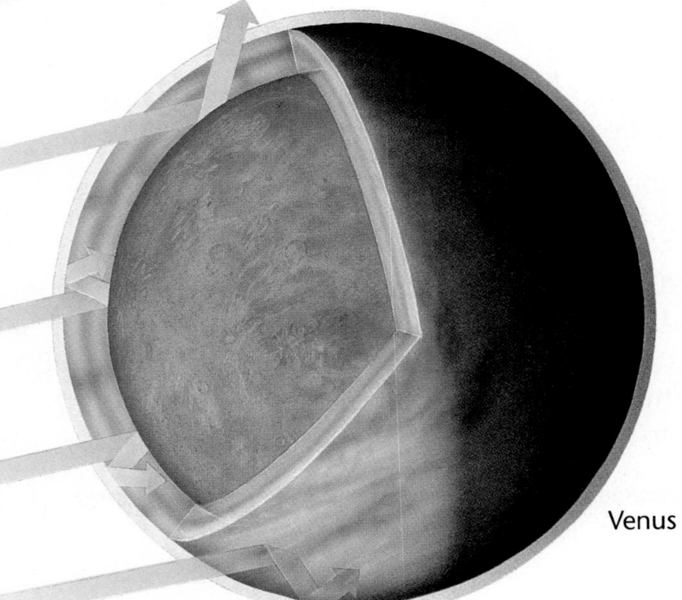

can escape. Venus is so hot that a living thing would instantly burn if it were put on Venus's surface.

In our solar system, there are nine planets circling the star we call the sun. Earth supports life, but the nearby planets, Mars and Venus, do not support life. How does the interaction between Earth's atmosphere and the sun provide an environment suitable for life?

The sun is the source of all energy in our atmosphere. When Earth receives energy from the sun, three different things happen to that energy. Some energy is reflected back into space, some is absorbed by the atmosphere, and some is absorbed by land and water surfaces. The balance among these three events controls the characteristics of our atmosphere and the life that it supports. Let's take a look at what happens to the energy that reaches Earth.

Radiation

Energy from the sun reaches our planet in the form of radiant energy, or radiation. **Radiation** is the transfer of energy by electromagnetic waves. Radiation from the sun travels through empty space as well as through our atmosphere. You experience radiation as light and heat. When the sun warms your face or when you sit by a fire and it warms the side of your body facing it, you experience radiant energy. You aren't in direct contact with the sun or the fire, but the energy still reaches you.

Mars

When radiation from the fire reaches you, the molecules of your skin absorb the energy and you feel heat. Heat is energy that flows from an object with a higher temperature to an object with a lower temperature. Once objects at Earth's surface, such as asphalt roads, rocks, houses, or ocean water, absorb radiation, they heat up. These heated surfaces then radiate energy, but the radiation they give off is at a longer wavelength than the energy coming from the sun. Much of the radiation coming from the sun can pass through the atmosphere, whereas most radiation coming from Earth's surface is absorbed and heats up our atmosphere.

The ozone layer absorbs ultraviolet radiation. When ozone and other gases absorb radiation, the temperature of the atmosphere rises, as you saw in **Figure 14-6** on page 398.

On Venus, even less radiation is able to escape back to space, making Venus hotter than Earth. On Earth, there is a delicate balance between energy received from the sun and energy escaping back to space. In the future weather forecast at the beginning of this chapter, high temperatures were forecast. This may be because smog and other pollutants in the

USING MATH

Most weather activity occurs in the airspace between the ground and an altitude of 11 km. Suppose the temperature in this airspace drops about 7°C for each kilometer of increase in altitude. If the ground temperature is 15°C, what is the temperature at an altitude of 3 km?

Figure 14-10

Pollution can change the energy balance in our atmosphere. Some types of pollution prevent radiation from escaping back into space, possibly causing Earth's temperature to rise.

atmosphere are preventing radiation from returning to space. **Figure 14-10** shows air pollution that can upset the balance of incoming and outgoing radiation on Earth.

Some radiation from the sun isn't absorbed by Earth's atmosphere or surface objects. Instead, it simply reflects off the atmosphere and surface, like a ball bouncing off a wall. **Figure 14-11** illustrates the percentages of radiation absorbed and reflected by Earth's surface and atmosphere.

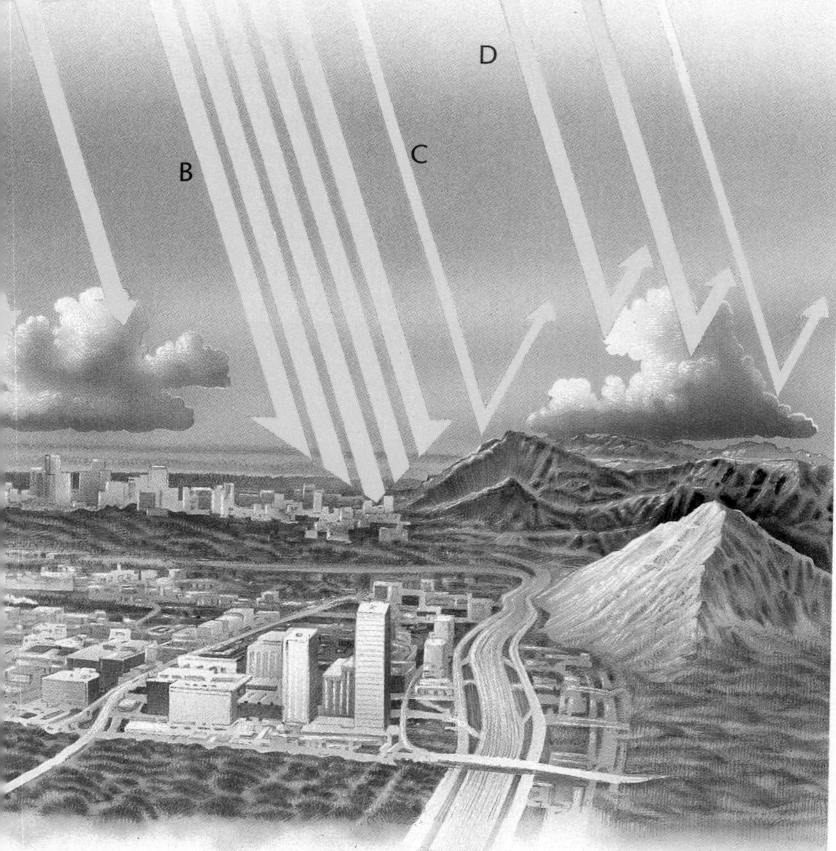

Conduction

If you walk barefoot on hot asphalt, your feet heat up because of conduction. Radiation from the sun heated the asphalt, but direct contact with the asphalt heated your feet. In a similar way, Earth's surface transfers energy directly to the atmosphere. As air moves over warm land or water, air molecules are heated by direct contact. A warm layer of molecules on Earth's surface comes in direct contact with a layer of air molecules and transfers energy.

Figure 14-11

The sun is the source of energy in our atmosphere. Thirty percent of incoming solar radiation is reflected back into space.
(A) Clouds and atmosphere absorb 20%
(B) Earth's surface absorbs 50%
(C) Surface reflects 5%
(D) Clouds and atmosphere reflect 25%

Conduction is the transfer of energy that occurs when molecules bump into one another. Molecules are always in motion, but molecules in hotter objects move more rapidly than those in cooler objects. When substances are in contact, energy is transferred from energized, fast-moving molecules to lower-energy molecules until all molecules are moving at about the same rate. **Figure 14-12** illustrates how the processes of heat transfer affect the atmosphere.

Convection

After the atmosphere is warmed by radiation or conduction, the heat is transferred throughout the atmosphere by a third process, convection. **Convection** is the transfer of heat by the flow of a heated material. Convection occurs in gases and liquids. Let's see how this works with air.

When air is warmed, the molecules move apart. This increases the volume of the air, which in turn reduces its density. With lower density, there is less air pressure because there are fewer molecules pressing in on each other. Cold temperatures affect the air density in just the opposite way. In cold air, molecules move closer together. Density and air pressure increase. Because cold air has a higher density, it sinks. As the cold air falls toward Earth, it pushes up less-dense, warm air. A circular movement of air, called a convection current, results.

CONNECT TO

PHYSICS

Infer which of the following would transfer heat by conduction the best: solids, liquids, or gases.

Figure 14-12

Heat is transferred within Earth's atmosphere by radiation, conduction, and convection.

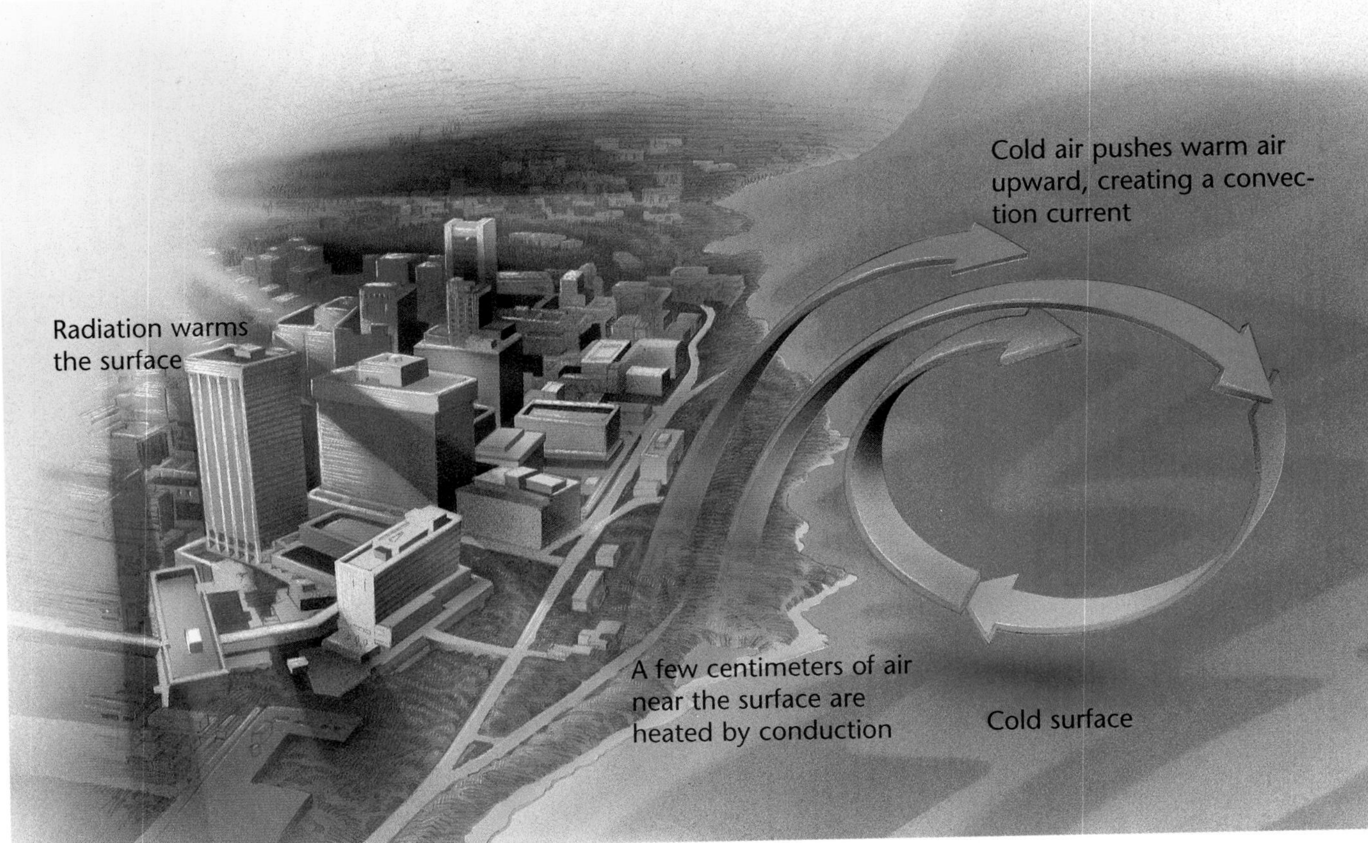

Science Journal

Mars and Venus have atmospheres very different from Earth's. In your Science Journal, write a report about how the atmospheres of these three planets differ. Include a discussion of the possibility of creating an Earthlike atmosphere on Mars.

Our Unique Atmosphere

Convection currents and other processes that transfer energy control our environment. As you have seen, radiation from the sun can escape back into space, be absorbed by the atmosphere, or be absorbed by bodies on Earth's surface. Once it's been absorbed, heat can be transferred by radiation, conduction, or convection. Just how much and what radiation is absorbed determines the type of life that can exist on this planet. Other planets in the solar system that are similar to Earth, such as Venus and Mars, don't absorb and lose the same amounts of radiation as Earth. Their atmospheres don't support life as we know it.

The Water Cycle

Another thing that allows our atmosphere to support life is water. All life as we know it needs water. Although most of Earth's water is in the oceans, it is also found in lakes, streams, rivers, groundwater, glaciers, and the atmosphere. All the water that occurs at Earth's surface is the **hydrosphere.** Although there's a lot of water on Earth, at any one time 97 percent is salt water; only three percent is fresh water. Two-thirds of the fresh water is frozen in ice caps at the North and South Poles. Therefore, the percentage of water moving through the atmosphere is low, but it is still very important because water is constantly moving between the atmosphere and Earth in the **water cycle,** as shown in **Figure 14-13.**

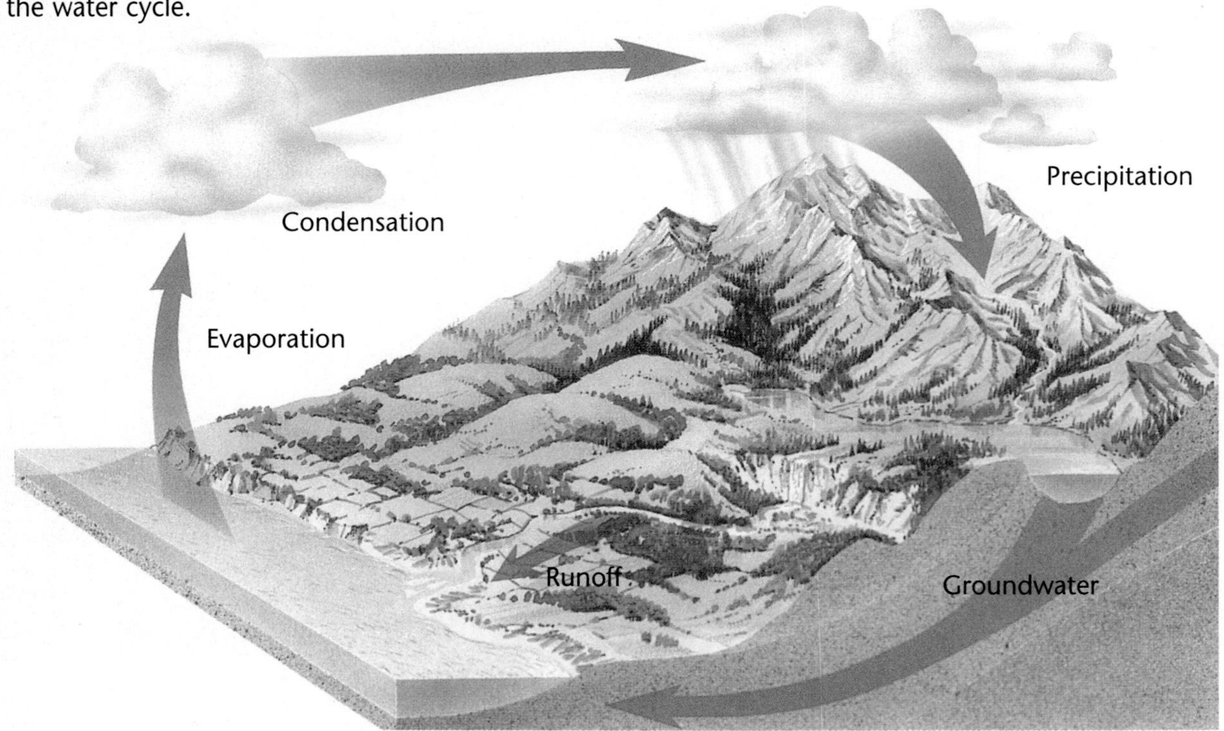

Figure 14-13

Water moves from Earth to the atmosphere and back to Earth again in the water cycle.

The sun provides the energy for the water cycle. Radiation from the sun causes water to change to a gas called water vapor. The process of water changing from a liquid state to a gas is called *evaporation.* Water evaporates from lakes, streams, and oceans and rises into Earth's atmosphere.

Figure 14-14

Rain is one form of precipitation. *Name another form.*

In the next step of the water cycle, water vapor rises in the atmosphere and cools. When it cools enough, it changes back into a liquid. This process of water vapor changing to a liquid is called *condensation.* When water vapor condenses, it forms clouds.

The third step in the water cycle is *precipitation.* Clouds are made up of millions of tiny water droplets that collide and form larger drops. When the drops grow so large that they can no longer stay suspended in the clouds, drops of water fall to Earth, and the water cycle continues. The moisture that falls from clouds is called precipitation. The desert in **Figure 14-14** is experiencing precipitation in the form of rain.

As you can see, many factors determine whether a planet will have an atmosphere capable of supporting life. How much energy is transferred to the atmosphere and how much energy escapes Earth's atmosphere is important to life on Earth. Learning about our atmosphere will help us protect it so it can continue to support life.

Section Wrap-up

Review

1. Pollution may be making our atmosphere more like that of Venus. How can that happen and how might it affect temperatures on Earth?
2. How does the sun transfer energy to Earth, and how does the atmosphere get heated?
3. **Think Critically:** Describe the role of the sun in the water cycle.

Skill Builder

Concept Mapping

Make a cycle concept map that explains what happens to energy that reaches Earth as radiant energy. If you need help, refer to Concept Mapping in the **Skill Handbook.**

USING MATH

Earth is about 150 000 000 km from the sun. The radiation coming from the sun travels at 300 000 km/second. About how long does it take for the radiation from the sun to reach Earth?

Activity 14-2

Design Your Own Experiment

The Heat Is On

Have you ever noticed how cool and refreshing a plunge in a pool or lake is on a hot summer day? Did you ever wonder why the land gets so hot when the water remains cool? At night the water feels warmer than the land. Let's explore how water and land absorb heat.

PREPARATION

Problem

Use soil and water to determine whether land and water absorb and release heat at different rates.

Form a Hypothesis

Form a hypothesis about how soil and water compare in their abilities to absorb and release heat. Write another hypothesis about how air temperatures above soil and above water differ during the day and night.

Objectives

- Design and carry out an experiment to compare the rates of heat absorption and of heat release of both soil and water.
- Observe how these differing rates of heat absorption and release affect the air above soil and above water.

Possible Materials

- ring stand
- soil
- metric ruler
- clear plastic boxes (2)
- overhead light with reflector
- thermometers (4)
- water
- masking tape
- colored pencils (4)

Safety Precautions

Be careful when handling the hot overhead light. Do not let the light or its cord make contact with water.

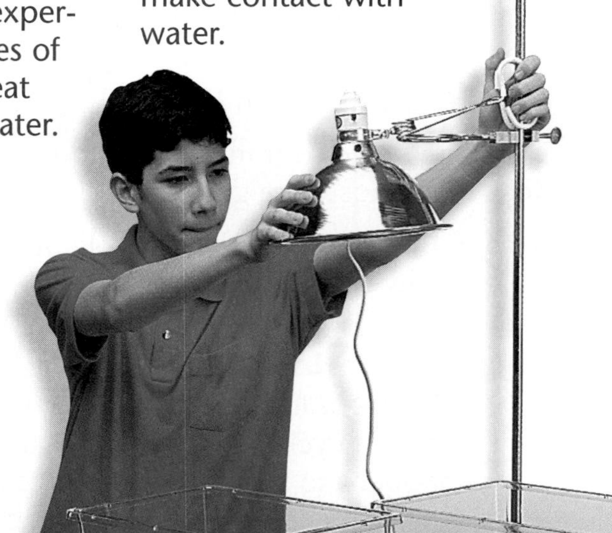

PLAN THE EXPERIMENT

1. As a group, agree upon and write out your hypotheses.
2. As a group, list the steps that you need to take to test your hypotheses. Include in your plan how you will use your equipment to (a) compare the rates of heat absorption of water and soil. Include in your measurements the temperatures of the air above the soil and the water, and (b) compare the rate of release of heat by water and soil. Include in your measurements the temperatures of the air above both substances. Do each test for 14 minutes.
3. Design a data table in your Science Journal for both parts of your experiment—when the light is on and energy can be absorbed and when the light is off and energy is released.

Check the Plan

1. Read over your entire plan to make sure that all steps are in a logical order.
2. Identify any constants and the variables of the experiment.
3. *Before you proceed, make sure your teacher approves your plan.*

DO THE EXPERIMENT

1. Carry out the experiment as planned.
2. During the experiment, record your observations and complete the data tables in your Science Journal.

Analyze and Apply

1. Use your colored pencils and the information in your data tables to make line graphs. Show the rate of energy absorption and energy release for both soil and water. (If you need help in making line graphs, refer to making and using graphs in the **Skill Handbook**.)
2. **Analyze** your graphs. When the light was on, which heated up faster, the soil or the water?
3. **Compare** how fast the air heated up over the water with how fast the air heated up over land.
4. **Infer** from your graphs which lost heat faster, the water or the soil.
5. **Compare** the temperatures of the air above the water and above the soil after the light was turned off.
6. Were your hypotheses supported or not? Explain.

Go Further

On a hot summer day at the beach, a cool breeze blows inland from the ocean. At night a cool breeze blows off the land toward the ocean. Based on what you have learned in your experiment, can you explain why the direction of the breeze changes?

14•4 Movement of Air

Science Words

Coriolis effect
jet stream
sea breeze
land breeze

Objectives

- Explain why different latitudes receive different amounts of solar energy.
- Explain the Coriolis effect, sea breezes, and land breezes.
- Locate the doldrums, trade winds, prevailing westerlies, polar easterlies, and jet streams.

Wind Formation

Have you ever watched a tree swaying in the breeze and wondered where wind comes from? Wind is caused by the uneven heating of Earth and its atmosphere. This uneven heating causes temperature differences that create areas of pressure differences in the atmosphere. Air moving from areas of high pressure to areas of lower pressure creates a general circulation of air around Earth. Wind is the movement of air from an area of high pressure to an area of lower pressure.

Temperature differences at Earth's surface are caused by Earth's tilt in its orbit around the sun and by Earth's curved surface. Areas of the Earth receive different amounts of solar radiation. **Figure 14-15** illustrates why more radiation is received at the equator than at other latitudes. Air above the equator is heated more than at any other place on Earth. As you know, heated air has a low density, so it is pushed upward by denser, cold air. **Figure 14-16** shows this general pattern of air circulation.

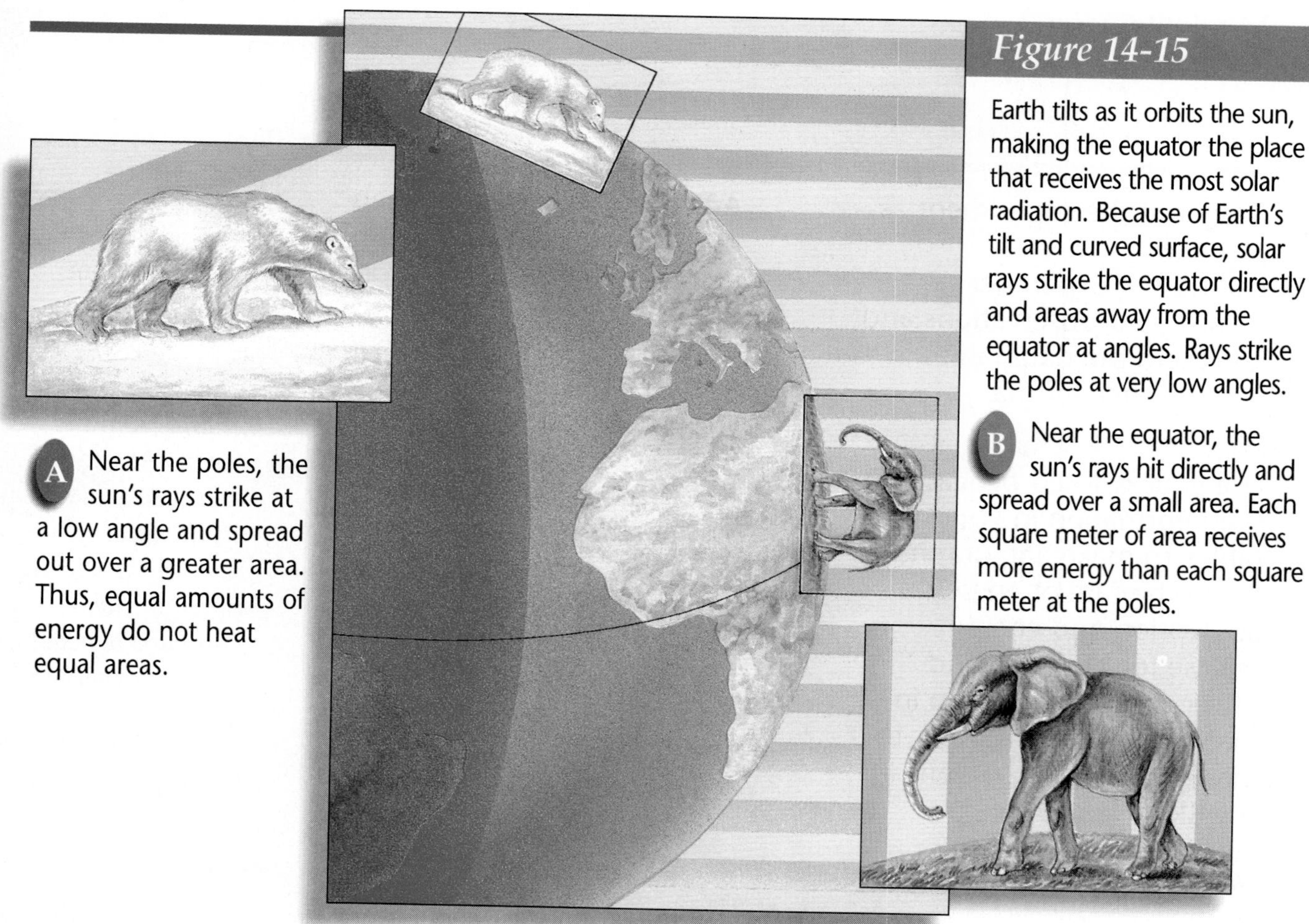

Figure 14-15

Earth tilts as it orbits the sun, making the equator the place that receives the most solar radiation. Because of Earth's tilt and curved surface, solar rays strike the equator directly and areas away from the equator at angles. Rays strike the poles at very low angles.

A Near the poles, the sun's rays strike at a low angle and spread out over a greater area. Thus, equal amounts of energy do not heat equal areas.

B Near the equator, the sun's rays hit directly and spread over a small area. Each square meter of area receives more energy than each square meter at the poles.

Where does this cold, denser air come from? It comes from the poles, which receive less radiation from the sun, making air at the poles much cooler. This dense, high pressure air sinks and moves along Earth's surface. Look at **Figure 14-17,** which shows the general wind systems around the globe. You'll see that cold, dense air pushing warmer, less-dense air upward cannot explain everything about wind.

The Coriolis Effect

The rotation of Earth creates the Coriolis effect. The **Coriolis effect** deflects all free-moving objects such as air and water to the right north of the equator and left to the south. It causes air moving south in the northern hemisphere to turn westward. To someone at the equator, southbound air appears to move to the west as Earth turns east. The diagram of Earth at the right shows this. The flow of air caused by differences in heating and by the Coriolis effect creates distinct wind patterns on Earth's surface. Not only do these wind systems influence the weather, but they also determine when and where ships and planes travel most efficiently.

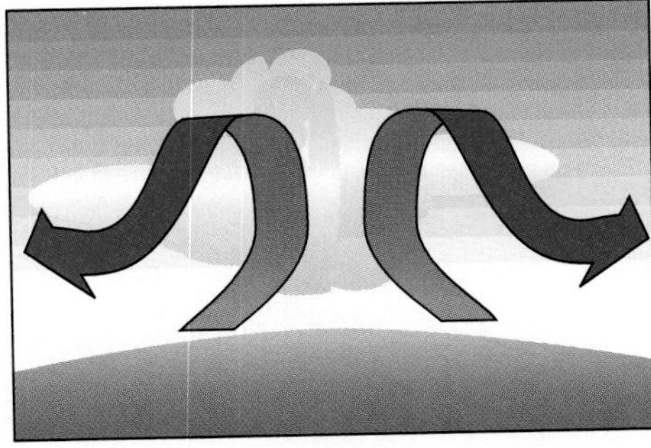

Figure 14-16

Wind develops from uneven heating on Earth. Cold air sinks and forces warm air upward.

The Coriolis-Go-Round

The Coriolis effect occurs with things like rockets shot from one place on Earth to another and with air and water—things that are not tied down to Earth. The Coriolis effect influences the general circulation of winds and storms. In some specific situations, you can notice its effect easily. Suppose your younger brother and sister are playing on a merry-go-round moving counterclockwise. Your sister sits in the middle of the merry-go-round. Your brother sits near the outer edge. She throws a ball to him. It misses. Why? Your sister threw the ball, but your brother had moved. To your sister, it appeared that she threw the ball straight. To your brother sitting at the edge, the ball appeared to curve to his left as he faced his sister.

Think Critically:

Does the ball actually curve as it travels from the center to the edge? Draw a diagram of the merry-go-round scenario to help you figure out the answers. What happens when your brother throws the ball back to your sister in the center?

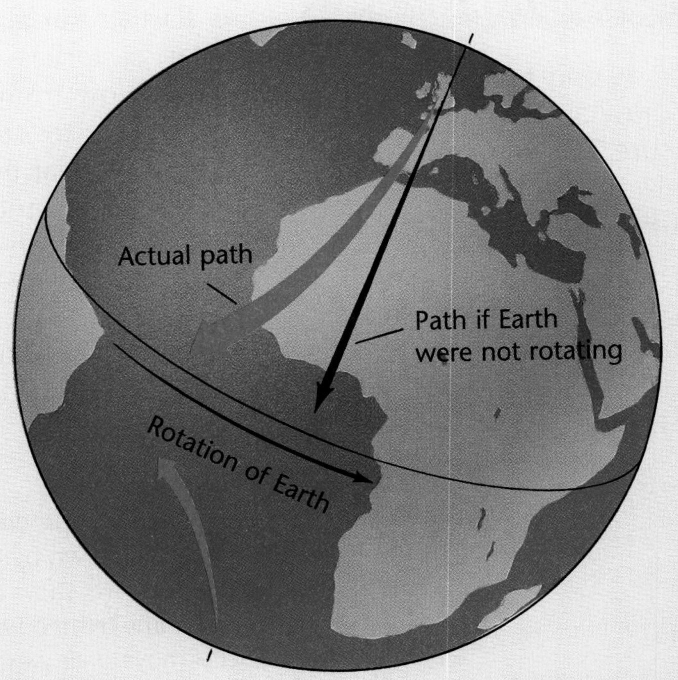

MiniLAB

What do convection currents look like?

Procedure

1. Fill a clear glass beaker with hot water.
2. Place ice cubes on the water along with a few drops of food coloring and observe the motion of the food coloring.

Analysis

1. Describe the motion of the food coloring.
2. Compare and contrast the convection current that you made with convection currents in the atmosphere.

Using a protractor, draw angles that measure 15°, 30°, and 60°.

Wind Systems

Let's venture into the past to imagine sailing the oceans during the time of the great sailing ships. No motors propel the ship. You depend entirely on the winds for energy. That means you must avoid getting into the doldrums, the windless zone at the equator. In the doldrums, the air seems motionless. Actually, the air is rising almost straight up. Do you remember why this happens?

Surface Winds

A better place to sail is between the equator and 30° latitude north or south. In that area, air descending to Earth's surface creates steady winds that blow to the southwest in the northern hemisphere. In the southern hemisphere, they blow toward the northwest. These are the trade winds. In the days of the great sailing ships, the northern trade winds provided a dependable route for trade. **Figure 14-17** shows the major wind systems on Earth, along with convection currents. Sailing ships like the one in **Figure 14-18** on page 414 were designed using knowledge of the winds.

Between 30° and 60° latitude north and south of the equator, winds blow in the opposite direction from the trade winds. These winds are called the prevailing westerlies. Sailors use the prevailing westerlies to sail from the Americas to Europe. The prevailing westerlies blow from the southwest to the northeast in the northern hemisphere. They are responsible for much of the movement of weather across the United States and Canada. In the southern hemisphere, the prevailing westerlies blow from the northwest to the southeast.

The last major wind systems at Earth's surface are the polar easterlies. These winds blow from the northeast to the southwest near the North Pole and from the southeast to the northwest near the South Pole.

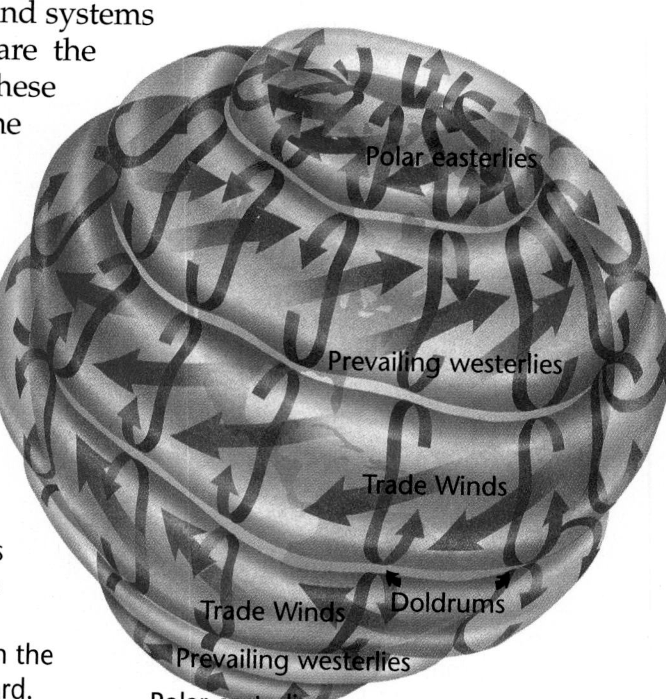

Figure 14-17

Uneven heating of the latitudes produces major convection currents, shown by the purple arrows. The blue arrows show the world's major wind systems created when the Coriolis effect deflects moving air westward.

People and Science

Tinho Dornellas, *Windsurfer*

On the Job

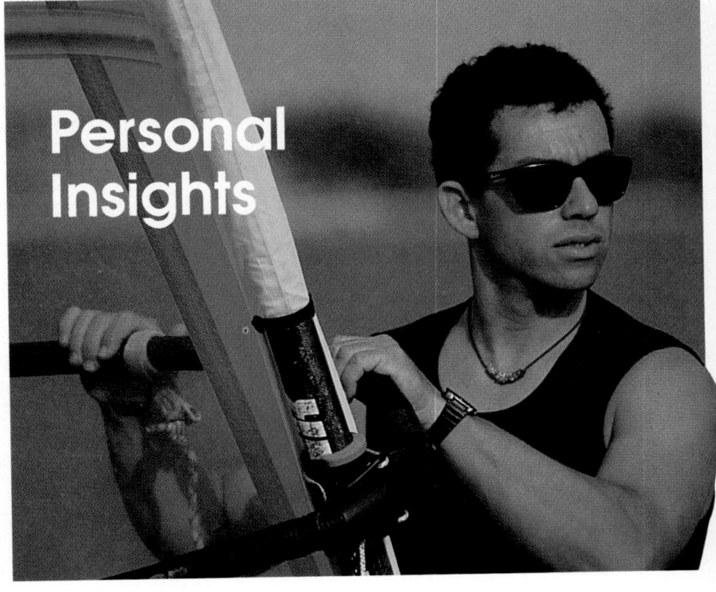

Q What do you need to know about wind to be a good windsurfer?

A You have to get a feel for where the wind is coming from and where it's going. You can picture the wind as a river, but it's a river that changes direction. You have to cross that river back and forth. The best way to find out where the wind is coming from is to look at signs around you—ripples in the water, smokestacks, flags. Another way to find out is to close your eyes and turn until you can feel the wind coming straight at your face.

Q Are certain places better for windsurfing than others?

A You can windsurf anywhere there is wind and water. Hawaii gets strong trade winds constantly. In the North Sea, you can have strong, but cold winds. Still, a lot of people windsurf there. In Washington, DC, I windsurfed close to the airport. The exhaust fumes from the airplanes were awful. Another bad place is close to cruise ships. The wind can blow toward you from the stacks of the ships.

Q Has windsurfing taught you any lessons that help you run your business?

A Running a business takes the same sort of perseverance that it takes to do well in windsurfing. You go out there and fall several times and try again. If the sail falls down, you have to have the willpower to pull that sail up and get going again. It's pretty much the same in business or in life in general—you take your falls, and you get up and keep going.

Career Connection

Modern windsurfing got its start when Jim Drake, an engineer, inventor, and sailing enthusiast (also known for codesigning the X-15 rocket research airplane, the B-70 bomber, and the cruise missile), met a surfer in the mid-1960s. The two started working together to design a surfboard with a sail. Engineers help design and build many things.

- **Engineer**
- **Meteorologist**

High-Altitude Winds

There are also winds at higher altitudes. Narrow belts of strong winds, called **jet streams,** blow near the top of the troposphere. There are two jet streams in each hemisphere that blow from west to east at the northern and southern boundaries of the prevailing westerlies. These streams of air resemble fast-moving, winding rivers. Their speeds average between 97 and 185 km/h. Their positions in latitude and altitude change from day to day and season to season. They have a major effect on our weather.

Just as sailors seek the trade winds, prevailing westerlies, and polar easterlies to help propel their ships, jet pilots take advantage of jet streams. When flying eastward, planes save time and fuel. Going west, planes fly at different altitudes to avoid the jet streams.

Daily and Seasonal Winds

The wind systems you've just read about determine the major weather patterns for the entire globe. Smaller wind systems determine local weather. If you live near a large body of water, you're familiar with two such wind systems—land breezes and sea breezes.

Convection currents over areas where the land meets the sea cause sea breezes and land breezes. **Sea breezes** are created during the day because solar radiation warms the land more than the water. Air over the land is heated

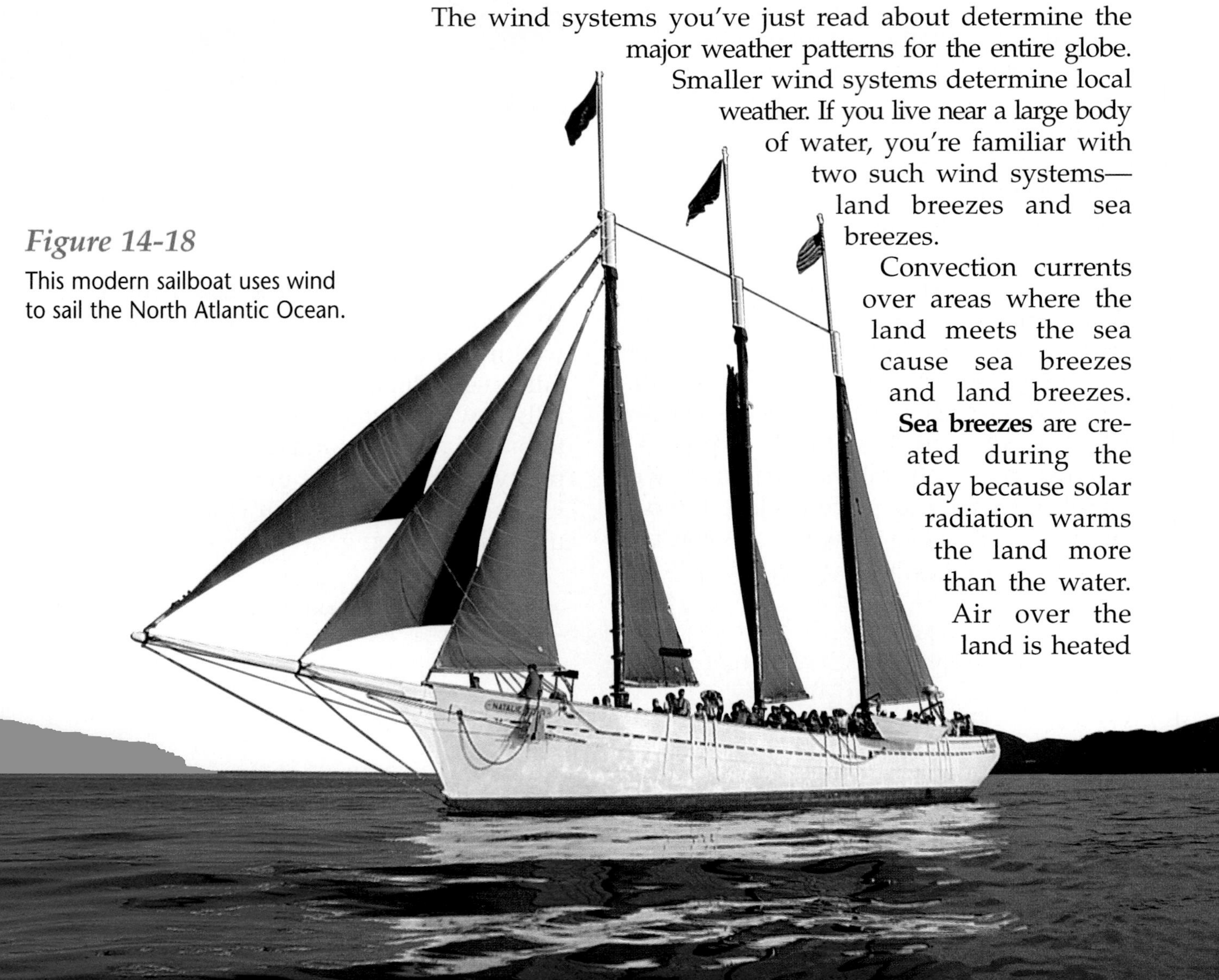

Figure 14-18

This modern sailboat uses wind to sail the North Atlantic Ocean.

Figure 14-19

These daily winds occur because a convection current changes its direction.

A During the day, cool air forces warm air over the land to rise, creating a sea breeze.

B At night, cold air over the land forces up the warmer air above the sea, creating a land breeze.

by conduction. This heated air becomes less dense and is forced upward by cooler, denser air moving inland from the ocean. A convection current results.

At night, the land cools much more rapidly than the ocean water. Air over the land becomes cooler than the air over the ocean. The cool, dense air from the land moves out over the water, pushing the warm air over the water upward. Movements of air toward the water are called **land breezes.** **Figure 14-19** can help you understand how sea breezes and land breezes occur.

Mountain-valley wind is another wind that has a daily cycle. In the mountains, about three hours after sunrise, a valley wind starts flowing from the valley upward along the slope of the mountain. A few hours after sunset, a mountain wind begins to blow down the slope into the valley. This mountain-valley wind circulation comes from heating of the mountainsides during the day. At night, the mountain slopes cool quickly and the cooler, denser air drains into the valley. At night, in deep, wide valleys, air temperatures are colder than temperatures on the mountain slopes.

Specific heat is the amount of heat required to change the temperature of a substance one degree. Substances with high specific heat values absorb a lot of heat for a small increase in temperature. ***Infer*** whether soil or water has the higher specific heat value.

Other winds change with the seasons. **Figure 14-20** shows monsoon winds that occur in tropical areas. During the winter, when land is cooler than the ocean, air flows away from the land. During the summer, when land is warmer, the air blows landward. Where the monsoon winds are strong, the summer monsoon blows moist, ocean air over the land and brings extremely heavy rain. The wind during the winter brings dry weather.

Wind is part of our weather. The general movement of air around Earth produces belts of prevailing winds and the flow of air masses and storms.

Figure 14-20

Monsoon winds are seasonal winds that occur in tropical areas.

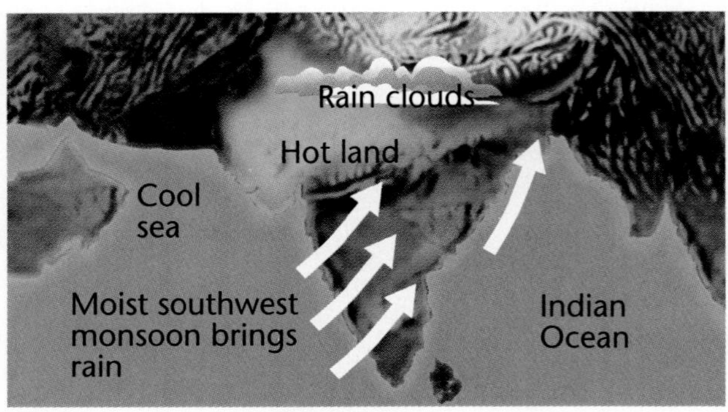

A When the land is intensely hot, the wet monsoon winds blow from the ocean onto the land, bringing rain.

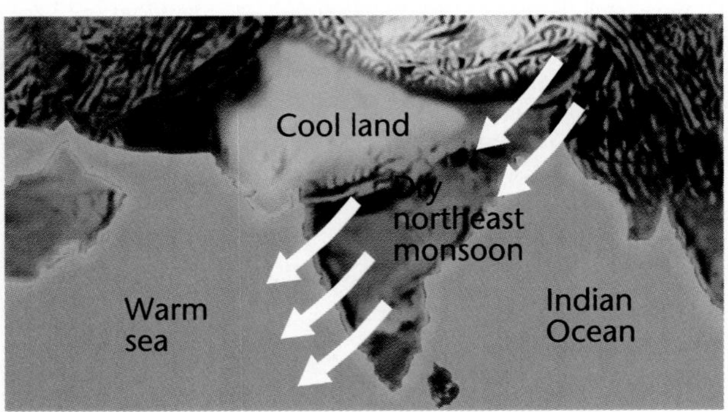

B When the sun is no longer shining directly over the land, the air above the ocean has a lower pressure. Dry winds flow from the land out over the ocean, bringing the dry season.

Section Wrap-up

Review

1. Why do latitudes differ in the amount of solar energy they receive?
2. How does the Coriolis effect influence the general wind circulation of Earth?
3. **Think Critically:** Explain why a jet that flies from South Carolina to Arizona uses more fuel and takes longer to complete its journey than a similar jet that flies from Arizona to South Carolina.

Skill Builder

Comparing and Contrasting

Compare and contrast land and sea breezes and seasonal winds versus daily winds. If you need help, refer to Comparing and Contrasting in the **Skill Handbook**.

Using Computers

Graphics Use a computer graphics package and Figure 14-17 on page 412 to draw the wind systems on Earth. Make separate graphics of major wind circulation cells shown by purple arrows. On another graphic, show major surface winds. On another, draw the jet streams. Print your graphs and share them with your class.

Chapter 14

Review

Summary

14-1: Earth's Atmosphere

1. Nitrogen and oxygen are the two most common gases in Earth's atmosphere.
2. Because the gases in Earth's atmosphere have mass, they push against one another, creating air pressure.
3. Earth's atmosphere is classified into layers based on temperature differences.

14-2: Science and Society: The Ozone Layer

1. Exposure to too much ultraviolet radiation can cause cancer and other health problems in living things.
2. When ozone reacts with chlorofluorocarbons, the molecules of ozone break apart, changing the ozone into oxygen.

14-3: Energy from the Sun

1. Some of the sun's energy that reaches Earth escapes back into space. Some energy is absorbed by Earth's air, land, and water.
2. Radiation transfers energy by electromagnetic waves. Conduction transfers heat when molecules bump into one another. Convection transfers heat due to density differences in the air.
3. Water cycles between the atmosphere and Earth's surface. The water cycle consists of evaporation, condensation, and precipitation.

14-4: Movement of Air

1. Earth's surface is curved; thus, not all areas receive the same amount of solar radiation.
2. Earth's rotation causes the Coriolis effect.
3. The doldrums are the windless zone at the equator. The trade winds lie between the equator and 30° north or south of the equator.

Key Science Words

a. chlorofluorocarbons
b. conduction
c. convection
d. Coriolis effect
e. hydrosphere
f. ionosphere
g. jet stream
h. land breeze
i. ozone layer
j. radiation
k. sea breeze
l. troposphere
m. ultraviolet radiation
n. water cycle

Reviewing Vocabulary

Match each phrase with the correct term from the list of Key Science Words.

1. layer of atmosphere closest to Earth
2. absorbs ultraviolet radiation
3. transfer of energy by electromagnetic waves
4. transfer of heat that occurs due to density differences in air
5. changes the direction of airflow and is caused by Earth's rotation
6. occurs when cold air over water moves inland
7. rays that can cause cancer
8. pollutant that breaks down the ozone layer
9. Earth's water
10. narrow wind belt at top of the troposphere

Chapter 14 Review

Checking Concepts

Choose the word or phrase that completes the sentence.

1. _______ is the most abundant gas in the air.
   a. Oxygen c. Argon
   b. Water vapor d. Nitrogen
2. Smog is a mixture of nitrogen oxide and _______.
   a. conduction c. hydrocarbons
   b. mud d. argon
3. The ______ is the uppermost layer of the atmosphere.
   a. troposphere c. exosphere
   b. stratosphere d. thermosphere
4. The warmest layer of air is the _______.
   a. troposphere c. mesosphere
   b. stratosphere d. thermosphere
5. _______ protects living things from too much ultraviolet radiation.
   a. Ozone layer c. Nitrogen
   b. Oxygen d. Argon
6. Air pressure is greatest in the ______.
   a. troposphere c. exosphere
   b. stratosphere d. thermosphere
7. When objects are in contact, energy is transferred by _______.
   a. trade winds c. radiation
   b. convection d. conduction
8. A barometer measures _______.
   a. temperature c. humidity
   b. air pressure d. wind speed
9. Movement of air toward water can be a _______.
   a. sea breeze c. land breeze
   b. doldrum d. barometer
10. _______ are near the top of the troposphere.
   a. Doldrums c. Polar easterlies
   b. Jet streams d. Trade winds

Understanding Concepts

Answer the following questions in your Science Journal using complete sentences.

11. Why do sailors avoid sailing into the doldrums?
12. Why have countries agreed to ban the use of chlorofluorocarbons?
13. How does radiation differ from conduction?
14. Explain why air rises at the equator and sinks near the poles.
15. Explain why Venus is so much hotter than Earth.

Thinking Critically

16. Why are there few or no clouds in the stratosphere?
17. It is thought that life could not have existed on land until the ozone layer formed, about 2 billion years ago. Why does life on land require an ozone layer?
18. Explain how soup in a pan on a stove is heated by conduction and convection.
19. What happens when water vapor rises and cools?
20. Why does air pressure decrease with an increase in altitude?

Developing Skills

If you need help, refer to the **Skill Handbook.**

21. Concept Mapping: Make a cycle concept map using the following phrases that explains how air moves to form a convection current: *air becomes cool and dense, warm air rises and cools, cool air pushes up warm air, cool air is warmed by conduction, cool air sinks.*

22. Making and Using Graphs: On the graph below, does air pressure increase more rapidly at high or low altitudes? Why doesn't the air pressure drop to zero on the graph? At what altitude would it drop to zero?

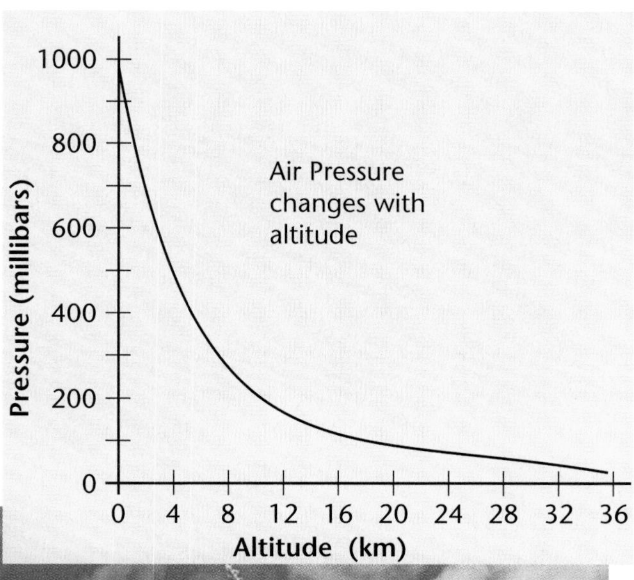

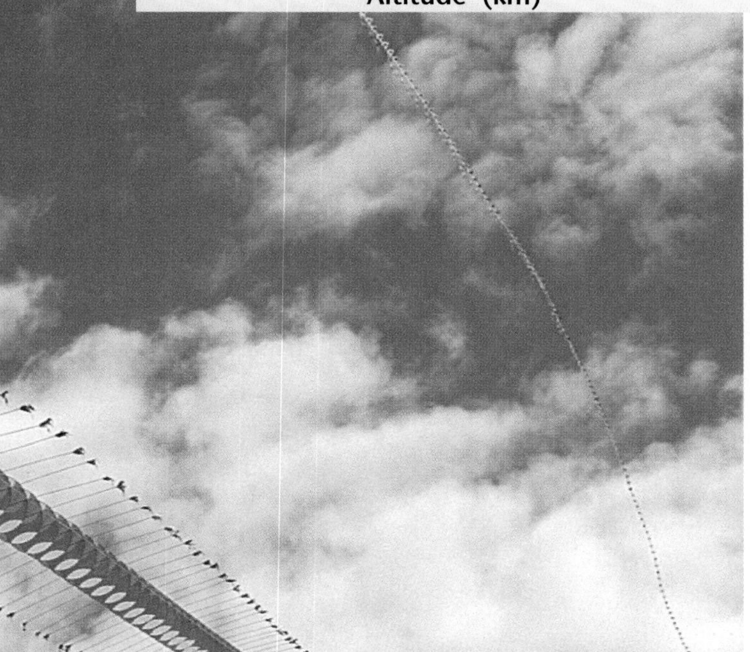

23. Observing and Inferring: In an experimen a student measured the air temperature on meter above the ground on a sunny afternoon and again one hour after sunset. The second reading was lower than the first. What can you infer from this?

24. Hypothesizing: Trees use carbon dioxide to photosynthesize. Carbon dioxide in the atmosphere prevents radiation from Earth's surface from escaping to space. Hypothesize how the temperature on Earth would change if many trees were cut down.

25. Using Variables, Constants, and Controls: Design an experiment to find out how plants are affected by differing amounts of ultraviolet radiation. In the design, use filtering film made for car windows. What is the variable you are testing? What are your constants? Your controls?

Performance Assessment

1. **Data Table:** Use the barometer you made in Activity 14-1 to make 24-hour forecasts. Record your barometer reading and your prediction on the data table. Later, compare your prediction with the actual weather that occurs. Keep a record for a week and find out how accurate your barometer predictions were overall.
2. **Poster:** Draw or find magazine photos of convection currents that occur in everyday life.
3. **Lab Report:** Design and conduct an experiment to find out how different surfaces such as asphalt, soil, sand, and grass absorb and reflect solar energy. Share the results of this with your class.

Previewing the Chapter

Chapter 15

Weather

Tornadoes are severe weather events. A tornado can roar through farms and cities, smashing everything in its path, leaving a trail of destruction. Winds in a tornado sometimes reach 500 km per hour, strong enough to flatten buildings. Updrafts in the center can act like a giant vacuum cleaner, sucking buildings, cars, and even animals high into the air.

EXPLORE ACTIVITY

Make a model of a tornado.

1. Obtain two 2-L plastic bottles.
2. Fill one about three-quarters full of water and add one drop of dishwashing soap to the water.
3. Put the empty bottle on top and tape the bottles securely, opening to opening.
4. Flip the bottles to put the one with water in it on top. Move the top bottle in a circular motion.

Observe: In your Science Journal, describe what happens in the bottles. Compare this model of a tornado with a real tornado.

Previewing Science Skills

- In the **Skill Builders**, you will **map concepts, recognize cause and effect**, and **compare and contrast.**
- In the **Activities**, you will **make a model, observe, infer**, and **interpret a scientific illustration.**
- In the **MiniLABs**, you will **observe, infer** and **make a model.**

15•1 What is weather?

Science Words

weather
humidity
relative humidity
saturated
dew point
fog
precipitation

Objectives

- Explain the role of water vapor in the atmosphere and how it affects weather.
- Describe how clouds form and how they are classified.
- Compare the development of rain, hail, sleet, and snow.

Factors of Weather

"What's the weather going to be today?" That's probably one of the first things you ask when you get up. Weather information can affect what you wear to school, how you get to and from school, and what you do after school.

Everyone discusses the weather. Can you explain what it is? **Weather** refers to the present state of the atmosphere and describes current conditions. One kind of weather is seen in **Figure 15-1.** Important factors determining the state of the atmosphere are air pressure, wind, temperature, and the amount of moisture in the air.

In Chapter 14, you learned about air pressure and how water moves around the hydrosphere in the water cycle. In the water cycle, the sun provides the energy to evaporate water into the atmosphere, where it forms clouds and eventually falls back to Earth.

The water cycle forms the basis of our weather. But the sun does more than just evaporate water. It also heats air, causing the winds that you learned about in Chapter 14. The interaction of air, water, and the sun causes weather.

Figure 15-1

The weather influences what you can do.

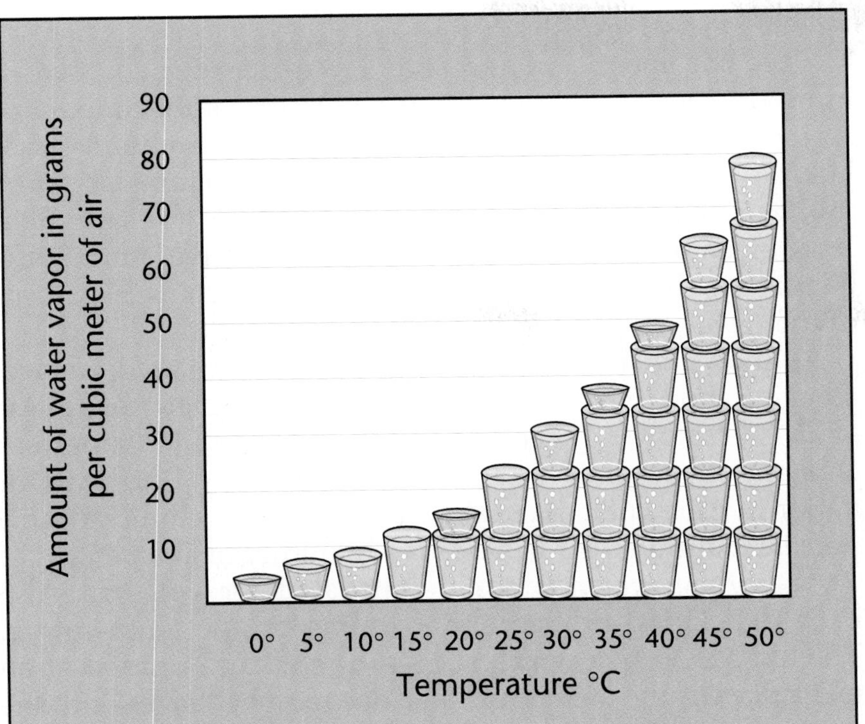

Figure 15-2

This graph shows that as the temperature of air increases, air can hold more water vapor.

CONNECT TO

LIFE SCIENCE

Birds and mammals maintain a constant internal temperature. When the temperature outside the bodies of birds and mammals changes, their body temperature is regulated to remain fairly constant. On the other hand, the internal temperature of fish and reptiles changes when the temperature around them changes. ***Infer*** from this which group is more likely to survive a quick change in the weather.

Humidity

The sun evaporates water into the atmosphere. How does this happen? How can the atmosphere hold water? The air of the atmosphere is somewhat like a sponge. The holes in a sponge enable it to hold water. The atmosphere holds water in a similar way. Water vapor molecules fit into spaces between the molecules that make up air. The amount of water vapor held in air is called **humidity.**

Humidity varies from day to day because the temperature of the air changes. The amount of water vapor that air can hold depends on the temperature. At cooler temperatures, molecules in air move more slowly. This slow movement in cool air allows water vapor molecules to join together (condense). At warmer temperatures, air and water vapor molecules move too quickly to join together. The higher energy of the molecules in warm air prevents condensation and allows more water molecules to remain as water vapor than in cooler air. If you look at **Figure 15-2,** you'll see that at 25°C, a cubic meter of air can hold a maximum of 22 g of water vapor. The same air cooled to 15°C can hold only about 13 g of water vapor.

On hot summer days when the air seems damp and sticky, people often comment on the high humidity. When they mention humidity, they are actually talking about the relative humidity.

Figure 15-3

When the air next to the glass cools to its dew point, condensation forms on the glass.

Relative Humidity

Have you ever heard a weather forecaster speak of relative humidity? **Relative humidity** is a measure of the amount of water vapor that air is holding compared to the amount it can hold at a specific temperature. When air contains as much moisture as possible at a specific temperature, it is **saturated.** If you hear a weather forecaster say the relative humidity is 50 percent, that means the air on that day contains 50 percent of the water needed for the air to be saturated.

As shown in **Figure 15-2,** air at 40°C is saturated when it contains about 50 g of water vapor per cubic meter of air. Air at 25°C is saturated when it contains 22 g of water vapor per cubic meter of air. If air at 25°C contains only 11 g of water vapor in each cubic meter of air, the relative humidity is 50 percent. Saturated air has relative humidity of 100 percent.

Additional water vapor in saturated air will condense back to a liquid or freeze, depending on the temperature. The temperature at which air is saturated and condensation takes place is the **dew point.** The dew point changes with the amount of moisture in the air.

You've probably seen water droplets form on the outside of a glass of ice water, as in **Figure 15-3.** The cold glass cooled the air next to it to its dew point. The water vapor in the air condensed and formed water droplets on the glass. Dew on grass in the early morning forms the same way. When air near the ground is cooled to its dew point, water vapor condenses and forms droplets on the grass.

Science Journal

Weather affects history. Research what happened to American colonial troops at Valley Forge during the War of Independence in the winter of 1777–1778. Imagine that you were a soldier at Valley Forge during that winter. In your Science Journal, describe your experiences.

Cloud Formation

Why are there clouds in the sky? Clouds form as warm air is forced upward, expands, and cools, as shown in **Figure 15-4.** As the air cools, the amount of water vapor needed for saturation decreases and the relative humidity increases. When the relative humidity reaches 100 percent, the air is saturated. Water vapor begins to condense in tiny drops around nuclei, small particles of dust, salt, and smoke in the atmosphere. These drops of water are so small they become suspended in the air. When millions of these drops collect, a cloud forms.

Cloud Classification

As you will see in **Figure 15-5** on page 426, there are many different types of clouds in the sky. Clouds are classified mainly by shape and height. Some clouds stack up vertically, reaching high into the sky, while others are low and flat. Some dense clouds bring rain or snow, while thin clouds appear on mostly sunny days.

Shape

The three main cloud types are stratus, cumulus, and cirrus. Stratus clouds form layers or smooth, even sheets in the sky. When layers of air cool below their dew point temperatures, stratus clouds appear. Stratus clouds usually form at low altitudes. Stratus clouds are associated with both fair weather and precipitation; sometimes they form a dull, gray blanket that hangs low in the sky and brings drizzle. When air is cooled to its dew point and condenses near the ground, it forms a stratus cloud called **fog. Figure 15-7** on page 430 shows a stratus cloud fog in San Francisco.

Figure 15-4

Clouds form when moist air is pushed high enough to reach its dew point. The water vapor condenses, forming water droplets that group together. Here are three ways clouds form.

A Clouds form when warm air is forced up in a convection current caused by solar radiation heating Earth's surface.

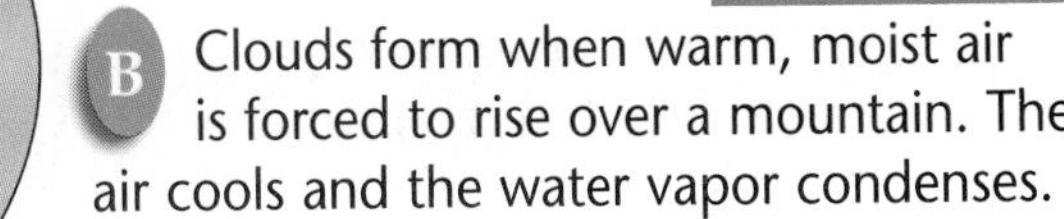

B Clouds form when warm, moist air is forced to rise over a mountain. The air cools and the water vapor condenses.

C Clouds form when two air masses meet. Warmer air is forced up over the cold air. As the warm air cools, the water vapor in it condenses.

How can dew point be determined?

Procedure

1. Partially fill a metal can with room-temperature water. Dry the outer surface of the can.
2. Slowly stir the water and add small amounts of ice.
3. In a data table in your Science Journal, note the exact water temperature at which a thin film of moisture first begins to form on the outside of the metal can.
4. Repeat steps 1-3 two more times.
5. The average of the three temperatures at which the moisture begins to appear is the dew point temperature of the air around the container.

Analysis

1. What factors determine the dew point temperature?
2. Will a change in air temperature also cause the dew point temperature to change? Explain.

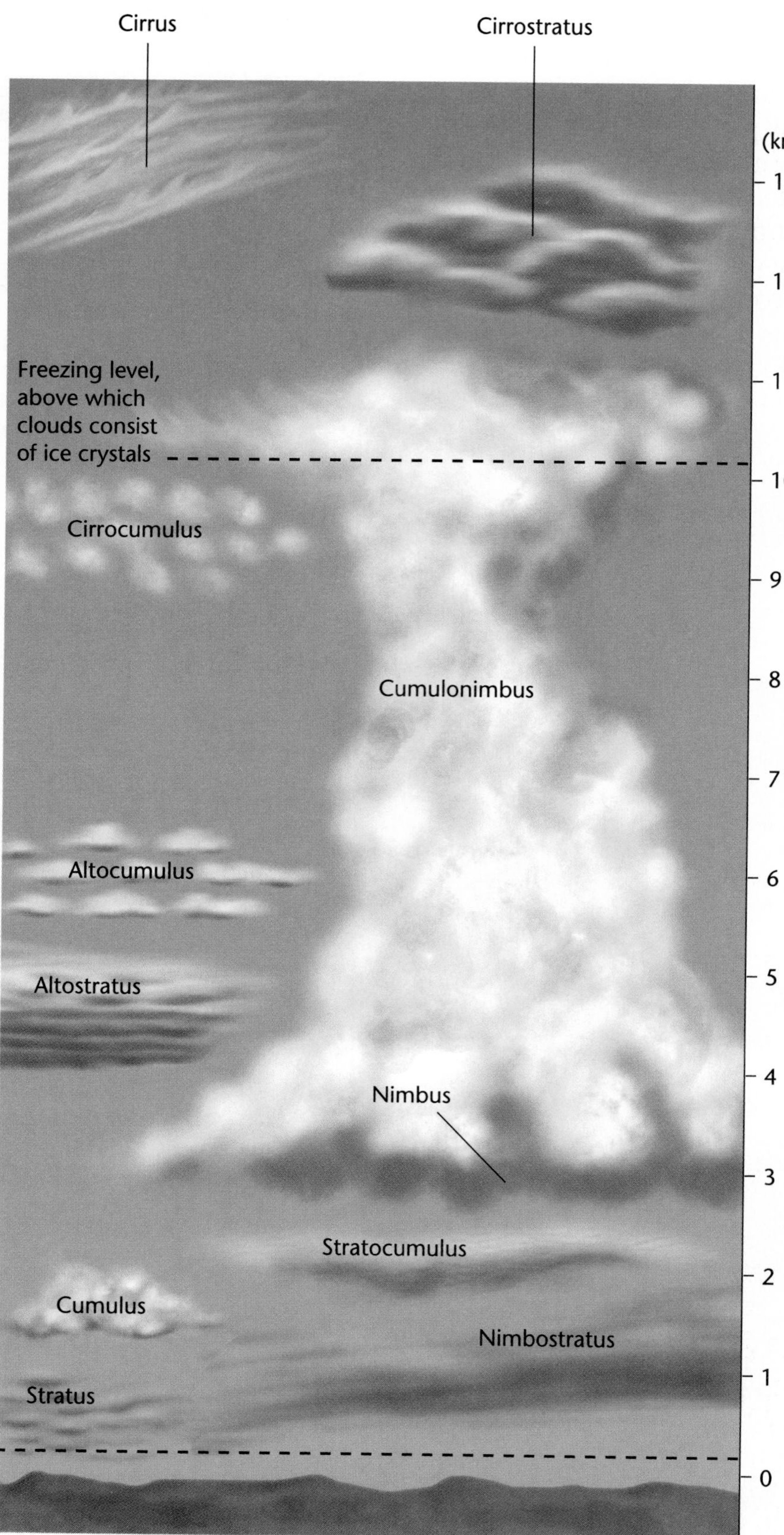

Figure 15-5

These are the most common types of clouds. *What do you think causes them to have different shapes?*

Cumulus clouds are masses of puffy, white clouds, often with flat bases. Some people refer to them as cauliflower clouds. They form when air currents rise. They may tower to great heights and can be associated with both fair weather and thunderstorms.

Cirrus clouds appear fibrous or curly. They are high, thin, white, feathery clouds containing ice crystals. Cirrus clouds are associated with fair weather, but they may indicate approaching storms.

Height

Some prefixes of cloud names describe the height of the cloud base. The prefix *cirro-* describes high clouds—clouds with a base starting above 6000 m; *alto-* describes middle elevation clouds—their base is between 2000 to 6000 m; and *strato-* refers to clouds below 2000 m. Some clouds' names combine the altitude prefix with the term *stratus* or *cumulus.*

Cirrostratus clouds are high clouds that look like fine veils. They are made of ice crystals that appear to form halos around the moon or sun. Usually cirrostratus clouds indicate fair weather, but they may also signal an approaching storm.

Altostratus clouds form at middle levels. They look like thick veils or sheets of gray or blue. If the clouds are not too thick, sunlight can filter through them. They produce light, continuous precipitation.

Rain Capacity

Nimbus clouds are dark clouds associated with precipitation. They are so full of water that no sunlight penetrates them. When a nimbus cloud is also a towering cumulus cloud, it is called a cumulonimbus cloud. Some cumulonimbus clouds grow huge, starting near Earth's surface and towering to nearly 18 000 m. Sudden, gigantic thunderstorms can be unleashed from them.

Nimbostratus clouds bring long, steady rain. They often have streaks that extend to the ground.

As long as the water drops in a cloud remain small, they stay suspended in the air. But when the water droplets combine and reach the size of 0.2 mm, they become too heavy and fall out of suspension in the cloud.

Figure 15-6

When water vapor in air collects on nuclei to form water droplets, the type of precipitation that is received on the ground depends on the temperature of the air.

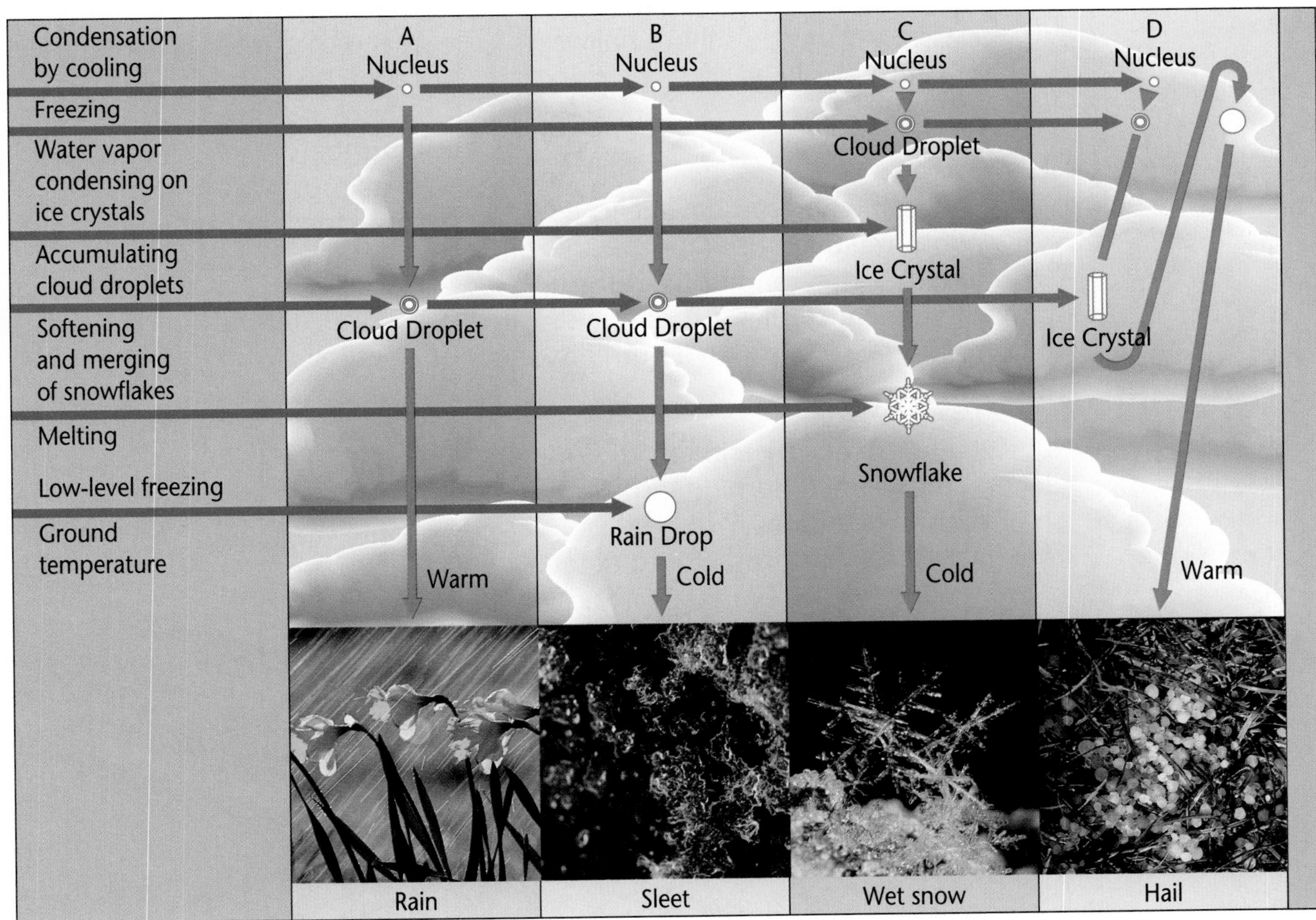

A When the air near the ground is warm, water vapor forms raindrops that fall as rain.

B When the air near the ground is cold, sleet, made up of many small ice pellets, falls.

C When the air is very cold, water vapor forms snowflakes.

D Hailstones are pellets of ice that form as the ice nuclei go up and down in the cloud.

Activity 15-1

Design Your Own Experiment

Is it humid or not?

Does it feel humid today or does the air feel dry? You can tell by examining your hair. On a day when the relative humidity is high, hair tends to be more curly than on a dry day. The relative humidity can be more accurately determined by using a psychrometer, a device with thermometers—one wet and one dry.

PREPARATION

Problem

Where is the relative humidity the highest? In your science classroom, in the school hallway, or on the walkway outside the school?

Form a Hypothesis

Recall what you have learned about relative humidity. Decide whether the conditions in the three locations lend themselves to high or low relative humidity.

Objectives

- Design an experiment to find the relative humidity in the three locations.
- Make a psychrometer to determine relative humidity at the three locations. On the psychrometer below, the wet bulb has material covering it.

Possible Materials

- identical Celsius thermometers (2)
- piece of gauze, 2-cm square
- tape
- string
- cardboard
- beaker of water
- table for determining relative humidity in Appendix J

Safety Precautions

Be careful when you handle glass thermometers, especially those containing mercury. If one breaks, have your teacher dispose of the glass and the mercury safely.

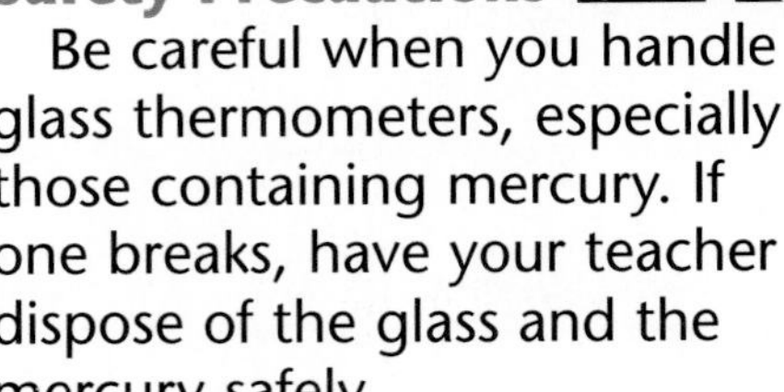

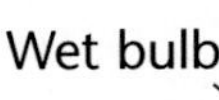

PLAN THE EXPERIMENT

1. As a group, write out your hypothesis statement and plan for the experiment.
2. Here is how a psychrometer works. Moisture from the wet thermometer uses heat energy as it evaporates. When the moisture evaporates, it lowers the temperature of its environment.
3. List the steps needed to measure the relative humidity of the three locations. Include how you will (a) saturate each wet bulb thermometer; (b) record the temperatures of the thermometers; and (c) determine the relative humidity using the relative humidity table provided by your teacher.
4. Design a data table in your Science Journal for your data.

Check the Plan

1. Make certain that you understand how to obtain accurate temperature readings.
2. Read through the concluding questions for more information about expected results of the experiment.
3. Identify the constant and the variable in the experiment.
4. How can you make certain that the tests at each location are exactly the same?
5. *Before you proceed, have your teacher approve your plan.*

DO THE EXPERIMENT

1. Carry out the experiment as planned.
2. While the experiment is going on, write down observations in your Science Journal. Complete your data table.

Analyze and Apply

1. From your results, **infer** why the molecules of water on the gauze of the wet bulb thermometer react differently in dry air and in saturated air.
2. **Conclude** why the relative humidity varied at your three test sites, if this happened.
3. **Hypothesize** about how the relative humidity in your classroom can be decreased.

Go Further

Continue your experiment at the three locations for a longer period and make a graph from your data. Discuss any patterns you see.

Evaporation and Condensation

Suppose you decide to make some spaghetti for lunch. You fill a large pot full of water and turn the burner to the highest temperature setting. After a few minutes of watching the pot, waiting for the water to boil, you get bored. You watch TV while you wait.

You get so interested in the show that you forget about the water. When you finally return to the kitchen, you see that the pot is now only half full of boiling water. On the wall above the stove are droplets of water.

Solve the Problem

1. **What happened to half of the water that was originally put into the pot?**
2. **How did water get on the wall?**

Think Critically:

If you get into a car and the windows begin fogging up, what can you do to make the moisture disappear—turn on the heater or turn on the air conditioner, or does it matter? Explain your answer.

Figure 15-7

Fog surrounds the Golden Gate Bridge, San Francisco. Fog is a stratus cloud near the ground.

Precipitation

Water falling from clouds is called **precipitation.** Air temperature determines whether the water droplets form rain, snow, sleet, or hail—the four main types of precipitation. **Figure 15-6** on page 427 shows how four types of precipitation form. Drops of water falling in temperatures above freezing fall as rain. Snow forms when the air temperature is so cold (at least below freezing) that water vapor changes directly to a solid. Sleet forms when snow passes through a layer of warm air, melts, and then refreezes near the ground.

Hail is precipitation in the form of lumps of ice. Hail forms in cumulonimbus clouds of a thunderstorm when drops of water freeze in layers around a small nucleus of ice. Hailstones grow larger as they're tossed up and down by rising and falling convection currents. Most hailstones are smaller than 2.5 cm but can

Figure 15-8

A large hailstone appears to have a layered structure much like an onion.

grow to the size shown in **Figure 15-8.** Of all forms of precipitation, hail produces the most damage immediately, especially if winds blow during a hailstorm. Falling hailstones can break windows and destroy crops.

By understanding the role of water vapor in the atmosphere, you can begin to understand weather. The relative humidity of the air helps determine whether a location will have a dry day or experience some form of precipitation. The temperature of the atmosphere determines the form of precipitation. Studying clouds can add to your ability to forecast weather.

MiniLAB

Can you make it rain?

Procedure

1. Pour a few centimeters of hot water into a tall, clear, wide-mouthed jar.
2. Put ice cubes in a small plastic bag. Suspend the bag from the top of the jar, and let it hang down inside.

Analysis

1. In your Science Journal, describe what you see.
2. Describe what is happening to the water vapor in the jar.

Section Wrap-up

Review

1. When does water vapor in air condense?
2. How do clouds form?
3. **Think Critically:** How can the same cumulonimbus cloud produce both rain and hail?

Skill Builder

Concept Mapping

Make a network tree concept map that compares four clouds. Use these terms: *cirrus, cumulus, stratus, nimbus, feathery, fair weather, puffy, layered, precipitation, clouds, dark,* and *steady precipitation.* If you need help, refer to Concept Mapping in the **Skill Handbook.**

USING MATH

Use the graph in **Figure 15-2** to determine the amount of water vapor air can hold when its temperature is 50°C.

15•2 Weather Patterns

Science Words

air mass
front
tornado
hurricane

Objectives

- Describe the weather associated with fronts and high and low pressure areas.
- Explain how low pressure systems form at fronts.
- Relate thunderstorms to tornadoes.

Changes in Weather

Why do you ask about the weather in the morning when you get up? Isn't it safe to assume that the weather is the same as it was the day before? Of course not! Weather is always changing because of the constant movement of air and moisture in the atmosphere. These changes are generally related to the development and movement of air masses.

Air Masses

An **air mass** is a large body of air that has the same properties as the surface over which it develops. For example, an air mass that develops over land is dry compared with one that develops over water. Also, an air mass that develops in the tropics is warmer than one that develops at a higher latitude. When you witness a change in the weather from one day to the next, it is due to the movement of air masses. **Figure 15-9** shows air masses that affect the United States.

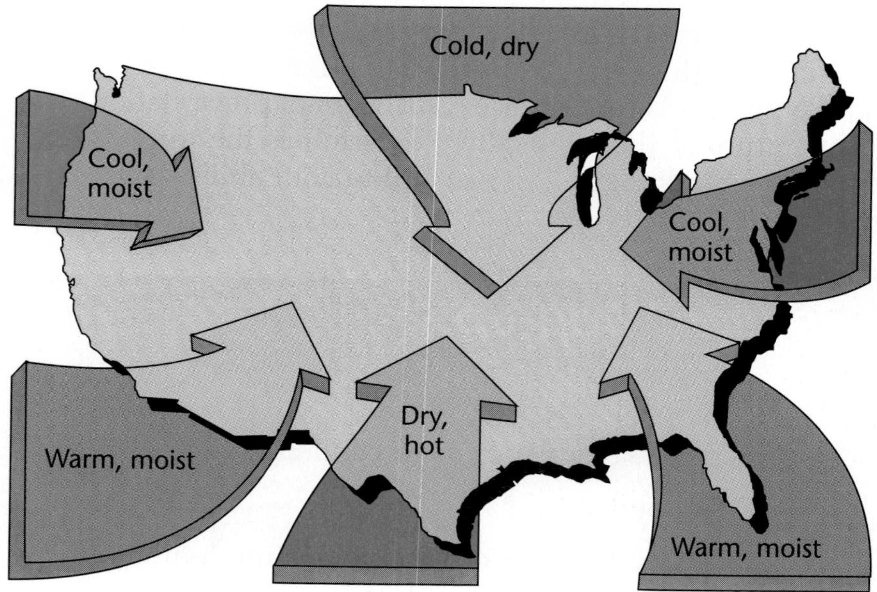

Figure 15-9

There are six major air masses that affect weather in the United States. Each air mass has the same characteristics of temperature and moisture content as the area over which it formed. *What air masses affect the weather in your region of the country?*

Pressure Systems

You have heard weather forecasters mention low and high pressure systems. What are they? In the atmosphere, great masses of air molecules push down from above, creating atmospheric pressure at Earth's surface. As you learned in the last chapter, atmospheric pressure at sea level varies over the surface of Earth. The temperature, the density, and the amount of water vapor of the air help determine the atmospheric pressure.

Variation in atmospheric pressure affects the weather. Areas of high pressure at Earth's surface are regions of descending air. Section 15-1 explained that clouds form when air rises and cools. The sinking motion in high pressure air masses makes it difficult for air to rise and clouds to form. That's why high pressure usually means good weather. Areas of low pressure usually have cloudy weather.

Fronts

In between regions of higher pressure will be areas of lower pressure. Low pressure systems form along the boundaries of air masses. The boundary between two different air masses is called a **front.** Storms and precipitation occur at these fronts.

At a front, air at the surface moves from the high pressure systems into the low pressure systems. As the air converges into the low pressure area, it flows under the less dense, warm air, forcing it upward. As the air in a low pressure system rises, it cools. At a certain elevation, the air reaches its dew point, and the water vapor in it condenses, forming clouds. **Figure 15-10** shows how a low pressure system can develop at the boundary between cold and warm air.

Figure 15-10

These diagrams show how a disturbance occurs along a front.

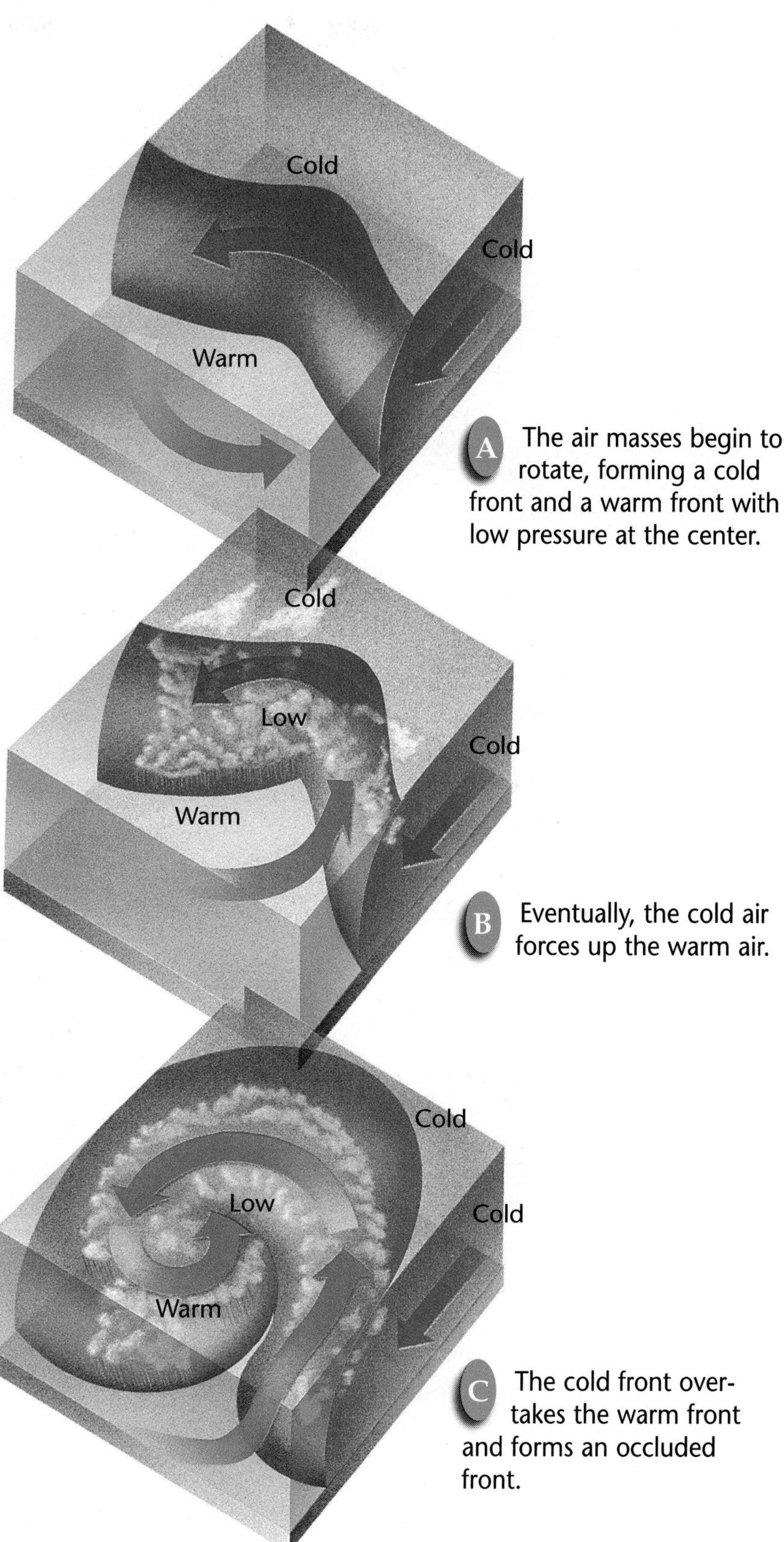

At fronts where two air masses with different characteristics meet, the air does not mix, but instead, the cold air mass moves under the warm air. The warm air rises. Winds begin. As surface winds blow from a high pressure area into a low pressure area, the Coriolis effect turns the winds and makes them circulate counterclockwise around the low pressure area.

Figure 15-11

These diagrams show the structure of a warm, a cold, an occluded, and a stationary front.

Most changes in weather occur at one of four types of fronts—warm, cold, occluded, or stationary, as illustrated in **Figure 15-11.** Fronts usually bring a change in temperature and always bring a change in wind direction.

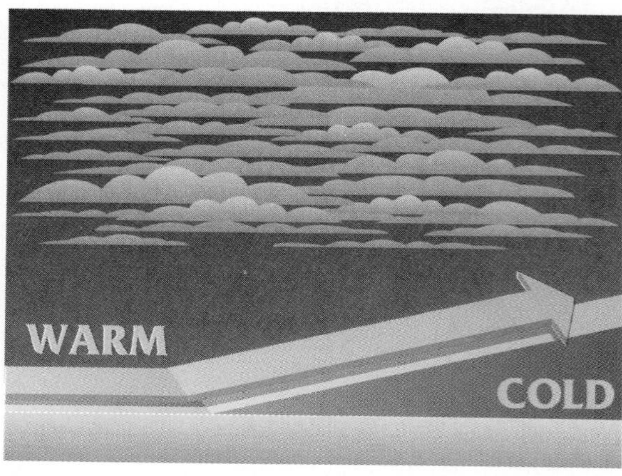

A A warm front develops when a less dense, warm air mass slides over a departing cold air mass. Precipitation occurs over a wide band. Look for high cirrus clouds to form as water vapor condenses. *What other clouds occur at a warm front?*

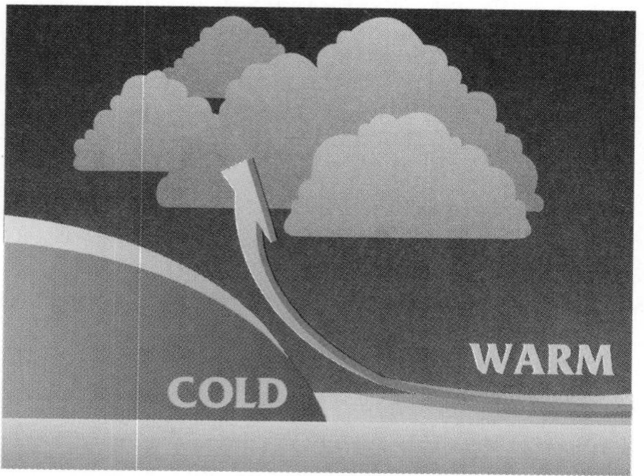

B In a cold front, a cold air mass pushes under a warm air mass and forces the warm air aloft along a steep front. There is a narrow band of violent storms. Cold fronts often move at twice the speed of warm fronts. Cumulus and cumulonimbus clouds form along the front. *What weather do these clouds bring?*

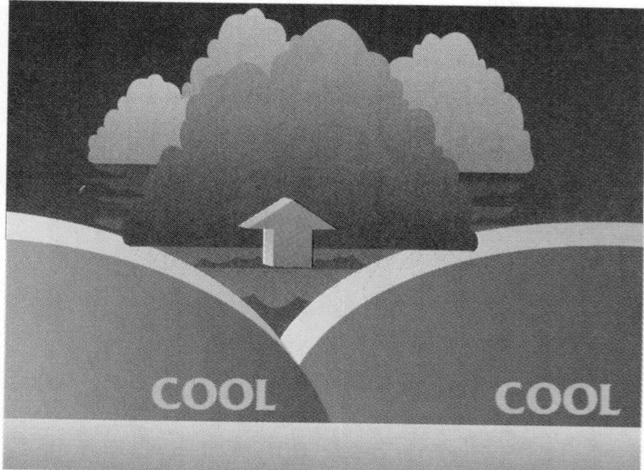

C An occluded front results from two cool air masses merging and forcing warmer air between them to rise. Strong winds and heavy precipitation may occur.

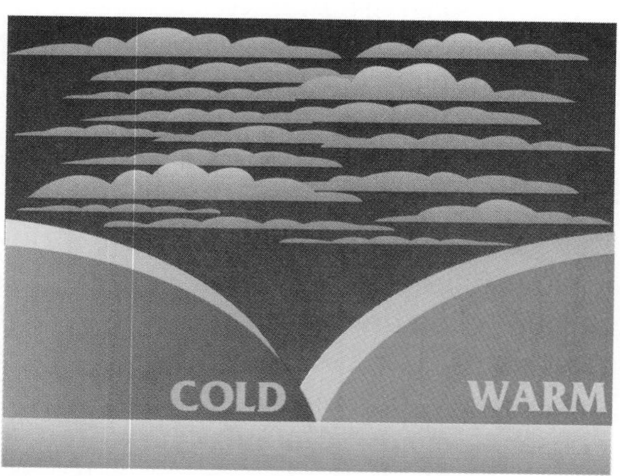

D A stationary front occurs when pressure differences cause a warm front or a cold front to stop moving. A stationary front may remain in the same place for several days. Weather conditions include light wind and precipitation across the entire frontal region.

Severe Weather

Weather affects you every day. Usually, you can still go about your business regardless of the weather. If it's raining, you can still go to school. Even if it snows a little, you can still get to school. But some weather conditions, such as those caused by blizzards, thunderstorms, and tornadoes, prevent you from going about your normal routine. Severe weather poses danger to people and animals.

Thunderstorms

INTEGRATION Physics

In a thunderstorm, heavy rain falls, lightning flashes, thunder roars, and maybe hail falls. What forces cause such extreme weather conditions? Thunderstorms occur inside warm, moist air masses and at fronts. They occur when warm, moist air moves upward rapidly, cools, condenses, and forms cumulonimbus clouds that can reach heights of 18 km. As the rising air reaches its dew point, water droplets and ice form and begin falling the long distance through the clouds toward Earth's surface. The falling droplets collide with other droplets and grow larger. The heavier raindrops fall, dragging down the air with them and creating downdrafts of air that spread out at Earth's surface. These downdrafts cause the strong winds associated with thunderstorms.

Thunderstorms also contain thunder and lightning. Lightning, like that in **Figure 15-12,** occurs when a rapid uplift of air builds up electric charges in the clouds. Some places in the clouds have a positive electrical charge and some have a negative electrical charge. When current flows between regions of opposite electrical charge, lightning flashes. Bolts of lightning can leap from cloud to cloud and from Earth to clouds.

Thunder results from the rapid heating of the air around a bolt of lightning. Lightning can reach temperatures of about 30 000°C, more than five times the temperature of the surface of the sun! This extreme heat causes the air around the lightning to expand rapidly. Then it cools quickly and contracts. The molecules, moving rapidly back and forth, form sound waves heard as thunder.

Figure 15-12
This time-elapsed photo shows a thunderstorm over Arizona.

USING MATH

The speed of light is 300 000 kilometers per second. The speed of sound in air is 332 meters per second. How much faster does light travel than sound? Explain why you do not usually see lightning and hear the thunder it creates at the same time.

Thunderstorms can cause a lot of damage. Their heavy rain sometimes causes flooding, and lightning can strike objects and set them on fire. Strong winds generated by thunderstorms can also cause damage. If a thunderstorm has winds traveling faster than 89 km per hour, it is classified as a severe thunderstorm. Hail from a thunderstorm can make dents in cars and the aluminum siding on houses. Although rain from thunderstorms helps crops grow, hail has been known to flatten and destroy a crop in a matter of minutes.

Tornadoes

Some of the most severe thunderstorms produce tornadoes. A **tornado** is a violent, whirling wind that moves in a narrow path over land, usually in a direction from southwest to northeast. Most tornadoes form along a front. In very severe thunderstorms, the wind at different heights blows in different directions and at different speeds. This difference in wind direction and speed is called *wind shear*. A strong updraft will tilt the wind shear and produce rotation inside the thunderstorm. A funnel cloud appears. **Figure 15-13** shows how a tornado funnel forms. Recall the tornado you made in the Explore activity.

Some tornado funnels do not reach Earth. Funnel clouds that touch down pick up dirt and debris from the ground, giving the funnels their dark gray or black color. Sometimes tornadoes

Figure 15-13

The diagram shows how wind forms a funnel cloud like the one to the right in a farm field. The destructive winds of a tornado can reach up to 500 km per hour.

strike Earth, go back up into the atmosphere, then dip down and strike another area.

When tornadoes touch the ground, their destructive winds rip apart buildings and trees. Winds of the tornado can reach up to 500 km per hour. High winds can blow through broken windows. When winds blow inside a house, they can lift off the roof and blow out the walls, making it look as though the building exploded. The updraft in the center of a powerful tornado can lift animals, cars, and even houses into the air. Although tornadoes rarely exceed 200 m in diameter and usually last only a few minutes, they are often extremely destructive.

Tornadoes occur worldwide, but most tornadoes touch down in the United States—about 700 per year. Tornadoes most frequently

USING TECHNOLOGY

Tailing a Tornado

A tornado can develop in less than an hour and can destroy property and kill people within a matter of minutes.

Next Generation Weather Radar, or NEXRAD, is a system of radar stations that uses Doppler radar to track severe weather storms such as tornadoes. Doppler radar sends out radio waves toward the storms. The waves reflect off the water droplets of storm clouds and are recorded at the radar station. The shift in frequency of the reflected signals allows meteorologists to determine the position, strength, and wind speed of the storm. Doppler radar helps scientists detect funnel clouds before they touch down. The weather service issues an advisory if the conditions are right for a tornado to form. If a tornado is spotted, the weather service issues a warning to people in its path.

Watching for a Tornado

Think Critically:

Suppose a tornado were moving toward your town at 100 km/h. If Doppler radar spotted the storm 160 km from your town, how much time would you have to prepare for the storm?

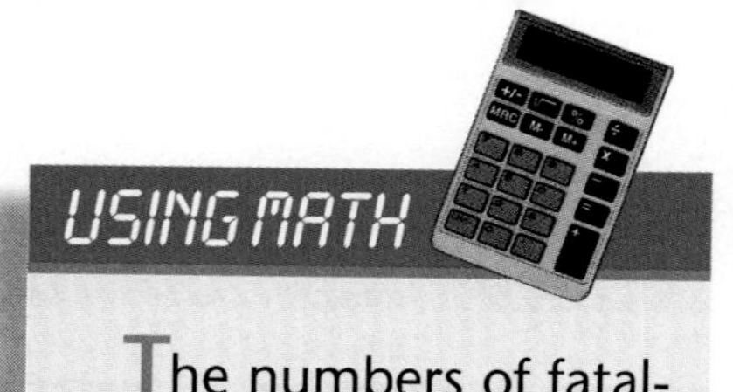

USING MATH

The numbers of fatalities due to hurricanes in the United States are as follows:

1940s	216
1950s	877
1960s	587
1970s	217
1980s	118

Make a bar graph of these data.

strike the Midwest and South, usually in spring or early summer. Texas, Oklahoma, and Kansas report the most tornadoes.

Hurricanes

The most powerful storm is the hurricane. A **hurricane** is a large, swirling, low pressure system that forms over tropical oceans. It is like a machine that turns heat energy from the ocean into wind. A storm must have winds of at least 120 km per hour to be called a hurricane.

Hurricanes are similar to low pressure systems on land, but they are much stronger. **Figure 15-14** illustrates the mechanics of the hurricane. In the North Atlantic, the southeast trade winds and the northeast trade winds sometimes meet. A low pressure area develops in the middle of the swirl and begins rotating counterclockwise in the northern hemisphere. This usually happens between 5° and 20° north latitude, where the water is quite warm. Around the middle of the low pressure area, warm, moist air is forced up. As it rises to higher elevations, it cools and moisture condenses.

Figure 15-15 shows a hurricane hitting land. When a hurricane strikes land, the high winds, tornadoes, heavy rains, and high waves of the storm surge cause a lot of damage. Floods from the heavy rains can cause additional damage. The weather of the hurricane can destroy crops, demolish buildings, and kill people and other animals. In 1992, Hurricane Andrew hit Florida and Louisiana, killing 14 people and causing more than 25 billion dollars in damage.

As long as a hurricane is over water, the warm, moist air rises and provides energy for the storm. When a hurricane reaches land, however, its supply of warm, moist air disappears and the storm loses power.

Figure 15-14

In this hurricane cross section, the small red arrows indicate rising warm, moist air. This rising air forms cumulus and cumulonimbus clouds in bands around the eye. The blue arrows indicate cool, dry air sinking in the eye and between the cloud bands. The large red arrows indicate the circular motion of the rising spiral cloud bands.

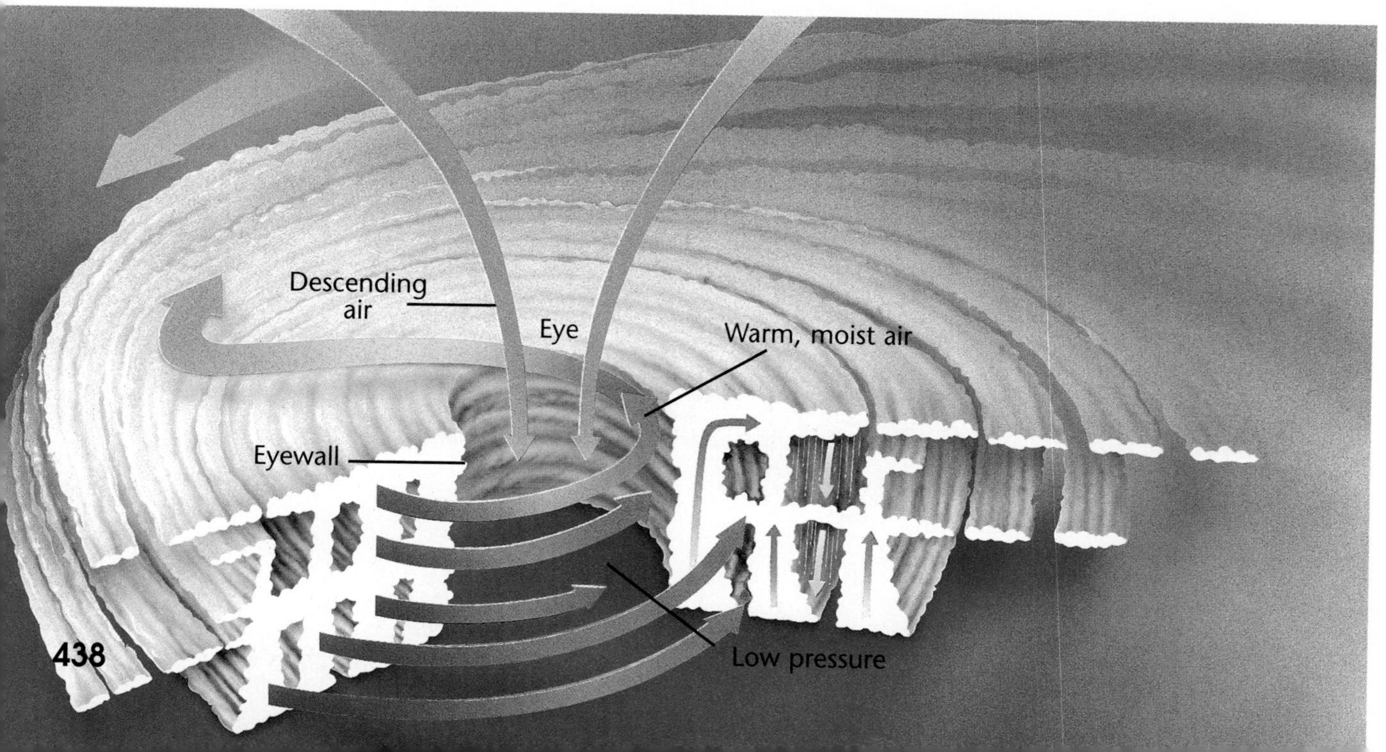

Most hurricanes in the United States strike along the Gulf of Mexico or along the Atlantic Coast. Mexico often sustains damage from hurricanes along its Pacific coast as well as its Atlantic coast. Find out what happens to the islands of the Caribbean Ocean during an average hurricane season.

Changes in weather affect your life. The interaction of air and water vapor cause constant change in the atmosphere. Air masses meet and fronts form, causing changes in weather. Severe weather can affect human lives and property.

Figure 15-15
Hurricanes can be very destructive, killing people and destroying property.

Section Wrap-up

Review

1. Why do high pressure areas usually have clear skies?
2. Explain how a tornado evolves from a thunderstorm.
3. **Think Critically:** How do two fronts form at a low pressure area? Which would bring the most severe weather?

Skill Builder

Recognizing Cause and Effect

Use your knowledge of weather to answer the following questions. If you need help, refer to Recognizing Cause and Effect in the **Skill Handbook.**

1. What effect does a warm, dry air mass have on the area over which it moves?
2. What causes a cold front? What effect does a cold front produce?
3. Describe the cause and effect of an occluded front that might form over your city.

Using Computers

Spreadsheet Make a spreadsheet comparing warm fronts, cold fronts, stationary fronts, and occluded fronts. Indicate what kind of clouds and weather systems form with each.

15•3 Forecasting Weather

Science Words

meteorologist
station model
isotherm
isobar

Objectives

- Explain the collection of data for weather maps and forecasts.
- Explain the symbols used in a weather station model.

Weather Observations

You can determine current weather conditions by observing the temperature and looking to see if clouds are in the sky. You know if it's raining. You also have a general idea of the weather because you are familiar with the typical weather where you live. If you live in Florida, you probably don't worry about snow in the forecast. But if you live in Maine, you assume it will snow every winter. What weather concerns do you have in your region?

A **meteorologist** studies the weather. Meteorologists take measurements of temperature, air pressure, winds, humidity, and precipitation. In Chapter 14, you learned about weather satellites. In addition to satellites, Doppler radar, computers, and instruments attached to balloons are used to gather data. Instruments for observing the weather improve meteorologists' ability to predict the weather. Meteorologists use the information provided by weather instruments to make weather maps. They use these maps to make weather forecasts.

Weather Forecasts

Storms such as hurricanes, tornadoes, blizzards, and thunderstorms can be dangerous. When conditions indicate that severe weather might occur or when dangerous weather is

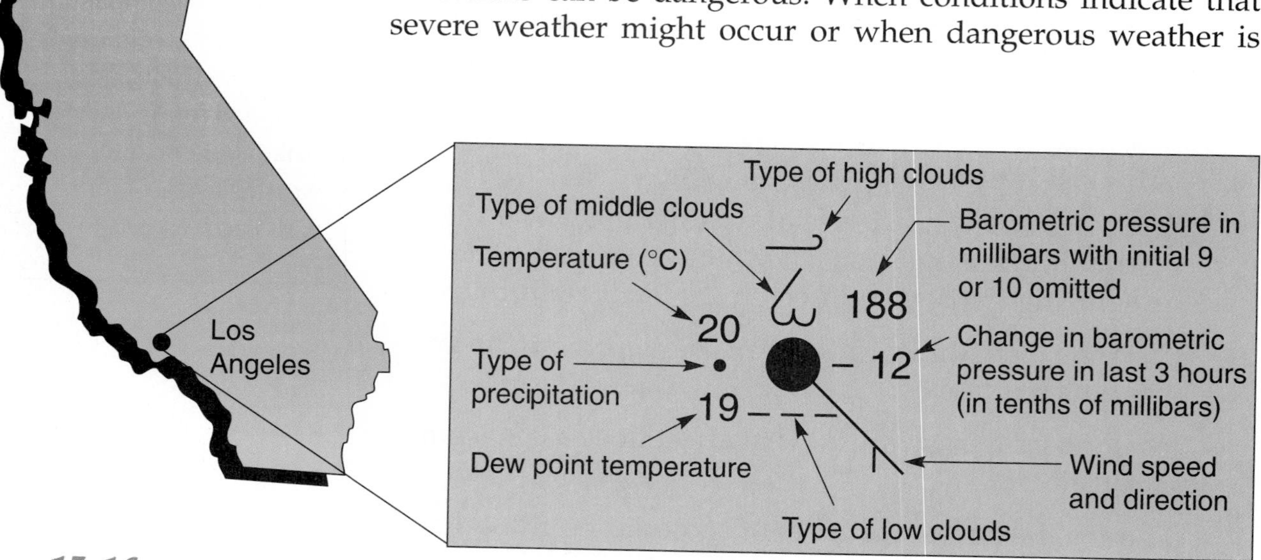

Figure 15-16

A station model shows the weather conditions at one specific location.

observed, meteorologists at the National Weather Service issue advisories. When they issue a weather watch, you should prepare for severe weather. Watches are issued for severe thunderstorms, tornadoes, floods, blizzards, and hurricanes. During a watch, stay tuned to a radio or television station reporting the weather. When a warning is issued, severe weather conditions already exist. You should take immediate action. During a severe thunderstorm warning, take shelter. During a tornado warning, go to the basement or a room in the middle of the house away from windows.

Weather Information

The National Weather Service depends on two sources for its information: meteorologists from around the world and satellites. Meteorologists take measurements in a specific location and give the data to the National Weather Service. The Weather Service uses this information to make weather maps. The Service records the information on maps with a combination of symbols, forming a **station model. Figure 15-16** shows information contained in a station model. Weather satellites provide cloud maps, surface temperatures, photos of Earth, and other data. All this information is used to forecast weather and to issue warnings about severe weather.

In addition to station models, weather maps have lines that indicate atmospheric pressure and temperature. These lines are like the contour lines you studied in Chapter 5, but instead of connecting points of equal elevation, they connect locations of equal temperature or pressure. A line that connects points of equal temperature is called an **isotherm.** *Iso* means "same" and *therm* means "temperature." You've probably seen isotherms on weather maps on TV or in the newspaper.

Figure 15-17

This tornado chaser uses a Doppler radar unit to obtain weather data about the severe storm.

An **isobar** is a line drawn to connect points of equal atmospheric pressure. You can tell how fast wind is blowing in an area by noting how closely isobars are spaced. Isobars close together indicate a large pressure difference over a small area. A large pressure difference causes strong winds. Isobars spread apart indicate a smaller difference in pressure. Winds in this area are more gentle. Isobars also indicate the locations of high and low pressure areas. On a weather map like the one in **Figure 15-18** on page 443, these areas are drawn as circles with a *High* or a *Low* in the middle.

Activity 15-2

Reading a Weather Map

Meteorologists use a series of symbols to provide a picture of local and national weather conditions. You've already learned about station models, isotherms, and isobars. Let's see how you can interpret weather information from weather map symbols.

Problem

How do you read a weather map?

Materials

- hand lens
- Figure 15-18 on page 443
- Appendix L

Procedure

Use the information provided in the questions below and Appendix L to learn how to read a weather map.

Analyze

1. Find the station models on the map for Tucson, Arizona, and Albuquerque, New Mexico. Find the dew point, cloud coverage, pressure, and temperature at each location.
2. After reviewing information about the spacing of isobars and wind speed in Section 15-3, determine whether the wind would be stronger at Roswell, New Mexico, or at Fort Worth, Texas. Record your answer. What is another way to judge the wind speed at these locations?
3. Determine the type of front near Key West, Florida. Record your answer.
4. The triangles or half-circles on the weather front symbol are on the side of the line towards the direction the front is moving. Determine the direction that the cold front located over Colorado and Kansas is moving. Record your answer.

Conclude and Apply

5. Locate the pressure system over Winslow, Arizona. After reviewing Section 15-3, **describe** what would happen to the weather of Wichita, Kansas, if this pressure system should move there.
6. Prevailing westerlies are winds responsible for the movement of much of the weather across the United States and Canada. Based on this, would you expect Charleston, South Carolina, to continue to have clear skies? **Explain** your answer.
7. The wind direction line on the station model indicates the direction from which the wind blows. The wind is named for that direction. **Infer** from this the name of the wind blowing at Jackson, Mississippi.

As you've learned so far in this chapter, weather is constantly changing. When you watch the weather forecasts on television, notice how weather fronts move across the United States from west to east. This is a pattern that meteorologists depend on in forecasting the weather. However, national and regional weather forecasters cannot always predict the exact weather because weather conditions change rapidly and local conditions also influence the weather. However, improved technologies enable forecasters to be increasingly more accurate.

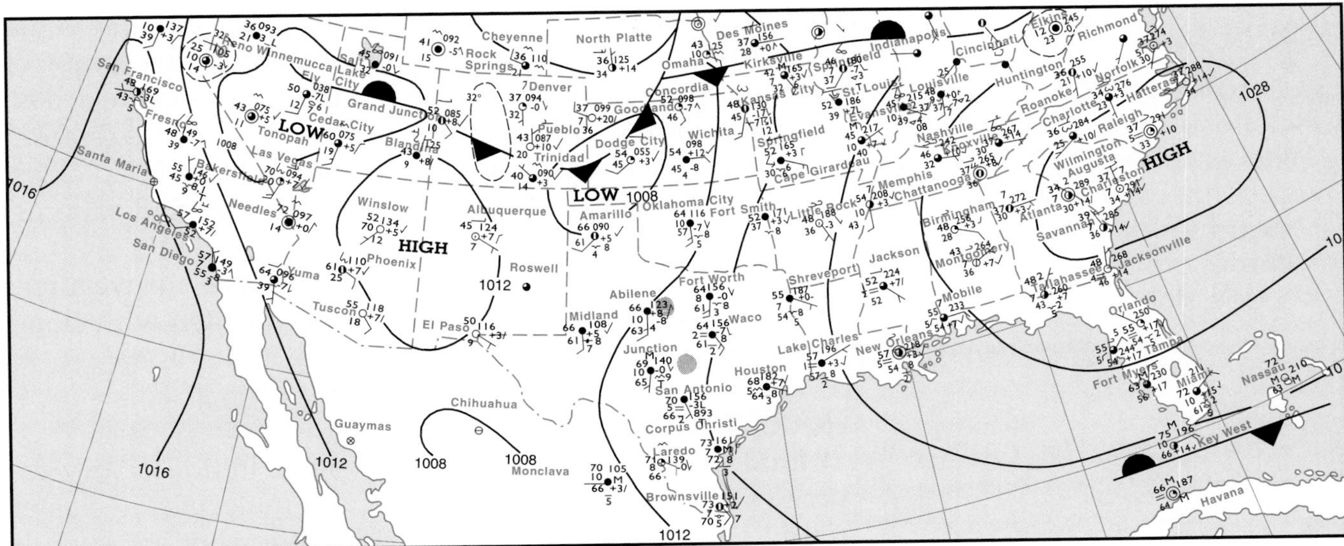

Figure 15-18

Highs, lows, isobars, and fronts on this weather map help meteorologists forecast the weather.

Section Wrap-up

Review

1. What symbols are used in a station model?
2. How do meteorologists collect the data that they need for making maps and forecasts?
3. **Think Critically:** Use Appendix L to analyze the station model shown in **Figure 15-16.** What is the temperature, type of clouds, wind speed and direction, and type of precipitation at that location?

Skill Builder

Comparing and Contrasting

Contrast a weather watch and a weather warning. If you need help, refer to Comparing and Contrasting in the **Skill Handbook.**

Science Journal

Analyze today's weather by looking at the cloud cover and a newspaper weather map. In your Science Journal, write a meteorological forecast for later today and tomorrow.

SCIENCE & SOCIETY

15•4 TECHNOLOGY: Changing the Weather

Science Words

cloud seeding

Objectives

- Describe human activities that alter the weather.
- Describe why many cloud seeding experiments are not successful.

Weather Modification

Mark Twain once said, "Everybody talks about the weather, but nobody does anything about it." Actually, that statement is not true. When we plant trees, we change the patterns of the wind. When we mulch around plants to reduce the escape of moisture from the soil or irrigate to add more moisture to soil, we change local weather conditions. When we release smoke, other particles, and gases into the atmosphere, or build cities, cut forests, and make artificial lakes, we change wind patterns, temperature, and air moisture.

What if we could have even more control of our weather? For nearly fifty years, people have experimented with cloud seeding as a way to modify weather.

Cloud Seeding

Cloud seeding involves placing silver iodide or dry ice particles into clouds as nuclei for water molecules to cluster around and freeze. The particles are either shot into clouds

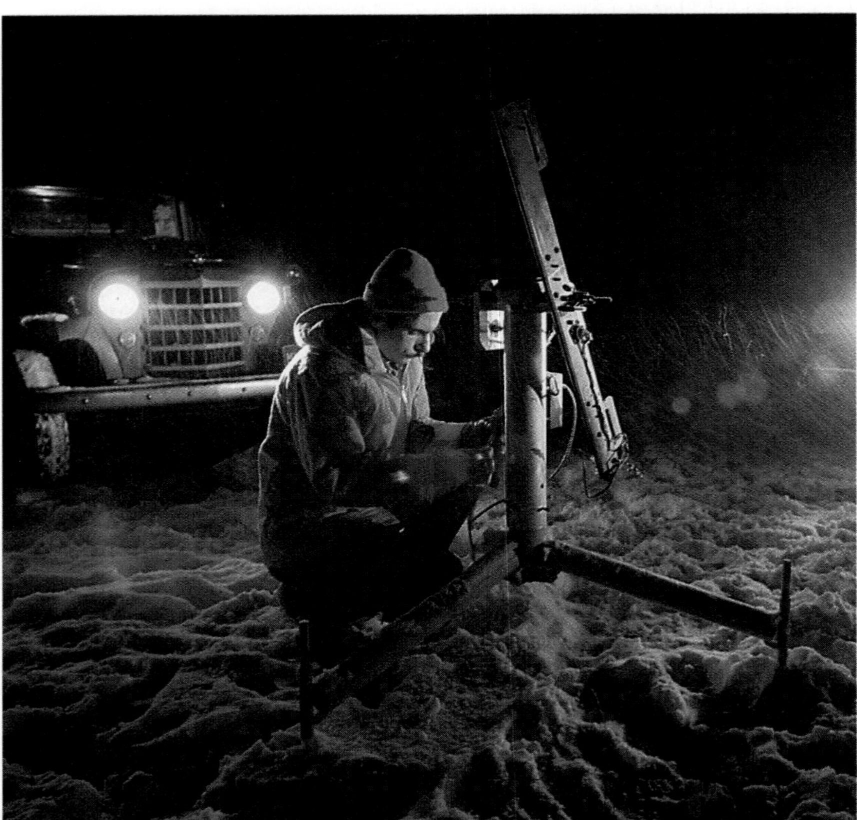

Figure 15-19

In 1973, the University of Colorado used a rocket in a cloud seeding experiment in Leadville, Colorado.

from the ground or dropped into the clouds from airplanes. One cloud seeding experiment is shown in **Figure 15-19.**

Once a particle and its added ice crystals reach 0.2 mm in size, it falls from the cloud as precipitation. Some scientists hypothesize that cloud seeding can be used to encourage precipitation wherever water vapor is present and the air temperature is near the dew point.

In some areas, cloud seeding has been successful. It has been used for clearing fog at airports. It has also been successful in increasing snowfall in some mountain resorts and rainfall in some areas of drought.

Throughout the years, a number of cloud seeding experiments have been unsuccessful because conditions that affect weather vary so much from place to place and from day to day. Experiments have been carried out to reduce winds, waves, and storm surges to make hurricanes less destructive. Although some studies claimed success, overall this cloud seeding did not work. Some studies were stopped because scientists feared cloud seeding might make hurricanes more powerful. Cloud seeding to prevent hailstorms has been successful in Russia, but the same techniques used in the United States failed to produce helpful results. In recent years, funding for cloud seeding experiments has disappeared. It appears at this time that our ability to modify the weather is quite limited.

CONNECT TO

PHYSICS

Because of the concrete and asphalt in a city, the air over a city is usually warmer than the air over rural communities. Because of this difference, convection currents develop at the boundary of cities and rural areas. Copy the drawing below into your Science Journal. ***Illustrate*** the drawing with arrows showing the direction the convection currents move.

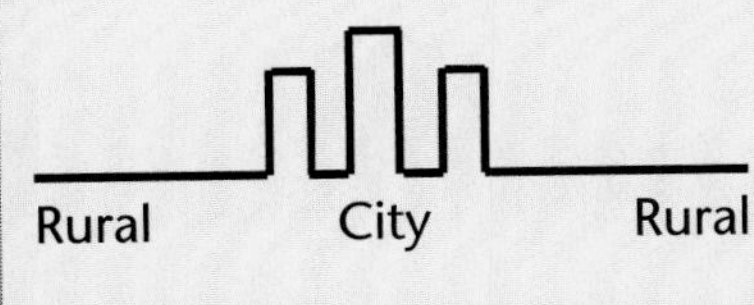

Section Wrap-up

Review

1. How can people modify local weather?
2. How could successful cloud seeding affect the water cycle in an area?

Explore the Technology

If cloud seeding becomes both successful and popular, conflict between communities could result. Community A might be accused of "cloud rustling" because they have taken moisture from clouds that normally would precipitate onto community B. Do you think clouds and their moisture are "owned" by the community beneath them? Should people intensify their efforts to find ways to modify the weather?

Science Journal

Find out how drought affected American history, especially between the years 1931 and 1938. Write an essay about your findings in your Science Journal.

How Weather Affected History

An article by Charles C. Hazewell in the May 1862 issue of the *Atlantic Monthly* expressed it well:

A hard frost, a sudden thaw, a "hot spell," a "cold snap," a long drought, a storm of sand—all these things have had their part . . . in the fate of nations. Leave weather out of history, and it is as if night were left out of day, and winter out of the year.

Examples from history illustrate Hazewell's idea:

Long droughts in West Africa during the 1970s and 1980s forced hungry people to migrate to cities. During the droughts, the dust from the dry areas in West Africa blew up and across the Atlantic Ocean and darkened the skies over the Caribbean.

Big storms also affect history. In 1279, after he conquered China, Kublai Khan set out to conquer Asia. When he sent a fleet of ships to conquer Japan, Japan had an ally: typhoons (hurricanes). These storms wrecked Kublai Khan's fleets.

In World War II, typhoons again aided Japan. In December 1944, a typhoon east of the Philippines sank three U.S. destroyers. Another typhoon struck Okinawa in 1945. It sank three U.S. naval vessels and damaged another 127 ships. Also in World War II, cold and snow affected Hitler's armies.

During the Revolutionary War, cold weather helped George Washington. In January 1777, General Washington was trapped at Princeton, New Jersey. His army was outnumbered. Muddy roads on one side and an ice-blocked river on the other prevented escape. The British general Cornwallis planned to attack the trapped army in the morning.

However, Washington knew that the wind was from the northwest and the temperature had not risen during the day. From past experiences, Washington expected a freeze that would harden the roads. By 9 P.M. the temperature was -3°C. Washington moved his army at 1 A.M. over frozen roads. The weather helped the Americans escape from a position from which they would have been defeated.

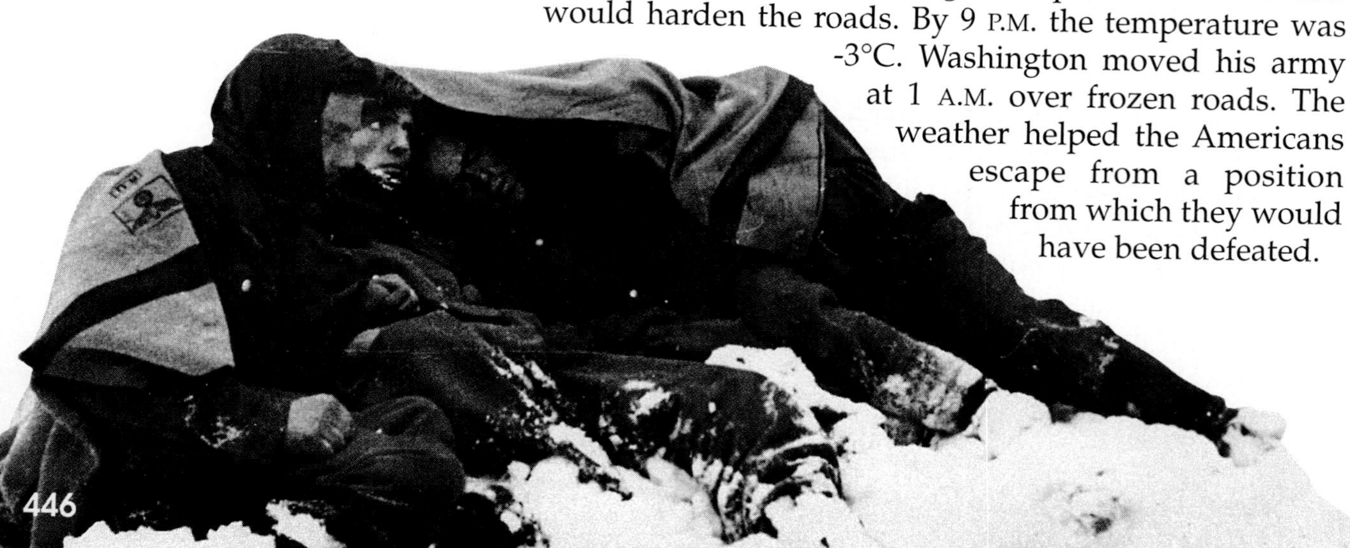

In WWII, cold weather stopped the German army's march into the U.S.S.R.

Chapter 15

Review

Summary

15-1: What is weather?

1. Water vapor forms most clouds and the precipitation that falls from them. Water vapor also determines the humidity.
2. Rain, hail, sleet, and snow are types of precipitation. The type of the precipitation depends on air temperature.
3. When air cools to its dew point, water vapor condenses and forms clouds.

15-2: Weather Patterns

1. Warm fronts may produce precipitation over a wide band. Cold fronts produce a narrow band of violent storms. A stationary front produces weak winds and precipitation. Occluded fronts are sometimes associated with high winds and heavy precipitation. High pressure brings clear skies and fair weather. Low pressure areas are cloudy.
2. Low pressure systems form when air at a front begins to swirl, rises, cools, and forms clouds.
3. Tornadoes are intense, whirling windstorms that can result from strong winds and low pressure in thunderstorms.

15-3: Forecasting Weather

1. Meteorologists use information from radar, satellites, computers, and other instruments to make weather maps and forecasts.
2. Symbols on a station model indicate the weather at a particular location.

15-4: Science and Society: Changing the Weather

1. Planting trees, irrigating, and building cities can modify wind patterns, temperature, and air moisture in a region.
2. Cloud seeding has been used to modify the weather. Cloud seeding changes regional precipitation patterns and may create ecological problems.

Key Science Words

a. air mass
b. cloud seeding
c. dew point
d. fog
e. front
f. humidity
g. hurricane
h. isobar
i. isotherm
j. meteorologist
k. precipitation
l. relative humidity
m. saturated
n. station model
o. tornado
p. weather

Reviewing Vocabulary

Match each phrase with the correct term from the list of Key Science Words.

1. temperature at which condensation occurs
2. falling water droplets
3. air that can't hold any more water vapor
4. boundary between air masses
5. violent swirling storm moving in a narrow path
6. large, swirling tropical storm
7. person who studies the weather
8. symbols that describe local weather conditions
9. line on a weather map that indicates points of equal pressure
10. stratus cloud near the ground

Chapter 15 Review

Checking Concepts

Choose the word or phrase that completes the sentence.

1. Water vapor will condense from air when the air is _______.
a. humid c. dry
b. temperate d. saturated

2. A large body of air that has the same properties as the area over which it formed is a(n) _______.
a. air mass c. front
b. station model d. isotherm

3. Water vapor in air condenses at its _______.
a. dew point c. front
b. station model d. isobar

4. _______ forms when water vapor changes directly into a solid.
a. Rain c. Sleet
b. Hail d. Snow

5. _______ clouds are high feathery clouds made of ice crystals.
a. Cirrus c. Cumulus
b. Nimbus d. Stratus

6. A(n) _______ front forms when two cool air masses merge.
a. warm c. stationary
b. cold d. occluded

7. _______ is used to seed clouds.
a. Water vapor c. Dew
b. Styrofoam d. Silver iodide

8. A _______ is issued when severe weather conditions exist and immediate action should be taken.
a. front c. station model
b. watch d. warning

9. The amount of water vapor in the air is the _______.
a. dew point c. humidity
b. precipitation d. relative humidity

10. A psychrometer is used to measure _______.
a. air pressure c. wind speed
b. relative humidity d. precipitation

Understanding Concepts

Answer the following questions in your Science Journal using complete sentences.

11. Explain how temperature affects relative humidity.

12. Describe the characteristics of an air mass that forms over central Canada in December.

13. Describe how water and the sun interact to cause our weather.

14. Describe several human activities that alter local weather.

15. Compare the structure and weather of a cold front to a warm front.

Thinking Critically

16. If you learn that there is 79 percent relative humidity, what does that mean?

17. If the temperature difference between a wet bulb thermometer and a dry bulb thermometer is great, what do you know about the relative humidity?

18. Why don't hurricanes form in polar regions?

19. If a barometer shows that the air pressure is dropping, what general weather prediction could you make?

20. What weather conditions would the very tall, thick clouds shown below indicate?

Developing Skills

If you need help, refer to the **Skill Handbook.**

21. **Comparing and Contrasting:** Compare and contrast tornadoes and severe thunderstorms.
22. **Interpreting Scientific Illustrations:** Describe the weather conditions shown on the station model below.

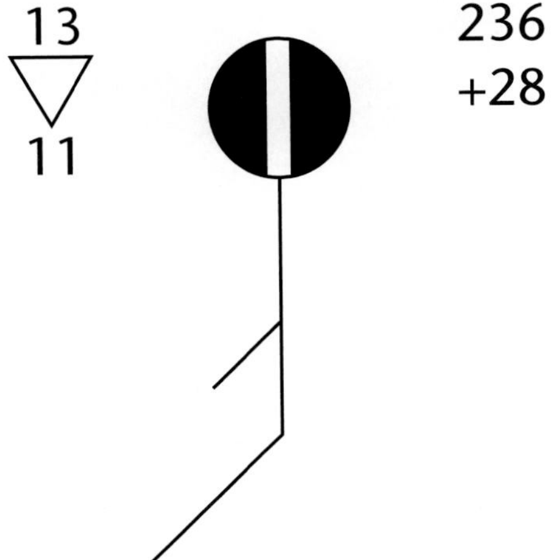

23. **Observing and Inferring:** You take a hot shower. The mirror in the bathroom "clouds up." Infer from this information what has happened.
24. **Interpreting Scientific Illustrations:** Use the cloud descriptions in **Figure 15-5** to describe the weather at your location today. Then try to predict tomorrow's weather.
25. **Concept mapping:** Construct sequence charts, one for each type of precipitation. Show the sequence from evaporation to falling precipitation.

Performance Assessment

1. **Graph:** In the MiniLAB on page 425, you learned how to find dew point temperature. Repeat this procedure for three days and graph your results.
2. **Prediction Display:** In Activity 15-2, you interpreted weather map symbols and predicted the weather. Check weather maps from the newspaper and predict the weather for your area for three days. Compare your predictions with the actual weather that occurs. Display the weather maps and your predictions.
3. **Models:** Design and construct an anemometer and a rain gauge. Use them for one week. Compare the accuracy of your instruments with reported data from radio or TV weather reports. Explain these weather instruments to your class.

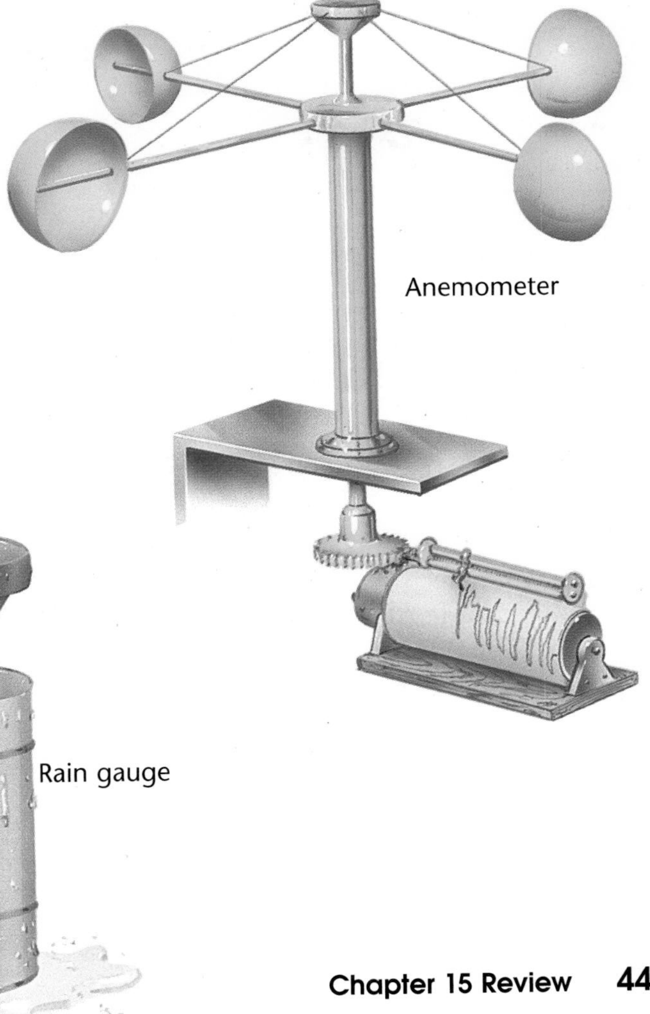

Previewing the Chapter

Section 16-1 What is climate?
- Climate
- Latitude Affects Climate
- Topographic Features Affect Climate

Section 16-2 Climate Types
- Climate Classification
- Organism Adaptations

Section 16-3 Climatic Changes
- Seasons
- El Niño
- Climatic Change
- Theories Explaining Major Climatic Change
- Climatic Changes Today

Section 16-4 Science and Society: Technology: How can global warming be slowed?

Chapter 16

Climate

Some places on Earth, such as this area of Iceland, receive so little solar radiation that glaciers form. Other places receive so little precipitation in a year that it is difficult to measure it. Why are some areas cold while others are hot? Why are some areas dry while others receive a lot of rain? How much solar radiation a place receives plays an important part in weather patterns. Wind plays a major role in distributing moisture and heat around the world.

EXPLORE ACTIVITY

Where are deserts found?

1. Obtain a world globe or atlas.
2. Locate several of the world's deserts.
3. Find the latitudes of these deserts.
4. Use **Figure 14-17** on page 412 to determine which winds affect these latitudes.

Observe: Many deserts are located next to mountain ranges. In your Science Journal, write about how you think mountains might affect precipitation patterns in different regions.

Previewing Science Skills

- In the **Skill Builders**, you will compare and contrast, and form hypotheses.
- In the **Activities**, you will design an experiment, use variables, constants, and controls, and make a model.
- In the **MiniLABs**, you will make a model, observe, measure in SI, and interpret data.

16•1 What is climate?

Science Words

tropics
polar zone
temperate zone

Objectives

- Describe what determines the climate of an area.
- Explain how latitude and topographic features affect the climate of a region.

Climate

When you travel around the world or around the United States, you experience a variety of climates. As you recall from Chapter 6, climate is the characteristic weather of a region. If you visit a rain forest, you'll find the climate there wetter than in a desert. The wettest rain forest averages 1168 cm of precipitation annually. A desert receives less than 25 cm of rain per year. Some places closer to the equator are much warmer than places near the poles. Temperatures on Earth range from -89.2°C to 57.8°C.

Climate is determined by averaging the weather over a long period of time, such as 30 years. Some weather conditions that are averaged include temperature, precipitation, air pressure, humidity, and days of sunshine. Factors that affect the climate of a region include latitude, topography, location of lakes and oceans, availability of moisture, global wind patterns, ocean currents, and location of air masses.

Latitude Affects Climate

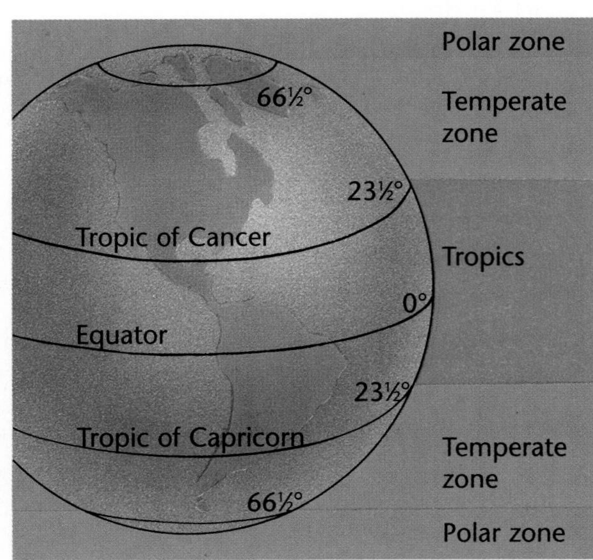

Figure 16-1

One factor that determines climate is the amount of solar energy received at different latitudes.

You learned in Chapter 14 that the amount of solar energy received at a particular location depends on the tilt of Earth on its axis. This tilt influences the angle at which sunlight strikes Earth and, therefore, the amount of radiation received by an area. The MiniLAB on the next page explains this concept.

Latitudes close to the equator receive the most radiation. As you can see in **Figure 16-1,** latitude affects climate. **Figure 16-2** shows a comparison of cities at different latitudes. The **tropics,** the region between latitudes 23½° north and 23½° south, receive the most solar radiation because the sun shines almost directly overhead. Year-round temperatures in the tropics are always hot, except at high elevations. The **polar zones** extend from the poles to 66½° north and south latitudes. Solar energy hits the polar zones at a low angle, spreading the same amount of energy over a larger area. Also, polar ice reflects some of this solar radiation. Therefore, polar regions are never very warm.

Between the tropics and the polar zones are the **temperate zones.** Temperatures in these zones are moderate. The continental United States is in a temperate zone.

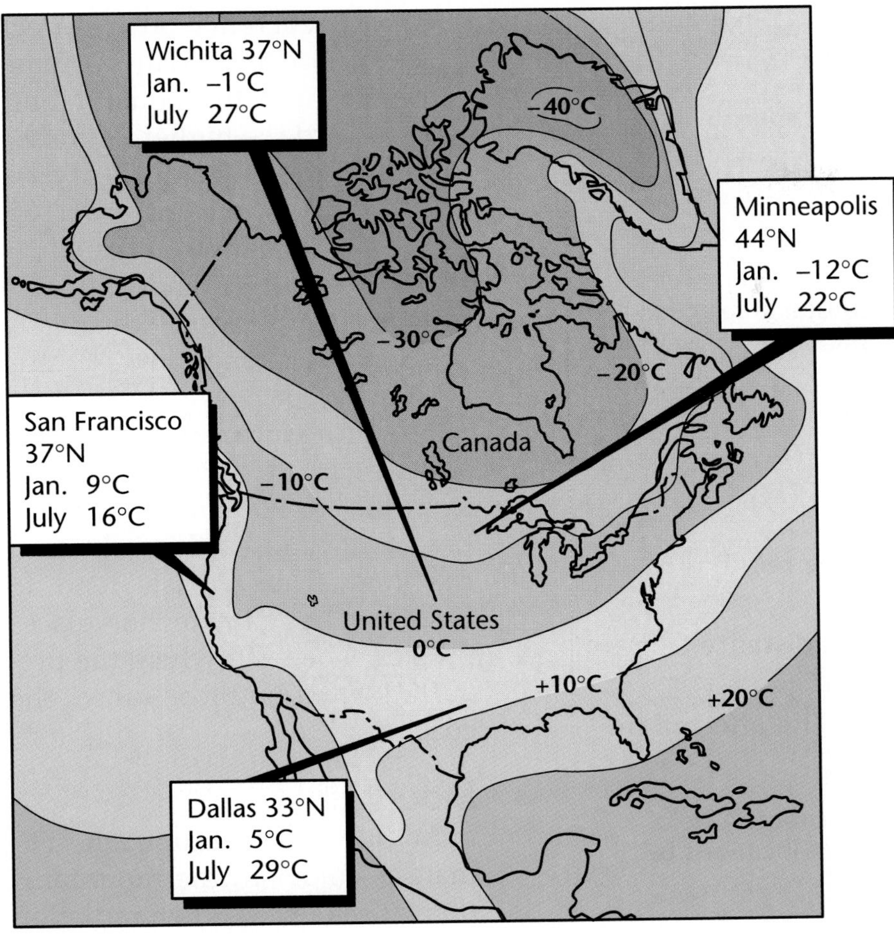

Figure 16-2

This map shows the normal daily minimum temperatures (°C) in January throughout the U.S.

A Minneapolis, Minnesota, at 44°N, receives less solar radiation than Dallas, Texas.

B San Francisco's climate is affected by the nearby ocean. Compare these temperatures to those of Wichita, Kansas on the same 37°N latitude line.

C Dallas, Texas, at 33°N, receives more solar radiation than Minneapolis.

Topographic Features Affect Climate

Climates are more complex than the general divisions of polar, temperate, and tropical. Within each zone, topographic features, such as mountains and large bodies of water, affect climate. Large cities also change weather patterns.

Large Bodies of Water

As you learned in Chapter 14, water heats up and cools down more slowly than land. Because of this, coastal areas experience sea breezes and land breezes. In a similar way, large bodies of water affect the climate of coastal areas. Many coastal regions are warmer in the winter and cooler in the summer than inland areas of similar latitude. **Figure 16-2** shows a comparison of July and January temperatures in a coastal city and a continental city, both at 37°N latitude.

MiniLAB

How does Earth's tilt affect radiation received?

Procedure

1. Hold a flashlight about 30 cm from a globe. Shine the light directly on the equator. With your finger, trace around the light.
2. Without moving the location of the light, tilt it to shine on 30°N latitude. The size of the illuminated area should increase.
3. Repeat at 60°N latitude.

Analysis

1. How did the size and shape of the light beam change as you directed the light toward higher latitudes?
2. How does the tilt of Earth affect the solar radiation received by different latitudes?

The Lake Effect

If oceans modify the climate of coastal areas near them, do you think large lakes do this as well? The following data were collected from four different Ohio cities near Lake Erie. Examine the data, and see if you can answer the questions.

Solve the Problem

1. **How is the distance from Lake Erie related to frost-free days?**
2. **What is the relationship between distance from the lake and annual precipitation?**
3. **Is the climate of a city near Lake Erie affected by the lake? Explain.**

Think Critically: Why would location A be a better location for growing grapes than locations B, C, and D?

Location	A	B	C	D
Distance from the Lake in Kilometers	0	1.6	48.3	80.5
Average Monthly Range of Temperature in °C	7.6	8.8	10.8	11.9
Frost-Free Days	205	194	162	154
Annual Precipitation in Centimeters	73.6	81.4	94.0	97.5

Ocean currents also affect coastal climate. Warm currents originate near the equator and flow toward the higher latitudes, warming the regions they pass by. When the currents cool off and flow back toward the equator, they cool the air and climates of land nearby. Some warm currents move along our Atlantic Coast.

The moisture available also affects climate. Winds blowing from the sea contain more moisture than those blowing from the land. Thus, coasts tend to have a wetter climate than places inland, especially where the prevailing winds blow onto the coast.

Mountains

At the same latitude, the climate is colder in the mountains than at sea level. When radiation absorbed by Earth's surface is emitted upward, there are less air molecules to absorb this heat at higher elevations.

Mountains also affect the climate of nearby areas, as shown in **Figure 16-3.** On the side of the mountain facing the wind, air rises, cools, and drops its moisture as precipitation. On the other side, the air descends, heats up, and dries out the land, often forming deserts. Deserts are common on the sides of mountains away from the wind.

Large Cities

Large cities affect local climates. When radiation strikes areas of vegetation, much of the energy is used in evaporating moisture. Radiation that strikes cities is absorbed by streets, parking lots, and buildings. They heat up and radiate heat into the atmosphere. Automobile exhaust and other pollutants in the air trap this heat, creating what some people call a "heat island" effect. Summer temperatures in a city can be 10° higher than in surrounding rural areas.

In addition to raising temperatures, cities affect the climate in other ways. Skyscrapers, being like small mountains, are barriers that change local wind and precipitation patterns. A study of St. Louis, Missouri found that 25 percent more rainfall, 45 percent more thunderstorms, and 31 percent more hailstorms occurred over the city than over the surrounding rural areas.

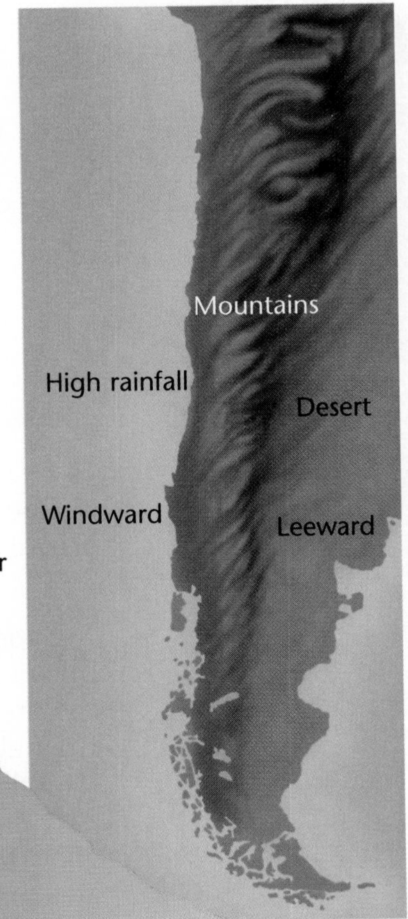

Figure 16-3

Climate differs on either side of a mountain. On the windward side, the air rises, cools, and drops its moisture. On the leeward side, the air descends and dries out the land. The map shows a comparison of the rainfall on either side of the Andes Mountains between Chile and Argentina.

Dry air

Moist air

Section Wrap-up

Review

1. What factors help determine the climate of a region?
2. How do mountains affect climate?
3. **Think Critically:** Explain why types of plants and animals found on different sides of the same mountain range might differ.

Skill Builder

Comparing and Contrasting

Contrast tropical, temperate, and polar climates. If you need help, refer to Comparing and Contrasting in the **Skill Handbook.**

Using the data from the Problem Solving Activity, predict the annual precipitation for a location 60 km from Lake Erie.

Activity 16-1

Design Your Own Experiment

Microclimates

A microclimate is a very localized climate that differs from the main climate surrounding it. Oceans, mountains, and cities affect climate. Do you think your school building affects the local climate, creating microclimates?

PREPARATION

Problem

Does your school building create microclimates?

Form a Hypothesis

Hypothesize how your school building affects the climate of the area surrounding it. Consider the height of the building, wind direction, and how the building may affect the temperature, wind speed, relative humidity, and precipitation above sidewalks, driveways, and grassy areas.

Objectives

- Design and carry out an experiment to study the microclimates around your school.
- Observe temperature, wind speed, relative humidity, and precipitation in areas outside your school. Infer how the building might affect these climate factors.

Possible Materials

- thermometers
- psychrometer (or materials to make one—see Activity 15-1)
- relative humidity chart, Appendix J
- paper strip or wind sock
- 4-5 large cans, beakers, or rain gauges
- piece of unlined paper

Safety Precaution

Be careful when you handle glass thermometers, especially if your school has only those that contain mercury. If a thermometer breaks, do not touch it. Have your teacher dispose of the glass and mercury safely.

PLAN THE EXPERIMENT

1. As a group, agree upon and write out your hypothesis statement.
2. As a group, list steps needed to test your hypothesis. Include in your plan how you will use your equipment to measure the temperature, wind speed, relative humidity, and precipitation at four or five sites around your school building. Select your sites and make certain the sites fit your hypothesis. Select a control site that is not affected by the building.
3. Sites around the school could be a sidewalk in front of the school, an area close to the building, one further away, one on the north side of the building, and one on the south side. Your control site should be selected carefully to make certain that buildings do not influence the microclimate.
4. Make a map of the school building and the test sites.
5. Design a data table in your Science Journal to use as your group collects data.

Check the Plan

1. Read over your entire experiment to make certain all steps are in a logical order.
2. Identify constants and variables in the experiment.
3. How important is a control to this experiment? How was it supported?
4. *Before you proceed, have your teacher approve your plan.*

DO THE EXPERIMENT

1. Carry out the experiment as planned.
2. While the experiment is going on, record your observations and complete the data table in your Science Journal.

Analyze and Apply

1. **Map** your data. Color code the areas to show which microclimates had the highest and lowest temperatures, the greatest and least wind speed, the greatest and least relative humidity, and the greatest and least precipitation.
2. **Analyze** your data to find patterns.
3. How did your test sites differ from your control site?
4. **Analyze** your hypothesis and the results of your experiment. Was your hypothesis supported?

Go Further

Repeat the experiment around your home. How does your home create different microclimates in its surroundings?

16•2 Climate Types

Science Words

adaptation
hibernation

Objectives

- Describe the Köppen Climate Classification System.
- Describe how organisms adapt to particular climates.

Climate Classification

If your job were to classify climates, where would you begin? Although there are several ways to classify climates, *climatologists* (persons who study climates) usually use a system developed in 1918 by Russian-German meteorologist and climatologist Wladimir Köppen. Köppen observed that the type of vegetation found in a region depended on the climate of the area. **Figure 16-4** shows differences Köppen might have observed. His classification system is based on his studies of temperature and precipitation.

Köppen Climate Types

In **Figure 16-5,** the Köppen Climate Classification System divides climates into six groups: tropical, mild, dry, continental, polar, and high elevation. These groups are further divided into types. For example, the dry climate classification is divided into *semiarid* and *arid.*

Examine the Köppen map and count how many different climates are found in the continental United States. Why do so many different types of climates exist here? What climate exists where you live?

Figure 16-4

The type of vegetation in a region depends on the climate.

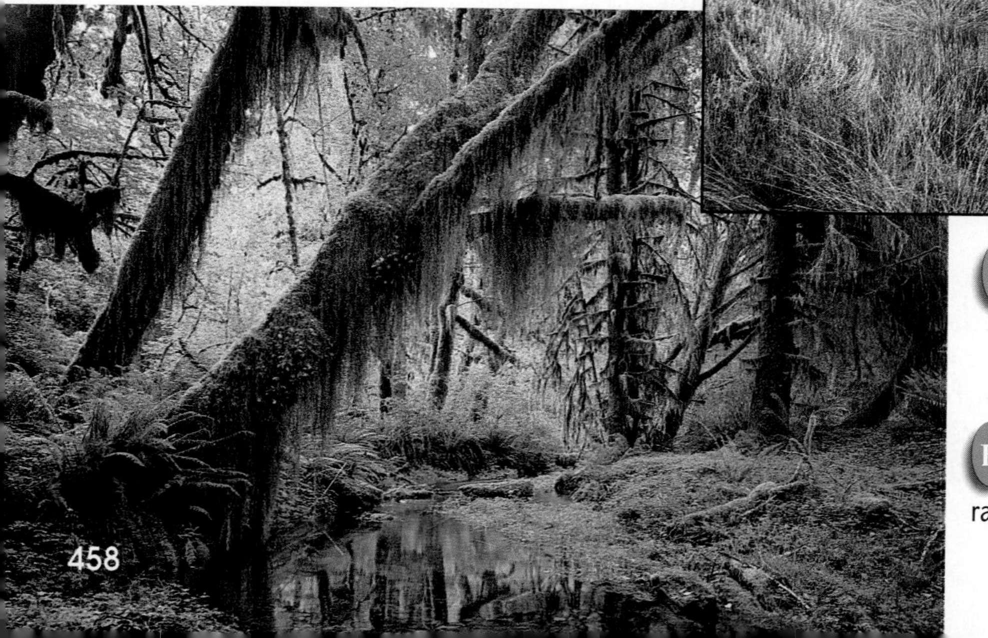

A Short grasses grow in fairly dry climates.

B Moss-draped trees and ferns grow in climates having heavy rainfall.

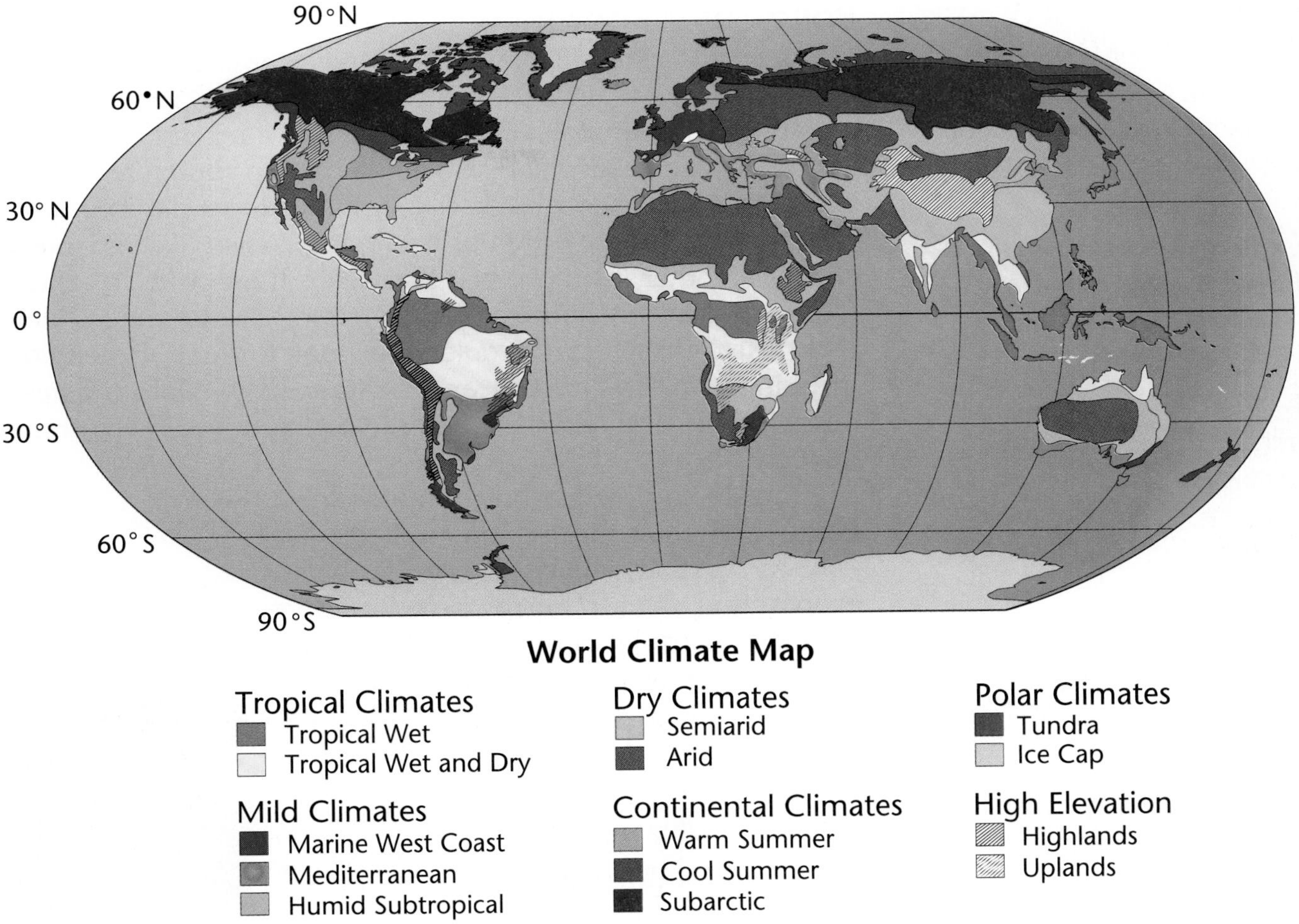

90°N
60°N
30°N
0°
30°S
60°S
90°S

World Climate Map

Tropical Climates
- Tropical Wet
- Tropical Wet and Dry

Mild Climates
- Marine West Coast
- Mediterranean
- Humid Subtropical

Dry Climates
- Semiarid
- Arid

Continental Climates
- Warm Summer
- Cool Summer
- Subarctic

Polar Climates
- Tundra
- Ice Cap

High Elevation
- Highlands
- Uplands

Figure 16-5

The Köppen Climate Classification System is shown on this map. Notice that most wet climates are located between latitudes 30° north and 30° south. *What other patterns do you see on this map?*

Organism Adaptations

As you have learned, climates vary around the world, and as Köppen noticed, the type of climate that exists in an area determines the vegetation found there. Fir trees aren't found in deserts, nor are cacti found in rain forests. In fact, all organisms have certain kinds of adaptations that allow them to survive in some climates, but not in others. An adaptation is any characteristic that develops in a population over a long time. An **adaptation** is any structure or behavioral feature that helps an organism survive in its environment. Some adaptations are shown in **Figure 16-6.** Climatic factors that may limit where an organism can live are temperature, moisture, and amount of daylight.

Structural Adaptations

Some organisms have body structures that enable them to survive in particular climates. The fur of mammals insulates them from cold temperatures. A cactus has a thick, fleshy stem. This adaptation helps a cactus hold water. The waxy skin covering the stem keeps water inside the plant from evaporating. Instead of broad leaves, cactus plants have spiny leaves that further reduce water loss.

Figure 16-6

Organisms have structural and behavioral adaptations that help them survive in particular climates.

Honeybees fan in fresh air to keep the hive cool.

Behavioral Adaptations

Some organisms display behavioral adaptations that help them survive in certain climates. For example, rodents and certain other mammals undergo a period of inactivity in winter called **hibernation.** During hibernation, body temperature drops and body processes are reduced to a minimum.

Other animals are differently adapted. When it's cold, bees cluster together in a tight ball that keeps them from freezing. During hot, sunny hours of the day, desert snakes hide under rocks. At night when it's cooler, they look for food. Instead of drinking water as turtles and lizards do in wet climates, desert turtles and lizards obtain the moisture they need from their food.

Lungfish survive periods of intense heat by going through an inactive state called *estivation.* As weather gets hot and water begins evaporating, the fish burrows into mud and secretes a capsule around itself. It lives this way until the warm, dry months pass.

B A bear's fur keeps it warm. Fur is hair. The spaces between the hairs trap air and heat and keep out the cold and rain.

C The needles and the waxy skin of a cactus reduce water loss.

Like other organisms, you have structural adaptations that help you cope with climate. You can maintain a fairly constant body temperature, regardless of the outside temperature. In hot weather, your sweat glands release water onto your skin. The water evaporates and as a result, you become cooler. What other adaptations help the people in **Figure 16-7** cope with climate?

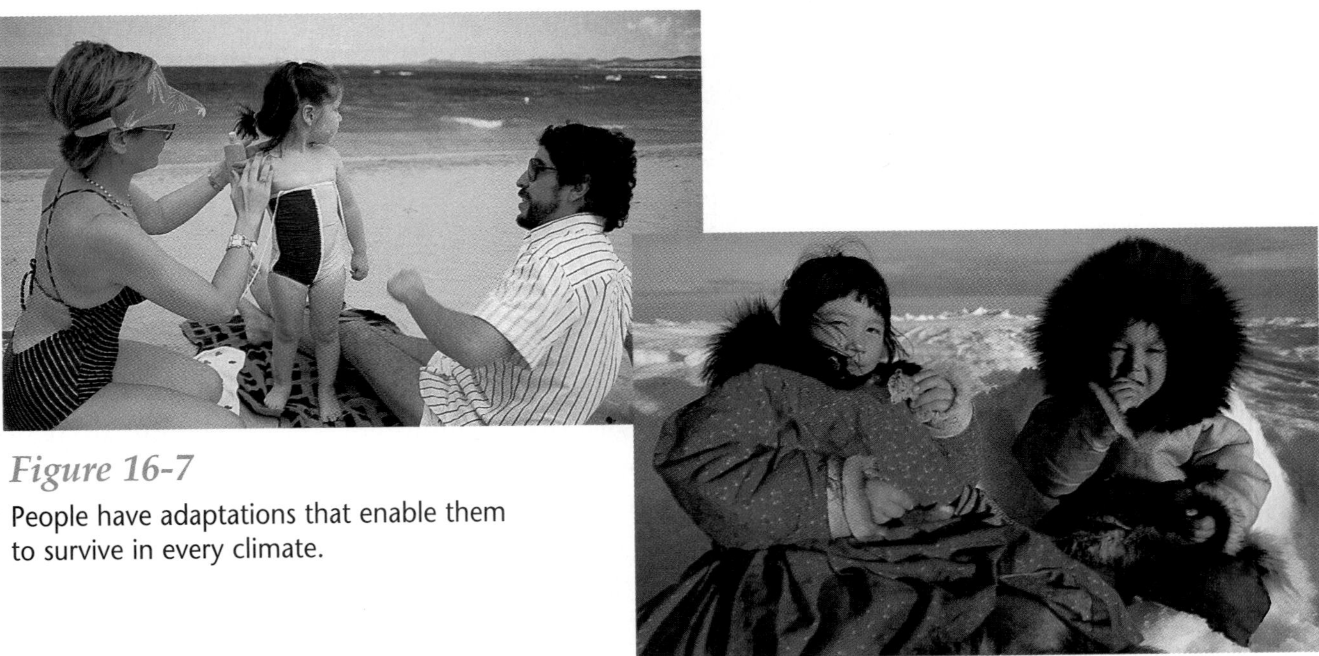

Figure 16-7
People have adaptations that enable them to survive in every climate.

Section Wrap-up

Review

1. Use **Figure 16-5** and Appendix G to identify the climate type for each of the following locations: Hawaiian Islands, North Korea, Egypt, Uruguay.
2. What are some behavioral adaptations that allow animals to stay warm?
3. **Think Critically:** What special adaptations must plants and animals have to live in the regions labeled as dry?

Skill Builder

Hypothesizing

Some scientists think Earth is becoming hotter. Suppose this is true. Form a hypothesis about which adaptations will allow some present-day organisms to survive this change. If you need help, refer to Hypothesizing in the **Skill Handbook.**

Science Journal

Research ways that people have adapted their behavior to survive in the six climate regions shown in **Figure 16-5.** Consider clothing, housing, and transportation. Write about these adaptations in your Science Journal.

16•3 Climatic Changes

Science Words

season
El Niño
greenhouse effect
global warming

Objectives

- Discuss what causes seasons.
- Describe how El Niño affects the climate.
- Describe past climatic changes.
- Explore possible causes of climatic change.
- Differentiate between the greenhouse effect and global warming.

Seasons

In temperate zones, weather generally changes with the season. **Seasons** are short-term periods of climate change caused by regular variation in daylight, temperature, and weather patterns. These variations are due to changes in the amount of solar radiation an area receives. **Figure 16-8** shows Earth revolving around the sun. Because Earth revolves at a tilt, different areas of Earth receive changing amounts of solar radiation. That affects wind patterns. In turn, wind patterns and topographic features help create seasonal climatic changes.

Effects of Latitude

In the middle latitudes or temperate zones, seasons often have warm summers and cool winters. Spring and fall are usually mild. Because of fairly constant solar radiation at the low latitudes near the equator, the tropics do not experience as much seasonal temperature change as the middle latitudes, but they do experience dry and rainy seasons.

During the year, the high latitudes near the poles experience great differences in temperature and in number of daylight hours. As shown in **Figure 16-8,** during summer in the northern hemisphere, the North Pole is tilted toward the sun. At the North Pole, there are 24 hours of daylight for six months. During that same time, the South Pole experiences 24-hour days of darkness. However, a little way from the North Pole, at 64° north latitude, Fairbanks, Alaska has more than five hours of daylight on the shortest day of the year and more than 18 hours of daylight on the longest day. At the equator, days are about the same length all year

Figure 16-8

During the year, as Earth moves around the sun, different areas of Earth tilt toward the sun, bringing different seasons. In the southern hemisphere, when the South Pole tilts away from the sun, the sun hangs low in the sky. The sun's rays hit at an angle, making the sunlight less concentrated. Days become short, bringing winter. At the same time in the northern hemisphere, the North Pole tilts toward the sun. The sun is high in the sky. Days are long. It's summer.

long. Recall from **Figure 16-2** on page 453 how latitude affects climate.

El Niño

Some climatic changes last longer than a season. **El Niño** is a climatic event that starts in the tropical Pacific Ocean and sets off changes in the atmosphere. El Niño used to occur every three to seven years but now happens more frequently. In El Niño, the Pacific Ocean warms along the equator. Near the equator, trade winds that blow east to west weaken and sometimes reverse. Instead of cold water rising off the coast of Peru, the change in the trade winds allows warm tropical water in the upper layers of the Pacific to flow eastward to South America. Ocean temperatures increase by 1°C to 7°C off the coast of Peru. Sea level rises.

El Niño does not directly cause unusual weather but instead affects the atmosphere and ocean to make stormy weather more likely. Warmer water brings more evaporation. Heavy rains fall over South America. Scientists also note that during El Niño, the jet stream often splits, changing the atmospheric pressure off California and wind and precipitation patterns around the world. It can cause drought in Australia and Africa, disruption of dependable monsoon rains, storms in California, and the weather events of **Figure 16-9.**

Scientists are not certain what causes El Niño, but many link it with global warming. Experts hypothesize that El Niño distributes the extra energy from global warming.

CONNECT TO LIFE SCIENCE

Some aquatic plants and animals living in waters off California are dying. ***Infer*** how El Niño may be the cause of their deaths.

Figure 16-9

The effects of El Niño are felt around the world. A strong El Niño occurred in 1983.

A severe drought brought dust storms to Australia.

A cyclone hit the Arutua Atoll in Polynesia.

MiniLAB

What can we learn from tree rings?

As a tree grows, it adds layers of growth rings. When conditions for growth are good, growth rings occur far apart. When conditions are poor, growth rings appear close together.

Procedure

1. Examine the growth rings of a cross section of wood. Note that the oldest wood is in the center.
2. Measure the thickness of several rings in your sample.

Analysis

1. How did the length of the growing seasons change as your tree grew older? Describe what evidence you found for this.
2. If scientists examine growth rings in petrified wood samples from different geologic times, what can they learn about climates of earlier geologic periods?

Climatic Change

Although some years are warmer, colder, drier, or wetter than others, Earth's climate remains fairly constant. However, in earlier geologic eras, Earth was sometimes much colder or much warmer than it is today.

Geological records show that in the past, the climate changed. Fossils of tropical plants and animals found in polar as well as temperate regions indicate warmer worldwide climates in the past. **Figure 16-10** illustrates how living things reflect variability in climate. Glacial erosion and deposition around the world indicate that in earlier eras extensive polar conditions existed on Earth.

In the past two million years, glaciers have covered large parts of Earth's surface. These periods of extensive glaciers are called ice ages. At least for the past three million years, ice ages have alternated with warm periods called interglacial intervals. We are now in an interglacial interval. Some interglacial intervals were warmer than now. Some ice ages lasted 60 000 years. Most interglacial periods lasted about 12 000 years. The present interglacial period began about 11 500 years ago. Cores drilled in Greenland ice give evidence of shorter variations in the climate. The evidence suggests that cold spells, lasting 1000 years or more, followed warm spells that lasted as long. Average temperatures in northern Europe changed as much as 10° C in a few years.

Theories Explaining Major Climatic Change

Research into the causes of climatic change suggests a variety of possibilities. Catastrophic events such as meteorite impacts and volcanic eruptions may have occurred, or perhaps the sun's output of energy isn't constant. It could be that when Earth's plates move, they change climate patterns, or perhaps Earth's movements in space cause climatic change.

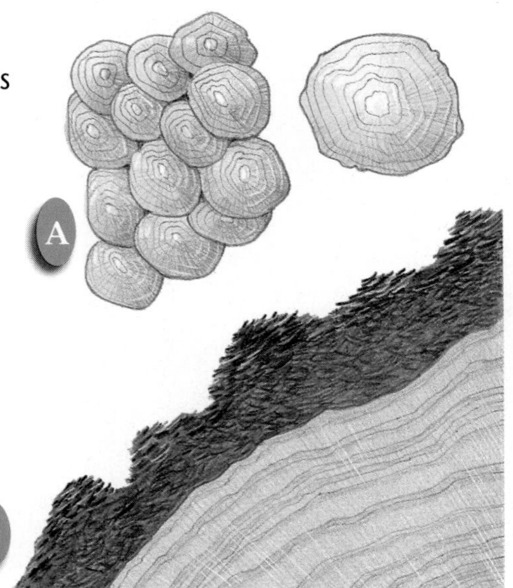

Figure 16-10

The length of growing seasons is recorded in the rings of fish scales (A) and tree rings (B).

Catastrophic events such as large meteorite impacts and volcanic eruptions put enormous volumes of dust, ash, and smoke into the atmosphere. These dust and smoke particles could have blocked so much solar radiation that they cooled the planet and changed the climate. **Figure 16-11** on the next page illustrates how a major volcanic eruption affects Earth's atmosphere.

Evidence suggests that an increase in clouds may lead to an increased greenhouse effect. It might also lead to less solar radiation reaching Earth's surface. Computer models have enabled scientists to conclude that clouds absorb more solar radiation than was previously thought. It is possible that an increase in clouds could lead to a cooling effect similar to that produced by volcanic ash.

In Chapter 14, you learned that solar radiation provides Earth's energy. If the output of the radiation from the sun varies, that could change Earth's climate.

Another possible explanation for major climatic change concerns the movement of the plates on Earth's crust. The movement of continents and oceans affects the transfer of heat on Earth's surface, which in turn affects

USING TECHNOLOGY

Ice Delivers Clues

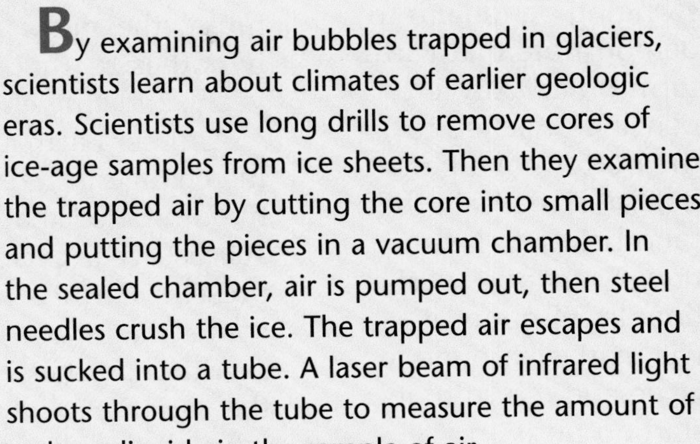

By examining air bubbles trapped in glaciers, scientists learn about climates of earlier geologic eras. Scientists use long drills to remove cores of ice-age samples from ice sheets. Then they examine the trapped air by cutting the core into small pieces and putting the pieces in a vacuum chamber. In the sealed chamber, air is pumped out, then steel needles crush the ice. The trapped air escapes and is sucked into a tube. A laser beam of infrared light shoots through the tube to measure the amount of carbon dioxide in the sample of air.

Scientists have discovered that the amount of carbon dioxide varies within ice cores. During warm periods in the past, when ice formed slowly, the concentrations of carbon dioxide ranged between 260 to 280 parts per million. In colder times, when ice layers formed quickly, concentrations of carbon dioxide ranged between 190 and 200 parts per million. No one knows the exact reason for these differences, but it is clear that carbon dioxide is associated with climate changes.

Glacial Ice Core

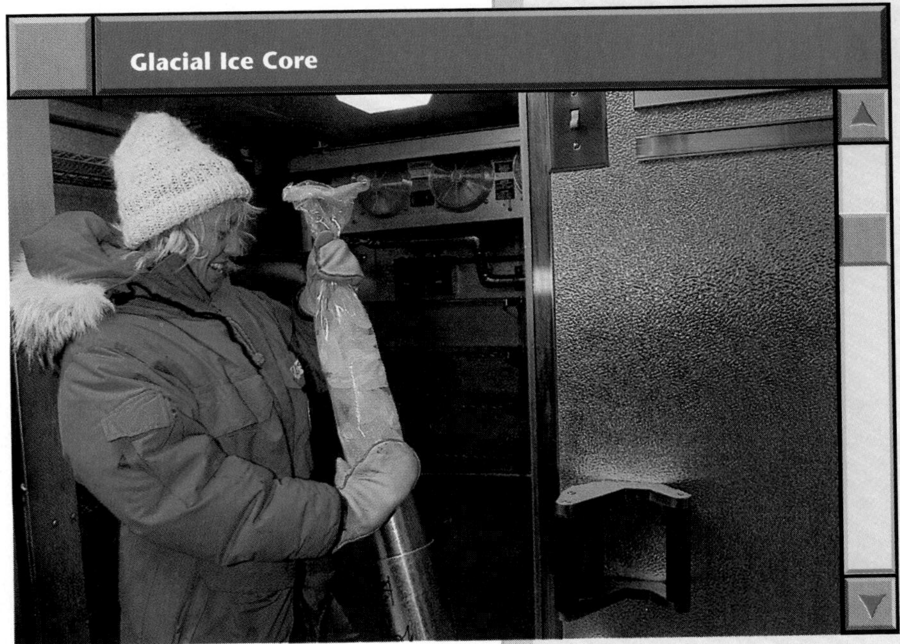

Think Critically:

Many experts theorize that we are living in an interglacial period and that another ice age will begin within 1000 years or so. Others think our climate will get warmer. Explain how measuring present-day atmospheric carbon dioxide levels may help predict the future climate.

CONNECT TO

PHYSICS

In 1851, the physicist Foucault used a free-swinging pendulum to demonstrate that Earth turns on its axis. If a pencil is attached to the bottom of a pendulum, it marks in the sand below as it continually changes position. In 24 hours, it will mark all around a circle and return to its starting position. ***Demonstrate*** this movement by making a model.

wind and precipitation patterns. Through time, these altered patterns may change the climate.

Another theory relates to Earth's movements in space. Earth is currently tilted on its axis at 23½° to the plane of its orbit around the sun. In the past, this tilt has increased to 25° and has decreased to 22°. When this tilt is at its maximum, the poles receive more solar energy. When the tilt is at its minimum, the poles receive less solar energy. Some scientists hypothesize that Earth's tilt changes about every 42 000 years, affecting climates.

Another Earth movement that may affect climatic change is the change in the shape of Earth's orbit around the sun. It's hypothesized that the shape of Earth's orbit changes over a 100 000 year cycle. When the orbit is more circular than at present, Earth is farther from the sun and temperatures are colder than those we are experiencing.

Although these movements of Earth may explain some of the variations in the most recent ice age, they do not explain why glaciers have occurred so rarely over the long span of geologic time.

As you've learned, there are many theories that attempt to answer questions about why Earth's climate has changed through the ages. Probably all of these factors play some role in changing climates. A great deal of research still needs to be done before we can understand all the interactions that affect climate.

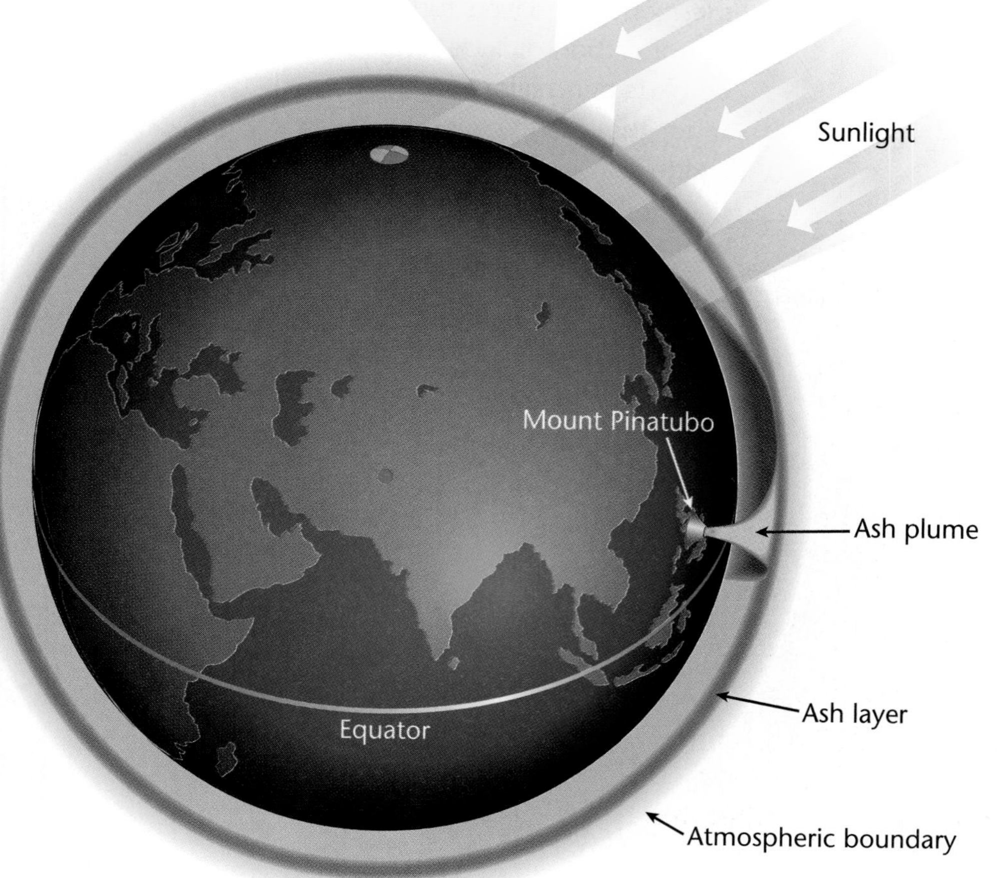

Figure 16-11

When Mt. Pinatubo in the Philippines erupted in 1991, its volcanic ash cooled temperatures around the world.

Figure 16-12

The sun's radiation travels through our atmosphere and heats Earth's surface. But the heat that is radiated from Earth can't escape back through the gases of the atmosphere into space. This is similar to how a greenhouse warms.

Climatic Changes Today

Today, many newspaper and magazine headlines warn us about the greenhouse effect and global warming. These two phenomena are related but are not the same thing.

The **greenhouse effect** is natural heating caused by gases in our atmosphere trapping heat. The greenhouse effect is illustrated in **Figure 16-12.** Carbon dioxide is the main greenhouse gas. Activity 16-2 demonstrates the greenhouse effect, which is not a bad thing. Without the greenhouse effect, life as we know it would not be possible on Earth. Like Mars, Earth would be too cold.

Global warming means global temperatures are rising. One reason for global warming is the increase of greenhouse gases in our atmosphere. An increase in greenhouse gases increases the greenhouse effect. In the last 100 years, the mean surface temperature on Earth has increased ½°C. This may be from global warming.

If the mean temperature continues to rise, ice caps will melt. Low-lying areas might experience increased flooding. Already some ice caps are beginning to break apart and sea level is rising in certain areas. Some people believe that these events are related to Earth's increased temperature. If you lived on an island or a river delta, would you need to be concerned?

Approximately 15 percent of the total land surface of Earth is covered by deserts. There are about 22 400 000 square km of deserts. Estimate the total land surface of Earth.

Activity 16-2

The Greenhouse Effect

Have you ever climbed into a car on a warm day and burned yourself on the seat? Why was it so hot inside the car, when it wasn't that hot outside? It was hotter in the car because the car functioned like a greenhouse. You experienced the greenhouse effect.

Problem

How can you demonstrate the greenhouse effect?

Materials

- identical empty aquariums (2)
- glass lid for one aquarium
- thermometers (3)

Safety Precaution 

Be careful when you handle glass thermometers, especially if your school has only those that contain mercury. If a thermometer breaks, do not touch it. Have your teacher dispose of the glass and mercury safely.

Procedure

1. Lay a thermometer inside each aquarium.
2. Place the aquariums next to each other by a sunny window. Lay the third thermometer between the aquariums.
3. Record the temperatures of the three thermometers. They should be the same.
4. Place the glass lid on one aquarium.
5. Record the temperatures of all three thermometers at the end of 5, 10, and 15 minutes.
6. Make a line graph that shows the temperatures of the three thermometers for the 15 minutes of the experiment.

Analyze

1. Explain why you placed a thermometer between the two aquariums.
2. What were the constants in this experiment? What was the variable?
3. Which thermometer experienced the greatest temperature change during your experiment? Why?

Conclude and Apply

4. **Analyze** what occurred in this experiment. How was the glass lid in this experiment like the greenhouse gases in the atmosphere?
5. **Infer** from this experiment why you should never leave a pet inside a closed car in warm weather.

You learned in section 16-2 that organisms are adapted to their environments. But when environments change, can organisms cope? In some tropical waters around the world, corals, like those in **Figure 16-13,** are dying. Are these deaths caused by warmer water? Many people think so.

Some climate models indicate that in the future, Earth's temperatures will increase faster than they have in the last 100 years. In Section 16-4, you will learn how people's activities may contribute to global warming, and you will find out what you can do to help lessen this problem.

USING MATH

When ice sheets melt, sea level rises. Between 1900 and 1970, sea level at New York City rose 23 cm. What was the average rise in sea level per year from 1900 to 1970? The present elevation of New York City is 396 cm above sea level. If sea level continues to rise at the same rate, when will New York City start to go underwater?

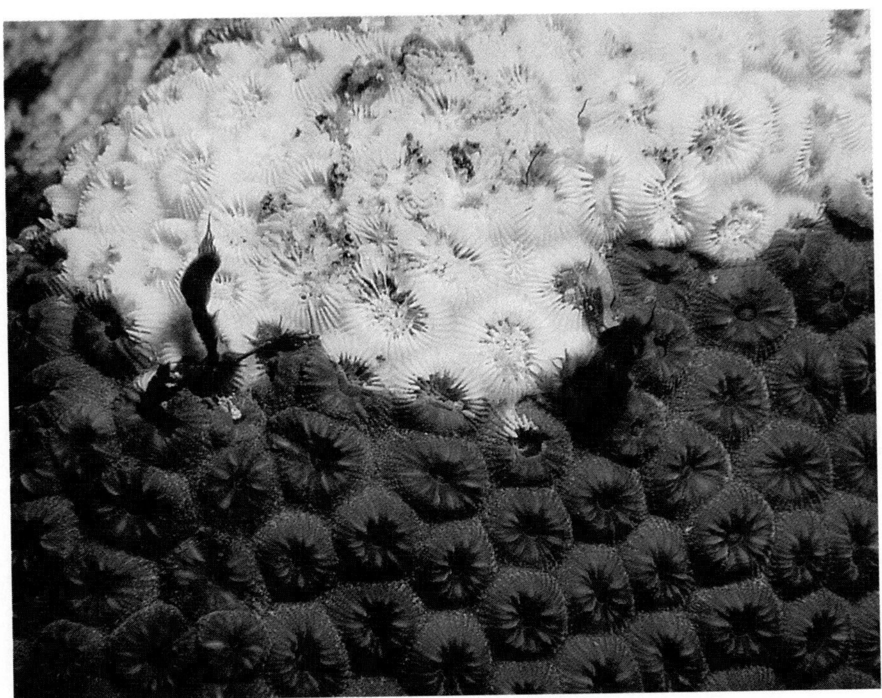

Figure 16-13

This coral is dying because water temperatures have risen in this area of the ocean.

Section Wrap-up

Review

1. What causes seasons?
2. In what way does El Niño change the climate?
3. *Think Critically:* How do we know that climates of earlier geologic eras were much different from today's climates?

Skill Builder

Comparing and Contrasting

Compare and contrast the greenhouse effect and global warming. If you need help, refer to Comparing and Contrasting in the **Skill Handbook.**

Using Computers

Word Processing
Design a pamphlet to inform people about why Earth's climate changes. What are your predictions for the future? On what evidence do you base these predictions? Include this in your pamphlet.

TECHNOLOGY:

16•4 How can global warming be slowed?

Science Words

deforestation
reforestation

Objectives

- Discuss some human activities that add carbon dioxide to the atmosphere.
- List ways we can reduce the amount of carbon dioxide in the atmosphere.

Human Activities and Carbon Dioxide

Human activities affect the air in our atmosphere. Burning fossil fuels and removing vegetation affect the atmosphere. Both activities add carbon dioxide to the atmosphere and contribute to global warming. Each year the amount of carbon dioxide in our atmosphere continues to increase by more than one part per million. By analyzing air bubbles trapped in glaciers, scientists have discovered that in about 1750, there were 280 parts per million of carbon dioxide in the atmosphere. Today there are more than 350 parts per million.

Burning of Fossil Fuels

When natural gas, petroleum, and coal are burned for energy, the carbon in these fossil fuels is combined with oxygen. This increases the amount of carbon dioxide (CO_2) in our atmosphere.

Deforestation

The mass removal of trees, called **deforestation,** also affects the amount of carbon dioxide in our atmosphere. Forests around the world are being cleared for mining, roads, buildings, grazing cattle, and drilling for oil. Forests are also dying from the effects of pollution.

As they grow, trees take in carbon dioxide. When trees are removed, the carbon dioxide they could have removed from the atmosphere is left. Cut-down trees are often burned. Burning produces more carbon dioxide.

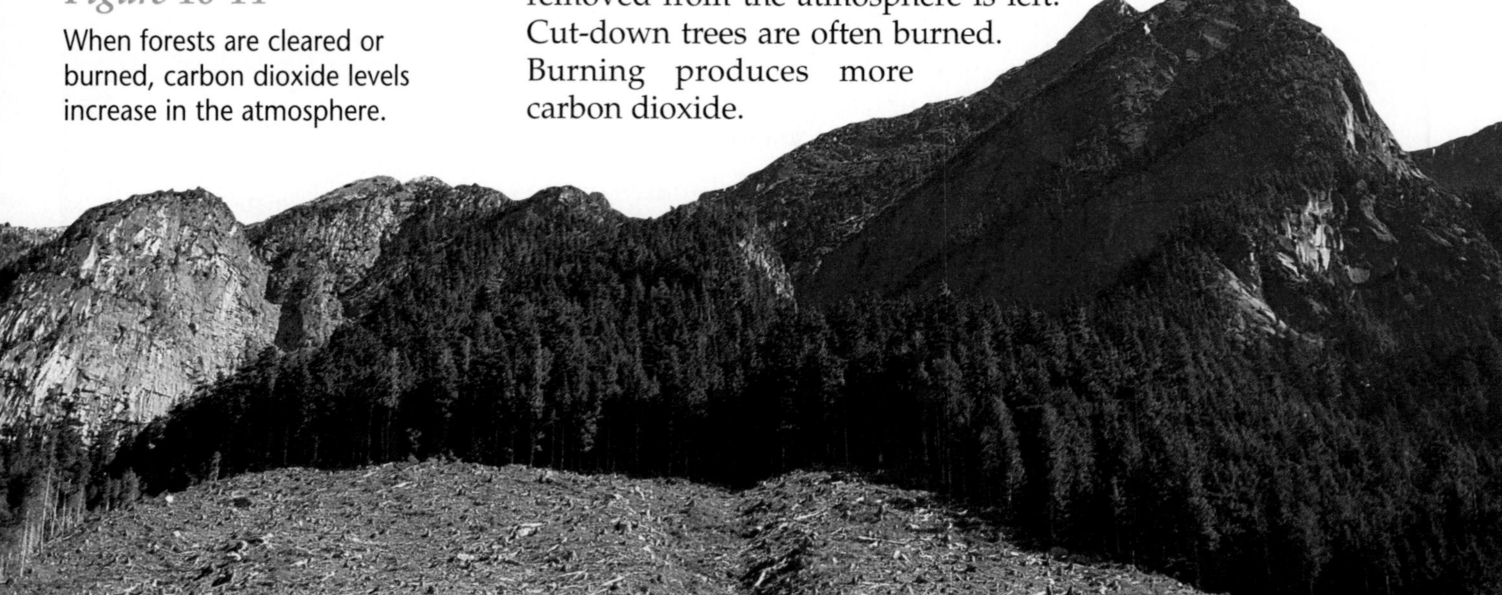

Figure 16-14

When forests are cleared or burned, carbon dioxide levels increase in the atmosphere.

Ways to Reduce CO_2

What can we each do to help reduce the amount of CO_2 in the atmosphere? Conserving electricity is one way. When we conserve electricity, we reduce the amount of fossil fuel that must be burned. One way to save fuel is to change your daily activities that rely on energy from burning fossil fuel. These activities might include car rides, watching television, and heating or cooling our homes.

Another way to reduce CO_2 is to plant vegetation. As you've learned, plants remove carbon dioxide from the atmosphere. Correctly planted vegetation can also shelter homes from cold winds or blazing sun and reduce the use of electricity.

Figure 16-15

This wind farm in California generates electricity without adding CO_2 to our atmosphere.

Futuristic Ideas to Cool Our Planet

Will our planet become as hot as Venus? A variety of ideas exist to cool Earth down. The planting of trees, called **reforestation,** would help. Investing in plants and equipment to produce electricity without generating more CO_2 would help. Wind farms, solar panels, and small water turbines generate electricity without producing CO_2. Or perhaps we could seed nutrients into polar oceans to encourage the growth of microscopic plants. These plants could remove much of the carbon dioxide. Perhaps releasing billions of aluminum balloons to reflect solar radiation would work. Or perhaps a giant screen 2000 square km in size could be put into space to decrease the amount of incoming radiation. Maybe dust from the moon could be used to make a cloud around Earth that would block solar radiation. What other things might help solve the problem?

Section Wrap-up

Review

1. How do burning fossil fuels and cutting down forests add carbon dioxide to the atmosphere?
2. What can you do to reduce the amount of carbon dioxide in the atmosphere? What are more global actions nations could undertake?

Explore the Technology

Nitrous oxide is another gas produced by human activity. Nitrous oxide also adds to global warming. Find out what activities produce nitrous oxide and how its output can be reduced.

SCIENCE & SOCIETY

People and Science

ELLEN MOSLEY-THOMPSON, *Geographer*

On the Job

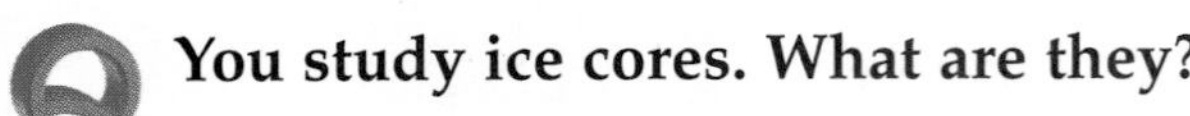

Q You study ice cores. What are they?

A In certain regions such as Antarctica in the southern hemisphere and Greenland in the northern hemisphere, the land is covered by ice sheets as much as 2.5 mi thick. Ice sheets form in places where snow falls every year and doesn't melt, and then more falls the next year. Gradually, the snow turns to firn (snow that has lain on the ground more than one year). With time, the firn becomes more dense and turns into ice. You end up with a stratigraphic sequence—layers deposited one on top of another like layers of a cake. By taking an ice core (a cylinder of ice), scientists can study these layers. The stratigraphic sequences contain different materials, such as dust that fell out of the atmosphere onto the ice sheet. Also, as firn turns into ice, air in the firn gets trapped as bubbles. By analyzing gases and other materials in the layers of ice, we track how dusty the atmosphere was in the past and what chemicals were present. Excess sulfate may indicate a volcanic eruption. Ice cores from mid-latitudes, such as those taken from the Tibetan Plateau in China, contain pollen grains. We can learn about vegetation in the past by tracking how pollen has changed.

Personal Insights

Q Why study ancient climate?

A From one perspective, just to know. Scientists are just naturally curious about the past. From a more practical view, climate is just one part of Earth's system, and it's important to understand that system–how it works, what processes are important, and how it will change in the future if such factors as atmospheric chemistry change. But it's so complex that humans can't sit down at a desk and figure it out. They have to use a computer model. One way to know whether your model is working is to try to simulate the climate of the past. If we can't "predict" what we already know has occurred, it means our model needs refinement.

Career Connection

To learn more about what researchers do in Antarctica, send for "Facts About the U.S. Antarctic Program" and "The United States Antarctic Program" from the National Science Foundation. To find out what other things geographers do, write to: Association of America Geographers, 1710 16th Street, N.W., Washington, DC 20009-3198.

- Geographer
- Climatologist

Chapter 16 Review

Summary

16-1: What is climate?

1. The climate of an area is the average weather over a long period, such as 30 years.
2. Higher latitudes experience cooler climates than lower latitudes do. Topographic features such as oceans, mountains, and large cities affect climate.

16-2: Climate Types

1. The Köppen Climate Classification System is based on temperature, precipitation, and vegetation.
2. Organisms have structural and behavioral adaptations that help them survive in particular climates.

16-3: Climatic Changes

1. Seasons are caused by Earth's tilt on its axis as it revolves around the sun.
2. El Niño disrupts the normal wind and precipitation patterns around the world.
3. Possible causes for major climatic changes include catastrophic events such as meteorite impacts and volcanic eruptions, variations in the sun's radiation, movements of Earth's plates, and Earth's changing movements as it rotates on its axis and revolves around the sun.
4. The greenhouse effect occurs because certain gases trap Earth's heat. Global warming occurs when global temperatures rise because of an increased greenhouse effect.

16-4: Science and Society: How can global warming be slowed?

1. When fossil fuels are burned and forests are cut down, carbon dioxide is added to the atmosphere.
2. When people plant vegetation and conserve electricity, they help reduce the amount of carbon dioxide in the atmosphere.

Key Science Words

a. adaptation
b. deforestation
c. El Niño
d. global warming
e. greenhouse effect
f. hibernation
g. polar zone
h. reforestation
i. season
j. temperate zone
k. tropics

Reviewing Vocabulary

Match each phrase with the correct term from the list of Key Science Words.

1. coldest area on Earth
2. climate zone between the tropics and the polar zones
3. replanting of a large number of trees
4. a characteristic that enables an organism to survive in a particular environment
5. occurs when an animal's temperature drops and body processes are reduced
6. regular change in the weather patterns
7. climatic change when trade winds weaken west of Peru
8. gases in the atmosphere trap Earth's heat
9. mass removal of trees
10. global temperatures rise

Chapter 16 Review

Checking Concepts

Choose the word or phrase that completes the sentence.

1. _______ are common where warm air crosses a mountain and descends.
   a. Lakes　　c. Deserts
   b. Rain forests　　d. Glaciers
2. The fossils of tropical plants and animals found in polar regions are used to tell _______.
   a. the temperature of earlier geologic eras
   b. the relative length of growing seasons
   c. behavioral adaptations
   d. the amount of carbon dioxide in the air in prehistoric times
3. The main greenhouse gas in our atmosphere is _______.
   a. helium　　c. hydrogen
   b. carbon dioxide　　d. oxygen
4. The latitude that receives the most direct rays of the sun year-round is _______.
   a. 60°N　　c. 30°S
   b. 90°　　d. 0°
5. As you climb a mountain, the _______.
   a. temperature decreases
   b. temperature increases
   c. air pressure increases
   d. air pressure remains constant
6. El Niño _______.
   a. occurs every ten to twenty years
   b. causes flooding in Australia
   c. cools the waters off California
   d. sometimes reverses the direction of the trade winds
7. Changes in Earth's orbit can affect Earth's _______.
   a. shape　　c. rotation
   b. temperatures　　d. tilt
8. The Köppen Climate Classification System is based on _______ and precipitation.
   a. temperature　　c. winds
   b. air pressure　　d. latitude
9. An example of structural adaptation is _______.
   a. hibernation　　c. fur
   b. migration　　d. estivation
10. We help reduce the global warming problem when we _______.
    a. conserve energy　　c. produce methane
    b. burn coal　　d. remove trees

Understanding Concepts

Answer the following questions in your Science Journal using complete sentences.

11. How do ocean currents affect climates of coastal regions?
12. Why are large cities usually warmer than the surrounding suburban and urban areas?
13. What are some adaptations of desert organisms that help them survive?
14. Compare and contrast the seasonal changes at low, middle, and high latitudes.
15. How could the eruption of a volcano affect climate?

Thinking Critically

16. Why will global warming lead to the extinction of some organisms?
17. What can you infer if you are digging in a desert and find fossils of tropical plants?
18. On a summer day, why would a Florida beach be cooler than an orange grove two kilometers away?
19. What would happen to Earth's climates if the sun became larger?
20. Why would you expect it to be cooler if you climb to a higher elevation in a desert?

Developing Skills

If you need help, refer to the **Skill Handbook.**

21. **Interpreting Scientific Illustrations:** Use Figure 16-5 and Appendix G to identify the climates of central Australia, Taiwan, and the east coast of Madagascar.
22. **Comparing and Contrasting:** Contrast the climate on windward and leeward sides of a mountain. Which side is shown in the photo on this page?
23. **Observing and Inferring:** It has rained for many days. You observe that mosses begin growing in a corner of your yard where grass used to grow. Infer from this which plant is better adapted to moist conditions. Explain your answer.
24. **Infer:** Explain how atmospheric pressure over the Pacific Ocean might affect the direction that the trade winds blow.
25. **Sequencing:** Make a chain-of-events chart to explain the effect of a major volcanic eruption on climate.

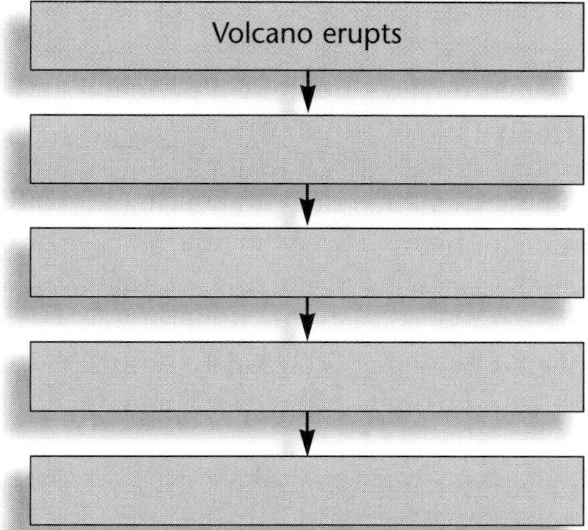

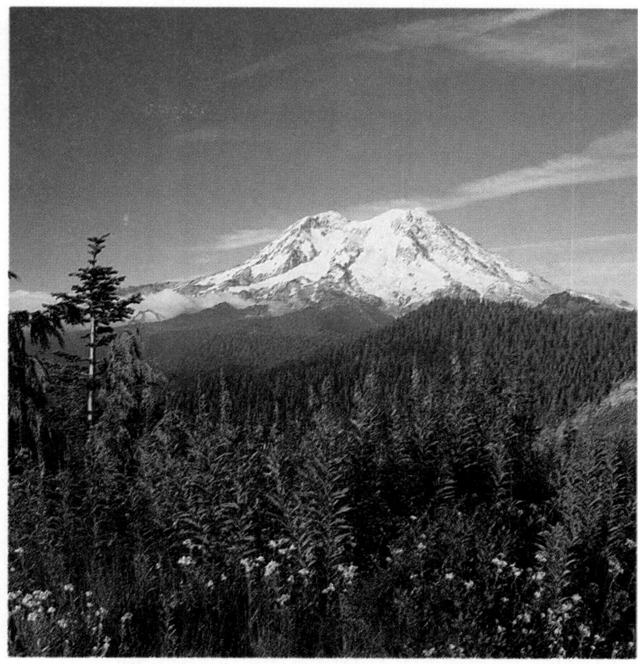

Performance Assessment

1. **Making a Graph:** In Activity 16-1, you studied the microclimates around your school building. Repeat these procedures at the same time of day for three days and graph the temperatures at the sites.
2. **Consumer Decision:** In Activity 16-2, you made a model of the greenhouse effect. Repeat your experiment, but instead of using glass, try aluminum foil, plastic wrap, construction paper, and cardboard lids for the aquarium. Compare your results with what you observed in Activity 16-2. Which lid creates the greatest greenhouse effect? What material would be best to use to cover a solar greenhouse? Which materials would be best to help keep a structure cool?
3. **Poster:** Use your local library to find the weather conditions of your area for the last thirty years. Graph and analyze your data. Does it appear that your climate has remained fairly constant, or do you see patterns that indicate climatic change? Present your findings to the class on a poster.

Previewing the Chapter

Ocean Motion

The ocean is constantly in motion. Waves like this one break against the shore with great energy. Wind causes most waves in the ocean, from small ripples to waves this size, to the giant waves of hurricanes—some more than 30 m high. Winds also cause general movements of ocean water in currents around the world. Other currents, caused by differences in energy and density, move through the ocean. In this chapter, you will learn more about these ocean motions and the interaction between the atmosphere and the oceans, as well as how waves, currents, and tides work.

EXPLORE ACTIVITY

Discover how currents work.

1. In a dish, melt ice to make ice water.
2. Fill a large beaker with warm water.
3. Add a few drops of food coloring to the ice water.
4. Use an eyedropper to place some of this ice water on top of the warm water. Repeat the experiment, placing the warm water on top of the ice water.

Observe: In which experiment was a current created? Infer why the current created is called a convection current.

Previewing Science Skills

- In the **Skill Builders**, you will map concepts, classify, and compare and contrast.
- In the **Activities**, you will design experiments, recognize cause and effect, measure in SI, observe, interpret, and hypothesize.
- In the **MiniLABs**, you will make a model and observe and infer.

17•1 Ocean Water

Science Words

basin
salinity

Objectives

- Learn the origin of the water in Earth's oceans.
- Explain how dissolved salts and other substances get into seawater.
- Describe the composition of seawater.

Oceans and You

Oceans affect your life, and your life affects oceans. No matter where you live, the ocean exerts an influence on you. When it rains, most of the water that falls comes from oceans. If the day is sunny, it is partly due to weather systems that develop over oceans. Oceans are a source of food and resources. **Figure 17-1** shows oil being drilled from the ocean and fish taken from the sea.

Oceans also act as barriers between continents. The price of clothing, cars, and gasoline may include the cost of shipping these materials across oceans. To go from the United States to Europe, a traveler must fly or take a ship across the ocean. But oceans also provide a means of transportation for plants and animals. Animals migrate with ocean currents. Plant seeds are washed from one coast to another by oceans. Humans have migrated across oceans to distant islands.

Human activity affects the oceans. If you live near a stream that is polluted, that pollution might travel to an ocean. Cities near oceans regularly dump sewage sludge into the ocean. Oil spills destroy beaches and marshes along the ocean.

A

B

Figure 17-1

We depend on the oceans. Some of our oil comes from wells drilled through rock layers under the oceans (A). Some food comes from the oceans (B).

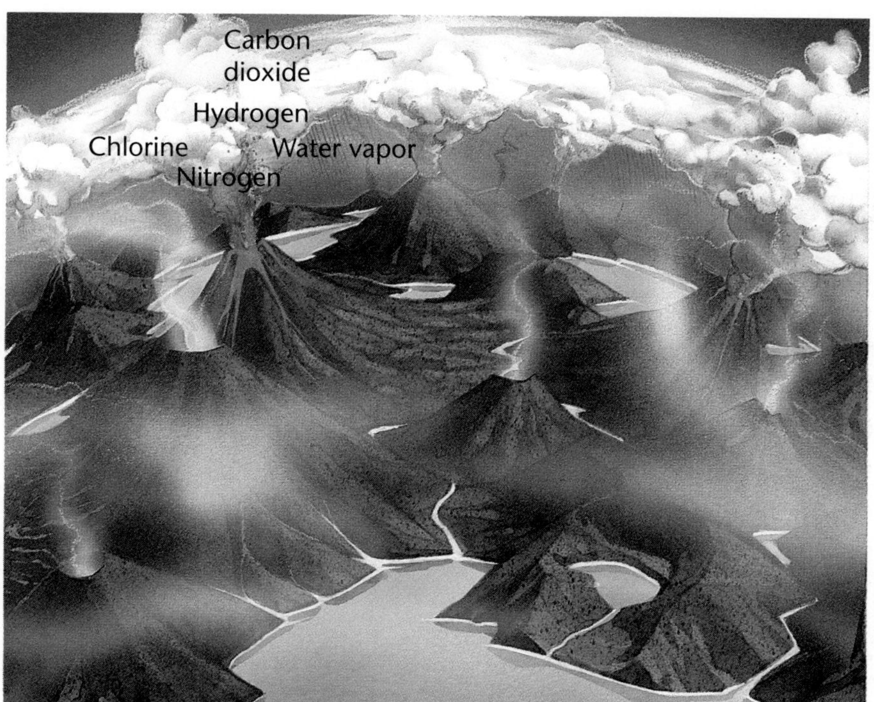

Figure 17-2

Earth's oceans may have formed from water vapor released into the atmosphere by volcanoes that also emitted chlorine, carbon dioxide, nitrogen, and hydrogen.

Origin of Oceans

In the first billion years after Earth formed, its surface was much more volcanically active than it is today. Volcanoes not only spew lava and ash, they give off water vapor along with carbon dioxide and other gases. Scientists hypothesize that about 4 billion years ago, this water vapor began to accumulate in Earth's early atmosphere. Over millions of years, it cooled enough to condense into storm clouds. Torrential rains began to fall. Oceans were formed as this water filled low areas on Earth called **basins.**

Composition of Oceans

We use a lot of water each day. In some areas, fresh water is limited. When water demand in these areas increases, too much water can be pumped from aquifers. This can cause problems.

Earth's surface is 70 percent ocean. It would help if we could use salt water to meet our water needs. But if you taste water from an ocean, immediately you'll know that it is different from the water you drink. It tastes salty because the ocean contains many dissolved salts. In the water, these salts are separated into ions of chloride, sodium, sulfur, magnesium, calcium, and potassium.

These salts come from rivers and groundwater slowly dissolving elements such as calcium, magnesium, and sodium from rocks and minerals. Rivers transport these elements to the oceans. Erupting volcanoes add elements, such as sulfur and chlorine, to the atmosphere and oceans.

USING MATH

Make a circle graph to show the percentage of Earth's surface that is ocean.

USING TECHNOLOGY

Desalting Ocean Water

In some areas that have little fresh water, salt is removed from ocean water to provide water for drinking and for other uses. Saudi Arabia, which borders the Red Sea and the Persian Gulf, uses a desalination system to make fresh water from salt water.

Desalting ocean water can be done in several ways. In one method, salt water is pumped into a large shallow tray covered by a sloping glass roof. Radiation from the sun heats the salt water. The water evaporates, rises to the glass, condenses, and drips down the glass into collecting troughs as fresh water. The salts are left behind.

In another method, an electric current passes through salt water. The positive and negative electrodes, as well as membranes placed between the electrodes, trap positive and negative salt ions. Desalted water is left behind.

In a third method of desalination, ocean water is frozen. Salt crystals are separated from the ice crystals by washing the salt from the ice with fresh water. The water is refrozen and washed several more times. Then, the ice is melted to produce fresh water.

Think Critically:

Why is it expensive to make fresh water from salt water?

A Desalinization Plant

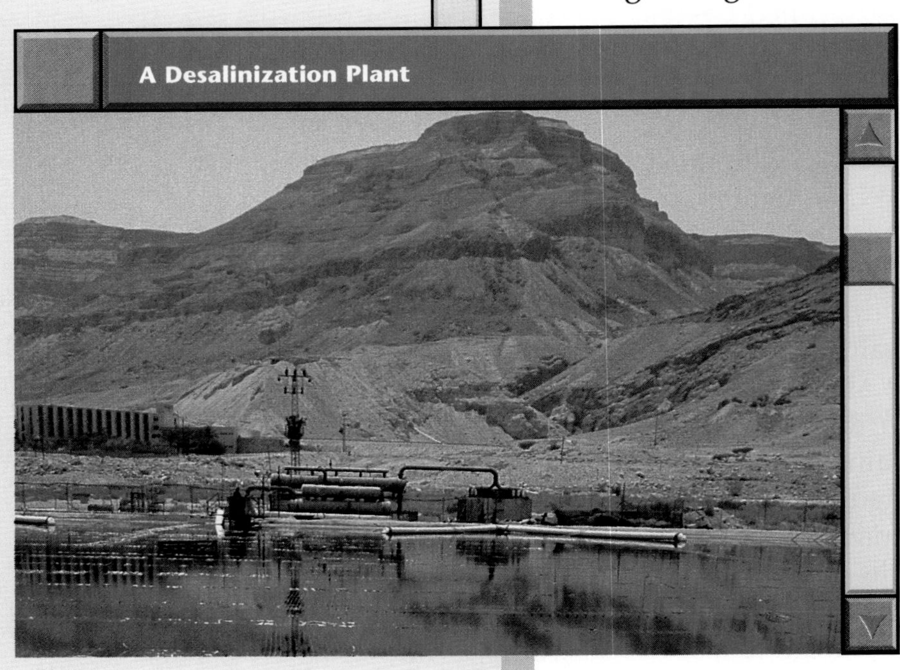

Salinity

In seawater, the most abundant elements are sodium and chlorine. Rivers that flow to the ocean dissolve sodium along the way. Volcanoes add chlorine gas. Most of the salt in seawater is made of sodium and chlorine. If seawater is evaporated, these sodium and chlorine ions combine to form a salt called halite. Halite is the salt you use to season food. It is this salt and similar ones that give ocean water its salty taste.

Salinity is a measure of the amount of solids dissolved in seawater. It is usually measured in grams of salt per kilogram. One kilogram of ocean water contains about 35 g of dissolved salts, or 3.5 percent. The graph in **Figure 17-3** lists the most abundant salts in ocean water. The proportions and amount of dissolved salts in seawater remain nearly constant and have stayed about the same for hundreds of millions of years. This tells us that the composition of the oceans is in balance. The ocean is not growing saltier.

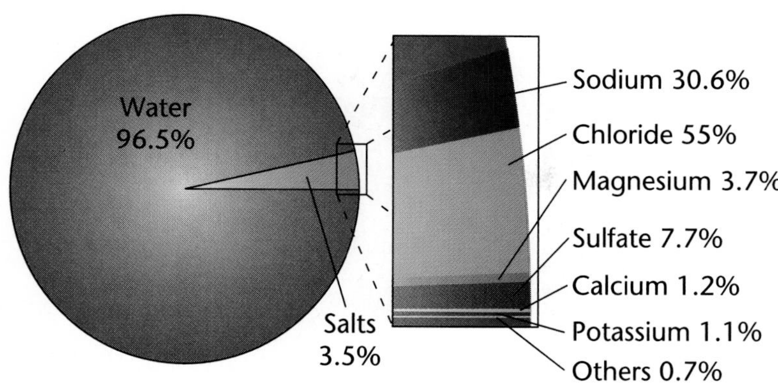

Figure 17-3

Ocean water contains about 3.5 percent salts. This graph lists the most abundant ions in seawater. Sodium and chloride make up nearly 86 percent of the substances in seawater.

Element Removal

Although rivers, volcanoes, and the atmosphere constantly add substances to the ocean, the oceans are considered to be in a steady state. As new substances come in, elements are removed from seawater by biological processes and by becoming sediment. Sea animals and algae use dissolved substances. Marine animals use calcium to form bones. Other animals, such as oysters and clams, use silica and calcium to form shells. Some algae, called diatoms, have silica shells. Because many organisms use calcium and silica, these ions are removed more quickly from seawater than are substances such as chloride or sodium. Iron is removed more quickly because iron forms solids that fall to the ocean floor.

INTEGRATION
Life Science

The oceans came from water vapor condensing and falling as rain, then collecting in basins. Volcanoes, groundwater, and rivers add salts to seawater.

Section Wrap-up

Review

1. According to scientific hypothesis, how were Earth's oceans formed?
2. Why does ocean water taste salty?
3. **Think Critically:** Why does the salinity of Earth's oceans remain constant?

Skill Builder

Concept Mapping

Make a concept map that shows how halite becomes dissolved in the oceans. Use the terms *rivers, volcanoes, halite, source of, sodium, chlorine,* and *combine to form.* If you need help, refer to Concept Mapping in the **Skill Handbook.**

Science Journal

Discuss what is happening in this excerpt from "The Rime of the Ancient Mariner" by Samuel Taylor Coleridge.

Day after day, day after day,
We stuck, nor breath nor motion;
As idle as a painted ship
Upon a painted ocean.
Water, water, everywhere,
And all the boards did shrink;
Water, water, everywhere,
Nor any drop to drink.

Activity 17-1

Design Your Own Experiment

Water, Water, Everywhere

Imagine being stranded on an island. You're thirsty, but you can't find any fresh water. There's plenty of salt water. How can you remove the salts from the water so that you can drink it?

PREPARATION

Problem

How can you make fresh water from salt water?

Form a Hypothesis

Remember what you have learned about evaporation and condensation in the water cycle. Form a hypothesis about how to make a simple desalination system to remove fresh water from salt water.

Objectives

- Design and carry out an experiment to remove fresh water from salt water.
- Recognize the cause-and-effect relationships of heating and cooling on evaporation and condensation.

Possible Materials

- Obtain from your teacher a rubber stopper with a plastic or glass right-angle tube inserted. A piece of rubber tubing should be attached to the end of the glass tube as shown in the photo below.
- pan balance
- table salt
- water
- 1000-mL flask
- 600-mL beaker
- towel
- plastic spoon
- hot plate
- scissors
- ice
- shallow pan
- goggles
- thermal mitts
- watch glasses (2)

Safety Precautions   

Be sure the glassware and spoon are clean before you begin your experiment.

Wear your goggles while you are boiling the water. Use the thermal mitts when handling hot objects.

PLAN THE EXPERIMENT

1. Agree upon and write out your hypothesis statement.
2. List the steps needed to test your hypothesis. Be specific, describing exactly what you will do. Include the following steps in your list: (a) dissolve 18 g of table salt in 500 mL of water; (b) mix the salt-water solution; (c) use the flask for heating and the beaker for collecting fresh water; and (d) use watch glasses on the hot plate to evaporate samples of water from both the beaker and the flask.

Check the Plan

1. Read over your entire experiment to make certain your experiment tests your hypothesis.
2. Make certain you measure and record carefully.
3. Have you included all safety precautions? *Pay attention to the boiling water. Do not leave it unattended. Do not touch the heated glassware. Let it cool before you touch it.*
4. Read over the concluding questions to see if you have included all the necessary steps in your experiment.
5. *Before you begin, have your teacher approve your plan.*

DO THE EXPERIMENT

1. Carry out the experiment as planned.
2. While the experiment is going on, write down any observations that you make in your Science Journal.

Analyze and Apply

1. **Analyze** what happened to the water in the flask as you boiled the solution. What remained in the flask? What happened to the water after it evaporated?
2. **Contrast** the samples on the watch glasses.
3. **Measure** the volume of water in the beaker and of the salt water remaining in the flask. **Compare** the combined volumes with the volume of the salt water at the beginning. Are they the same? Explain.
4. **Compare** your results with those of other laboratory teams. Did they have similar results?
5. **Explain** your results by drawing a diagram of the experiment. Label the processes involved.

Go Further

How could you use the desalination process to collect valuable minerals from seawater?

17 • 2 Ocean Currents

Science Words

surface current
upwelling
density current

Objectives

- Determine how winds, the Coriolis effect, and continents influence surface currents.
- Explain the temperatures of coastal waters.
- Describe density currents.

Surface Currents

The ocean is always moving. A mass movement or flow of ocean water is called a current. When you stir chocolate flavoring into milk or stir a pot of soup, you make currents with the spoon. The currents are what mix the liquid. An ocean current is like a river within the ocean. It moves masses of water from place to place.

Ocean currents carry water in various directions. **Surface currents** move water horizontally—parallel to Earth's surface. Surface currents are powered by wind. Friction between wind and water causes the water to move, driving ocean water in huge circular patterns all around the world. **Figure 17-4** shows these major surface currents. From looking at the map of surface currents, you may have guessed that the surface currents in the ocean are related to the general circulation of wind on Earth.

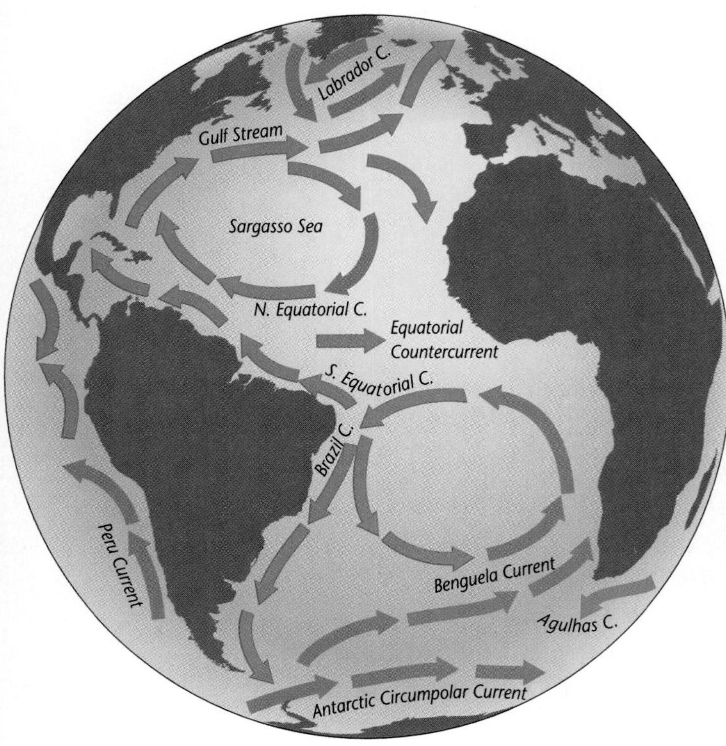

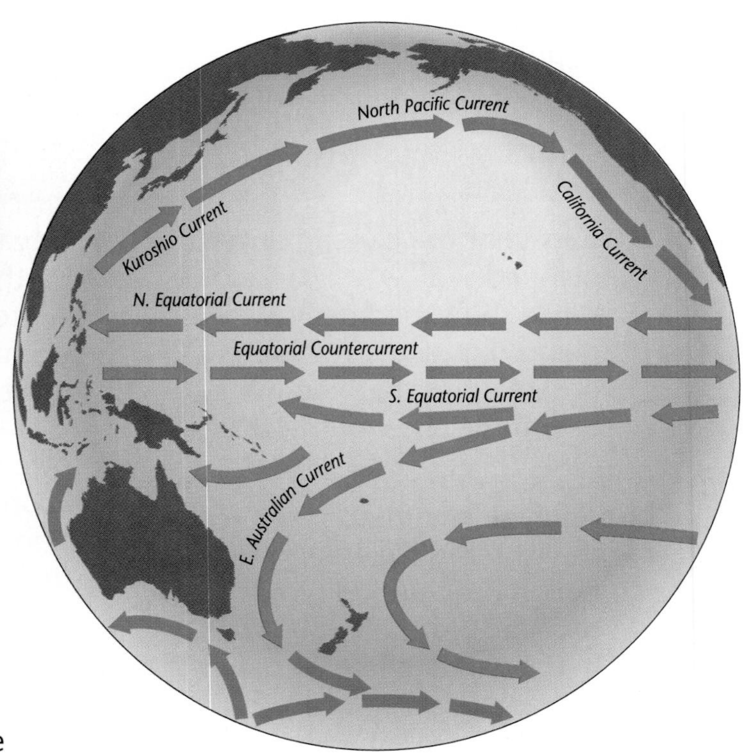

Figure 17-4

These are the major surface currents of Earth's oceans. Red arrows indicate a warm current. Blue arrows indicate a cold current.

Surface currents are called surface currents because they move only the upper few hundred meters of seawater, although sometimes they extend down to 1000 m. Surface currents carry seeds and plants between continents and float bottles with messages from one continent to another. Sailors now and in the past have used these currents in conjunction with winds to sail more efficiently from place to place. The Antarctic Circumpolar Current is the strongest current in the ocean. It is the only one that circles Earth.

Figure 17-5

Captain Timothy Folger drew this map of the Gulf Stream around 1770.

The Gulf Stream

Although satellites now provide information about ocean movements, much of what we know about surface currents comes from records kept by sailors of the 19th century. For as many years as ships have sailed, sailors have used surface currents to help them travel quickly. Sailing ships depended on one surface current to carry them to the west and another to carry them to the east. During the American colonial era, ships floated on the 100-km-wide Gulf Stream current to go quickly from North America to England. On the map, find the Gulf Stream current in the Atlantic Ocean.

We learn about ocean currents in a variety of ways, both scientific and nonscientific. In the late 1600s, a British explorer named William Dampier first wrote about the Gulf Stream in his *Discoveries on the Trade Winds.* A century later, Benjamin Franklin published a map of the Gulf Stream that his cousin Captain Timothy Folger, a Nantucket whaler, had drawn. This map is shown in **Figure 17-5.**

An unscientific way we learn about currents is from things that wash up on beaches. In May 1991, thousands of sneakers washed up on Oregon beaches. The North Pacific Drift Current and the California Current had carried the shoes from the site of a shipping accident a year earlier in the mid-Pacific and deposited 80 000 shoes on the western edge of the United States.

CONNECT TO

LIFE SCIENCE

When marine animals such as eels, turtles, and whales migrate, they often use surface ocean currents. ***Infer*** what would happen to animal migration without surface currents.

Factors Influencing Surface Currents

Surface ocean currents, like surface winds, are influenced by the Coriolis effect. The Coriolis effect deflects currents north of the equator, such as the Gulf Stream, to the right. Currents south of the equator are deflected to the left. Look again at the map of surface currents in **Figure 17-4** on page 484.

The continents also influence ocean currents. Notice that currents moving toward the west in the Pacific are deflected by Asia and Australia. Then currents, influenced by the Coriolis effect, move eastward until North and South America deflect them. On the map, what other surface currents are deflected by continents?

On the same map, note that currents on the western coast of continents are usually cold, whereas currents on eastern coasts are warm. Warm currents on eastern coasts originate near the equator.

Importance of Surface Currents

Surface currents distribute heat from equatorial regions to other areas of Earth. Warmer waters flowing away from the equator transfer heat to the atmosphere. The atmosphere is warmed. Heat moving in the ocean and into the atmosphere influences climate.

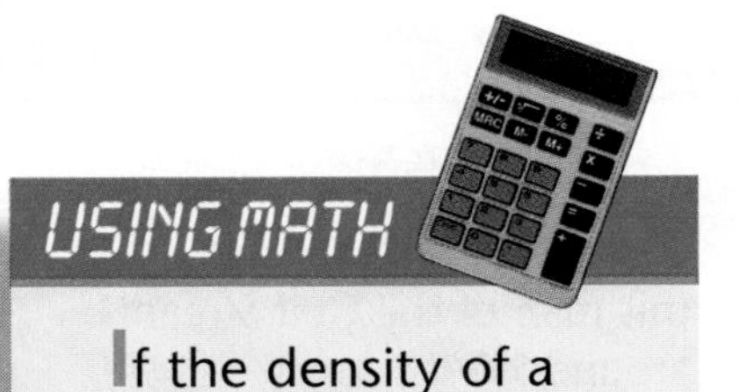

If the density of a sample of seawater is 1.02716 g/mL, what would be the mass of 4 mL of the sample?

As an example of how surface currents affect the climate of land they pass, find Iceland on the map. It is located so far north you would think it has a frigid climate. But part of the Gulf Stream flows past Iceland. The current's warm water heats the surrounding air. Because of the Gulf Stream, Iceland has a surprisingly mild climate. How do warm and cold surface currents affect the east and west coasts of the United States? **Figure 17-6** shows warm waters of the Gulf Stream.

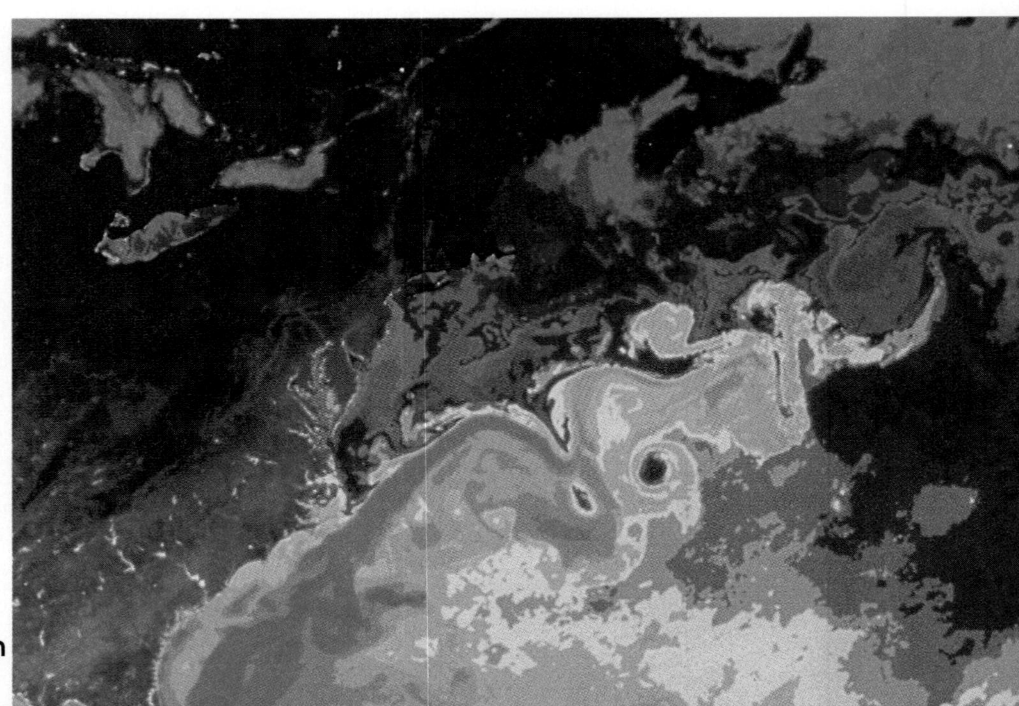

Figure 17-6

Ocean-temperature data collected by a satellite were used to make this surface-temperature image of the Atlantic Ocean. The warm Gulf Stream waters appear as orange and red, and cooler waters appear as blue and green.

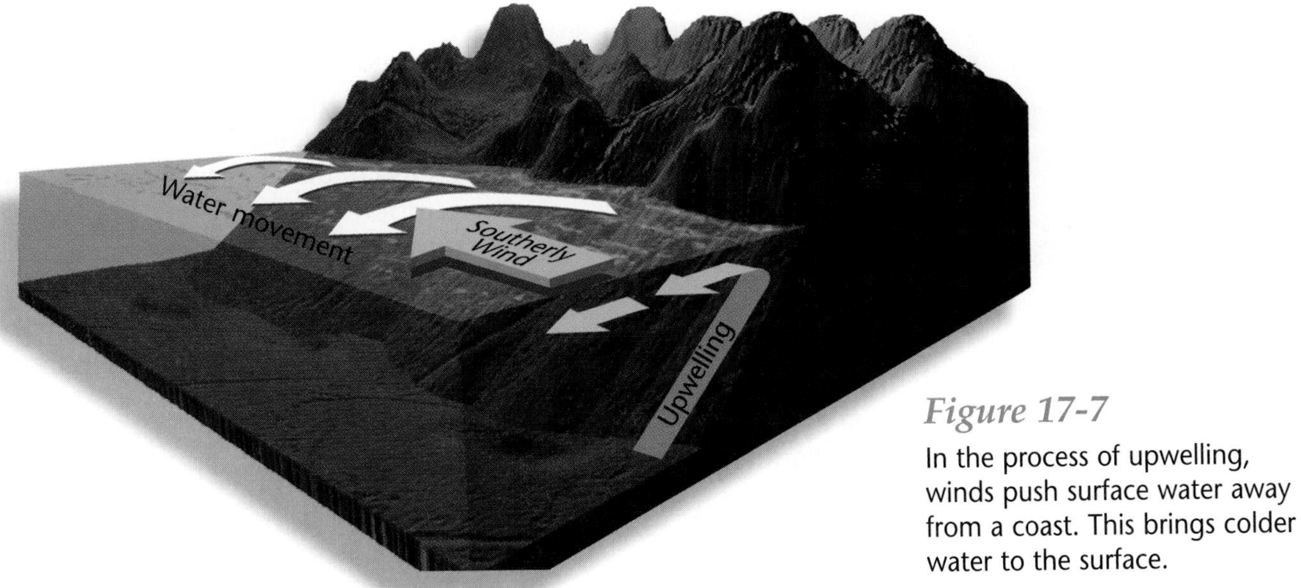

Figure 17-7

In the process of upwelling, winds push surface water away from a coast. This brings colder water to the surface.

Upwelling

Along some coasts, upwelling occurs. **Upwelling** is a circulation in the ocean that brings deep, cold water to the ocean surface. **Figure 17-7** illustrates upwelling. In upwelling, wind blowing offshore or parallel to the coast carries water away from the land. When surface water is pushed away, cold water from deep in the ocean rises. This cold, deep water contains high concentrations of nutrients from organisms that decayed, sank, and then rose with the deep water. These nutrients bring fish to areas of upwelling along the coasts of Oregon, Washington, and Peru, which are known as good fishing areas. Upwelling also affects the climate of adjacent coastal areas. In the United States, upwelling contributes to San Francisco's cool summers and famous fogs.

Density Currents

Wind-driven surface currents affect only the upper layers of Earth's oceans. Upwelling is a vertical mixing of ocean waters in specific areas. In the depths of the ocean, waters circulate not because of wind but because of density differences.

Differences in the density of different masses of seawater cause density currents. A **density current** occurs when more dense seawater sinks under less dense seawater, mixing surface ocean water farther into the ocean.

You made a density current in the Explore activity at the beginning of this chapter. The cold water was more dense than the warm water in the beaker. The cold water sank to the bottom. This created a density current that moved the food coloring. In the MiniLAB on this page, you modeled a density current with salt water and fresh water.

MiniLAB

How can you make a density current model?

Procedure

1. Fill a clear plastic storage box with room-temperature water.
2. Mix several teaspoons of table salt into a beaker of water at room temperature.
3. Add a few drops of food coloring to the saltwater solution. Pour the solution slowly into the fresh water in the large container.

Analyze

1. Explain what happened when you added salt water to fresh water.
2. How does this lab relate to density currents?

Ocean Temperatures and Density

Imagine that your job is to analyze the data shown in the chart below. The data were collected from three different water masses in the North Atlantic Ocean. You know that one sample came from near the surface, one came from a depth of 750 m, and a third came from the ocean floor. The samples are not labeled and became mixed up in shipment.

Sample #	Temperature (°C)	Density (g/mL)
1	6	1.02716
2	3	1.02781
3	14	1.02630

Solve the Problem

1. **Which sample came from near the surface? How do you know?**
2. **Which sample of water came from the bottom? How do you know?**
3. **Hypothesize how less dense seawater can be colder than the most dense seawater.**

Think Critically:

During El Niño, the surface winds blow toward South America. How do you think the usual upwelling near Peru is affected?

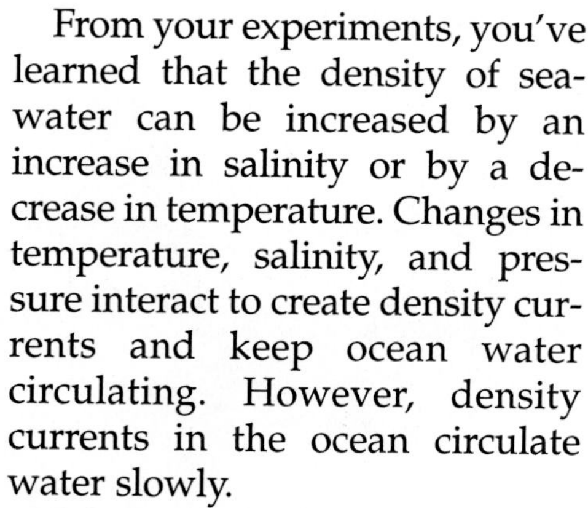

From your experiments, you've learned that the density of seawater can be increased by an increase in salinity or by a decrease in temperature. Changes in temperature, salinity, and pressure interact to create density currents and keep ocean water circulating. However, density currents in the ocean circulate water slowly.

An important density current occurs in Antarctica where the most dense ocean water forms during the winter. As ice forms, seawater freezes, but the salt is left in the unfrozen water. The extra salt not incorporated into the ice increases the density of the ocean water until it is very dense. This dense water sinks and slowly spreads along the ocean bottom toward the equator, forming a density current. In the Pacific, this water may take 1000 years to reach the equator. In the Atlantic Ocean, dense surface waters that sink into the depths circulate more quickly. Atlantic density currents may bring water back to the surface in just 275 years.

A density current also occurs in the Mediterranean, a nearly enclosed sea. Warm temperatures evaporate water from the surface, increasing the salinity and the density of the water. This dense water from the Mediterranean flows outward into the less dense water of the Atlantic Ocean. When it reaches the Atlantic, it flows downward in a density current.

Figure 17-8

Surface currents in the ocean carry objects from long distances. This couple gathered up these sport shoes from a beach in Oregon. The shoes fell into the Pacific Ocean from five boxcar-sized containers a year earlier. An oceanographer from Seattle is studying Pacific Ocean currents with information from the drifting sneakers.

Our knowledge has come a long way since Benjamin Franklin investigated the Gulf Stream. Today, we can track ocean currents by objects that wash up on shore as shown in **Figure 17-8** and more scientifically by satellite. This information helps ships navigate, fishing fleets locate upwellings, and us to understand climate and track hurricanes.

Section Wrap-up

Review

1. How does the energy from wind affect surface currents?
2. How do density currents affect the circulation of water in deep parts of the oceans?
3. **Think Critically:** The latitudes of San Diego, California and Charleston, South Carolina are exactly the same. However, the average yearly water temperature in the ocean off Charleston is much higher than the water temperature off San Diego. Explain why.

Skill Builder

Predicting

A river flows into the ocean. Predict what will happen to this layer of fresh water. Explain your prediction. If you need help, refer to Predicting in the **Skill Handbook.**

Using Computers

Spreadsheet Make a spreadsheet that compares surface and density currents. Focus on characteristics such as wind, horizontal and vertical movement, temperature, density, and anything else that you think helps describe each current.

17•3 Ocean Waves and Tides

Science Words

wave
crest
trough
breaker
tide
tidal range

Objectives

- Define a *wave*.
- Differentiate between the movement of water particles in a wave and the movement of wave energy.
- Describe how the energy of wind and the gravitational force of the moon and sun create tides.

Waves

If you've been to the seashore or seen a beach on TV, you've watched waves roll in. Waves in water are caused by winds, earthquakes, and the gravitational force of the moon and sun. You learned about ocean waves caused by earthquakes. But what is an ocean wave? A **wave** is a rhythmic movement that carries energy through matter or space. In the ocean, waves move through seawater.

Several terms are used to describe waves. **Figure 17-9** shows that the **crest** is the highest point of the wave. The **trough** is the lowest point. Wave height is the vertical distance between crest and trough. Wavelength is the horizontal distance between the crests or between the troughs of two successive waves.

Wave Movement

When you watch an ocean wave, it looks as though the water is moving forward. But unless the wave is breaking onto shore, the water does not move forward. Each molecule of water stays in about the same place as the wave passes. **Figure 17-10** shows this. If you want to see another example of a wave carrying energy, tie a rope to a tree. Wiggle the rope until a wave starts across it. Notice that as the energy passes along the rope, the rope moves, but each piece of rope is still in the same place; it is the same distance from the tree.

Figure 17-9

The crest, trough, wavelength, and wave height are parts of a wave. *What other types of waves have you learned about?*

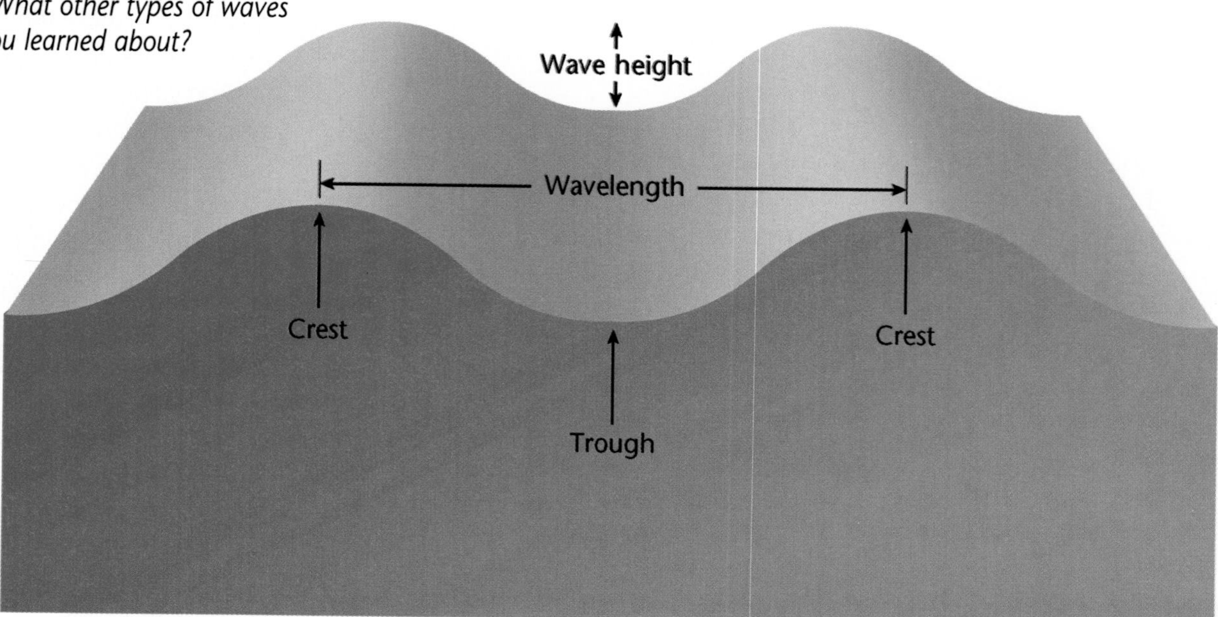

Name ______________ Class ______________ Date ______________

Assessment

Quiz

Section: Health and Wellness

Listed below are the components of total health and the various actions people can take to improve them. On the line next to each phrase, write the letter of the matching health component. Components may be used more than once or not at all.

a. physical health

b. social health

c. mental health

d. emotional health

e. environmental health

f. spiritual health

_______ **1.** Stop smoking.

_______ **2.** Take steps to prevent the spread of sexually transmitted diseases.

_______ **3.** Talk to trusted family members or friends about concerns you have.

_______ **4.** Get involved in volunteer work

Name Dewayne Pulley Class ____________ Date 9-3-03

Assessment

Quiz

Section: Health and Teens

Write the letter of the correct answer in the blank.

X A b **1.** Something that increases the likelihood of injury, disease, or other health problems is

a. a lifestyle disease.
b. a risk factor.
c. unhealthy.
d. uncontrollable.

✓ D **2.** Health risk behaviors include

a. sedentary lifestyle and poor diet.
b. alcohol or drug use and smoking.
c. sexual activity and behaviors that can cause injury.
d. All of the above

X D C **3.** Paul plays soccer three afternoons a week, does not smoke, eats a health-promoting diet, and maintains a recommended weight. Paul

a. will not develop a lifestyle disease.
b. cannot change the fact that he will develop heart disease, because his father has it.

a lifestyle disease.

d. has an increased chance of developing a lifestyle disease.

D ____ **4.** Prior to the 20th century, the primary cause of death was

a. infectious disease.

b. lifestyle disease.

c. a high-fat diet.

d. smoking.

A ____ **5.** Uncontrollable risk factors include

a. age.

b. race.

c. gender.

d. All of the above

_____ 5. Do things that will make you feel good about yourself and increase your sense of self-worth.

The MiniLAB on page 492 will help explain this further. In the same way, an object floating on water will rise and fall as a wave passes, but the object will not move forward. Only the energy moves forward while the water particles remain in the same place.

When a wave breaks against a shore, it is different. In a shallow area near shore, a wave changes shape. Friction with the ocean bottom slows water at the bottom of the wave. As the wave slows, its crest and trough come closer together. The wave height increases. The top of a wave, not slowed by friction, moves faster than the bottom. Eventually, the top of the wave outruns the bottom and collapses. The wave crest falls; water tumbles over on itself. The wave breaks onto the shore. This collapsing wave is a **breaker.** It is the collapse of this wave that propels a surfer and surfboard onto shore. After a wave breaks onto shore, gravity pulls the water back into the sea.

Now let's look at two different types of waves: (1) the common sea waves caused by wind and (2) the long waves of the tides.

CONNECT TO PHYSICS

The *period* of a wave is the time it takes for successive wave crests to pass a fixed point. If the period (T) is 2 s and the wavelength (L) is 6 m, you can find the velocity (C) of the wave by using the formula C = L/T. ***Calculate*** the velocity of this wave.

Figure 17-10

In a wave, individual particles of water move in circles as the energy passes through.

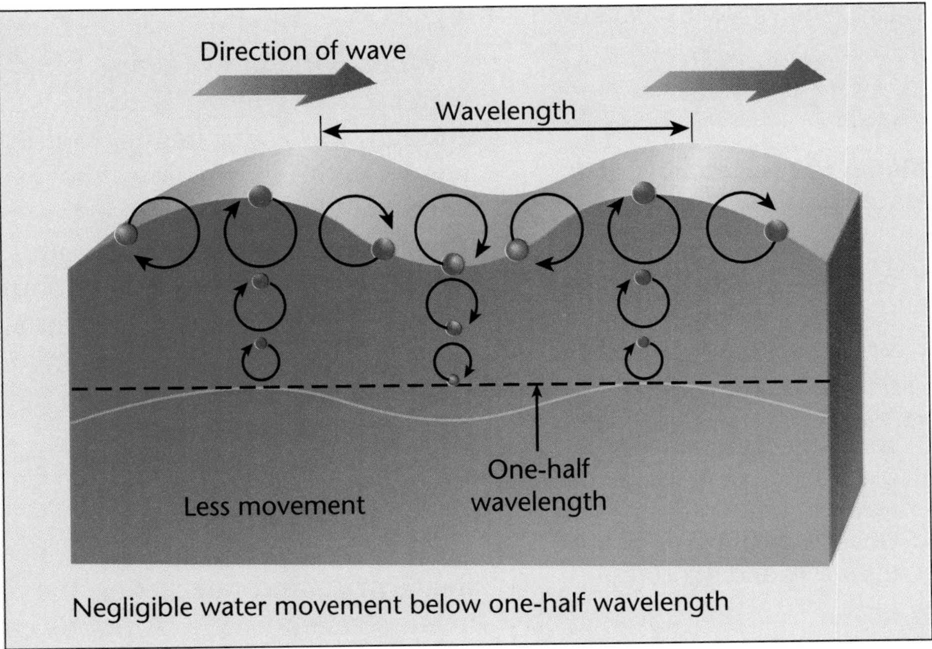

A Wave energy moves forward somewhat like the motion of falling dominoes. Like water particles, individual dominoes remain near where they were standing as they fall and transfer energy to the next domino.

B Particles of water in a wave move around in circles. The farther below the surface, the smaller the circles are. At a depth about equal to half the wavelength, particle movement stops.

Figure 17-11

Wavelength decreases and wave height increases as waves approach the shore. This leads to the formation of breakers.

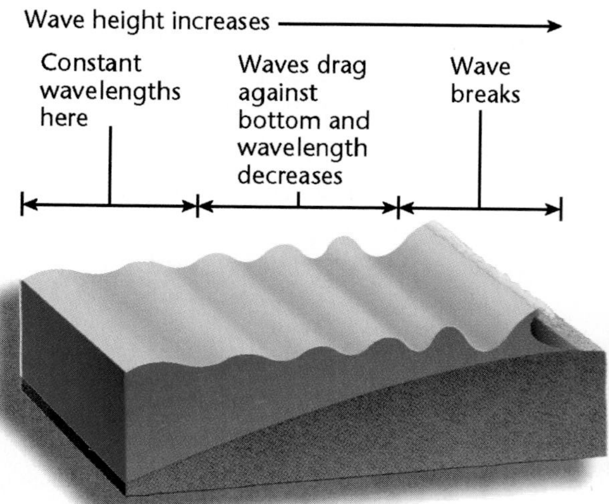

MiniLAB

Model the movement of a water particle in a wave.

Procedure

1. Put a piece of tape on the bottom of a clear, rectangular plastic storage box. Fill the box with water.
2. Float a cork in the container above the piece of tape.
3. Use a spoon to make gentle waves in the container.
4. Observe the movement of the waves and the cork.

Analysis

1. Describe the movement of the waves and the motion of the cork.
2. Did the cork move from its original location in the container?
3. Compare the movement of the cork with the movement of water particles in a wave.

Waves Caused by Wind

When wind blows across a body of water, friction causes the water to move along with the wind. If the wind speed is great enough, the water begins to pile up, forming a wave. As the wind continues to blow, the wave increases in height. The height of waves depends on the speed of the wind, the distance over which the wind blows, and the length of time the wind blows. You will learn more about this in Activity 17-2. When the wind stops blowing, waves stop forming. But, once set in motion, waves continue moving for long distances. Waves at a seashore may have originated halfway around the world.

Tides

Have you ever been to the beach and noticed the level of the sea rise and fall during the day? This rise and fall in sea level is called a **tide.** A tide is caused by a giant wave. This wave is only 1 m or 2 m high but thousands of kilometers long. As the crest of this wave approaches the shore, sea level appears to rise. This is called high tide. Then a few hours later, as the trough of the wave approaches, sea level appears to drop. This is referred to as low tide.

One low-tide-high-tide cycle takes 12 hours and 25 minutes. A daily cycle of two high and low tides takes 24 hours and 50 minutes, slightly more than a day. The **tidal range** is the difference between the level of the ocean at high tide and low tide. Notice the tidal range in the photos of **Figure 17-12** on page 494.

Activity 17-2

Making Waves

As you have learned, wind helps generate some waves. The energy of motion is transferred from the wind to the surface water of the ocean. The force of the wind affects the waves created. What other factors influence the generation of waves?

Problem

How does the speed of the wind and the length of time it blows affect the height of a wave?

Materials

- white paper
- electric fan (3-speed)
- light source
- clock or watch
- a rectangular, clear plastic storage box
- ring stand with clamp
- water
- metric ruler

Procedure

1. Position the box on white paper beside the ring stand.
2. Clamp a light source on the ring stand or use a gooseneck lamp. Direct the light onto the box. Almost fill the plastic box with water. **CAUTION:** *Do not allow any part of the light or cord to come in contact with the water.*
3. Place the fan at one end of the box to create waves. Start on slow. Keep the fan on during measuring. **CAUTION:** *Do not allow any part of the fan or cord to come in contact with the water.*
4. After three minutes, measure the height of the waves caused by the fan. Record your observations in a table similar to the one shown. Through the plastic box, observe the shadows of the waves on the white paper.
5. After five minutes, measure the wave height and record your observations.
6. Repeat steps 3 to 5 with the fan on medium, then on high.
7. Turn off the fan. Observe what happens.

Analyze

1. **Analyze** your data to determine whether the wave height is affected by the length of time that the wind blows. Explain.
2. **Analyze** your data to determine whether the height of the waves is affected by the speed of the wind. Explain.
3. How does an increase in fan speed affect the pattern of the wave shadows on the white paper?
4. What caused shadows to appear on the paper?
5. **Compare** what happened with the fan turned on with what happened when you turned off the fan.

Conclude and Apply

6. **Hypothesize** what three factors cause wave height to vary in the ocean.
7. **Infer** how wave energy could be used to reduce Earth's dependence on its decreasing supply of fossil fuels.

Data and Observations

Fan speed	Time	Wave height	Observations
Low			
Medium			
High			

The Gravitational Effect of the Moon

For the most part, these tides or giant waves are caused by the interaction of gravity in the Earth-moon system. The moon and Earth are relatively close together, so the moon's gravity exerts a strong pull on Earth. The water in the oceans responds to this pull. The water is thrown outward as Earth and moon revolve around a common center of mass between them. These events are explained in **Figure 17-13.**

Two bulges of water form, one directly under the moon and one on the exact opposite side of Earth. As Earth spins, these bulges follow the moon on its daily transit. To the earth-bound observer, these bulges appear as waves traveling across the oceans. The crests of the waves are the high tides.

As Earth rotates, different locations on Earth's surface pass through high and low tide. Many coastal locations, such as the Atlantic and Pacific coasts of the United States, experience two high tides and two low tides each day. Ocean basins vary in size and shape, so some coastal locations, such as many along the Gulf of Mexico, have only one high and one low tide each day.

The Gravitational Effect of the Sun

The sun also affects tides. The sun can strengthen or weaken the moon's effects. The moon, Earth, and the sun lined up together cause a greater effect called spring tides.

Figure 17-12

The large difference between high and low tide is seen at Mt. Saint Michel off the northwestern coast of France on a summer day.

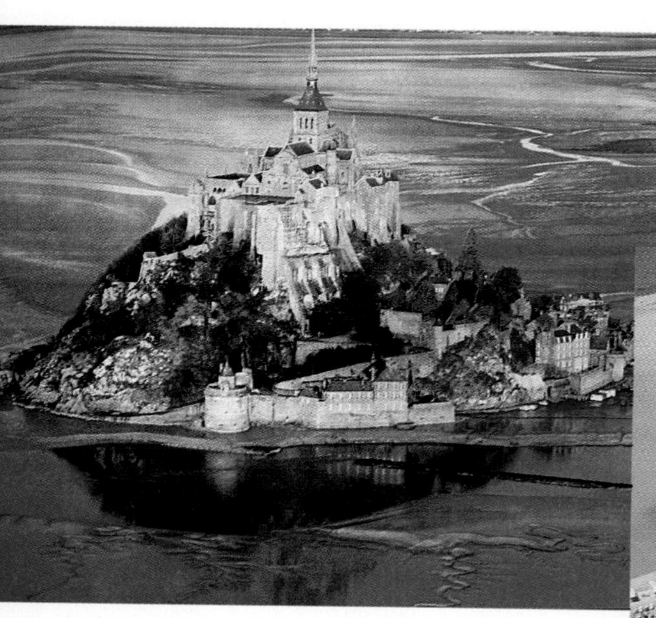

A High tide at 6 A.M.

B Low tide at 6:25 P.M.

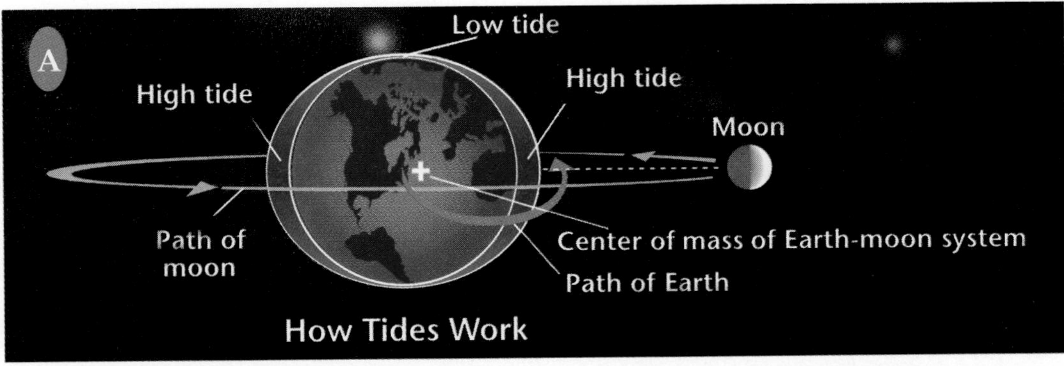

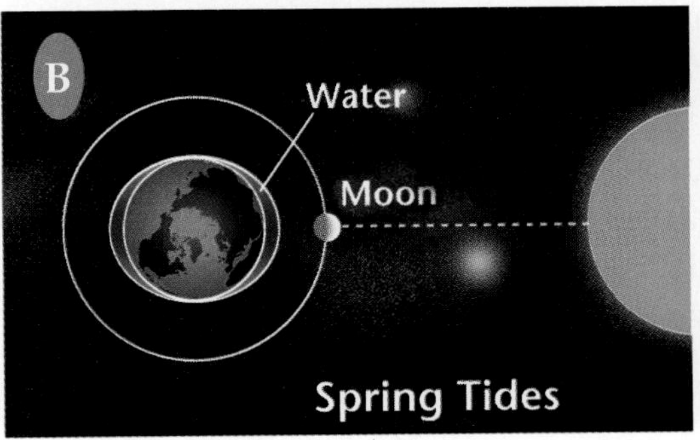

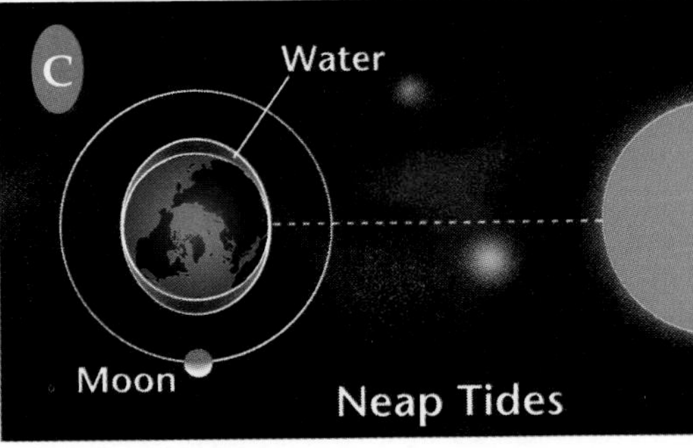

Figure 17-13

(A) The moon doesn't really revolve around Earth. They both revolve around a common center of mass called the barycentre. The action of the moon and Earth around the barycentre causes the bulge of water on the side of Earth opposite the moon. (B) When the sun, moon, and Earth are aligned, spring tides occur. (C) When the sun, moon, and Earth form a right angle, neap tides occur.

During spring tides, high tides are higher and low tides are lower than normal. When the sun, Earth, and the moon form a right angle, high tides are lower and low tides are higher than normal. These are called neap tides.

In this section, you've learned how the energy of the wind and gravity causes waves. In the next lesson, you'll see a way that people use the energy of these waves.

Section Wrap-up

Review

1. Describe the parts of an ocean wave.
2. What causes high tides?
3. **Think Critically:** At the ocean, you spot a wave about 200 m from shore. A few seconds later, the wave breaks on the beach. Explain why the water in the breaker is not the same water that was in the wave 200 m away.

Skill Builder

Comparing and Contrasting

Compare and contrast the effects of the sun and moon on Earth's tides. If you need help, refer to Comparing and Contrasting in the **Skill Handbook.**

Science Journal

In the waters off Southern California, the reproductive cycle of the grunion fish follows the tide schedule. Find out about the egg-laying habits of the grunion and then write about it in your Science Journal.

SCIENCE & SOCIETY

17 • 4 ISSUE: Tapping Tidal Energy

Objectives

- Explain the advantages of a tidal power plant in the Bay of Fundy.
- Describe how a tidal dam converts energy from tides into electricity.
- Consider the consequences of a power plant at the Bay of Fundy.

Tidal Power

Imagine having a constant supply of nonpolluting energy that you could use to generate electricity. No burning fossil fuels. No polluting the air. No worry about running out of electricity for heat, air conditioning, and appliances. In areas where there is a large tidal range, this can happen. Tides can be used to generate electricity.

Electric power-generating plants that use tidal energy already have been built in France, Russia, and China. In the Bay of Fundy in Nova Scotia, the tidal range reaches 16 m, the greatest tidal range on Earth. It is a perfect location for tapping the energy of the tides. At the Bay of Fundy, the water moves through a narrow opening into a funnel-shaped bay. At high tide, the water is trapped in the bay by the force of the incoming tide. Then, at low tide, the bay flushes out.

In the late 1980s, the Canadian government planned to build a tidal dam like the one in **Figure 17-14** in the Bay of Fundy. The large gates of the dam would let water enter the bay at high tide. Then the gates would close. Water would be stored behind the dam. As the tide went out, the dammed water would pass over turbines in the dam, generating electricity. A turbine is similar to a fan blade. When the water pushes against the blades, a turbine spins and generates electricity.

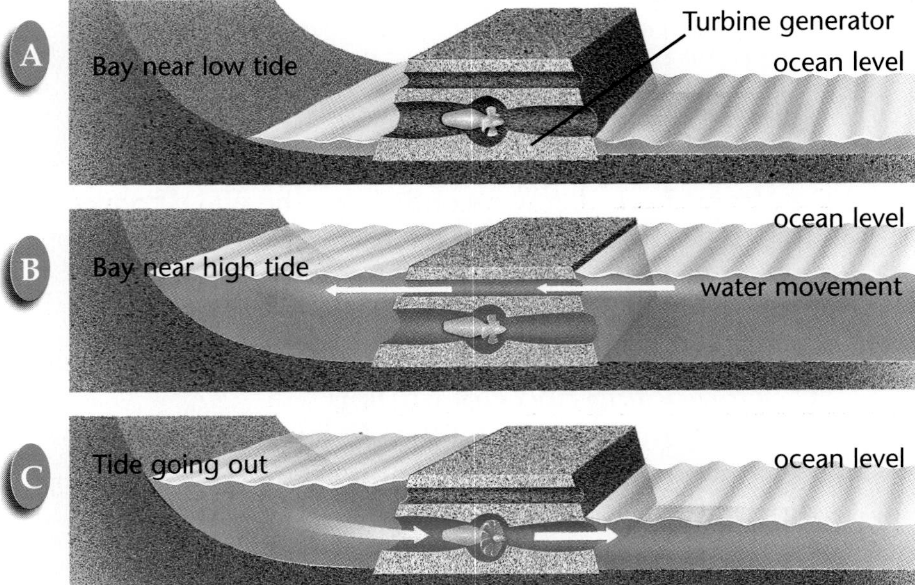

Figure 17-14

This diagram shows the basic concept of a tidal-dam power plant. As the tide starts to rise (A), water flows into the bay and becomes trapped by the dam (B). Then as the tide goes down, the water flows through the dam (C) and spins the turbine generator, making electricity.

2 Points of View

A Tidal Dam Works Well

There would be no expenses for fuel, pollution control, or waste disposal. The cost of the power would be relatively low. Tidal energy will not run out.

A Tidal Dam Creates Problems

So far, this Canadian tidal dam has not been built. There are many reasons for this. It will be expensive, and unfortunately, tidal dams produce electricity at times not related to peak demand. The dam also may create environmental problems. Salt water could contaminate water wells. Valuable salt-marsh habitats could be lost. If these habitats were destroyed, many shorebirds and marine organisms would die. At the dam site, fish might be killed by the turbines. Also, the mud flats at the Bay of Fundy, where many migrating birds feed on mud shrimp, would be flooded by the water retained behind the dam.

Shores as far south as Boston, Massachusetts, 600 km away, could be affected. At low tide, some Boston harbor channels could become too shallow to use.

Figure 17-15

These houses on stilts along the Bay of Fundy indicate how much the water rises from low tide to high tide.

Section Wrap-up

Review

1. Why was the Bay of Fundy picked as a possible site to build a tidal power plant?
2. How does a tidal dam convert tidal energy into electricity?

Explore the Issue

When we make changes in our environment, we must consider the consequences. This certainly is the case with the building of a tidal dam at the Bay of Fundy. Consider the benefits and problems involved with the building of this dam. Do you think this tidal dam should be built? Explain your answer.

Science & History

Polynesian Art of Navigation

Captain James Cook, a British navigator, landed on the Hawaiian Islands in 1778. There he learned about the theory that the Polynesians originally migrated from the East Indies or Indonesia. At first, Cook didn't think this could be possible. To make such a journey, the Polynesians would have had to sail against the trade winds that blow from the east. Later on, Cook learned that during the months of November, December, and January, westerly winds with rain blow from the west, so this was possible.

In this chapter, you learned how ocean currents vary around the world. Early Polynesians used winds, waves, and currents to steer them from one Pacific island to another in double-hulled canoes *without instruments.* They did use maps showing star positions and currents.

Some Polynesians still use this method of navigation. Anthropologist Ben Finney and others have learned how ancient navigators sailed without instruments. The researchers built a double-hulled canoe like the kind used by ancient Polynesians to sail the ocean. The canoe was called *Hokule`a,* which is Hawaiian for "Arcturus," a star seen directly over Hawaii. Anthropologists hoped to find out whether the Polynesians could have sailed across currents to settle the Pacific islands.

The *Hokule`a* was sailed from Hawaii to Tahiti and back. The first navigator of *Hokule`a* was a Polynesian who learned to sail from his father and grandfather. He steered by zenith stars, those vertically above him, chiefly the North Star. He also learned to determine latitude by the stars and to understand the currents and waves. To reach Hawaii from Tahiti, the canoe had to go north in the Pacific and cross two great west-flowing ocean currents, the north equatorial and south equatorial. Between them is the east-flowing equatorial countercurrent. This is shown on the map at the left.

The *Hokule`a* made the round-trip from Hawaii to Tahiti to Hawaii, navigating the way Polynesians would have in ancient times.

Science Journal

Imagine how the Polynesians on the *Hokule`a* must have felt as they re-created a voyage made by their ancestors. Write a poem or an essay in your Science Journal that the Polynesians might have written about *Hokule`a.*

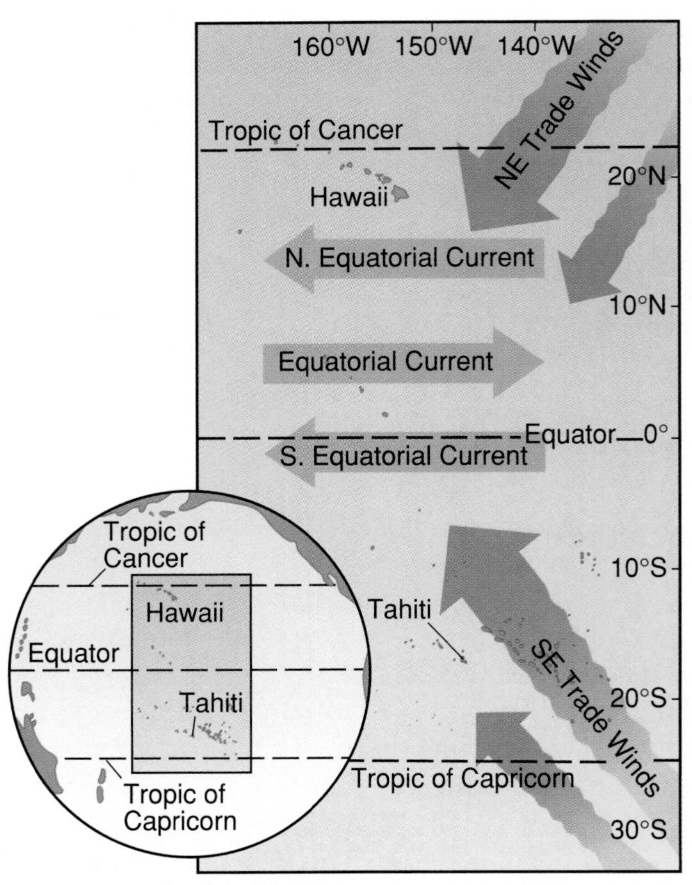

Chapter 17

Review

Summary

17-1: Ocean Water

1. Water that fills Earth's oceans may have started as water vapor released from volcanoes. Over millions of years, the water vapor condensed and rain fell, forming the oceans.
2. Groundwater and rivers weather rocks and cause some elements to dissolve. Both carry the dissolved elements to the oceans. Volcanoes also add elements to seawater.
3. Nearly 90 percent of the salt in seawater is halite. Ocean water also contains calcium, sulfate, and many other substances.

17-2: Ocean Currents

1. Friction between air and the ocean's surface causes surface currents. Surface currents are affected by the Coriolis effect. Currents are also deflected by landmasses.
2. Cool currents off western coasts originate far from the equator. Warmer currents along eastern coasts begin near the equator.
3. Differences in temperatures and densities between water masses in the oceans set up circulation patterns called density currents.

17-3: Ocean Waves and Tides

1. A wave is a rhythmic movement that carries energy. The crest is the highest point of a wave. The trough is the lowest point.
2. Energy moves forward while water particles move in place.
3. Wind causes water to pile up and form waves. Attraction among Earth, the moon, and the sun causes long waves called tides.

17-4: Science and Society: Tapping Tidal Energy

1. Because of its high tidal range, the Bay of Fundy is a good place for tapping tidal energy to generate electricity.
2. A tidal dam can generate electricity.
3. Energy from tides is essentially nonpolluting but can cause problems such as changes in tidal ranges, flooding, and contamination.

Key Science Words

a. basin
b. breaker
c. crest
d. density current
e. salinity
f. surface current
g. tidal range
h. tide
i. trough
j. upwelling
k. wave

Reviewing Vocabulary

Match each phrase with the correct term from the list of Key Science Words.

1. low area that collects water
2. nutrient-rich water that comes to the surface
3. a horizontal current at the top of the ocean
4. the lowest point on a wave
5. movement as energy passes through water
6. a collapsing wave
7. mixes surface water deep into the ocean
8. difference between high and low tide
9. rise and fall of the ocean related to gravitational pull
10. amount of solids dissolved in seawater

Chapter 17 Review

Checking Concepts

Choose the word or phrase that completes the sentence.

1. Ocean water may have developed from _______.
   a. salt marshes
   b. volcanoes
   c. basins
   d. surface currents
2. Chlorine gas enters the oceans from _______.
   a. volcanoes
   b. rivers
   c. density currents
   d. groundwater
3. _______ is the most common ocean salt.
   a. Chloride
   b. Calcium
   c. Boron
   d. Sulfate
4. Most surface currents are caused by _______.
   a. density differences
   b. the Gulf Stream
   c. salinity
   d. wind
5. The highest point on a wave is the _______.
   a. wave height
   b. trough
   c. crest
   d. wavelength
6. Ocean _______ are movements in which water alternately rises and falls.
   a. currents
   b. waves
   c. crests
   d. upwellings
7. A problem with tidal power plants is _______.
   a. fuel expense
   b. pollution
   c. they flood some environments
   d. they only work at high tide
8. The Coriolis effect causes currents in the northern hemisphere to turn _______.
   a. eastward
   b. southward
   c. left
   d. right
9. Tidal power plants use energy from _______.
   a. Earth's oceans
   b. salt marshes
   c. fossil fuels
   d. floods
10. Surface currents are affected by _______.
    a. crests
    b. upwellings
    c. the Coriolis effect
    d. calcium

Understanding Concepts

Answer the following questions in your Science Journal using complete sentences.

11. How do spring tides differ from high tides?
12. In the Mediterranean Sea, a density current forms because of the high rate of evaporation of water from the surface. Explain how evaporation can lead to the formation of a density current.
13. Describe the effect of temperature and salinity on density currents.
14. What happens to a wave when it enters shallow water?
15. How can surface currents affect climate?

Thinking Critically

16. Describe the position of the moon and sun when a low tide is higher than normal.
17. Halite makes up nearly 90 percent of the salt in ocean water. How much halite is found in 1000 kg of ocean water?
18. Why do silica and calcium remain in seawater for a shorter time than sodium?
19. Describe the Antarctic density current.
20. How could a tidal power plant in the Bay of Fundy cause flooding?

Developing Skills

If you need help, refer to the **Skill Handbook.**

21. **Recognizing Cause and Effect:** What causes upwelling? What effect does it have? When upwelling disappears from an area, what effect can this have?
22. **Comparing and Contrasting:** Compare and contrast waves and currents. How are they related?

23. **Comparing and Contrasting:** Compare the pros and cons of tidal power.
24. **Interpreting Data:** Based on the information in this chapter, is there more sodium or calcium dissolved in the oceans? Explain your answer.
25. **Graphing:** Plot the data from the table below on graph paper. Title your graph "Monthly Tides." Use a scale from −0.6 m to 2.4 m. The horizontal axis should show days 1 to 30.

Performance Assessment

1. **Design an Experiment:** Find out how a current moves a floating object. Base the test on the MiniLAB on page 492.
2. **Invention:** In Activity 17-1, you made fresh water. How could you desalinate seawater without electricity? Design a method, draw it, and display it for your class.
3. **Make a Lab Report:** Create an experiment to test the density of water at different temperatures.

Day	Height of High Tide (meters)	Height of Low Tide (meters)	Day	Height of High Tide (meters)	Height of Low Tide (meters)
1	1.4	0.5	16	1.6	0.2
2	1.5	0.4	17	1.7	0.1
3	1.7	0.2	18	1.7	−0.1
4	1.8	−0.1	19	1.8	−0.2
5	2.1	−0.3	20	1.9	−0.2
6	2.2	−0.5	21	1.9	−0.2
7	2.3	−0.6	22	1.9	−0.2
8	2.3	−0.6	23	1.9	−0.2
9	2.3	−0.6	24	1.8	−0.2
10	2.1	−0.5	25	1.7	−0.1
11	1.9	−0.2	26	1.6	0.0
12	1.6	−0.1	27	1.4	0.1
13	1.6	0.2	28	1.3	0.3
14	1.6	0.4	29	1.5	0.4
15	1.6	0.4	30	1.5	0.5

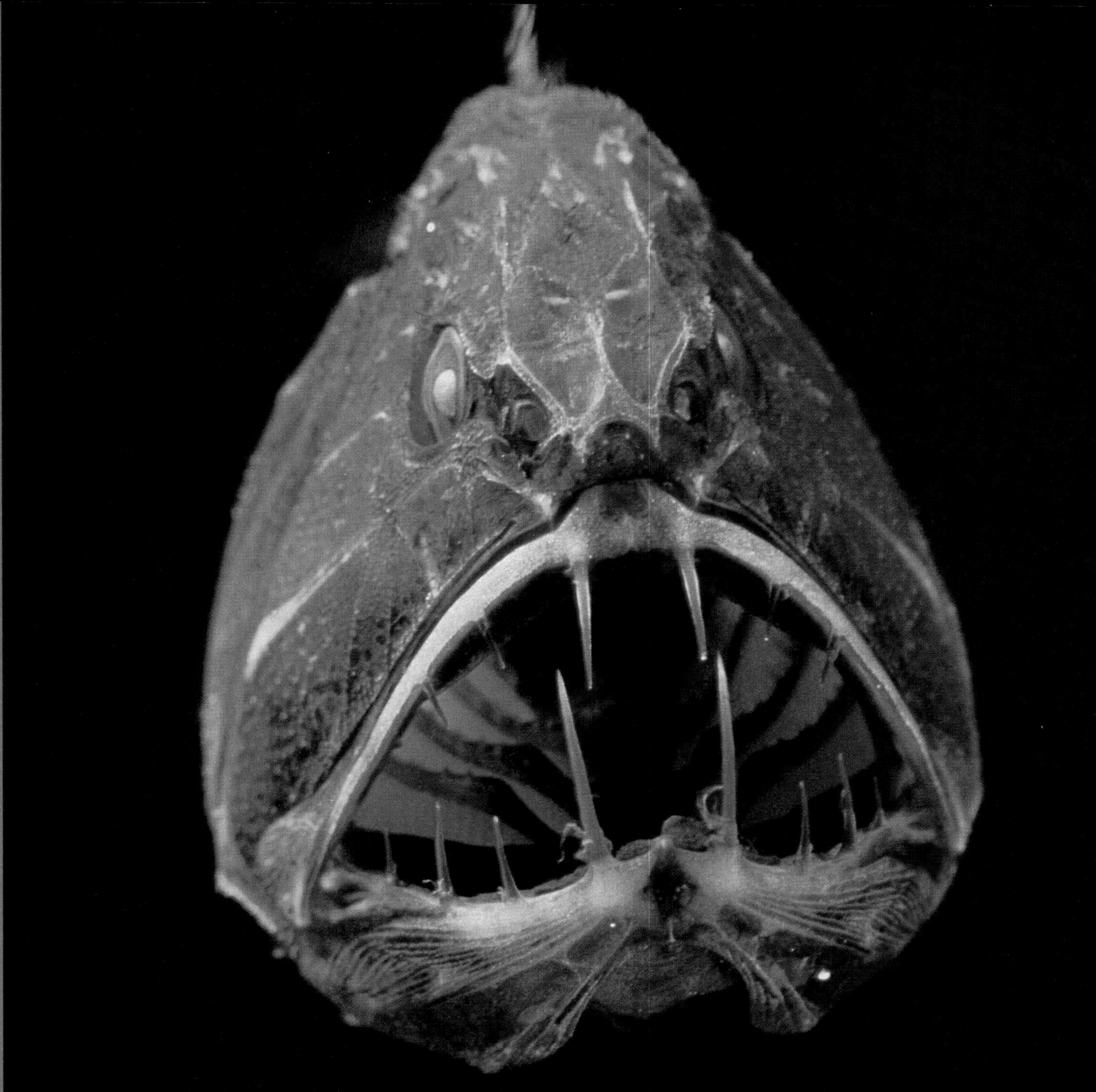

Previewing the Chapter

Chapter

18 Oceanography

In the Pacific Ocean Basin, ocean floor is being destroyed. In the Atlantic Basin, new seafloor is forming. Where the seafloor is spreading apart, communities of organisms live deep in the ocean. This bony, hard-bodied fish, called an anoplogaster, swims in the deep waters of the Pacific Ocean, west of Peru. We were not aware of some communities of organisms until a submersible called *Alvin* explored the ocean bottom in 1977. How do submersibles like *Alvin* dive and surface in the oceans?

EXPLORE ACTIVITY

Make a model of a submersible.

1. Fill a 2-L, plastic soda bottle three-quarters full of water.
2. Draw water into an eyedropper. Leave some air in the bulb of the dropper and place the dropper in the plastic bottle, bulb end up. The dropper should float.
3. Put a cap on the bottle, then gently squeeze the sides of the bottle.

Observe: What happens to the level of the water inside the eyedropper and what happens to the eyedropper? What happens when pressure on the bottle is released? Answer these questions in your Science Journal.

Previewing Science Skills

- In the **Skill Builders**, you will **compare and contrast** and **use variables, constants, and controls.**
- In the **Activities**, you will **make and use a graph, interpret data, design an experiment,** and **hypothesize.**
- In the **MiniLABs**, you will **interpret a scientific illustration, make inferences, observe,** and **classify.**

18•1 The Seafloor

Science Words

continental shelf
continental slope
abyssal plain
mid-ocean ridge
trench

Objectives

- Differentiate the continental shelf from the continental slope.
- Describe a mid-ocean ridge, an abyssal plain, and an ocean trench.
- Describe seafloor mining.

Ocean Basin Features

You can find the biggest mountains, the deepest valleys, and the flattest plains on Earth not on land but at the bottom of the ocean. In this chapter, you'll learn how the mountains, valleys, and plains on the ocean bottom differ from those on the continents. **Figure 18-1** shows major features of the ocean basins.

Beyond the ocean shoreline is the continental shelf. The **continental shelf** is a gradually sloping end of a continent that extends out under the ocean. Along some coasts, the continental shelf extends a long distance. On North America's Atlantic and Gulf coasts, it extends 100 to 350 km into the sea.

But on the Pacific Coast, where mountains are close to the shore, the shelf is only 10 to 30 km wide. The ocean covering the shelf varies from a few centimeters to 150 m deep. As you will see, the continental shelf illustrated in **Figure 18-2** resembles a flat shelf.

At the end of the shelf is the continental slope. The **continental slope** is the end of the continent extending from the outer edge of the continental shelf down to the ocean floor. The continental slope is steeper than the continental shelf. Beyond the continental slope lie the trenches, valleys, plains, and ridges of the ocean basin.

In the ocean, currents flow along the continental shelves and down the slopes, and deposit sediment on the seafloor. These deposits fill in valleys. Deep in the ocean, this creates flat seafloor areas called **abyssal plains.** Abyssal plains are from 4000 to 6000 km deep in the ocean and can be seen in various regions of the map below.

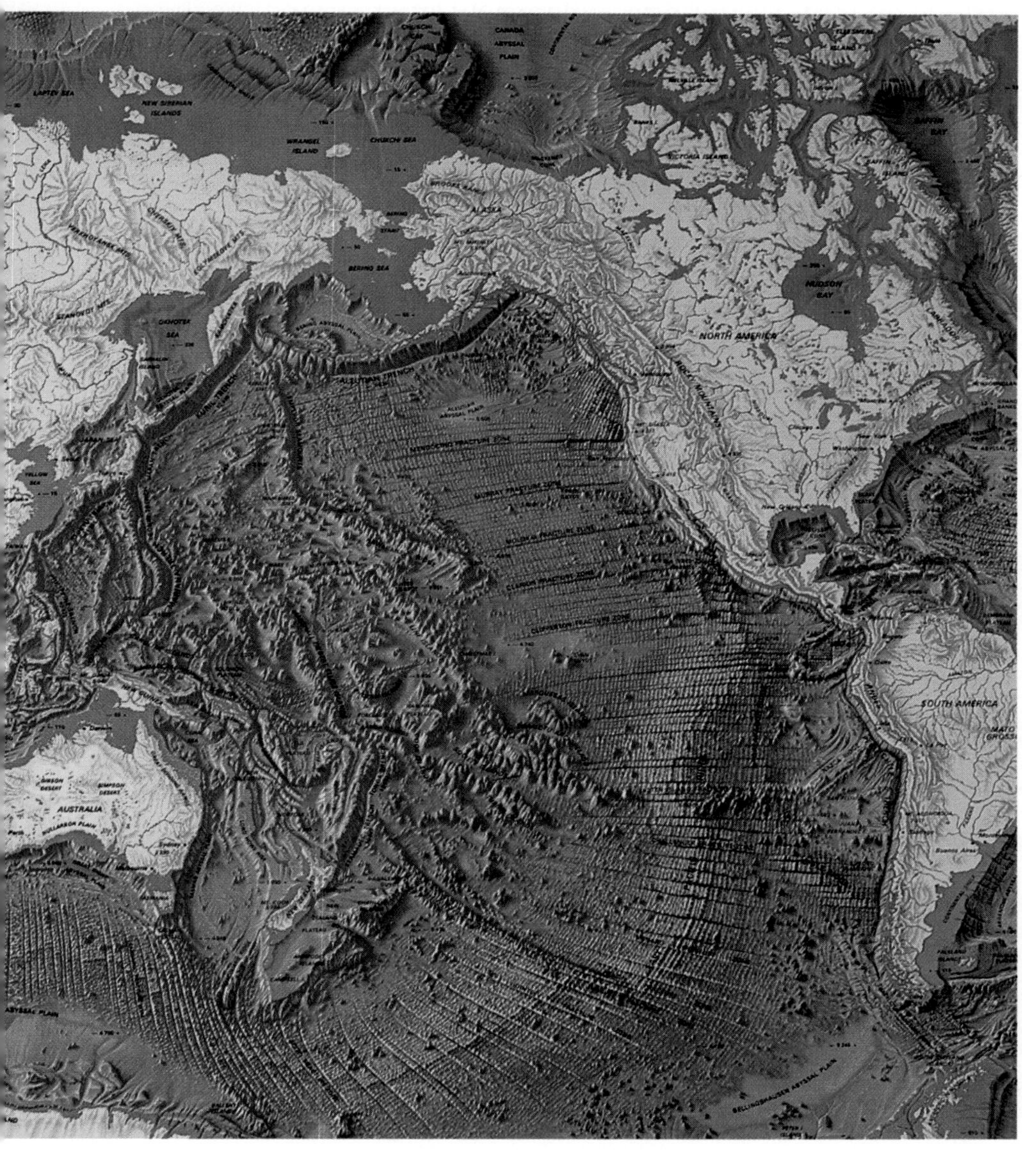

Figure 18-1

This map shows the features of the ocean basins: the continental shelf, mountain ranges, trenches, valleys, and plains. *Find a trench and the Mid-Atlantic Ridge.*

CONNECT TO

LIFE SCIENCE

Marine plants need minerals and sunlight to survive. ***Infer*** from this why many marine plants live on the continental shelves.

Plate Boundary Structures

The ocean floor map in **Figure 18-1** shows structures related to plate movements occurring on the seafloor. In Chapter 11, you learned that some crustal plates under the ocean are diverging. Where these plates diverge, new crust forms. Some plates on the ocean floor are converging. At convergent boundaries, old ocean crust descends beneath other crust and is destroyed.

The place where new ocean floor forms is called a **mid-ocean ridge.** A mid-ocean ridge resembles a chain of mountains. New crust forms along the mid-ocean ridge as a result of volcanic eruptions and the upwelling of molten lava through cracks. Magma from Earth's interior oozes from these cracks. When the lava hits the water, it cools quickly into solid rock and forms new seafloor.

Seamounts are inactive volcanic cones found on the ocean floor. Some scientists theorize that these old volcanoes erupted on mid-ocean ridges where new crust formed. As the new crust moved further from the ridge, these seamounts moved with it. After millions of years, the volcanoes became inactive. Seamounts that extend up out of the ocean are volcanic islands and are shown in **Figure 18-2.**

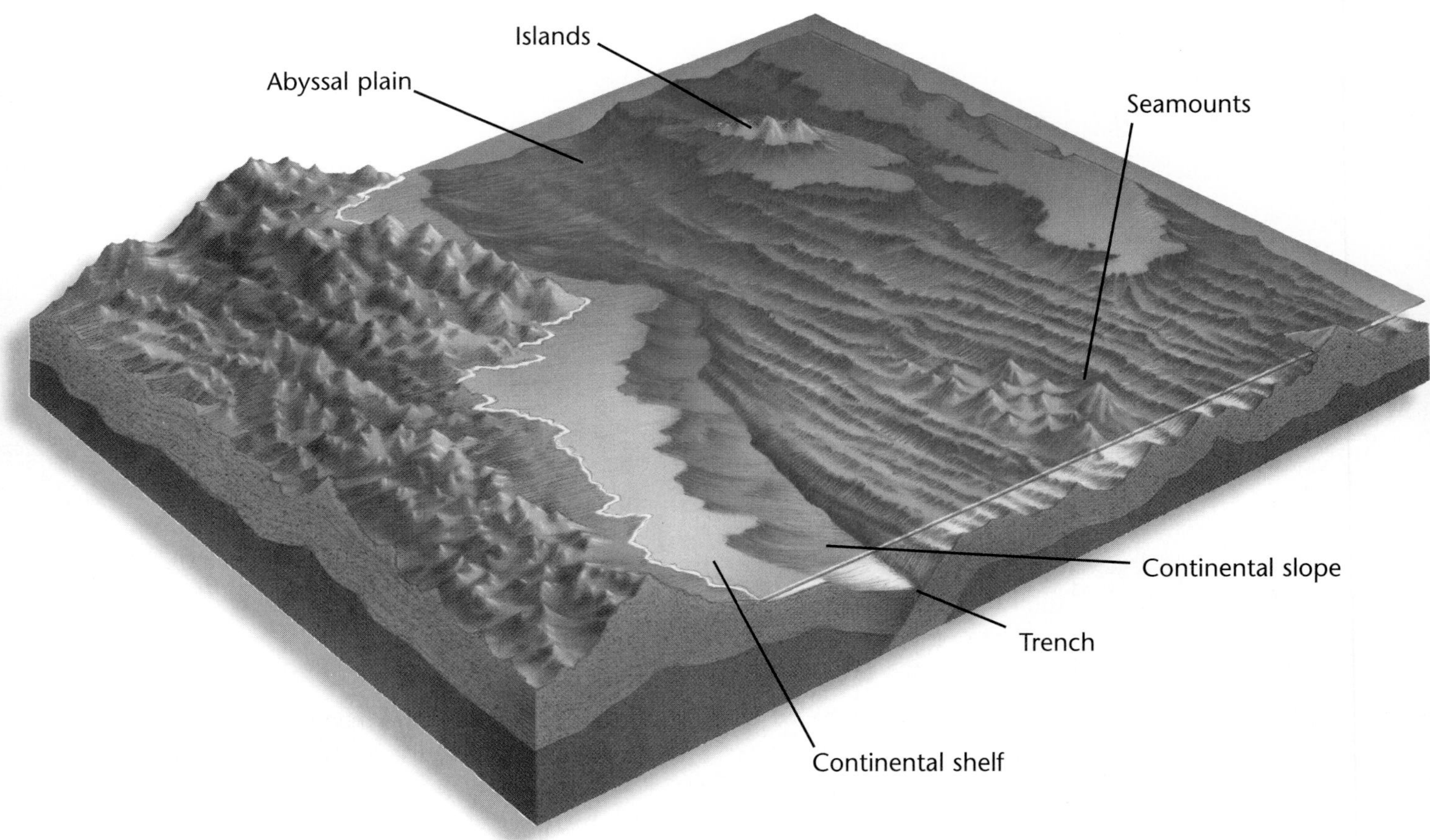

Figure 18-2

This illustration shows some features found in ocean basins. (Features in this diagram are not to scale.)

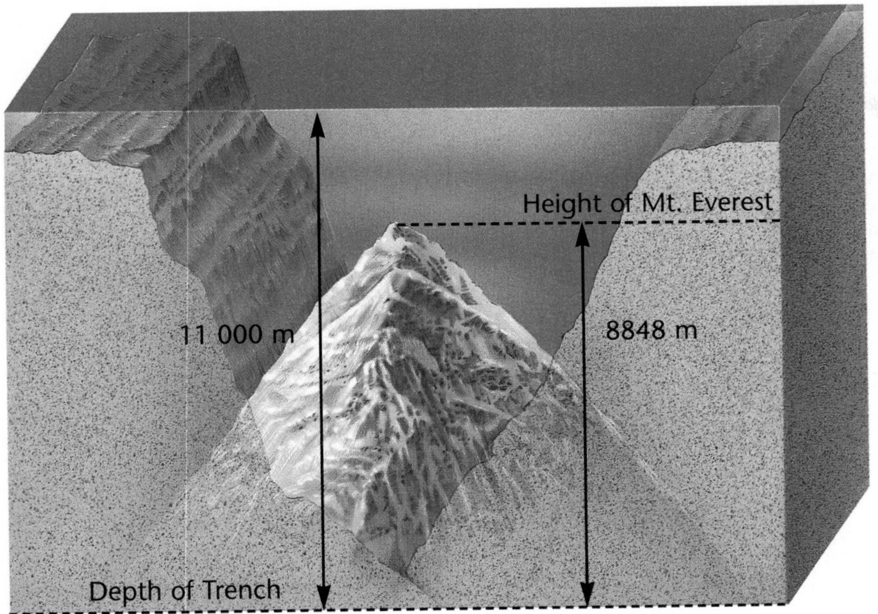

Figure 18-3

If Earth's tallest mountain, Mt. Everest, were set in the bottom of the Marianas Trench of the Pacific Basin, it would be covered with more than 2000 m of water.

On the ocean floor, crustal plates converge at trenches. A **trench** is a long, narrow, steep-sided depression in the ocean floor where one crustal plate is forced beneath another. Most trenches are found in the Pacific Basin. Oceanic trenches are often longer and deeper than any valley on any continent. The Marianas Trench, the deepest place in the Pacific, is 11 km deep. The Grand Canyon is only 1.6 km deep. A comparison is shown in **Figure 18-3.**

Not only are trenches the deepest places on Earth, they are also the most geologically active. Seaquakes and chains of volcanoes often occur along trenches.

Mining the Seafloor

People have always collected salt from ocean water. About one third of all common salt used to season food comes from the sea. Now we also obtain oil and gas from under the sea.

Continental Shelf Deposits

There are many petroleum and natural gas deposits in continental shelf sediments. To pump these substances to the surface, drills are attached to floating vessels and fixed platforms. Approximately 20 percent of the world's oil comes from under the seabed.

Other deposits on the continental shelf are also mined. In Chapter 8, you learned that when a river loses its velocity, it can no longer carry as much sediment. When this occurs, some of the sediment is deposited. These deposits, called *placer deposits,* are mined around the world. Placer deposits can occur in coastal regions where rivers entering the ocean suddenly slow down. Sand, metals, and even diamonds are mined from placer deposits in some coastal regions.

MiniLAB

What does a topographic map of the ocean basins show?

Procedure

1. Use the map in **Figure 18-1.**
2. Trace a mid-oceanic ridge with your finger.
3. Observe the width of the continental shelves at various locations around the world.

Analysis

1. Locate a trench and give its approximate longitude and latitude.
2. Describe the continental shelves of Australia and Alaska.

Figure 18-4

These manganese nodules lie on the floor of the Pacific Ocean.

Deep-Water Deposits

Valuable minerals are being deposited on the ocean floor. As molten substances and hot water are forced into cool ocean water through holes and cracks along ridges, "chimneys" of mineral deposits sometimes form. Today, no one is trying to mine these minerals from the depths, but in the future, these deposits may become important.

In addition to these deposits, other minerals in solution precipitate out and form solids on the ocean floor. Manganese nodules are small black lumps strewn across 20 to 50 percent of the Pacific Basin. **Figure 18-4** shows examples of these nodules. Manganese nodules form in a chemical process not fully understood. They concentrate around nuclei such as discarded sharks' teeth. They grow slowly, perhaps as little as 1 mm to 10 mm per million years. These nodules are rich in manganese, nickel, and cobalt, all of which are used in the manufacture of steel, paint, and batteries. Most of the nodules lie thousands of meters deep in the ocean and are not currently being mined by industry, although suction devices similar to huge vacuum cleaners have been tested to collect them.

In this section, you've learned about the features of ocean basins and how they relate to seafloor spreading and seafloor disappearance. You've also been introduced to the idea of seafloor mining. In the next section, you'll learn about organisms that live in the oceans.

USING MATH

Using the data from Activity 18-1 on page 509, calculate the average depth of the Atlantic Ocean floor along the 39° north latitude line.

Section Wrap-up

Review

1. Compare and contrast ocean ridges and trenches.
2. How do placer deposits form?
3. **Think Critically:** If sea level rises significantly because of global warming, how will island and continental shelf regions be affected?

Skill Builder

Comparing and Contrasting

Compare and contrast the continental shelf, the continental slope, and the abyssal plain. If you need help, refer to Comparing and Contrasting in the **Skill Handbook.**

Science Journal

Nations disagree about mining rights to mineral deposits. Since 1958, nations have met to decide international policy at the United Nations Law of the Sea Conferences. Research these conferences and find out why nations cannot agree. Summarize information in your Science Journal.

Activity 18-1

The Ups and Downs of the Ocean Floor

In Chapter 5, you graphed a profile of the United States. It is not easy to see the ocean floor, but data on depth of the ocean have been collected. In this activity, you will use these data to profile the Atlantic Ocean at 39° north latitude.

Problem

What does the ocean floor look like?

Materials

- graph paper

Procedure

1. Set up a graph as shown below.
2. Examine the data listed in the table. These data were collected at 29 ocean locations along 39° north latitude from New Jersey to Portugal.
3. Plot each data point and connect the points with a smooth line.
4. Color the ocean bottom brown and the water blue.

Analyze

1. What ocean-floor structures occur between 160 and 1050 km east of New Jersey? Between 2000 and 4500 km? Between 5300 and 5600 km?
2. When a profile of a feature is drawn to scale, both the horizontal and vertical scales must be the same. What is the vertical scale of your profile? What is the horizontal scale?
3. Does your profile give an accurate picture of the ocean floor? Explain.

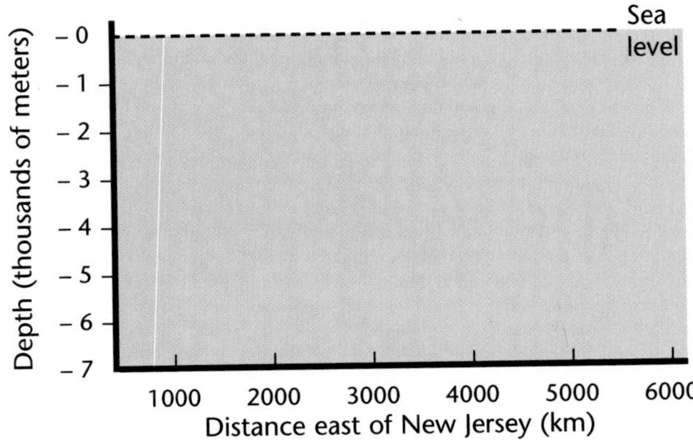

Data and Observations

Station Number	Distance from New Jersey (km)	Depth to Ocean Floor (m)
1	0	0
2	160	165
3	200	1800
4	500	3500
5	800	4600
6	1050	5450
7	1450	5100
8	1800	5300
9	2000	5600
10	2300	4750
11	2400	3500
12	2600	3100
13	3000	4300
14	3200	3900
15	3450	3400
16	3550	2100
17	3600	1330
18	3700	1275
19	3950	1000
20	4000	0
21	4100	1800
22	4350	3650
23	4500	5100
24	5000	5000
25	5300	4200
26	5450	1800
27	5500	920
28	5600	180
29	5650	0

18•2 Life in the Ocean

Science Words

photosynthesis
chemosynthesis
plankton
nekton
benthos
reef

Objectives

- List five things that the oceans provide for organisms.
- Describe how photosynthesis transfers energy into the food chains in the ocean.
- List the key characteristics of plankton, nekton, and benthos.

Life Processes in the Ocean

You live on land. Your basic needs for water, energy, and food are met in your environment. This is also true of organisms that live in the ocean. A saltwater environment provides the vital necessities of life for marine organisms, such as the eel shown in **Figure 18-5.**

Water

All life on Earth is water-based. Life began in the oceans. Organisms need water for processes that take place in cells. Water is used for all the basic processes of living things, such as digestion and cellular respiration.

Energy Relationships

Nearly all the energy used by organisms in the ocean ultimately comes from the sun. Radiant energy from the sun penetrates seawater and is trapped by chlorophyll-containing organisms in the ocean. Like chlorophyll-containing land organisms, these organisms capture the sun's energy and make food using light energy, carbon dioxide, and water. This process of food making is called **photosynthesis.** Organisms that undergo photosynthesis are called producers. Those that feed on them are called consumers.

Figure 18-5

A moray eel swims through the warm waters of Hawaii.

Figure 18-6

There are numerous food chains in the ocean. Some food chains are simple and some are complex.

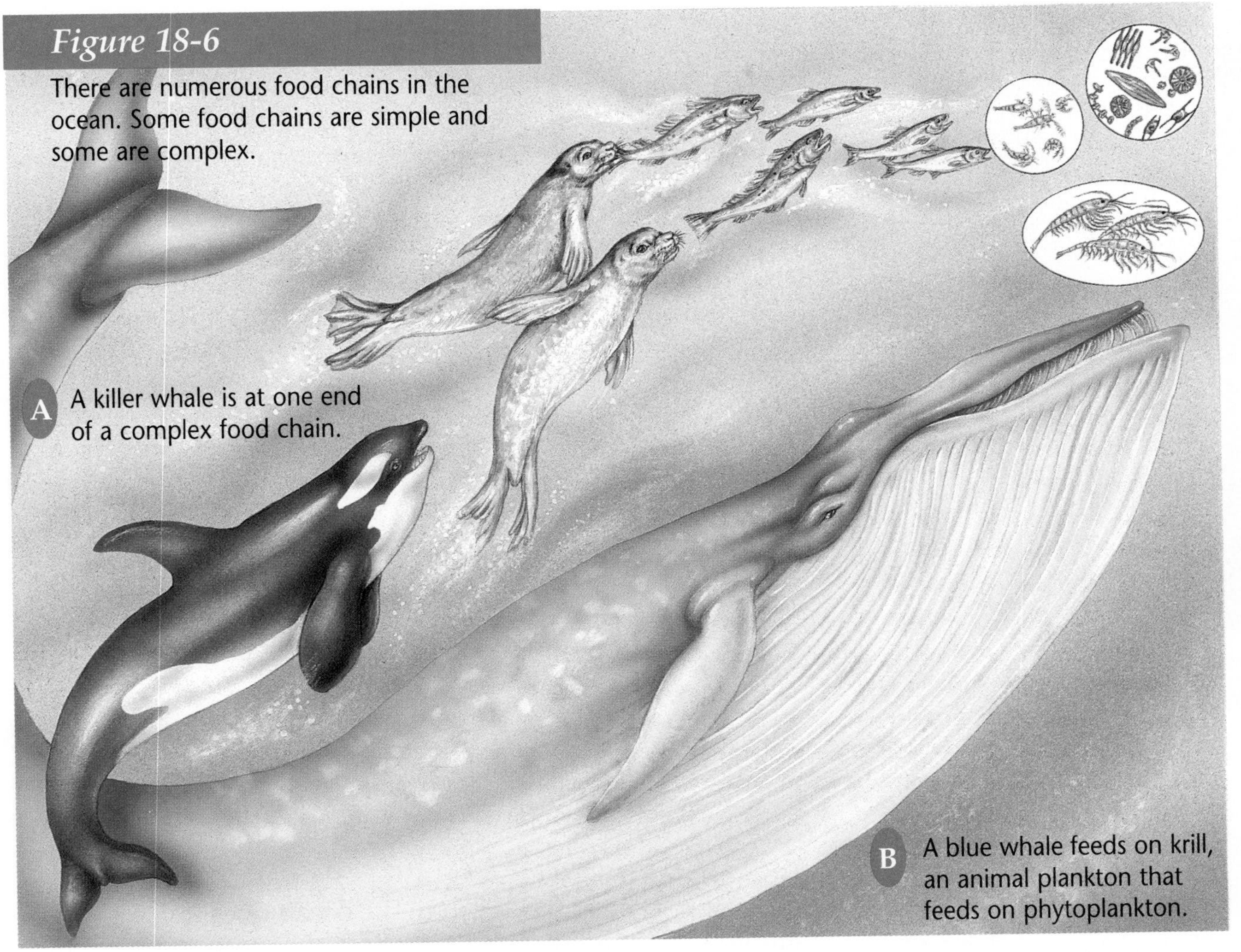

A A killer whale is at one end of a complex food chain.

B A blue whale feeds on krill, an animal plankton that feeds on phytoplankton.

Throughout the ocean, energy from the sun is transferred through food chains. Although the organisms of the ocean capture only a small part of the sun's energy, this energy is passed from producer to consumer, to other consumers. In **Figure 18-6,** notice that in one food chain, a large blue whale consumes small organisms, krill, as its basic food. In the other chain, phytoplankton are eaten by copepods that are, in turn, eaten by a herring. Cod eat the herring, seals eat the cod, and eventually killer whales eat the seals. At each stage in the food chain, energy obtained from one organism is used by other organisms to move, to grow, to repair cells, to reproduce, and to eliminate waste. Energy not used in these life processes is transferred along the food chain as organisms feed on other organisms.

A food chain is a model of energy transfer. In an ecosystem, there are many complex feeding relationships because most organisms depend on more than one species for food. For example, herring eat more than copepods, and in turn are eaten by animals other than cod. In an actual ecosystem, food chains are interconnected to form highly complex systems called food webs.

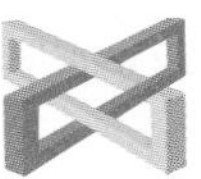

INTEGRATION
Life Science

CONNECT TO
PHYSICS

Buoyant force is the upward force of fluid on an object within the fluid. If an object is floating in the ocean, ***infer*** which is greater: the buoyant force being exerted on it or the force of gravity upon it.

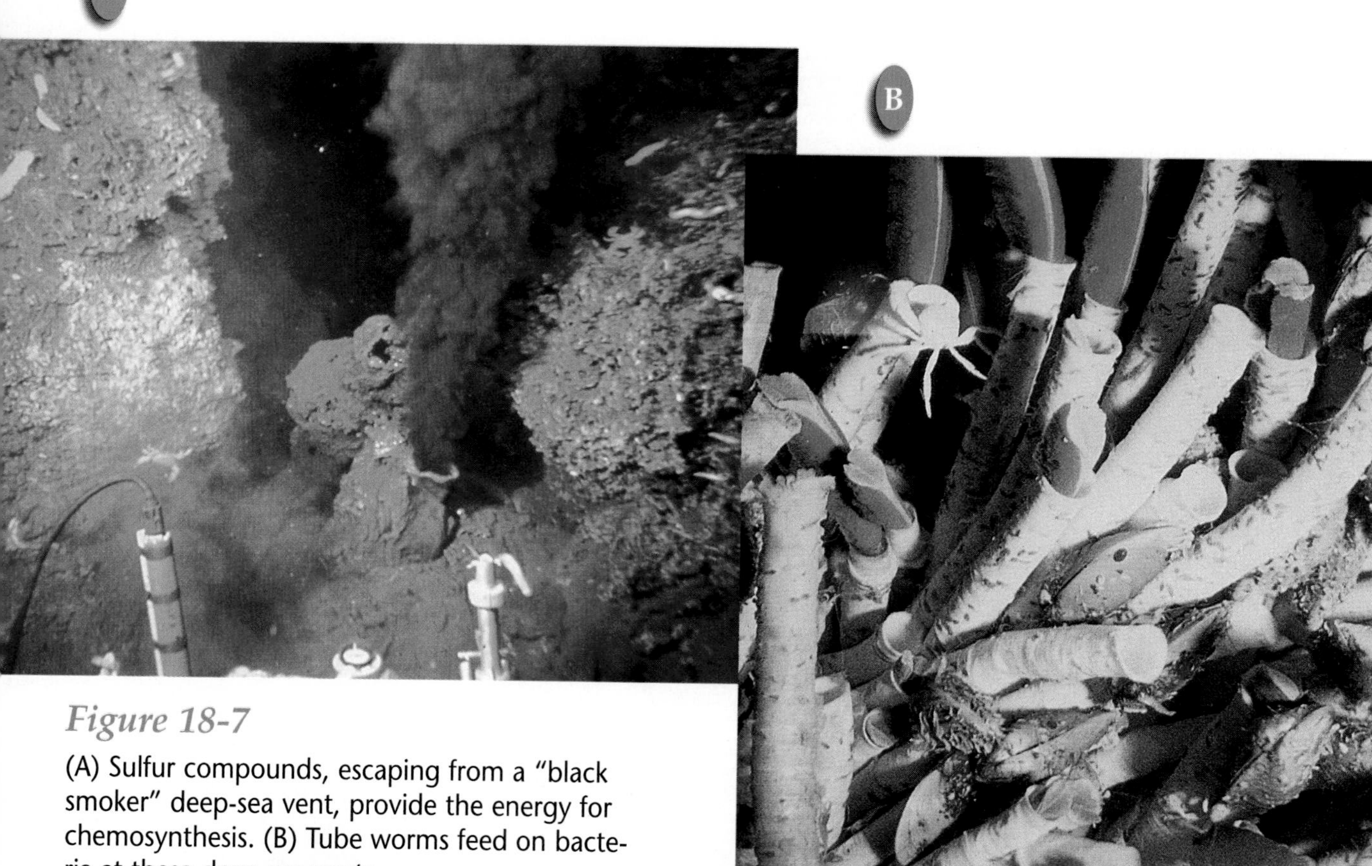

Figure 18-7

(A) Sulfur compounds, escaping from a "black smoker" deep-sea vent, provide the energy for chemosynthesis. (B) Tube worms feed on bacteria at these deep-sea vents.

USING MATH

Blue whales can consume more than 8 metric tons of krill a day. Adults average 26 m in length and have a mass of 137 metric tons. Use Appendices A and B to answer the following questions. How many *grams* of krill can a blue whale consume in one day? How many *centimeters* long is an average blue whale? What is the *mass* in kilograms of an average adult blue whale?

Chemosynthesis

Another type of food web in the ocean does not depend on sunlight. This food web depends on bacteria performing a process called **chemosynthesis.** *Chemo* means "chemical," and as you have learned, *synthesis* means "to make." This process takes place along mid-ocean ridges where in some places superheated water blasts from the crust and in some places seeps. In these areas, bacteria produce food and oxygen by using dissolved sulfur compounds that escape from magma. Organisms then feed on the bacteria. The ocean creatures in **Figure 18-7** receive their energy from chemosynthesis.

Other Life Processes

Another vital life process is reproduction. Ocean water provides an easy way for organisms to reproduce. Many ocean organisms release reproductive cells into the water where they unite to form new organisms.

Seawater also provides a thermally stable environment for organisms. That means ocean water protects organisms from sudden temperature changes. Because the ocean has a large volume, the temperature of the ocean changes slowly. This slow change of temperature means that marine organisms do not experience the sudden changes in temperature that some organisms on land endure.

Ocean Life

Many varieties of plants, animals, and algae live in the ocean. Most exist in the sunlight zone of the ocean. Sufficient sunlight to promote photosynthesis can penetrate about 140 m down into ocean water. This means that many marine organisms live on the continental shelf because this is where they can obtain food.

Most organisms in the ocean are adapted to obtain food in one of three ways: they make it by photosynthesis; they move through the water to obtain food; or they filter water to obtain nutrients.

Plankton

Plankton are tiny marine algae and animals that drift with currents. Most plant plankton are one-celled organisms that float in the upper layers of the ocean where light needed for photosynthesis is found. To see most plankton, you need a microscope. A one-celled, golden alga, called a diatom, is a tiny but abundant form of plankton. Diatoms are the most important source of food for animal plankton, the next step in the food chain of the oceans. Examples of animal plankton include eggs, very young fish, jellyfish and crabs, and tiny adults of some organisms. Most animal plankton depend on surface currents to move them, but some can swim.

Nekton

In the ocean, animals that actively swim, rather than drift with the currents, are **nekton.** Nekton include all swimming forms of fish and other animals, from tiny herring to huge whales. Shrimps, squid, and seals are also nekton. In **Figure 18-8**, the swimming turtle and angler fish are nekton. Nekton move from one depth to another easily. This ability to swim reduces the effects of surface currents on them. For nekton, buoyancy is important. In the water, organisms less dense than water float. By changing how buoyant or how easily they float, organisms can change depth in the ocean. Changing depth allows animals to search more areas for food. Some nekton live in cold water, whereas others exist in warm regions. Some, like whales, roam the entire ocean. Many nekton come to the surface at night to feed on plankton, while others remain in deeper parts of the ocean where sunlight does not reach.

MiniLAB

What do plankton look like?

Procedure

1. Place one or two drops of water that contains ocean or freshwater plankton onto a microscope slide.
2. With light coming through the sample from beneath, use a microscope to observe your sample.
3. Find at least three different types of plankton.

Analysis

1. Draw detailed pictures of three types of plankton.
2. Classify the plankton as plant or animal.

Figure 18-8

Swimming animals, such as this turtle (A) and this angler fish (B), are nekton.

USING TECHNOLOGY

Drugs from the Sea

Did you know that half of all drugs that are developed come from living organisms? Until recently, nearly all of the organisms used in drug making lived on land. Now, scientists are looking at organisms in the oceans as sources of drugs to treat asthma, arthritis, cancers, and AIDS.

Scientists are especially interested in studying soft-bodied organisms that live in the sea. These organisms seem defenseless, and yet other organisms do not eat them. These organisms use chemical defense systems. Scientists suspect that we could develop drugs from these chemicals. Slugs, soft corals, sponges, and even some sea worms are among the organisms being researched.

Many different kinds of scientists work together to develop new drugs. Biologists and ecologists collect and separate the species. Chemists extract substances from the organisms. Biologists and pharmaceutical specialists test and purify the ingredients. Finally, researchers perform clinical trials to see whether a chemical is effective in the treatment of certain diseases.

One drug from the ocean, in use since 1982, is acyclovir. It fights herpes infection. Acyclovir is derived from a Caribbean sponge.

Sponges

Think Critically:

Once a chemical is isolated from an organism and found to be an effective drug, scientists synthesize the chemical in the laboratory instead of harvesting the organisms. Why do you think this is done?

Some of these deep-dwelling organisms have special light-generating organs for attracting live food in the dark depths. The angler fish, shown in **Figure 18-8,** dangles a luminous lure over its forehead. When small animals bite at the lure, the angler fish swallows them whole. The viper fish has luminous organs in its mouth. It swims with its mouth open, and small organisms swim right into the viper fish's mouth.

Bottom-Dwellers

Plants, algae, and animals also live on the seafloor. These plants and animals living in the lowest levels of the ocean are the **benthos.** Some benthos live on the bottoms of the continental shelf in fairly shallow water. Others live on the ocean bottom hundreds of meters deep. Some are attached to the bottom, others

Figure 18-9
Many organisms live on this coral reef in the South Pacific.

burrow into the bottom sediments, and others walk or swim over the bottom.

Forests of kelp grow in sunlit areas of the continental shelf, in cool water near shore. Did you know that kelp is the fastest growing organism in the world? It can grow up to 30 cm each day and reach lengths of 25 m. Kelp is anchored to the ocean bottom and is part of the benthos. A special organ holds each kelp to the bottom. Gas-filled bladders keep the plant upright in the water.

Other benthos organisms include crabs, corals, snails, clams, sea urchins, and bottom-dwelling fish. How do you suppose these animals get food? Many of them eat the partially decomposed matter that sinks to the ocean bottom. Some prey on other benthos, eating them whole. Those that are permanently attached to the bottom filter food particles from water currents. Still others have specialized organs that sting prey that comes near. Coral is an attached benthic organism that stings its prey.

Problem Solving

Washed Ashore

A chemical plant and a nuclear power plant are both located along Barney Beach. The chemical plant produces its electricity by burning coal. The nuclear power plant generates power using nuclear fission, a process that produces vast amounts of energy. The chemical plant is located about 20 km north of Barney Beach. The nuclear power plant is about 10 km south of Barney Beach. Recently, many of the fish along Barney Beach died and washed ashore.

Solve the Problem

1. **How could a chemical plant pollute Barney Beach?**
2. **How could a nuclear power plant pollute Barney Beach?**

Think Critically:

What questions would you want to ask about the Barney Beach fish kill that would help you determine the source of the pollution?

Activity 18-2

Design Your Own Experiment

Cleaning Up Oil Spills

Oil spills can be destructive to life in the oceans. Oil forms thick layers that smother organisms and clog their feeding and breathing mechanisms. Organisms exposed to oil spills frequently die. Food webs are disrupted.

PREPARATION

Problem

How can an oil spill be cleaned up?

Form a Hypothesis

Consider the list of materials at right. Hypothesize which items will be most effective in cleaning up oil spills.

Objectives

- Design and carry out an experiment to study the effectiveness of different materials in cleaning up oil spills.
- Observe the effectiveness of different materials in cleaning up oil spills.

Possible Materials

- olive oil
- pan
- gravel
- water
- toothpicks
- sponge
- paper towels
- cotton balls
- feathers
- string
- liquid soap
- cardboard
- eyedropper

Safety Precautions

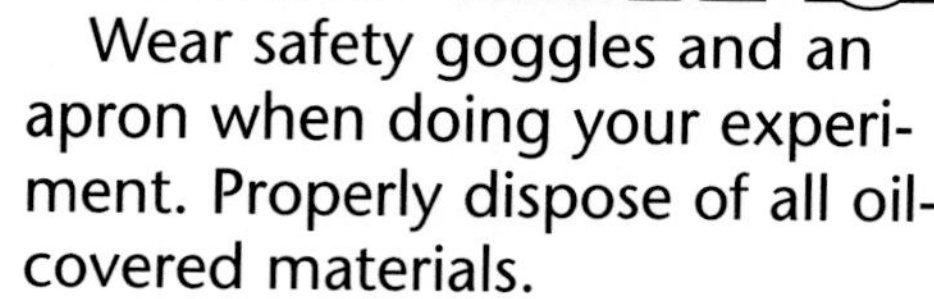

Wear safety goggles and an apron when doing your experiment. Properly dispose of all oil-covered materials.

PLAN THE EXPERIMENT

1. As a group, agree upon and write out your hypothesis statement.
2. As a group, list the steps that you need to take to test your hypothesis. Include in your plan how you will (a) make the oil spills using the pan, water, olive oil, and gravel (to simulate a rocky shore); (b) conduct your tests using the different items from the materials list; and (c) determine which material is the most effective in cleaning up an oil spill.
3. Design a data table in your Science Journal so that it is ready to use as your group collects data.

Check the Plan

1. Read over your experiment to make sure that all steps are in a logical order.
2. Identify any constants and the variables of the experiment.
3. Make certain that your experiment tests one variable at a time.
4. *Make sure your teacher approves your plan before you proceed.*

DO THE EXPERIMENT

1. Carry out the experiment as planned.
2. While the experiment is going on, write your observations in the data table in your Science Journal.

Analyze and Apply

1. **Compare** your results with those of other groups.
2. Which materials were the most effective in cleaning up the oil from the water's surface? **Explain** your criteria for judging effectiveness.
3. **Explain** why certain materials were the most effective in cleaning up the oil from the gravel.
4. **Compare and contrast** characteristics of the most effective and the least effective materials.
5. Why should different materials be used to clean up oil spills in different locations?

Go Further

Oil is really a mixture of different compounds. You have tested different methods of cleaning up oil that floats. Now, design a way to clean up oil that sinks to the ocean bottom.

Figure 18-10

A diver enjoys looking at the variety of organisms on a coral reef.

Corals

Corals are organisms that live attached to the seafloor. Corals obtain food by stinging their prey. They need clear, warm water and a lot of light. This means that generally they form no farther than 30 m from the surface in the warm waters near the equator.

Each coral builds a hard capsule around its body from the calcium it removes from seawater. Each capsule is cemented to others to form a large colony called a reef. A **reef** is a rigid, wave-resistant structure built from skeletal materials and calcium carbonate. As a coral reef forms, other benthos and nekton begin living on it.

If you could swim through the coral reef shown in **Figure 18-10,** you'd see many ocean organisms interacting. Fish, crabs, sea urchins, and many others congregate in reefs. Nutrients, food, and energy are cycled among these organisms in complex food webs. Plankton, nekton, and benthos all depend on each other for survival, not only in coral reefs, but throughout the entire ocean.

Section Wrap-up

Review

1. How do organisms in the ocean obtain food?
2. List the key characteristics of plankton, nekton, and benthos. How do they depend on one another?
3. **Think Critically:** When whales die and their carcasses sink to the bottom, sometimes sulfides are produced. Bacteria use the sulfides for food, and giant clams and worms eat the bacteria. Does this food web depend on photosynthesis or chemosynthesis? Explain.

Skill Builder

Using Variables, Constants, and Controls

Describe how you could test the effects of salinity on marine organisms. If you need help, refer to Using Variables, Constants, and Controls in the **Skill Handbook.**

Using Computers

Graphics Design a poster that illustrates the relationship between photosynthesis and energy in a food chain.

TECHNOLOGY:

Pollution and Marine Life

18 • 3

Science Words

pollution
pollutant

Objectives

- List seven human activities that pollute the oceans.
- Explain how ocean pollution affects the entire world.
- Determine how we can live on this planet without destroying the oceans.

Ocean Pollution

How would you feel if someone came into your bedroom; spilled oil on your carpet; littered your room with plastic bags, cans, bottles, and newspapers; sprayed insect killer all over; scattered sand; and then poured hot water all over you? Organisms in the ocean experience these very things when people pollute seawater.

Pollution is the introduction by humans of harmful waste products, chemicals, and substances into an environment. A **pollutant** is a substance that causes damage to organisms by interfering with biochemical processes. **Figure 18-11** illustrates how pollutants from the land can eventually flow to the oceans and pollute them. Most ocean pollution caused by humans is concentrated along the coasts of continents.

Figure 18-11

Ocean pollution comes from many sources. *At your school, what are the sources of pollution that could eventually pollute the oceans?*

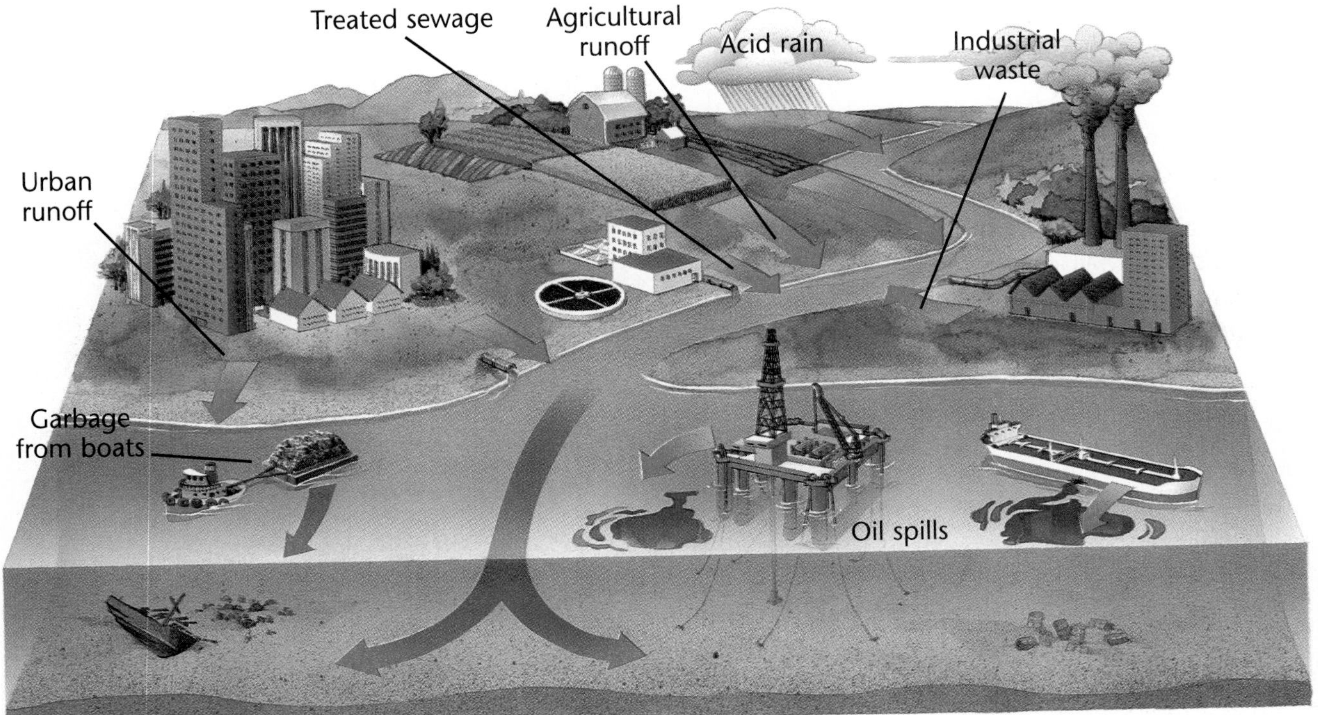

CONNECT TO

PHYSICS

Oil is a mixture of different compounds. When oil is spilled on water, various things happen to these compounds. Some of them evaporate, some float on the water, and some sink. ***Infer*** what this indicates about the density of these different compounds.

Chemical and Solid-Waste Pollution

Industrial waste sometimes gets into seawater. Often, this waste contains concentrations of metals and chemicals that harm organisms. Solid wastes, such as plastic bags and fishing line left lying on beaches, can entangle animals. Illegally dumped medical waste such as needles, plastic tubing, and bags are a threat to both humans and animals.

Pesticides, including herbicides (weed killers) and insecticides (insect killers), used in farming and on lawns run off and reach the ocean. In the ocean food chains, these pesticides become concentrated in the tissues of organisms.

Crop fertilizers and human sewage create a different kind of problem. Because they fertilize the water, fertilizers cause some types of algae to reproduce rapidly. When these algae die, they're decomposed by huge numbers of bacteria. The problem is that the bacteria use up much of the oxygen in the water during respiration. Therefore, other organisms such as fish can't get the oxygen they need, and they die.

Oil spills also pollute the ocean. You've heard in the news about major oil pollution caused by tanker collisions and leaks at offshore oil wells. Another source of oil pollution is oil mixed with wastewater that is pumped out of ships. In addition to these sources, oil washes from cars, from parking lots, and from streets and flows into streams. Eventually, this oil can reach the ocean. Part of this cycle is shown in **Figure 18-12.**

Human activities such as agriculture, cutting down forests, and construction tear up the soil. Rain washes the soil into streams and eventually into an ocean. This causes huge amounts of silt to accumulate in some coastal areas. In silted areas, the filter-feeding systems of benthos such as oysters and clams become clogged. The organism can die. Also, when saltwater marshes are filled for land development, marine habitats are destroyed.

Figure 18-12

A ship discharges waste into Pearl Harbor, Hawaii. *How does this affect the ocean environment?*

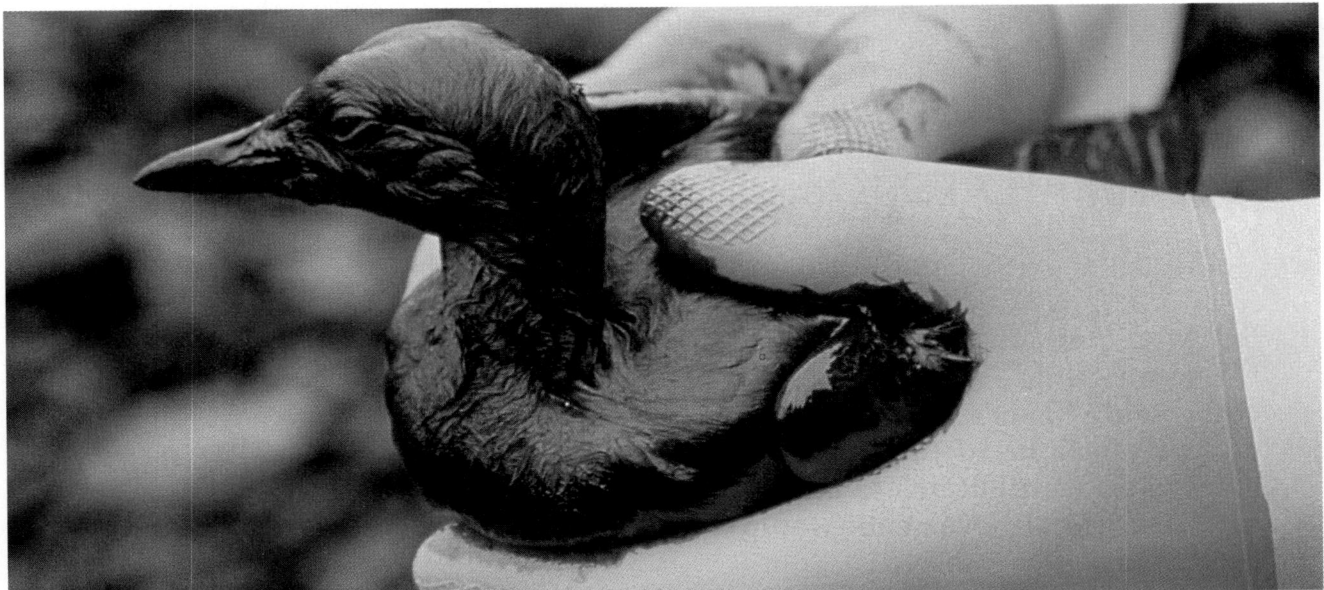

Some pollutants cause instant death to some organisms. Other pollutants, such as fertilizer, inflict damage slowly as they change the environment. Some pollutants remain in the ocean only a short time before they are broken down. However, many synthetic pollutants such as Styrofoam resist being broken down.

When an ocean becomes polluted, it isn't just the resident animals, plants, and algae that are affected. As food chains in the ocean are disrupted, Earth's oxygen supply is affected because most of the world's oxygen is produced by plankton. If an ocean's plankton were to die, this could result in the death of many organisms.

Figure 18-13

Pollution doesn't just look bad. It hurts living things. It can threaten our food supply.

Section Wrap-up

Review

1. List seven human activities that pollute the oceans. Suggest a solution to each.
2. How does pollution of the oceans affect the entire world?

Explore the Technology

In this section, you learned that human activities can severely pollute our oceans. However, many of these activities are part of the systems that provide your food, home, energy, transportation, and recreation. As Earth's population increases, the problem may grow worse. How can we maintain our quality of life without destroying life in the oceans? What are your ideas?

SCIENCE & SOCIETY

People and Science

ROBERT BALLARD, *Oceanographer*

On the Job

Q After you discovered the sunken ocean liner RMS *Titanic*, your mother said, "That's too bad," because you'll be remembered for that discovery, not for your contributions to science. For what scientific achievements would you most like to be remembered?

A Without a doubt, the first manned exploration of the mid-ocean ridge in the summers of 1973 and 1974. That was really the first time anyone went into the largest feature on the planet Earth. What we found there helped confirm the theory of plate tectonics. The second thing would be the discovery in 1977 of the hydrothermal vents in the Galápagos Rift. The chemosynthetic ecosystem we discovered there has had a lot to do with understanding the origins of the ocean's chemistry and the origins of life on the planet itself.

Q Of all the amazing undersea sights you've seen, what stands out most?

A Seeing a black smoker for the first time. It looked like a fire hydrant underwater, belching out this black plume.

Personal Insights

Q Through the JASON Project, you have gotten a lot of young people excited about ocean exploration. What would you say to kids who still think science is boring?

A A lot of young people think that it's all done—that all the exploration is over with, and they're just picking up the pieces. I want to tell them that's not true at all. We've just begun. Their generation will probably explore more of Earth than all the previous generations combined. We're not going to live in space in their lifetime, at least not in any numbers. So it's the ocean that is the frontier of exploration.

Career Connection

At the library, look up information on the underwater research vehicles *Alvin, Jason,* and *Argo.* Design your own system of submersible vehicles for deep-sea exploration. Then make a report about an ocean career that interests you, perhaps

- **Marine Geologist**
- **Marine Biologist**

Chapter 18 Review

Summary

18-1: The Seafloor

1. The continental shelf is a gently sloping part of the continent that extends out into the oceans. The continental slope extends beyond the continental shelf and down to the ocean floor. The abyssal plain is a flat area of the ocean floor.
2. Cracks form where the seafloor is spreading apart. Along the mid-ocean ridges, new crust appears. Seafloor slips beneath another section of seafloor at a trench.
3. Petroleum, natural gas, and placer deposits are mined from continental shelves. In the future, perhaps manganese nodules will be mined from the deep ocean.

18-2: Life in the Ocean

1. A water habitat provides water, nutrients, waste disposal, constant temperatures, and a means of reproduction for marine organisms.
2. Photosynthesis is the basis of food chains in the ocean. Chemosynthesis is a special process of food making near thermal vents.
3. Microscopic algae and animals called plankton drift in ocean currents. Nekton are marine organisms that swim. Benthos are plants, algae, and animals that live on or near the ocean floor.

18-3: Science and Society: Pollution and Marine Life

1. Industrial, solid, and medical wastes; pesticides, herbicides, and fertilizers; and oil spills, hot water, and silt all pollute oceans.
2. Ocean pollution can disrupt food webs and threaten Earth's oxygen supply. It is a serious worldwide problem.

Key Science Words

a. abyssal plain
b. benthos
c. chemosynthesis
d. continental shelf
e. continental slope
f. mid-ocean ridge
g. nekton
h. photosynthesis
i. plankton
j. pollutant
k. pollution
l. reef
m. trench

Reviewing Vocabulary

Match each phrase with the correct term from the list of Key Science Words.

1. damages biochemical processes
2. forms when deposits fill in valleys in the deep ocean
3. sharp drop-off beyond the continental shelf
4. mountains formed where seafloor is spreading apart
5. process using sunlight that produces food
6. drifting marine plants, algae, and animals
7. animals, algae, and plants that live on and near the ocean floor
8. colony made of many corals
9. occurs whenever harmful substances are introduced to the environment
10. where ocean crust is destroyed

Chapter 18 Review

Checking Concepts

Choose the word or phrase that completes the sentence.

1. The flattest feature of the ocean floor is the _______.
   a. rift valley c. seamount
   b. continental slope d. abyssal plain
2. _______ may be found where rivers enter oceans.
   a. Manganese nodules c. Abyssal plains
   b. Rift valleys d. Placer deposits
3. When organisms _______, they produce food.
   a. photosynthesize c. benthos
   b. excrete wastes d. respire
4. _______ are formed along diverging plates.
   a. Mid-ocean ridges c. Continental slopes
   b. Trenches d. Density currents
5. Ocean organisms that drift in ocean currents are called _______.
   a. nekton c. benthos
   b. fish d. plankton
6. The seafloor is spreading apart at _______.
   a. trenches c. abyssal plains
   b. mid-ocean ridges d. continental shelves
7. Some deep-water bacteria in the ocean make food through _______ .
   a. photosynthesis c. respiration
   b. chemosynthesis d. rifting
8. Ocean swimmers are _______.
   a. benthos c. kelp
   b. nekton d. coral
9. Most ocean pollution _______.
   a. is near shore c. is thermal pollution
   b. is in the ocean depths d. harms only plants
10. Fertilizers cause some _______ to reproduce rapidly, resulting eventually in a lack of oxygen.
    a. fish c. algae
    b. animal plankton d. corals

Understanding Concepts

Answer the following questions in your Science Journal using complete sentences.

11. Why might some industries be interested in mining manganese nodules?
12. Where would you expect to find the most marine organisms—closer to continents or in deeper waters? Explain.
13. Why is pollution an international problem?
14. Discuss how agricultural chemicals can kill marine organisms.
15. How do plate tectonics affect the ocean floor?

Thinking Critically

16. Some areas of the oceans are so polluted that organisms are dying. Some species are becoming extinct. What effect does pollution have on our ability to discover new drugs?
17. Animal plankton are disappearing in waters off the coast of California. How will this disappearance affect the food web in that area?
18. To widen their beaches, some cities pump offshore sediments onto beaches. How does this affect organisms that live in coastal waters?
19. Would you expect to find coral reefs growing around the bases of volcanoes off the coast of Alaska? Explain your answer.
20. Some oil leaks in the ocean are natural. Where does the oil come from?

Developing Skills

If you need help, refer to the **Skill Handbook.**

21. **Using Scientific Illustrations:** Use **Figure 18-6** to determine which organisms would starve to death if phytoplankton became extinct.
22. **Comparing and Contrasting:** Compare and contrast mid-ocean ridges and trenches.
23. **Classifying:** Classify each of the following sea creatures as plankton, nekton, or benthos: shrimp, dolphins, starfish, krill, coral, algae.
24. **Concept Mapping:** Make an events chain concept map that describes how crop fertilizers can harm marine fish.
25. **Measuring in SI:** If kelp, as pictured below, grows at a steady rate of 30 cm/day, how long would it take to reach a length of 25 m?

Performance Assessment

1. **Graph a Profile:** In Activity 18-1, you made a profile of the ocean bottom along the 39º north latitude line, but it was not drawn to scale. Make a profile to scale of the area between 3600 and 4100 km from New Jersey. Use the scale 1 mm = 1000 m.
2. **Design an Experiment:** In Activity 18-2, you tried different methods to clean up an oil spill. Design an experiment using different methods to clean up solid-waste pollution.
3. **Poster:** Choose a particular sea animal to research. Find out where it lives, how it moves, what it eats, how it gets rid of wastes, and how it reproduces. Classify it as plankton, nekton, or benthos. Draw its food web on a poster and share this with your class.

UNIT PROJECT 5

Wet and Windy Inventions

Air and water have unique characteristics that we all enjoy. Many of you have thrown paper and balsa airplanes, boomerangs, and Frisbees into the air and watched them glide. You may have run through a field, pulling on a kite until air currents finally carried it high into the sky, or blown soap bubbles and watched them float and drift in the air. Some of you have even played air hockey. Over the years, people have invented many ways to play in the wind. You've played in the water too, with beach balls, surf boards, water skis, boogie boards, and inner tubes. This is your chance to create a new air or water toy. Your goal is to design and make a new toy. Where should you begin?

Procedures

1. First, research the types of water and air toys that are already on the market.
2. Next, make a list of at least ten of your ideas for new toys.
3. Select several of your most original ideas from your list, and consider the following about each one. What materials are needed to make it? What will it cost to make? Are the materials for building it available? Is it really possible for you to make this toy? Will it be fun to play with? Will it be durable? Will it be safe?
4. Select the toy from your list that you feel best fits the criteria in step 3, and sketch your idea on paper.
5. Obtain the materials you need to make your toy, and build it.
6. Test your toy to be sure that it does what you intended it to do, and if there are problems, make corrections.
7. What will you name it? The name could include your name, or it could include the components or ingredients you used. Whatever you decide to call your toy, try to think of a name people will remember.

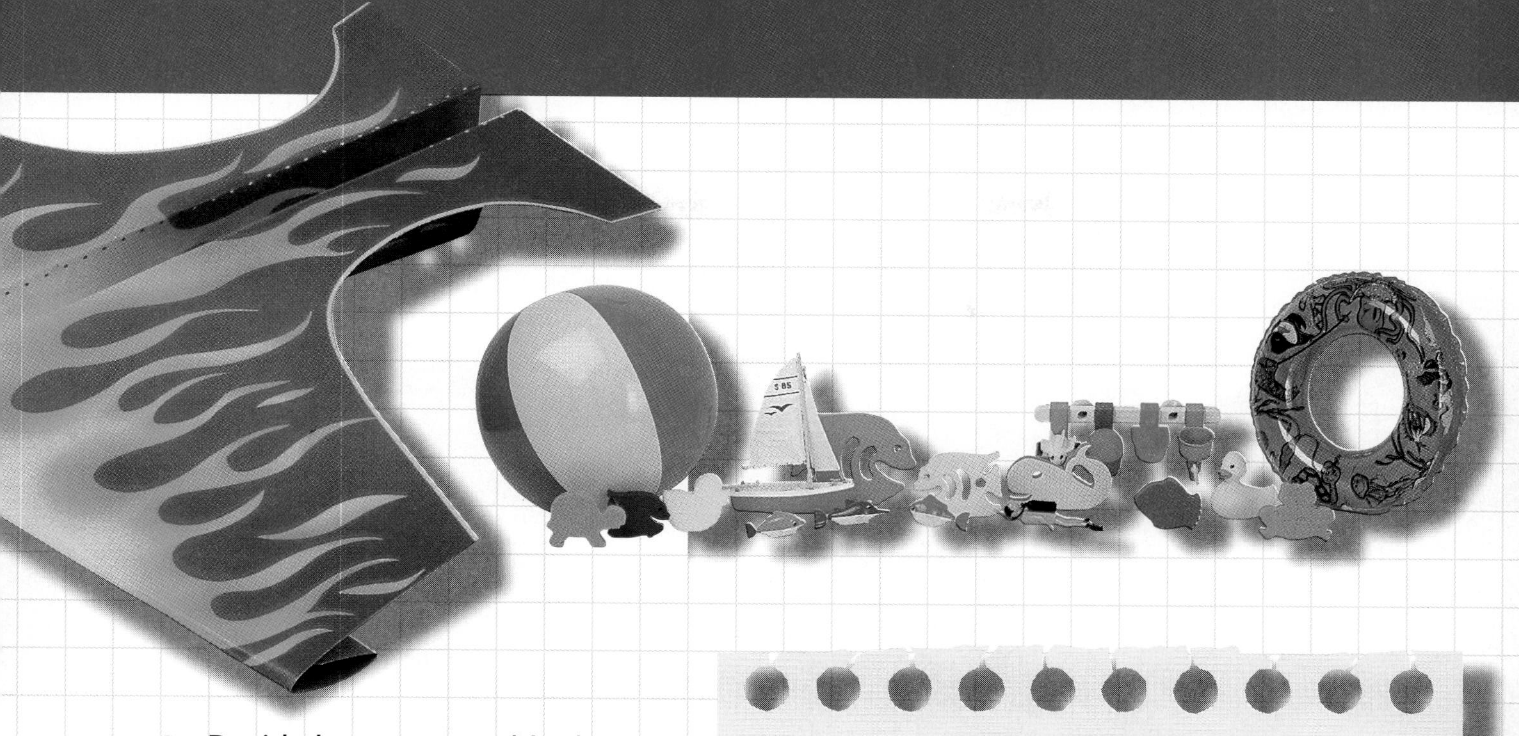

8. Decide how you would advertise. Is there an age group of people who would be likely to play with your toy?
9. How much would your toy cost? Determine the price by calculating your cost to make it, how much profit you would want to make on each toy, and how much you think people would be willing to pay for the toy.

Going Further

Enter your toy in a class Toy Invention Fair. Make a colorful poster that shows a scale model drawing of your toy, its name, the age of the people who would play with it, and how much it would cost if you were selling it. Demonstrate how your toy works to others at the fair.

UNIT

6

Unit Contents

You and the Environment

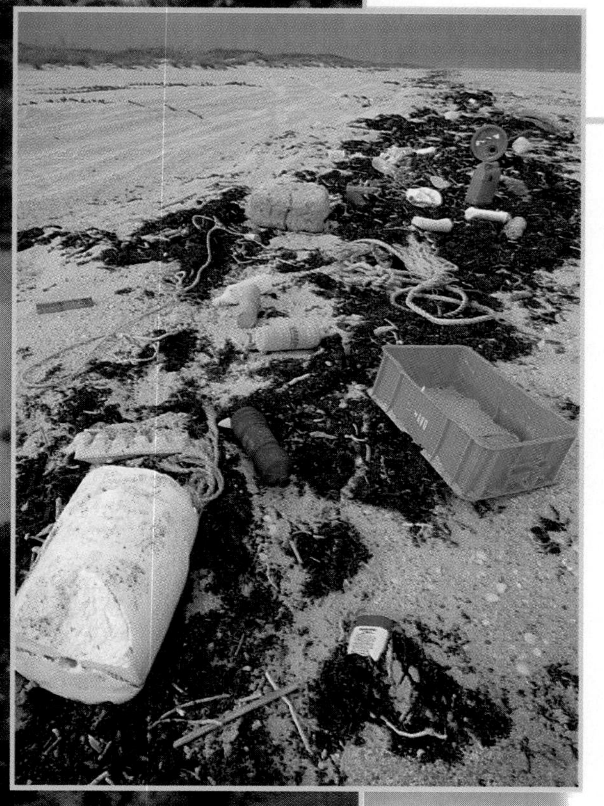

What's Happening Here?

Life abounds in the nutrient-rich waters of the tidal pool, as these students are discovering. Before the day is through, they will likely spot burrowing shellfish, foraging birds, a dozen species of aquatic plants, and much more. They will also learn that the tidal pool is a fragile environment that can be easily disrupted by human activities such as illegal dumping of trash. Today, more people realize that clean air, land, and water are precious resources, and they are working to restore polluted areas to their natural state. In this unit, you will learn how humans impact the environment, and ways to make your own personal impact a wise one.

Science Journal

Contact your local newspaper for information on a recent environmental problem in your area. In your Science Journal, describe how people in the community have responded to the problem.

Previewing the Chapter

Chapter 19

Our Impact on Land

Did you ever consider the impact that every one of us has on the land? Each person on Earth competes for space and resources. Each one generates trash, consumes products made in factories, uses natural resources such as water, soil, and energy, and creates pollution. Do you think our impact on the land is very significant?

EXPLORE ACTIVITY

Draw a population growth model.

1. Use a piece of paper and pencil to draw a square that is 10 cm on each side. This represents one square kilometer of Earth's land surface.
2. In 1965, the average number of people for every square kilometer of land was 21. Draw 21 small circles inside your square to represent this.
3. In 1990, the average rose to 35. Add 14 circles to illustrate this increase.
4. In 2025, there will be an estimated 67 people per square kilometer. Add circles to represent the average number of people per square kilometer of land in 2025.

Observe: Based on the numbers above, is human population increasing more rapidly as time goes on? In your Science Journal, hypothesize how population growth affects the environment.

Previewing Science Skills

- In the **Skill Builders,** you will use graphs, and recognize cause and effect.
- In the **MiniLABs,** you will collect data, measure in SI, and hypothesize.
- In the **Activities,** you will experiment, make models, analyze data, make and use tables, make and use graphs, and draw conclusions.

19 • 1 Population Impact on the Environment

Science Words

carrying capacity
population
population explosion

Objectives

- Interpret data from a graph that shows human population growth.
- List reasons for Earth's rapid increase in human population.
- List several ways each person in an industrialized nation affects the environment.

The Human Population Explosion

At one time, people thought of Earth as a world with unlimited resources. They thought the planet could provide them with whatever materials they needed. Earth seemed to have an endless supply of metals, fossil fuels, and rich soils. Today, we know this isn't true. Earth has a carrying capacity. The **carrying capacity** is the maximum number of individuals of a particular species that the planet will support. Thus, Earth's resources are limited. Unless we treat those resources with care, they will disappear.

Many years ago, there were few people on Earth. Fewer resources were used and less waste was produced than today. But in the last 200 years, the number of people on Earth has increased at an alarming rate. The increase in the world population has changed the way we must view our world and how we care for it for future generations, like the babies in **Figure 19-1.**

Figure 19-1

The human population is growing at an alarming rate. *In your Science Journal, discuss how this impacts Earth's limited resources.*

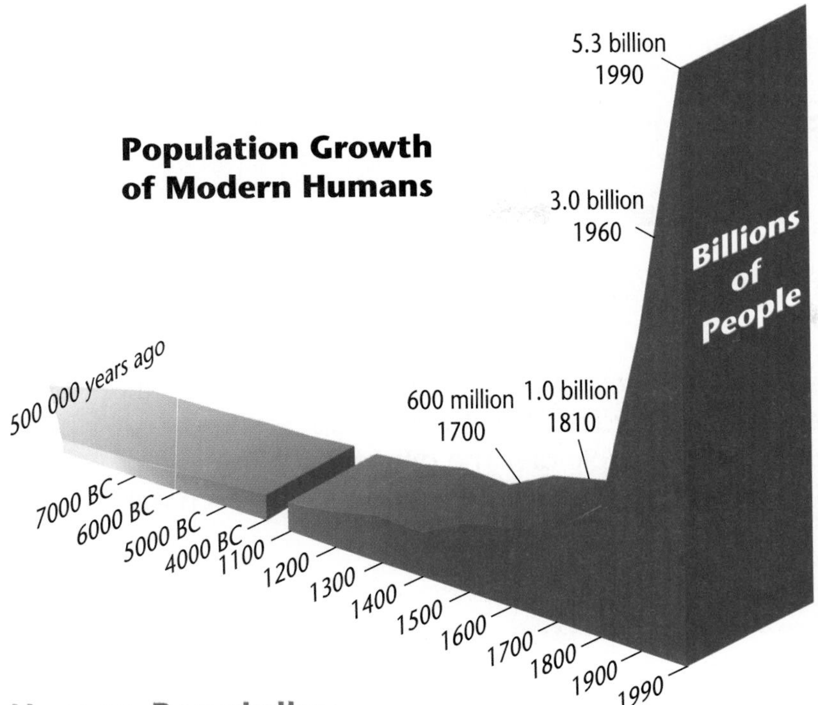

Figure 19-2

The human species, *Homo sapiens*, may have appeared about 500 000 years ago. Our population numbers remained relatively steady until about 200 years ago. Since that time, we have experienced a sharp increase in growth rate.

Human Population

A **population** is the total number of individuals of a particular species in a particular area. The area can be small or large. For example, we can talk about the human population of one particular community, such as Los Angeles, or about the human population of the entire planet.

Have you ever wondered how many people live on Earth? The global population in 1995 was 5.7 billion. Each day, approximately 242 000 people are added. Earth is experiencing a **population explosion.** The word *explosion* is used because the rate at which the population is growing has rapidly increased in recent history.

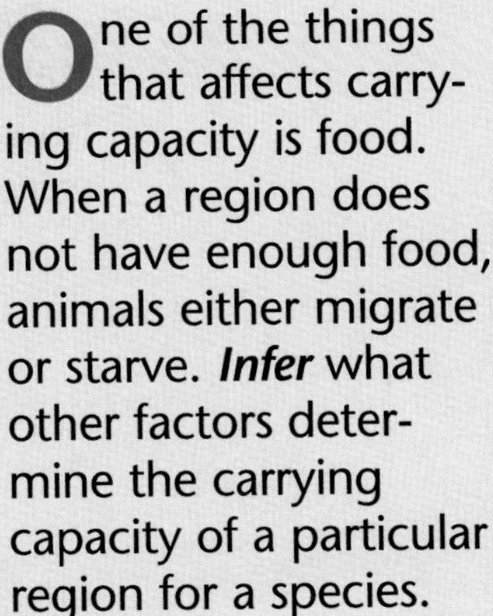

CONNECT TO

LIFE SCIENCE

One of the things that affects carrying capacity is food. When a region does not have enough food, animals either migrate or starve. ***Infer*** what other factors determine the carrying capacity of a particular region for a species.

Why Our Population Is Increasing

Look at **Figure 19-2.** You can see that it took hundreds of thousands of years for Earth's population to reach 1 billion people. After that, the population increased much faster. Population has increased so rapidly in recent years because the death rate has been slowed by modern medicine, better sanitation, and better nutrition. This means that more people are living longer and remaining in the population. Also, births have increased because more people have reached the age at which they can have children.

The population explosion has seriously affected the environment. Scientists predict even greater changes as more people use Earth's limited resources. The population is predicted to be 14 billion sometime in the next century, nearly three times what it is now. We need to be aware of the effect such a large human population will have on our environment. We need to ask ourselves whether we have enough natural resources to support such a large population.

Activity 19-1

Design Your Own Experiment

A Crowded Encounter

Think about the effects of our rapidly increasing population. One of these is overcrowding. Every minute approximately 265 people are born and 97 people die. The difference in these two numbers, 168, is the net increase in human population each minute.

PREPARATION

Problem

How does the net increase in human population affect Earth's limited land and resources?

Form a Hypothesis

Based on the numbers above, state a hypothesis that shows how population increases in ten minutes' time can be modeled.

Objectives

- Design and carry out an experiment to model human population growth in ten minutes' time.
- Observe the effects of human population increase on space.

Possible Materials

- many small objects, such as dried beans or paper clips
- 250-mL beaker
- clock or watch

PLAN THE EXPERIMENT

1. As a group, agree upon and write out your hypothesis statement.
2. Your group should list the steps needed to test your hypothesis. Include in your plan how you will (a) let the empty beaker represent the unoccupied space left on Earth at the moment you begin the experiment; (b) let each of your small objects represent five people; and (c) design a table with two columns in your Science Journal showing time (1 through10 minutes) and population at the designated time.

Check the Plan

1. Review your plan to make certain that all steps are clear and logical.
2. *Have your teacher approve the plan and include any suggested changes before beginning.*

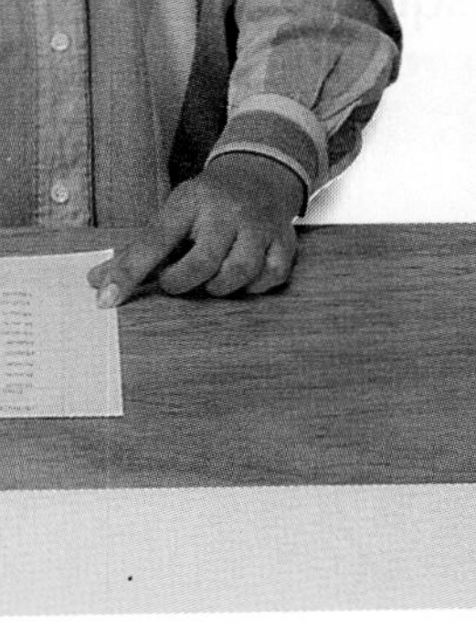

DO THE EXPERIMENT

1. Carry out the experiment as planned.
2. While the experiment is going on, write down any observations that you make and complete the data table in your Science Journal.
3. Make a graph that shows the results of your experiment. Plot time in minutes on the horizontal axis and population on the vertical axis.

Analyze and Apply

1. At the end of ten minutes, what is the net increase in human population?
2. **Explain** why the model you made was not a scale model.
3. Today there are approximately 5.7 billion people on Earth. That number will double in about 45 years. Assuming there is no change in that rate, **predict** what the population will be in 35 years. In 135 years.

Go Further

If you gathered all the humans on Earth into one place, we would take up only a very small portion of Earth's surface. Why then is human population growth a problem?

19•2 Using the Land

Land Usage

You may not think of land as a natural resource. Yet, it is as important to people as oil, clean air, and clean water. Through agriculture, logging, refuse disposal, and urban development, we use land—and sometimes abuse it.

Science Words

landfill
sanitary landfill
hazardous waste
conservation
composting

Objectives

- List ways that we use land.
- Discuss environmental problems created because of land use.
- List things you can do to help protect the environment.

Farming

Earth's total land area is 145 million square km. We use 15 million square km as farmland. Even so, about 20 percent of the people living in the world are hungry. Millions starve to death each year. To meet this need, farmers work to increase the productivity of croplands by using higher-yield seeds and chemical fertilizers. Herbicides and pesticides are also used to reduce weeds and insects that can damage crops.

Poor use of these chemicals, however, can contaminate the environment. For this reason, some farmers rely on organic farming techniques to lessen the environmental impact on the land. **Figure 19-5** shows an organic farm in China that has been farmed for 20 centuries.

Figure 19-5

Organic farming techniques can rebuild topsoil, rather than deplete it. Organic farmers use natural fertilizers, crop rotations, and biological pest controls to help their crops thrive.

Whenever vegetation is removed from an area, such as in construction sites or in agricultural tilling, the soil can become easily eroded. With plants gone, there is nothing to prevent the soil from being carried away by running water and wind. Several centimeters of topsoil may be lost in one year. In some places, it can take more than 1000 years for new topsoil to develop and replace the eroded topsoil. **Figure 19-6** shows one way farmers work to reduce this type of erosion.

Figure 19-6

Contour plowing is one way farmers reduce erosion. Water follows the contour of the land, along the plowed rows, instead of running straight downslope.

Grazing Livestock

Land is also used for grazing livestock. Animals such as cattle eat vegetation and are then often used as food for humans. In the United States, the majority of land used for this purpose is unsuitable for crops. However, about 20 percent of the cropland in our country is used to grow feed for livestock.

A square kilometer of vegetable crops can feed many more people than a square kilometer used to raise livestock. Some people argue that a more efficient use of the land would be to grow crops directly for human consumption, rather than for livestock. For many, however, meat and dairy products are an important part of their diet. They also argue that livestock have ecological benefits that justify livestock production.

Figure 19-7

Corn is used for human food, for livestock feed, and for industrial products, such as ceramics, textiles, ethanol, and paint. About half the corn grain raised in the United States is fed to livestock.

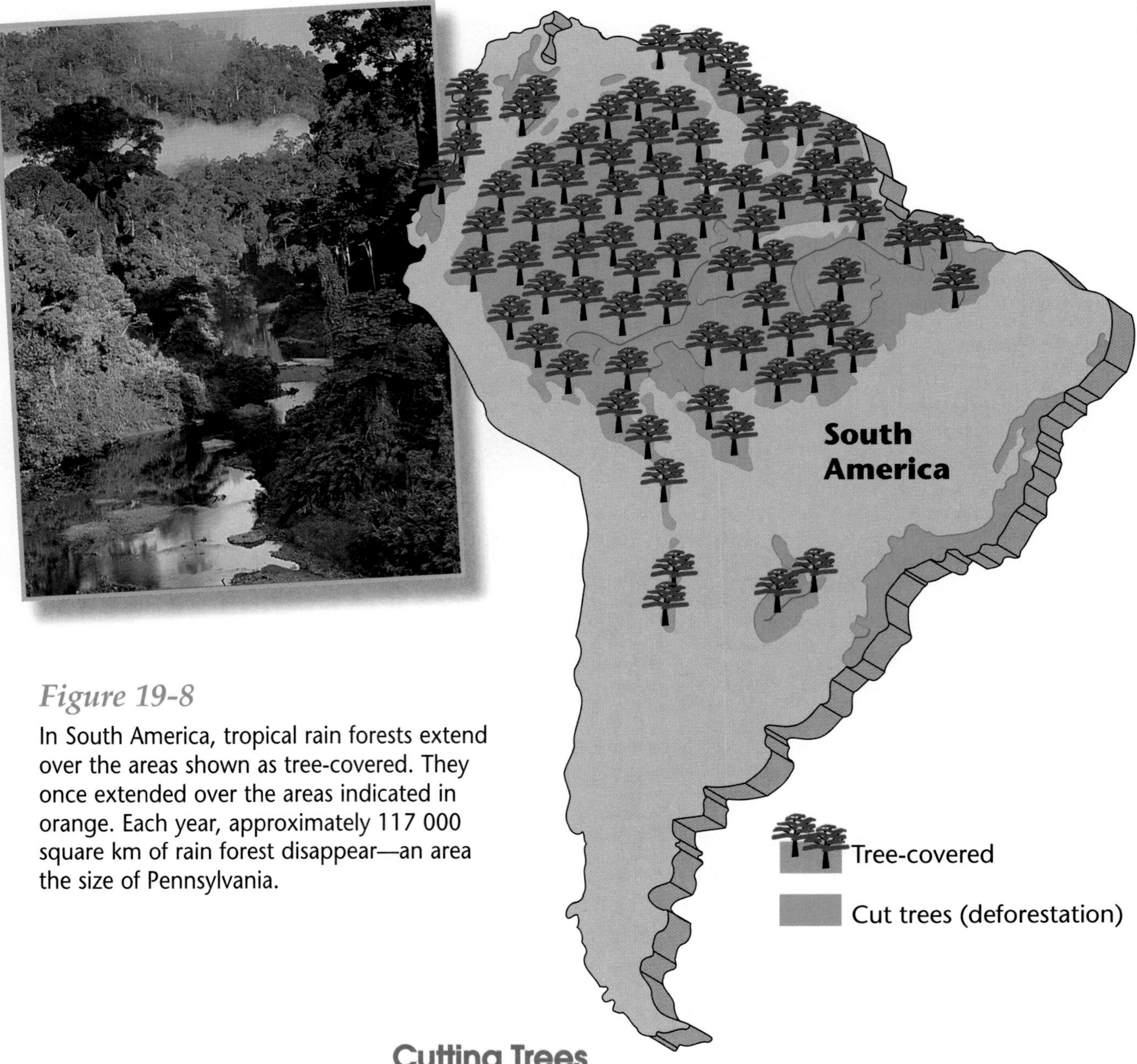

Figure 19-8

In South America, tropical rain forests extend over the areas shown as tree-covered. They once extended over the areas indicated in orange. Each year, approximately 117 000 square km of rain forest disappear—an area the size of Pennsylvania.

Cutting Trees

Some land is used as a source of wood. Trees are cut down and used for lumber, fuel, and paper. Often, new trees are planted to take their places. In some cases, especially in the tropical regions shown in **Figure 19-8,** whole forests are cut down without being replaced. Each year an area of rain forest the size of Pennsylvania disappears. It is difficult to estimate, but evidence suggests that worldwide up to 50 000 species may become extinct each year due to loss of habitat.

Organisms living outside of the tropics also suffer because of the lost vegetation. Plants remove carbon dioxide from the air, and when they photosynthesize, they produce oxygen that organisms need to breathe. Reduced vegetation also results in higher levels of carbon dioxide in the atmosphere. Carbon dioxide is one of the gases that contribute to the greenhouse effect.

Landfills

Land is also used when we dispose of the products we consume. Eighty percent of our garbage goes into landfills. A **landfill** is an area where waste is deposited. In a **sanitary landfill,** such as the one illustrated in **Figure 19-9,** each day's deposit is covered with dirt. The dirt prevents it from blowing away, and it reduces the odor produced by the decaying waste. Sanitary landfills are also designed to prevent liquid wastes from draining into the soil and groundwater below. A sanitary landfill is lined with plastic or concrete, or it's located in clay-rich soils that trap the liquid waste.

Hazardous Wastes

Some of the wastes we put into landfills are dangerous to organisms. Poisonous, cancer-causing, or radioactive wastes are called **hazardous wastes.** Hazardous wastes are put into landfills by industries. But individuals contribute to the problem, too, when they throw away insect sprays, batteries, drain cleaners, bleaches, medicines, and paints.

Sanitary landfills greatly reduce the chance that hazardous substances will leak into the surrounding soil and groundwater. However, some still find their way into the environment.

Another problem is that we're filling up our landfills and running out of acceptable areas to build new ones. Many materials placed into landfills decompose slowly.

It may seem that when we throw something in the garbage can, it is gone and we don't need to be concerned with it anymore. But our garbage does not disappear. It can simply remain in the landfill for hundreds of years. In the case of radioactive waste, it may be around for thousands of years, creating problems for future generations. Fortunately, industries and people are becoming more aware of the problems associated with landfills, and are disposing of their wastes in a more responsible manner. Later in this chapter, you will learn about what you can do personally to help reduce these problems.

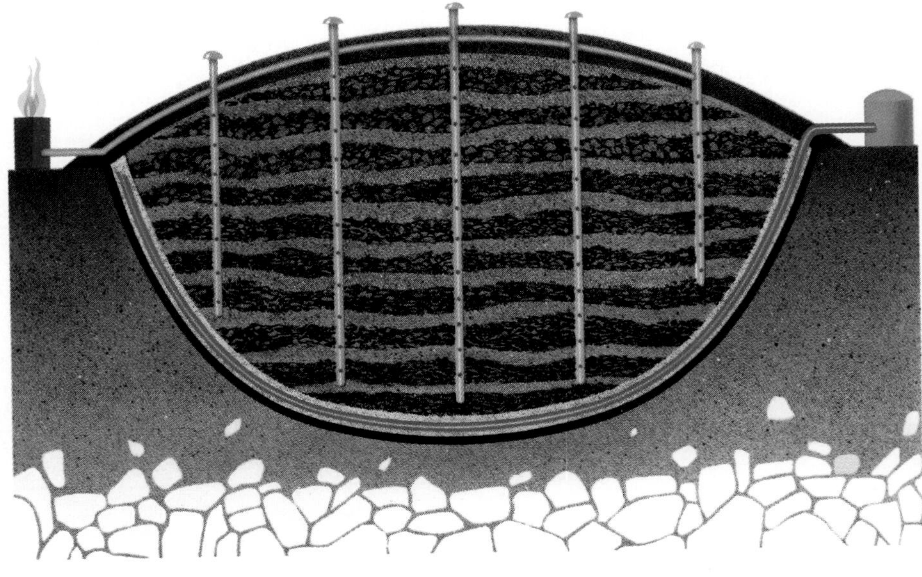

Figure 19-9

The vast majority of our garbage is deposited in landfills. *In your Science Journal, list some problems associated with landfill disposal.*

MiniLAB

How much trash do you generate in one day?

Determine the total mass and volume of your trash.

Procedure

1. Carry a garbage bag around with you all day, and place all of your trash, including food wastes, into it.
2. At the end of the day, determine the mass and volume of the contents.

Analysis

1. Calculate the percentage of the mass and volume of each of the following wastes: plastics, metals, food wastes, paper, and glass.
2. Compare your mass percentages with those of your classmates.
3. Hypothesize how you can reduce the amount of trash you generate each day.

Where does our trash go?

In the early days of the United States, population was more sparse and few people considered the impact of their actions on the environment. They threw their trash into rivers, buried it, or burned it. Given our increased population and output of trash, such methods are no longer feasible. Use the graph below to solve the following problem.

Solve the Problem:

1. **Since 1960, what three things have we done with most of our trash? Where does most of our trash go?**
2. **In your Science Journal, describe how our use of landfills, incinerators, and recycling has changed since 1960.**

Think Critically:

More than half of the states in our country are running out of landfill space. New landfills will have to be made, but most people have a NIMBY attitude. NIMBY means "not in my backyard." People don't want to live near a landfill.

1. What should we do about our trash problem?
2. What are some ways that we can generate less trash?

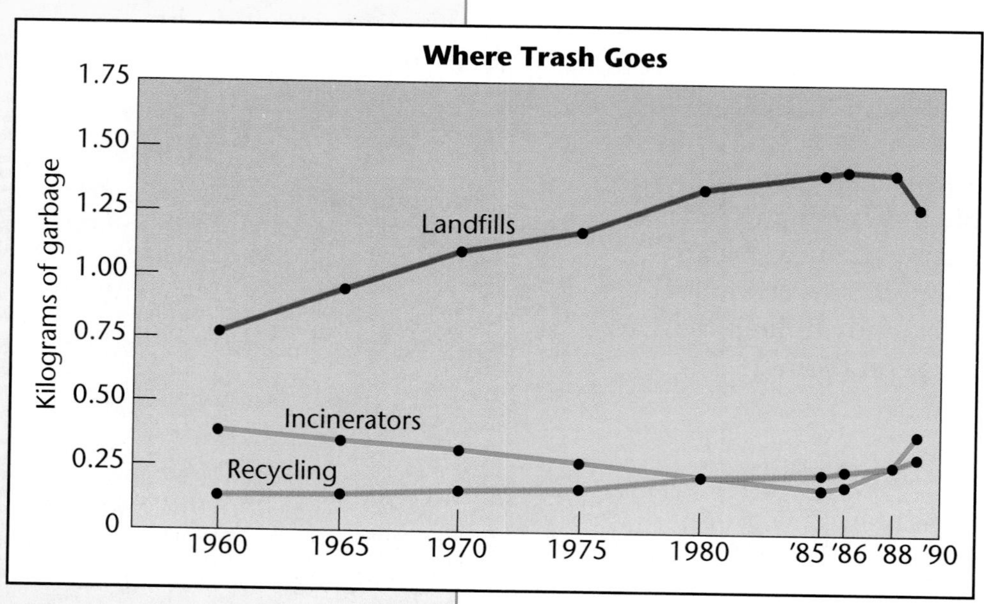

Structures, Mines, and Natural Environments

Concrete and asphalt are quickly replacing grass and woodlands in our communities. The impact on the environment, particularly in urban areas such as the one shown in **Figure 19-10,** is easy to observe. Asphalt and concrete absorb a lot of solar radiation. The atmosphere is then heated by conduction and convection, which causes the air temperature to rise. You may have observed this if you've ever traveled from a rural area to the city and noticed a rise in temperature.

Another effect of paving over the land is that less water is able to soak into the soil. Instead, it runs off into sewers or streams. During heavy rainstorms in urban areas, sewer pipes can overflow or become clogged with debris. Some of the water that does not soak into the soil evaporates. This reduces the amount of water in groundwater aquifers. Many communities rely on groundwater for drinking

Figure 19-10

Our cities are covered by concrete sidewalks and paved streets. *How does this impact the environment?*

water. But covering more and more of the land with roads, sidewalks, and parking lots prevents the water from reaching aquifers. In previous chapters, you've read about mining and how it can adversely affect the environment. As more people populate Earth, increased demands for fossil fuels and mineral ores will result in more mining operations.

Natural Preserves

Not all land on Earth is being used to produce usable materials or for storing waste. Look at **Figure 19-11.** Some land remains uninhabited by people. National parks in the United States are protected from much of the pollution and destruction that you've read about in this section. In many countries throughout the world, land is set aside as natural preserves. As the world population continues to rise, the strain on our environment is likely to increase. Let's hope that we will be able to continue preserving some land as natural environments.

MiniLAB

Can one person make a difference?

Find out what you can do to help conserve resources and protect the environment.

Procedure

1. Based on what you have learned in this chapter, select at least one way to help the environment.
2. As a class, agree upon a length of time in which to complete your tasks.
3. Report back to the class on what you did to help the environment.

Analysis

1. Do you think your actions made a difference? Explain.
2. Hypothesize what changes might occur if each person in the United States did his or her part to save planet Earth.

Figure 19-11

Many countries set aside land as natural preserves. This means that the land will not be used to produce goods, but will remain in its natural state.

USING TECHNOLOGY

Recycling Paper

What happens to the newspaper you recycle? After the paper is taken to a plant where it is to be recycled, the paper is sorted and put into a device called a pulper. The pulper contains water and other substances that remove ink from the paper. The paper then becomes part of a soggy mixture called pulp.

Pulp is run through a machine that removes any solid objects such as rubber bands, paper clips, or staples that may have been fastened to the paper. Another device then squeezes the water from the paper mixture. The final stage of processing includes sifting and washing the pulp to remove any unwanted debris.

Water is added to the clean pulp to make a thick, pasty substance. This substance is rolled into thin sheets and dried to form sheets of recycled paper.

Think Critically:

If 17 trees are saved when one metric ton of paper is recycled, how many trees could be saved in a year if a community recycled 150 000 kg of paper each month?

Recycling Paper Saves Trees

Conserving Resources

In the United States and other industrialized countries, people have a throwaway lifestyle. When we are done with something, we throw it away. This means more products have to be produced to replace what we've thrown away, more land is used, and landfills overflow. You can help by conserving resources. **Conservation** is the careful use of resources to reduce damage to the environment.

Reduce, Reuse, Recycle

The United States makes up only 5 percent of the world's human population, yet it consumes 25 percent of the world's natural resources. Each of us can reduce our consumption of materials in simple ways, such as using both sides of notebook paper or carrying lunch to school in a nondisposable container. Ways to conserve resources include reducing our use of materials and reusing and recycling materials.

Reusing an item means finding another use for it instead of throwing it away. You can reuse old clothes by giving them to someone else, or by cutting them up into rags. The rags can be used in place of paper towels for cleaning jobs around your home.

Reusing plastic and paper bags is another way to reduce waste. Some grocery stores even pay a few cents when you return and reuse paper grocery bags. Out-of-doors, there are things you can do, too. If you cut grass or rake leaves, you can compost these items instead of putting them into the trash. **Composting** means piling yard wastes where they can gradually decompose. The decomposed matter can be used in gardens or flower beds to fertilize the soil. Some cities no longer pick up yard waste to take to the landfills. In these places, composting is common. If everyone in the United States composted, it would reduce the trash put into landfills by 20 percent.

USING MATH

You use 16 L of water if you let the water run while brushing your teeth. Only 2 L are used if the water is turned off until it's time to rinse. How much water could you save per year by not letting the water run? How much water could be saved by the entire U.S. population (260 million people)?

The Population Outlook

The human population explosion has already had devastating effects on the environment and the organisms that inhabit Earth. It's unlikely that the population will begin to decline in the near future. To compensate, we must use our resources wisely. Conserving resources by reducing, reusing, and recycling is an important way that you can make a difference.

Section Wrap-up

Review

1. In your Science Journal, list six ways that we use land.
2. Discuss environmental problems that are sometimes created by agriculture, mining, and trash disposal.
3. **Think Critically:** Choose one of the following items, and list three ways it can be reused: an empty milk carton, vegetable scraps, used notebook paper, or an old automobile tire. Be sure your uses are environmentally friendly.

Skill Builder

Recognizing Cause and Effect

Review Section 19-2 and answer the questions below. If you need help, refer to Recognizing Cause and Effect in the **Skill Handbook.**

1. How many uses of land are described in this section?
2. How can farming negatively impact the environment?
3. How can grazing livestock negatively impact it?

Using Computers

Word Processing

Suppose that a new landfill is needed in your community. Where do you think it should be located? Now, suppose that you want to convince people that you've selected the best place for the landfill. Use your word processing skills to write a letter to the editor of the local newspaper, listing reasons in favor of your choice.

SCIENCE & SOCIETY

19•3 ISSUE: Should recycling be required?

Science Words

recyclable

Objectives

- List the advantages of recycling.
- List the advantages and disadvantages of mandatory recycling.
- Express your feelings about whether recycling should be required.

Recycling

Did you know that any object is **recyclable** if it can be processed and used again? Look at **Figure 19-12A.** Paper and aluminum are two of the many things that can be recycled.

Paper makes up about 40 percent of the mass of our trash. If you recycle it, you save landfill space and trees. Also, the production of recycled paper takes 61 percent less water and generates 71 percent fewer air pollutants than the production of brand-new paper made from trees.

How much energy do you think is saved when you recycle one aluminum can? Answer: enough to keep a TV running for about three hours. Twenty aluminum cans can be recycled with the energy needed to produce a single brand-new can from ore.

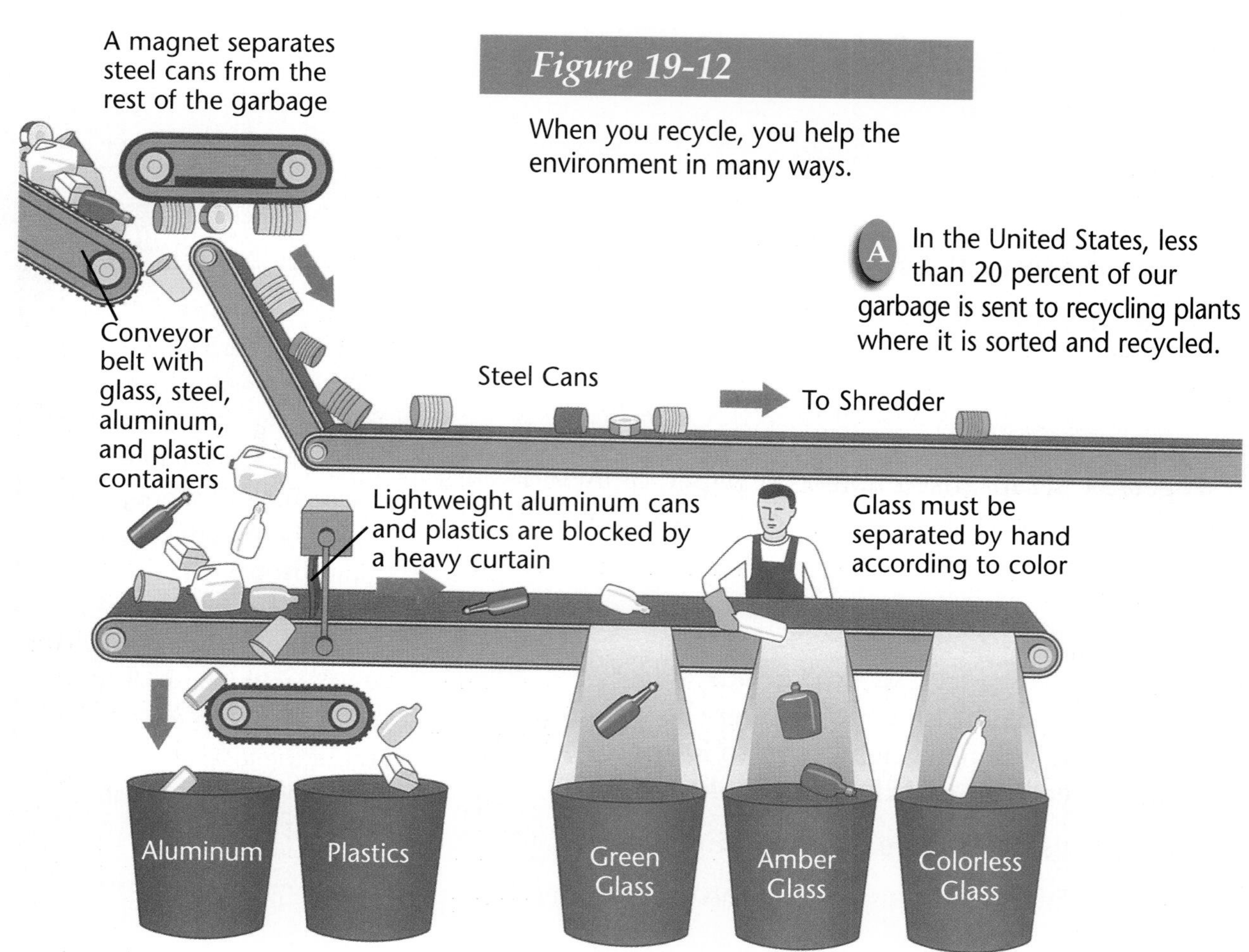

Figure 19-12

When you recycle, you help the environment in many ways.

A In the United States, less than 20 percent of our garbage is sent to recycling plants where it is sorted and recycled.

B Recycling plastics, aluminum, and glass saves landfill space, energy, and natural resources.

2 Points of View

Everyone agrees that recycling is good for the environment. It saves landfill space, energy, and natural resources. Recycling also helps reduce the damage caused by mining, cutting trees, and manufacturing. Did you know that if you recycle, you will reduce the trash you generate in your lifetime by 60 percent? If you don't recycle, you'll generate trash equal to at least 600 times your mass. But the question is, should recycling be required?

Mandatory Recycling

Many things aren't recycled because some people haven't gotten into the habit of recycling. In the United States, much less garbage is recycled than in countries with mandatory recycling, such as Japan and Germany. Mandatory recycling requires people to participate and creates new jobs in "reuse" industries, such as the production of items from recycled plastics.

C Many civic groups "adopt" sections of a highway. They pick up trash and recycle the salvageable part. *In your Science Journal, write a letter to your local chamber of commerce suggesting ways to beautify your community.*

In 1993, 17 states already had some form of recycling laws. People in these states comply with the laws because they benefit directly in some way. For example, in some places people who recycle pay lower trash-collection fees. In other places, garbage is not collected if it contains items that should have been recycled.

Some states have container laws whereby a five-cent refundable deposit is made on all beverage containers. This means paying five cents extra at the store for a drink. But you get your nickel back if you return the container to the store for recycling.

There are those who believe that if we had a national container law, we could save enough energy to light up a large city for four years.

CONNECT TO

PHYSICS

Iron and aluminum have different properties. Using **Figure 19-12A**, ***determine*** which property allows recycling centers to separate iron from aluminum. (Hint: Steel is made from iron.)

Voluntary Recycling

Today, many people already recycle voluntarily because their cities provide curbside collection or convenient drop-off facilities. In this case, people have the freedom to decide whether to recycle or not, without government intervention.

Some people will argue that the cost of recycling outweighs the benefits. Recycling requires money to pay for workers, trucks, and buildings. Also, some workers, such as miners and manufacturers who make brand-new containers, might lose their jobs.

Another problem is what to do with all of the recyclable items. Recycling businesses have to make a profit or they can't exist. The only way to make a profit in recycling is to sell the recycled material. There must be a market for the material. People or businesses must want to purchase the recycled products. In some cities, from time to time, old newspapers can't be recycled because there is no market for them. New paper is cheaper than recycled paper.

Section Wrap-up

Review

1. List at least four advantages of recycling.
2. What are the advantages and disadvantages of mandatory recycling?

Explore the Issue

Do you think the government should pass laws requiring people to recycle?

Activity 19-2

A Model Landfill

When garbage is put into landfills, it is buried under other trash and soil; therefore, it isn't exposed to sunlight and other things that help decomposition. When examined by a researcher, one landfill was found to contain grass clippings that were still green, and bread that had not molded.

Problem

At what rates do different materials decompose in a landfill?

Materials

- two 2-L bottles
- thermometer
- graph paper
- soil
- plastic wrap
- rubber band
- trash (including fruit and vegetable scraps, a plastic item, a metal item, a foam cup, and notebook paper or newsprint)

Procedure

1. Cut off the top of two 2-L bottles.
2. Add soil to each bottle until it is half-filled.
3. On graph paper, trace the outline of all the garbage items that you will place into each bottle. Label each outline. Keep the graph paper as a record of the original sizes of the items.
4. Place the items, one at a time, in each bottle. Completely cover each item with soil.
5. Add water to the "landfill" until the soil is slightly moist. Place a thermometer in each bottle and seal the bottle with the plastic wrap and a rubber band. Store one bottle at a cold temperature and place the other on a shelf.
6. Check the temperature of the "landfill" on the shelf each day for two weeks. Record the temperatures in a data table that you design.

7. After two weeks, remove all of the items from the soil. Trace the outlines of each on a new sheet of graph paper. Compare the sizes of the items with their original sizes.
8. Wash your hands thoroughly after cleaning up your lab space. Be sure to properly dispose of each item as instructed by your teacher.

Analyze

1. Which items decomposed the most? Which showed the least decomposition?
2. Most decomposition in a landfill is due to the activity of microorganisms. The organisms can live only under certain temperature and moisture conditions. Why was it necessary to add moisture to the soil? **Explain** how the decomposition rates would have differed if the soil had been completely dry.
3. **Compare** your results with the results of the bottle that was stored at a cold temperature. **Explain** the differences you observe.

Conclude and Apply

4. Why do some items decompose more rapidly than others?
5. What problems are created in landfills by plastics?

People and Science

LEIGH STANDINGBEAR, Horticulturist, City of Tulsa

On the Job

Q Describe how you transformed a city park into a habitat for native waterfowl.

A Swan Lake Park, one of Tulsa's most popular parks, has a two-acre lake right in the middle of an urban neighborhood. One of the many native plants we used is button bush—a woody shrub that provides roosting places for birds and cascades out over the water to provide good cover. Rushes and sedges do a very good job of holding the soil together at water's edge to prevent erosion.

Q You've also created ethnobotanical gardens, which display plants that were important to two groups of Native American people. Tell us about them.

A In the Osage Precolumbian Garden, one of the plants on display is the Osage orange tree. The strong, beautiful wood from that tree was used to make bows. The Creek Nation Council Oak Park garden, right on the edge of downtown Tulsa, has plants that were used by the Creeks for food, fiber, medicinal and ceremonial purposes in their ancestral homelands in Alabama.

Personal Insights

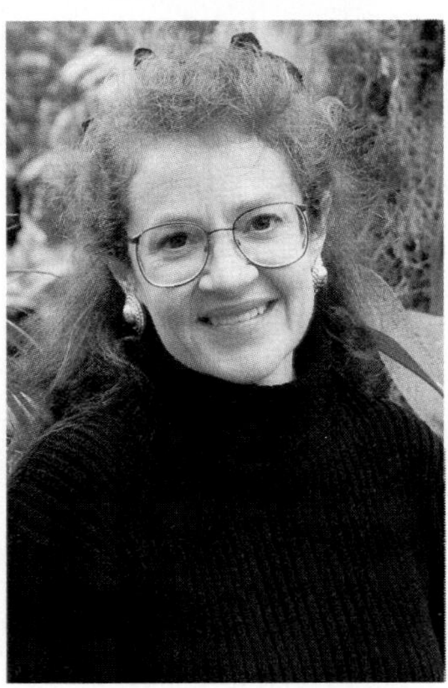

Q You feel a personal responsibility to care for the land. What in your background has led you to that feeling?

A Although I wasn't around at the time, I know from Oklahoma history that the famous Dust Bowl changed the thinking, technology, and methodology used in agriculture to prevent erosion and to protect marshlands and grasslands. Today, people are beginning to understand that protected places in urban settings are a valuable, important part of life.

Career Connection

Visit a park near your home or school, taking along a notebook or sketchpad. List or sketch five ways the park could be improved to make it a better place for plants, animals, and people.

- **Biologist**
- **Ethnobotanist**

Chapter 19 Review

Summary

19-1: Population Impact on the Environment

1. The rapid increase in Earth's human population in recent years is due to an increase in the birthrate, advances in medicine, better sanitation, and better nutrition.
2. People in industrial nations strongly impact the environment when they use electricity, burn fossil fuels, contaminate water, and use food that's been grown with pesticides and herbicides.

19-2: Using the Land

1. Land is used for farming, grazing livestock, cutting trees, and mining coal and mineral ores. We also build structures and landfills on the land. Some land is preserved as natural environments.
2. Land becomes polluted by hazardous wastes thrown away by industries and individuals. Fertilizers and pesticides used by farmers pollute groundwater and soil. When trees are cut down, soil is destroyed and more carbon dioxide remains in the atmosphere.
3. Recycling, reducing, and reusing materials are important ways we can conserve natural resources.

19-3: Science and Society: Should recycling be required?

1. Recycling saves energy, natural resources, and much-needed space in landfills.
2. If recycling were required, more people would recycle, new jobs would be created, and additional landfill space and energy would be saved. But there would be government control instead of free choice. Also, recycling is expensive, some people would lose their jobs, and we would have some recyclable items that no one wants.

Key Science Words

a. carrying capacity
b. composting
c. conservation
d. hazardous waste
e. landfill
f. population
g. population explosion
h. recyclable
i. sanitary landfill

Reviewing Vocabulary

Match each phrase with the correct term from the list of Key Science Words.

1. total number of individuals of a particular species in an area
2. describes the rapid increase in birthrate and decrease in death rate
3. area used to deposit garbage
4. trash that's dangerous to organisms
5. piling up organic material to decompose
6. careful use of resources
7. area lined with plastic, concrete, or clay where garbage is dumped
8. items that can be processed and used again
9. maximum number of individuals of a particular type that the planet will support

Chapter 19 Review

Checking Concepts

Choose the word or phrase that completes the sentence.

1. Most of the trash in the United States is disposed of in _______.
   a. recycling centers c. hazardous waste sites
   b. landfills d. old mine shafts
2. Between 1960 and 1995, world population increased by _______ billion people.
   a. 5.7 c. 1.0
   b. 3.2 d. 2.7
3. The United States uses about _______ percent of Earth's resources.
   a. 5 c. 25
   b. 10 d. 50
4. About _______ of U.S. land is used to graze livestock.
   a. 100 percent c. 50 percent
   b. one percent d. 25 percent
5. When an object is _______, it can be used again.
   a. trash c. disposable
   b. recyclable d. hazardous
6. _______ makes up about 40 percent of the mass of our trash.
   a. Glass c. Yard waste
   b. Aluminum d. Paper
7. In a _______, trash is covered with soil.
   a. recycling center c. sanitary landfill
   b. surface mine d. coal mine
8. _______ of people starve to death each year.
   a. Hundreds c. Millions
   b. Thousands d. Billions
9. Organisms living outside the tropics suffer when rain forests are cut down because the trees are no longer there to produce _______.
   a. carbon dioxide c. water
   b. methane d. oxygen
10. An example of a hazardous waste is a _______.
    a. piece of glass c. steel can
    b. plastic jug d. can of paint

Understanding Concepts

Answer the following questions in your Science Journal using complete sentences.

11. On average, how many people are added to Earth's population each hour?
12. Why is raising vegetable crops more efficient than raising livestock?

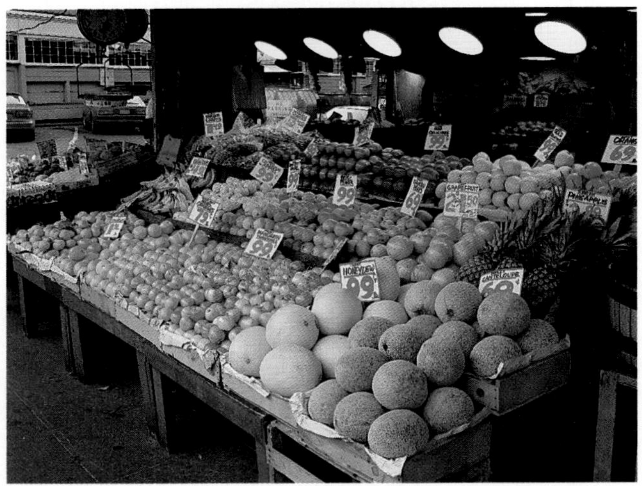

13. Discuss why recycling paper is good for the environment.

Thinking Critically

14. A ten-minute shower uses about 190 L of water. If a person takes one shower a day and reduces the time to five minutes, how much water would be saved in a year?
15. Renewable resources are those resources that can be replenished by nature in the foreseeable future. Nonrenewable resources cannot. Which kind of resource is oxygen? Explain.
16. Although land is farmable in many developing countries, hunger is a major problem in many of these countries. Give some reasons why this might be so.
17. Forests in Germany are dying due to acid rain. What effects might this loss have on the environment?
18. Describe how you could encourage your neighbors to recycle their newspapers.

Chapter 19 Review

Developing Skills

If you need help, refer to the **Skill Handbook.**

19. **Recognizing Cause and Effect:** Suppose a city decided to pave over a large, grassy lot. What effects will this have on the local environment?
20. **Measuring in SI:** If each person in your class produced 47 174 kg of trash in a lifetime, how much trash would be produced by this small population?
21. **Classifying:** Analyze the garbage you throw away in one day. Classify each piece of garbage.
22. **Making and Using Graphs:** In a population of snails, each snail produces two offspring each month. Each offspring also produces two offspring. Using the graph below, determine how many snails would be present after five months if the initial population was only two snails.
23. **Interpreting Scientific Illustrations:** Why does the curve of the line graph in Question 22 change its slope over time? Suppose half of the snails died after six months. Draw a new graph to illustrate the effect.

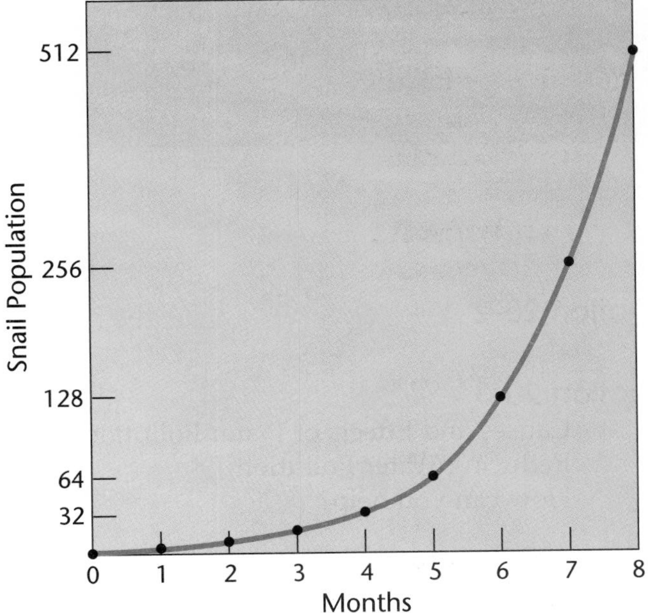

Performance Assessment

1. **Using Math in Science:** Collect your family's junk mail for one week. Determine the mass of the accumulated mail. Divide the mass by the number of persons living in your home. Using this number as an average for all people living in the U.S. (260 million), calculate the total mass of junk mail received each week in the U.S. Calculate the amount per year. If 17 trees are killed to make each metric ton of paper, calculate how many trees are killed each year to make junk mail.
2. **Evaluating a Hypothesis:** Design an experiment to determine the factors that decrease the time it takes for newspapers or yard wastes to decompose.

Previewing the Chapter

Chapter 20

Our Impact on Air and Water

Did you know that most sources of drinking water in the United States are polluted? The water has to be treated with chemicals or filters before it's safe to use. Air pollution is another problem. Factories such as the paper pulp mill shown here release dust and chemicals into the air. In the United States, 12 percent of all toxic chemicals we produce end up in the atmosphere. Is the air in your community polluted?

EXPLORE ACTIVITY

Study your air.

1. Find a high shelf or the top of a tall cabinet—someplace that hasn't been cleaned for a while.
2. Run a cloth over this area and observe the dirt under a microscope.
3. In your Science Journal, describe and sketch what you see.

Observe: Where did these particles come from? What do you suppose happens when you breathe in these particles?

Previewing Science Skills

- In the **Skill Builders,** you will map concepts and interpret scientific illustrations.
- In the **Activities,** you will collect and analyze data, make and use a data table, make and use a graph, calculate, infer, and draw conclusions.
- In the **MiniLABs,** you will experiment, observe, compare and contrast, and infer.

20•1 Air Pollution

Science Words

photochemical smog
sulfurous smog
acid rain
pH scale
acid
base
Clean Air Act

Objectives

- Identify the sources of pollutants that cause photochemical smog, sulfurous smog, holes in the ozone layer, and acid rain.
- Describe how air pollution affects people and the environment.
- Explain how air pollution can be reduced.

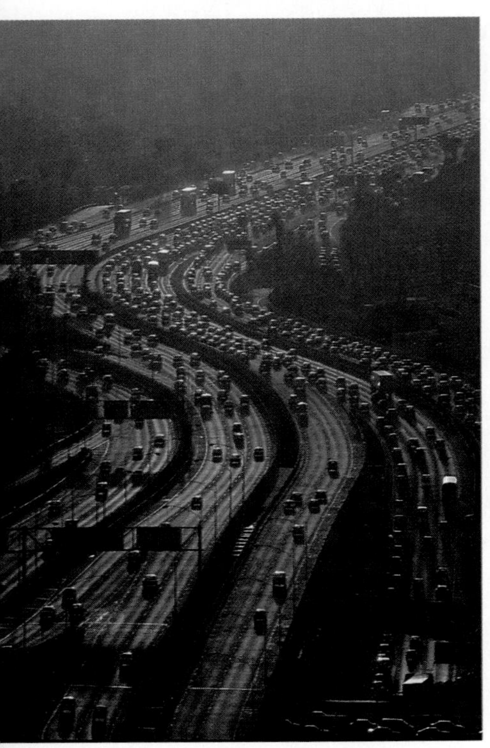

What causes air pollution?

Have you ever noticed that the air looks hazy on some days? Do you know what causes this haziness? Some industries generate dust and chemicals. Thus, the more industries there are in a region, the more dust and chemicals there are in the air. Other human activities add pollutants to the air, too. Look at **Figure 20-1.** Cars, buses, trucks, trains, and planes all burn fossil fuels for energy. Their exhaust—the waste products from burning the fossil fuels—adds polluting chemicals to the air. Other sources include smoke from burning trash and dust from plowed fields, construction sites, and mines.

Natural sources add pollutants to the air, too. Examples are volcanic eruptions, forest fires, and grass fires.

Smog

Pollutants produced by human activities and by natural processes cause the haze you sometimes see in the air. **Figure 20-2** shows the major sources of air pollution. Around cities, polluted air is called smog, a word made by combining the words *smoke* and *fog*. Two types of smog are common—photochemical smog and sulfurous smog.

Photochemical Smog

In areas such as Los Angeles, Denver, and New York City, a hazy, brown blanket of smog is created when sunlight reacts with pollutants in the air. This brown smog is called **photochemical smog** because it forms with the aid of light. The pollutants get into the air when fossil fuels are burned. Coal, natural gas, and gasoline are burned by factories, airplanes, and cars. Burning fossil fuels causes nitrogen and oxygen to chemically combine to form nitrogen compounds. These compounds react in the presence of sunlight and produce other substances. One of the substances produced is ozone. Recall that ozone in the stratosphere protects us from the sun's ultraviolet radiation. But ozone that forms in smog near Earth's surface causes health problems.

Figure 20-1

Cars are one of the main sources of air pollution in the United States. *In your Science Journal, list five other sources of air pollution.*

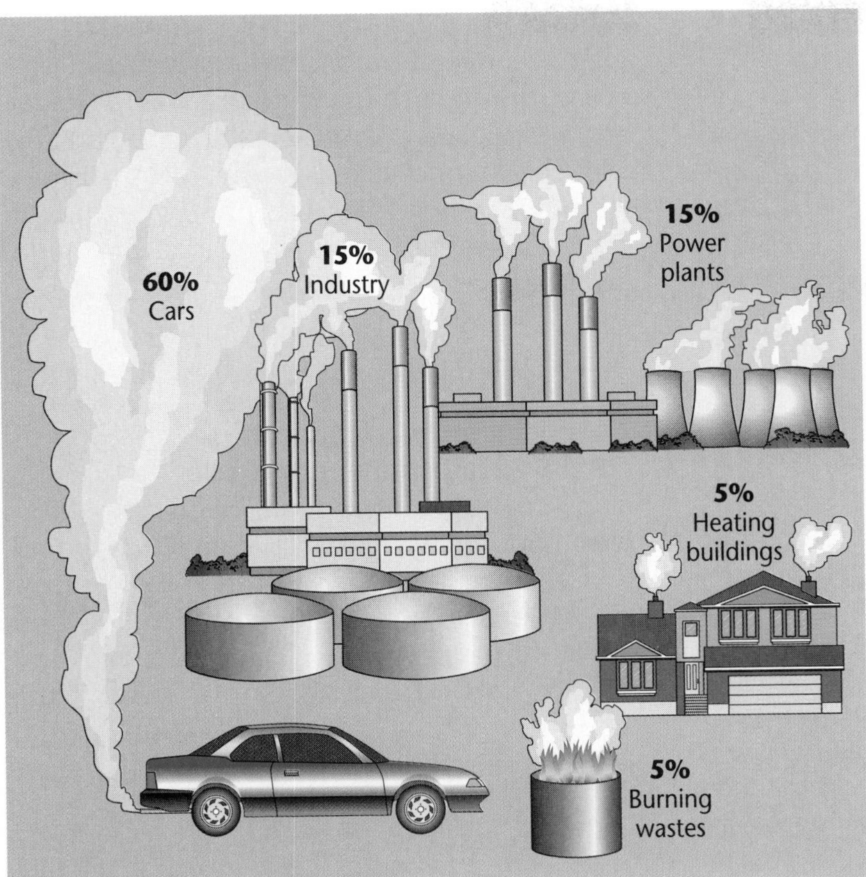

Figure 20-2

Proportions of smog caused by the major sources of human-created air pollution.

Sulfurous Smog

A second type of smog is called **sulfurous smog.** It's created when fossil fuels are burned in electrical power plants and home furnaces. The burning releases sulfur compounds, dust, and smoke particles into the air. Sulfurous smog forms when these substances collect in an area where there's little or no wind. A blanket of gray smog may hang over a city for several days and be hazardous to breathe.

Nature and Smog

Nature plays an important role in creating smog. Sunlight helps form photochemical smog. Sulfurous smog forms when weather systems are calm and the air is not being moved around. Also, sometimes a layer of warm air lies on top of cooler air. Normally, the warmer air is near Earth's surface. But in cases where warm air is above, it becomes a barrier that prevents cool air under it from rising. As a result, pollutants are trapped near the ground. Eventually, the weather changes and cleaner air is blown in, dispersing the pollutants in the air.

Landforms also affect smog development. For example, mountains may help cause smog by restricting the movement of air. Los Angeles' problem with smog is made worse by nearby mountains. Also, some cities are in valleys, where dense, dirty air tends to collect.

MiniLAB

Do you have acid rain in your community?

Procedure

1. The next time it rains or snows, use a glass or plastic container to collect a sample of the precipitation.
2. Use pH paper to determine the acidity level of your sample. If you have collected snow, melt it before measuring its pH.
3. Record the indicated pH of your sample and compare it with the results of other classmates who have followed the same procedure.

Analysis

1. What is the average pH of the samples obtained from this precipitation?
2. Compare and contrast the level of acidity in your samples with those of the substances shown on the pH scale in **Figure 20-4.**

Figure 20-3

These trees are dying because acids in the soil have removed essential nutrients and lowered the trees' resistance to diseases, insects, and bad weather. Acid rain also increases the acidity of streams, rivers, and lakes, killing fish that live in the water. It even damages the surfaces of buildings and cars.

Smog isn't the only air pollution problem we have. Recall from Chapters 14 and 16 that chlorofluorocarbons from air conditioners and refrigerators are thought to be destroying the ozone layer in the stratosphere. Carbon dioxide from burning coal, oil, natural gas, and forests is increasing the greenhouse effect.

Acid Rain

Another major pollution problem is acid rain. Acid rain is created when sulfur dioxide from coal-burning power plants combines with moisture in the air to form sulfuric acid. It is also created when nitrogen oxides from car exhausts combine with moisture in the air to form nitric acid. The acidic moisture falls to Earth as rain or snow. We call this **acid rain.**

What would happen if you watered a plant with lemon juice instead of water? As you might guess, the plant would die. Lemon juice is an acid that is toxic to tender shoots and roots. Acid rain sometimes is as acidic as lemon juice. Thus, acid rain can kill plants and trees, as in **Figure 20-3.** To understand this, let's learn how acidity is measured.

The pH Scale

We describe how acidic a solution is by using the **pH scale.** **Figure 20-4** illustrates the pH scale. Substances with a pH lower than seven are considered **acids.** The lower the pH number, the greater the acidity. The higher the number, the lower the acidity. Substances with a pH above seven are considered **bases.**

Figure 20-4

The natural pH of rainwater is about 5.6. Acid rain is precipitation with a pH below 5.6.

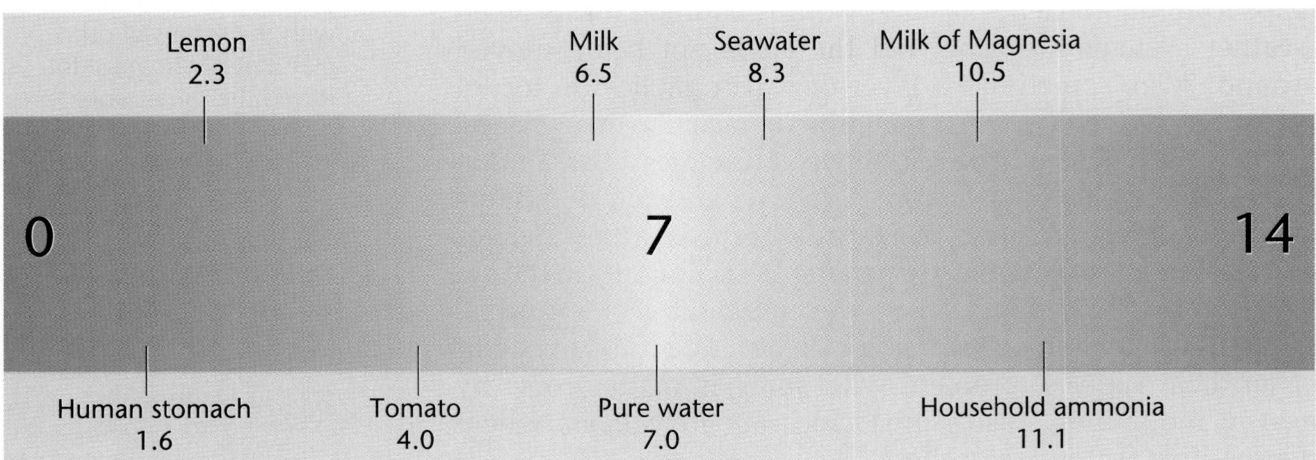

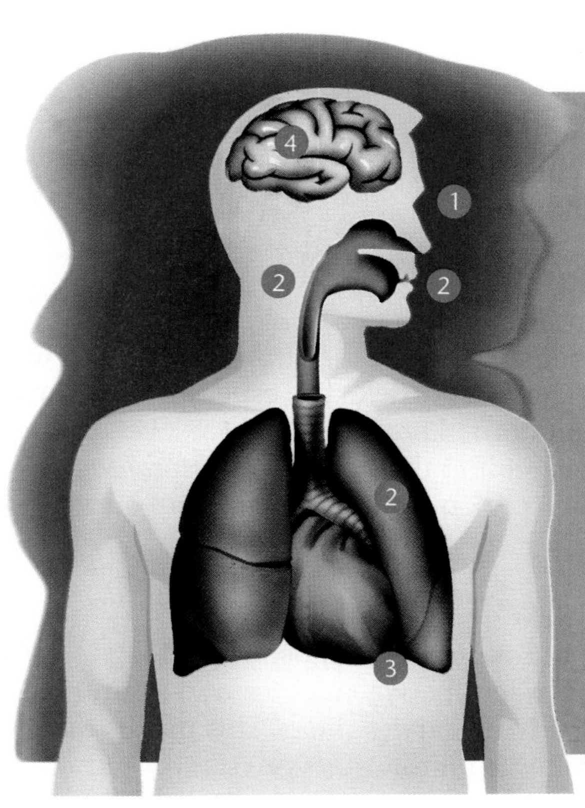

Effects of Air Pollution on the Body

1 Eyes
Compounds found in smog cause the eyes to water and sting. If conditions are bad enough, vision may be blurred.

2 Nose, throat, and lungs
Ozone irritates the nose and throat, causing burning. It reduces the ability of the lungs to fight infections.

3 Heart
Inhaled carbon monoxide is absorbed by red blood cells, rendering them incapable of transporting oxygen throughout the body. Chest pains result because of low oxygen levels.

4 Brain
Motor functions and coordination are impaired because oxygen levels in the brain are reduced when carbon monoxide is inhaled.

How Air Pollution Affects Our Health

Suppose you're an athlete in a large city and you're training for a big, upcoming competition. You have to get up at 4:30 A.M. to exercise. Later in the day, the smog levels will be so high that it won't be safe for you to do strenuous exercise. In southern California, in Denver, and in other areas, athletes adjust their training schedules to avoid exposure to ozone and other smog. Schools schedule football games for Saturday afternoons when smog levels are low. Parents are warned to keep their children indoors when smog levels exceed certain levels. Breathing dirty air, especially taking deep breaths of it when you are actively exercising, can cause health problems.

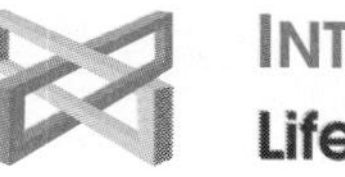

INTEGRATION
Life Science

Health Disorders

How hazardous is dirty air? Approximately 250 000 people in the United States suffer from pollution-related breathing disorders. About 60 000 deaths each year in the U.S. are blamed on air pollution. Ozone damages lung tissue, making people more susceptible to diseases such as pneumonia and asthma. Less severe symptoms of breathing ozone include a stinging chest, burning eyes, dry throat, and headache.

Carbon monoxide also contributes to smog. A colorless, odorless gas, carbon monoxide makes people ill, even in small concentrations.

The pH scale is a logarithmic scale. This means that there is a tenfold difference for each pH unit. For example, pH 4 is ten times more acidic than pH 5 and 100 times more acidic than pH 6. ***Calculate*** how much more acidic pH 1 is than pH 4.

CONNECT TO

LIFE SCIENCE

Air pollution causes many people to develop bronchitis. What organs are inflamed when a person has bronchitis? ***Describe*** the symptoms of bronchitis.

What do you suppose happens when you inhale the humid air from acid rain? Acid is deposited deep inside your lungs. This causes irritation, reduces your ability to fight respiratory infections, interferes with oxygen absorption, and puts stress on your heart.

Particulate Pollution

Particulate pollution also harms people. Particulate pollution is caused by airborne solids that range from large grains to microscopic particles. The fine particles are especially dangerous because they disrupt normal breathing and can even cause lung disease when they get into the lungs. Some of these particles are produced when coal, oil, and petroleum are burned.

Reducing Air Pollution

Pollutants moving through the atmosphere don't stop when they reach the borders between states and countries. They float wherever the wind carries them. This makes them very difficult to control. Even if one state or country reduces its air pollution, pollutants from another state or country can blow across the border.

When states and nations cooperate, pollution problems can be reduced. Diplomats from around the world have met on several occasions since 1990 to try to eliminate some kinds of air pollution. Of particular concern are chlorofluorocarbons and carbon dioxide.

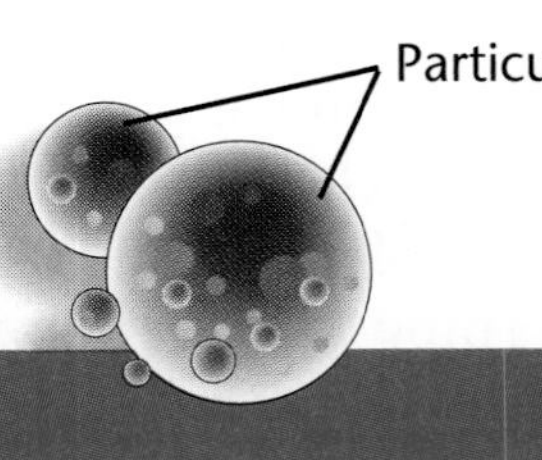

Figure 20-5

Particulate pollution is caused by solids in the air that are produced, in part, by burning fossil fuels. *Why is particulate pollution difficult to control?*

Table 20-1

Clean Air Regulations

Smog	Acid Rain	Airborne Toxins

Car emissions of nitrogen oxides had to be reduced by 60 percent and hydrocarbons by 35 percent of 1990 levels in all new cars as of 1996.	"Clean-coal" technologies must reduce sulfur dioxide emissions by 14 million tons by the year 2000 and nitrogen oxides by several million tons immediately.	Starting in 1995, industries had to limit the emission of 200 compounds that cause cancer and birth defects.

Air Pollution in the United States

The Congress of the United States has passed several laws to protect the air. The 1990 **Clean Air Act** attacked the problems of smog, chlorofluorocarbons, particulates, and acid rain by regulating car manufacturers, coal technologies, and other industries. In **Table 20-1,** you can read about some of these regulations.

The good news is that since the passage of the Clean Air Act, the quality of the air in some regions of the United States has improved. The bad news is that one of four U.S. citizens still breathes unhealthy air.

It is the role of the federal Environmental Protection Agency to monitor progress toward the goals of the Clean Air Act. However, it is the consumers who must pay increased prices and taxes, and change their habits in order to really help protect the environment. The Clean Air Act can work only if we all cooperate. In Chapter 19 you discovered several ways you can conserve energy and reduce trash. When you do these things, you are also reducing air pollution. We all must do our share to clean up the air.

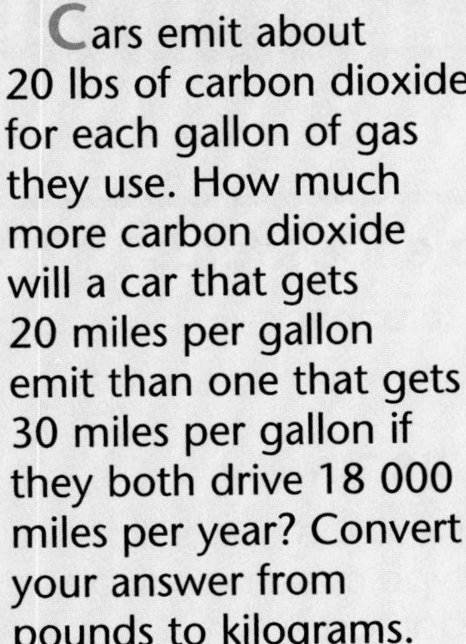

USING MATH

Cars emit about 20 lbs of carbon dioxide for each gallon of gas they use. How much more carbon dioxide will a car that gets 20 miles per gallon emit than one that gets 30 miles per gallon if they both drive 18 000 miles per year? Convert your answer from pounds to kilograms.

Section Wrap-up

Review

1. In what ways does air pollution affect the health of people?
2. How can you help reduce air pollution?
3. **Think Critically:** An earlier Clean Air Act, passed in 1970, required that coal-burning power plants use very tall smokestacks so that air pollutants would be ejected high into the sky, where high-altitude winds would disperse them. Power plants in the Midwestern states complied with that law, and people in eastern Canada began complaining about acid rain. Explain the connection.

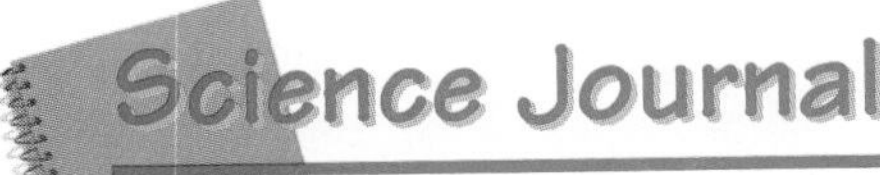

In your Science Journal, make a crossword puzzle using at least 12 important terms found in this section.

Make a concept map that explains how sulfurous smog forms and how weather affects how long it persists. If you need help, refer to Concept Mapping in the **Skill Handbook.**

20 • 2 TECHNOLOGY: Acid Rain

Science Words

scrubber

Objectives

- Describe how soil type affects acid rain damage.
- Describe activities that help reduce acid rain.
- Decide who should pay the cost of reducing sulfur dioxide emissions.

Acid Rain Damage

You may live in an area where there is very little acid rain. The amount of acid rain in an area depends on the number of factories and cars in the area. They are the sources of the sulfur and nitrogen gases that cause acid rain. Whether acid rain falls where you live also depends on your area's pattern of precipitation and wind direction. Even if industry and cars are producing a lot of pollution near you, much of it may be carried away by the wind before it falls back down as acid rain.

Soil Type Affects Acid Rain Damage

It's not just the volume of acid rain that falls that determines how much damage is done in an area. Different types of soil react differently to acid rain. Some soils already are acidic, and acid rain makes them even more so. But other soils are basic. If you mix an acidic solution with a basic solution, they neutralize each other. Therefore, if acid rain falls on basic soil, the acid rain becomes partly neutralized and causes less damage.

Soils in the midwestern states are basic. Thus, when acid rain falls in the Midwest, it's neutralized in the soil. But the northeastern states and eastern Canada don't have basic soils. In these places, acid rain makes the ground and groundwater even more acidic, causing harm to plants and fish. **Figure 20-6** shows the pH levels of rainfall for the United States.

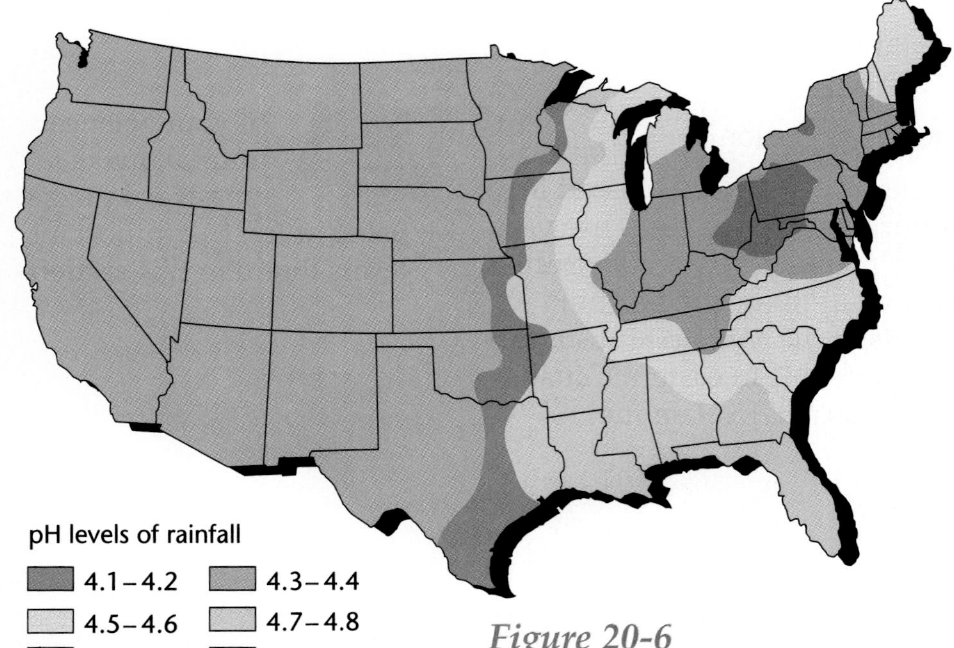

Figure 20-6

The pH levels of rainfall in the eastern states and Canada are dangerously low. *How does the soil type of this region affect acid rain damage?*

What can be done about acid rain?

The main source of the nitric acid in acid rain is car exhaust. Better emission control devices on cars will help reduce acid rain. So will car pooling and public transportation, because they reduce the number of trips and thus the amount of fuel used.

Reducing Emissions

Coal-burning power plants can help correct the problem, too. Some coal has a lot of sulfur in it. When the coal is burned, the sulfur combines with moisture in the air to form sulfuric acid. Power plants can wash coal to remove some sulfur before the coal is burned. This way, burning produces less sulfur in the smoke. Power plants also can run the smoke through a scrubber. A **scrubber** lets the gases in the smoke dissolve in water, as they would in nature, until the smoke's pH increases to a safe level.

USING TECHNOLOGY

Solar Grids

Some scientists think many of our energy and air pollution problems could be solved if we began collecting solar energy in space and beaming its power to Earth. One idea is to use huge orbiting grids of solar cells to collect the sun's power. These sunsats (sun satellites) would convert the solar energy into electricity. Then the electricity would be converted into microwaves that would be transmitted to Earth. Antennas on Earth would convert the microwaves into direct-current electricity, which could then be used for powering our factories, homes, and even cars.

Other scientists suggest that instead of using satellites, grids of solar cells could be placed on the moon. The energy would be collected and sent to Earth in the same way as with sunsats. There is a problem that needs to be worked out in both methods, however. Sending the energy to Earth in the form of microwaves could be dangerous for organisms living here.

Sunsats Beaming Power to Earth

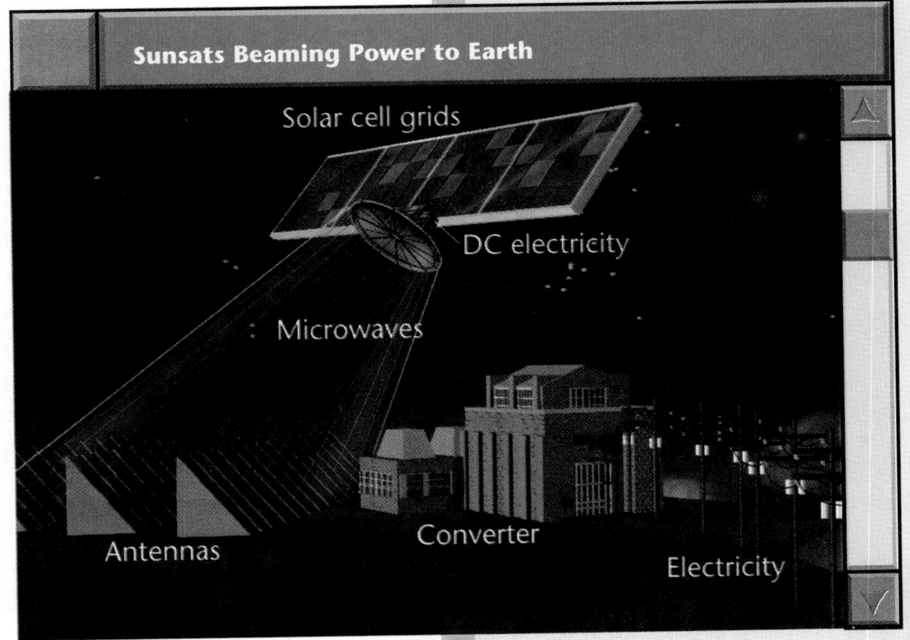

Think Critically: Once perfected, how would sunsats or grids of solar cells on the moon help reduce acid rain and other forms of air pollution?

Figure 20-7

This solar power plant in California provides a clean, renewable source of fuel. Other alternative fuel sources include nuclear, wind, and geothermal power.

Alternative Fuels

Another thing we could do to reduce air pollution is switch to other fuel sources such as solar, nuclear, and geothermal power. **Figure 20-7** shows a solar power plant in California that provides a nonpolluting source of fuel. However, these alternative fuel sources have disadvantages, too. And even if everyone agreed to make fuel changes, it would take years for some areas to change. This is because many people, especially in the midwestern states, depend on coal-burning power plants for home heating and electricity. Changing the kind of fuel used would be costly.

As you read in the last section, the 1990 Clean Air Act requires great reductions in auto exhaust and sulfur dioxide emissions. This will cost all of us billions of dollars. It especially affects the Midwest, where thousands of people will lose their jobs. These include miners, workers in coal-burning power plants, and people who work for factories.

Section Wrap-up

Review

1. How does soil type affect acid rain damage?
2. How can acid rain be reduced?

Explore the Technology

Much of the acid rain problem comes from coal-burning power plants in the Midwest. Some people think that these midwestern states should pay to clean up the problem. But others propose that everyone in the nation should pay for the cleanup. This would reduce the financial burden on the midwestern states. They also propose a job-protection program to help coal miners and others who might lose their jobs. Is this fair to everyone? Why or why not?

SCIENCE & SOCIETY

Science & History

The Aswan High Dam and Its Effects on Egypt's Soil and Water

For many centuries, the Nile River overflowed during each year's rainy season, leaving behind moist, fertile soil. This cycle of flooding usually meant that only one crop could be planted each year. Water was so scarce during the dry season that additional crops could not be grown.

The Aswan High Dam was built during the 1960s to control flooding along the portion of the Nile that flows through Egypt. During the rainy season, water is stored in a huge artificial lake, or reservoir, called Lake Nasser. The dam provides water to irrigate close to a million hectares of land year round. The stored water also operates a hydroelectric power plant that generates much of Egypt's electricity. However, the Aswan High Dam has also created some unexpected pollution problems.

No More Natural Fertilizer

The silt left behind when the Nile was allowed to flood was a natural fertilizer. It replaced soil nutrients used up by crops grown the previous year and replaced eroded topsoil. When the dam was built, flooding stopped and silt was no longer deposited. Farmers began using chemical fertilizers. Over the years, valuable topsoil has been lost to erosion.

Too Much Salt

Irrigation water from Lake Nasser contains small amounts of salts that are left behind in the soil as the water evaporates. The salt levels have become so high in some fields that they can no longer be used to grow crops. In addition, wells and streams are polluted by runoff containing residues of fertilizers, pesticides, and salts.

An Increase in Disease

Many disease-carrying organisms, such as water snails, used to die in the annual floods of the Nile. Now, the calm waters of Lake Nasser serve as a perfect breeding ground for some of these organisms, and more people are becoming ill from diseases such as dysentery.

Science Journal

People in many regions of the world, including Africa, India, and Asia, have soil and water pollution problems much like Egypt's. In your Science Journal, research possible solutions for these problems.

Activity 20-1

Design Your Own Experiment

What's in the air?

Have you ever gotten a particle of dust in your eye? Before it got into your eye, it was one of the many pieces of particulate matter in the air. In the Explore Activity at the beginning of the chapter, you observed some of the dust particles that settled out of the air. Just imagine how many pieces of particulate matter the air must hold.

PREPARATION

Problem

What kinds of particulate matter are in your environment? Are some areas of your environment more polluted with particulates than others?

Form a Hypothesis

Based on the observations you made in the Explore Activity on page 555 and your knowledge of your neighborhood, hypothesize what kinds of particulate matter you will find in your environment. Will all areas in your community contain the same types and amounts of particulate matter?

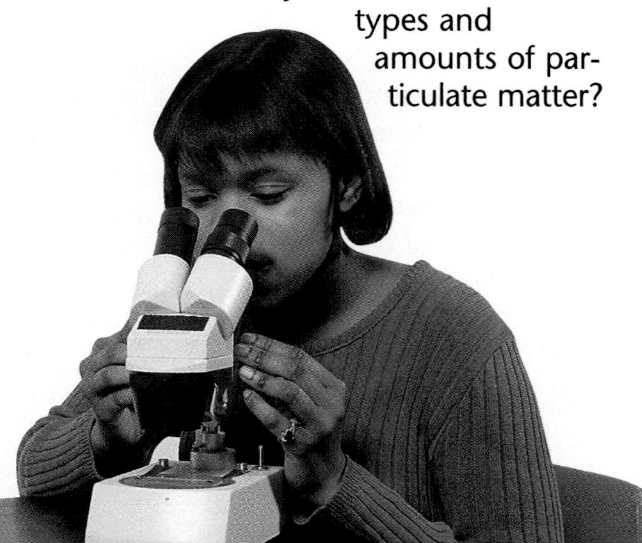

Objectives

- Design an experiment to collect and analyze particulate matter in the air in your community.
- Observe and describe the particulate matter you collect.

Possible Materials

- small box of plain gelatin
- hot plate
- pan or pot
- water
- safety goggles
- thermal mitt
- marker
- refrigerator
- plastic lids (4)
- stereomicroscope

Safety Precautions

Wear a thermal mitt and safety goggles while working with a hot plate and while pouring the gelatin from the pan or pot into the lids.

PLAN THE EXPERIMENT

1. As a group, agree upon and write out your hypotheses statements.
2. As a group, list the steps you need to take to test your hypotheses. Be specific, describing exactly what you will do at each step.
3. List your materials.
4. Design a data table in your Science Journal so that it is ready to use as your group collects data.
5. Mix the gelatin according to the directions on the box. Carefully pour a thin layer of gelatin into each lid. Use this to collect air particulate matter.
6. Decide where you will place each plastic lid in order to collect particulate matter in the air.

Check the Plan

1. Read over your entire experiment to make sure that all steps are in a logical order.
2. Identify any constants, variables, and controls of the experiment.
3. *Make sure your teacher approves your plan before you proceed.*

DO THE EXPERIMENT

1. Carry out the experiment as planned.
2. While the experiment is going on, write down any observations that you make and complete the data table in your Science Journal.

Analyze and Apply

1. **Describe** the types of materials you collected in each lid.
2. **Graph** your results using a bar graph. Place the number of particulates on the *y*-axis and the test-site location on the *x*-axis.
3. Why do you think some lids had more particles and larger particles than the rest?
4. Which of the materials that you collected might be the result of human activities?

Go Further

Do you think any of the materials you collected might be harmful to humans? Explain your answer. How does your body filter solid particles from the air you breathe?

20•3 Water Pollution

Science Words

Safe Drinking Water Act
Clean Water Act

Objectives

- List five water pollutants and their sources.
- Describe ways that international agreements and U.S. laws are designed to reduce water pollution.
- Relate ways that you can help reduce water pollution.

Causes and Effects of Water Pollution

Suppose you were hiking along a stream or lake and became very thirsty. Do you think it would be safe to drink the water? In most cases, it wouldn't. Many streams and lakes in the United States are quite polluted, such as the one shown in **Figure 20-8A.** Even streams that look clear and sparkling are not safe for drinking.

You learned in Chapter 18 how pollutants get into the oceans. Groundwater, streams, and lakes are polluted by similar sources. There is strong reason to believe that these pollutants cause health problems such as cancer, dysentery, birth defects, and liver damage in humans and other animals.

Pollution Sources

How do you think pollutants get into the water? Bacteria and viruses get into the water because some cities illegally dump raw sewage directly into the water supply. Underground septic tanks can leak, too. Radioactive materials can get into the water from leaks at nuclear power plants and radioactive waste dumps.

Pesticides, herbicides, and fertilizers from farms and lawns are picked up by rainwater and carried into streams. Some people dump motor oil into sewers after they've changed the oil in their cars. Water running through mines also carries pollutants to streams and underground aquifers. Industrial

Figure 20-8

Clean water is one of our most precious resources. Before we can use water for our daily activities, we must purify it. After we use it, we must clean it again before returning it to the environment.

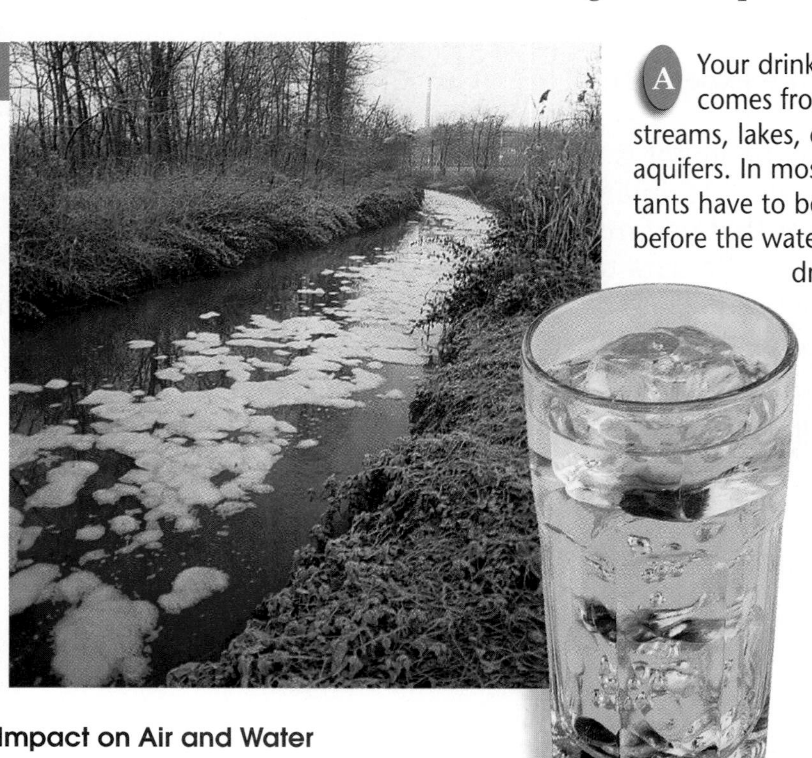

A Your drinking water comes from nearby streams, lakes, or underground aquifers. In most cases, pollutants have to be removed before the water is safe to drink. *Name some sources of water pollution.*

B The plant known as water hyacinth removes many pollutants from water. Water hyacinths can be used at wastewater treatment facilities to clean up water before it flows back into streams and aquifers.

chemicals get into the water because some factories illegally dump toxic materials directly into the water. Waste from landfills and hazardous waste dumps leaks into the surrounding soil and groundwater.

But most water pollution is caused by legal, everyday activities. Water is polluted when we flush our toilets, wash our hands, brush our teeth, and water our lawns. Nitrogen and phosphorus are difficult to remove from sewage and wastewater. Large amounts of these elements are released into rivers and streams from household detergents, soaps, and other cleaning agents. Water is also polluted when oil and gasoline run off of pavement, down storm sewers, and into streams.

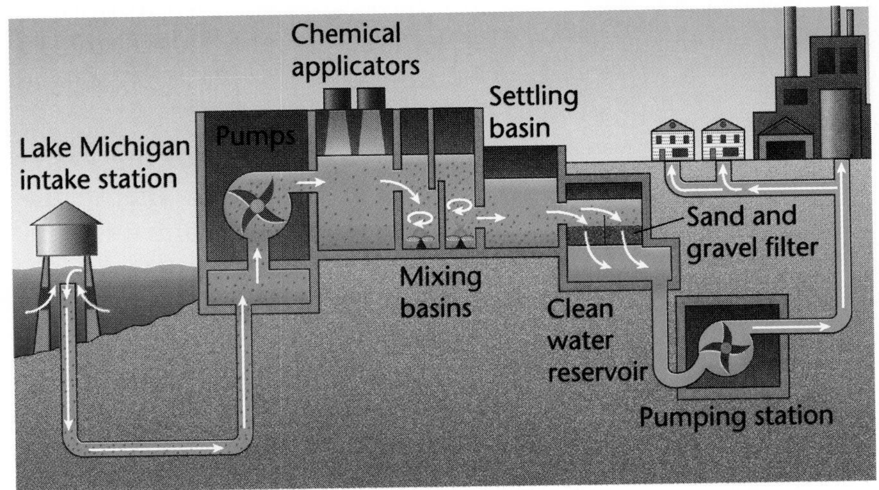

C This water purification plant in Chicago provides drinking water for millions of people. Water taken from Lake Michigan is pumped into a tank where alum, chlorine, lime, and other compounds are added to kill microorganisms. It is thoroughly mixed and the large particles of matter settle out. Some smaller particles are filtered by sand and gravel. Clean water is then pumped to consumers.

Problem Solving

What's happening to the fish?

Suppose you've been fishing the same pond since you were only seven years old. You used to catch a lot of fish there. But lately, you haven't been able to catch many fish at all, and those that you do catch seem sick. You also notice that the algae layer covering the surface of the water seems to be getting thicker each year. You remember that high levels of phosphates and nitrates in water can cause an increase in algae populations. You test the water and find that the dissolved oxygen content is very low.

Solve the Problem:

1. **Why might you suspect that crop fertilizers or sewage could be polluting the pond?**
2. **What could be the sources of these pollutants?**

Think Critically:

1. As the algae populations die, what happens to the population of decomposers in the water?
2. If the decomposers use up oxygen in the water, what happens to the fish population?
3. If the pond is being polluted by phosphates and nitrates, what could be done to decrease the pollution?

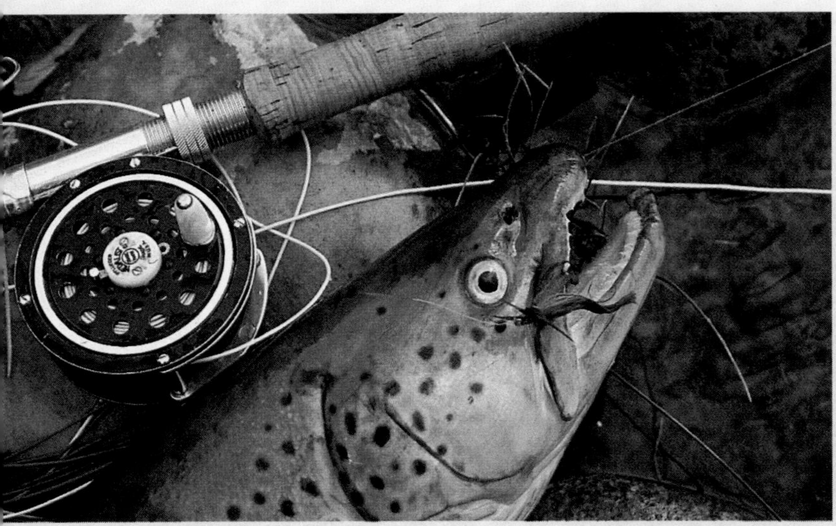

Reducing Water Pollution

Several countries have worked together to reduce water pollution. Let's look at one example. Lake Erie is on the border between the United States and Canada. In the 1960s, Lake Erie was so polluted by phosphorus from sewage, soaps, and fertilizers that it was turning into a green, soupy mess. Large areas of the lake bottom had no oxygen and, therefore, no life.

International Cooperation

In the 1970s, the United States and Canada made two water quality agreements. The two countries spent $15 billion to stop the sewage problem. Today, the green slime is gone, and the fish are back. However, more than 300 human-made chemicals can still be found in Lake Erie, and some of them are very hazardous. The United States and Canada now are studying ways to get them out of the lake.

Pollution Control in the United States

The U.S. Congress also has reduced water pollution by passing several laws. Two important laws are the 1986 Safe Drinking Water Act and the 1987 Clean Water Act.

The 1986 **Safe Drinking Water Act** is a law to ensure that drinking water in our country is safe. This act has not been very successful. Today, there are 200 000 public water systems in the United States, and many still have not met the standards set by the 1986 Safe Drinking Water Act.

Figure 20-9

Sewage and industrial wastes pollute many rivers and lakes. However, these sources of pollution can be controlled, and the water made safe for recreational purposes. *Explain how the 1987 Clean Water Act encourages communities to clean up their water.*

Clean Water Act

The 1987 **Clean Water Act** gives money to the states for building sewage and wastewater treatment facilities. The money also is for controlling runoff from streets, mines, and farms. Runoff caused up to half of the water pollution in the United States before 1987. This act also requires states to develop quality standards for all their streams.

The U.S. Environmental Protection Agency (EPA) makes sure that cities comply with both the Safe Drinking Water Act and the Clean Water Act. Most cities and states are working hard to clean up their water. Look at **Figure 20-9.** Many streams that once were heavily polluted by sewage and industrial wastes are now safe for swimming and fishing.

However, there is still much to be done. For example, the EPA discovered that 30 percent of the nation's rivers, 42 percent of its lakes, and 32 percent of its estuaries are still polluted because of silt and nutrients from farm runoff and municipal sewage overflows from cities. An estimated 40 percent of the country's freshwater supply is still not usable because of pollution. While Americans have spent $260 billion in the past 20 years on sewage treatment plants, water treatment is still unsatisfactory in 1100 U.S. cities.

The average shower uses 19 L of water per minute. If you take a five-minute shower each day, how much water do you use in one year by showering?

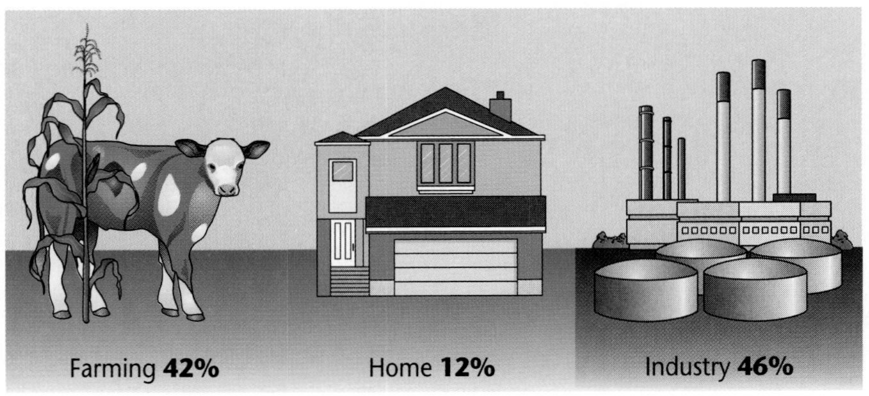

Figure 20-10

The industries that produce the products you use each day consume nearly half of the fresh water used in the United States.

MiniLAB

How hard is your water?

When minerals such as calcium carbonate, magnesium carbonate, or sulfates are dissolved in water, the water is said to be *hard.* This is a type of natural pollution that occurs in some areas.

Procedure

1. Use small containers such as baby food jars to collect samples of water from the tap, a nearby pond or well, and a local stream.
2. Add one drop of liquid soap to each jar.
3. Cap each container tightly and shake it rapidly.
4. Observe how many soapsuds are produced in each container.

Analysis

1. Which of your water samples had the hardest water? (Hint: The container with the most suds contains the softest water.)
2. Infer what problems might be caused by having a hard water supply in your home or community.

How can you help?

As you discovered in Chapter 19, we are often the cause of our environmental problems. But we are also the solution. What can you do to help?

Safely Dispose of Wastes

If you dispose of household chemicals such as paint and motor oil, don't pour them down the drain or onto the ground. Also, don't put them out with your other trash.

Hazardous wastes poured directly onto the ground move through the soil and eventually reach the groundwater below. When you pour them down the drain, they flow through the sewer, through the wastewater treatment plant, and into wherever the wastewater is drained, usually into a stream. This is how rivers become polluted. If you put hazardous wastes out with the trash, they end up in landfills, where they may leak out.

What should you do with these wastes? First, read the label on the container for instructions on disposal. Don't throw the container into the trash if the label specifies a different method of disposal. Recycle if you can. Many cities have recycling facilities for metal, glass, and plastic containers. Store chemical wastes so that they can't leak. If you live in a city, call the sewage office, water office, or garbage disposal service and ask them how to safely dispose of the others.

Conserve Energy and Water

Another way you can reduce water pollution is to conserve energy. With less use of fuels, there will be less acid rain falling into forests and streams. And with less nuclear power, there will be a reduced risk of radioactive materials leaking into the environment.

Another way you can help is to conserve water. Look at **Figure 20-10.** How much water do you use every day? Think of all the ways you depend on water. How many times do you flush a toilet each day? How much water do you use every day for taking a bath or cleaning your clothes? How much water do you use when you wash dishes, wash a car, or use a hose or lawn sprinkler? Typical U.S. citizens like the one shown in **Figure 20-11** use about 397 L every day.

All of this water must be purified before it reaches your home. Then it must be treated as wastewater after you use it. It takes a lot of energy to treat this water and pump it to your home. Remember, when you use energy, you add to the pollution problem. Therefore, when you reduce the amount of water you use, you help prevent pollution.

Water pollution is everybody's problem, and we all must do our part to help reduce it. Consider some changes you can make in your life that will make a difference.

Figure 20-11

Simple things you can do to use less water include taking a shower instead of a bath and turning off the water while brushing your teeth.

Section Wrap-up

Review

1. What are three things you can do to help reduce water pollution?
2. How have the United States and Canada joined forces to clean up Lake Erie?
3. **Think Critically:** Southern Florida is home to nearly 5 million people, many dairy farms, and sugarcane fields. It is also the location of Everglades National Park—a shallow river system with highly polluted waters. What kinds of pollutants do you think are in the Everglades? How do you think they got there?

Skill Builder

Interpreting Scientific Illustrations

Use **Figure 20-8C** to answer the following questions: *Why are mixing basins needed? What's the purpose of the sand and gravel filter? Gravity forces the water through the system until it reaches the reservoir. Why is a pump needed after this point?* If you need help, refer to Interpreting Scientific Illustrations in the **Skill Handbook.**

Science Journal

In your Science Journal, make a list of things that will reduce the amount of water you use.

Activity 20-2

Water Use

How much water goes down the drain at your house? Did you know that up to 75 percent of the water used in homes is used in the bathroom? Flushing the toilet accounts for 50 percent of that water; the rest is for bathing, showering, washing hands, and brushing teeth. By learning to read a water meter, you can find out how much water your family uses.

Problem
How much water does your family use?

Materials
- home water meter

Background
There are several different types of water meters. Meter A has six dials. As water moves through the meter, the pointers on the dials rotate. To read a meter similar to A, find the dial with the lowest denomination indicated. The bottom dial is labeled 10. Record the last number that the pointer on that dial has passed. Continue this process for each dial. Meter A shows 28 853 gallons. Meter B is read like a digital watch. It indicates 1959.9 cubic feet. Meter C is similar to meter B, but indicates water use in cubic meters.

Procedure
1. Design a data table and record your home water meter reading at the same time of the day for eight days.
2. Subtract the previous day's reading to determine the amount of water used each day.
3. Record how much water is used in your home each day. Also, record the activities in your home that used water each day.
4. Plot your data on a graph like the one shown below. Label the vertical axis with the units used by your meter.

Analyze
1. During which day was the most water used? Why?
2. **Calculate** the total amount of water used by your family during the week.
3. **Calculate** the average amount of water each person used during the week by dividing the total amount of water used by the number of persons. **Calculate** a monthly average.

Conclude and Apply
4. Why is your answer to question 3 only an estimate of the amount of water used?
5. **Infer** how the time of year might affect the rate at which your family uses water.
6. What are some things your family could do to conserve water?

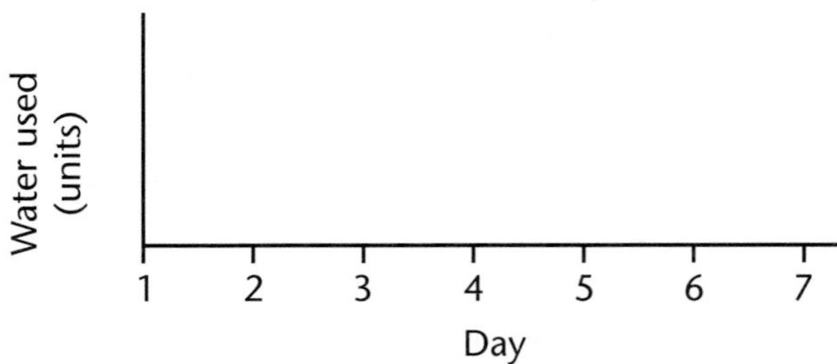

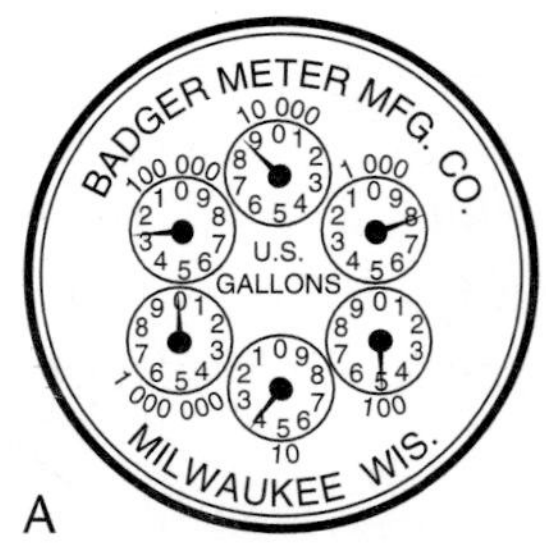

A

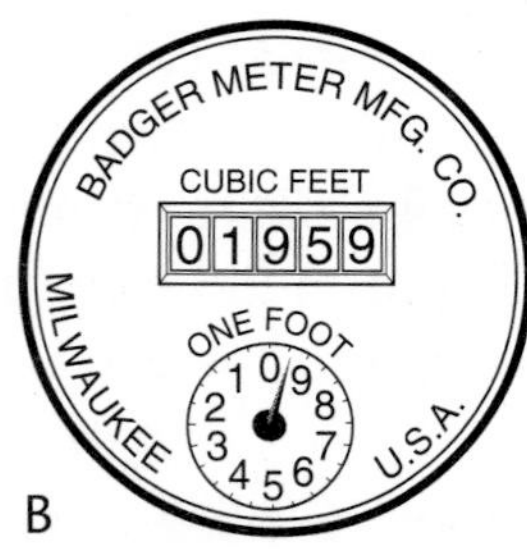

B

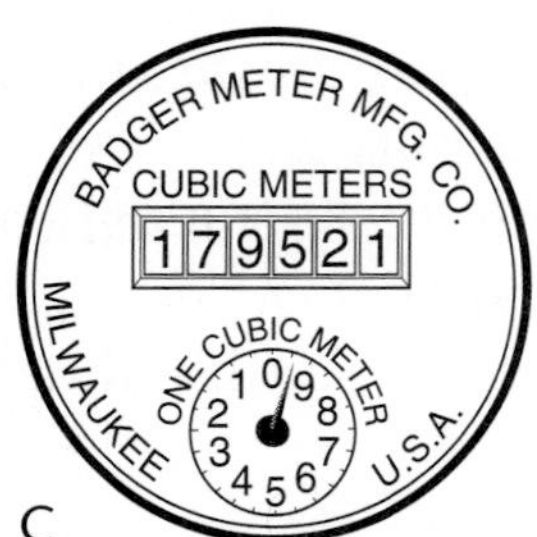

C

Chapter 20 Review

Summary

20-1: Air Pollution

1. Photochemical smog forms when fossil fuels are burned and the pollutants react with sunlight. Sulfurous smog is created when fuels are burned in electrical power plants and home furnaces.
2. Chlorofluorcarbons cause holes in the ozone layer. Acid rain is the result of burning coal and gasoline, which produce gases that react to form acids in the air.
3. Air pollution causes health problems in people and other organisms.
4. Recycling and conservation of Earth's resources reduce pollution.

20-2: Science and Society: Acid Rain

1. If soils are acidic, acid rain causes a lot of damage. Basic soils neutralize acid rain, resulting in less damage.
2. Better emission control devices on cars, car pooling, and the use of public transportation can help reduce acid rain. Washing coal that contains a lot of sulfur and using scrubbers can also reduce acid rain.
3. The cost of reducing sulfur dioxide emissions could be shared by everyone or could be limited to the areas that produce the emissions.

20-3: Water Pollution

1. Bacteria and viruses from animal wastes, radioactive waste, pesticides and herbicides used in agriculture, fossil fuels from mines and wells, and industrial chemicals from manufacturing processes all pollute freshwater bodies.
2. The passing of laws and the enforcement of these laws by the EPA and other agencies has helped to reduce water pollution.
3. Safely disposing of hazardous wastes is one way to reduce water pollution. Recycling and conservation are also important.

Key Science Words

a. acid rain
b. acid
c. base
d. Clean Air Act
e. Clean Water Act
f. pH scale
g. photochemical smog
h. Safe Drinking Water Act
i. scrubber
j. sulfurous smog

Reviewing Vocabulary

Match each phrase with the correct term from the list of Key Science Words.

1. smog that forms with the aid of light
2. smog that forms when pollutants mix with a layer of stagnant air
3. acidic rain, snow, sleet, or hail
4. law passed to protect air in the [illegible]
5. a measure of the acidity or [illegible] a solution
6. device that lowers [illegible] coal-burning [illegible]
7. law ensur[illegible]
8. law [illegible]
9. [illegible]

Chapter 20 Review

Checking Concepts

Choose the word or phrase that completes the sentence.

1. _______ cause more smog pollution than any other source.
   a. Power plants c. Industries
   b. Burning wastes d. Cars
2. _______ forms when chemicals react with sunlight.
   a. pH
   b. Photochemical smog
   c. Sulfurous smog
   d. Acid rain
3. A solution with a pH of 3 is more _______ than a solution with a pH of 4.
   a. neutral c. dense
   b. acidic d. basic
4. Acid rain can form when _______ combines with moisture in the air.
   a. ozone c. carbon dioxide
   b. sulfur dioxide d. oxygen
5. The industries that produce the products you use each day consume nearly _______ of the fresh water used in the United States.
   a. one-tenth c. one-third
   b. half d. two-thirds
6. One goal of the _______ Act is to reduce the level of car emissions.
   a. Clean Water c. Safe Drinking Water
   b. Clean Air d. Hazardous Waste
7. Acid rain has a pH of_______.
   a. less than 5.6 c. greater than 7.0
   b. between 5.6 & 7.0 d. greater than 9.5
8. The damage done by acid rain depends on _______.
   a. the pH scale c. soil type
   b. ozone emissions d. sunlight
9. Most water pollution is caused by _______.
   a. illegal dumping
   b. industrial chemicals
   c. everyday water use in the home
   [illegible]tewater treatment facilities
10. The _______ Act gives money to local governments to treat wastewater.
   a. Clean Water c. Safe Drinking Water
   b. Clean Air d. Hazardous Waste

Understanding Concepts

Answer the following questions in your Science Journal using complete sentences.

11. Compare and contrast photochemical smog and sulfurous smog.
12. Explain two ways in which acid rain can form.
13. Discuss the factors that affect the severity of the damage done by acid rain.
14. Describe natural sources of air pollution.
15. Compare and contrast the Safe Drinking Water Act with the Clean Water Act.

Thinking Critically

16. How might cities with smog problems lessen the dangers to people who live and work in the cities?
17. How are industries both helpful and harmful to humans?
18. How do trees help to reduce air pollution?
19. Thermal pollution occurs when heated water is dumped into a nearby body of water. What effects does this type of pollution have on organisms living in the water?
20. What steps might a community in a desert area take to cope with water supply problems?

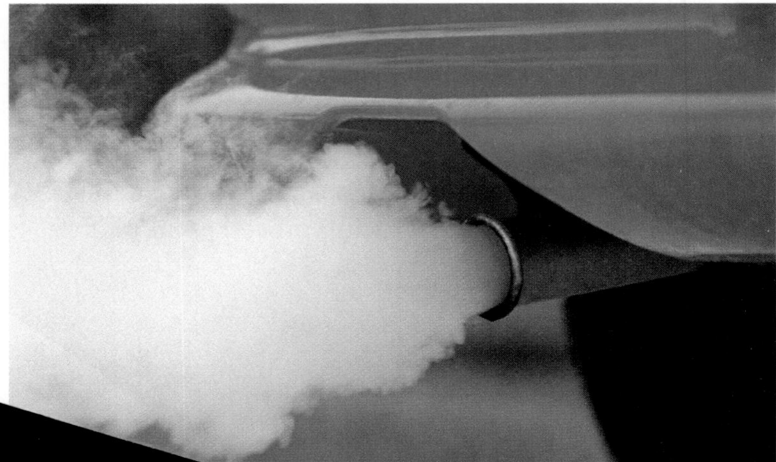

Developing Skills

If you need help, refer to the **Skill Handbook.**

21. **Hypothesizing:** Earth's surface is nearly 75 percent water. Yet much of this water is not available for many uses. Explain.
22. **Recognizing Cause and Effect:** What effect will an increase in the human population have on the need for fresh water?
23. **Classifying:** If a smog is brownish in color, is it sulfurous or photochemical smog?
24. **Concept Mapping:** Complete a concept map of the water cycle. Indicate how humans interrupt the cycle. Use the following phrases: *evaporation occurs, purified, drinking water, atmospheric water, wastewater, precipitation falls,* and *groundwater* or *surface water.*
25. **Outlining:** Make an outline that summarizes what you personally can do to reduce air pollution.

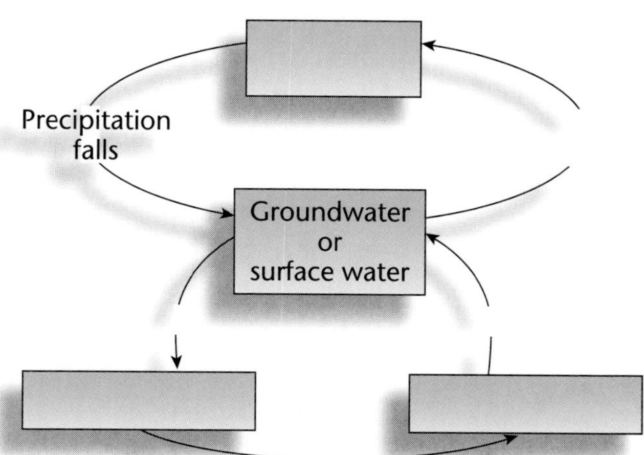

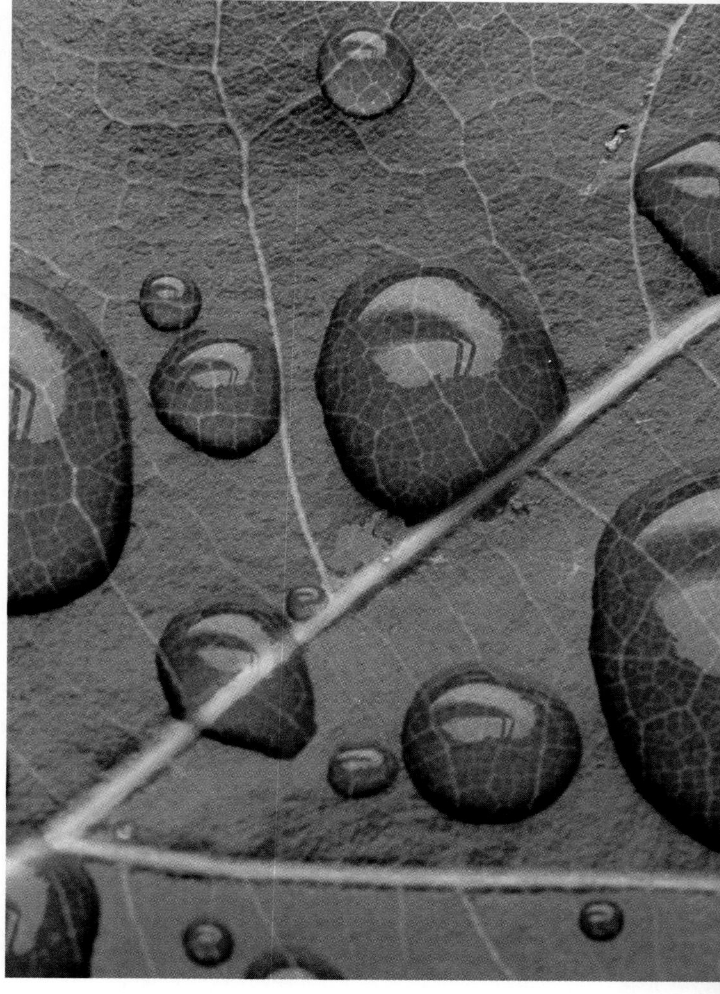

Performance Assessment

1. **Carrying Out a Strategy and Collecting Data:** Design an experiment to test the effects of acid rain on various kinds of vegetation.
2. **Interpreting and Using Data:** Design an experiment to determine which kind of sediment—gravel, sand, or clay—is most [illegible]tive in filtering pollutants from w[illegible]
3. **Design:** In the MiniLAB on [illegible] observed the hardness [illegible] Review the proce[illegible] 17-1 on page[illegible] way to [illegible]

578 Unit 6 Project

UNIT PROJECT 6

Become an Advocate for the Environment

Endangered species, acid rain, ozone depletion, global warming, loss of wildlife habitat, and many other environmental topics make the headlines of our newspapers every day. Do these things bother you? As you do this project, you'll learn about issues that deal with energy and conservation, air pollution, water pollution, soil pollution, and human population growth and their effects on the environment. What can you do about these things?

Procedures

1. The first thing you can do is learn more about these problems. Choose an environmental problem to study in depth. Use the library and local and national organizations to get information.
2. Make a scrapbook of the information you collect. Include the following in your scrapbook: a table of contents; all research you have done including a background report, interviews, pictures, copies of letters you've sent and received, summaries of newspaper and magazine articles that you've read; and a bibliography.
3. After collecting your information, become an advocate for the environment. You can do this in a variety of ways:
   - Write letters to the editors of newspapers, your local government leaders, your representatives in Congress, and even the President of the United States.
   - Become a pen pal with other students who live in the United States, or in other countries, and tell them about what you've learned.
   - Teach young children. Arrange to teach an elementary class, a Boy Scout or Girl Scout troop, or a church group.
   - Make pamphlets and hand them out to your neighbors and/or classmates.
   - Make posters and display them around your school and/or the local library.
   - Have a bake sale or car wash and send the money you collect to an environmental group of your choice.

- Adopt a stream or section of a roadway and clean it up.
- Organize a tree planting celebration.
- Begin an antilitter campaign in your school.
- Adopt an endangered species.
- Be creative. Think of other things you can do.

4. At the end of Unit 6, share your project with your classmates. During your presentation, use one or more visual aids. These might include videotapes, audiocassettes, filmstrips, slides, overhead transparencies, models, and posters. Be sure to answer the following questions during your presentation.
   - What do you know now that you didn't know when you started this project?
   - What did you do to help correct the environmental problem that you studied?
   - Do you think you really made a difference in helping to correct the problem? Explain.
   - What would you do differently if you could do this project again?
   - If you had unlimited money and other resources, what would you do to help correct the problem you studied?
   - Do you think the problem will get worse or better in the next ten years?
   - How has this project changed your attitude/outlook about the problem you chose to study?
   - Will you continue to be involved in helping correct the problem? If your answer is yes, how will you do this?

Going Further

Participate in an environmental fair. Create a display of information about the problem you studied, and describe what you did to help correct the problem. Be ready to answer questions from people who attend the fair.

Be a Part of the Solution, Not the Problem!

UNIT 7

Unit Contents

Astronomy

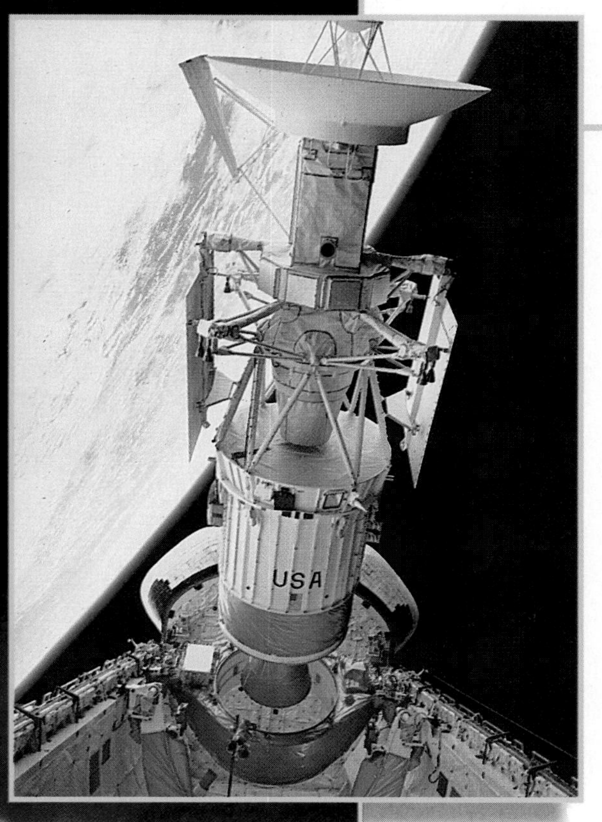

What's Happening Here?

Deep in the blackness of space, particles of dust and gas are swirling together in the first step of a wondrous process that will climax in the birth of a star. With new technology, such as the *Hubble Space Telescope,* we can witness this spectacular event through photographs like the one shown here. Space telescopes are only one of the tools scientists use to explore space. Space probes, such as *Magellan,* shown in the smaller photo, have traveled throughout our solar system, sending back data to scientists on Earth. Read on to learn about space and the technology we use to observe it.

Science Journal

In the 1600s, Galileo Galilei turned his telescope on Saturn and saw what appeared to be a triple planet—one large planet with two smaller planets hidden behind it. In your Science Journal, research what Galileo made of his discovery, and how this compares to what we know about Saturn today.

Previewing the Chapter

Section 21-1 Radiation from Space
- ▶ Electromagnetic Waves
- ▶ Optical Telescopes
- ▶ Radio Telescopes

Section 21-2 Science and Society: Issue: Light Pollution

Section 21-3 Artificial Satellites and Space Probes
- ▶ The First Steps into Space
- ▶ Space Probes
- ▶ The Race to the Moon

Section 21-4 The Space Shuttle and the Future
- ▶ The Space Shuttle
- ▶ Space Stations
- ▶ Cooperation in Space

Chapter 21

Exploring Space

The first space exploration didn't occur in a spaceship or a satellite. Instead, it was done by a person simply looking upward, studying countless points of shimmering light. Over time, people devised more and more accurate ways of keeping track of the stars and their apparent movements. Today we have powerful telescopes, spacecraft, and computers to help us observe what lies beyond Earth. But as old as astronomy is, it seems we've barely scraped the surface of what there is to discover.

EXPLORE ACTIVITY

Is white light more than it seems?

1. Cover the end of a flashlight with a green gelatin (or Plexiglas) filter.
2. Cover another flashlight with a blue filter and a third with a red filter.
3. In a darkened room, experiment with the three lights by shining different combinations of two lights on the same spot on a sheet of white paper. Record your observations in your Science Journal.

Observe: *Observe* all colors shining on the same spot on the paper and *infer* what is happening. Record your observations and inferences in your Science Journal.

Previewing Science Skills

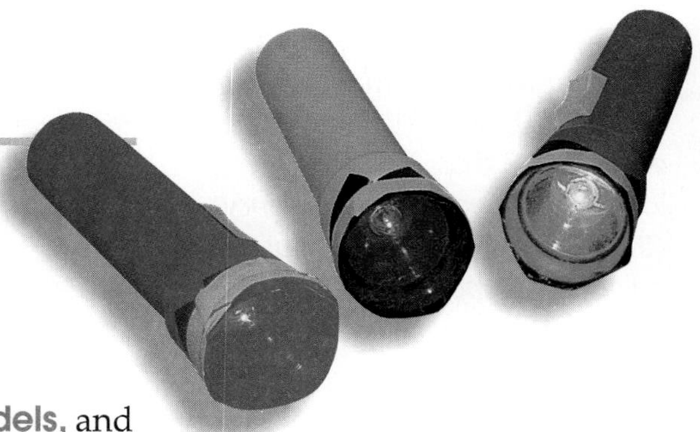

- In the **Skill Builders**, you will **sequence events**, **make concept maps**, and **outline**.
- In the **Activities**, you will **observe**, **analyze data**, **make inferences**, and **draw conclusions**.
- In the **MiniLABs**, you will **experiment**, **construct models**, and **draw conclusions**.

21•1 Radiation from Space

Science Words

electromagnetic spectrum
refracting telescope
reflecting telescope
observatory
radio telescope

Objectives

- Define the electromagnetic spectrum.
- Compare and contrast refracting and reflecting telescopes.
- Compare and contrast optical and radio telescopes.

Electromagnetic Waves

On a crisp autumn evening, you take a break from your homework to gaze out the window at the many stars filling the night sky. Looking up at the night sky, it's easy to imagine future spaceships venturing through space and large space stations circling above Earth where people work and live. But when you look into the night sky, what you're really seeing is the distant past, not the future.

Light from the Past

When you look at a star, you see light that left it many years ago. The light that you see travels very fast, but at a finite speed. The distances across space are so great that it takes years for the light to reach Earth—sometimes millions of years.

The light and other energy leaving a star are forms of radiation. Recall that radiation is energy that's transmitted from one place to another by electromagnetic waves. Because of the electric and magnetic properties of this radiation, it's called electromagnetic radiation. Electromagnetic waves carry energy through space as well as through matter.

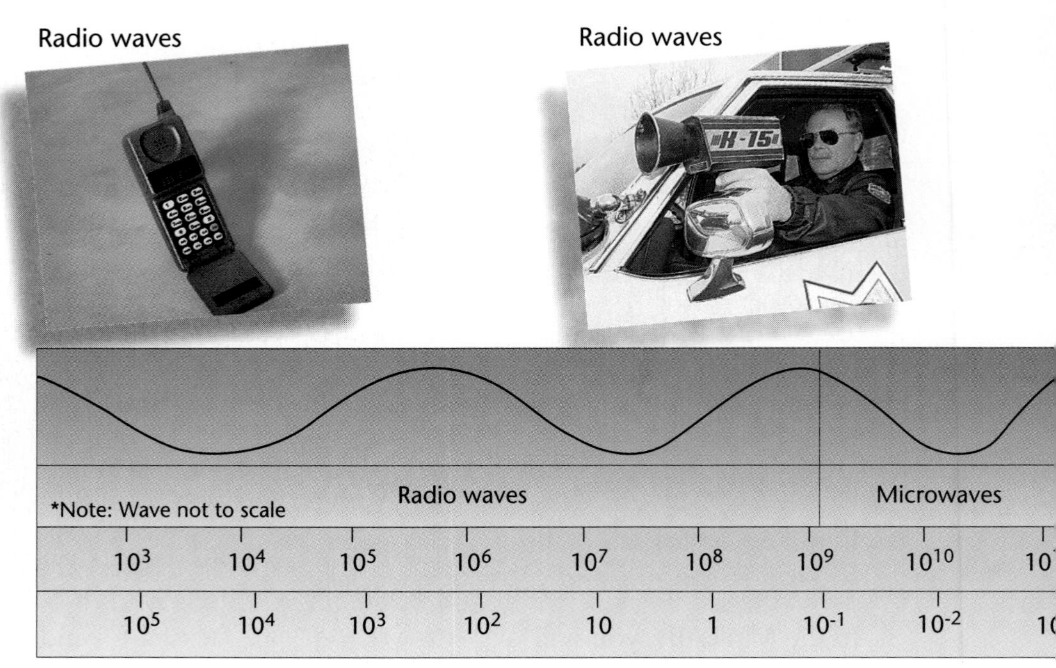

Figure 21-1

The electromagnetic spectrum ranges from gamma rays with wavelengths of less than 0.000 000 000 01 m to radio waves more than 100 000 m long. *What happens to frequency (the number of waves that pass a point per second) as wavelength shortens?*

Electromagnetic Radiation

Sound waves, a type of mechanical wave, can't travel through empty space. How do we hear the voices of the astronauts while they're in space? When they speak into a microphone, the sound is converted into electromagnetic waves called radio waves. The radio waves travel through space and through our atmosphere. They are then converted back into sound by electronic equipment and audio speakers.

Radio waves and visible light from the sun are just two types of electromagnetic radiation. The other types include gamma rays, X rays, ultraviolet waves, infrared waves, and microwaves. **Figure 21-1** shows these forms of electromagnetic radiation arranged according to their wavelengths. This arrangement of electromagnetic radiation is called the **electromagnetic spectrum.**

Although the various electromagnetic waves differ in their wavelengths, they all travel at the speed of 300 000 km/s in a vacuum. You're probably more familiar with this speed as the "speed of light." Visible light and other forms of electromagnetic radiation travel at this incredible speed, but the universe is so large that it takes millions of years for the light from some stars to reach Earth.

Once electromagnetic radiation from stars and other objects reaches Earth, we can use it to learn about the source of the electromagnetic radiation. What tools and methods do scientists use to discover what lies beyond our planet? One tool for observing electromagnetic radiation from distant sources is a telescope.

Pass a beam of white light through a prism. Note that different colors of light are bent, forming a spectrum. ***Infer*** how the white light and prism form a spectrum with violet on one end and red on the other.

Infrared

Visible light

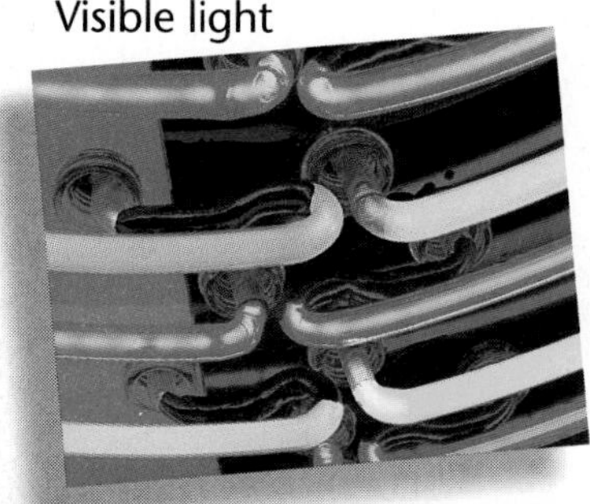

X rays

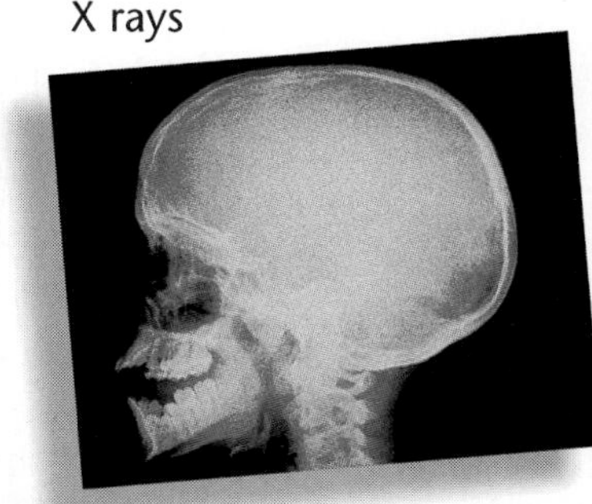

rared radiation		Visible light	Ultraviolet radiation		X rays		Gamma rays				
10^{13}	10^{14}	10^{15}	10^{16}	10^{17}	10^{18}	10^{19}	10^{20}	10^{21}	10^{22}	10^{23}	Frequency (hertz)
10^{-5}	10^{-6}	10^{-7}	10^{-8}	10^{-9}	10^{-10}	10^{-11}	10^{-12}	10^{-13}	10^{-14}	10^{-15}	Wavelength (meters)

Optical Telescopes

Optical telescopes produce magnified images of objects. Light is collected by an objective lens or mirror, which then forms an image at the focal point of the telescope. The eyepiece lens then magnifies the image. The two types of optical telescopes are shown in **Figure 21-2.**

In a **refracting telescope,** the light from an object passes through a convex objective lens and is bent to form an image on the focal point. The image is then magnified by the eyepiece.

A **reflecting telescope** uses a mirror as an objective to focus light from the object being viewed. Light passes through the open end of a reflecting telescope and strikes a concave mirror at its base. The light is then reflected to the focal point to form an image. A smaller mirror is often used to reflect the light into the eyepiece lens so the magnified image can be viewed.

Figure 21-2

These diagrams show how each type of optical telescope collects light and forms an image.

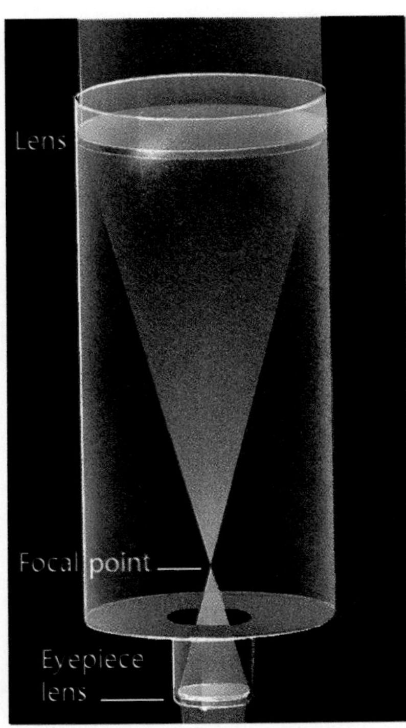

A In a refracting telescope, a convex lens focuses light to form an image at the focal point.

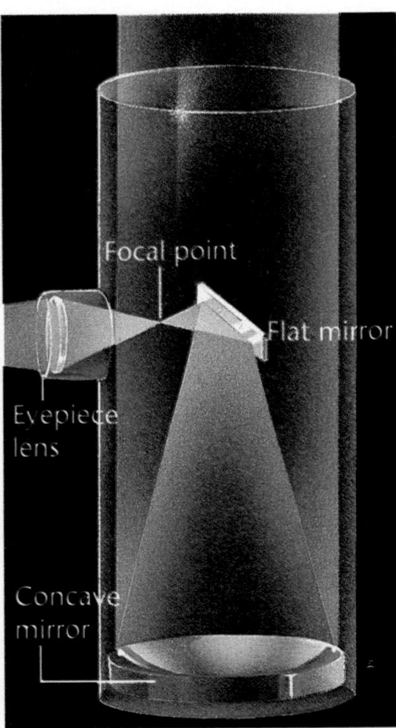

B In a reflecting telescope, a concave mirror focuses light to form an image at the focal point.

Using Optical Telescopes

Most optical telescopes used by professional astronomers are housed in buildings called **observatories.** Observatories often have a dome-shaped roof

C *Which type of optical telescope is this student using?*

that opens up to let light in. **Figure 21-4,** on the next page, shows the twin Keck telescopes on Mauna Kea in Hawaii. However, not all telescopes are in observatories.

The *Hubble Space Telescope* was launched in 1990 by the space shuttle *Discovery*. Earth's atmosphere absorbs and distorts some of the energy we receive from space. Because *Hubble* didn't have to view space through our atmosphere, it should have produced very clear images. However, when the largest mirror of this reflecting telescope was shaped, there was a mistake. Images obtained by the telescope were not as clear as expected. In December 1993, a team of astronauts repaired *Hubble's* optics and other equipment. Now the clear images obtained by *Hubble* are changing scientists' ideas about space.

Radio Telescopes

As you know, stars and other objects radiate energy throughout the electromagnetic spectrum. A **radio telescope** is used to study radio waves traveling through space. Unlike visible light, radio waves pass freely through Earth's atmosphere. Because of this, radio telescopes are useful under most weather conditions and at all times of day and night.

Radio waves reaching Earth's surface strike the large, curved dish of a radio telescope. This dish reflects the waves to a focal point where a receiver is located. The information allows us to detect objects in space, to map the universe, and to search for intelligent life.

A Homemade Antenna

Suppose that you, a classmate, and your cousin had been looking forward to a camping trip for several weeks. You have just set up camp when some large clouds begin to roll in and the wind begins to stir. As your classmate looks for wood to use in the campfire, you pull out your radio to listen to the weather forecast. You think a storm might be approaching and you might need to seek better shelter. The radio reception is poor and you can barely hear the weather report.

Your cousin digs through the camping supplies and finds some aluminum foil wrap, an umbrella, and some string. She suggests that these items could be used to improve the radio reception.

Solve the Problem:

1. **What type of radiation is the radio capable of receiving?**
2. **How could you and the others improve the radio reception with the supplies described above?**

Think Critically:

How are the parts of your radio and antenna like those of a radio telescope?

Figure 21-3

The Hubble Space Telescope was deployed from the cargo bay of the space shuttle *Discovery* on April 25, 1990. It's now orbiting Earth, sending back images and data about distant space objects.

Figure 21-4

The twin Keck telescopes on Mauna Kea in Hawaii can be used in combination, effectively more than doubling the resolving power. Each individual telescope has an objective mirror 10 m in diameter. To cope with the difficulty of building such a large mirror, this telescope design used several smaller mirrors positioned to work as one. *Although the Keck telescopes are much larger than the* Hubble Space Telescope, *why is the* Hubble *able to achieve better resolution?*

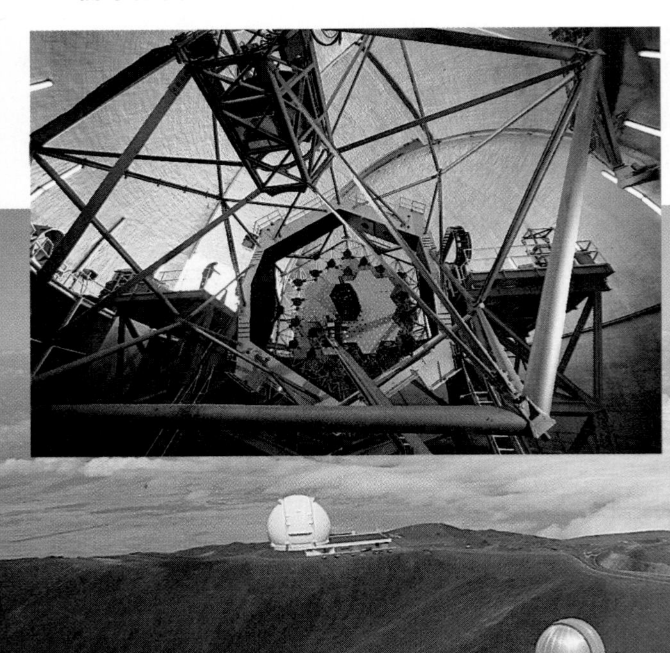

Since the early 1600s when the Italian scientist Galileo Galilei first turned a telescope toward the stars, people have been searching for better ways to study what lies beyond our atmosphere. Today, the largest reflector has a segmented mirror 10 m wide, and the largest radio telescope is 300 m wide. The most recent innovations in optical telescopes involve active and adaptive optics. With active optics, a computer is used to compensate for changes in temperature, mirror distortions, and bad viewing conditions. Even more ambitious is adaptive optics, which uses a laser to probe the atmosphere and relate information to a computer about air turbulence. The computer then adjusts the telescope's mirror thousands of times per second, thus reducing the effects of atmospheric turbulence.

Even with active optics and adaptive optics, there is still a need to travel into space for exploration. In the remainder of this chapter, you'll learn about the instruments that travel into space and send back information not obtainable through telescopes on Earth's surface.

Section Wrap-up

Review

1. How are radio telescopes and optical telescopes different from one another?
2. Why should telescopes such as *Hubble* produce more detailed images than Earth-based telescopes?
3. The frequency of electromagnetic radiation is the number of waves that pass a point in a specific amount of time. If red light has a longer wavelength than blue light, which would have a greater frequency?
4. **Think Critically:** It takes light from the closest star to Earth (other than the sun) about four years to reach us. If there were intelligent life on a planet circling that star, how long would it take for us to send them a radio transmission and for us to receive their reply?

Skill Builder

Sequencing

Sequence these electromagnetic waves from longest wavelength to shortest wavelength: *gamma rays, visible light, X rays, radio waves, infrared waves, ultraviolet waves,* and *microwaves.* If you need help, refer to Sequencing in the **Skill Handbook.**

USING MATH

The magnifying power (Mp) of a telescope is determined by dividing the focal length of the objective (FL_{obj}) by the focal length of the eyepiece (FL_{eye}) using the following equation:

$$Mp = \frac{FL_{obj}}{FL_{eye}}$$

If $FL_{obj} = 1200$ mm and $FL_{eye} = 6$ mm, what is the telescope's magnifying power?

Activity 21-1

Telescopes

You have learned that optical telescopes use lenses and mirrors as objectives to collect light from an object. They use eyepiece lenses to magnify images of that object. Try this activity to see how the paths of light differ in reflecting and refracting telescopes.

Problem

In what way are paths of light affected by the lenses and/or mirrors in refracting and reflecting telescopes?

Materials  

- candle
- flashlight
- hand lens
- concave mirror
- masking tape
- empty paper towel tube
- cardboard, white, 50 cm × 60 cm
- glass of water
- plane mirror
- convex mirror

Procedure

1. Observe your reflection in a plane, convex, and concave mirror.
2. Hold an object in front of each of the mirrors. Compare the size and position of the images.
3. Darken the room and hold the convex mirror in front of you at a 45° angle, slanting downward. Direct the flashlight beam toward the mirror. Note the size and position of the reflected light.
4. Repeat step 3 using a plane mirror. Draw a diagram to show what happens to the beam of light.
5. Attach the empty paper towel tube to the flashlight with masking tape so that the narrow beam of light will pass through the tube. Direct the light into a large glass of water, first directly from above, then from an angle 45° to the water surface. Observe the direction of the light rays when viewed from the side of the glass.
6. Light a candle and set it up some distance from the vertically held cardboard screen. **CAUTION:** *Keep hair and clothing away from the flame.* Using the hand lens as a convex lens, move it between the candle and the screen until you have the best possible image.
7. Move the lens closer to the candle. Note what happens to the size of the image. Move the cardboard until the image is in focus.

Analyze

1. What is the purpose of the concave mirror in a reflecting telescope?
2. How did you determine the position of the focal point of the hand lens in step 6? What does this tell you about the position of all the light rays?
3. The eyepiece of a telescope is convex. **Infer** its purpose.
4. **Compare and contrast** the effect the three types of mirrors had on your reflection.

Conclude and Apply

5. Discuss your observations of the relationship of the distance between the object and lens and the clearest and largest image you could obtain in steps 6 and 7.
6. **Compare and contrast** the path of light in refracting and reflecting telescopes.

ISSUE:

Light Pollution

SCIENCE & SOCIETY

21 • 2

Light Pollution and Stargazing

When you gaze out your window at the night sky, what do you see? Chances are, if you live in or near a city, you don't see a star-filled sky. Instead, you see only a few of the brightest stars scattered throughout a hazy, glowing sky. You're looking through a sky full of light pollution.

City lights cause a glow in the sky called **light pollution.** Light pollution makes the sky glow bright enough that dim stars can't be seen. What effect does light pollution have on your ability to stargaze? How do you think this affects astronomers working near large cities? Many people feel that their right to a dark night sky has been taken away.

Observing objects through Earth's atmosphere can be difficult even without light pollution. As you learned in Chapter 14, the atmosphere absorbs some of the electromagnetic radiation entering it. Visible light from objects in space can't pass directly through clouds or smog. Light pollution makes observing even more difficult as shown in **Figure 21-5.** If an object is faint, it's difficult to distinguish between visible light from the object in space and visible light from the city.

Science Word

light pollution

Objectives

- Explain how light pollution affects our ability to see dim objects in the sky.
- Discuss the issue of light pollution as it relates to security and safety lighting.

Figure 21-5

These two photographs were taken of the same area of night sky, from two different places. *Which photo was taken within a city and which was taken in a more rural setting?*

CONNECT TO

CHEMISTRY

Why is light from low-pressure sodium lights easier for astronomers to filter out? If possible, ***compare*** the spectrum of a low-pressure sodium light with the spectra of other types of security lights. ***Describe*** the importance of low-pressure sodium lights to astronomers.

Most people agree that lights are needed on city streets and parking lots for safety and security. What can be done to reduce light pollution without reducing public safety and security?

In several cities in the United States, work has begun to reduce light pollution. As shown in **Figure 21-6,** bad lighting can cause lots of glare. Instead of increasing security, such lighting provides places for criminals to hide. Tucson, Arizona, located only 80 km from Kitt Peak National Observatory, has replaced its street lights with low-pressure sodium lamps. These lights shine at wavelengths that can be filtered out by astronomers. After upgrading to low-pressure sodium lights, Tucson's lighting improved throughout the city. In some cases, the amount of light increased by a factor of 10. These low-pressure sodium lights produce better lighting for the streets and even cost less to operate. Another solution is to put hoods on billboards, parking-lot lights, and floodlights so they illuminate the object or the ground rather than the sky.

Figure 21-6

The type of lighting can make a big difference in the amount of light pollution.

A The main mall at the University of Arizona in Tucson had been lit by 400-watt mercury vapor lights. There was enough light but there were also many dark, shadowy areas and glare.

B After changing to low-pressure sodium lights, the lighting on the mall improved everywhere and glare was eliminated. Also, the university has saved money because the new 135-watt sodium lights use 67 percent less electricity.

2 Points of View

The Right to Have Security Lighting

The problem of light pollution is not limited to the big cities. Just as in urban and suburban areas, crime and safety is becoming a more important issue in rural areas. To deal with this, security lighting is being used more extensively in all areas of the country. Property owners feel they have the right to protect themselves and their property by installing security lights. To many, the safety that comes with illuminating their property is well worth the problems caused by light pollution.

The Right to a Dark Sky for Observation

On the other hand, imagine you have invited a few friends over to sit outside and watch for meteors, or so-called shooting stars. You set up your chairs and sit down for an evening of searching the skies. As dusk slowly turns to night, your neighbor's security lights turn on, flooding your backyard with bright, glaring light. Soon you realize the sky is too bright to observe meteors. Your neighbor's security lights are casting a glow into the sky. You feel you have the right to darkness so you can observe the night sky.

MiniLAB

Compare the effects of light pollution.

Procedure:

1. Obtain a cardboard tube from an empty roll of paper towels.
2. Select a night when clear skies are predicted. Go outside about two hours after sunset, and look through the cardboard tube at a specific constellation decided upon ahead of time.
3. Count the number of stars you are able to see without moving the observing tube. Repeat this three times.
4. Determine the average number of observable stars at your location.

Analysis:

1. Compare and contrast the number of stars visible from other students' homes.
2. Recognize the cause and effect of differences in your observations. Explain.

Section Wrap-up

Review

1. Why would it be difficult to pass laws that ban light pollution?
2. What happens to city lights within Earth's atmosphere that makes stargazing difficult?

Explore the Issue

Should your neighbors be required to turn off their lights whenever you are observing? If yes, how can they protect their property if they must darken it? If not, how can they use security lights and still enable you and your friends to observe dark skies? To help solve this problem, refer to **Figure 21-6** and the work of one group, the International Dark-Sky Association.

21•3 Artificial Satellites and Space Probes

Science Words

satellite
orbit
space probe
Project Mercury
Project Gemini
Project Apollo

Objectives

- Compare and contrast natural and artificial satellites.
- Differentiate between an artificial satellite and a space probe.
- Trace the history of the race to the moon.

The First Steps into Space

If you had your choice of watching your favorite sports team on television or from the stadium, which would you prefer? You would probably want to be as close as possible so you wouldn't miss any of the action. Scientists feel the same way. Even though telescopes have taught them a great deal about the moon and planets, they want to learn more by actually going to those places, or by sending spacecraft where they can't go.

Satellites

Space exploration began in 1957 when the former Soviet Union used a rocket to send *Sputnik I* into space. It was the first artificial satellite. A **satellite** is any object that revolves around another object. When an object enters space, it will travel in a straight line unless a force such as gravity deflects it. When Earth's gravity pulls on a satellite, it falls toward Earth. The result of the satellite traveling forward while at the same time falling toward Earth is a curved path, or **orbit** around Earth. This is shown in **Figure 21-7.**

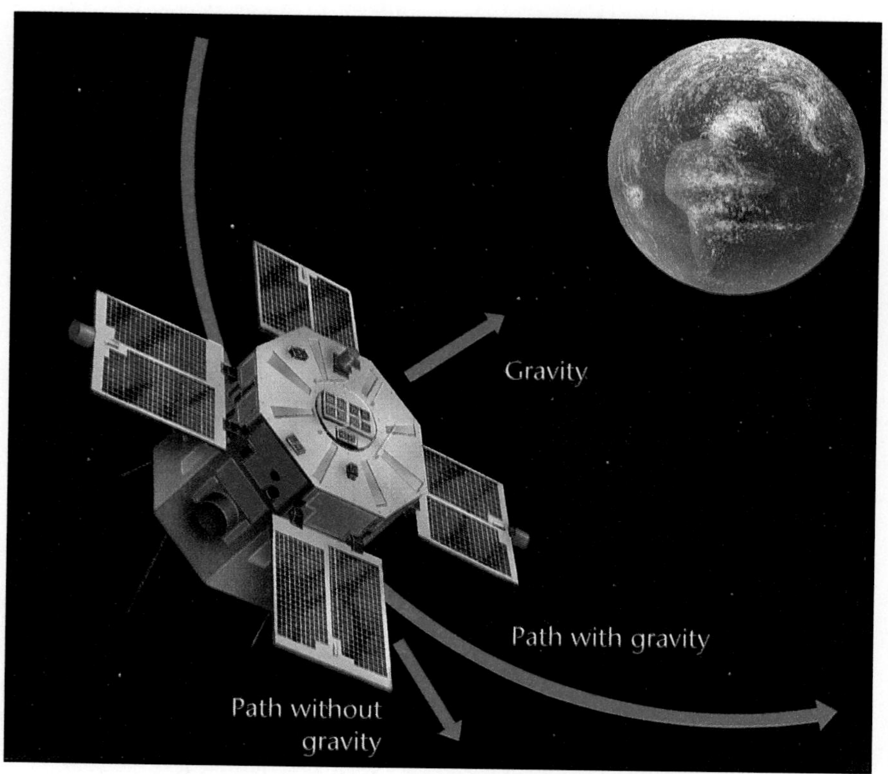

Figure 21-7

The combination of the satellite's forward movement and the gravitational attraction of Earth causes the satellite to travel in a curved path, or orbit. *What would happen if the forward speed of the satellite decreased?*

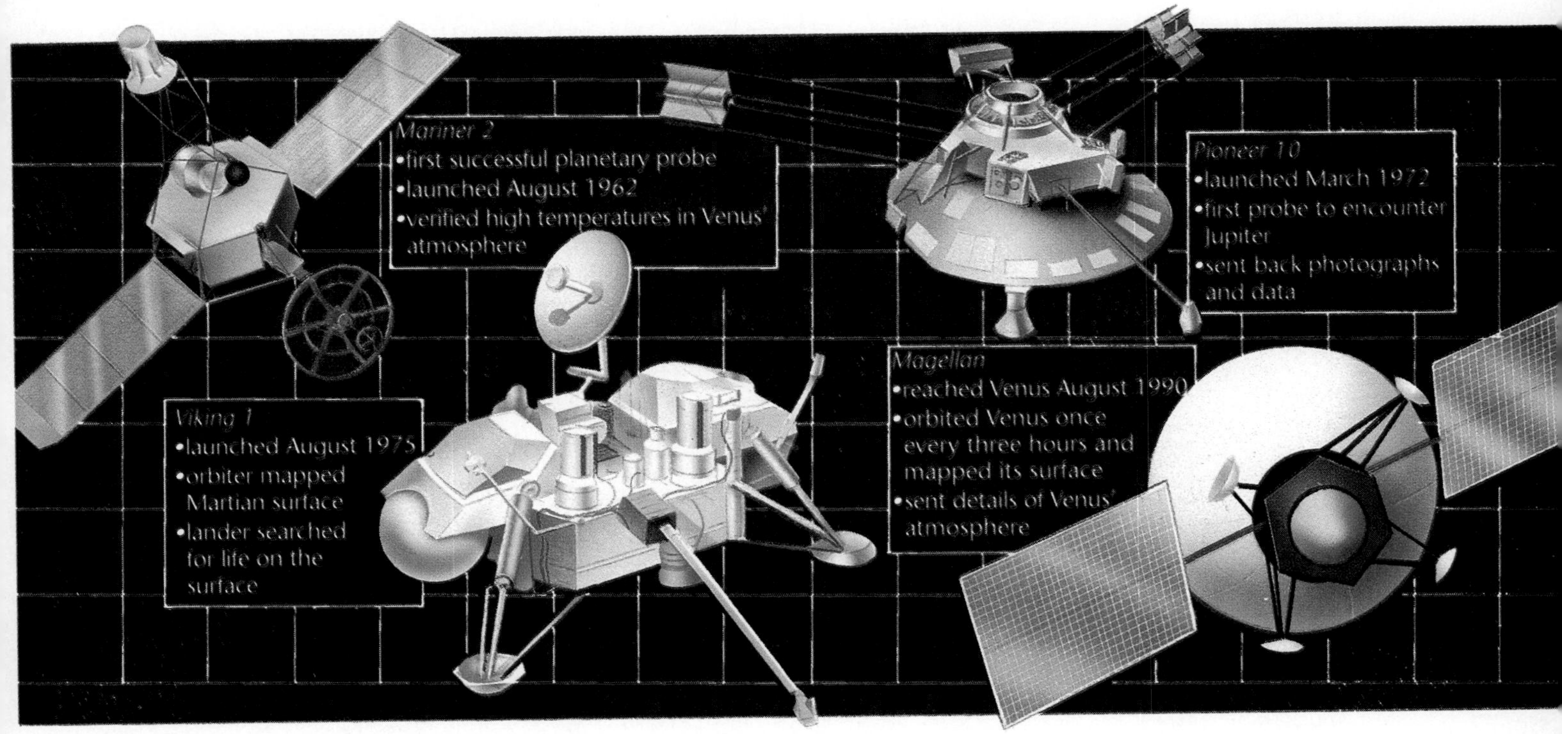

Mariner 2
- first successful planetary probe
- launched August 1962
- verified high temperatures in Venus' atmosphere

Pioneer 10
- launched March 1972
- first probe to encounter Jupiter
- sent back photographs and data

Viking 1
- launched August 1975
- orbiter mapped Martian surface
- lander searched for life on the surface

Magellan
- reached Venus August 1990
- orbited Venus once every three hours and mapped its surface
- sent details of Venus' atmosphere

Figure 21-8

Some U.S. space probes and their missions.

The moon is a natural satellite of Earth. It completes one orbit every month. *Sputnik I* orbited Earth for three months before gravity pulled it back into the atmosphere, where it burned up. *Sputnik I* was an experiment to show that artificial satellites could be made. Today, thousands of artificial satellites are in orbit around Earth.

Present-day communication satellites transmit radio and television programs to locations around the world. Other satellites gather scientific data that can't be obtained from Earth, and weather satellites constantly monitor Earth's global weather patterns.

Space Probes

Not all objects carried into space by rockets become satellites. Rockets can also be used to send instruments into space. A **space probe** is an instrument that gathers information and sends it back to Earth. Unlike satellites that orbit Earth, space probes travel far into the solar system. Some have even traveled out of the solar system. Space probes, like many satellites, carry cameras and other data-gathering equipment as well as radio transmitters and receivers that allow them to communicate with scientists on Earth. **Figure 21-8** is a time line showing some of the space probes launched by NASA (National Aeronautics and Space Administration).

You've probably heard of the space probes *Voyager 1* and *Voyager 2*. These two probes were launched in 1977 and are

USING MATH

Suppose a spacecraft is launched at a speed of 40 200 kilometers per hour. Express this speed in kilometers per second.

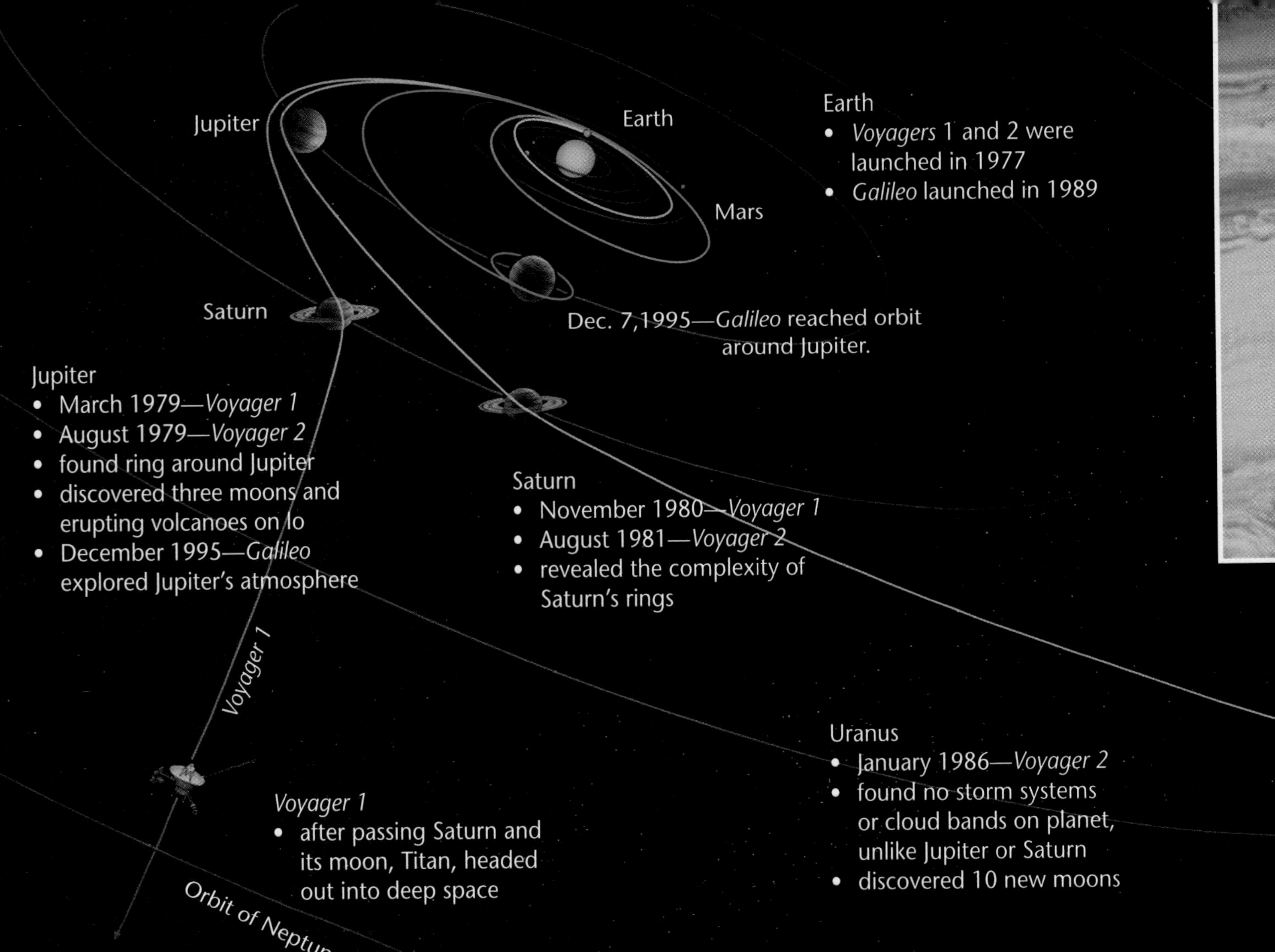

Figure 21-9

The *Voyager* and *Galileo* spacecraft helped make many major discoveries.

now heading outward toward deep space. *Voyager 1* flew past Jupiter and Saturn. *Voyager 2* flew past Jupiter, Saturn, Uranus, and Neptune. **Figure 21-9** describes a portion of what we've learned from the *Voyager* probes. Scientists expect these probes to continue to transmit data to Earth from beyond the solar system for at least 20 more years.

The fate of a probe is never certain and not all are successful. *Mars Explorer* was in orbit around Mars when communication with the probe was lost. It has been silent ever since.

Galileo, launched in 1989, reached Jupiter in 1995. In July 1995, *Galileo* released a smaller probe that began a five-month approach to Jupiter. The small probe took a parachute ride through Jupiter's violent atmosphere in December 1995. Before being crushed by the atmospheric pressure, it transmitted information about Jupiter's composition, temperature, and pressure to the ship orbiting above. *Galileo* studied Jupiter's moons, rings, and magnetic fields and then relayed this information back to scientists eagerly awaiting it on Earth.

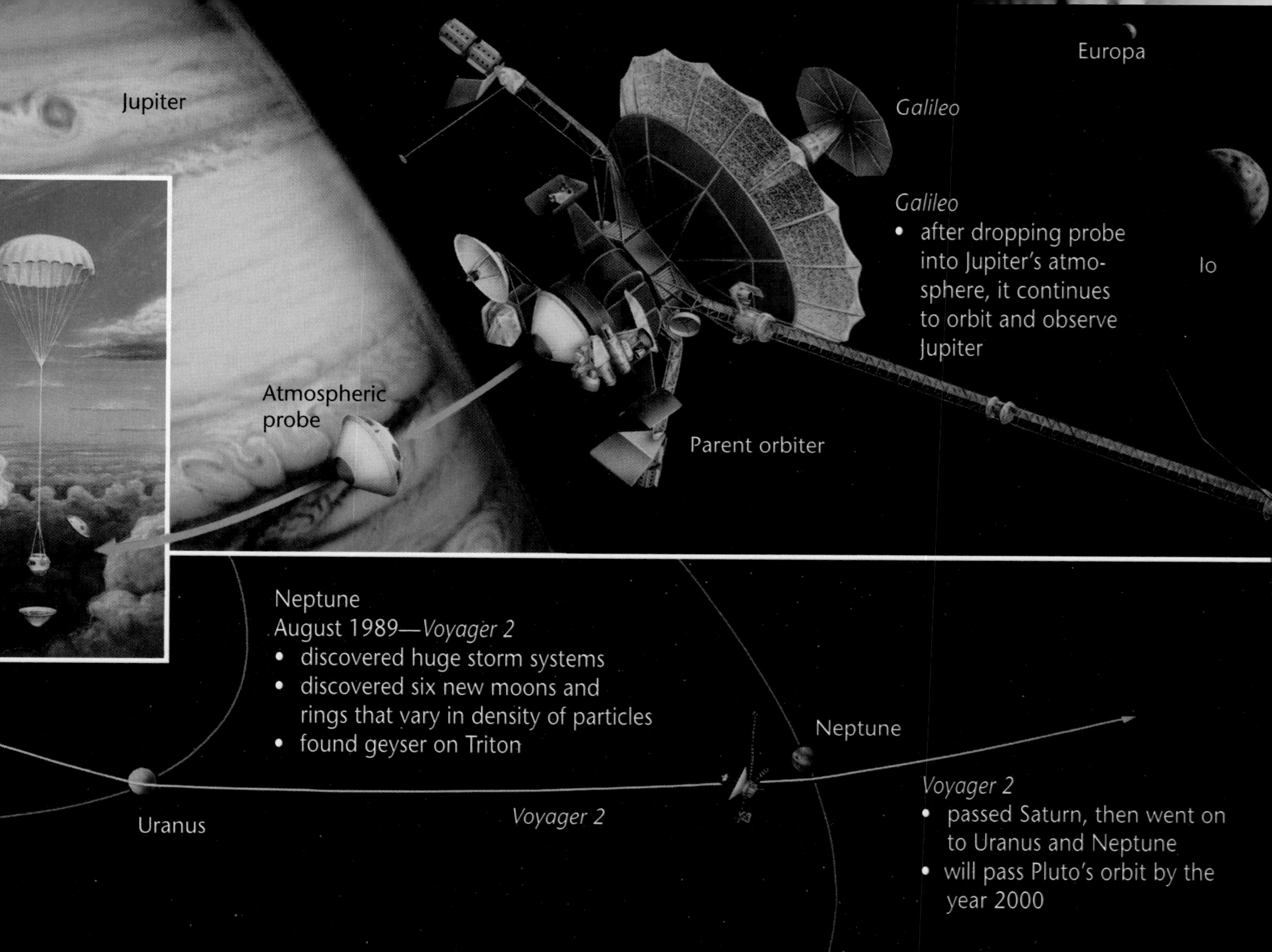

The Race to the Moon

It was quite a shock to people throughout the world when they turned on their radio and television sets in 1957 and heard the radio transmissions from *Sputnik I* as it orbited over their heads. All that *Sputnik I* transmitted was a sort of beeping sound, but people quickly realized that putting a human into space wasn't far off.

In 1961, the Soviet cosmonaut Yuri A. Gagarin became the first human in space. He orbited Earth and then returned safely. Soon, President John F. Kennedy called for the United States to place people on the moon and return them to Earth by the end of that decade. The "race for space" had begun.

The U.S. program to reach the moon began with **Project Mercury.** The goals of Project Mercury were to orbit a piloted spacecraft around Earth and to bring it safely back. The program provided data and experience in the basics of space flight. On May 5, 1961, Alan B. Shepard became the first U.S. citizen in space. In 1962, *Mercury* astronaut John Glenn became the first U.S. citizen to orbit Earth.

CONNECT TO

LIFE SCIENCE

Scientists are very careful when building spacecraft so as not to contaminate equipment with microorganisms from Earth. ***Explain*** in your Science Journal why this is so important for craft sent into space or to other worlds.

USING TECHNOLOGY

Spin-Offs

The technology developed by NASA to achieve its goals in space has been remarkable. Much of it is now being used by people throughout the world. The technologies developed by NASA that are later used by the general public are called spin-offs.

NASA had to develop lightweight, compact breathing systems for the astronauts to carry as they ventured out of their spacecraft and onto the moon. Today, firefighters use these breathing systems as well as fire-resistant uniforms originally designed as flight suits for NASA pilots. The lightweight material in the suits won't burn or crack.

Another material, designed for boots worn by astronauts on the moon, is now found in some athletic shoes. Other materials have been incorporated into ski goggles, blankets, and bicycle seats.

Persons who are visually impaired have benefited from spin-offs too. They are able to use a device that vibrates ink on a printed page. This enables them to read materials that aren't in Braille. Another device determines the denomination of currency and generates an audible signal.

Other spin-offs include pens that write without the help of gravity and sunglasses that adjust to various light levels.

Think Critically:

How has the space program affected your life? You may need to look no farther than your wrist!

NASA Spin-Off

Project Gemini

Project Gemini was the next step in reaching the moon. Teams of two astronauts in the same *Gemini* spacecraft orbited Earth. One *Gemini* team met and connected with another spacecraft in orbit—a skill that would be needed on a voyage to the moon.

Along with the *Mercury* and *Gemini* programs, a series of robotic probes was sent to the moon. *Ranger* proved we could get spacecraft to the moon. *Surveyor* landed gently on the moon's surface, indicating that the moon's surface could support spacecraft and humans. The mission of *Lunar Orbiter* was to take pictures of the moon's surface that would be used to determine the best landing sites on the moon.

Project Apollo

The final stage of the U.S. program to reach the moon was **Project Apollo.** On July 20, 1969, *Apollo 11* landed on the lunar surface. Neil Armstrong was the first human to set foot on the moon. His first words as he stepped onto its surface were: "That's one small step for a man, one giant leap for mankind." Edwin Aldrin, the second of the three *Apollo 11* astronauts, joined Armstrong on the moon and they explored its surface for two hours. Michael Collins remained in the Command Module orbiting the moon, where Armstrong and Aldrin returned before beginning the journey home. A total of six lunar landings brought back more than 2000 samples of moon rock and soil for study before the program ended in 1972.

Figure 21-10

The Lunar Rover Vehicle was first used during the *Apollo 15* mission. Riding in the "moon buggy," *Apollo 15, 16,* and *17* astronauts explored large areas of the lunar surface.

During the past three decades, most missions in space have been carried out by individual countries, often competing to be the first or the best. Today, there is much more cooperation among countries of the world to work together and share what each has learned. Projects are now being planned for cooperative missions to Mars and elsewhere. As you read the next section, you'll see how the U.S. program has progressed since the days of Project Apollo, and where it may be going in the future.

Section Wrap-up

Review

1. Currently, no human-made objects are orbiting Neptune, yet Neptune has eight satellites. Explain.
2. *Galileo* was considered a space probe as it traveled to Jupiter. Once there, however, it became an artificial satellite. Explain.
3. **Think Critically:** Is Earth a satellite of any other body in space? Explain your answer.

Skill Builder

Concept Mapping

Make an events-chain concept map that lists the events in the U.S. space program to place people on the moon. If you need help, refer to Concept Mapping in the **Skill Handbook.**

Using Computers

Spreadsheet Use the spreadsheet feature on your computer to generate a table of recent successful satellites and space probes launched by the United States. Include a description of the craft, the date launched, and the mission.

People and Science

GIBOR BASRI, *Astronomer*

On the Job

Q What is it like to use the Keck Telescopes on Mauna Kea, Hawaii?

A One of my most exciting moments as an astronomer was the first night I was able to operate the telescopes. They're amazing telescopes to use; each one is a complex, fascinating machine. The summit of Mauna Kea, Hawaii, is the world's premier site for observing. It's almost 4270 m high, so when you go up there you are somewhat oxygen deprived, and some people get sick. But you're on top of the mountain peak, you can see Maui in the distance, there's ocean all around, and the air is particularly clear and stable. It's kind of the pinnacle of astronomy.

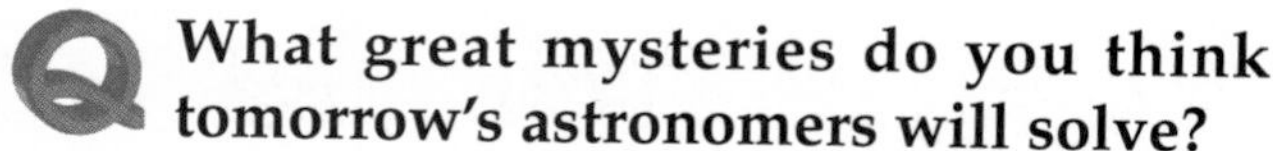

Q What great mysteries do you think tomorrow's astronomers will solve?

A One of the big ones that the next generation will solve is the question of other planets out there. A great deal of work and new technology will be needed to answer the questions, "Is there life out there?" "How many planets are there around other stars, what are they like, and how do they form?" and "Are there a lot of Earth-like planets out there?"

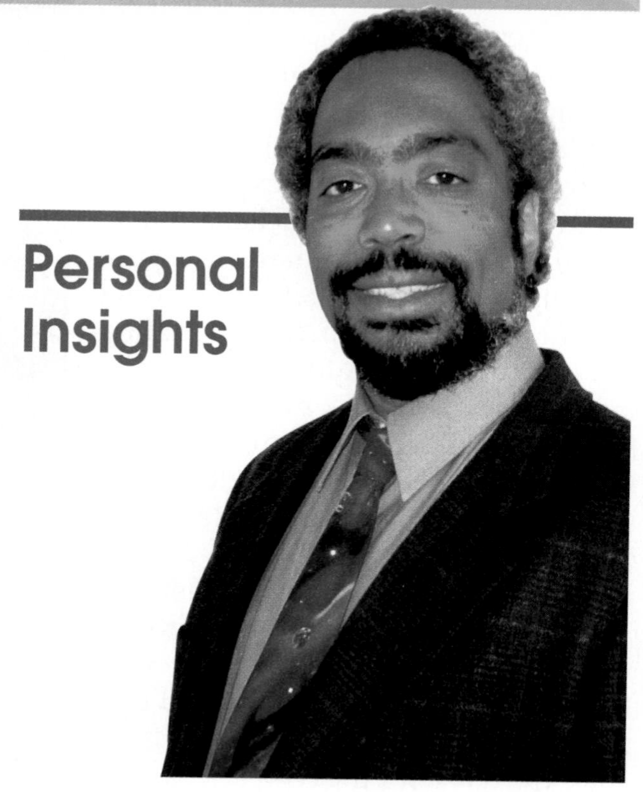

Personal Insights

Q Did your interest in science fiction play a part in your becoming an astronomer?

A I think so. I was fascinated with all aspects of outer space, the cosmos, and time. Science fiction was one outlet for my fascination. It opens your mind to thinking about things in different ways—not accepting things as they seem to be, but imagining how they could be different. It's a broadening experience in that sense.

Career Connection

Contact a local planetarium or astronomy club, or search the Worldwide Web, to find out more about what astronomers do and about careers in astronomy.

Make a profile describing one of the following careers in astronomy:

- **Observatory Scientist**
- **Planetarium Director**

Activity 21-2

Design Your Own Experiment

Probe to Mars

Before attempting a new mission in space, NASA plans, organizes, and practices the operation of the mission many times on Earth. In this way, NASA can be ready to solve many problems that otherwise might have canceled or hindered a successful mission.

PREPARATION

Problem

You and your classmates will conduct a simulated mission to successfully land an exploration probe on the surface of Mars, retrieve samples, and return those samples to Earth for analysis. To achieve this goal, you must do what NASA does before, during, and after a mission.

Mission Planning

The entire class will plan the mission. Include methods to be used for observing, mapping, choosing a landing site, operating the robot surface probe (a battery-operated remote-control car), collecting samples, and analyzing the success of the mission.

Possible Materials

- portable screen
- remote-controlled car
- bag to hold rock and mineral samples
- large rocks (30 cm)
- camera (instant), film
- stopwatch

Activity 21-2 (continued)

PLAN THE EXPERIMENT

1. **Orbital Photographic Study of the Martian Surface:** Two students will photograph (from above) all parts of the Martian surface to be studied.
2. **Mapping the Martian Surface:** Three students will use photographs from the orbital study and draw a map of the Martian surface. Be sure to identify large boulders, craters, flat areas, locations of samples to be retrieved, and any other key surface features.
3. **Landing Site Selection:** Three students will study the maps of the surface and decide on a landing site and route for the robot surface probe.
4. **Design of RSR (Remote Sample Retrieval) System:** Three students will design a method of retrieving a container of rock samples from Mars' surface. Rock samples will be encased in a baglike container, so the retrieval system must be able to pick up and carry the bag of samples.
5. **Mission Control (part 1):** A team of five students will conduct the countdown, blast off, flight, and landing of the robot surface probe.
6. **Mission Control (part 2):** A mission specialist and assistant will operate remote controls that will enable the robot surface probe to move around on the Martian surface. A second mission specialist will operate the RSR System and collect samples for study. A successful mission is one in which the lander explores all parts of the Martian surface within range, without colliding with boulders or falling into craters. A successful mission also involves retrieving rock samples from the planet's surface and returning them to Earth for study. As the mission specialist operates controls from

Data and Observations

Time of Mission		Accuracy of Mission	
Mission start		Were all sites visited?	
Mission stop		Was the mission completed?	
Mission length		List the number of times the mission was delayed by surface features.	

PLAN THE EXPERIMENT (continued)

behind a screen, the assistant will give directions on which way to move the surface probe. The assistant does the job that sensors or cameras would perform on an actual mission.

7. **Sample Analysis:** A team of eight students will conduct an analysis of the rocks brought back from Mars's surface. This team of students is to use methods of identifying minerals and rocks learned in Chapters 3 and 4.
   - Look for any evidence of life on the surface of Mars by searching for fossils in the rocks brought back from the planet.
8. **Mission Debriefing:** A team of three students will:
   - Observe the probe mission.
   - Record the time required to conduct the surface probe mission.
   - Judge when the mission is complete.
   - Record the number of times the probe is delayed or hindered by various features.
   - Review the reports of the sample analysis teams for general conclusions regarding the content of rocks on Mars's surface.

DO THE EXPERIMENT

1. Carry out the mission as planned.
2. While the mission is going on, record any observations of the mission and complete the data table in your Science Journal.
3. Write a final report summarizing the success of the mission and including suggestions for improvement.

Analyze and Apply

1. What part of the mission did all other parts depend on?
2. What feature of the total project brought about a successful mission?
3. In an actual mission, how would photographs of possible landing sites be obtained?
4. What was the most common mineral found on the "Martian" surface?
5. What types of rocks were found?

Go Further

What conclusion can be drawn from the mission about the past or present existence of life on Mars? Why was the mission specialist placed behind a screen? What task did the assistant perform when working with the mission specialist? From your observations, list some likely sources of problems for an actual mission of this type.

21•4 The Space Shuttle and the Future

Science Words

space shuttle
space station

Objectives

- Describe the benefits of the space shuttle.
- Evaluate the usefulness of orbital space stations.

The Space Shuttle

Imagine spending millions of dollars to build a machine, sending it off into space, and watching its 3000 metric tons of metal and other materials burn up after only a few minutes of work. That's exactly what NASA did for many years. The early rockets lifted into orbit a small capsule holding the astronauts. Sections of the rocket separated from the rest of the rocket body and burned as they reentered the atmosphere.

A Reusable Spacecraft

NASA administrators, like many people, realized that it can be less expensive and less wasteful to reuse resources. The reusable spacecraft that transports astronauts, satellites, and other materials to and from space is the **space shuttle.**

At launch, the space shuttle orbiter stands on end and is connected to an external liquid-fuel tank and two solid-fuel booster rockets. When the shuttle reaches an altitude of about 45 km, the emptied solid-fuel booster rockets drop off and parachute back to Earth. They are recovered and used again. The larger external liquid-fuel tank eventually separates and falls back to Earth. It isn't recovered.

Once the space shuttle orbiter reaches space, it begins to orbit Earth. There, astronauts perform many different tasks. The cargo bay can carry a self-contained laboratory where astronauts conduct scientific experiments and determine the effects of space flight on the human body. On missions in which the cargo bay isn't used as a laboratory, the shuttle can launch, repair, and retrieve satellites.

To retrieve a satellite, a large mechanical arm in the cargo bay is extended. An astronaut inside the shuttle orbiter moves the arm by remote control. The arm grabs the satellite and pulls it back into the cargo bay. The doors are closed and it can then be returned to Earth.

Similarly, the mechanical arm can be used to lift a satellite or probe out of the cargo bay and place it into space. In some cases, a defective satellite can be pulled in by the mechanical

Figure 21-11

The space shuttle is designed to make many trips into space.

arm, repaired while in the cargo bay, and then placed into space once more.

After the completion of each mission, the space shuttle orbiter glides back to Earth and lands like an airplane. A very large landing field is needed because the gliding speed of the orbiter is 335 km/hr.

Space Stations

Astronauts can spend only a short time in space in the space shuttle orbiter. Its living space is very small, and the crew needs more space to live, exercise, and work. A **space station** has living quarters, work and exercise space, and all the equipment and support systems needed for humans to live and work in space.

The United States had such a station in the past. The space station *Skylab* was launched in 1973. Crews of astronauts spent up to 84 days in it performing experiments and collecting data on the effects on humans living in space. In 1979, the abandoned *Skylab* fell out of orbit and burned up as it entered Earth's atmosphere.

Crews from the former Soviet Union have spent the most time in space aboard the space station *Mir.* Two cosmonauts spent a record 365 days on board.

Science Journal

The *Galileo* craft rendezvoused with Jupiter after it first flew by Venus, circled the sun, and then flew past Earth. Research newspaper and magazine articles to find out why *Galileo* took such an indirect path to Jupiter. Research NASA's method of gravity-assisted space travel.

MiniLAB

How can gravity be simulated in a space station?

Procedure:

1. Locate an LP record album you can use for this activity.
2. Fold an 8-cm-wide strip of construction paper in half, then unfold it.
3. Wrap the paper along the fold around the circumference of the record, so there is a 4-cm wall around the outside edge of the disc.
4. Securely tape the rest underneath the record.
5. Place the record on a turntable and place three marbles at its center.
6. Switch the turntable on.

Analysis:

1. What do you observe about the movements of the marbles?
2. Hypothesize how what you've observed could be useful for simulating the effects of gravity on a space station.

Cooperation in Space

In 1995, the United States and Russia ushered in a time of cooperation and trust in exploring space. Early in the year, Dr. Norman Thagard was launched into orbit aboard the Russian *Soyuz* spacecraft along with two Russian cosmonaut crewmates. Dr. Thagard was the first United States astronaut launched into space by a Russian or Soviet booster. He then became the first American resident of the Russian space station *Mir.* Dr. Thagard was studying the long-term effects of space travel on humans.

In June 1995, Russian cosmonauts Anatoli Solovyev and Nikolai Budarin rode into orbit aboard the space shuttle *Atlantis,* America's 100th crewed launch. The mission of *Atlantis* involved, among other studies, a rendezvous and docking, shown in **Figure 21-13,** with space station *Mir.* Upon completion of its mission, *Atlantis* returned the *Mir 18* team, Dr. Thagard and cosmonauts Vladimir Dezhurov and Gennadiy Strekalov, to Earth. Both nations hope the

Figure 21-12

The proposed International Space Station is shown with the space shuttle docked.

cooperation that existed on this mission will continue through at least seven future space shuttle-*Mir* docking missions, an important step leading to the building and operating of the International Space Station set to begin in 1997.

International Space Station

The International Space Station will be a permanent laboratory designed for long-term research. A diverse selection of research topics will be studied. One example is the plans for research and growth of protein crystals. This project will help scientists determine protein structure and function. This could enhance work on drug design and the treatment of diseases.

The space station will draw on the resources of more than a dozen nations. Various nations will build modules for the space station that will then be transported into space aboard the space shuttle. The space station itself will be constructed in space. **Figure 21-12** shows what the completed station will look like with a space shuttle attached. NASA is planning the

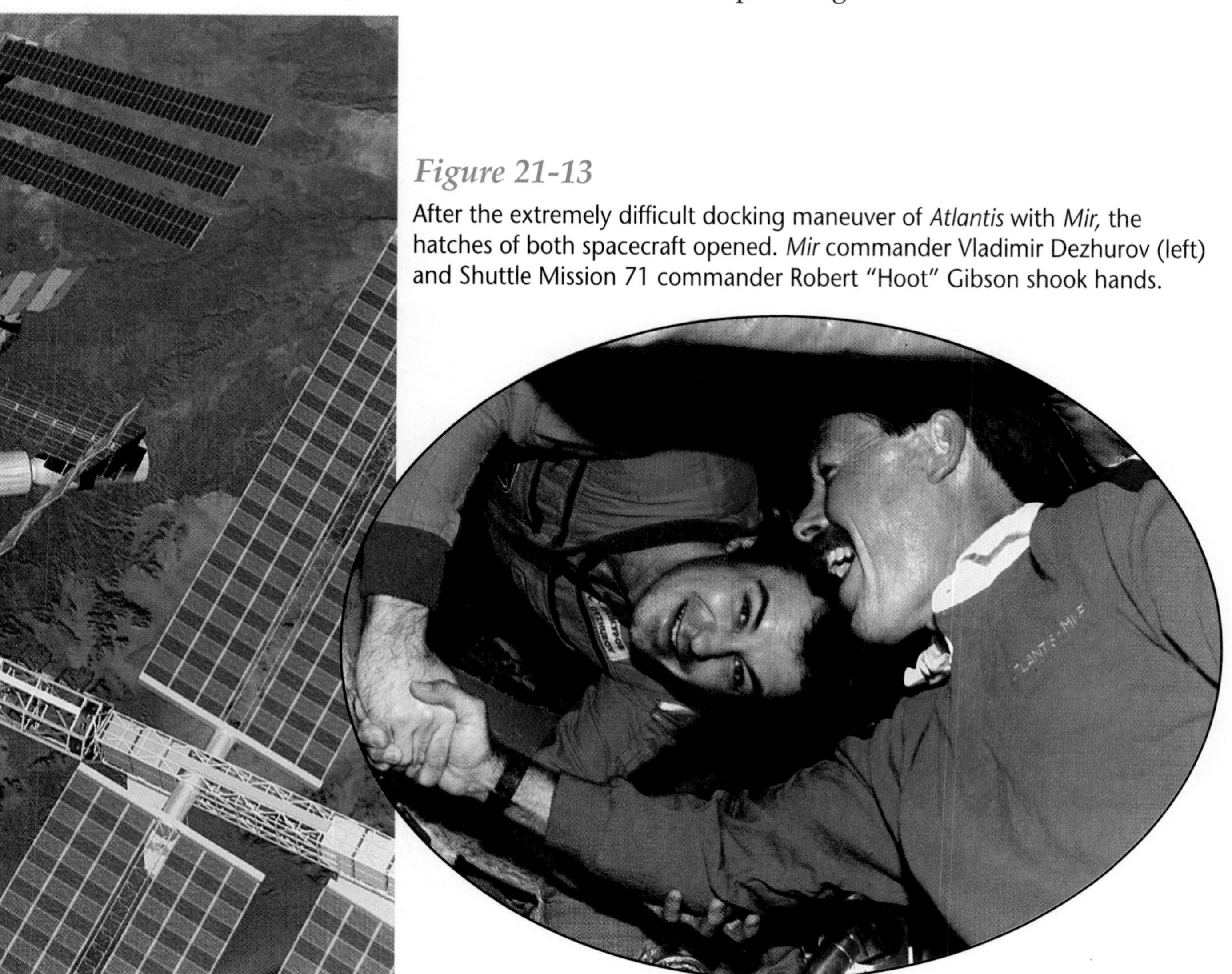

Figure 21-13

After the extremely difficult docking maneuver of *Atlantis* with *Mir*, the hatches of both spacecraft opened. *Mir* commander Vladimir Dezhurov (left) and Shuttle Mission 71 commander Robert "Hoot" Gibson shook hands.

space station program in three phases. Phase One involves the space shuttle-*Mir* docking missions. In Phase Two, NASA and RSA (Russian Space Agency) will assemble the bilateral station Alpha. In Phase Three, a crew will permanently operate a multinational space station.

NASA plans for crews of astronauts to stay on board the station for several months at a time. While there, researchers will make products that are returned for use on Earth. One day, the station could be a construction site for ships to the moon and Mars.

Figure 21-14

Using the space shuttle, scientists have already performed extensive experiments in the weightlessness of space.

Section Wrap-up

Review

1. What is the main advantage of the space shuttle?
2. Why are the space shuttle-*Mir* docking missions so important?
3. **Think Critically:** Why is the space shuttle more versatile than earlier spacecraft?

Skill Builder

Making and Using Tables

Make a table outlining the possible uses of the mechanical arm of the space shuttle's cargo bay. If you need help, refer to Making and Using Tables in the **Skill Handbook.**

Science Journal

Suppose you're in charge of assembling a crew for a new space station. Select 100 people you want for the station. Remember, you will need people to do a variety of jobs, such as farming, maintenance, scientific experimentation, and so on. In your Science Journal, relate whom you would select and why.

Chapter 21 Review

Summary

21-1: Radiation from Space

1. The arrangement of electromagnetic waves according to their wavelengths is the electromagnetic spectrum.
2. Optical telescopes produce magnified images of objects. A refracting telescope bends light to form an image. A reflecting telescope uses mirrors to focus light to produce an image.
3. Radio telescopes collect and record radio waves given off by some space objects.

21-2: Science and Society: Light Pollution

1. Lights are needed in towns and cities for safety and security reasons. However, light pollution can cause problems for stargazers as well as astronomers.

21-3: Artificial Satellites and Space Probes

1. A satellite is an object that revolves around another object. The moons of planets are natural satellites. Artificial satellites are those made by people.
2. An artificial satellite collects data as it orbits a planet. A space probe travels out into the solar system, gathers data, and sends it back to Earth.
3. Early American piloted space programs included the Mercury, Gemini, and Apollo projects.

21-4: The Space Shuttle and the Future

1. The space shuttle is a reusable spacecraft that carries astronauts, satellites, and other payloads to and from space.
2. Space stations provide the opportunity to conduct research not possible on Earth. The International Space Station will be constructed in Earth orbit with the cooperation of over a dozen nations.

Key Science Words

a. electromagnetic spectrum
b. light pollution
c. observatory
d. orbit
e. Project Apollo
f. Project Gemini
g. Project Mercury
h. radio telescope
i. reflecting telescope
j. refracting telescope
k. satellite
l. space probe
m. space shuttle
n. space station

Reviewing Vocabulary

Match each phrase with the correct term from the list of Key Science Words.

1. the arrangement of electromagnetic waves according to their wavelengths
2. uses lenses to bend light toward a focal plane
3. uses mirrors to collect light and form an image
4. glow in the night sky caused by city lights
5. an object that revolves around another object
6. the path traveled by a satellite
7. the first piloted U.S. space program
8. space program that reached the moon
9. carries people and tools to and from space
10. a place in space to live and work

Chapter 21 Review

Checking Concepts

Choose the word or phrase that completes the sentence.

1. The _______ spacecraft has sent back images of Venus.
   a. *Voyager* c. *Apollo 11*
   b. *Viking* d. *Magellan*
2. _______ telescopes use mirrors to collect light.
   a. Radio c. Refracting
   b. Electromagnetic d. Reflecting
3. A(n) _______ telescope can be used during the day or at night and during bad weather.
   a. radio c. refracting
   b. electromagnetic d. reflecting
4. _______ reduce light pollution.
   a. Radio telescopes
   b. Observatories
   c. Low-pressure sodium lamps
   d. Security lights
5. *Sputnik I* was the first _______.
   a. telescope c. observatory
   b. artificial satellite d. U.S. space probe
6. Goals of _______ were to put a spacecraft in orbit and bring it safely back.
   a. Project Mercury c. Project Gemini
   b. Project Apollo d. *Viking I*
7. The _______ of the space shuttle are reused.
   a. liquid-fuel tanks c. booster engines
   b. *Gemini* rockets d. *Saturn* rockets
8. The _______ of the space shuttle can place a satellite into space and retrieve it.
   a. liquid-fuel tank c. mechanical arm
   b. booster rocket d. cargo bay
9. *Skylab* was a(n) _______ that fell from its orbit.
   a. space probe c. space shuttle
   b. space station d. optical telescope
10. A natural satellite of Earth is _______.
    a. *Skylab* c. the sun
    b. the space shuttle d. the moon

Understanding Concepts

Answer the following questions in your Science Journal using complete sentences.

11. How do electromagnetic waves differ from mechanical waves? Give an example to support your answer.
12. Compare and contrast two types of optical telescopes.
13. List one natural and two artificial sources that prevent clear observations of the night sky.
14. Explain what two motions keep a satellite in orbit around Earth.
15. If given the chance, would you choose to fly on a shuttle mission? Give reasons for your answer.

Thinking Critically

16. How would a moon-based telescope have advantages over the Earth-based telescopes being used today?
17. Would a space probe to the sun's surface be useful? Explain.
18. Suppose NASA had to choose between continuing either the space flight programs with people aboard or the robotic space probes. Which do you think is the more valuable program? Explain your choice.
19. Suppose two astronauts were outside of the space shuttle orbiter while orbiting Earth. The audio speaker in the helmet of one of the astronauts quits working. The other astronaut is only one meter away, so she shouts a message to him. Can he hear her? Explain.
20. No space probes have visited the planet Pluto, the outermost planet of our solar system. Nevertheless, probes have crossed Pluto's orbit. Explain how this is possible.

Developing Skills

If you need help, refer to the **Skill Handbook.**

21. **Sequencing:** Arrange these events in order from earliest to the most recent: Galileo Galilei discovered four moons orbiting Jupiter, *Sputnik I* orbited Earth, humans landed on the moon, *Galileo* began its journey to Jupiter, Yuri Gagarin orbited Earth, Project Apollo began, *Discovery* launched the *Hubble Space Telescope,* astronauts and cosmonauts worked together on *Mir.*
22. **Measuring in SI:** Explain whether or not each of the following pieces of equipment could be used aboard the space shuttle as it orbits Earth: a balance, a graduated cylinder, a meterstick, and a thermometer.
23. **Concept Mapping:** Make an events chain map that explains what happens to different parts of the space shuttle, including the orbiter, liquid-fuel tank, and solid-fuel booster engines, from takeoff to landing.
24. **Making and Using Tables:** Copy the table below. Use information in the chapter as well as news articles and other resources to complete your table.

U.S. Space Probes

Probe	Launch Date	Destinations	Planets or Objects Visited
Mariner 4			
Vikings 1 & 2			
Pioneers 10 & 11			
Voyagers 1 & 2			
Magellan			
Galileo			

25. **Classifying:** Classify each of the following as a satellite or a space probe: Mercury spacecraft, *Sputnik I,* the *Hubble Telescope,* the space shuttle orbiter, and *Voyager 2.*

Performance Assessment

1. **Experiment:** Use several different-sized concave mirrors and determine which mirror concentrates more light into the image.
2. **Formulating a Hypothesis:** In Activity 21-2, you planned, carried out, and analyzed a robot surface probe mission to Mars. Based on how successful your mission was, what type of problems might hamper the success of a true robot mission to Mars? Would a crewed mission be more or less likely to be hampered by the same problems? Explain.
3. **Model:** Design and build a model of a space station. Include a way for people and equipment to be moved into and out of the station.

Previewing the Chapter

Chapter 22

The Sun-Earth-Moon System

The position of Earth and the moon in space affects our lives in many ways. For instance, as Earth revolves around the sun on its tilted axis, we experience the changing of the seasons. But seasons are not the same on different parts of Earth. When it's winter in the northern hemisphere, it's summer in the southern hemisphere, and vice versa. Why is this the case?

EXPLORE ACTIVITY

Determine what causes seasons.

1. Use a lamp without a shade to represent the sun.
2. Turn the lamp on, and hold a globe of Earth about 2 m from the lamp.
3. Tilt the globe slightly so the northern half points toward the sun.
4. Keeping the globe tilted in the same direction, walk halfway around the sun. Take care not to turn or twist the globe as you walk.

Observe: In which direction is the northern hemisphere pointing relative to the sun in step 3? In step 4? In your Science Journal, describe which seasons these positions represent for the northern hemisphere.

Previewing Science Skills

- In the **Skill Builders**, you will recognize cause and effect and measure in SI.
- In the **Activities**, you will make models, analyze and interpret data, and make inferences.
- In the **MiniLABs**, you will measure in SI, analyze data, interpret graphs, and draw conclusions.

22•1 Planet Earth

Science Words

sphere
axis
rotation
revolution
ellipse
equinox
solstice

Objectives

- Describe Earth's shape and list physical data about Earth.
- Compare and contrast the rotation and revolution of Earth.
- Demonstrate how Earth's revolution and tilt cause seasons to change on Earth.

Planet Earth Data

You rise early in the morning while it's still dark outside. You sit by the window and watch the sun come up. Finally, day breaks, and the sun begins its journey across the sky. But is the sun moving, or are you?

Today, we know that the sun appears to move across the sky because Earth is spinning as it travels around the sun. But it wasn't long ago that people believed Earth was the center of the universe. They believed Earth stood still and the sun traveled around it.

As recently as the days of Christopher Columbus, some people also believed Earth was flat. They thought that if you sailed far out to sea, you would eventually fall off the edge of the world. How do you know this isn't true? How have scientists determined Earth's shape?

Earth's Shape

Space probes and artificial satellites have sent back images that show Earth is sphere-shaped. A **sphere** is a round, three-dimensional object whose surface at all points is the same distance from its center. Tennis balls and basketballs are examples of spheres. But people had evidence of Earth's true shape long before cameras were sent into space.

Around 350 B.C., the Greek astronomer and philosopher Aristotle reasoned that Earth was spherical because it always

Figure 22-1

If Earth were flat, its shadow would look like this on the moon during an eclipse.

Table 22-1

Physical Properties of Earth	
Diameter (pole to pole)	12 714 km
Diameter (equator)	12 756 km
Circumference (poles)	40 008 km
Circumference (equator)	40 075 km
Mass	5.98×10^{27} g
Density	5.52 g/cm^3
Average distance to the sun	149 600 000 km
Period of rotation (1 day)	23 hr, 56 min
Period of revolution (1 year)	365 day, 6 hr, 9 min

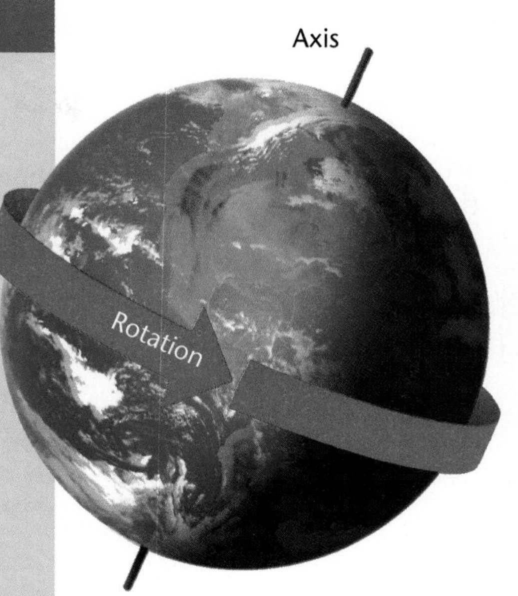

casts a round shadow on the moon during an eclipse. Only a spherical object always produces a round shadow. If Earth were flat, it would cast a shadow like the one in **Figure 22-1.**

Other evidence of Earth's shape was observed by early sailors. They watched ships approach from across the ocean and saw that the top of the ship would come into view first. As they continued to watch the ship, more and more of it would come into view. As the ship moved over the curved surface of Earth, they could see all of it.

Today, we know that Earth is sphere-shaped, but it is not a perfect sphere. It bulges slightly at the equator and is somewhat flattened at the poles. The poles are located at the north and south ends of Earth's axis. Earth's **axis** is the imaginary line around which Earth spins. The spinning of Earth on its axis, called **rotation,** causes day and night to occur.

Earth's Rotation

As Earth rotates, the sun comes into view at daybreak. Earth continues to spin, making it seem that the sun moves across the sky until it appears to set at night. During night, your area of Earth has spun away from the sun. Because of this, the sun is no longer visible. Earth continues to steadily rotate, and the sun eventually comes into view the next morning. One complete rotation takes about 24 hours, or one day. How many rotations does Earth complete during one year? As you can see in **Table 22-1,** it completes about 365 rotations during its journey around the sun.

MiniLAB

What is the shape of Earth?

Compare Earth's shape with other spheres.

Procedure

1. Use a long piece of string to measure the circumference of a basketball or volleyball.
2. Measure the circumference of the ball at a right angle to your first measurement.
3. Determine the roundness ratio by dividing the larger measurement by the smaller one.
4. Compare this data with the data about Earth's circumference provided in **Table 22-1.**

Analysis

1. How round is Earth compared with the ball?
2. Is Earth larger through the equator or through the poles?
3. Explain how your observations support your answer.

Earth rotates through an angle of 360° in one day. Through how many degrees does Earth rotate in one hour?

Earth's Magnetic Field

Recall from Chapter 11 that convection currents inside Earth's mantle power the movement of tectonic plates. Scientists hypothesize that movement of material inside Earth also generates a magnetic field.

The magnetic field of Earth is much like that of a bar magnet. Earth has a north and a south magnetic pole, just as a bar magnet has opposite magnetic poles at its ends. **Figure 22-2** illustrates the effects of sprinkling iron shavings over a bar magnet. The shavings align with the magnetic field of the magnet. Earth's magnetic field is similar, almost as if Earth had a giant bar magnet in its core.

Magnetic North

When you observe a compass needle pointing toward the north, you are seeing evidence of Earth's magnetic field. Earth's magnetic axis, the line joining its north and south magnetic poles, does not align with its rotational axis. The magnetic axis is inclined at an angle of 11.5° to the rotational axis. If you followed a compass needle pointing north, you would end up at the magnetic north pole rather than the geographic (rotational) north pole.

Earth's magnetic field and other physical properties affect us every day. What occurrences can you explain in terms of Earth's physical properties and movement in space?

Figure 22-2

Particles in the solar wind streaming through space from the sun distort Earth's magnetic field. As a result, it doesn't have the same shape as a magnetic field surrounding a bar magnet.

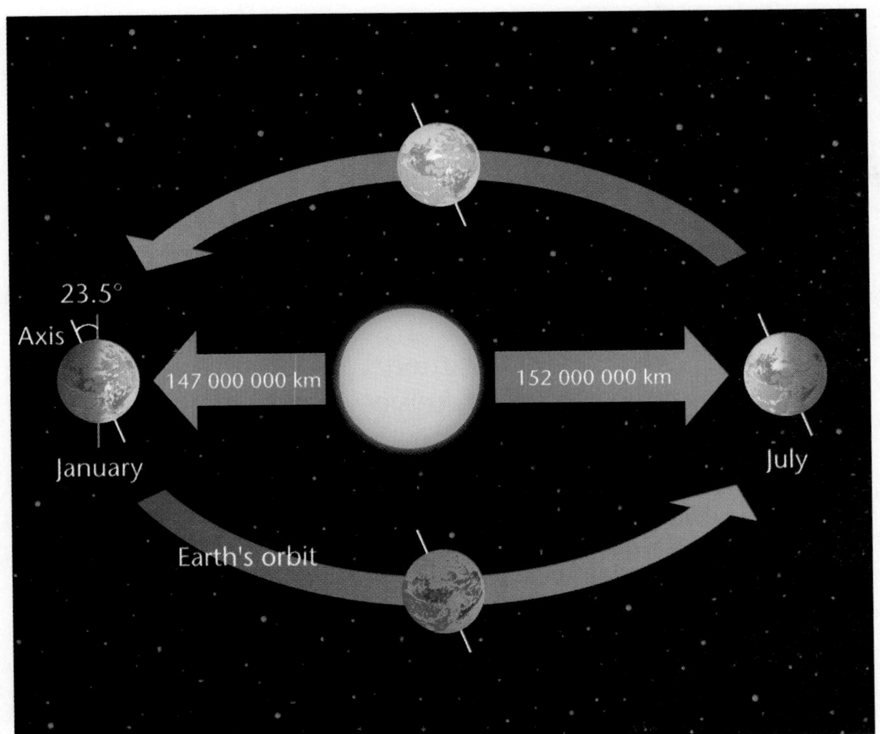

Figure 22-3

The northern hemisphere experiences summer when Earth is farthest from the sun. It experiences winter when Earth is closest to the sun. *Is the change of seasons caused by Earth's elliptical orbit? Explain your answer.*

Seasons

INTEGRATION Life Science

Autumn is coming and each day it gets colder outside. The sun rises later each morning and is lower in the sky. A month ago, it was light enough to ride your bike at 8:00 P.M. Now it's dark at 6:30 P.M. What is causing this change?

Earth's Revolution

You learned earlier that Earth's rotation causes day and night to occur. Another important motion of Earth is its **revolution,** or yearly orbit around the sun. Just as the moon is a satellite of Earth, Earth is a satellite of the sun. If Earth's orbit were a circle, and the sun were at the center of the circle, Earth would maintain a constant distance from the sun. However, this is not the case. Earth's orbit is an **ellipse,** an elongated closed curve. As **Figure 22-3** shows, the sun is offset from the center of the ellipse. Because of this, the distance between Earth and the sun changes during Earth's yearlong orbit. Earth gets closest to the sun—about 147 million km away—on January 3. The farthest point in Earth's orbit is about 152 million km away from the sun and is reached on July 4. Is this elliptical orbit causing the changing temperatures on Earth? If it were, you would expect the warmest days in January. You know this isn't the case in the northern hemisphere. Something else is causing the change.

Even though Earth is closest to the sun in January, the overall amount of energy Earth receives from the sun changes little throughout the year. However, the amount of energy any one place on Earth receives can vary quite a bit.

USING TECHNOLOGY

Clocks: Old and New

Atomic clocks measure time by recording the frequency of electromagnetic waves given off by atoms. Unlike conventional clocks, atomic clocks are not affected by changes in temperature or the wearing of their parts. As a result, they gain or lose less than one second in 200 000 years.

Using atomic clocks, scientists have found that each successive day on Earth is getting longer. Evidence indicates that Earth's rotation is slowing down.

Apparently, Earth's rotation has been slowing down for millions of years. By studying the growth lines on 375-million-year-old corals, scientists have determined that there were 440 days in a year at the time these corals were growing. Corals deposit monthly growth lines on their shells in much the same way trees develop yearly growth rings.

Atomic clocks and ancient corals indicate that Earth's rotation is slowing down. Scientists hypothesize that it is being dragged on by the gravitational attraction of the moon.

Think Critically:

As this drag continues, what will happen to the length of a day on Earth? To the length of a year?

Atomic Clock

Earth's Tilted Axis

Earth's axis is tilted 23.5° from a line perpendicular to its orbit. This tilt causes the seasons. Daylight hours are longer for the hemisphere tilted toward the sun. Think of how early it gets dark in the winter compared with the summer. The hemisphere tilted toward the sun receives more hours of sunlight than the hemisphere tilted away from the sun.

Another effect of Earth's tilt is that the sun's radiation strikes the hemisphere tilted toward it at a higher angle than it does the other hemisphere. Because of this, the hemisphere tilted toward the sun receives more electromagnetic radiation per unit area than the hemisphere tilted away. In other words, if you measured the amount of radiation received in a one-square-kilometer area in the northern hemisphere and, at the same time, measured it for one square kilometer in the southern hemisphere, you would find a difference. The hemisphere tilted toward the sun would be receiving more energy.

A summer season results when the sun is in the sky longer and its electromagnetic radiation strikes Earth at a higher angle. Just the opposite occurs during winter.

Science Journal

Suppose that Earth's rotation took twice the time that it presently does. In your Science Journal, write a report on how conditions such as global temperatures, work schedules, plant growth, and other factors might be different.

Equinoxes and Solstices

Because of the tilt of Earth's axis, the sun's position relative to Earth's equator constantly changes. Most of the time, the sun is north or south of the equator. Two times during the year, however, the sun is directly over the equator.

Equinox

Look at **Figure 22-4.** When the sun reaches an **equinox,** it is directly above Earth's equator, and the number of daylight hours equals the number of nighttime hours all over the world. At that time, neither the northern nor the southern hemisphere is tilted toward the sun. In the northern hemisphere, the sun reaches the spring equinox on March 20 or 21 and the fall equinox on September 22 or 23. In the southern hemisphere, the equinoxes are reversed. These dates are the first days of spring and fall.

Figure 22-4

At the northern hemisphere's summer solstice, the sun's rays directly strike the Tropic of Cancer, 23.5° north latitude. The sun is directly over the Tropic of Capricorn, 23.5° south latitude, at the northern hemisphere's winter solstice. At both fall and spring equinoxes, the sun is directly over the equator.

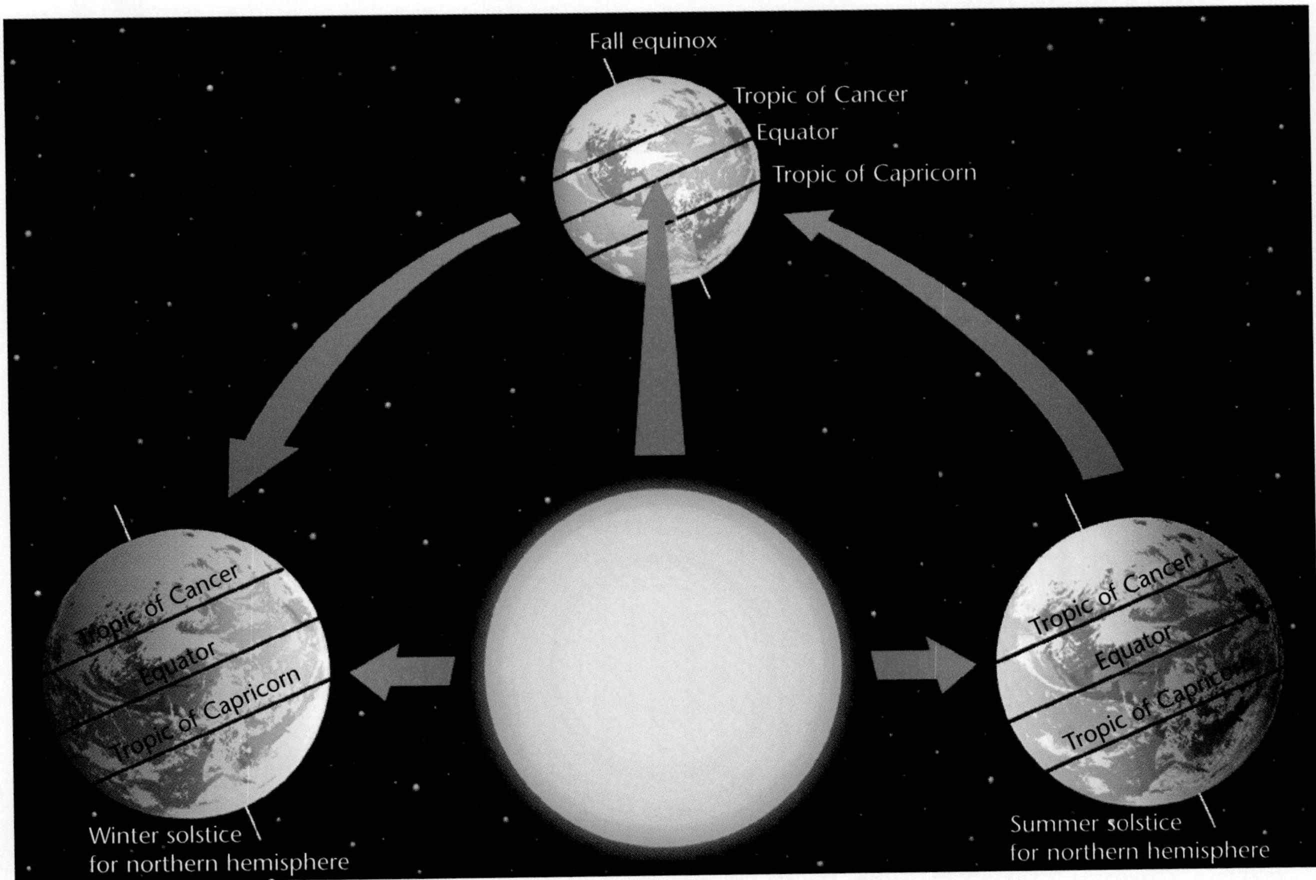

Activity 22-1

Design Your Own Experiment

Tilt and Temperature

Have you ever noticed how hot the surface of a blacktop driveway can get during the day? The sun's rays hit Earth more directly as the day progresses. Now, consider the fact that Earth is tilted on its axis. How does this affect the amount of heat an area on Earth receives from the sun?

PREPARATION

Problem

How is the angle at which light strikes an area on Earth related to the changing of the seasons?

Form a Hypothesis

State a hypothesis about how the angle at which light strikes an area affects the amount of heat energy received by that area.

Objectives

- Measure the amount of heat generated by a light as it strikes a surface at different angles.
- Describe how light striking a surface at different angles is related to the changing of the seasons on Earth.

Possible Materials

- tape
- black construction paper
- gooseneck lamp with 75-watt bulb
- Celsius thermometer
- watch
- protractor

Safety Precautions

Do not touch the lamp. The lightbulb and shade can be hot even when the lamp has been turned off. Handle the thermometer carefully. If it breaks, do not touch anything. Inform your teacher immediately.

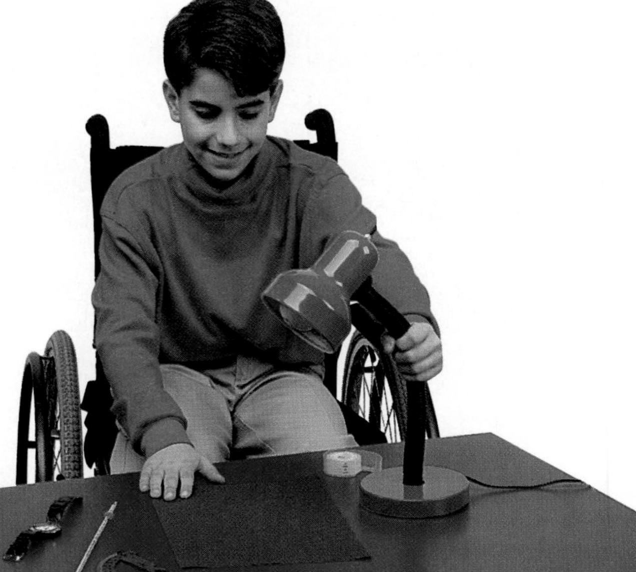

PLAN THE EXPERIMENT

1. As a group, agree upon and write out your hypothesis statement.
2. As a group, list the steps you need to take to test your hypothesis. Be specific, describing exactly what you will do at each step. List your materials.
3. Make a list of any special properties you expect to observe or test.

Check the Plan

1. Read over your entire experiment to make sure that all steps are in a logical order.
2. Identify any constants, variables, and controls in the experiment.
3. Will you summarize data in a graph, table, or some other form of organization?
4. How will you determine whether the length of time the light is turned on affects heat energy?
5. How will you determine whether the angle at which light strikes an area causes changes in heat and energy?
6. *Make sure your teacher approves your plan before you proceed.*

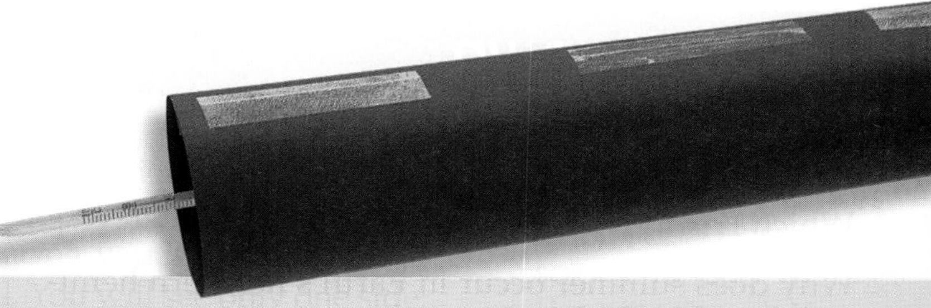

DO THE EXPERIMENT

1. Carry out the experiment as planned.
2. While the experiment is going on, write down any observations that you make and complete the data table in your Science Journal.

Analyze and Apply

1. **Describe** your experiment, including how you used independent variables to test your hypothesis.
2. What happened to the temperature of the area being measured as you modified your variables?
3. **Identify** the dependent variable in your experiment.
4. Did your experiment support your hypothesis? If not, **determine** how you might change the experiment in order to retest your hypothesis. How might you change your hypothesis?

Go Further

Predict how the absorption of heat would be affected by changing your independent variables. Try your experiment with different values for your independent variables.

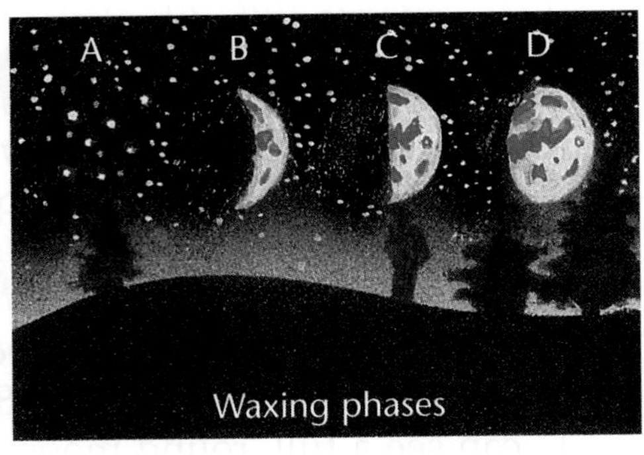

Figure 22-6

The phases of the moon: (A) new moon, (B) waxing crescent, (C) first quarter, (D) waxing gibbous, (E) full moon, (F) waning gibbous, (G) third quarter, and (H) waning crescent.

Why the Moon Shines

The moon shines by reflecting sunlight from its surface. Just as half of Earth experiences day as the other half experiences night, half of the moon is lighted while the other half is dark. As the moon revolves around Earth, you see different portions of its lighted side, causing the moon's appearance to change. **Moon phases** are the changing appearances of the moon as seen from Earth. The phase you see depends on the relative positions of the moon, Earth, and the sun.

USING MATH

The formula for the circumference of a circle is $2\pi r$, where r is the radius (one half the diameter) of the circle and π is approximately equal to 3.14. The moon's diameter at its equator is 3476 km. Find the circumference of the moon at its equator.

Phases of the Moon

A new moon occurs when the moon is between Earth and the sun. During a **new moon,** the lighted half of the moon is facing the sun and the dark side faces Earth. The moon is in the sky, but it cannot be seen.

Waxing Phases

Shortly after a new moon, more and more of the moon's lighted side becomes visible—the phases are **waxing.** About 24 hours after a new moon, you can see a thin slice of the side of the moon that is lighted by the sun. This phase, shown in **Figure 22-6,** is called the waxing crescent. About a week after a new moon, you can see half of the lighted side, or one-quarter of the moon's surface. This phase is **first quarter.**

The phases continue to wax. When more than one-quarter is visible, it is called waxing gibbous. A **full moon** occurs when all the moon's surface facing Earth is lit up.

Waning Phases

After a full moon, the amount of the moon's lighted side that can be seen becomes smaller. The phases are said to be **waning.** Waning gibbous begins just after a full moon. When you can see only half of the lighted side, the **third quarter** phase occurs. The amount of the moon that can be seen continues to become smaller. Waning crescent occurs just before another new moon. Once again, you can see a small slice of the lighted side of the moon.

The complete cycle of the moon's phases takes about 29.5 days. Recall that it takes about 27.3 days for the moon to revolve around Earth. The discrepancy between the time it takes for the moon to revolve around Earth and the time it takes for the moon to complete its phases is due to Earth's revolution. It takes the moon about two days to "catch up" with Earth's advancement around the sun.

Eclipses

Imagine yourself as one of your ancient ancestors, living 50 000 years ago. You are out foraging for nuts and other fruit in the bright afternoon sun. Gradually, the sun disappears from the sky, as if being swallowed by a giant creature. But the darkness lasts only a short time, and as quickly as the sun disappeared, it returns to full brightness. You realize something unusual has happened, but you don't know what caused it. It will be almost 48 000 years before anyone can explain the event that you just experienced.

The event just described was a total solar eclipse. Today, we know what causes such eclipses; but for our early ancestors, they must have been terrifying events. Many animals act as if night has come: cows return to their barns, and chickens go to sleep. What causes the day to suddenly change into night and suddenly back into day?

MiniLAB

How does the moon affect the ocean?

Find out how the phases and the position of the moon relate to tides.

Procedure

1. Use the graph you constructed on tidal ranges in the Chapter 17 Review. If you did not make the graph, you should do so now.
2. Review what you know about the position of the moon during neap and spring tides.

Analysis

1. On what days in your graph can you infer that the moon was at new or full phase?
2. On what days can you infer it was in first or third quarter phase?
3. How are the phases and position of the moon related to tides?

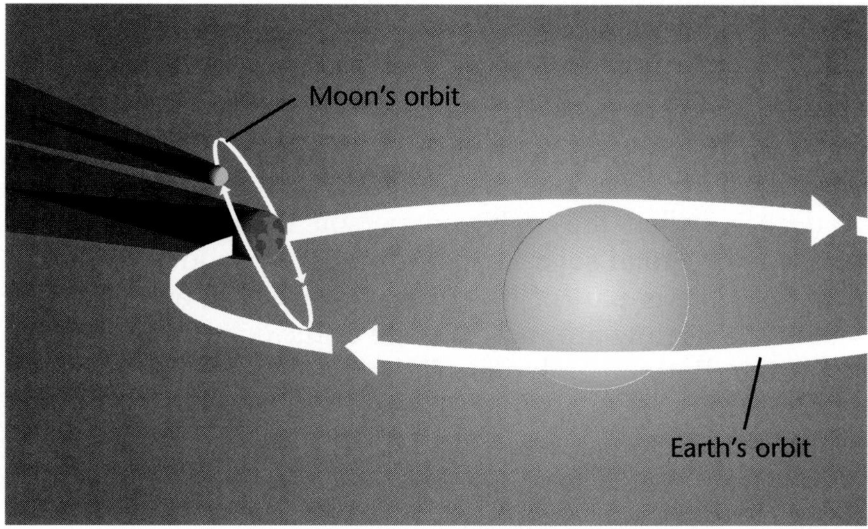

Figure 22-7

The orbit of the moon is not in the same plane as Earth's orbit around the sun. If it were, we would experience a solar eclipse each month during the new moon. The plane of the moon's orbit is tilted about five degrees to the plane of Earth's orbit. This illustration exaggerates the tilt to make it clear.

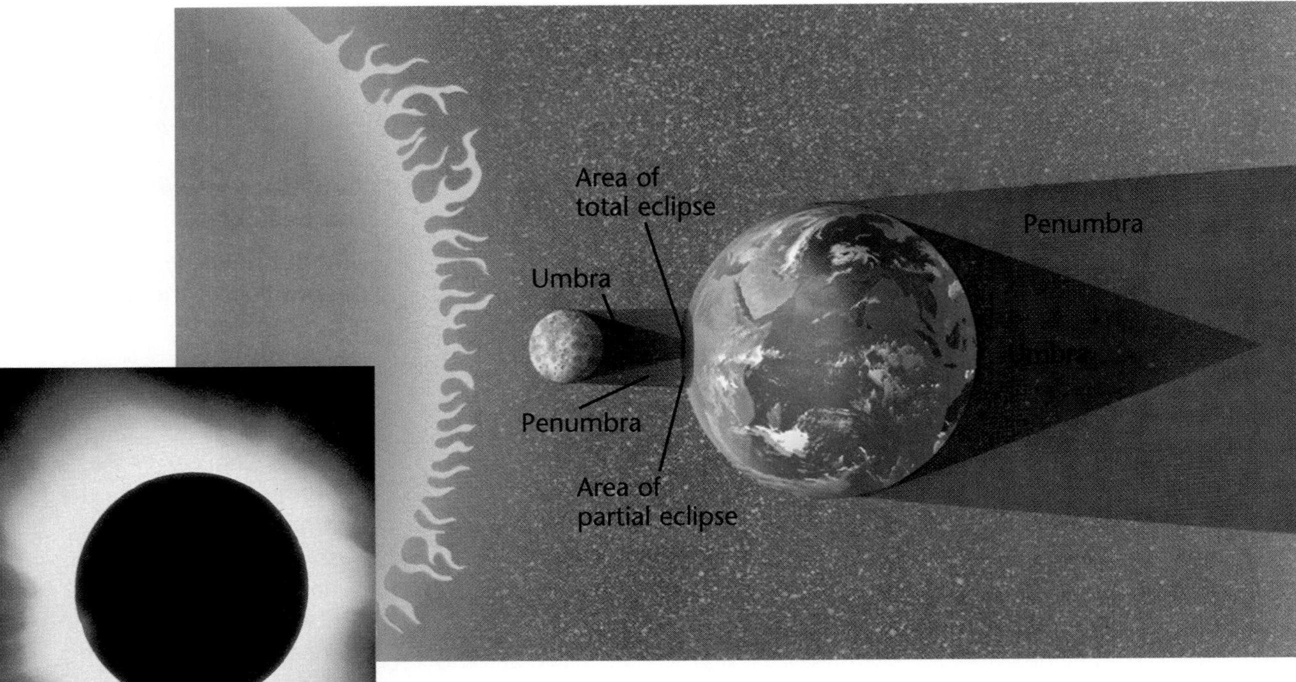

Figure 22-8

Only a small area of Earth experiences a total solar eclipse during the eclipse event. Only the outer portion of the sun's atmosphere is visible during a total solar eclipse. Distances are not drawn to scale.

The Cause of Eclipses

Revolution of the moon causes eclipses. Eclipses occur when Earth or the moon temporarily blocks the sunlight reaching the other. Sometimes during a new moon, a shadow cast by the moon falls on Earth and causes a solar eclipse. During a full moon, a shadow of Earth can be cast on the moon, resulting in a lunar eclipse.

Eclipses can occur only when the sun, the moon, and Earth are perfectly lined up. Because the moon's orbit is not in the same plane as Earth's orbit around the sun, eclipses happen only a few times each year.

CONNECT TO

PHYSICS

When a total solar eclipse occurs, Earth's moon completely blocks out the sun, which is about 400 times larger in diameter. In order for this to happen, ***describe*** what can be inferred about the relative distances to the two objects.

Solar Eclipses

A **solar eclipse,** such as the one in **Figure 22-8,** occurs when the moon moves directly between the sun and Earth and casts a shadow on part of Earth. The darkest portion of the moon's shadow is called the *umbra.* A person standing within the umbra sees a total solar eclipse. The only portion of the sun that is visible is part of its atmosphere, which appears as a pearly white glow around the edge of the eclipsing moon.

Surrounding the umbra is a lighter shadow on Earth's surface called the *penumbra.* Persons standing in the penumbra see a partial solar eclipse. **CAUTION:** *Regardless of where you are standing, never look directly at a solar eclipse. The light will permanently damage your eyes.*

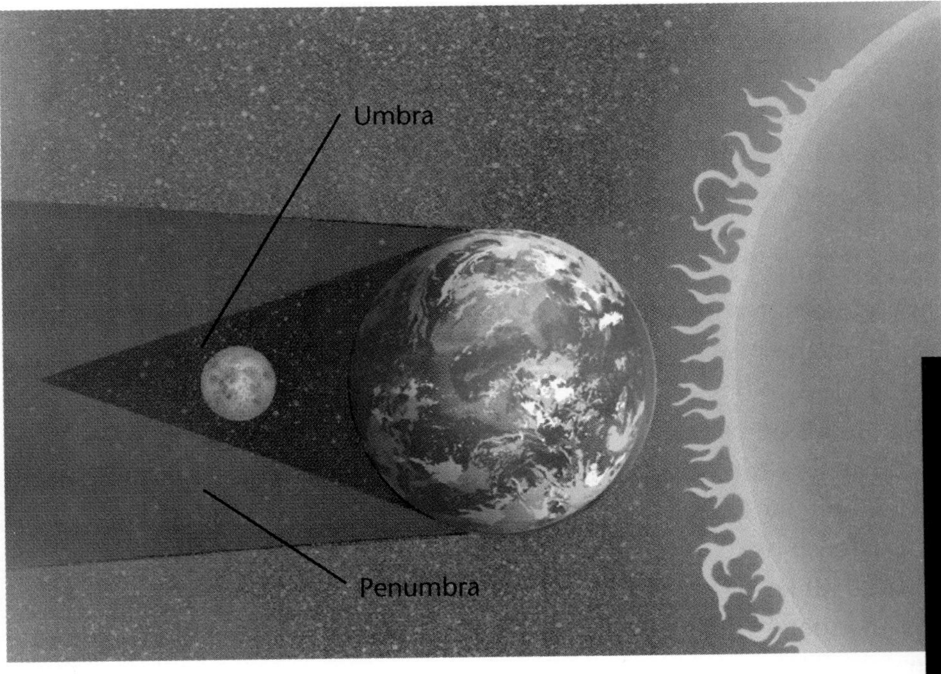

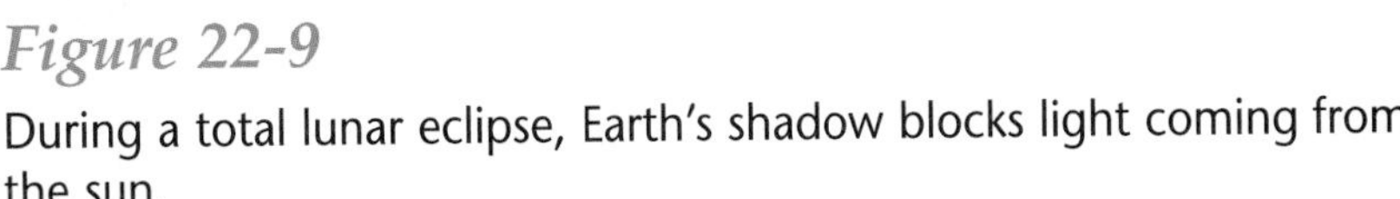

Figure 22-9

During a total lunar eclipse, Earth's shadow blocks light coming from the sun.

Lunar Eclipses

When Earth's shadow falls on the moon, a **lunar eclipse** like the one shown in **Figure 22-9** occurs. A lunar eclipse begins with the moon moving into Earth's penumbra. As the moon continues to move, it enters Earth's umbra and you see a curved shadow on the moon's surface. It was from this shadow that Aristotle concluded that Earth's shape was spherical. When the moon moves completely into Earth's umbra, the moon becomes dark red because light from the sun is refracted by Earth's atmosphere onto the moon. A total lunar eclipse has occurred.

A partial lunar eclipse occurs when only a portion of the moon moves into Earth's umbra. The remainder of the moon is in Earth's penumbra and, therefore, receives some direct sunlight.

A total solar eclipse occurs up to two times every year, yet most people live their entire lives never witnessing one. You may not be lucky enough to see a total solar eclipse, but it is almost certain you will have a chance to see a total lunar eclipse in your lifetime. The reason it is so difficult to view a total solar eclipse is that only those people in the small region where the moon's umbra strikes Earth can witness one. In contrast, anyone on the nighttime side of Earth can see a total lunar eclipse.

USING MATH

The moon's orbit is tilted at an angle of about 5° to Earth's orbit around the sun. Using a protractor, draw an angle of 5°.

Problem Solving

Survival on the Moon

You and your crew have crash-landed on the moon, far from your intended landing site at the moon colony. It will take one day to reach the colony on foot. The side of the moon that you are on will be facing away from the sun during your entire trip back. You manage to salvage the following items from your wrecked ship: food, rope, solar-powered heating unit, battery-operated heating unit, three 70-kg oxygen tanks, map of the constellations, magnetic compass, oxygen-burning signal flares, matches, 8 L of water, solar-powered radio receiver and transmitter, three flashlights and extra batteries, signal mirror, and binoculars. Keep in mind that the moon's gravity is about ⅙ that of Earth's, and it lacks a magnetic field.

Solve the Problem:

1. **Determine which items will be of no use to you.**
2. **Determine which items to take with you on your journey to the colony.**

Think Critically:

Based on what you have learned about the moon, describe why each of the salvaged items is useful or not useful.

1. How did the moon's physical properties affect your decisions?
2. How did the lack of sunlight affect your decisions?

Structure of the Moon

When you look at the moon, you can see many of its larger surface features. The dark-colored, relatively flat regions are called **maria.** Maria formed when ancient lava flows from the moon's interior filled large basins on the moon's surface. The basins formed early in the moon's history.

Craters

Many depressions on the moon were formed by meteorites, asteroids, and comets, which strike the surfaces of planets and their satellites. These depressions are called craters. During impact, cracks may have formed in the moon's crust, allowing lava to reach the surface and fill in the large craters. The igneous rocks of the maria are 3 to 4 billion years old. They are the youngest rocks found on the moon thus far.

The Moon's Interior

Seismographs left on the moon by *Apollo* astronauts have enabled scientists to study moonquakes. Recall from Chapter 9 that the study of earthquakes allows scientists to map Earth's interior. Likewise, the study of moonquakes has led to a model of the moon's interior. One model of the moon shows that its crust is about 60 km thick on the side facing Earth and about 150 km thick on the far side. Below the crust, a solid mantle may extend to a depth of 1000 km. A partly molten zone of the mantle extends farther down. Below this may be an iron-rich solid core.

Activity 22-2

Moon Phases and Eclipses

You know that moon phases and eclipses result from the relative positions of the sun, the moon, and Earth. In this activity, you will demonstrate the positions of these bodies during certain phases and eclipses. You will also see why only people on a small portion of Earth's surface see a total solar eclipse.

Problem

Can a model be devised to show the positions of the sun, the moon, and Earth during various phases and eclipses?

Materials 

- light source (unshaded)
- polystyrene ball on pencil
- globe
- **Figure 22-6**

Procedure

1. Study the positions of the sun, the moon, and Earth in **Figure 22-6.**
2. Use a polystyrene ball on a pencil as a model moon. Move the model moon around the globe to duplicate the exact position that would have to occur for a lunar eclipse to take place.
3. Move the model moon to the position that would cause a solar eclipse.
4. Place the model moon at each of the following phases: first quarter, full moon, third quarter, and new moon. Identify which, if any, type of eclipse could occur during each phase. Record your data.
5. Place the model moon at the location where a lunar eclipse could occur. Move it slightly toward, then away from Earth. Note the amount of change in the size of the shadow causing the eclipse. Record this information.
6. Repeat step 5 with the model moon in a position where a solar eclipse could occur.

Analyze

1. During which phase(s) of the moon is it possible for an eclipse to occur?
2. **Describe** the effect that a small change in the distance between Earth and the moon has on the size of the shadow causing the eclipse.
3. As seen from Earth, how does the apparent size of the moon **compare** with the apparent size of the sun? How can an eclipse be used to confirm this?

Conclude and Apply

4. **Infer** why a lunar and solar eclipse do not occur every month.
5. Suppose you wanted to more accurately model the movement of the moon around Earth. **Explain** how your model moon moves around the globe. Would it always be in the same plane as the light source and the globe?
6. Why have only a few people experienced a total solar eclipse?

Data and Observations

Moon Phase	Observations
first quarter	
full	
third quarter	
new	

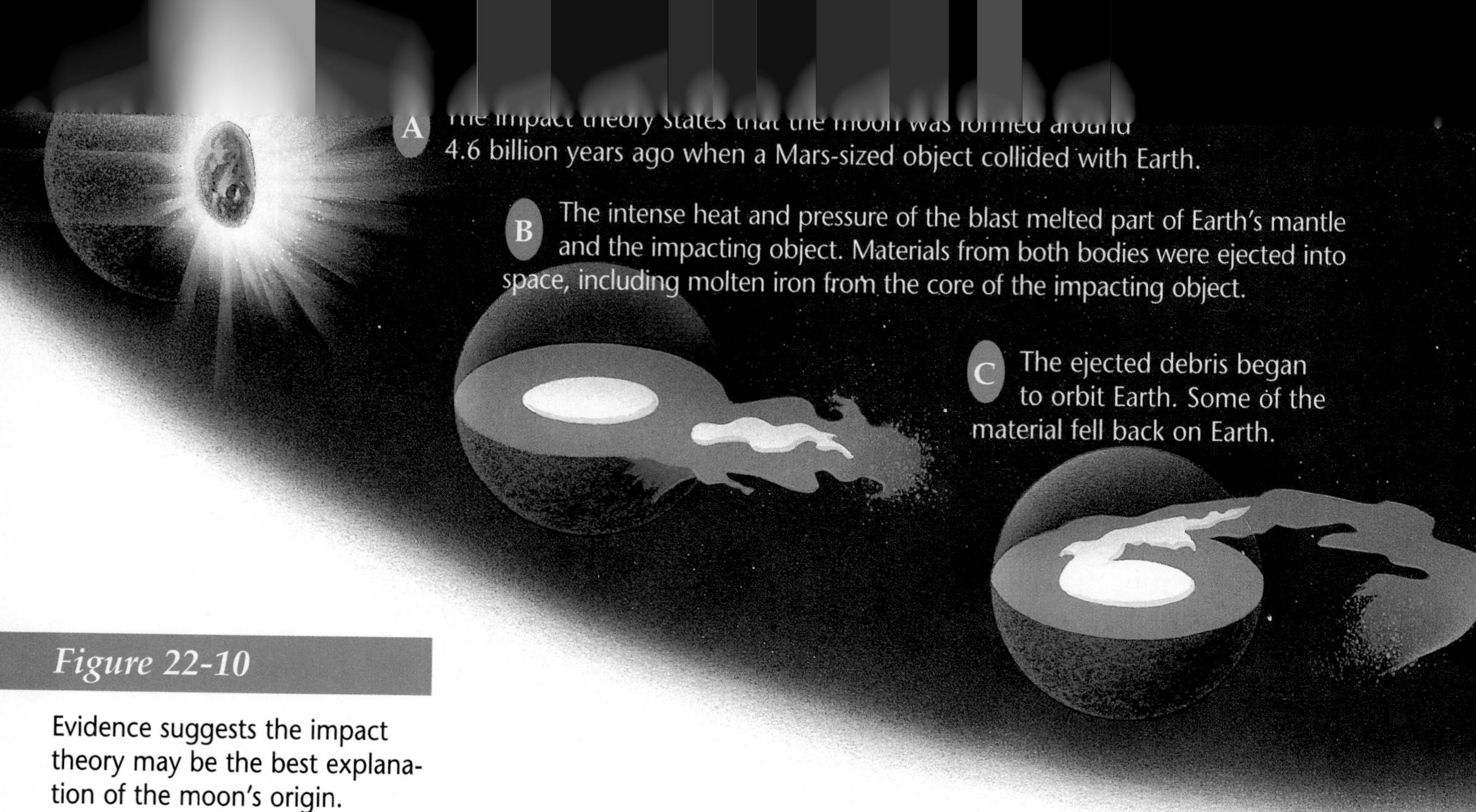

Figure 22-10

Evidence suggests the impact theory may be the best explanation of the moon's origin.

Origin of the Moon

Prior to the data obtained from the *Apollo* space missions, there were three theories about the moon's origin. The first is that the moon was captured by Earth's gravity. It formed elsewhere and wandered into Earth's vicinity. The second model is that the moon condensed from loose material surrounding Earth during the early formation of the solar system. The last theory is that a blob of molten material was ejected from Earth while Earth was still in its early molten stage.

Impact Theory

The data gathered by the *Apollo* missions have led many scientists to support the impact theory. According to the impact theory, the moon was formed about 4.6 billion years ago when a Mars-sized object collided with Earth, throwing gas and debris into orbit. The gas and debris then condensed into one large mass, forming the moon. **Figure 22-10** illustrates the impact theory.

Regardless of the moon's true origin, it has played an important role in our history. It was a source of curiosity for many early astronomers. Studying the phases of the moon and eclipses led people to conclude that Earth and the moon were in motion around the sun. Earth's shadow on the moon proved that Earth's shape was spherical. When Galileo first turned his telescope to the moon, he found a surface scarred by craters and maria. Before that time, many people believed

that all planetary bodies were "perfect," without surface features.

By studying the moon, we can learn about ourselves and the planet we live on. As you will read in the next section, not only is the moon important as an object from our past, but it is important to our future as well.

Section Wrap-up

Review

1. What are the relative positions of the sun, the moon, and Earth during a full moon?
2. Why does a lunar eclipse occur only during a full moon?
3. **Think Critically:** What provides the force necessary to form craters on the moon?

Skill Builder
Measuring in SI

The moon's mass is $1/81$ of Earth's mass. The moon's density is 3.3 g/cm^3. Calculate the moon's volume using the following formula.

volume = mass/density

If you need help, refer to Measuring in SI in the **Skill Handbook.**

Science Journal

Research the moon's origin in astronomy books and magazines. In your Science Journal, write a report about the various theories, including the theory about a Mars-sized object colliding with Earth. Make a drawing of each theory.

SCIENCE & SOCIETY

22•3 TECHNOLOGY: Exploration of the Moon

Science Words

mascon

Objectives

- List and discuss new information about the moon discovered by the *Clementine* spacecraft.
- List two facts about the moon's south pole that may be important to future space travel.

Return to the Moon

At this time, there are no ongoing, long-term missions planned for a study of the moon. However, one spacecraft was placed into lunar orbit in February 1994 to conduct a two-month survey of the moon's surface. Its name is *Clementine.*

The *Clementine* Spacecraft

The *Clementine* spacecraft has been successful in obtaining scientific data. *Clementine's* mission was to test new sensors for tracking cold objects in space. Cold objects are items such as satellites, warheads, or even near-Earth asteroids.

In addition, *Clementine* was placed in lunar orbit to take high-resolution photographs intended for use in compiling a detailed map of the moon's surface. **Figure 22-11** shows a photograph taken by *Clementine. Clementine's* four cameras were able to resolve features as small as 200 m across, enhancing our knowledge of the moon's surface.

The Moon's South Pole

The South Pole-Aitken Basin is the oldest identifiable impact feature on the moon's surface. It is also the largest and deepest impact basin or depression found thus far anywhere in the solar system, measuring 12 km in depth and 2500 km in diameter. Data returned by *Clementine* have given scientists the first set of high-resolution photographs of this area of the

Figure 22-11

This false-color photograph, taken by cameras on the *Clementine* spacecraft, shows the moon, the sun, and the planet Venus.

moon. Much of this depression stays in shadow throughout the moon's rotation, providing an area where ice deposits from impacting comets may have collected. Also, a large plateau that is always in sunlight was discovered in this area. With the possibility of water from the ice, this would be an ideal location for a moon colony powered by solar energy.

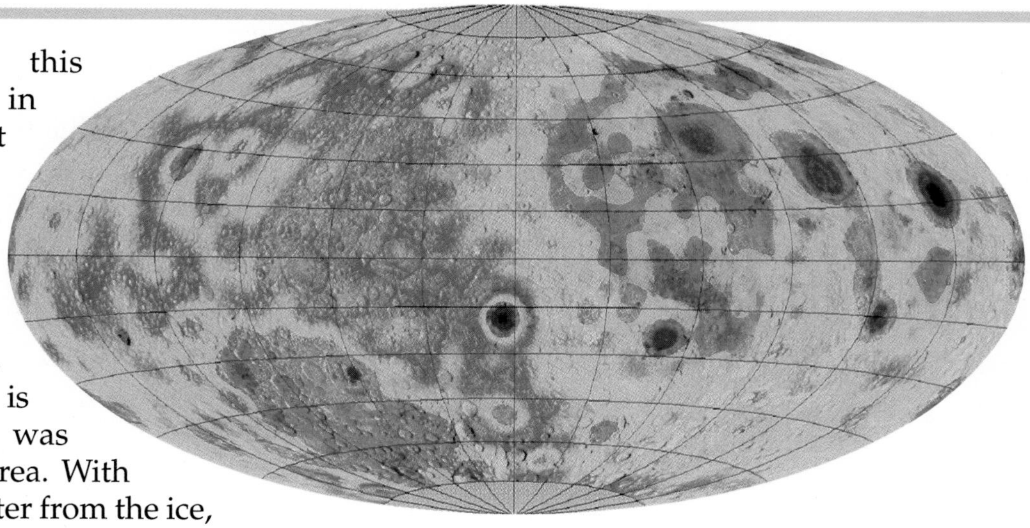

Figure 22-12

This computer-enhanced map based on *Clementine* data indicates the thickness of the moon's crust. It shows that the crust on the side of the moon facing Earth is thinner than the crust on the far side of the moon.

Figure 22-12 is a global map showing crustal thickness based on *Clementine* data. It shows that the crust thins under impact basins and that the moon's crust on the side facing Earth is much thinner than that on the far side. Such maps show the location of **mascons** or concentrations of mass. Mascons are located under impact basins. Data collected by *Clementine* also provided information on the composition or mineral content of moon rocks. In fact, this part of its mission was instrumental in naming the spacecraft. *Clementine* was the daughter of a miner in the ballad "My Darlin' Clementine."

Section Wrap-up

Review

1. List two discoveries about the moon made by the *Clementine* spacecraft.
2. Why would the discovery of ice at the moon's south pole be so important?

Explore the Technology

One of the most important contributions of the technology that placed *Clementine* in lunar orbit was the program's ability to accomplish 90 percent of its mission without going over budget. The spacecraft was built to budget, built on schedule, and used new technology only when it was cost efficient. Research how this compares to earlier space missions. Were they launched on schedule? Did they exceed projected budgets?

SCIENCE & SOCIETY

Science & Literature

The Sun and the Moon: A Korean Folktale

Science Journal

Use the library to research information about the history of astronomical science. In your Science Journal, make a time line showing the history of discoveries in astronomy dating back to its origins in ancient Egypt. Also try to find out when astronomical science began independently in various cultures around the world. Are there any similarities or differences in the way these cultures studied astronomy?

In this chapter, you have learned about the interactions among the sun, the moon, and Earth, and how these relationships relate to the changing seasons on Earth. Due to advances in technology such as satellites, advanced telescopes, and computer-imaging systems, scientists have learned more about the sun, the moon, and Earth in the last 50 years than they have throughout history.

Astronomy is our oldest science, having been studied for at least 4000 years. In fact, astronomy was the first science to be studied in most cultures around the world, from the ancient Egyptians to the native peoples of North America. People have always been curious about the objects they see in the sky and how they affect life on Earth.

Korea has long had an interest in science, particularly astronomy. Many scientists believe that early Korean astronomers built the first astronomical observatory in the Far East in 647 A.D. The Korean folktale "The Sun and the Moon" illustrates their keen interest in astronomy, as well as their knowledge about the sun and moon. This excerpt from the story describes the origin of the sun and moon.

. . . The two children lived peacefully in the Heavenly Kingdom, until one day the Heavenly King said to them, "We do not allow anyone to sit here and idle away the time. So I have decided on duties for you. The boy shall be the Sun, to light the world of men, and the girl shall be the Moon, to shine by night." Then the girl answered, "Oh King, I am not familiar with the night. It would be better for me not to be the Moon." So the King made her the Sun instead, and made her brother the Moon.

It is said that when she became the Sun, people used to gaze up at her in the sky. But she was modest, and greatly embarrassed by this. So she shone brighter and brighter, so that it was impossible to look at her directly. And that is why the Sun is so bright, that her modesty might be forever respected.

Can you see how this folktale shows an interest in and an understanding about certain astronomical facts? Which sentences in the story describe facts about the sun? The moon?

Chapter 22

Review

Summary

22-1: Planet Earth

1. Earth is a sphere that is slightly flattened at its poles. Earth's mass is nearly 6.0×10^{27} g and its density is 5.52 g/cm^3. Earth's magnetic field is due to convection currents deep inside Earth.
2. Earth rotates once each day and revolves around the sun in a little more than 365 days.
3. Seasons on Earth are due to the amount of solar energy received by a hemisphere at a given time. The tilt of Earth on its axis causes the amount of solar energy to vary.

22-2: Earth's Moon

1. Earth's moon goes through phases that depend on the relative positions of the sun, the moon, and Earth.
2. Eclipses occur when Earth or the moon temporarily blocks out the sunlight reaching the other. A solar eclipse occurs when the moon moves directly between the sun and Earth. A lunar eclipse occurs when Earth's shadow falls on the moon.
3. The moon's maria are the result of ancient volcanism. Craters on the moon's surface formed from impacts with meteorites, asteroids, and comets.

22-3: Science and Society: Exploration of the Moon

1. The *Clementine* spacecraft took detailed, high-resolution photographs of the moon's surface.
2. The moon's South Pole-Aitken Basin may contain ice deposits that could supply water for a moon colony.

Key Science Words

a. axis
b. ellipse
c. equinox
d. first quarter
e. full moon
f. lunar eclipse
g. maria
h. mascon
i. moon phase
j. new moon
k. revolution
l. rotation
m. solar eclipse
n. solstice
o. sphere
p. third quarter
q. waning
r. waxing

Reviewing Vocabulary

Match each phrase with the correct term from the list of Key Science Words.

1. Earth's approximate shape
2. causes day and night to occur on Earth
3. Earth's movement around the sun
4. shape of Earth's orbit
5. occurs when the sun's position is directly above the equator
6. the moon cannot be seen during this phase
7. moon phase in which all of the lighted side of the moon is seen
8. eclipse that occurs when the moon is between Earth and the sun
9. relatively flat region on the moon
10. concentration of mass on the moon located under an impact basin

Chapter 22 Review

Checking Concepts

Choose the word or phrase that completes the sentence.

1. The sun appears to rise and set because _______.
   a. Earth revolves
   b. the sun moves in space
   c. Earth rotates
   d. Earth orbits the sun

2. Earth's circumference at the _______ is greater than it is at the _______.
   a. equator, poles   c. poles, equator
   b. axis, mantle   d. mantle, axis

3. When the sun reaches equinox, the _______ is facing the sun.
   a. southern hemisphere
   b. northern hemisphere
   c. equator
   d. pole

4. The moon rotates once every _______.
   a. 24 hours   c. 27.3 hours
   b. 365 days   d. 27.3 days

5. The moon revolves once every _______.
   a. 24 hours   c. 27.3 hours
   b. 365 days   d. 27.3 days

6. As the lighted portion of the moon appears to get larger, it is said to _______.
   a. wane   c. rotate
   b. wax   d. be crescent-shaped

7. During a _______ eclipse, the moon is directly between the sun and Earth.
   a. solar   c. full
   b. new   d. lunar

8. The _______ is the darkest part of the moon's shadow during a solar eclipse.
   a. waxing gibbous   c. waning gibbous
   b. umbra   d. penumbra

9. _______ are depressions on the moon.
   a. Eclipses   c. Phases
   b. Moonquakes   d. Craters

10. Data gathered from the *Clementine* spacecraft support the fact that the moon _______.
   a. rotates once in 29.5 days
   b. has a thinner crust on the side facing Earth
   c. revolves once in 29.5 days
   d. has a thicker crust on the side facing Earth

Understanding Concepts

Answer the following questions in your Science Journal using complete sentences.

11. Compare and contrast rotation and revolution.

12. Why have observers on Earth never seen craters on one side of the moon?

13. How do equinoxes and solstices differ?

14. What causes umbras and penumbras?

15. Do you think the *Clementine* mission was worth the cost of funding the project? Explain.

Thinking Critically

16. How would the moon appear to an observer in space during its revolution? Would phases be observable? Explain.
17. Would you weigh more at Earth's equator or at the North Pole? Explain.
18. Recall from Chapter 17 that tides occur due to the gravitational attraction among the sun, the moon, and Earth. During which phases of the moon are tides the highest? Explain.
19. If you were lost on the moon's surface, why would it be more beneficial to have a star chart rather than a compass?
20. Which of the moon's motions are real? Which are apparent? Explain why each occurs.

Developing Skills

If you need help, refer to the **Skill Handbook.**

21. **Hypothesizing:** Hypothesize why locations near Earth's equator travel faster during one rotation than do places near the poles.
22. **Using Variables, Constants, and Controls:** Describe a simple activity to show how the moon's rotation and revolution work to keep one side facing Earth at all times.
23. **Inferring:** The moon does not produce its own light. Why can we see it in the night sky?
24. **Comparing and Contrasting:** Compare and contrast a waning moon with a waxing moon.
25. **Concept Mapping:** Copy and complete the cycle map shown on this page. Show the sequences of the moon's phases.

Performance Assessment

1. **Design an Experiment:** In Activity 22-1, you learned how the angle at which light strikes a surface determines the amount of heat absorbed. Design a new experiment that will enable you to measure the amount of heat absorbed by a paved surface from the sun at different times of the day. Explain what causes any differences in heat absorption that you notice.
2. **Model:** In Activity 22-2, you used models of Earth, the sun, and the moon to demonstrate moon phases and eclipses. Repeat the activity using yourself as the model of Earth. Have a classmate move the moon model around you until you observe all phases and eclipses.
3. **Oral Presentation:** Write a report describing how your life might be different if you lived in the southern hemisphere. Be sure to discuss the seasons and how they compare with the northern hemisphere's.

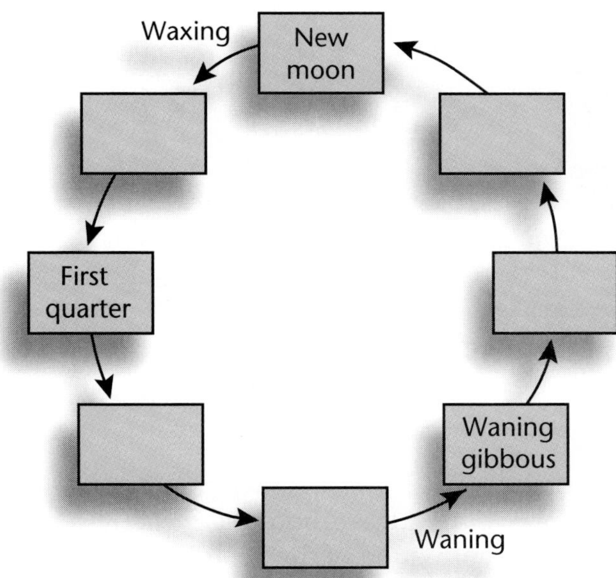

Previewing the Chapter

Chapter 23

The Solar System

The planets of our solar system are our nearest neighbors in space. But to us on Earth, they look like tiny points of light among the thousands of others visible on a clear night. With the help of telescopes and space probes, the points of light become giant colorful spheres, some with rings, others pitted with countless craters. This false-color image of the surface of Venus was made using space probe data and computers. In this chapter, explore our wandering neighbors as we look at the solar system.

EXPLORE ACTIVITY

What might happen to one of the solid-surface planets if a comet collided with it?

1. Place fine white flour into a cake pan to a depth of 3 cm, completely covering the bottom of the pan.
2. Cover the flour with 1 cm of fine, gray, dry cement mix, or try different colors of gelatin powder.
3. From varying heights, drop various-sized objects into the pan. Use marbles, lead weights, bolts, and nuts.

Observe: In your Science Journal, draw what happens to the surface of the powder in the pan as each object is dropped from different heights.

Previewing Science Skills

- In the **Skill Builders**, you will **map concepts**, **interpret data**, **recognize cause and effect**, and **interpret a scientific illustration.**
- In the **Activities**, you will **measure**, **predict**, **interpret**, and **use numbers.**
- In the **MiniLABs**, you will **make a model** and plan for a trip to Mars.

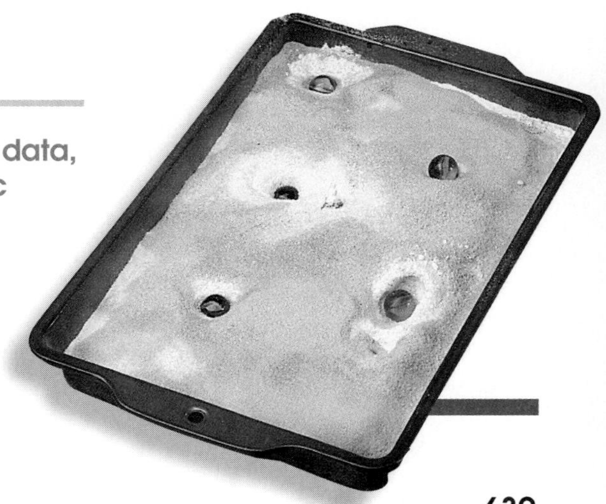

23•1 The Solar System

Science Words

solar system
inner planet
outer planet

Objectives

- Compare and contrast the sun-centered and Earth-centered models of the solar system.
- Describe current models of the formation of the solar system.

Early Ideas About the Solar System

Imagine yourself on a warm, clear summer night lying in the grass and gazing at the stars and the moon. The stars and the moon seem so still and beautiful. You may even see other planets in the solar system, thinking they are stars. Although the planets are very different from stars, they blend in with the stars and are usually hard to pick out.

The Earth-Centered Model

As you learned in Chapter 22, the sun and the stars appear to move through the sky because Earth is moving. This wasn't always an accepted fact. Many early Greek scientists thought the planets, the sun, and the moon were embedded in separate spheres that rotated around Earth. The stars were thought to be embedded in another sphere that also rotated around Earth. Early observers described moving objects in the night sky using the term *planasthai,* which means "to wander." This model is called the Earth-centered model of the solar system. To the astronomers who believed in this model of the solar system, there were Earth, the moon, the sun, five planets—Mercury, Venus, Mars, Jupiter, and Saturn—and the sphere of stars.

The Sun-Centered Model

The idea of an Earth-centered solar system was held for centuries until the Polish astronomer Nicholas Copernicus published a different view in 1543. He proposed that the moon revolved around Earth, which was a planet. Earth, along with

Figure 23-1
This instrument, called an astrolabe, was used for a variety of astronomical calculations.

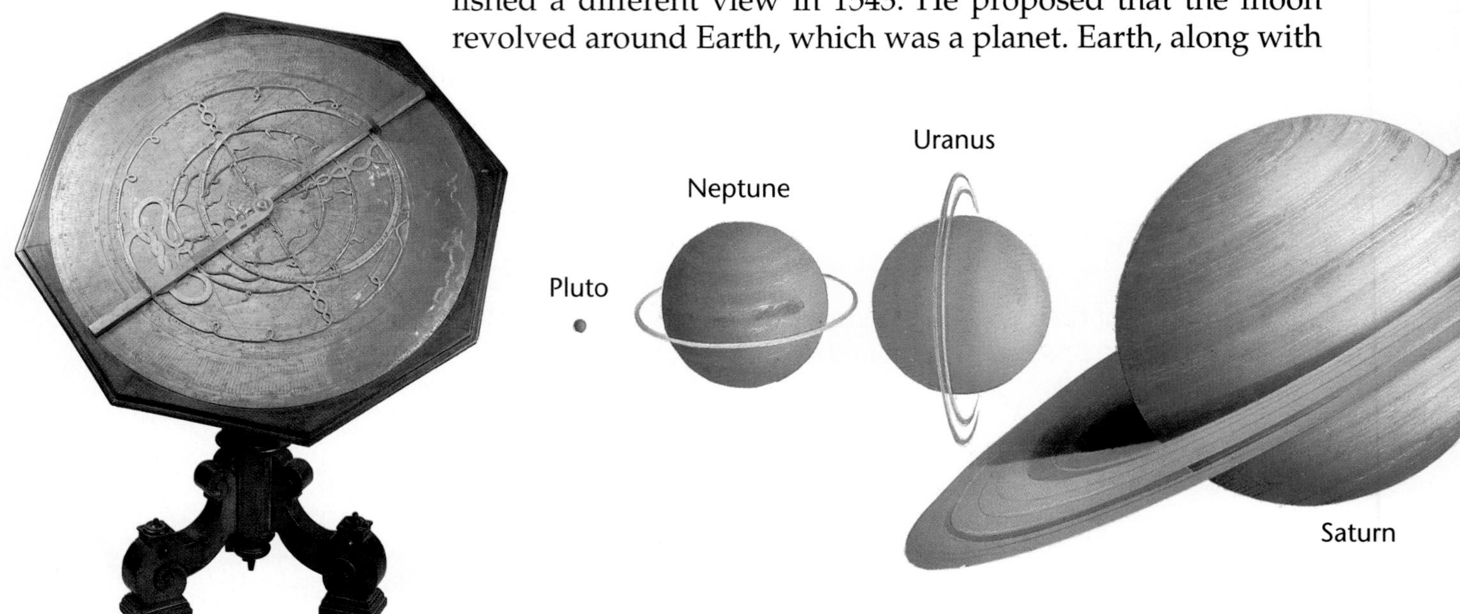

the other planets, revolved around the sun. He also stated that the daily movement of the planets and the stars was due to Earth's rotation. This is the sun-centered model of the solar system.

Using his telescope, the Italian astronomer Galileo Galilei found evidence that supported the ideas of Copernicus. He discovered that Venus went through phases like the moon's. These phases could be explained only if Venus were orbiting the sun. From this, he concluded that Venus revolves around the sun, and the sun is the center of the solar system.

Modern View of the Solar System

We now know that Earth is one of nine planets and many smaller objects that orbit the sun and make up the **solar system.** The sizes of the nine planets and the sun are shown to scale in **Figure 23-2;** however, the distances between the planets are not to scale. The dark areas on the sun are sunspots, which you will learn about in Chapter 24. Notice how small Earth is compared with some of the other planets and the sun, which is much larger than any of the planets.

The solar system includes a vast territory extending billions of kilometers in all directions from the sun. The sun contains 99.86 percent of the mass of the whole solar system. Because of its gravitational pull, the sun is the central object around which other objects of the solar system revolve.

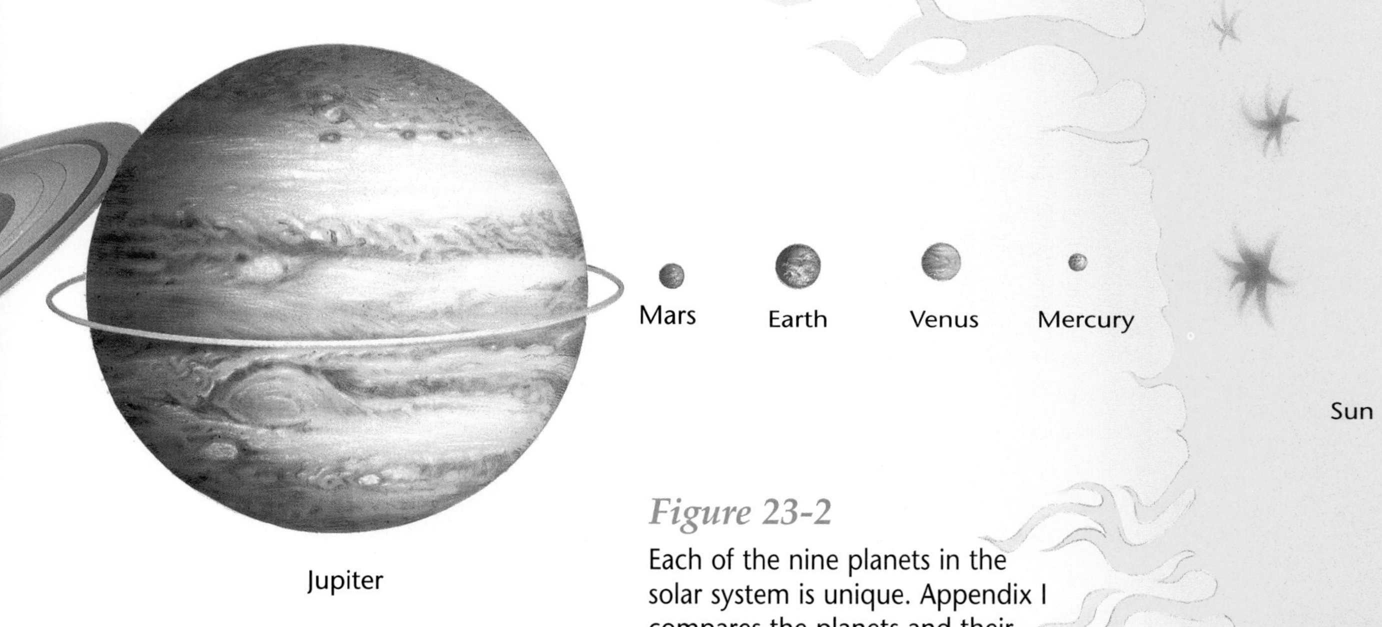

Figure 23-2

Each of the nine planets in the solar system is unique. Appendix I compares the planets and their characteristics.

Formation of the Solar System

Scientists hypothesize that the sun and the solar system formed from a cloud of gas, ice, and dust about 5 billion years ago. **Figure 23-3** illustrates how this may have happened. This cloud was slowly rotating in space. A nearby star may have exploded and the shock waves from this event may have caused the cloud to start contracting. Initially the cloud was rotating very slowly. But as it contracted and the density became greater, increased gravitational attraction pulled more gas and dust toward the cloud center. This caused the cloud to rotate faster, which in turn caused it to flatten into a disk with a dense center.

As the cloud contracted, the temperature began to increase. Eventually the temperature in the core of the cloud reached about 10 million °C and nuclear fusion began. This was the beginning of the sun becoming a star. Nuclear

Figure 23-3

Through careful observations, astronomers have found clues that help explain how our solar system may have formed.

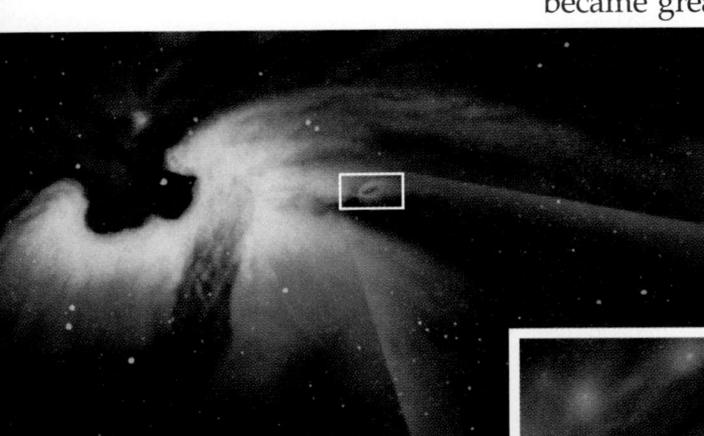

A About 5 billion years ago, a large cloud of gas, ice, and dust occupied our place in space.

B As gravity pulled matter inward, the cloud began to contract and spin. The densely packed matter grew extremely hot.

C The center of the rotating disk continued to heat. Meanwhile, gas and dust particles in the outer rim clumped together, forming larger objects.

D The larger clumps continued to grow through more collisions.

E Eventually, the larger clumps gathered enough matter to become the nine planets. The core of the disk grew even denser and hotter.

fusion occurs when atoms with low mass, such as hydrogen, combine to form heavier elements, such as helium. The new heavy element contains slightly less mass than the sum of the light atoms that formed it. The lost mass is converted into energy.

F Nuclear fusion began in the core and the sun became a star. Smaller objects became moons and rings around the planets.

Not all of the nearby gas, ice, and dust were drawn into the core of the cloud. Remaining gas, ice, and dust particles collided and stuck together, forming larger objects that in turn attracted more particles because of the stronger gravitational pull. Close to the sun the temperature was quite hot, and easily vaporized elements could not condense into solids. This accounts for the fact that light elements are more scarce in the planets closer to the sun than in planets further out in the solar system. Instead, the inner solar system is dominated by small, rocky planets with iron cores.

The **inner planets** of Mercury, Venus, Earth, and Mars are the solid, rocky planets closest to the sun. The **outer planets** of Jupiter, Saturn, Uranus, Neptune, and Pluto are those furthest from the sun. Except for Pluto, which is rocky, the outer planets are composed mostly of lighter elements such as hydrogen and helium.

Figure 23-4
This device is an early model of the solar system.

Motions of the Planets

When Nicholas Copernicus developed his sun-centered model of the solar system, he thought that the planets orbited the sun in circles. In the early 1600s, the German mathematician Johannes Kepler began studying the orbits of the planets. He discovered that the shapes of the orbits are not circular, but elliptical. He also calculated that the sun is not at the center of the ellipse, but is offset from the center.

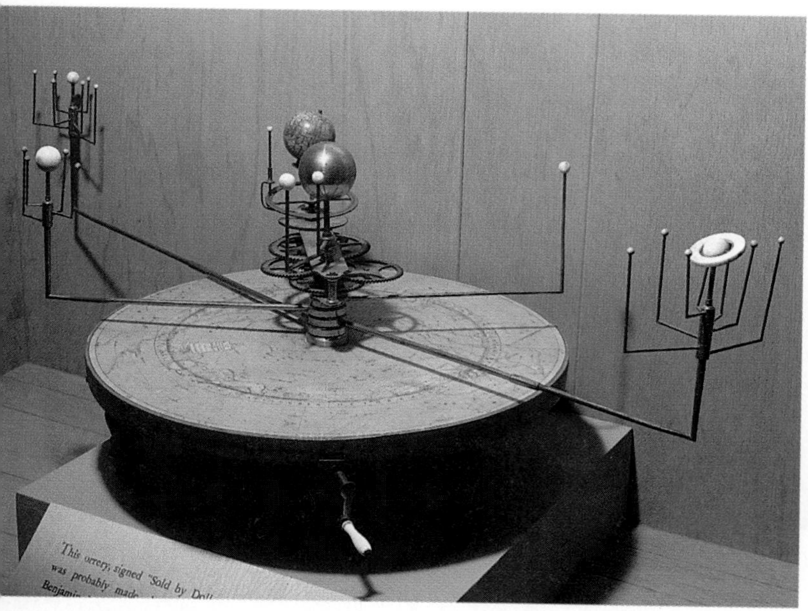

Kepler also discovered that the planets travel at different speeds in their orbits around the sun. By studying Appendix I, you can see that the planets closer to the sun travel faster than planets farther away from the sun. As a result, the outer planets take much longer to orbit the sun than the inner planets do.

Copernicus's ideas, considered radical at the time, led to the birth of modern astronomy. Early scientists didn't have technology such as space probes to learn about the planets. Nevertheless, they developed theories about the solar system that we still use today.

Section Wrap-up

Review

1. What is the difference between the sun-centered and the Earth-centered models of the solar system?
2. How do scientists hypothesize the solar system formed?
3. The outer planets are rich in water, methane, and ammonia—the materials needed for life. Yet life is unlikely on these planets. Explain.
4. **Think Critically:** Would a year on the planet Uranus be longer or shorter than an Earth year? Explain.

Skill Builder
Concept Mapping

Make a concept map that compares and contrasts the Earth-centered model with the sun-centered model of the solar system. If you need help, refer to Concept Mapping in the **Skill Handbook.**

Assuming that the planets travel in nearly circular orbits, use the information in Appendix I to determine how much faster (in km/s) Mercury travels in its orbit than Earth travels in its orbit.

Activity 23-1

Planetary Orbits

You have learned that planets travel around the sun along fixed paths called orbits. Early theories about the solar system stated that planetary orbits were perfect circles. Do this activity to construct a model of a planetary orbit. You will observe that the shape of planetary orbits is an ellipse, not a circle.

Problem

Can a model be constructed that will show planetary orbits to be elliptical?

Materials

- thumbtacks or pins
- string
- cardboard (21.5 cm × 28 cm)
- metric ruler
- pencil
- paper

Procedure

Part A

1. Place a blank sheet of paper on top of the cardboard and place two thumbtacks or pins about 3 cm apart.
2. Tie the string into a circle with a circumference of 15 to 20 cm. Loop the string around the thumbtacks. With someone holding the tacks or pins, place your pencil inside the loop and pull it taut.

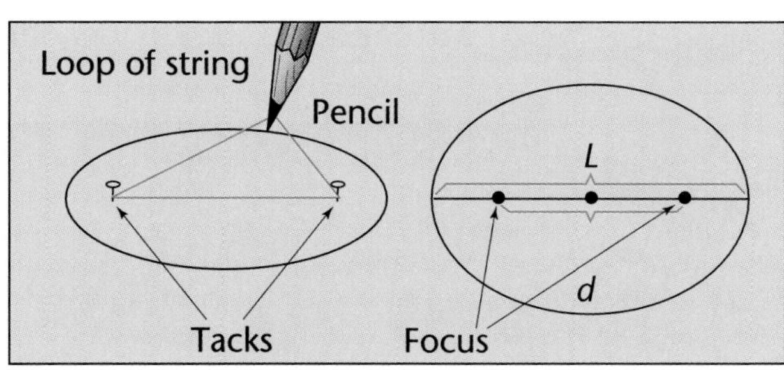

3. Move the pen or pencil around the tacks, keeping the string taut, until you have completed a smooth, closed curve, or ellipse.
4. Repeat steps 1 through 3 several times. First vary the distance between the tacks and then vary the length of the string. However, change only one of these each time. Note the effect on the size and shape of the ellipse with each of these changes.
5. Orbits are usually described in terms of eccentricity (*e*). The eccentricity of any ellipse is determined by dividing the distance (*d*) between the foci (here, the tacks) by the length of the major axis (*L*). See the diagram at left.
6. Calculate and record the eccentricity of the ellipses that you constructed.

Part B

7. Refer to Appendix I to determine the eccentricities of planetary orbits.
8. Construct an ellipse with the same eccentricity as Earth's orbit.
9. Repeat step 8 with the orbit of either Pluto or Mercury.

Analyze

1. **Analyze** the effect a change in the length of the string or the distance between the tacks has on the shape of the ellipse.
2. **Hypothesize** what must be done to the string or placement of tacks to decrease the eccentricity of a constructed ellipse.

Conclude and Apply

3. **Describe** the shape of Earth's orbit. Where is the sun located within the orbit?
4. **Identify** the planets that have the most eccentric orbits.
5. **Describe** the path of an orbit with an eccentricity of zero.

23•2 The Inner Planets

Science Words

Mercury
Venus
Earth
astronomical unit
Mars

Objectives

- List the inner planets in their relative order from the sun.
- Identify important characteristics of each inner planet.
- Compare and contrast Venus and Earth.

Inner Planets

We have learned much about the solar system since the days of Copernicus and Galileo. Advancements in telescopes allow astronomers to observe the planets from Earth. And space probes have explored much of our solar system, adding greatly to the knowledge we have about the planets. Let's take a tour of the solar system through the "eyes" of the space probes.

Mercury

The closest planet to the sun is **Mercury.** It is also the second-smallest planet. The first and only American spacecraft mission to Mercury was in 1974-1975 by *Mariner 10,* which flew by the planet and sent pictures back to Earth. *Mariner 10* imaged only 45 percent of Mercury's surface—we do not know what the other 55 percent looks like. What we have seen is that the surface of Mercury has many craters and looks much like our moon. It also has cliffs as high as 3 km on its surface, as seen in **Figure 23-5.** These cliffs may have formed when Mercury apparently shrank about 2 km in diameter.

Why did Mercury apparently shrink? Scientists think the answer may lie inside the planet. *Mariner 10* detected a weak magnetic field around Mercury, indicating that the planet has a large iron core. Some scientists hypothesize that the crust of Mercury solidified while the iron core was still hot and molten. Then, as the

Figure 23-5

Giant cliffs on Mercury, like the one marked by the arrow, suggest that the planet might have shrunk.

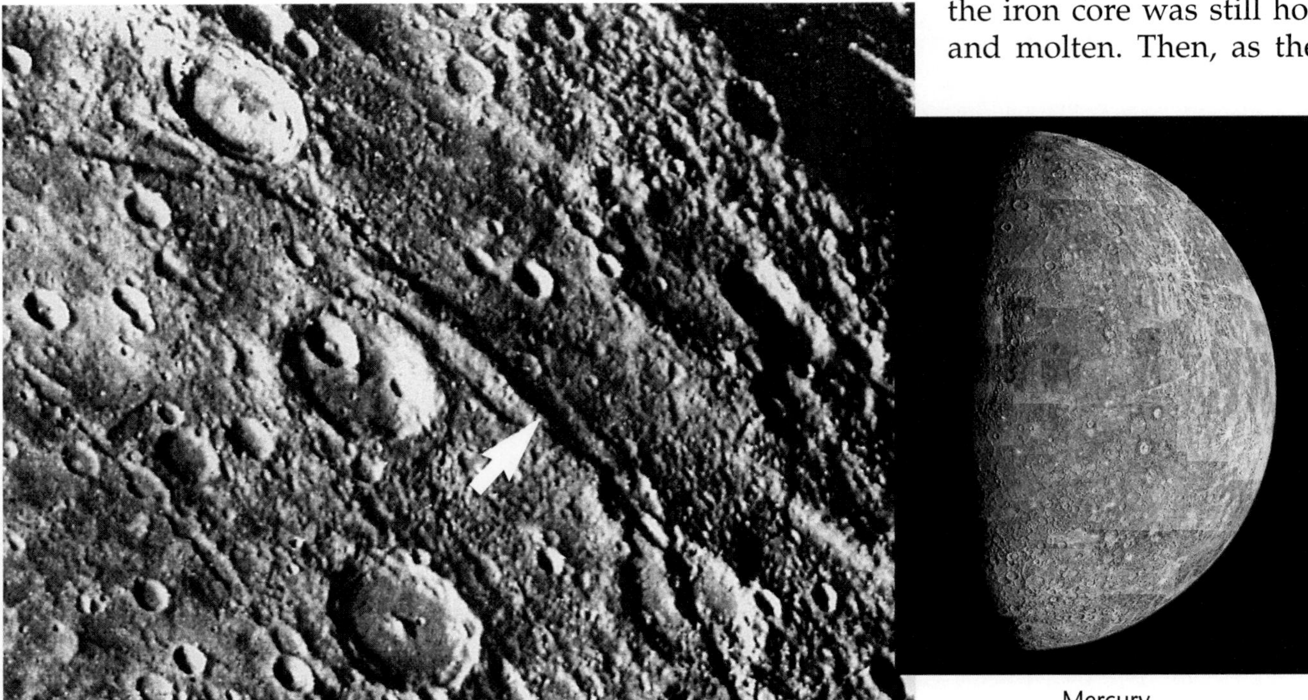

Mercury

core cooled and solidified, it contracted, causing the planet to shrink. The large cliffs may have resulted from breaks in the crust caused by this contraction, similar to what happens when an apple dries out and shrivels up.

Because of Mercury's small size and low gravitational pull, most gases that could form an atmosphere escape into space. Mercury's thin atmosphere is composed of hydrogen and helium (possibly trapped from the solar wind) and sodium and potassium. The sodium and potassium may diffuse upward through the crust. The thin atmosphere and the proximity of Mercury to the sun cause this planet to have great extremes in temperature. Mercury's surface temperature can reach 450°C during the day and drop to −170°C at night.

USING MATH

The average distance from the sun to Earth is 150 million km. How many minutes does it take light traveling at 300 000 km/s to reach Earth?

Venus

The second planet outward from the sun is **Venus.** Venus is sometimes called Earth's twin because its size and mass are similar to Earth's. One major difference is that the entire surface of Venus is blanketed by an atmosphere with dense clouds. The atmosphere of Venus, which has 90 times the surface pressure of Earth's at sea level, is mostly carbon dioxide. The clouds in the atmosphere contain droplets of sulfuric acid, which gives them a slightly yellow color, as seen in **Figure 23-6.**

Clouds on Venus are so dense that only two percent of the sunlight that strikes the top of the clouds reaches the planet's surface. This solar energy is trapped by the carbon dioxide gas and causes a greenhouse effect similar to Earth's greenhouse effect, discussed in Chapter 16. Due to this intense greenhouse effect, the temperature on the surface of Venus is 470°C.

The former Soviet Union led the exploration of Venus. Beginning in 1970 with the first *Venera* probe, the Russians have photographed and mapped the surface of Venus using radar and several surface probes. Between 1990 and 1994, the United States' *Magellan* probe used its radar to make the most detailed maps yet of the surface of Venus. *Magellan* revealed huge craters, faultlike cracks, and volcanoes with visible lava flows.

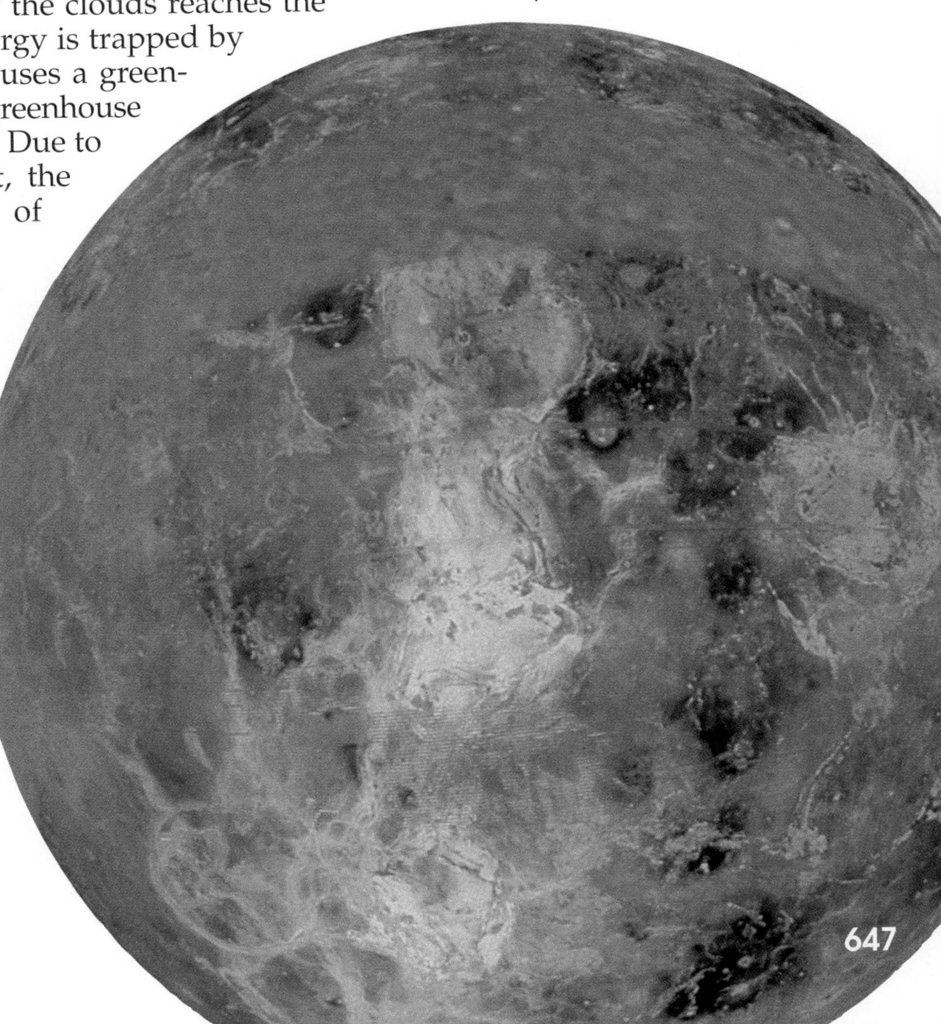

Figure 23-6

Although Venus is similar to Earth, there are important differences. *How could studying Venus help us learn more about Earth?*

Which planet is hotter?

Your teacher asks you to determine which planet's surface is hotter, Mercury or Venus. She also asks you to explain why one of them is hotter than the other. You decide that this assignment is going to be very easy. Of course Mercury has to be hotter than Venus because it is much closer to the sun. Venus is almost twice as far away as Mercury. You write your answer and turn in your paper. Later, when you receive your paper back, you find out that your assumptions were evidently wrong. Your teacher suggests that you research the question further.

Solve the Problem:

1. **Research both planets and determine their surface temperatures and atmospheric density and content.**
2. **Consider how a greenhouse works to keep it warmer inside than outside and relate this to what might happen to a planet with a thick atmosphere.**

Think Critically:

What causes Venus to have a higher surface temperature than Mercury? Explain.

Earth

Earth, shown in **Figure 23-7,** is the third planet from the sun. The average distance from Earth to the sun is 150 million km, or one **astronomical unit** (AU). Astronomical units are used to measure distances to objects in the solar system.

Surface temperatures on Earth allow water to exist as a solid, liquid, and gas. Earth's atmosphere causes most meteors to burn up before they reach the surface. The atmosphere also protects life from the sun's intense radiation.

Mars

Mars is the fourth planet from the sun. It's referred to as the red planet because iron oxide in the weathered rocks on its surface gives it a reddish color, as seen in **Figure 23-8.** Other features of Mars visible from Earth are its polar ice caps, which get larger

Figure 23-7

More than 70 percent of Earth's surface is covered by liquid water, causing our planet to appear blue from space.

during the Martian winter and shrink during the summer. The southern polar ice cap is made mostly of frozen carbon dioxide. The northern polar ice cap is made of water in the form of ice.

Most of the information we have about Mars came from the *Mariner 9* and *Viking* probes. *Mariner 9* orbited Mars in 1971–1972. It revealed long channels on the planet that may have been carved by flowing water. Because liquid water no longer exists on Mars's surface, the existence of the channels suggests that climatic conditions on Mars may have been much different in the past. *Mariner 9* also discovered the largest volcano in the solar system, Olympus Mons. Large rift zones that formed in the Martian crust were discovered as well. One such rift, Valles Marineris, is shown in **Figure 23-9.**

In 1976, the *Viking 1* and *2* probes arrived at Mars. Each spacecraft consisted of an orbiter and a lander. The *Viking 1* and *2* orbiters photographed the entire surface of Mars from orbit, while the *Viking 1* and *2* landers touched down on the planet's surface to conduct meteorological, chemical, and biological experiments. The biological experiments found no evidence of life in the soil. The *Viking* landers also sent back pictures of a reddish-colored, barren, rocky, and windswept surface.

Figure 23-8

Mars has many features in common with Earth.

MiniLAB

How would construction be different on Mars?

Procedure

1. Suppose you are a crane operator who is sent to Mars to help build a Mars colony.
2. You know that your crane can lift 44 500 newtons on Earth, but the gravity on Mars is only 0.4 of Earth's gravity.
3. Using numbers, determine how much mass your crane could lift on Mars.

Analysis

1. How can what you have discovered be an advantage over construction on Earth?
2. In what ways might construction advantages change the overall design of the Mars colony?

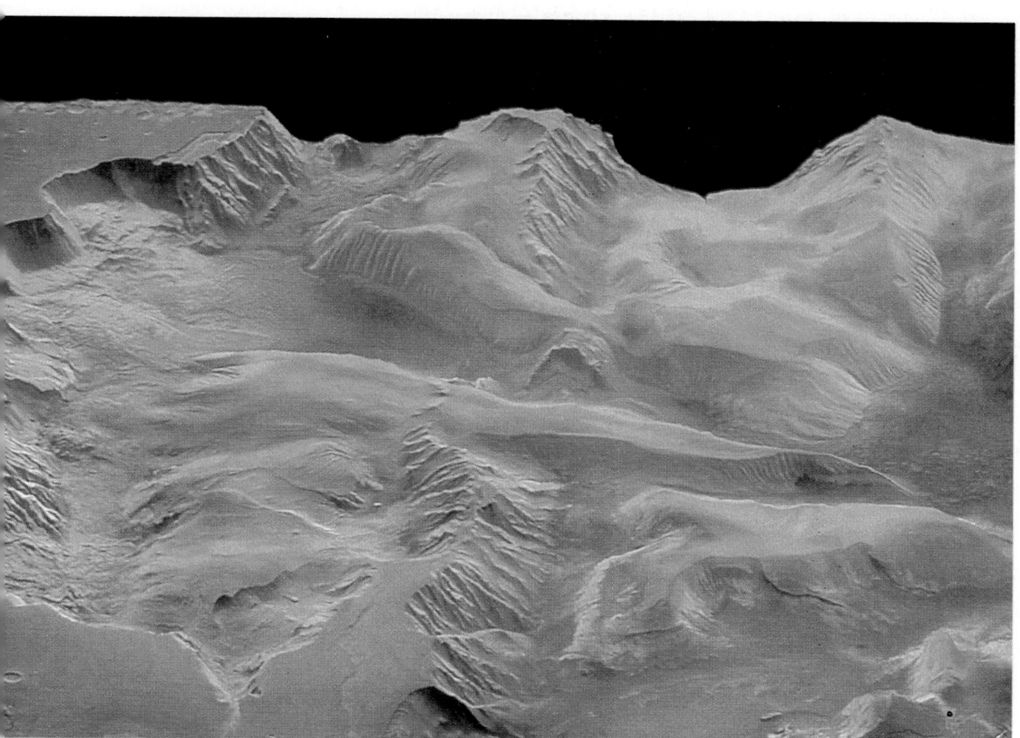

Figure 23-9

Valles Marineris is more than 4000 km long, up to 240 km wide, and more than 6 km deep.

Figure 23-10

These photos show two features of Mars.

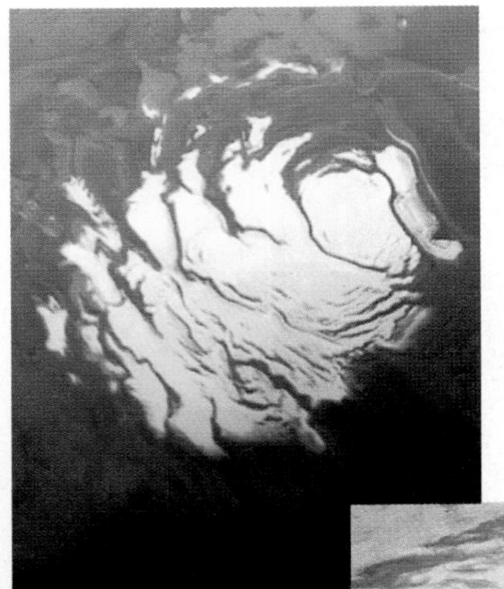

Polar ice caps

Olympus Mons

The Martian atmosphere is much thinner than Earth's and is composed mostly of carbon dioxide, with some nitrogen and argon. Surface temperatures range from 35°C to −170°C. The temperature difference between day and night also sets up strong winds on the planet, which can cause global dust storms during certain seasons.

Mars has two small, heavily cratered moons. Phobos is only 25 km in diameter, and Deimos is 13 km in diameter. Phobos is in an orbit that is slowly spiraling inward towards Mars. It is expected to impact the Martian surface in about 50 million years.

As you toured the inner planets using the "eye" of the space probes, you saw how each planet is unique. Mercury, Venus, Earth, and Mars are quite different from the outer planets that you'll tour in the next section.

Section Wrap-up

Review

1. How are Mercury and Earth's moon similar?
2. List one important characteristic of each inner planet.
3. Although Venus is often called Earth's twin, why would life as we know it be unlikely on Venus?
4. **Think Critically:** Why is the surface temperature of Venus higher than Mercury's, even though Mercury is closer to the sun?

Skill Builder

Interpreting Data

Using the information in this section and Appendix I, explain how Mars is like Earth. If you need help, refer to Interpreting Data in the **Skill Handbook.**

Science Journal

Use textbooks and NASA materials to investigate the *Viking* mission to Mars. In your Science Journal, report on the possibility of life on Mars and the tests that have been conducted to see whether life is there. Include your own opinion as to whether there is or was life on Mars.

Science & Literature

A Brave and Startling Truth by Maya Angelou

In this chapter, you have learned some of the physical characteristics—scientific facts and figures—of our solar system. Find out how one poet, Maya Angelou, uses Earth and its motion through space to describe people and the quest for world peace. Below are several excerpts, or parts, from her poem *A Brave and Startling Truth.*

We, this people, on a small and lonely planet
Traveling through casual space
Past aloof stars, across the way of indifferent suns
To a destination where all signs tell us
It is possible and imperative that we learn
A brave and startling truth.

When we come to it
Then we will confess that not the Pyramids
With their stones set in mysterious perfection
Nor the Gardens of Babylon
Hanging as eternal beauty
In our collective memory
Not the Grand Canyon
Kindled into delicious color
By Western sunsets

Nor the Danube, flowing its blue soul into Europe
Not the sacred peak of Mount Fuji
Stretching to the Rising Sun
Neither Father Amazon nor Mother Mississippi
who, without favor,
Nurture all creatures in the depths and on the shores
These are not the only wonders of the world

When we come to it
We, this people, on this miniscule and kithless globe
Who reach daily for the bomb, the blade and the dagger
Yet who petition in the dark for tokens of peace
We, this people, on this mote of matter

When we come to it
We, this people, on this wayward, floating body
Created on this earth, of this earth
Have the power to fashion for this earth
A climate where every man and every woman
Can live freely without sanctimonious piety
Without crippling fear

When we come to it
We must confess that we are the possible
We are the miraculous, the true wonder of this world
That is when, and only when
We come to it.

Science Journal

Use your knowledge of Earth's features and/or its place in the solar system to write a short poem that describes a social issue important to you.

SCIENCE & SOCIETY

23•3 ISSUE: Mission to Mars

Objectives

- Recognize problems that astronauts will encounter during a trip to Mars.
- Decide whether a piloted mission to Mars is necessary or whether it could be done by robotics.

How should Mars be explored?

You learned about exploring space in Chapter 21. Scientists are currently developing plans to further explore Mars. Because Mars is 55 million km away from Earth at the closest, it will take about three years to get to Mars and back. Because of the long duration of the flight, astronauts would face much more danger than they currently do in space shuttle missions. How should Mars be explored?

2 Points of View

Should humans explore Mars?

Mars may once have been able to support life. One type of information that Mars explorers would be looking for is evidence of present or past life. Mars may not have life on it now, but early in its history it may have been a warm and wet planet with a much thicker atmosphere than it has now. If evidence of life is found, it would show that life elsewhere in the universe is also possible.

The search for evidence of life would not be easy. To properly search for fossils on Mars would require many years of work for trained Mars geologists. To study just one geologic formation would require field trips by humans to many different sites.

Figure 23-11

Robotic explorers such as this may one day help us explore Mars.

Because of the near-zero gravity in outer space, bones lose calcium and weaken. Bones might fracture more easily once astronauts land on Mars and return to Earth. Also, muscles get weak because they don't have to hold the body up as they do under Earth's gravity.

In addition to these problems, body fluids move upward because no gravity is pulling them down. The movement of fluids could signal the kidneys to excrete more fluids, causing dehydration. Also, astronauts would be exposed to more radiation from the sun than they are during space shuttle flights.

Should robots explore Mars?

For the reasons just stated, some scientists are suggesting that advanced robots currently under development could operate equipment and carry out scientific experiments in space or on Mars. This new technology uses video and artificial touch sense. The artificial senses of the robot are connected to the real senses of a human. The robot has video cameras for eyes and special touch sensors in its limbs. The human operator looks at tiny video screens worn as goggles and sees exactly what the robot sees. Sometimes data gloves worn by the operator have sensors connected by radio signal to the robot's hands. Any movement performed by the operator is duplicated by the robot. **Figure 23-12** shows one system for exploring potentially hazardous environments.

One problem involved in using robots to do this type of in-depth work is the long distance from Earth to Mars. Radio signals from the operator to a robot on Mars would take up to 20 minutes to transmit—too long to be practical.

Researchers would like to develop robot technology for use on the moon or on a space station first, then work on developing artificial intelligence for robots on Mars. A robot with artificial intelligence would be programmed to do a task and would have some ability to "think" on its own.

Figure 23-12

Using virtual reality, hazardous environments can be explored safely. The video screen allows humans to see what a remote probe senses.

Section Wrap-up

Review

1. How could we benefit from a piloted mission to Mars?
2. Why would robots on Mars need to be able to "think" on their own?

Explore the Issue

Do you think humans or robots should explore Mars? If you think humans should, explain how they would be protected from the hazards of extended time in space. If you think robots should, explain how they could be used to search for evidence of past life on Mars.

SCIENCE & SOCIETY

23•4 The Outer Planets

Science Words

Jupiter
Great Red Spot
Saturn
Uranus
Neptune
Pluto

Objectives

- List the major characteristics of Jupiter, Saturn, Uranus, and Neptune.
- Recognize how Pluto differs from the other outer planets.

Outer Planets

You have learned that the inner planets are small, solid, rocky bodies in space. By contrast, the outer planets, except for Pluto, are very large, gaseous objects.

You first read of the *Voyager* probes in Chapter 21. Although they were not the first probes to the outer planets, they have uncovered a wealth of new information about Jupiter, Saturn, Uranus, and Neptune. Let's follow the *Voyager* probes on their journeys to the outer planets of the solar system.

Jupiter

In 1979, *Voyager 1* flew past **Jupiter,** the largest planet and the fifth planet from the sun. *Voyager 2* flew past Jupiter later that same year. The major discoveries of the probes include new information about the motions of Jupiter's atmosphere and the discovery of three new moons. *Voyager* probes also discovered that Jupiter has faint dust rings around it and that one of its moons has volcanoes on it.

Jupiter is composed mostly of hydrogen, helium, and some ammonia, methane, and water vapor. Scientists believe the atmosphere of hydrogen and helium gradually changes to a planetwide ocean of liquid hydrogen and helium toward the middle of the planet. Below this liquid layer, there may be a solid rocky core. The extreme pressure and temperature, however, make the core different from any rock on Earth.

You've probably seen pictures from the *Voyager* probes of Jupiter's colorful clouds. Its atmosphere has bands of white, red, tan, and brown clouds, as shown in **Figure 23-13.** Continuous storms of swirling, high-pressure gas have been observed on Jupiter. The **Great Red Spot** is the most spectacular of these storms. Lightning has also been observed within Jupiter's clouds.

Figure 23-13

Jupiter is the largest planet in our solar system, containing more mass than all of the other planets combined. The Great Red Spot is a giant storm that is about 12 000 km from top to bottom.

Table 23-1

Large Moons of Jupiter			
Io	**Europa**	**Ganymede**	**Callisto**
The most volcanically active object in the solar system. Sulfur lava gives it its distinctive red and orange color. Possesses a thin, temporary sulfur-dioxide atmosphere.	Rocky interior is covered by a 100-km-thick ice crust, which has a network of cracks, indicating tectonic activity. Possesses a thin oxygen atmosphere.	Has an ice crust about 100 km thick, covered with grooves. Crust may surround a 900-km-thick slushy mantle of water and ice. Has a rocky core.	Has a heavily cratered ice-rock crust several hundred km thick. Crust may surround a water or ice mantle around a rocky core.

Moons of Jupiter

In orbit around Jupiter are 16 moons. The four largest, shown in **Table 23-1,** were discovered by Galileo in 1610. Io is the closest large moon to Jupiter. Jupiter's tremendous gravitational force and the gravity of Europa pull on Io. This force heats up Io, causing it to be the most volcanically active object in the solar system. The next moon out is Europa. It is composed mostly of rock with a thick crust of ice. Next is Ganymede, which is the largest satellite in the solar system. It's larger than the planet Mercury. Callisto is the fourth moon out, and is composed of ice and rock.

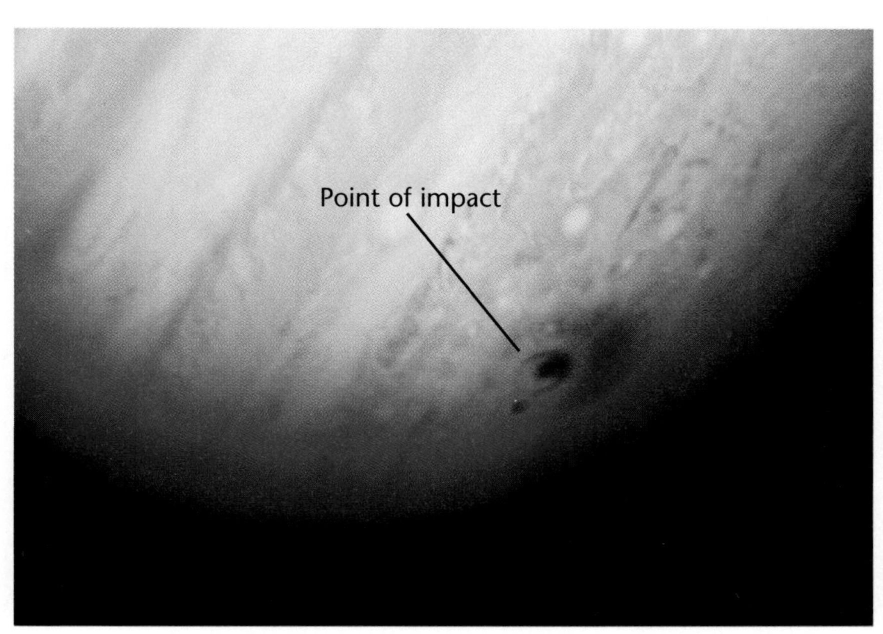

Figure 23-14

In 1994, a comet named Shoemaker-Levy 9 collided into Jupiter with a series of spectacular explosions. Information from this impact gives us clues about what might happen if such an impact occured on Earth.

MiniLAB

Can you draw planets to scale?

Procedure

1. To determine how the sizes of the planets in the solar system compare with each other, use the information on planet diameters in Appendix I.
2. Select a scale diameter of Earth.
3. Formulate a model by drawing a circle with this diameter on paper.
4. Using Earth's diameter as 1.0, draw each of the other planets to scale also.

Analysis

1. At this scale, how far would your model Earth have to be located from the sun?
2. What would 1 AU be equal to in this model?
3. Using a scale of 1 AU = 2 m, how large would the sun and Earth models have to be to remain in scale?

Saturn

The next planet surveyed by the *Voyager* probes was Saturn, in 1980 and 1981. **Saturn** is the sixth planet from the sun, and is also known as the ringed planet. Saturn is the second-largest planet in the solar system, but has the lowest density. Its density is so low that the planet would float on water.

Similar to Jupiter, Saturn is a large, gaseous planet with a thick outer atmosphere composed mostly of hydrogen and helium. Saturn's atmosphere also contains ammonia, methane, and water vapor. As you go deeper into Saturn's atmosphere, the gases gradually change to liquid hydrogen and helium. Below its atmosphere and liquid ocean, Saturn may have a small rocky core.

The *Voyager* probes gathered new information about Saturn's ring system and its moons. The *Voyager* probes showed that Saturn has several broad rings, each of which is composed of hundreds of thin ringlets. Each ring is composed of countless ice and rock particles ranging in size from a speck of dust to tens of meters across, as shown in **Figure 23-15**. This makes Saturn's ring system the most complex of all the outer gaseous planets'.

At least 18 moons orbit Saturn. That's more than any other planet in our solar system has. The largest of these, Titan, is larger than Mercury. It has an atmosphere of nitrogen, argon, and methane. Thick clouds of smog prevent us from seeing the surface of Titan.

Figure 23-15

Saturn's rings are composed of pieces of rock and ice.

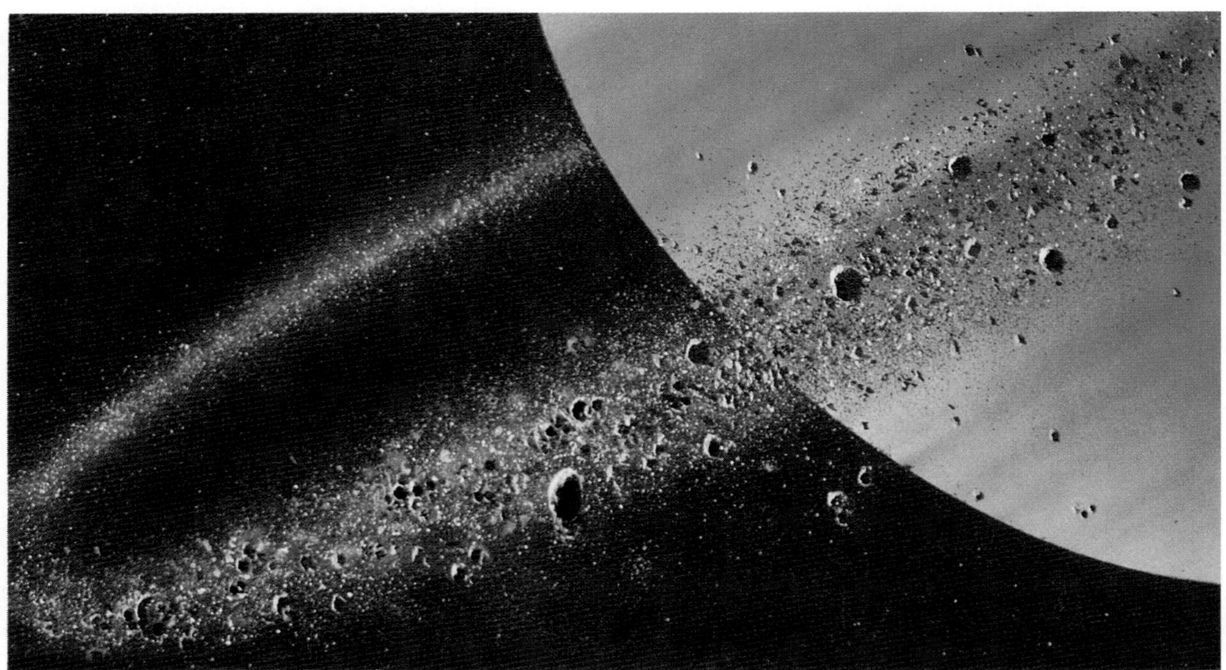

Uranus

After touring Saturn, *Voyager 2* flew by Uranus in 1986. **Uranus** is the seventh planet from the sun and wasn't discovered until 1781. It is a large, gaseous planet with 15 satellites and a system of thin, dark rings.

Voyager revealed numerous thin rings and ten moons that had not been seen earlier. *Voyager* also detected that the magnetic field is tilted 60° from its rotational poles. The field is also offset from the planet's center by one-third of Uranus's radius. Only Neptune's magnetic field is offset more than this.

The atmosphere of Uranus is composed of hydrogen, helium, and some methane. The methane gives the planet its blue-green color. Methane absorbs the red and yellow light, and the clouds reflect the green and blue. No cloud bands and few storm systems are seen on Uranus. Evidence suggests that under its atmosphere, Uranus has a mantle of liquid water, methane, and ammonia surrounding a rocky core.

One of the most unique features of Uranus is that its axis of rotation is tilted on its side compared with the other planets. The axes of rotation of the other planets, except Pluto, are nearly perpendicular to the planes of their orbits. Uranus, however, has a rotational axis nearly parallel to the plane of its orbit, as shown in **Figure 23-17.**

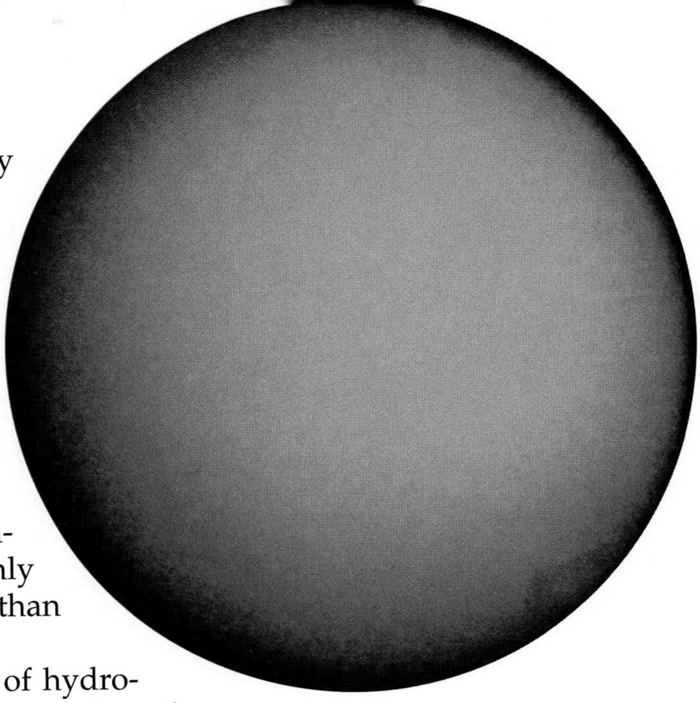

Figure 23-16

The atmosphere of Uranus gives the planet its distinct blue-green color.

CONNECT TO

CHEMISTRY

Recall what you've learned about the chemical and physical properties of hydrogen, helium, methane, and ammonia. ***Describe*** the properties of these substances that caused them to be more abundant in the outer planets than in the inner planets.

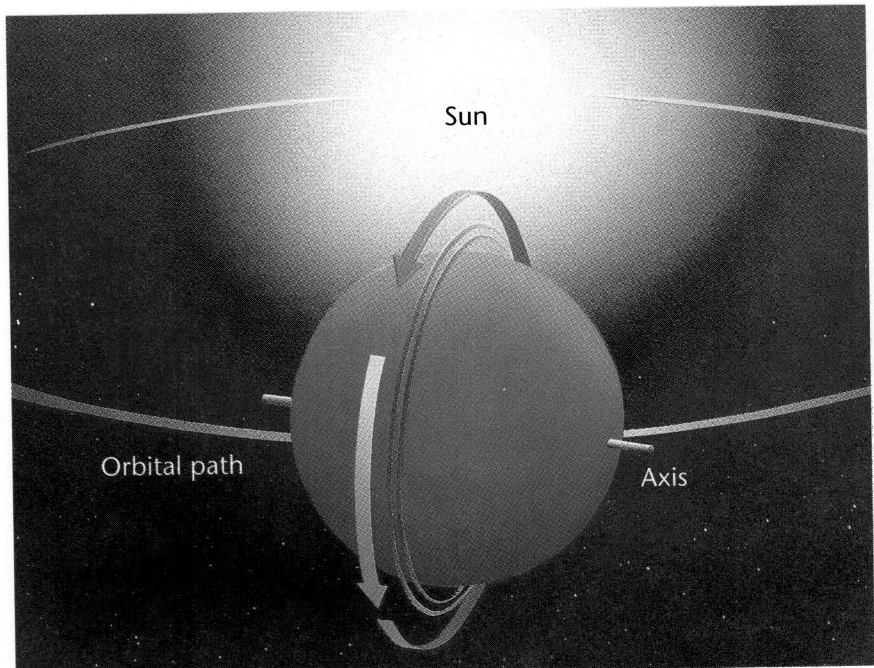

Figure 23-17

Uranus rotates on an axis nearly parallel to the plane of its orbit. At some point during its revolution around the sun, one of the poles points almost directly at the sun.

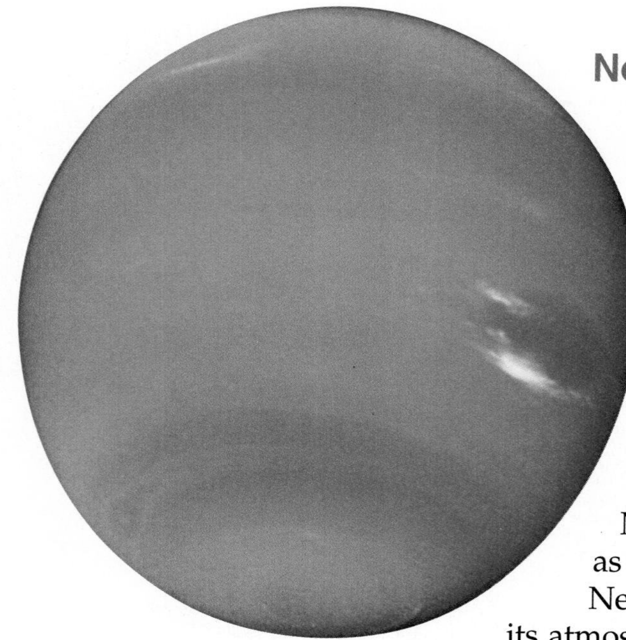

Figure 23-18

Neptune has an atmosphere similar to Uranus's, but it is much bluer due to the higher methane content.

Neptune

From Uranus, *Voyager 2* traveled on to **Neptune,** a large, gaseous planet. Discovered in 1846, Neptune is the eighth planet from the sun most of the time. However, Pluto's orbit crosses inside Neptune's during a part of its voyage around the sun. Since 1979, Pluto has been closer to the sun than Neptune, and it will remain closer to the sun until 1999.

Neptune's atmosphere is very similar to that of Uranus. The methane content gives Neptune its distinctive blue-green color just as it does for Uranus.

Neptune has dark-colored, stormlike features in its atmosphere that are similar to the Great Red Spot on Jupiter. One that was discovered by *Voyager,* called the Great Dark Spot, is shown in **Figure 23-18.**

Under its atmosphere, Neptune is thought to have liquid water, methane, and ammonia. Neptune probably has a rocky core.

With *Voyager 2*'s detection of six new moons, the total number of Neptune's known moons is now eight. Of these, Triton is the largest. Triton, shown in **Figure 23-19,** has a diameter of 2700 km and has a thin atmosphere composed mostly of nitrogen. *Voyager* detected methane geysers erupting on Triton. *Voyager* also detected that Neptune has rings that are thin in some places and thick in other places. Neptune's magnetic field is tilted 55° from its rotational axis and offset from center by half Neptune's radius.

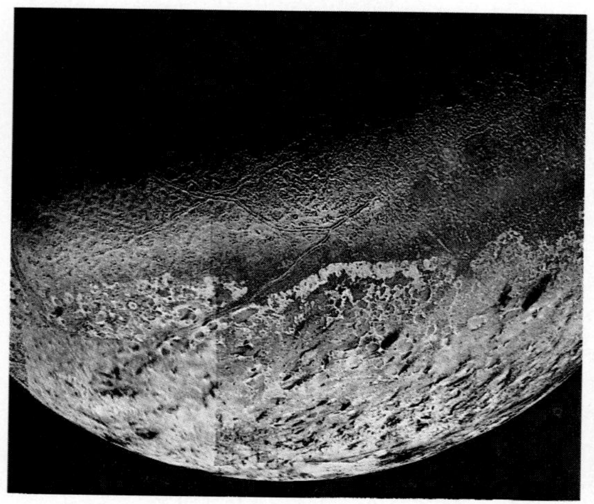

Figure 23-19

Triton is Neptune's largest moon.

Voyager ended its tour of the solar system with Neptune. Both *Voyager* probes are now beyond the orbits of Pluto and Neptune. They will continue into space sensing the extent of the sun's effect on charged particles.

Pluto

The smallest planet in our solar system, and the one we know the least about, is Pluto. Because **Pluto** is farther from the sun than Neptune during most of its orbit around the sun, it is considered the ninth planet from the sun. Pluto is not like the other outer planets. It's surrounded by only a thin atmosphere, and it's the only outer planet with a solid, icy-rock surface.

Pluto's only moon, Charon, has a diameter about half of Pluto's. Charon orbits very close to Pluto. Because of their close size and orbit, Charon and Pluto are sometimes considered to be a double planet.

Recent data from the *Hubble Space Telescope* indicate the presence of a vast disk of icy comets near Neptune's orbit, called the Kuiper belt. Some of the ice comets are hundreds of kilometers in diameter. Are Pluto and Charon members of this belt, are they escaped moons of one of the larger gaseous giants, or did they simply form at the distance they are? Maybe planets at that distance from the sun should be small and composed of icy rock. We may not find out until we send a probe to Pluto.

With the *Voyager* probes, we entered a new age of knowledge about the solar system. The space probe *Galileo*, which arrived at Jupiter in 1995, and the upcoming *Cassini* probe to Saturn will continue to extend our understanding of the solar system.

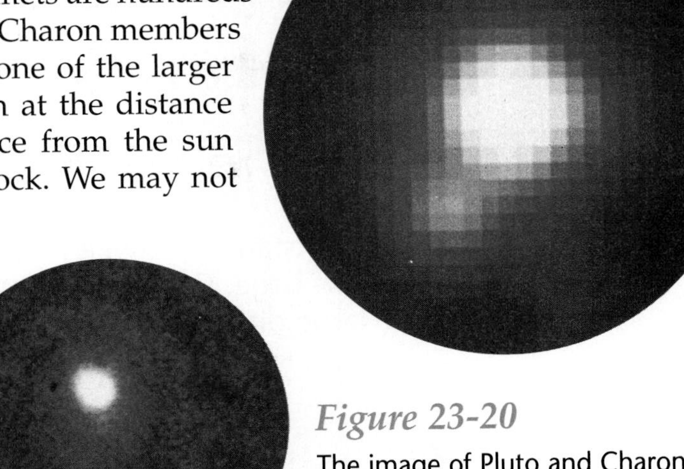

Figure 23-20

The image of Pluto and Charon (above) was taken by an Earth-based telescope. The *Hubble Space Telescope* gave astronomers their first clear view of Pluto and Charon as distinct objects (left) .

Section Wrap-up

Review

1. What's the difference between the outer planets and the inner planets?
2. Are there moons in the solar system that are larger than planets? If so, what are they?
3. How does Pluto differ from the other outer planets?
4. **Think Critically:** Why is Neptune sometimes the farthest planet from the sun?

Using Computers

Spreadsheet Use the spreadsheet or table-making features of your computer to produce a table of the nine planets that compares their characteristics.

Skill Builder

Recognizing Cause and Effect

Answer the following questions about Jupiter. If you need help, refer to Recognizing Cause and Effect in the **Skill Handbook.**

1. What causes Jupiter's surface color?
2. How is the Great Red Spot affected by Jupiter's atmosphere?
3. How does Jupiter's mass affect its gravitational force?

Activity 23-2

Design Your Own Experiment

Solar System Distance Model

Distances between the planets of the solar system are very large. Can you design a model that will demonstrate the large distances between and among the sun and planets in the solar system?

PREPARATION

Problem

How can a model be designed that will show the relative distances between and among the sun and planets of the solar system?

Form a Hypothesis

State a hypothesis about how a model with scale dimensions of the solar system can be constructed.

Objectives

- Make a table of scale distances that will represent planetary distances to be used in a model of the solar system.
- Make a model of the distances between the sun and planets of the solar system.

Possible Materials

- meterstick
- pencil
- paper
- scissors
- string

Safety Precautions 

Take care when handling scissors.

Data and Observations

Planet	Distance to Sun (km)	Distance to Sun (AU)	Scale Distance (1 AU =)	Scale Distance (1 AU = 2m)
Mercury	5.8×10^7			
Venus	1.08×10^8			
Earth	1.50×10^8			
Mars	1.50×10^8			
Jupiter	7.80×10^8			
Saturn	1.43×10^8			
Uranus	2.88×10^8			
Neptune	4.51×10^8			
Pluto	5.92×10^8			

PLAN THE EXPERIMENT

1. As a group, agree upon and write out your hypothesis statement.
2. As a group, list the steps that you need to take in making your model to test your hypothesis. Be very specific, describing exactly what you will do at each step.
3. Make a list of the materials that you will need to complete your model.
4. Make a table of scale distances you will use in your model.
5. Write a description of how you will build your model, explaining how it will demonstrate relative distances between and among the sun and planets of the solar system.

Check the Plan

1. Read over your entire model description to make sure that all steps are in a logical order.
2. Do any parts of your model need to be constructed before the overall model is produced?
3. Will you summarize data used in construction, as well as data obtained from your model, in a graph, table, or some form of organization other than the one on page 660?
4. How will you determine whether your model successfully demonstrates the large distances between and among the sun and planets in the solar system?
5. *Make sure your teacher approves your plan before you proceed.*

DO THE EXPERIMENT

1. **Construct the model** as planned using your scale distances.
2. While **constructing the model,** write down any observations that you or other members of your group make and complete the data table in your Science Journal.
3. **Calculate** the scale distance that would be used in your model if 1 AU = 2 m.

Analyze and Apply

1. **Explain** how a scale distance is determined.
2. Was it possible to work with your scale?
3. **Explain** why or why not.
4. How much string would be required to construct a model with a scale distance 1 AU = 2 m?

Go Further

Proxima Centauri, the closest star to our sun, is about 270 000 AU from the sun. Based on your original scale, how long would a piece of string need to be in order to include this star on your model? Can you translate this to a distance from your school to some feature in your town?

23•5 Other Objects in the Solar System

Science Words

comet
Oort Cloud
meteor
meteorite
asteroid

Objectives

- Explain where a comet comes from and describe how a comet develops as it approaches the sun.
- Differentiate among comets, meteoroids, and asteroids.

Comets

Although the planets and their satellites are the most noticeable members of the sun's family, there are many other objects that orbit the sun. Comets, meteoroids, and asteroids are other objects in the solar system.

You've probably heard of Halley's comet. It was last seen from Earth in 1986. English astronomer Edmund Halley realized that comet sightings that had taken place about every 76 years were really sightings of the same comet. This comet, which takes about 76 years to orbit the sun, was named after him. Halley's comet is just one example of the many other objects in the solar system besides the planets. **Figure 23-21** illustrates comets.

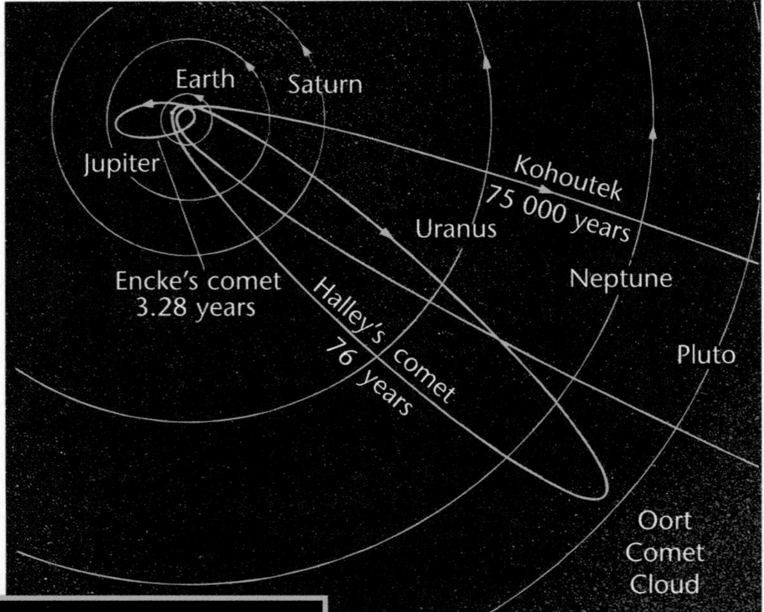

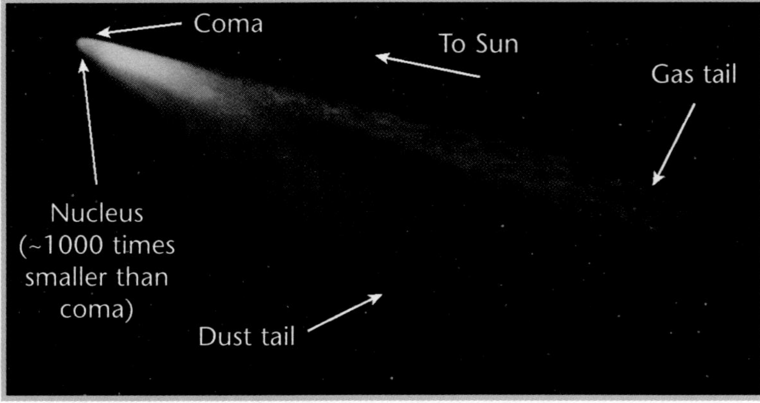

Figure 23-21

A comet consists of a nucleus, a coma, and a tail, as shown at left. The orbits of three famous comets are shown above.

A **comet** is composed of dust and rock particles mixed in with frozen water, methane, and ammonia. The Dutch astronomer Jan Oort proposed the idea that a large collection of comets lies in a cloud that completely surrounds the solar system. This cloud is located beyond the orbit of Pluto and is called the **Oort Cloud.** Evidence suggests that the gravity of nearby stars or another planet interacts with comets in the Oort Cloud. The sun's gravity pulls the comet toward it. The comet then either escapes from the solar system or gets captured into a much smaller orbit. As mentioned earlier, another belt of comets, called the Kuiper belt, may exist near the orbit of Neptune.

An exciting discovery made by two backyard astronomers on July 23, 1995 is a new comet heading toward the sun. This comet, Comet Hale-Bopp, is larger than most that approach the sun and may turn out to be the brightest comet visible from Earth in 20 years. It will be at its brightest in March and April 1997.

CONNECT TO

PHYSICS

Research Kepler's laws of planetary motion. ***Describe*** how his second law explains the speed of Halley's comet at various positions within its orbit.

Structure of Comets

The structure of a comet is like a large, dirty snowball, or a mass of frozen ice and rock. But as the comet approaches the sun, it develops a very distinctive structure. Ices of water, methane, and ammonia begin to vaporize because of the heat from the sun. The vaporized gases form a bright cloud called a coma around the nucleus, or solid part, of the comet. The solar wind pushes on the gases in the coma. These particles form a tail that always points away from the sun.

After many trips around the sun, most of the frozen ice in a comet has vaporized. All that is left are small particles that spread out in the orbit of the original comet.

Figure 23-22

Comet West was visible in March 1976.

Meteoroids, Meteors, and Meteorites

You learned that comets tend to break up after they have passed close to the sun many times. The small pieces of the comet nucleus spread out into a loose group within the original orbit of the broken comet. These small pieces of rock moving through space are then called meteoroids.

When the path of a meteoroid crosses the position of Earth, it enters our atmosphere at between 15 and 70 km/s. Most meteoroids are so small that they are completely vaporized in Earth's atmosphere. A meteoroid that burns up in Earth's atmosphere is called a **meteor.** People often see these and call them shooting stars.

Each time Earth passes through the loose group of particles within the old orbit of a comet, many small particles of rock and dust enter the atmosphere. Because more meteors than usual are seen, this is called a meteor shower.

Figure 23-23

When a meteorite collides with an object such as a moon or a planet, it can leave huge craters such as these. The crater on the left is in Arizona. The crater on the right is on the moon.

If the meteoroid is large enough, it may not completely burn up in Earth's atmosphere. When it strikes Earth, it is called a **meteorite.** Meteor Crater in Arizona, shown in **Figure 23-23,** was formed when a large meteorite struck Earth. Most meteorites are probably debris from asteroid collisions, but some are from the moon and Mars.

Asteroids

An **asteroid** is a piece of rock similar to the material that later formed into the planets. Most asteroids are located in an area between the orbits of Mars and Jupiter called the asteroid belt, shown in **Figure 23-24.**

Why are they located there? The gravity of Jupiter may have kept a planet from forming in the area where the asteroid belt is now located. In addition, some asteroids may have been gravitationally thrown out of the belt and are probably scattered

USING TECHNOLOGY

Space Junk

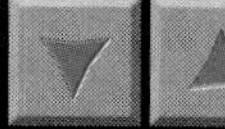

Since the 1950s, people have been exploring space. And since then, more than 3.5 million kg of garbage—from wrenches to satellites—have been dumped in space! Much of the space trash is only a few centimeters in size, but because there is almost no friction in the near-vacuum environment of space, such particles could move at speeds that exceed 24 000 km per hour and cause much damage to spacecraft!

NASA scientists are currently working on designs that would protect spacecraft and the International Space Station from space debris. One proposed system is a collision warning system that would alert astronauts and ground controllers to potentially dangerous debris. The system would use infrared sensors mounted like headlights onto a spacecraft or space station to detect objects. Space-based radars would then determine the path of the object and allow the astronauts time to either avoid or prepare for the collision.

Satellite Damage Caused by Debris

Think Critically:

One estimate projects that there are 3.5 million particles of garbage in space. What is the average mass of each particle?

throughout the present-day solar system. Some may have since been captured as moons around other planets.

The sizes of the asteroids in the asteroid belt range from tiny particles to 940 km for Ceres, the largest and first discovered. The next three in size are Pallas (523 km), Vesta (501 km), and Juno (244 km). Two asteroids, Gaspra and Ida, were photographed by *Galileo* on its way to Jupiter. *Galileo* discovered that Ida has a small moon orbiting it.

Comets, meteoroids, and asteroids are probably composed of material that formed early in the history of the solar system. Scientists study the structure and composition of these space objects in order to better understand what the solar system may have been like long ago. Understanding what the early solar system was like could help scientists to better understand the formation of Earth and its relationship to other objects in the solar system.

Figure 23-24

The asteroid belt lies between the orbits of Mars and Jupiter.

Section Wrap-up

Review

1. How does a comet's tail form as it approaches the sun?
2. What type of feature might be formed on Earth if a large meteorite reached its surface?
3. Describe differences among comets, meteoroids, and asteroids.
4. **Think Critically:** What is the chemical composition of comets? Are comets more similar to the inner or the outer planets?

Skill Builder

Interpreting Scientific Illustrations

Identify the coma and tail of the comet shown in **Figure 23-22.** In which direction is the sun relative to the comet? If you need help, refer to Interpreting Scientific Illustrations in the **Skill Handbook.**

Science Journal

The asteroid belt contains many objects ranging in size from tiny particles to objects almost 1000 km in diameter. In your Science Journal, describe how mining the asteroids for valuable minerals might be accomplished. Include your opinion as to whether we have the technology to attempt this type of space mission.

Chapter 23

Review

Summary

23-1: The Solar System

1. The Earth-centered model proposed that the planets, the moon, the sun, and the stars were embedded in separate spheres that rotated around Earth. The sun-centered model states that the sun is the center of the solar system.
2. Our solar system may have formed about 5 billion years ago from a cloud of gas, ice, and dust.

23-2: The Inner Planets

1. The inner planets, in increasing distance from the sun, are Mercury, Venus, Earth, and Mars.
2. The moonlike Mercury has craters and cliffs on its surface. Venus has a dense atmosphere of carbon dioxide and sulfuric acid. On Earth, water exists in three states. Mars appears red due to the iron oxide content of its weathered rocks.
3. Venus and Earth are similar in size and mass. Both have greenhouse effects.

23-3: Science and Society: Mission to Mars

1. Problems that astronauts to Mars would face include muscle and bone weakness and exposure to dangerous levels of radiation.
2. Robots may be a better alternative than sending humans to Mars. However, robots may not be able to study Mars's surface in enough detail.

23-4: The Outer Planets

1. Faint rings and 16 moons orbit the gaseous Jupiter. Jupiter's Great Red Spot is a storm. Saturn is made mostly of gas and has rings. Uranus is a large, gaseous planet with many moons and several rings. Neptune is similar to Uranus in composition and has stormlike features.
2. Pluto has a thin, changing atmosphere, and its surface is icy rock.

23-5: Other Objects in the Solar System

1. As a comet approaches the sun, vaporized gases form a bright coma around the comet's nucleus. A tail that points away from the sun is formed by solar wind.
2. Meteoroids are small pieces of rock moving through space. An asteroid is a piece of rock usually found in the asteroid belt.

Key Science Words

a. asteroid
b. astronomical unit
c. comet
d. Earth
e. Great Red Spot
f. inner planet
g. Jupiter
h. Mars
i. Mercury
j. meteor
k. meteorite
l. Neptune
m. Oort Cloud
n. outer planet
o. Pluto
p. Saturn
q. solar system
r. Uranus
s. Venus

Reviewing Vocabulary

Match each phrase with the correct term from the list of Key Science Words.

1. any of the four solid, rocky planets close to the sun
2. planet most like Earth in size and mass
3. planet with carbon dioxide ice caps
4. Ganymede and Io are two of its moons
5. planet that could float on water

Chapter 23 Review

6. most are large, gaseous planets
7. blue-green planet with cloud bands and storm systems
8. large group of comets beyond Pluto's orbit
9. a rock that enters Earth's atmosphere
10. a meteoroid that strikes Earth

Checking Concepts

Choose the word or phrase that completes the sentence.

1. ____ proposed a sun-centered solar system.
   a. Ptolemy c. Galileo
   b. Copernicus d. Oort
2. _______ produces energy in the sun.
   a. Magnetism c. Nuclear fusion
   b. Nuclear fission d. The greenhouse effect
3. Planets orbit the sun _______.
   a. in circles c. by rotation
   b. in ellipses d. at the same speed
4. _______ has very extreme temperatures because it has essentially no atmosphere.
   a. Earth c. Mars
   b. Jupiter d. Mercury
5. Water is a solid, liquid, and gas on _______.
   a. Pluto c. Saturn
   b. Uranus d. Earth
6. The largest volcano in the solar system is on _______.
   a. Earth c. Mars
   b. Jupiter d. Uranus
7. A problem with living in space is _______.
   a. bones gain calcium
   b. body tissue gains water
   c. muscles become stronger
   d. bones lose calcium
8. _______ has a very complex ring system made of hundreds of ringlets.
   a. Pluto c. Uranus
   b. Saturn d. Mars
9. The magnetic pole of _______ is tilted 60°.
   a. Uranus c. Jupiter
   b. Earth d. Pluto
10. The tail of a comet always points _______.
   a. toward the sun c. toward Earth
   b. away from the sun d. away from the Oort Cloud

Understanding Concepts

Answer the following questions in your Science Journal using complete sentences.

11. Contrast Copernicus's model of the solar system with Kepler's model.
12. Describe the general characteristics of the inner and outer planets.
13. Describe how the structure of a comet changes as it nears the sun.
14. How is Uranus different from the other eight planets?
15. Compare and contrast Mars and Earth. Use the photographs on the next page to help you.

Thinking Critically

16. Why is the surface temperature on Venus so much higher than that on Earth?
17. Describe the relationship between the mass of a planet and the number of satellites it has.
18. Why are probe landings on Jupiter or Saturn unlikely events?
19. What evidence suggests that water is or once was present on Mars?
20. An observer on Earth can watch Venus go through phases much like Earth's moon does. Explain why this is so.

Developing Skills

If you need help, refer to the **Skill Handbook.**

21. **Concept Mapping:** Make a concept map that shows how a comet changes as it travels through space.
22. **Hypothesizing:** Mercury is the closest planet to the sun, yet it does not reflect much of the sun's light. What can you say about Mercury's color?
23. **Measuring in SI:** The Great Red Spot of Jupiter is about 40 000 km long and about 12 000 km wide. What is its approximate area in km^2?
24. **Sequencing:** Arrange the following planets in order from the planet with the most natural satellites to the one with the least: Earth, Jupiter, Saturn, Neptune, Uranus, and Mars.
25. **Making and Using Tables:** Make a table that summarizes the main characteristics of each planet in the solar system.

Performance Assessment

1. **Display:** Mercury, Venus, Mars, Jupiter, and Saturn can be observed by the unaided eye. Research where in the sky these planets can be observed in the next year. Construct a display with your findings. Include time of day, day of the year, and locations with respect to known landmarks in your area on your display.
2. **Scientific Drawing:** You learned that planetary orbit shapes are ellipses, not circles. Based on the models of planetary orbits that you formulated, how could you draw a circular orbit? How could you draw an orbit that is extremely elliptical?
3. **Model:** Build a three-dimensional model of the solar system. Include features such as the asteroid belt.

Surface of Mars

Surface of Earth

Previewing the Chapter

Stars and Galaxies

This photo may look like science fiction, but it is a real place. It is a photo of the Andromeda galaxy. From our point in the universe, billions of galaxies are visible, each containing billions of stars. By studying deep space, astronomers have observed that the universe is expanding in all directions. In the following activity, you can model how the universe might be expanding.

EXPLORE ACTIVITY

Make a model of the expansion of the universe.

1. Partially inflate a balloon. Clip the neck shut with a clothespin.
2. Draw six evenly spaced dots on the balloon with a felt-tip marker. Label the dots A through F.
3. Use a string and ruler to measure the distance, in millimeters, from dot A to each of the other dots.
4. Remove the clothespin and inflate the balloon some more.
5. Measure the distance of each dot from A again.
6. Inflate the balloon again and take the new measurements.

Infer: If each dot represents a cluster of galaxies, describe the motion of the galaxy clusters relative to one another. Is the universe expanding?

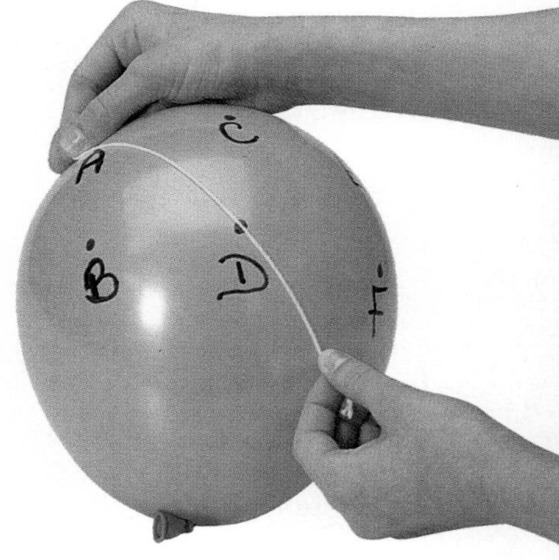

Previewing Science Skills

- In the **Skill Builders**, you will **recognize cause and effect, sequence events,** and **interpret scientific illustrations.**
- In the **Activities,** you will **collect and analyze data** and **draw conclusions.**
- In the **MiniLABs,** you will **formulate models, interpret scientific illustrations,** and **make inferences.**

24•1 Stars

Science Words

constellation
absolute magnitude
apparent magnitude
parallax
light-year

Objectives

- Explain why the positions of the constellations change throughout the year.
- Compare and contrast absolute magnitude and apparent magnitude.
- Describe how parallax is used to determine distances.

Constellations

Have you ever watched clouds drift by on a summer day? It's fun to look at the clouds and imagine they have shapes familiar to you. One may look like a face. You might see a cloud that resembles a rabbit or a bear. People long ago did much the same thing with patterns of stars in the sky. They named certain groups of stars, called **constellations,** after animals, characters in mythology, or familiar objects. Two constellations are shown in **Figure 24-1.**

From Earth, a constellation looks like a group of stars that are relatively close to one another. In most cases, the stars in a constellation have no relationship to each other in space.

The position of a star in the sky can be given as a specific location within a constellation. For example, you can say that the star Betelgeuse is in the right shoulder of the mighty hunter Orion. Orion's faithful companion is his dog, Canis Major. The brightest star in the sky, Sirius, is in the constellation Canis Major.

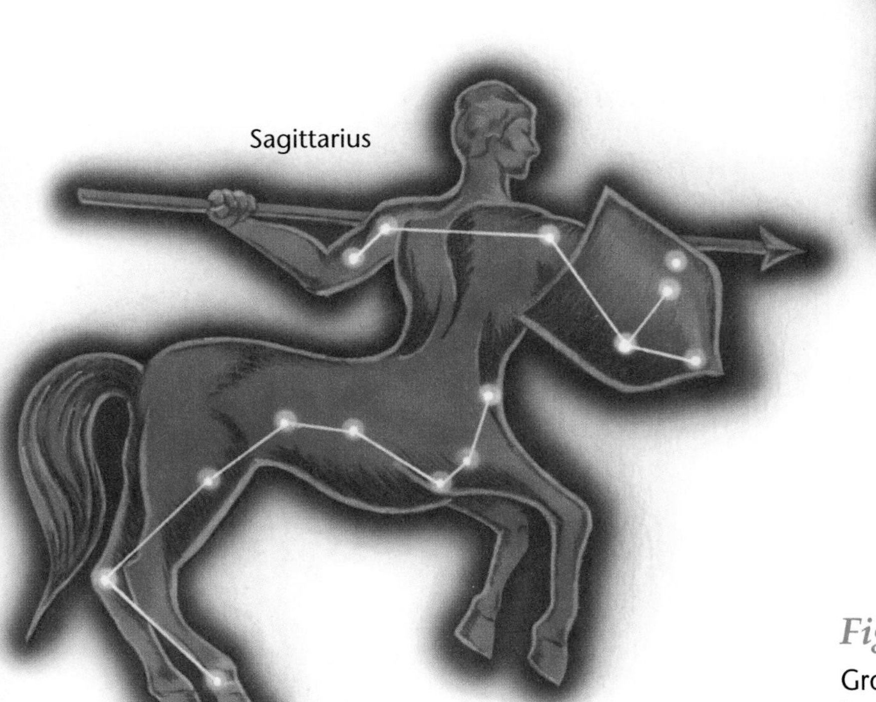

Figure 24-1

Groups of stars can form patterns that look like familiar objects or characters. More constellations are shown in Appendix K.

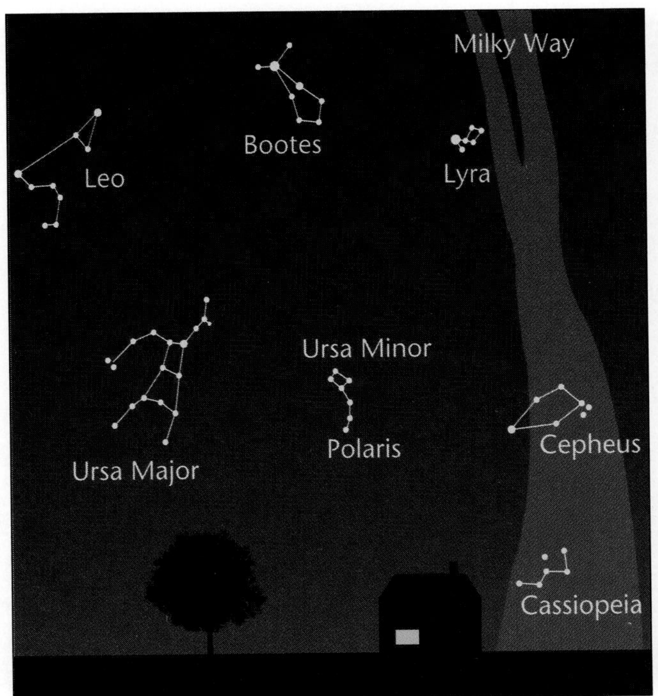

Summer

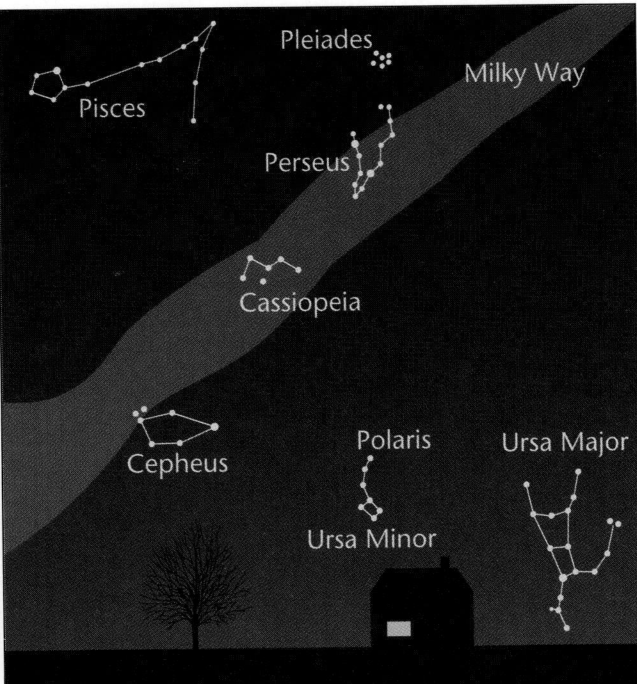

Winter

Figure 24-2

Some constellations are visible only during certain seasons of the year. Others, such as those close to Polaris, are visible year-round.

Early Greek astronomers named many constellations, and modern astronomers have divided the sky into 88 constellations. You may already know some of them. Have you ever tried to find the Big Dipper? It's part of the constellation Ursa Major. Notice how the front two stars of the Big Dipper point directly at the star Polaris. Polaris, also known as the North Star, is located at the end of the Little Dipper in the constellation Ursa Minor. Polaris is almost directly over Earth's north pole.

Circumpolar Constellations

As Earth rotates, you can watch Ursa Major, Ursa Minor, and other constellations in the northern sky circle around Polaris. Because these constellations circle Polaris they are called circumpolar constellations.

All of the constellations appear to move because Earth is moving, as shown in **Figure 24-3.** The stars appear to complete one full circle in the sky in just under 24 hours as Earth rotates on its axis. The stars also appear to change positions in the sky throughout the year as Earth revolves around the sun.

Figure 24-3

This photograph shows the path of circumpolar stars over several hours. Polaris is almost directly over the north pole and doesn't appear to move much as Earth rotates.

Circumpolar constellations are visible all year long, but other constellations are not. As Earth orbits the sun, different constellations come into view while others disappear. Orion, which is visible in the winter, can't be seen in the summer because the daytime side of Earth is facing it.

Star Light, Star Bright

Mary conducted an experiment to determine the relationship between distance and the brightness of stars. She used a meterstick, a light meter, and a light bulb. The bulb was mounted at the zero end of the meterstick. Mary placed the light meter at the 20-cm mark on the meter-stick and recorded the distance and the light-meter reading in the data table below. Readings are in luxes, which are units for measuring light intensity. Mary doubled and tripled the distance and took more readings.

Solve the Problem:

1. **Does the light meter measure absolute or apparent brightness?**
2. **What happened to the amount of light recorded when the distance was increased from 20 cm to 40 cm? From 20 cm to 60 cm?**

Think Critically:

What is the relationship between light intensity and distance? What would the light intensity be at 100 cm?

Distance (cm)	Meter Reading (luxes)
20	4150
40	1037.5
60	461.1
80	259.4

Absolute and Apparent Magnitudes

When you look at constellations, you'll notice that some stars are brighter than others. Sirius looks much brighter than Rigel. But is Sirius actually a brighter star, or is it just closer to Earth, which makes it appear to be brighter? As it turns out, Sirius is 100 times closer to Earth than Rigel. If Sirius and Rigel were the same distance from Earth, Rigel would appear much brighter in the night sky than would Sirius.

When you refer to the brightness of a star, you can refer to either its absolute magnitude or its apparent magnitude. The **absolute magnitude** of a star is a measure of the amount of light it actually gives off. A measure of the amount of light received on Earth is called the **apparent magnitude.** A star that's actually rather dim can appear quite bright in the sky if it's close to Earth. A star that's actually bright can appear dim if it's far away. If two stars are the same distance away, what factors might cause one of them to be brighter than the other?

You can experience the effect of distance on apparent magnitude when driving in a car at night. Observe the other cars' headlights as they approach. Which car's headlights are brighter, those that are closer to you or those that are further away?

Determining the Distances to Stars

How do we know when a star is close to our solar system? One way is to measure its parallax. **Parallax** is the apparent shift in the position of an object when viewed from two different positions. You can easily observe parallax. Hold your hand at arm's length and look at one finger first with your left eye and then with your right eye. Your finger appears to change position with respect to the background. Now try the same experiment with your finger closer to your face. What do you observe? The nearer an object is to the observer, the greater its parallax.

We can measure the parallax of relatively close stars to determine their distances from Earth, as shown in **Figure 24-4.** When astronomers first realized how far away stars actually are, it became apparent that a new unit of measure would be needed to record their distances. Measuring star distances in kilometers would be like measuring the distance between cities in millimeters.

Distances in space are measured in light-years. A **light-year** is the distance that light travels in one year. Light travels at 300 000 km/s, or about 9.5 trillion kilometers in one year. The nearest star to Earth, other than the sun, is Proxima Centauri. Proxima Centauri is 4.2 light-years away, or about 40 trillion kilometers.

MiniLAB

What patterns do you see in the stars?

Procedure

1. On a clear night go outside after dark and study the stars. Take an adult with you and see if you can help each other find constellations.
2. Let your imagination go to work and try to see any patterns of stars in the sky that look like something with which you are familiar.
3. Draw the stars you see, where they are in the sky, and include a drawing of what you think the star pattern resembles.

Analysis

1. How do your constellations compare with those observed by your classmates?
2. How do you think recognizing star patterns could be useful?

A
Background stars
Star
P
Earth's orbit
January
1 AU
Sun
1 AU
July
Baseline

Figure 24-4

Parallax can be seen if you observe the same star while Earth is at two different points during its orbit around the sun (A). The star's position relative to the more-distant background stars will appear to change (B and C).

B

As seen in January

C

As seen in July

Activity 24-1

Design Your Own Experiment

Measuring Parallax

You have learned that parallax is the apparent shift in the position of an object when viewed from two locations. You have also learned that the nearer an object is to the observer, the greater its parallax. Do this activity to design a model and use it in an experiment that will show how distance affects the amount of observed parallax.

PREPARATION

Problem

Can a model be built that will show how distance affects the amount of observed parallax?

Form a Hypothesis

State a hypothesis about how a model must be built in order for it to be used in an experiment to show how distance affects the amount of observed parallax.

Objectives

- Design a model to show how the distance from an observer to an object affects the object's parallax shift.
- Using your model, carry out an experiment that shows how distance affects the amount of observed parallax.

Possible Materials

- meterstick
- metric ruler
- masking tape
- object to observe (pencil)

PLAN THE EXPERIMENT

1. As a group, agree upon and write out your hypothesis statement.
2. As a group, list the steps that you need to take to build your model. Be very specific, describing exactly what you will do at each step.
3. Devise a method to test how distance from an observer to an object, such as a pencil, affects the relative position of the object.
4. As a group, list the steps of your experiment that will test your hypothesis. Be very specific, describing exactly what you will do at each step.
5. Make a list of the materials that you will need to complete your experiment.

Check the Plan

1. Read over your plan for the model to be used in this experiment.
2. Read over your entire experiment to make sure that all steps are in a logical order.
3. Will you summarize data in a graph, table, or some other form of organization?
4. How will you determine whether observed parallax is changed because of the distance from the observer to the object?
5. How will the position from which observations are made differ?
6. How will you measure distances accurately and compare relative position shift?
7. *Make sure your teacher approves your plan before you proceed.*

DO THE EXPERIMENT

1. Construct the model your team has planned.
2. Carry out the experiment as planned.
3. While conducting the experiment, write down any observations that you or other members of your group make in your Science Journal.

Analyze and Apply

1. **Compare** what happened to the object when it was viewed from two locations.
2. At what distance from the observer did the object appear to shift the greatest distance?
3. **Infer** what happened to the apparent shift of the object's location as the distance from the observer was increased or decreased.

Go Further

Based on your hypothesis and on the tests you have performed, how does distance from an observer to an object affect the parallax observed for that object? How might astronomers use parallax?

Figure 24-5

A triangle-shaped glass called a prism can be used to separate different wavelengths of light into a spectrum.

Determining a Star's Temperature and Composition

The color of a star indicates its temperature. For example, very hot stars are a blue-white color. A relatively cool star looks orange or red. Stars the temperature of our sun have a yellow color.

Astronomers learn about other properties of stars by studying their spectra. They use spectrographs to break visible light from a star into its component colors. If you look closely at the spectrum of a star, such as the ones shown in **Figure 24-23** on page 695, you will see dark lines in it. The lines are caused by elements in the star's atmosphere.

As light radiated from a star passes through the star's atmosphere, some of it is absorbed by elements in the atmosphere. The wavelengths of visible light that are absorbed appear as dark lines in the spectrum. Each element absorbs certain wavelengths, producing a certain pattern of dark lines. The patterns of lines can be used to identify which elements are in a star's atmosphere.

Section Wrap-up

Review

1. Explain how Earth's revolution affects constellations that are visible throughout the year.
2. If two stars give off the same amount of light, what might cause one to look brighter than the other?
3. If the spectrum of another star shows the same absorption lines as the sun, what can be said about its composition?
4. **Think Critically:** Only about 700 stars have large enough parallaxes that their distances can be determined using parallax. Most of them are invisible to the naked eye. What does this indicate about their absolute magnitudes?

Skill Builder

Recognizing Cause and Effect

Suppose you viewed Proxima Centauri through a telescope. How old were you when the light that you see left Proxima Centauri? Why might Proxima Centauri look dimmer than the star Betelgeuse, a very large star 489 light-years away? If you need help, refer to Recognizing Cause and Effect in the **Skill Handbook.**

Using Computers

Graphics Use drawing software on a computer to make a star chart of major constellations visible from your home during the current season. Include reference points to help someone using the chart to find the constellations.

The Sun

24•2

The Layers of the Sun

More than 99 percent of all of the matter in our solar system is in the sun. It is the center of our solar system, and it makes life possible on Earth. To you and everyone else on Earth, the sun is a special object in the sky and one of the most important objects in your life. Nevertheless, our sun is just like many other stars in our galaxy.

The sun is an average middle-aged star. Its absolute magnitude is about average and it shines with a yellow light. The sun is an enormous ball of gas, producing energy by fusing hydrogen into helium in its core. **Figure 24-6** is a model of the sun's interior.

Science Words

photosphere
chromosphere
corona
sunspot
binary system

Objectives

- Describe how energy is produced in the sun.
- Recognize that sunspots, prominences, and solar flares are related.
- Explain why our sun is considered an average star and how it differs from stars in binary systems.

The Sun's Atmosphere

The lowest layer of the sun's atmosphere and the layer from which light is given off is the **photosphere.** The photosphere is often called the surface of the sun. Temperatures there are around 6000°C. Above the photosphere is the **chromosphere.** This layer extends upward about 2000 km. A transition zone occurs between 2000 and 10 000 km above the photosphere. Above the transition zone is the **corona.** This is the largest layer of the sun's atmosphere and extends millions of kilometers into space. Temperatures in the corona are as high as 2 000 000°C. Charged particles continually escape from the corona and move through space as the solar wind.

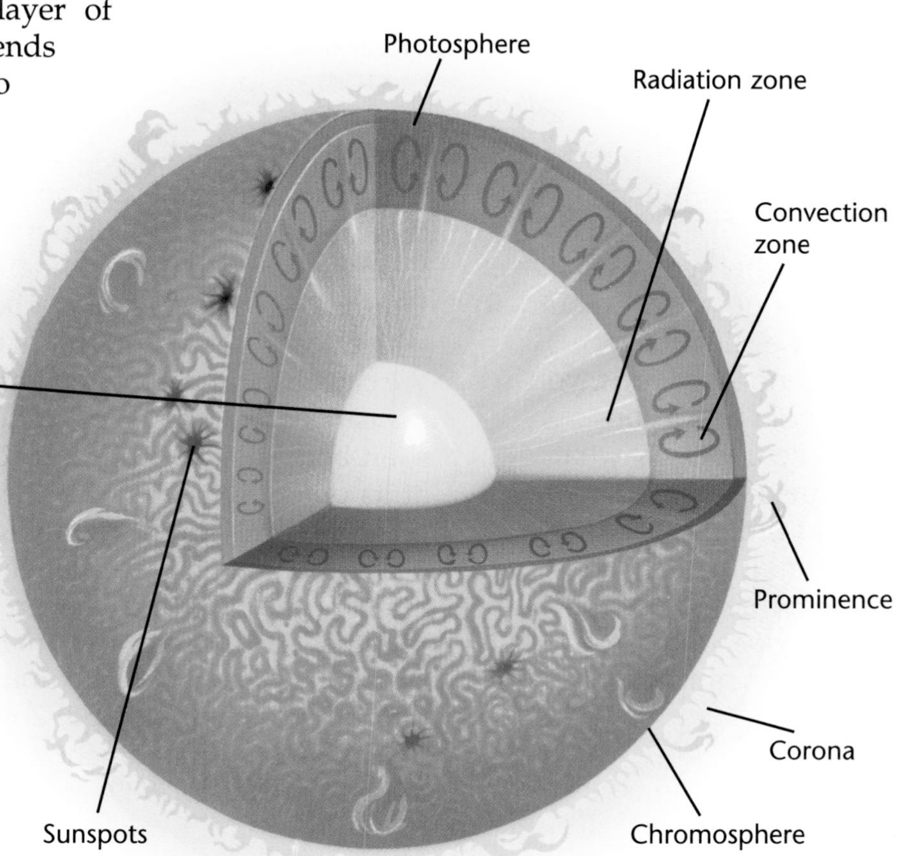

Figure 24-6

Energy produced by fusion in the sun's core travels outward by radiation and convection. The sun's atmosphere, composed of the photosphere, the chromosphere, and the corona, is illuminated by the energy produced in the core.

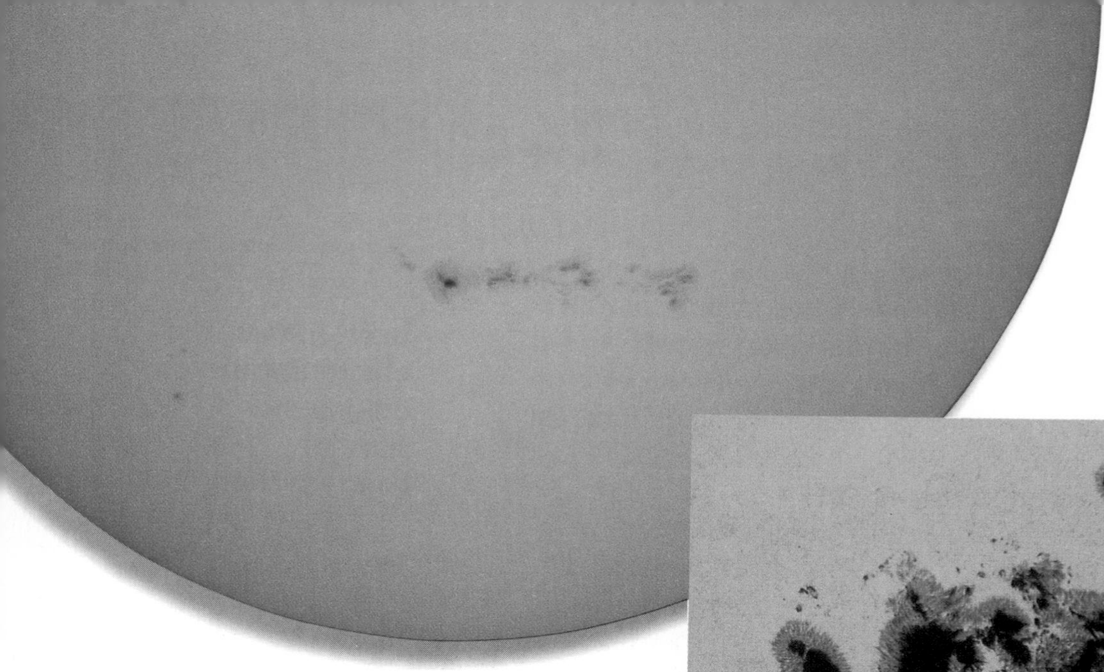

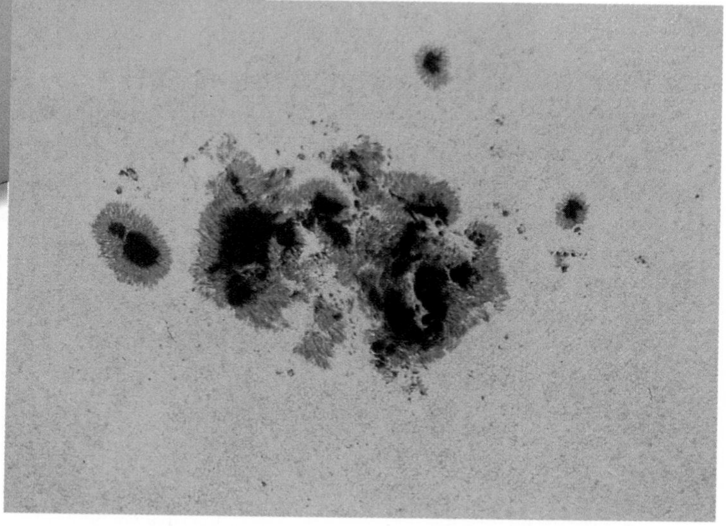

Figure 24-7

Sunspots are very bright, but when viewed against the rest of the photosphere, they appear dark. The photo on the right is a close-up of a sunspot.

Surface Features of the Sun

Because the sun is a hot ball of gas, it's hard to imagine its surface as anything but a smooth layer of gas. But there is much more that we can see. There are many features that can be studied, including sunspots, prominences, and flares.

CONNECT TO

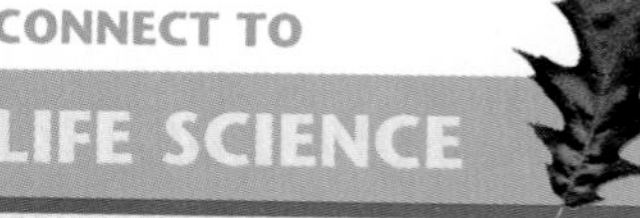

LIFE SCIENCE

When you are going to be out in the sunlight for extended periods, you should apply sunscreen. ***Identify*** what type of solar radiation this protects you from. If little or no sunscreen is used, ***describe*** what problems might occur.

Sunspots

Areas of the sun's surface that appear to be dark because they are cooler than surrounding areas are called **sunspots.** Ever since Galileo first identified sunspots like those in **Figure 24-7**, scientists have been studying them. One thing we've learned by studying sunspots is that the sun rotates. We can observe the movement of individual sunspots as they are carried by the sun's rotation. The sun doesn't rotate as a solid body, as does Earth. It rotates faster at its equator than at its poles. Sunspots near the equator take about 27 days to go around the sun; at higher latitudes, they take 31 days.

Sunspots aren't permanent features on the sun. They appear and disappear over a period of several days or months. Also, there are times when there are many large sunspots—a sunspot maximum—and times when there are only a few small sunspots or none at all—a sunspot minimum. Sunspot maxima occur about every 11 years. The last maximum was in 1989.

Prominences and Flares

Sunspots are related to several features on the sun's surface. The intense magnetic field associated with sunspots may cause prominences, huge arching columns of gas. Some prominences are so eruptive that material from the sun is blasted into space at speeds ranging from 600 to over 1000 km/s.

Gases near a sunspot sometimes brighten up suddenly, shooting gas outward at high speed. These violent eruptions from the sun, shown in **Figure 24-8**, are called solar flares.

Ultraviolet light and X rays from solar flares can reach Earth and cause disruption of radio signals. This makes communication by radio and telephone very difficult at times. High energy particles emitted by solar flares are captured by Earth's magnetic field. These particles interact with Earth's atmosphere near the polar regions and create light. This light is called the aurora borealis, or northern lights, when it occurs in the northern hemisphere. In the southern hemisphere this light is called the aurora australis.

MiniLAB

How do the sizes of stars compare?

Procedure

1. Find an area where you can make a chalk mark on pavement or another surface.
2. Tie a piece of chalk to one end of a string that's 125 cm long.
3. Hold the other end of the string to the pavement.
4. Have a friend pull the string tight and walk around you, leaving a mark on the pavement as he or she circles you.
5. Draw a circle with a 1-cm diameter in the middle of the large circle.

Analysis

1. The small circle represents a star the size of our sun, and the larger circle represents a supergiant, the largest kind of star. How big is the supergiant compared to our solar system?
2. Formulate a model of our sun and of a white dwarf star by drawing two new circles. Our sun is about 100 times larger than a white dwarf.
3. What are the diameters of your two new circles?

Our Sun: A Typical Star?

Although our sun is an average star, it is somewhat unusual in one way. Most stars are in systems in which two or more stars orbit each other. When two stars orbit each other, it is called a **binary system.**

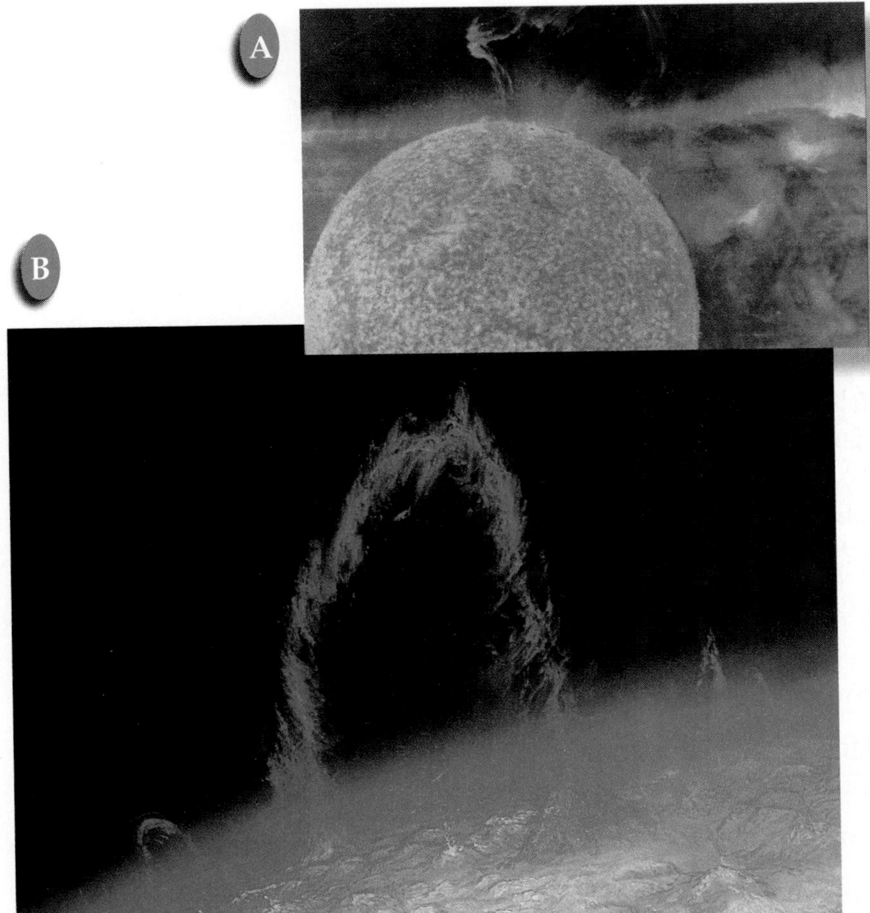

Figure 24-8

Features such as solar flares (A) and solar prominences (B) can reach tens of thousands of kilometers into space. *How big is this compared with the size of Earth?*

Figure 24-9

Pleiades is a cluster of stars gravitationally bound to each other.

In some cases, astronomers can detect binary systems because one star occasionally eclipses the other. The total amount of light from the star system becomes dim and then bright again, on a regular cycle. Algol in Pleiades is an example of this.

In many cases, stars move through space together as a cluster. In a star cluster, many stars are relatively close to one another and are gravitationally attracted to each other. The Pleiades star cluster can be seen in the constellation of Taurus in the winter sky. On a clear, dark night, you may be able to make out seven of the stars of this cluster. Most star clusters are far from our solar system, and appear as a fuzzy patch in the night sky.

Section Wrap-up

Review

1. How are sunspots, prominences, and solar flares related?
2. What properties does the sun have in common with other stars? What property makes it different from most other stars?
3. **Think Critically:** Since most stars are in multiple-star systems, what might explain why the sun is a single star?

Skill Builder

Interpreting Scientific Illustrations

Use **Figure 24-6** to answer the questions below. If you need help, refer to Interpreting Scientific Illustrations in the **Skill Handbook.**

1. Compare **Figure 24-6** with **Figure 22-8** on page 626, which shows a total solar eclipse. What part of the sun is visible in **Figure 22-8?**
2. Which layers make up the sun's atmosphere?
3. What process occurs in the sun's convection zone that enables energy produced in the core to reach the surface?

Science Journal

Write a brief description in your Science Journal that explains how the sun generates energy. Hypothesize what might happen to the sun when it exhausts the supply of hydrogen in its core.

Sunspots

Sunspots are dark, relatively cool areas on the surface of the sun. They can be observed moving across the face of the sun as it rotates. Do this activity to measure the movement of sunspots and use your data to determine the sun's period of rotation.

Problem

Can sunspot motion be used to determine the sun's period of rotation?

Materials 

- several books
- clipboard
- small refracting telescope
- cardboard
- drawing paper
- small tripod
- scissors

Procedure

1. Find a location where the sun may be viewed at the same time of day for a minimum of five days. **CAUTION:** *Do not look directly at the sun. Do not look through the telescope at the sun. You could damage your eyes.*
2. Set up the telescope with the eyepiece facing away from the sun, as shown below. Align so that the shadow cast by the telescope on the ground is a minimum size. Set up the clipboard with the drawing paper attached.
3. Use the books to prop the clipboard upright. Point the eyepiece at the drawing paper.
4. If the telescope has a small finder scope attached, remove the finder scope or keep it covered.
5. Move the clipboard back and forth until you have the largest possible image of the sun on the paper. Adjust the telescope to form a clear image.
6. Trace the outline of the sun on the paper.
7. Trace any sunspots that appear as dark areas on the sun's image. Repeat this at the same time each day for a week.
8. Using the sun's diameter as approximately 1 390 000 km, estimate the size of the largest sunspots that are observed.
9. Calculate how many kilometers any observed sunspots appear to move each day.
10. At the rate determined in step 9, predict how many days it will take for the same group of sunspots to return to about the same position in which you first observed them.

Analyze

1. Which part of the sun showed up in your image?
2. What was the average number of sunspots observed each day?

Conclude and Apply

3. **Infer** how sunspots can be used to determine that the sun's surface is not solid like Earth's.

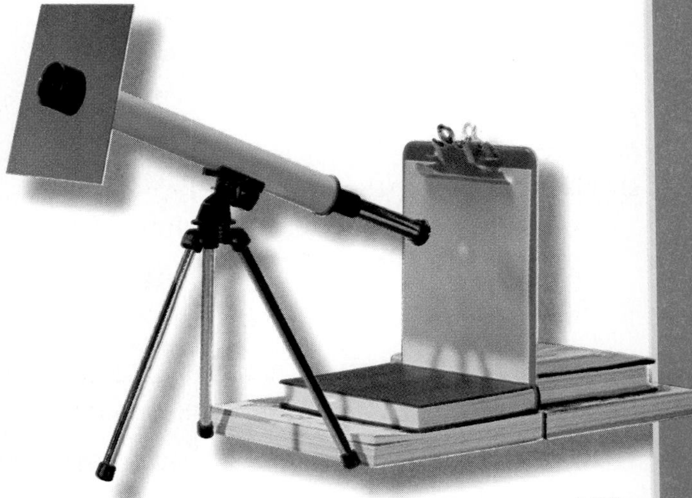

24•3 Evolution of Stars

Science Words

main sequence
nebula
giant
white dwarf
supergiant
neutron star
black hole

Objectives

- Diagram how stars are classified.
- Relate the temperature of a star to its color.
- Outline the evolution of a star through all stages of its development.

The H-R Diagram

In the early 1900s, Ejnar Hertzsprung and Henry Russell noticed that for most stars, the higher their temperatures, the brighter their absolute magnitudes. They developed a graph to show this relationship.

Hertzsprung and Russell placed the temperatures of the stars across the bottom of the graph and the absolute magnitudes of the stars up one side. A graph that shows the relationship of a star's temperature to its absolute magnitude is called a Hertzsprung-Russell (H-R) diagram. **Figure 24-10** shows a typical H-R diagram.

The Main Sequence

As you can see, stars seem to fit into specific areas of the chart. Most stars fit into a diagonal band that runs from the upper left to the lower right of the chart. This band, called the **main sequence,** contains hot, blue, bright stars in the upper left and cool, red, dim stars in the lower right. Yellow, medium-temperature, medium-brightness stars fall in between. The sun is a yellow main sequence star.

About 90 percent of all stars are main sequence stars, most of which are small, red stars found in the lower right of the

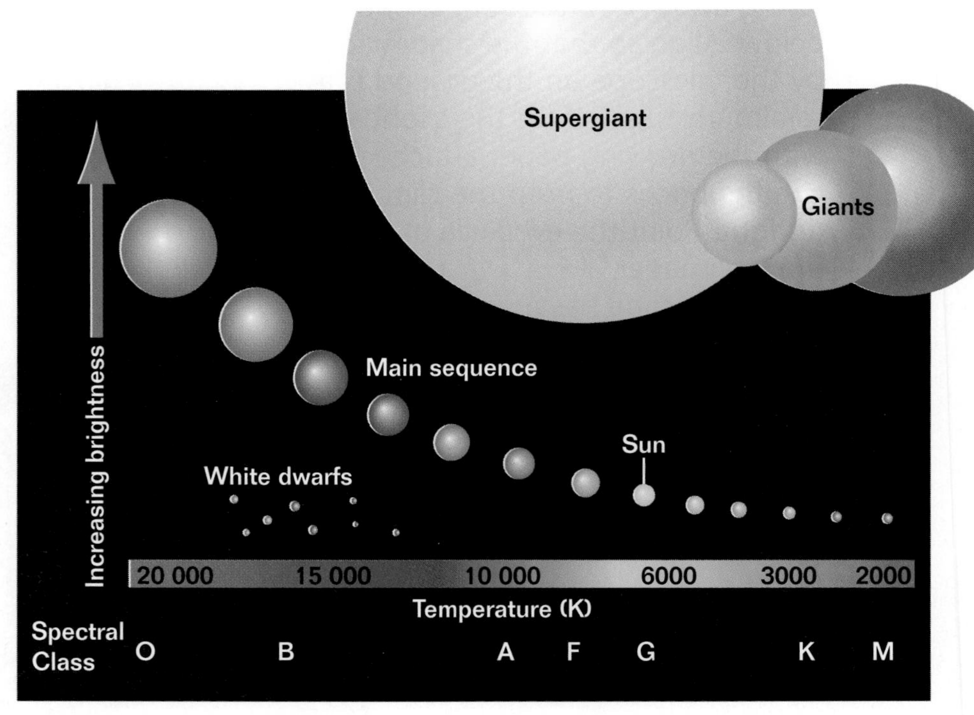

Figure 24-10
The Hertzsprung-Russell diagram shows the relationships among a star's color, temperature, and brightness. Stars in the main sequence run from hot, bright stars in the upper left corner of the diagram to cool, faint stars in the lower right corner. *What type of star shown in the diagram is the coolest, brightest star?*

H-R diagram. Among main sequence stars, the hottest stars generate the most light and the coolest generate the least. But what about the remaining ten percent? Some of these stars are hot but not very bright. These small stars are located on the lower left of the H-R diagram and are called white dwarfs. Other stars are extremely bright but not very hot. These large stars on the upper right of the H-R diagram are called giants, or red giants because they are usually red in color. The largest giants are called supergiants.

Fusion

When the H-R diagram was developed, scientists didn't know what caused stars to shine. Hertzsprung and Russell developed their diagram without knowing what produced the light and heat of stars.

For centuries, people had been puzzled by the question of what stars were and what made them shine. It wasn't until the early part of this century that scientists were forced to explain how a star could shine for billions of years. Until that time, many had estimated that Earth was only a few thousand years old. The sun could have been made of coal and shined for that long. But what material could possibly burn for billions of years?

In 1920, A. S. Eddington suggested that temperatures in the center of the sun would be very high. Robert Atkinson then suggested that with these very high temperatures, hydrogen could fuse into helium in a reaction that would release tremendous amounts of energy. **Figure 24-12** on page 686 illustrates how four hydrogen nuclei could combine to create one helium nucleus. The mass of one helium nucleus is less than the mass of four hydrogen nuclei, so some mass is lost in the reaction. In the 1930s, Hans Bethe suggested how carbon could be used as a catalyst in fusion reactions that explain the energy production in hotter stars.

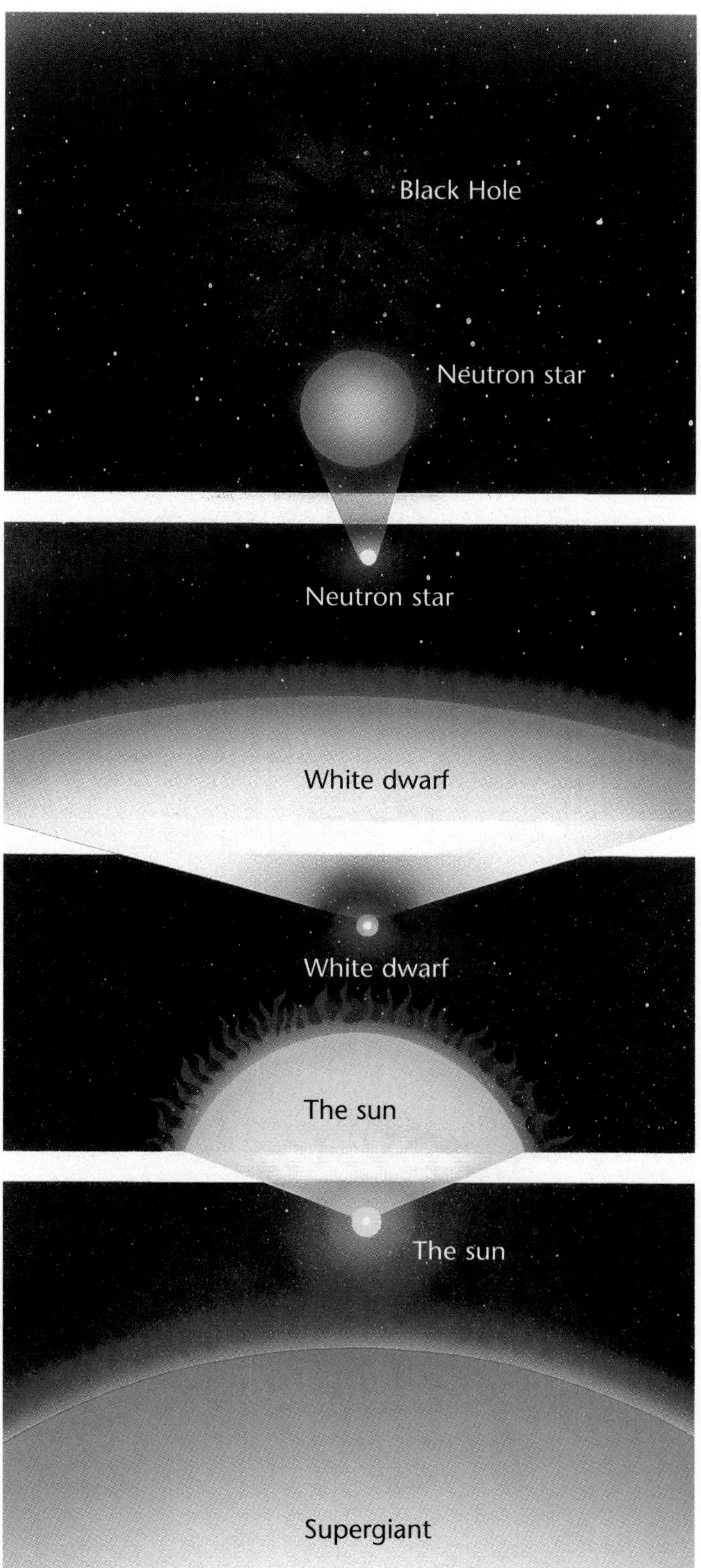

Figure 24-11

The relative sizes of stars range from supergiants as much as 800 times larger than the sun, to neutron stars and black holes possibly 30 km or less across. The relative sizes of a supergiant, the sun, a white dwarf, a neutron star, and a black hole are shown.

Years earlier, in 1905, Albert Einstein had proposed a theory stating that mass can be converted into energy. This was stated as the famous equation $E = mc^2$, where E is the energy produced, m is the mass, and c is the speed of light. The small amount of mass "lost" when hydrogen atoms fuse to form a helium atom is converted to a large amount of energy.

Fusion occurs in the cores of stars. Only there are temperatures and pressures high enough to cause atoms to fuse. Normally, they would repel each other, but in the core of a star, atoms are forced close enough together that their nuclei fuse together.

INTEGRATION
Physics

The Evolution of Stars

The H-R diagram and Eddington's and Einstein's theories explained a lot about stars. But they also led to more questions. Many wondered why some stars didn't fit in the main sequence group and what happened when a star exhausted its supply of hydrogen fuel. Today, we have a theory of how stars evolve, what makes them different from one another, and what happens when they die. **Figure 24-13** illustrates the lives of different types of stars.

Nebula

Stars begin as a large cloud of gas and dust called a **nebula.** The particles of gas and dust exert a gravitational force on each other, and the nebula begins to contract. Gravitational forces cause instability within the nebula. The nebula can fragment into smaller pieces. Each will eventually collapse to form a star.

Figure 24-12

In a star's core, fusion begins as two hydrogen nuclei (protons) are forced together.

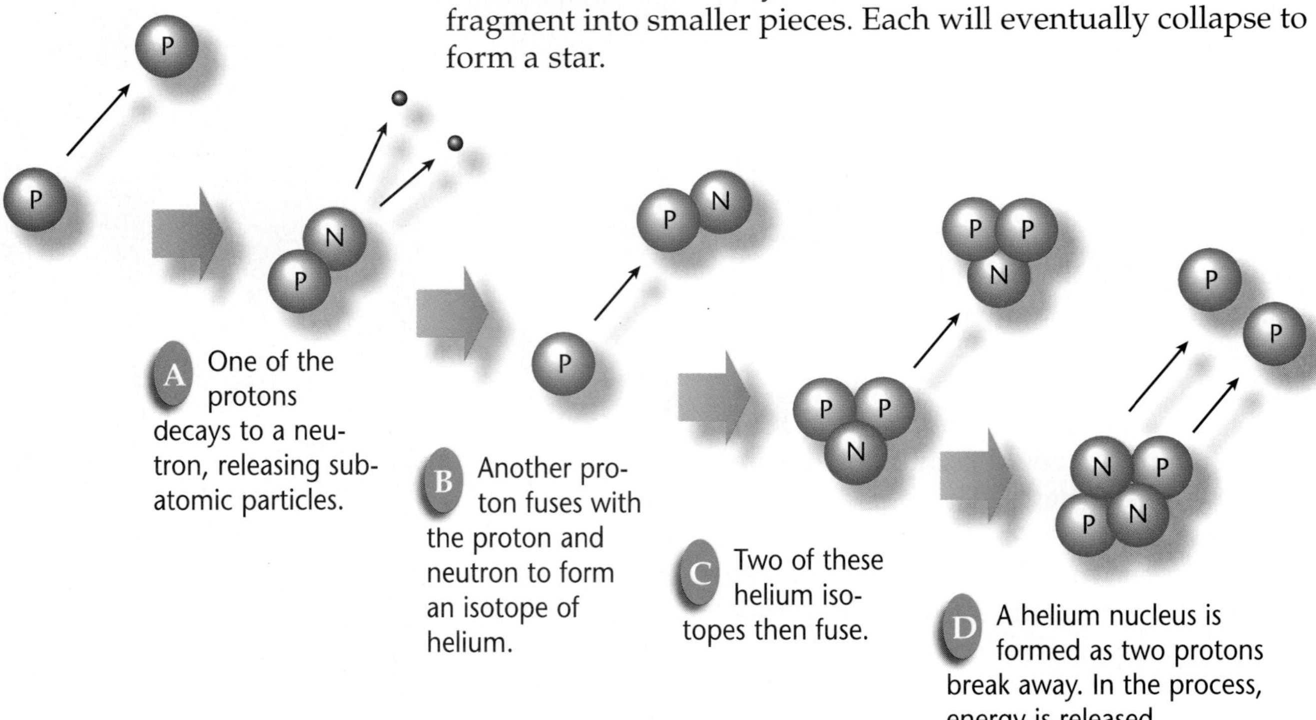

As the particles in the smaller clouds move closer together, the temperatures in each nebula increase. When temperatures inside each nebula reach 10 000 000 K, fusion begins. The energy released radiates outward through the condensing ball of gas. As the energy radiates into space, stars are born.

Main Sequence to Giant Stars

The heat from the fusion causes pressure that balances the attraction due to gravity, and the star becomes a main sequence star. It continues to use up its hydrogen fuel.

When the hydrogen in the core of the star is exhausted, there is no longer a balance between pressure and gravity. The core contracts, and the temperatures inside the star increase. This causes the outer layers of the star to expand. The star has evolved into a **giant.**

Once the core temperature reaches 100 000 000 K, helium nuclei fuse to form carbon in the giant's core. By this time, the star has expanded to an enormous size, and its outer layers are much cooler than they were when it was a main sequence star. In about 5 billion years, our sun will become a giant.

CONNECT TO

CHEMISTRY

If the spectrum of a star shows absorption lines of helium and hydrogen and is very bright in the blue end, ***describe*** as much as you can about the star's composition and surface temperature.

Figure 24-13

The life of a star depends greatly on its mass. Massive stars eventually become neutron stars, or possibly black holes. *What happens to stars the size of our sun?*

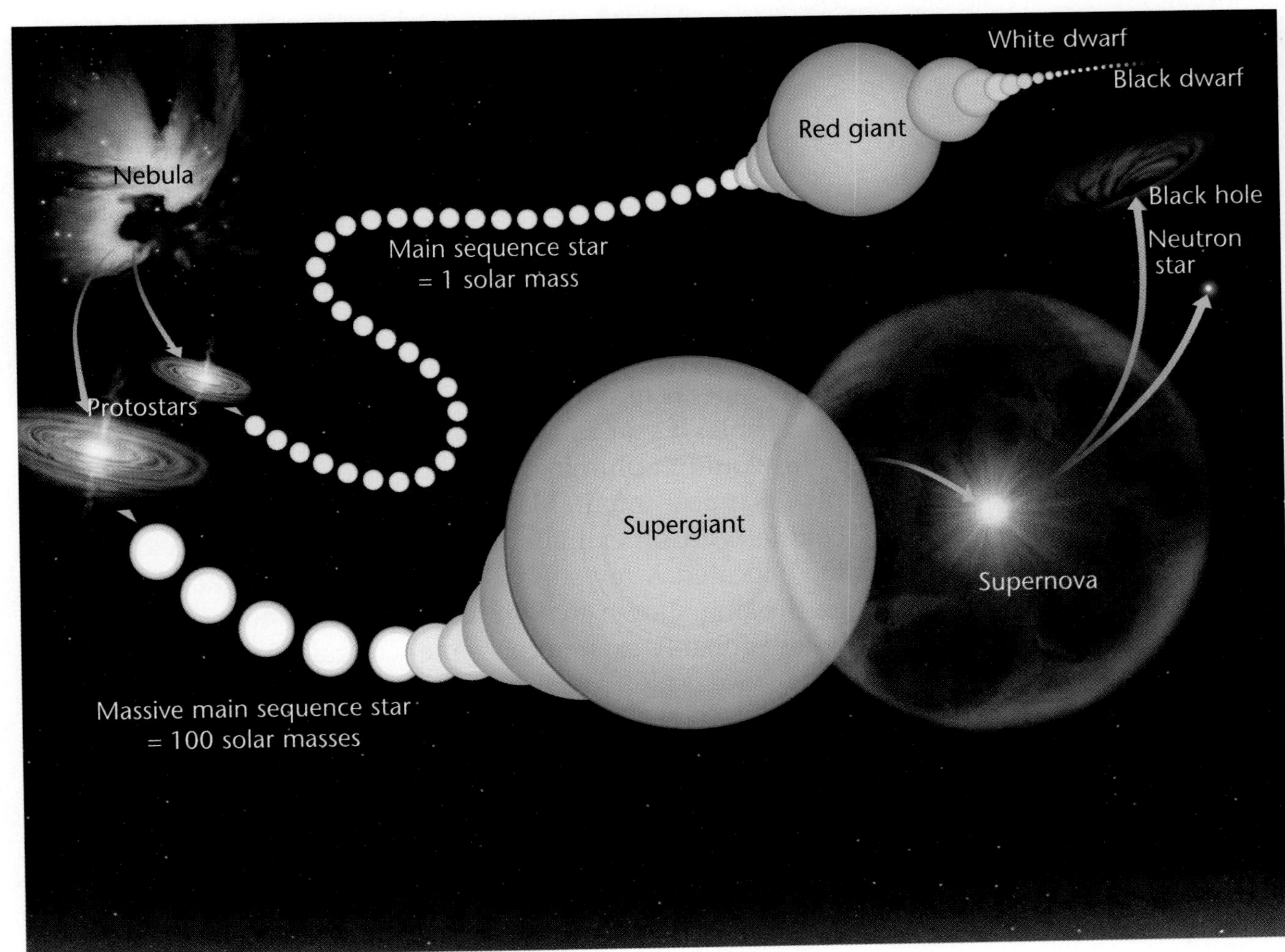

White Dwarfs

After the star's core uses up its supply of helium, it contracts even more. As the core of a star like the sun runs out of fuel, the outer layers escape into space. This leaves behind the hot, dense core. The core contracts under the force of gravity. At this stage in a star's evolution, it is a **white dwarf.** A white dwarf is about the size of Earth.

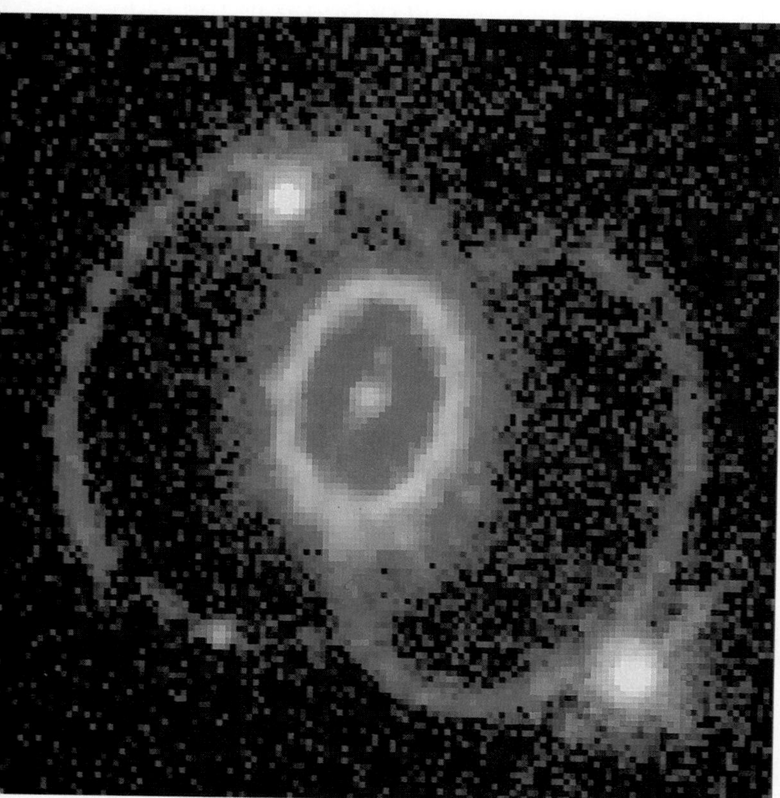

Figure 24-14

This image, taken by the *Hubble Space Telescope,* is of a supernova. The two white objects are other stars in the field of view. The two large rings are somewhat of a mystery. There may be an invisible, hourglass-shaped bubble of gas blown outward by the supernova. The rings may form when a high-energy beam of radiation sweeps across the gas, much like a searchlight sweeping across clouds.

Supergiants and Supernovas

In stars that are over ten times more massive than our sun, the stages of evolution occur more quickly and more violently. The core heats up to much higher temperatures. Heavier and heavier elements form by fusion. The star expands into a **supergiant.** Eventually, iron forms in the core. Fusion can no longer occur once iron forms. The core collapses violently, sending a shock wave outward through the star. The outer portion of the star explodes, producing a supernova like the one shown in **Figure 24-14.** A supernova can be millions of times brighter than the original star.

Neutron Stars and Black Holes

The collapsed core of a supernova shrinks to about 10-15 km. Only neutrons can exist in the dense core, and the supernova becomes a **neutron star.**

If the remaining dense core is over three times more massive than the sun, probably nothing can stop the core's collapse. It quickly evolves into a **black hole**—an object so dense, nothing can escape its gravity field. In fact, not even light can escape a black hole. If you could shine a flashlight on a black hole, the light wouldn't illuminate the black hole. The light would simply disappear into it. Matter being pulled into a black hole can collide with other material, generating X rays. Astronomers have located X-ray sources around possible black holes, such as the one shown in **Figure 24-15.** Extremely massive black holes probably exist in the centers of galaxies.

What are nebulas?

A star begins its life as a nebula, but where does the matter in a nebula come from? Nebulas form partly from the matter that was once in other stars. A star ejects enormous amounts of matter during its lifetime. This matter can be incorporated into other nebulas, which can evolve into new stars. The matter in stars is recycled many times.

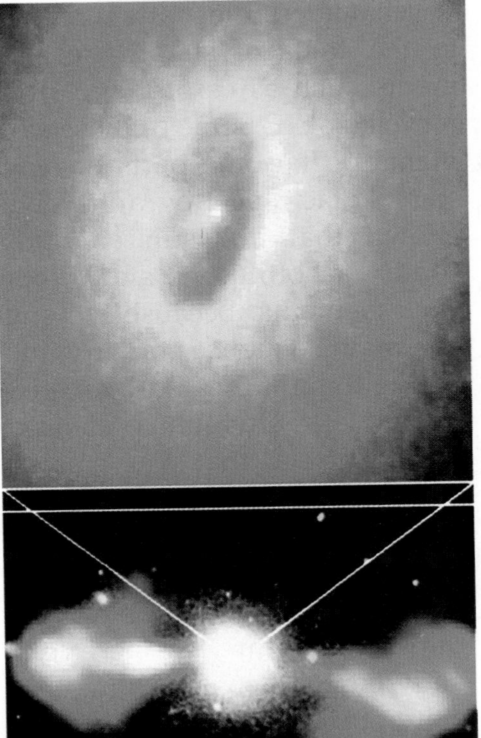

Figure 24-15

The bottom photo shows radio jets shooting from the core of the galaxy NGC 4261. These jets are thought to be associated with black holes. The dark ring in the top photo surrounds a possible black hole in the galaxy's nucleus.

USING TECHNOLOGY

Studying Supernovas

Near Cleveland, Ohio, and in Kamioka, Japan, large tanks of water and sensitive radiation-detecting instruments rest underground. The instruments are capable of detecting the radiation given off when subatomic particles called neutrinos strike protons or electrons in the water. In February 1987, the instruments recorded the presence of neutrinos. The records of the neutrinos striking the tanks went unnoticed until astronomers in the southern hemisphere observed a supernova.

Astronomers had previously hypothesized that neutrinos are emitted when a star evolves into a supernova. When the supernova was spotted, astronomers asked researchers at the underground tanks to check records of activity in the tanks. Records from Cleveland and Kamioka verified that neutrinos had been emitted during the explosion of the star. The neutrinos had traveled at the speed of light and arrived at Earth at about the same time as the visible light from the supernova.

Radiation Detection Tanks

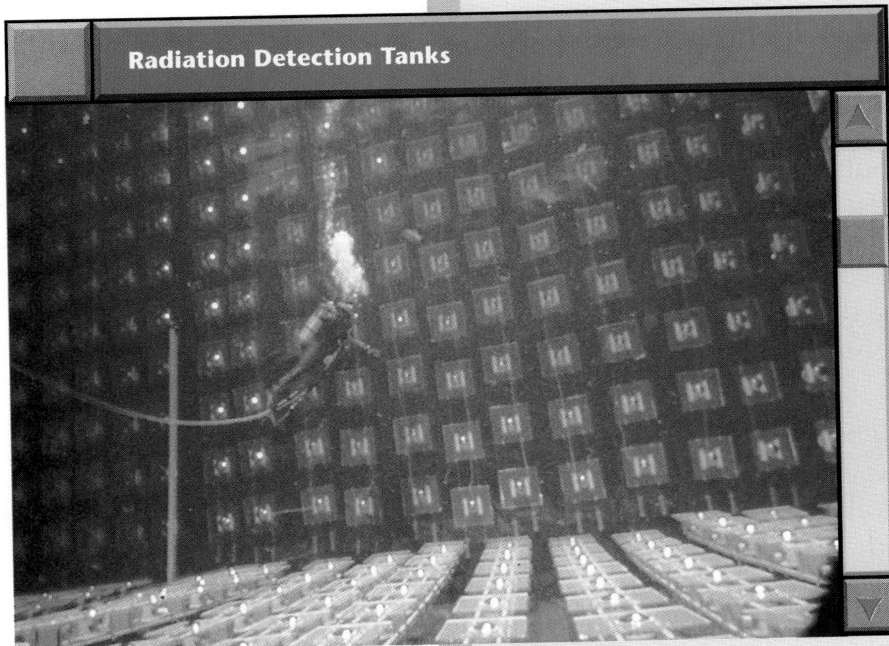

Think Critically:

The neutrinos that traveled through Earth in 1987 were produced about 169 000 years ago when Sanduleak –69°202 evolved into a supernova. How far away was Sanduleak –69°202?

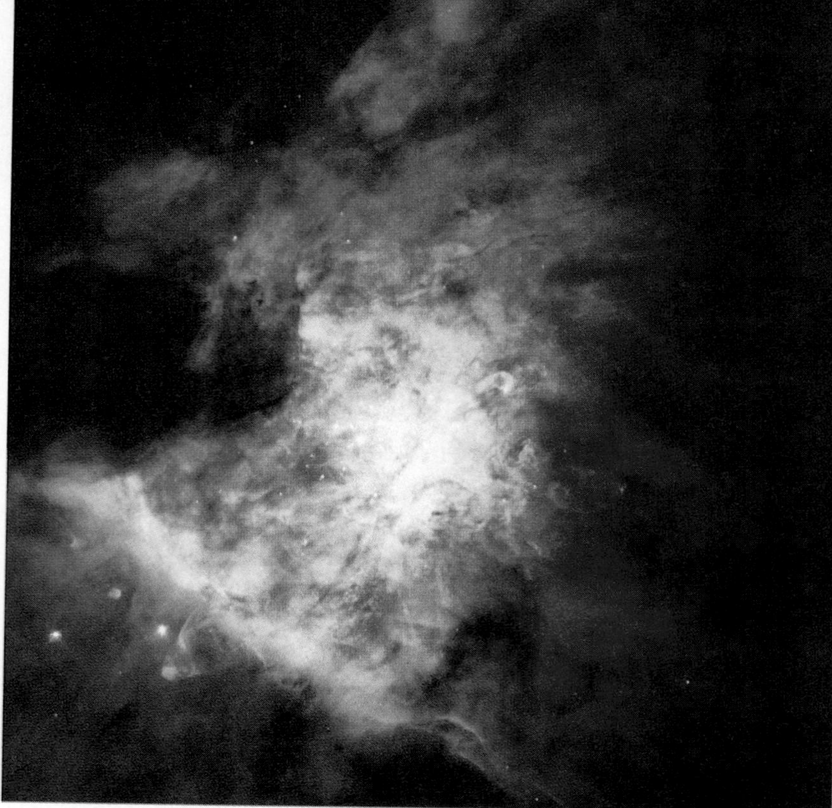

Figure 24-16

These stars in the Great Nebula of Orion are some of the youngest stars observed from Earth, although their formation began over 6 million years ago.

What about the matter created in the cores of stars? Are elements such as carbon and iron recycled also? Some of these elements do become parts of new stars. In fact, spectrographs have shown that our sun contains some carbon, iron, and other such elements. Because the sun is a main sequence star, it is too young to have created these elements itself. Our sun condensed from material that was created in stars that died many billions of years ago.

Some elements condense to form planets and other bodies rather than stars. In fact, your body contains many atoms that were fused in the cores of ancient stars. Evidence suggests that the first stars formed from hydrogen and helium and that all the other elements have formed in the cores of stars.

Section Wrap-up

Review

1. Explain why giants are not in the main sequence on the H-R diagram. How do their temperatures and absolute magnitudes compare with those of main sequence stars?
2. What can be said about the absolute magnitudes of two equal-sized stars whose colors are blue and yellow?
3. Outline the history and probable future of our sun.
4. **Think Critically:** Why doesn't the helium currently in the sun's core undergo fusion?

Skill Builder

Sequencing

Sequence the following in order of most evolved to least evolved: *main sequence star, supergiant, neutron star,* and *nebula.* If you need help, refer to Sequencing in the **Skill Handbook.**

Science Journal

Write a brief description in your Science Journal explaining what evidence exists that our sun evolved from material that existed in a previous star.

Galaxies and the Expanding Universe

24•4

Science Words

galaxy
big bang theory

Objectives

- Describe a galaxy and list the three main types of galaxies.
- Identify several characteristics of the Milky Way Galaxy.
- Explain how the big bang theory explains the observed Doppler shifts of galaxies.

Galaxies

One reason to study astronomy is to learn about your place in the universe. Long ago, people thought they were at the center of the universe and everything revolved around Earth. Today, you know this isn't the case. But do you know where you are in the universe?

You are on Earth, and Earth orbits the sun. But does the sun orbit anything? How does it interact with other objects in the universe? The sun is one star among many in a galaxy. A **galaxy** is a large group of stars, gas, and dust held together by gravity. Our galaxy, the Milky Way, contains about 200 billion stars, including the sun. Galaxies are separated by huge distances—often millions of light-years.

Figure 24-17

There may be more than 100 billion galaxies in the universe, and nearly all of them seem to be organized into clusters.

Figure 24-19

NGC 2997 is a spiral galaxy similar to our own. The scattered stars in the picture are in the foreground and belong to the Milky Way.

Figure 24-18

M87 is an example of an elliptical galaxy.

Just as stars are grouped together within galaxies, galaxies are grouped into clusters. The cluster the Milky Way belongs to is called the Local Group. It contains about 30 galaxies of various types and sizes.

Elliptical Galaxies

There are three major classes of galaxies: elliptical, spiral, and irregular. Probably the most common type of galaxy is the elliptical galaxy, shown in **Figure 24-18.** These galaxies are shaped like large, three-dimensional ellipses. Many are football-shaped, whereas some are spherical. Some elliptical galaxies are quite small, while others are so large that the entire Local Group of galaxies would fit inside one of them.

Spiral Galaxies

Spiral galaxies, as shown in **Figure 24-19**, have spiral arms winding outward from inner regions. These spiral arms are made up of bright stars and dust. The fuzzy patch you can see in the constellation of Andromeda is actually a spiral galaxy. It's so far away that you can't see its individual stars. Instead, it appears as a hazy spot in our sky. The Andromeda Galaxy is a member of the Local Group. It is about 2.2 million light-years away and is shown on page 670.

Arms in a normal spiral start close to the center of the galaxy. Barred spirals have spiral arms extending from a large bar that passes through the center of the galaxy.

Irregular Galaxies

The third class of galaxies, irregulars, includes most of those galaxies that don't fit into the other classifications. Irregular galaxies have many different shapes and are smaller and less common than the other types. Two irregular galaxies called the Clouds of Magellan orbit the Milky Way. The Large Magellanic Cloud is shown in **Figure 24-20**.

The Milky Way Galaxy

The Milky Way contains more than 200 billion stars. The visible disk of stars is about 100 000 light-years across, and the sun is located about 30 000 light-years out from its center. In our galaxy, all stars orbit around a central region. Based on a distance of 30 000 light-years and a speed of 235 kilometers per second, the sun orbits around the center of the Milky Way once every 240 million years.

A diagram of our galaxy is shown in **Figure 24-21.** The Milky Way is usually classified as a normal spiral galaxy. However, recent evidence suggests that it might be a barred spiral. It is difficult to know for sure because we can never see our galaxy from the outside.

You can't see the normal spiral or barred shape of the Milky Way because you are located within one of its spiral arms. You can see the Milky Way stretching across the sky as a faint band of light. All of the stars you can see in the night sky belong to the Milky Way Galaxy.

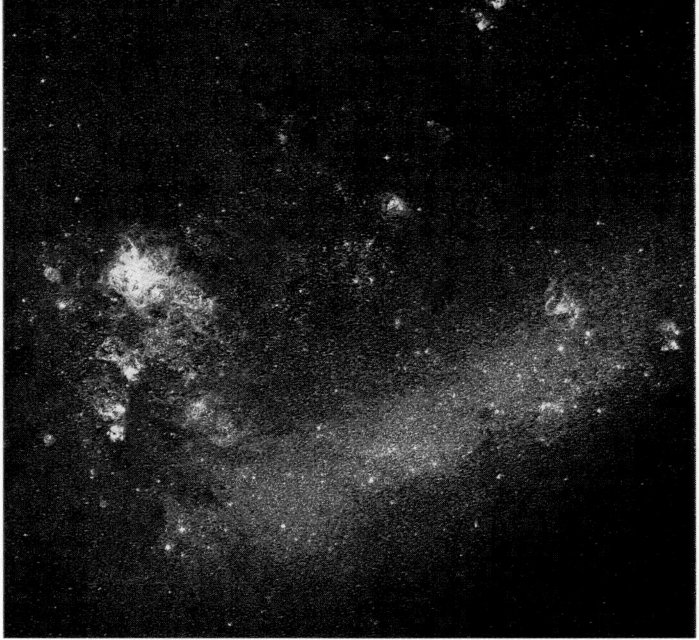

Figure 24-20

The Large Magellanic Cloud is an irregular galaxy. It's a member of the Local Group, and it orbits our own galaxy.

Expansion of the Universe

What does it sound like when a car is blowing its horn while it drives past you? The horn has a high pitch as the car approaches you, then the horn seems to drop in pitch as the car drives away. This effect is called the Doppler shift. The Doppler shift occurs with light as well as with sound.

Figure 24-21

The Milky Way Galaxy is usually classified as a normal spiral galaxy. Its spiral arms, composed of stars and gas, radiate out from an area of densely packed stars, the nucleus.

CONNECT TO

PHYSICS

Have you ever heard how the pitch of a train's horn changes as it first approaches and then recedes from you? Relate this to the changes shown in **Figure 24-22.** This figure shows how the Doppler shift causes changes in the light coming from distant galaxies. ***Infer*** how these changes relate to whether a galaxy is approaching or receding from Earth.

The Doppler Shift

Look at the spectrum of a star containing sodium in **Figure 24-23A.** Note the position of the dark lines. How do they compare with the lines in **Figures 24-23B** and **C?** They have shifted in position. What caused this shift? When a star is moving toward you, its wavelengths of light are pushed together, just as the sound waves from the car's horn are. The dark lines in the spectrum shift toward the blue-violet. A red shift in the spectrum occurs when an object is moving away from you. In a red shift, the dark lines shift toward the longer-wavelength red end of the spectrum.

In 1924, Edwin Hubble noticed an interesting fact about the light coming from most galaxies. When a spectrograph is used to study light from galaxies beyond the Local Group, there is a red shift in the light. What does this red shift tell you about the universe?

Because all galaxies beyond the Local Group show a red shift in their spectra, they must be moving away from Earth. If all galaxies outside the Local Group are moving away from you, this indicates that the entire universe must be expanding. Think of the Explore activity at the beginning of

Figure 24-22

The Doppler shift causes the wavelengths of light coming from galaxies to be compressed or stretched.

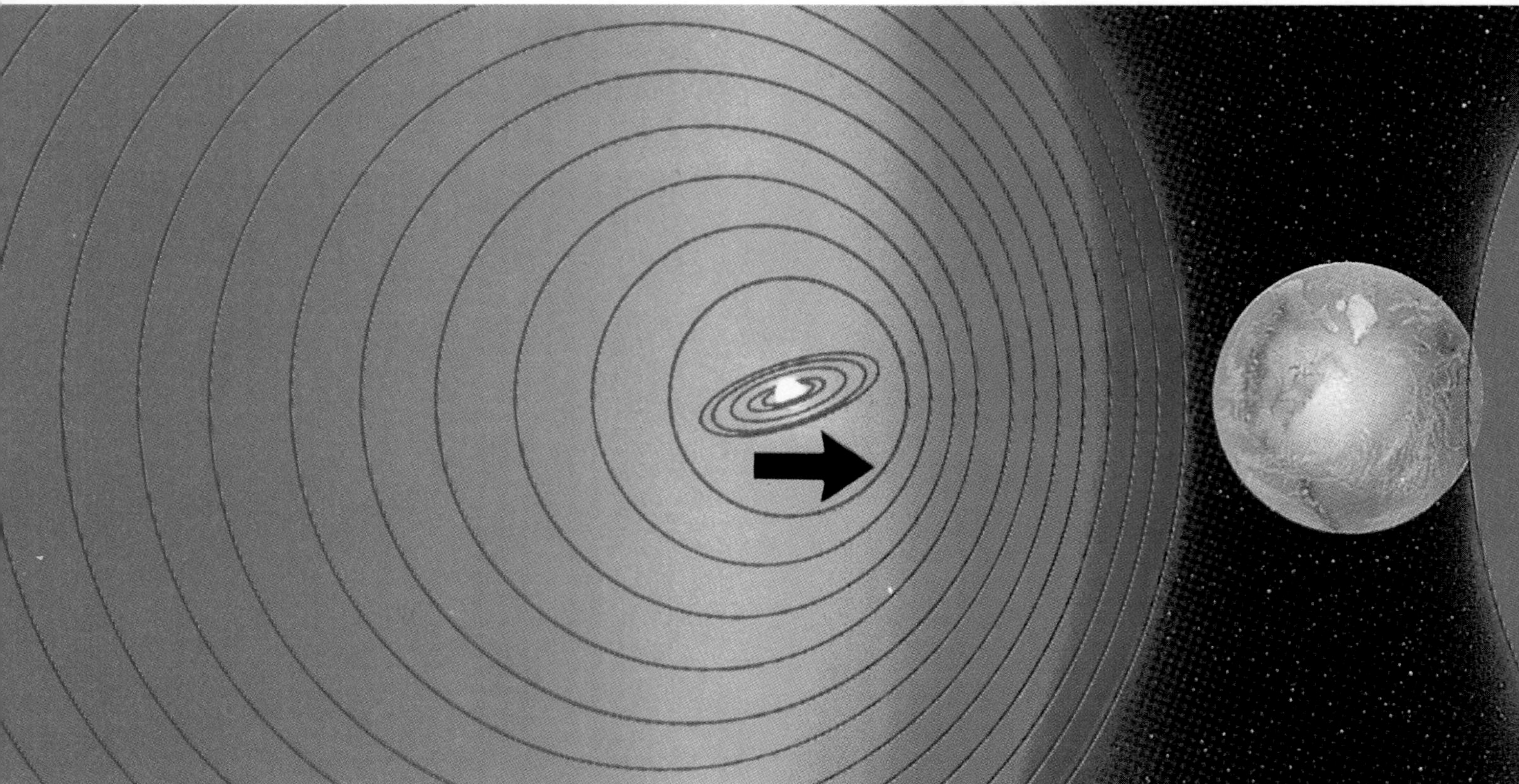

Figure 24-23

The dark lines in the spectra (A) are shifted toward the blue-violet end when a star is moving toward Earth (B). A red shift (C) indicates a star is moving away from Earth.

the chapter. The dots on the balloon moved apart as the model universe expanded. Regardless of which dot you picked, all the other dots moved away from it. Galaxies beyond the Local Group move away from us just as the dots moved apart on the balloon.

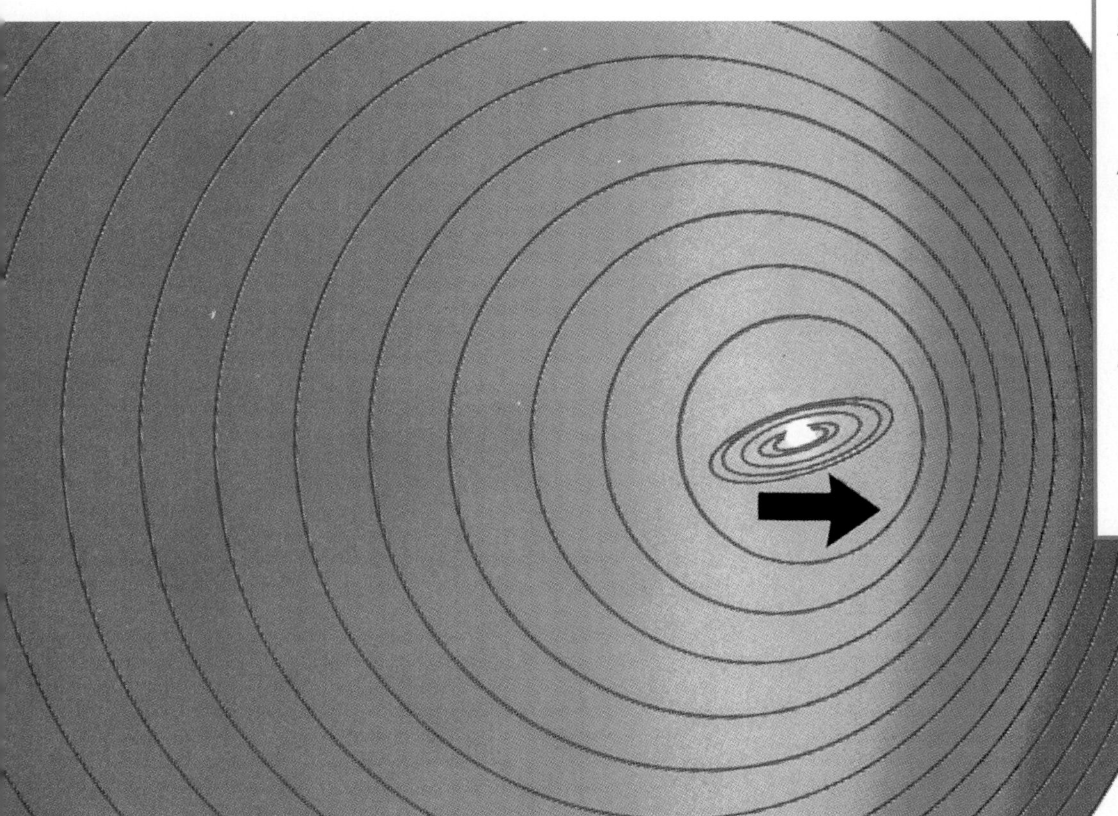

MiniLAB

How far can we "see" into space?

Procedure

1. On a large sheet of paper, formulate a model of the Milky Way Galaxy as it would appear looking down on it from somewhere out in space.
2. Mark the appropriate location of the solar system, about 2/3 of the way out on one of the spiral arms.
3. Draw a circle around the sun indicating the 4.2 light-year distance of the next closest star to the sun, Proxima Centauri.
4. One of the most distant stars that can be seen on a clear night without the aid of a telescope is Epsilon Aurigae in the constellation Auriga at a distance of 3300 light-years.
5. On your diagram, draw a circle with a radius of about 3300 light-years with the sun at the center.

Analysis

1. If we assume that Epsilon Aurigae is the farthest star visible without the aid of a telescope, interpret what this circle represents.
2. At this scale interpret how far away the Andromeda Galaxy (the next closest spiral galaxy) would be located.

Figure 24-24

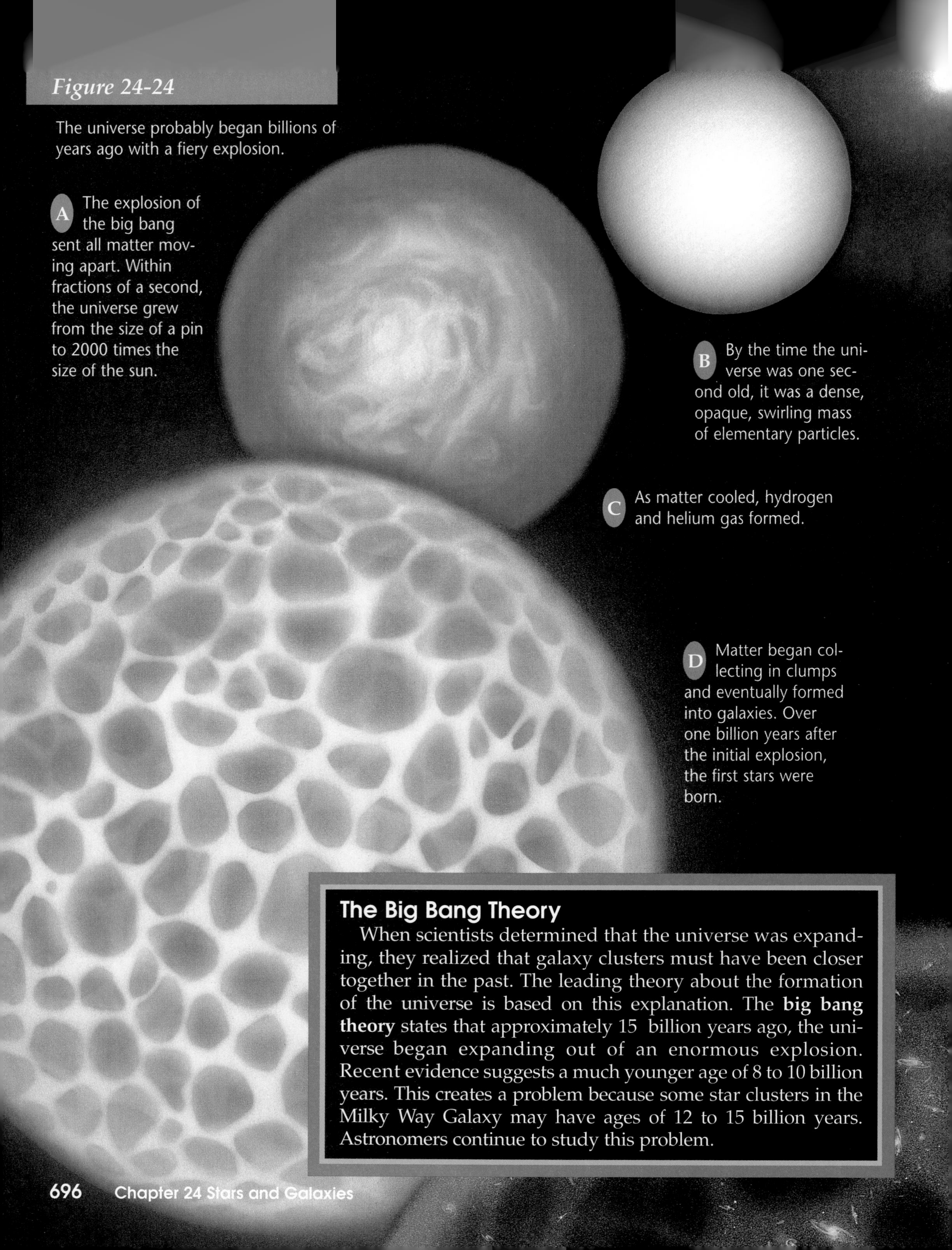

The universe probably began billions of years ago with a fiery explosion.

A The explosion of the big bang sent all matter moving apart. Within fractions of a second, the universe grew from the size of a pin to 2000 times the size of the sun.

B By the time the universe was one second old, it was a dense, opaque, swirling mass of elementary particles.

C As matter cooled, hydrogen and helium gas formed.

D Matter began collecting in clumps and eventually formed into galaxies. Over one billion years after the initial explosion, the first stars were born.

The Big Bang Theory

When scientists determined that the universe was expanding, they realized that galaxy clusters must have been closer together in the past. The leading theory about the formation of the universe is based on this explanation. The **big bang theory** states that approximately 15 billion years ago, the universe began expanding out of an enormous explosion. Recent evidence suggests a much younger age of 8 to 10 billion years. This creates a problem because some star clusters in the Milky Way Galaxy may have ages of 12 to 15 billion years. Astronomers continue to study this problem.

A time-lapse photograph taken in December 1995 by the *Hubble Space Telescope* shows over 1500 galaxies at a distance of more than 10 billion light-years. These galaxies may date back to when the universe was no more than 1 billion years old. The galaxies are in various stages of development. One astronomer indicates that we may be looking back to a time when our own galaxy was forming. Studies of this nature will eventually enable astronomers to determine the approximate age of the universe.

Whether the universe expands forever or stops depends on how much matter is in the universe. All matter exerts a gravitational force. If there's enough matter, gravity will halt the expansion, and the universe will contract until everything comes to one point.

Astronomers continue to study the structure of the known universe in hopes of learning a more exact age, how it has evolved, and how it might end. You will learn in the next section that astronomers also search the galaxy in hopes of finding evidence that life exists elsewhere in our solar system and galaxy.

USING MATH

The mass of a hydrogen nucleus is 1.008 atomic mass units (amu). A helium nucleus' mass is 4.004 amu. Calculate the mass lost when four hydrogen nuclei fuse to form one helium nucleus. What happens to this mass?

Section Wrap-up

Review

1. List the three major classifications of galaxies. What do they all have in common?
2. What is the name of the galaxy that you live in? What motion do the stars in this galaxy exhibit?
3. **Think Critically:** All galaxies outside the Local Group show a red shift in their spectra. Within the Local Group, some galaxies show a red shift and some show a blue shift. What does this tell you about the galaxies in the Local Group and outside the Local Group?

Skill Builder

Recognizing Cause and Effect

Current measurements and calculated densities of the observable universe show it's not dense enough to collapse on itself. Scientists are trying to prove the existence of so-called dark matter, which is not directly observable. If the universe contains an abundance of dark matter (as much as 90 percent of its total mass), what could you infer about the true density and the future of the universe? If you need help, refer to Recognizing Cause and Effect in the **Skill Handbook.**

Science Journal

Research and write a report in your Science Journal about the most recent evidence supporting or disputing the big bang theory. Describe how the big bang theory explains observations of galaxies made with spectrometers.

SCIENCE & SOCIETY

TECHNOLOGY:

24•5 The Search for Extraterrestrial Life

Objectives

- Name locations in our solar system where life may have existed in the past or could exist now.
- Describe the methods used by astronomers to search for extraterrestrial life within and beyond our solar system.
- Decide if methods used by scientists might cause contamination of other worlds.

Searching for Life on Other Worlds

On July 20, 1976, the first *Viking* lander touched down on Mars. It conducted three different types of tests on Martian soil in its search for evidence of extraterrestrial life. Extraterrestrial life is life that exists beyond Earth. It can be microorganisms living in the rocks of a barren world or intelligent life living on a planet orbiting a distant star.

Some of the tests made by the *Viking* probes produced results that could have been caused by life or by chemical reactions with the Martian soil. Analysis of the soil showed it lacked the types of molecules that appear to be necessary for life. It appears Mars doesn't contain life at the present time. But what if life had been found? Would the space probe have contaminated it?

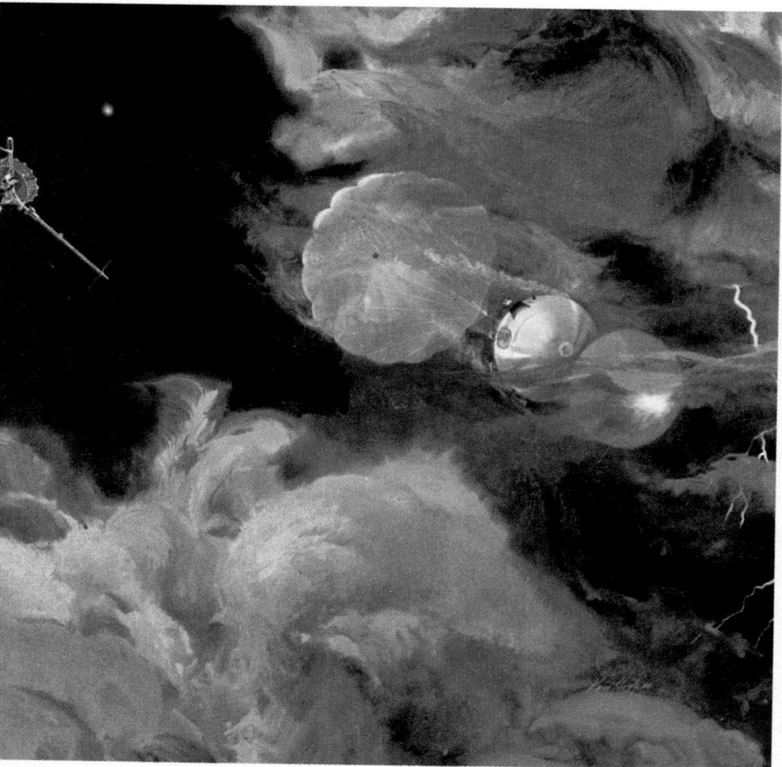

Figure 24-25

The *Galileo* spacecraft dropped a small probe into Jupiter's atmosphere. Before it was crushed by the intense pressure, it transmitted data back to the orbiter above.

Searching with Space Probes

The *Viking* spacecraft was carefully sterilized so the Martian soil would not be contaminated. The *Galileo* spacecraft arrived at Jupiter in 1995. It sent a probe down into the clouds of the planet as shown in **Figure 24-25.** It was not sterilized before being launched. Is there a danger that microbes from Earth could contaminate the environment of Jupiter's clouds? Probably not. *Galileo's* probe was crushed and sterilized in the depths of Jupiter's atmosphere, eliminating any chance of planetary contamination.

Titan, one of the satellites of Saturn, contains organic molecules. The molecules themselves aren't considered to be life, but they are thought to resemble the molecules from which life evolved on Earth. Given enough time, millions or billions of years, it's possible that the molecules will evolve into life as they did on Earth. Scientists hope to land a probe on Titan in the near future to further investigate the organic molecules there.

Cracks in the icy crust of Europa, one of Jupiter's moons, may allow sunlight into the ocean underneath. Although it is unlikely, the sunlight could have helped start biological activity. Could a space probe be used to study this possibility? What types of precautions should be taken before a probe is sent to Europa?

Are there ways to search for life on other worlds without the possibility of contamination? One method is to search the universe for sources of radio signals from intelligent life. This program is called SETI, the Search for Extraterrestrial Intelligence.

In October 1992, NASA began a program that, in addition to current programs by the University of California, the Planetary Society, and the Ohio State University, was to search for radio signals that could be evidence of extraterrestrial life. However, less than one year later, Congress voted to stop funding for NASA's SETI program. Since then, part of the project has received private funding. In October 1995, Project BETA, funded by the Planetary Society, began searching for extraterrestrial life. META II, operated by Argentine scientists, has recorded several signals from the direction of the Milky Way. As yet they have not repeated the discovery. The signals may have been interference.

Figure 24-26

This 305-m dish at Arecibo, Puerto Rico, has been used in an attempt to communicate with extraterrestrial life. In 1974, astronomers used this radio telescope to beam a radio message to a star cluster more than 26 000 light-years away.

Section Wrap-up

Review

1. Where is life known to exist in the solar system? Name two other places in the solar system where life might be evolving.
2. Which program searches for evidence of intelligent life using radio telescopes?
3. Why were the *Viking* probes sterilized, but the *Galileo* probe was not?

Explore the Technology

If you were in charge of studying the possibility of life on other worlds, how would you proceed? If space probes are used, how could they be used with no danger of contaminating present or future life? If your study is by radio telescopes alone, what evidence of life would you expect to find?

Science Journal

What kind of message would you place on a spacecraft? In your Science Journal, write a short essay describing the content of the message and the technology or technologies you would choose as a medium. Explain your reasoning.

Is anyone out there?

In 1977, two unmanned space probes–*Voyager 1* and *Voyage 2* –were launched from Earth on a mission to Jupiter, Saturn, Uranus, and Neptune. During the 1980s, the *Voyagers* broadcast back to Earth close-up photographs and other valuable data about our solar system's outer planets. Once this work was completed, both spacecraft continued on, and they are now hurtling past the orbit of Pluto and out into the depths of outer space.

The *Voyagers* have completed their original mission. But astronomers predict that these spacecraft will continue speeding through the frictionless vacuum of deep space for thousands of years. Just in case they should someday come into contact with intelligent beings, both *Voyagers* carry identical messages from Earth.

The *Voyager* Record

The message from Earth is in the form of a gold-plated copper disc that operates like a long-playing phonograph record. Included are a stylus and cartridge that can be used to play the recording. The message contains 90 minutes of music from around the world, greetings in dozens of human languages, sounds of nature, mathematical equations, and 118 photographs.

The photographs include pictures of children, families, classrooms, men and women at work, highways, trains, buses, airplanes, a rocket launch, an astronaut walking in space, musicians and instruments, and a variety of nature scenes and animals, including mountains, deserts, forests, beaches, seashores, elephants, gorillas, dogs, dolphins, birds, and insects. There is also a star map to show the location of Earth and a diagram of our solar system.

Nature sounds include volcanic eruptions, earthquakes, thunder, wind, waves, rain, crickets and frogs, footsteps, laughter, and heartbeats.

The recording includes the music of Bach, Mozart, Beethoven, a Javanese Gamelan, Pygmies of the Ituri Forest, percussion from Senegal, Chuck Berry, Louis Armstrong, bagpipes, a Navajo chant, Japanese flute, and music from China, India, Australia, and New Guinea.

Chapter 24

Review

Summary

24-1: Stars

1. The constellations seem to move because Earth rotates on its axis and revolves around the sun.
2. The magnitude of a star is a measure of the star's brightness. Absolute magnitude is a measure of the light emitted. Apparent magnitude is the amount of light received on Earth.
3. Parallax is the apparent shift in the position of an object when viewed from two different positions.

24-2: The Sun

1. The sun produces energy by fusing hydrogen into helium in its core.
2. Sunspots, prominences, and flares are all caused by the intense magnetic field of the sun.
3. The sun is a main sequence star.

24-3: Evolution of Stars

1. A main sequence star uses hydrogen as fuel. When the hydrogen is used up, the core collapses and the star's temperature increases. The star becomes a giant or a supergiant, which uses helium as fuel. As the star evolves, the outer layers escape into space. The core has no fuel left and the star becomes a white dwarf. Stars containing high amounts of mass can explode and evolve into neutron stars or black holes.

24-4: Galaxies and the Expanding Universe

1. A galaxy is a large group of stars, gas, and dust held together by gravity. Galaxies can be elliptical, spiral, or irregular in shape.
2. The Milky Way is a spiral galaxy.
3. Spectral line shifts toward red light suggest that all galaxies beyond the Local Group are moving away from our galaxy.

24-5: Science and Society: The Search for Extraterrestrial Life

1. Organic molecules, similar to those that evolved into life on Earth, have been found on Titan, a satellite of Saturn.
2. Space probes and radio telescopes are used to determine whether or not life exists in other parts of the universe.

Key Science Words

- **a.** absolute magnitude
- **b.** apparent magnitude
- **c.** big bang theory
- **d.** binary system
- **e.** black hole
- **f.** chromosphere
- **g.** constellation
- **h.** corona
- **i.** galaxy
- **j.** giant
- **k.** light-year
- **l.** main sequence
- **m.** nebula
- **n.** neutron star
- **o.** parallax
- **p.** photosphere
- **q.** sunspot
- **r.** supergiant
- **s.** white dwarf

Reviewing Vocabulary

Match each phrase with the correct term from the list of Key Science Words.

1. a group of stars that resembles an object, animal, or mythological character
2. a measure of the amount of light given off by a star
3. distance light travels in one year

Chapter 24 Review

4. the beginning of a star's life cycle
5. star that uses helium as fuel
6. dense object that allows nothing to escape its field of gravity
7. our sun belongs to this group of stars
8. layer of the sun's atmosphere that emits light
9. two stars that orbit each other
10. supported by observed red shifts in the spectra of galaxies

Checking Concepts

Choose the word or phrase that completes the sentence.

1. The stars of a constellation ______.
   a. are in the same cluster
   b. are all giants
   c. are equally bright
   d. form a pattern
2. ______ is a measure of the amount of a star's light received on Earth.
   a. Absolute magnitude
   b. Apparent magnitude
   c. Fusion
   d. Parallax
3. The closer an object is to an observer, the greater its ______.
   a. absolute magnitude c. parallax
   b. red shift d. size
4. Once a nebula contracts and temperatures increase to 10 000 000°C, ______ begins.
   a. main sequencing c. fusion
   b. a supernova d. a white dwarf
5. A ______ is about 10 km in size.
   a. giant c. black hole
   b. white dwarf d. neutron star
6. Our sun fuses ______ into ______.
   a. hydrogen/carbon c. carbon/helium
   b. helium/hydrogen d. hydrogen/helium
7. Loops of matter flowing from the sun are ______.
   a. sunspots c. coronas
   b. auroras d. prominences
8. Groups of galaxies are called ______.
   a. clusters c. giants
   b. supergiants d. binary systems
9. ______ galaxies are sometimes shaped like footballs.
   a. Spiral c. Barred
   b. Elliptical d. Irregular
10. A shift of wavelengths of light toward the red end of the ______ indicates that galaxies are moving away from the Local Group.
   a. spectrum c. corona
   b. surface d. chromosphere

Understanding Concepts

Answer the following questions in your Science Journal using complete sentences.

11. How can parallax be used to determine which of two stars is closer to Earth?
12. What variables determine a star's position on the H-R diagram?
13. Compare and contrast sunspots, solar flares, and prominences.
14. How do red shifts support the big bang theory?
15. What are the advantages and disadvantages of using radio telescopes and probes to search for life in the universe?

Thinking Critically

16. What is significant about the 1995 discovery by the *Hubble Space Telescope* of over 1500 galaxies at a distance of more than 10 billion light-years?
17. How do scientists know that black holes exist if these objects don't emit any visible light?
18. Use the autumn star chart in Appendix K to determine which constellation is directly overhead at 8 P.M. on November 21 for an observer in North America.
19. How are radio waves used to detect objects in space?
20. What kinds of reactions produce the energy emitted by stars?

Chapter 24 Review

Developing Skills

If you need help, refer to the **Skill Handbook.**

21. **Making and Using Tables:** Astronomical objects are given numbers to represent their absolute and apparent magnitudes. The lower the number, the greater the object's brightness. What object listed in the table below has the brightest absolute magnitude? What object is the brightest as seen from Earth?
22. **Making and Using Tables:** How does the table below show that apparent magnitude is dependent on both absolute magnitude and distance from an observer?
23. **Concept Mapping:** Make a concept map that shows the evolution of a main sequence star with a mass similar to that of the sun.
24. **Comparing and Contrasting:** Compare and contrast the sun with other stars on the H-R diagram.
25. **Measuring in SI:** The Milky Way Galaxy is 100 000 light-years in diameter. What scale would you use if you were to construct a scale model of the Milky Way with a diameter of 20 cm?

Star Magnitudes

Star	Absolute Magnitude	Apparent Magnitude	Light-Years from Earth
The Sun	4.9	−26.7	0.000 002
Sirius	1.5	−1.5	8.80
Arcturus	−0.3	−0.1	35.86
Alpha Centauri	4.4	0.0	4.34
Betelgeuse	−5.5	0.8	489.0
Deneb	−6.9	1.3	1401.80

Performance Assessment

1. **Model:** Design and construct scale models of a spiral and a barred spiral Milky Way Galaxy. Show the approximate position of the sun in each.
2. **Carrying Out a Strategy and Collecting Data:** Design and carry out an experiment that uses sunspot locations to compare rotational periods of different latitudes of the sun.
3. **Using Math in Science:** Find two distant objects in opposite directions from a single viewing point. Using a protractor and what you know about parallax, determine which object is closer.

UNIT PROJECT 7

Where? Out There!

What would it be like for humans to colonize space? What would they need to survive far from Earth? What problems might people encounter? These are just a few of the many questions that humans would face if they moved out into near-Earth space, to the moon once again, and on to Mars.

Suppose you were responsible for planning a permanent station for studying both deep space and nearby planets. Would you place this station in near-Earth orbit, on the Moon, or on Mars? What supplies would you need to build the station? How would you get them to the station? As you do this project you will make technical drawings, build a model, and describe your plan to build a space station. You'll also write an instruction manual for how your space station works.

Procedures

1. Conduct a brainstorming session with your group to decide the best location for your space station. Keep in mind your research goals, as outlined above.
2. Next, determine what materials you'd need to transport into space and how you'd get them there. Also decide where people building the station will live while they work on the project. How will your workers get air to breathe and food to eat?

   You'll also need to think about how the space station will get food and air. Will you use the same process that you used during construction, or will you need to use a different process?
3. Write a summary of your decisions and answers. Be certain that you say why you chose the location you did for your space station. Describe the basic form of your station (in pieces or in one large complex), what materials you would need, and how you intend to transport them. Tell where people working on the station will live and how they will get needed supplies such as food and air.
4. After you've finished your report, draw the overall plan of your space station and where each part of the space station will be located. Remember that this is a research station, so you'll need to have an area designed for research and study.

5. Decide what materials you'll need to construct a model of your space station. Assign each member of your team to work on building a specific part of the space station.
6. Examine your design closely and see what safety features might have to be built in. What hazards might there be in your space station's location? What kinds of things can you build in to help space colonists cope with emergency situations?
7. When you've completed your model and report, prepare a 10–15-minute oral presentation to describe your space station to the class. Tell why you made the choices you did, and what hazards people living in a space station might face. Also describe how your space station design helps make living comfortable for space colonists.

Going Further

To help get people interested in the space station, describe the benefits your space station will have for humans on Earth. Also describe how working on the space station will help humans prepare for future missions farther out into the solar system.

Appendices

Appendix A

International System of Units

The International System (SI) of Units is accepted as the standard for measurement throughout most of the world. Three base units in SI are the meter, kilogram, and second. Frequently used SI units are listed below.

Table A-1

Frequently Used SI Units	
Length	1 millimeter (mm) = 1000 micrometers (µm) 1 centimeter (cm) = 10 millimeters (mm) 1 meter (m) = 100 centimeters (cm) 1 kilometer (km) =1000 meters (m) 1 light-year = 9 460 000 000 000 kilometers (km)
Area	1 square meter (m^2) = 10 000 square centimeters (cm^2) 1 square kilometer (km^2) = 1 000 000 square meters (m^2)
Volume	1 milliliter (mL) = 1 cubic centimeter (cc) (cm^3) 1 liter (L) = 1000 milliliters (mL)
Mass	1 gram (g) = 1000 milligrams (mg) 1 kilogram (kg) = 1000 grams (g) 1 metric ton = 1000 kilograms (kg)
Time	1 s = 1 second

Temperature measurements in SI are often made in degrees Celsius. Celsius temperature is a supplementary unit derived from the base unit kelvin. The Celsius scale (°C) has 100 equal graduations between the freezing temperature (0°C) and the boiling temperature of water (100°C). The following relationship exists between the Celsius and Kelvin temperature scales:

$$K = °C + 273$$

Several other supplementary SI units are listed below.

Table A-2

Supplementary SI Units

Measurement	Unit	Symbol	Expressed in base units
Energy	Joule	J	$kg \cdot m^2/s^2$ or $N \cdot m$
Force	Newton	N	$kg \cdot m/s^2$
Power	Watt	W	$kg \cdot m^2/s^3$ or J/s
Pressure	Pascal	Pa	$kg/(m \cdot s^2)$ or $N \cdot m$

Appendix B

Table B-1

SI/Metric to English Conversions

	When you want to convert:	Multiply by:	To find:
Length	inches	2.54	centimeters
	centimeters	0.39	inches
	feet	0.30	meters
	meters	3.28	feet
	yards	0.91	meters
	meters	1.09	yards
	miles	1.61	kilometers
	kilometers	0.62	miles
Mass and Weight*	ounces	28.35	grams
	grams	0.04	ounces
	pounds	0.45	kilograms
	kilograms	2.20	pounds
	tons	0.91	tonnes (metric tons)
	tonnes (metric tons)	1.10	tons
	pounds	4.45	newtons
	newtons	0.23	pounds
Volume	cubic inches	16.39	cubic centimeters
	cubic centimeters	0.06	cubic inches
	cubic feet	0.03	cubic meters
	cubic meters	35.31	cubic feet
	liters	1.06	quarts
	liters	0.26	gallons
	gallons	3.78	liters
Area	square inches	6.45	square centimeters
	square centimeters	0.16	square inches
	square feet	0.09	square meters
	square meters	10.76	square feet
	square miles	2.59	square kilometers
	square kilometers	0.39	square miles
	hectares	2.47	acres
	acres	0.40	hectares
Temperature	Fahrenheit	5/9 (°F − 32)	Celsius
	Celsius	9/5 °C + 32	Fahrenheit

*Weight as measured in standard Earth gravity

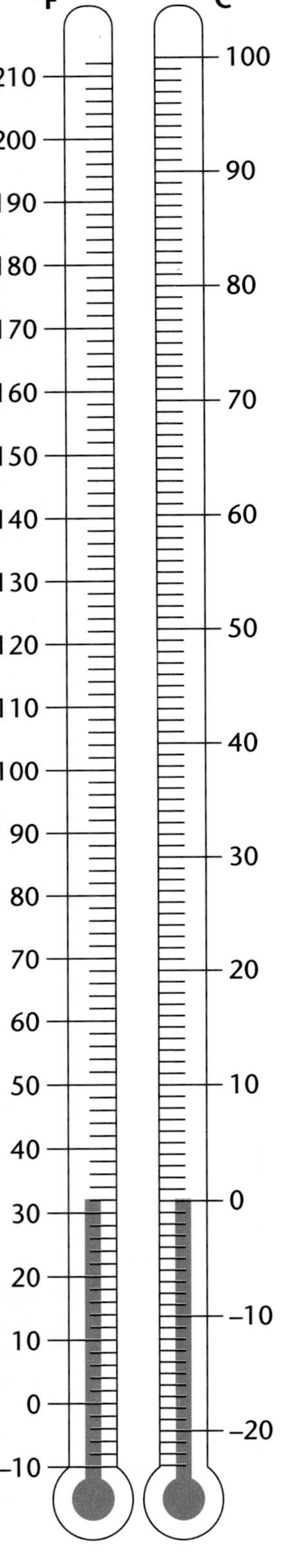

Appendix C

Safety in the Science Classroom

1. Always obtain your teacher's permission to begin an investigation.
2. Study the procedure. If you have questions, ask your teacher. Understand any safety symbols shown on the page.
3. Use the safety equipment provided for you. Goggles and a safety apron should be worn when any investigation calls for using chemicals.
4. Always slant test tubes away from yourself and others when heating them.
5. Never eat or drink in the lab, and never use lab glassware as food or drink containers. Never inhale chemicals. Do not taste any substances or draw any material into a tube with your mouth.
6. If you spill any chemical, wash it off immediately with water. Report the spill immediately to your teacher.
7. Know the location and proper use of the fire extinguisher, safety shower, fire blanket, first aid kit, and fire alarm.
8. Keep materials away from flames. Tie back hair and loose clothing.
9. If a fire should break out in the classroom, or if your clothing should catch fire, smother it with the fire blanket or a coat, or get under a safety shower. **NEVER RUN.**
10. Report any accident or injury, no matter how small, to your teacher.

Follow these procedures as you clean up your work area.

1. Turn off the water and gas. Disconnect electrical devices.
2. Return all materials to their proper places.
3. Dispose of chemicals and other materials as directed by your teacher. Place broken glass and solid substances in the proper containers. Never discard materials in the sink.
4. Clean your work area.
5. Wash your hands thoroughly after working in the laboratory.

Table C-1

First Aid

Injury	Safe response
Burns	Apply cold water. Call your teacher immediately.
Cuts and bruises	Stop any bleeding by applying direct pressure. Cover cuts with a clean dressing. Apply cold compresses to bruises. Call your teacher immediately.
Fainting	Leave the person lying down. Loosen any tight clothing and keep crowds away. Call your teacher immediately.
Foreign matter in eye	Flush with plenty of water. Use eyewash bottle or fountain.
Poisoning	Note the suspected poisoning agent and call your teacher immediately.
Any spills on skin	Flush with large amounts of water or use safety shower. Call your teacher immediately.

Appendix D

Safety Symbols

Table D-1

	**DISPOSAL ALERT** This symbol appears when care must be taken to dispose of materials properly.		**ANIMAL SAFETY** This symbol appears whenever live animals are studied and the safety of the animals and the students must be ensured.
	**BIOLOGICAL HAZARD** This symbol appears when there is danger involving bacteria, fungi, or protists.		**RADIOACTIVE SAFETY** This symbol appears when radioactive materials are used.
	**OPEN FLAME ALERT** This symbol appears when use of an open flame could cause a fire or an explosion.		**CLOTHING PROTECTION SAFETY** This symbol appears when substances used could stain or burn clothing.
	**THERMAL SAFETY** This symbol appears as a reminder to use caution when handling hot objects.		**FIRE SAFETY** This symbol appears when care should be taken around open flames.
	**SHARP OBJECT SAFETY** This symbol appears when a danger of cuts or punctures caused by the use of sharp objects exists.		**EXPLOSION SAFETY** This symbol appears when the misuse of chemicals could cause an explosion.
	**FUME SAFETY** This symbol appears when chemicals or chemical reactions could cause dangerous fumes.		**EYE SAFETY** This symbol appears when a danger to the eyes exists. Safety goggles should be worn when this symbol appears.
	**ELECTRICAL SAFETY** This symbol appears when care should be taken when using electrical equipment.		**POISON SAFETY** This symbol appears when poisonous substances are used.
	**PLANT SAFETY** This symbol appears when poisonous plants or plants with thorns are handled.		**CHEMICAL SAFETY** This symbol appears when chemicals used can cause burns or are poisonous if absorbed through the skin.

Appendix E

Topographic Map Symbols

Primary highway, hard surface
Secondary highway, hard surface
Light-duty road, hard or improved surface
Unimproved road
Railroad: single track and multiple track
Railroads in juxtaposition

Buildings
Schools, church, and cemetery — cem
Buildings (barn, warehouse, etc.)
Wells other than water (labeled as to type) — o oil o gas
Tanks: oil, water, etc. (labeled only if water) — water
Located or landmark object; windmill
Open pit, mine, or quarry; prospect

Marsh (swamp)
Wooded marsh
Woods or brushwood
Vineyard
Land subject to controlled inundation
Submerged marsh
Mangrove
Orchard
Scrub
Urban area

Spot elevation — ×7369
Water elevation — 670

Index contour
Supplementary contour
Intermediate contour
Depression contours

Boundaries: National
State
County, parish, municipal
Civil township, precinct, town, barrio
Incorporated city, village, town, hamlet
Reservation, National or State
Small park, cemetery, airport, etc.
Land grant
Township or range line, United States land survey
Township or range line, approximate location

Perennial streams
Elevated aqueduct
Water well and spring
Small rapids
Large rapids
Intermittent lake
Intermittent streams
Aqueduct tunnel
Glacier
Small falls
Large falls
Dry lake bed

Appendix F

Periodic Table of the Elements

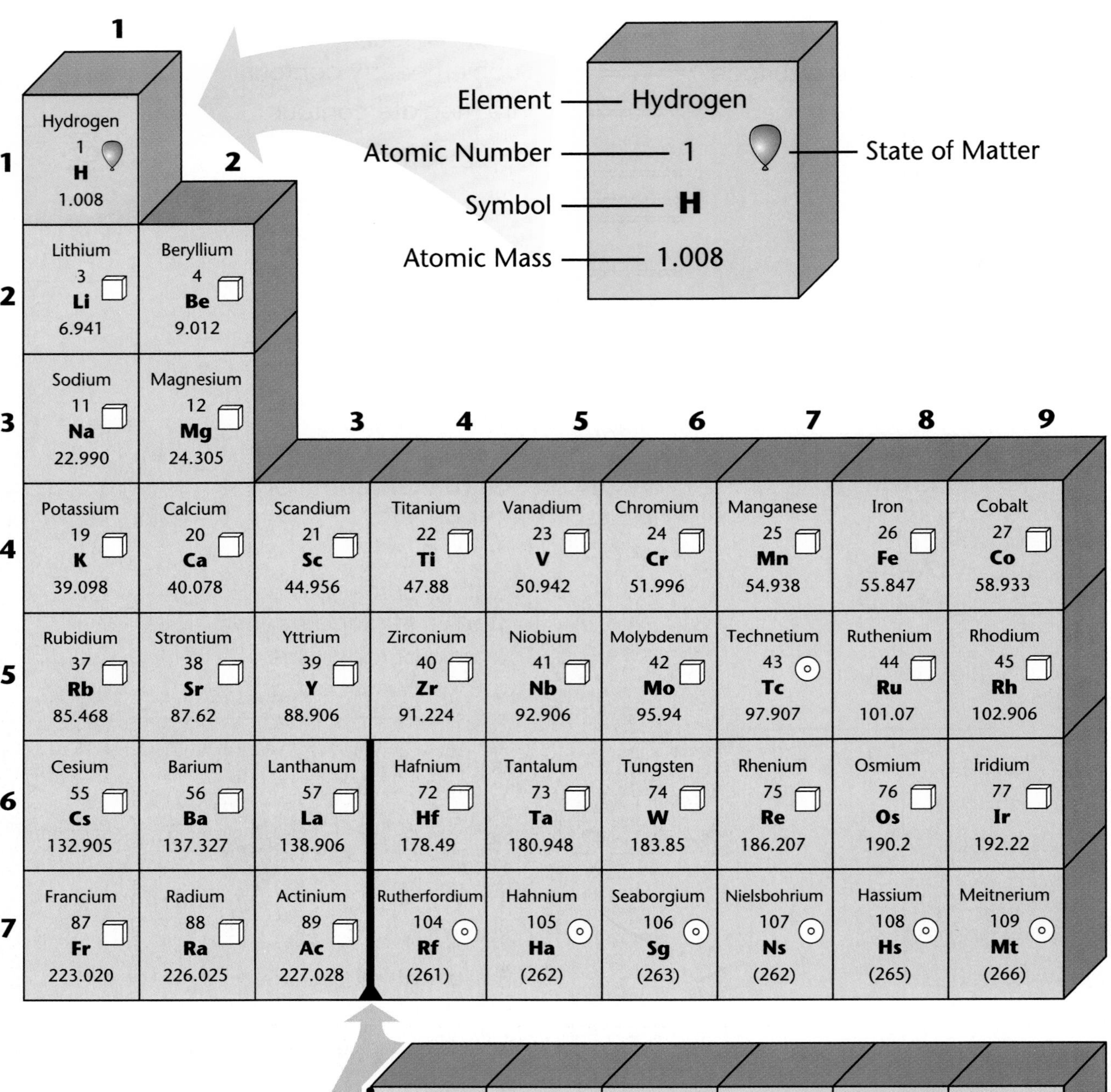

	1	2	3	4	5	6	7	8	9
1	Hydrogen 1 **H** 1.008								
2	Lithium 3 **Li** 6.941	Beryllium 4 **Be** 9.012							
3	Sodium 11 **Na** 22.990	Magnesium 12 **Mg** 24.305							
4	Potassium 19 **K** 39.098	Calcium 20 **Ca** 40.078	Scandium 21 **Sc** 44.956	Titanium 22 **Ti** 47.88	Vanadium 23 **V** 50.942	Chromium 24 **Cr** 51.996	Manganese 25 **Mn** 54.938	Iron 26 **Fe** 55.847	Cobalt 27 **Co** 58.933
5	Rubidium 37 **Rb** 85.468	Strontium 38 **Sr** 87.62	Yttrium 39 **Y** 88.906	Zirconium 40 **Zr** 91.224	Niobium 41 **Nb** 92.906	Molybdenum 42 **Mo** 95.94	Technetium 43 **Tc** 97.907	Ruthenium 44 **Ru** 101.07	Rhodium 45 **Rh** 102.906
6	Cesium 55 **Cs** 132.905	Barium 56 **Ba** 137.327	Lanthanum 57 **La** 138.906	Hafnium 72 **Hf** 178.49	Tantalum 73 **Ta** 180.948	Tungsten 74 **W** 183.85	Rhenium 75 **Re** 186.207	Osmium 76 **Os** 190.2	Iridium 77 **Ir** 192.22
7	Francium 87 **Fr** 223.020	Radium 88 **Ra** 226.025	Actinium 89 **Ac** 227.028	Rutherfordium 104 **Rf** (261)	Hahnium 105 **Ha** (262)	Seaborgium 106 **Sg** (263)	Nielsbohrium 107 **Ns** (262)	Hassium 108 **Hs** (265)	Meitnerium 109 **Mt** (266)

Lanthanide Series	Cerium 58 **Ce** 140.115	Praseodymium 59 **Pr** 140.908	Neodymium 60 **Nd** 144.24	Promethium 61 **Pm** 144.913	Samarium 62 **Sm** 150.36	Europium 63 **Eu** 151.965
Actinide Series	Thorium 90 **Th** 232.038	Protactinium 91 **Pa** 231.036	Uranium 92 **U** 238.029	Neptunium 93 **Np** 237.048	Plutonium 94 **Pu** 244.064	Americium 95 **Am** 243.061

Metal
Metalloid
Nonmetal

Gas
Liquid
Solid
Synthetic Elements

10	11	12	13	14	15	16	17	18
								Helium 2 **He** 4.003
			Boron 5 **B** 10.811	Carbon 6 **C** 12.011	Nitrogen 7 **N** 14.007	Oxygen 8 **O** 15.999	Fluorine 9 **F** 18.998	Neon 10 **Ne** 20.180
			Aluminum 13 **Al** 26.982	Silicon 14 **Si** 28.086	Phosphorus 15 **P** 30.974	Sulfur 16 **S** 32.066	Chlorine 17 **Cl** 35.453	Argon 18 **Ar** 39.948
Nickel 28 **Ni** 58.693	Copper 29 **Cu** 63.546	Zinc 30 **Zn** 65.39	Gallium 31 **Ga** 69.723	Germanium 32 **Ge** 72.61	Arsenic 33 **As** 74.922	Selenium 34 **Se** 78.96	Bromine 35 **Br** 79.904	Krypton 36 **Kr** 83.80
Palladium 46 **Pd** 106.42	Silver 47 **Ag** 107.868	Cadmium 48 **Cd** 112.411	Indium 49 **In** 114.82	Tin 50 **Sn** 118.710	Antimony 51 **Sb** 121.757	Tellurium 52 **Te** 127.60	Iodine 53 **I** 126.904	Xenon 54 **Xe** 131.290
Platinum 78 **Pt** 195.08	Gold 79 **Au** 196.967	Mercury 80 **Hg** 200.59	Thallium 81 **Tl** 204.383	Lead 82 **Pb** 207.2	Bismuth 83 **Bi** 208.980	Polonium 84 **Po** 208.982	Astatine 85 **At** 209.987	Radon 86 **Rn** 222.018
(unnamed) 110 **Uun**	(unnamed) 111 **Uuu**							

Gadolinium 64 **Gd** 157.25	Terbium 65 **Tb** 158.925	Dysprosium 66 **Dy** 162.50	Holmium 67 **Ho** 164.930	Erbium 68 **Er** 167.26	Thulium 69 **Tm** 168.934	Ytterbium 70 **Yb** 173.04	Lutetium 71 **Lu** 174.967
Curium 96 **Cm** 247.070	Berkelium 97 **Bk** 247.070	Californium 98 **Cf** 251.080	Einsteinium 99 **Es** 252.083	Fermium 100 **Fm** 257.095	Mendelevium 101 **Md** 258.099	Nobelium 102 **No** 259.101	Lawrencium 103 **Lr** 260.105

Appendix G

THE WORLD

- ● World's most populous cities
- ——— International boundary
- - - - - - Disputed boundary
- .... Undefined boundary

0 1000 2000 Miles
0 1000 2000 Kilometers

Projection: Robinson

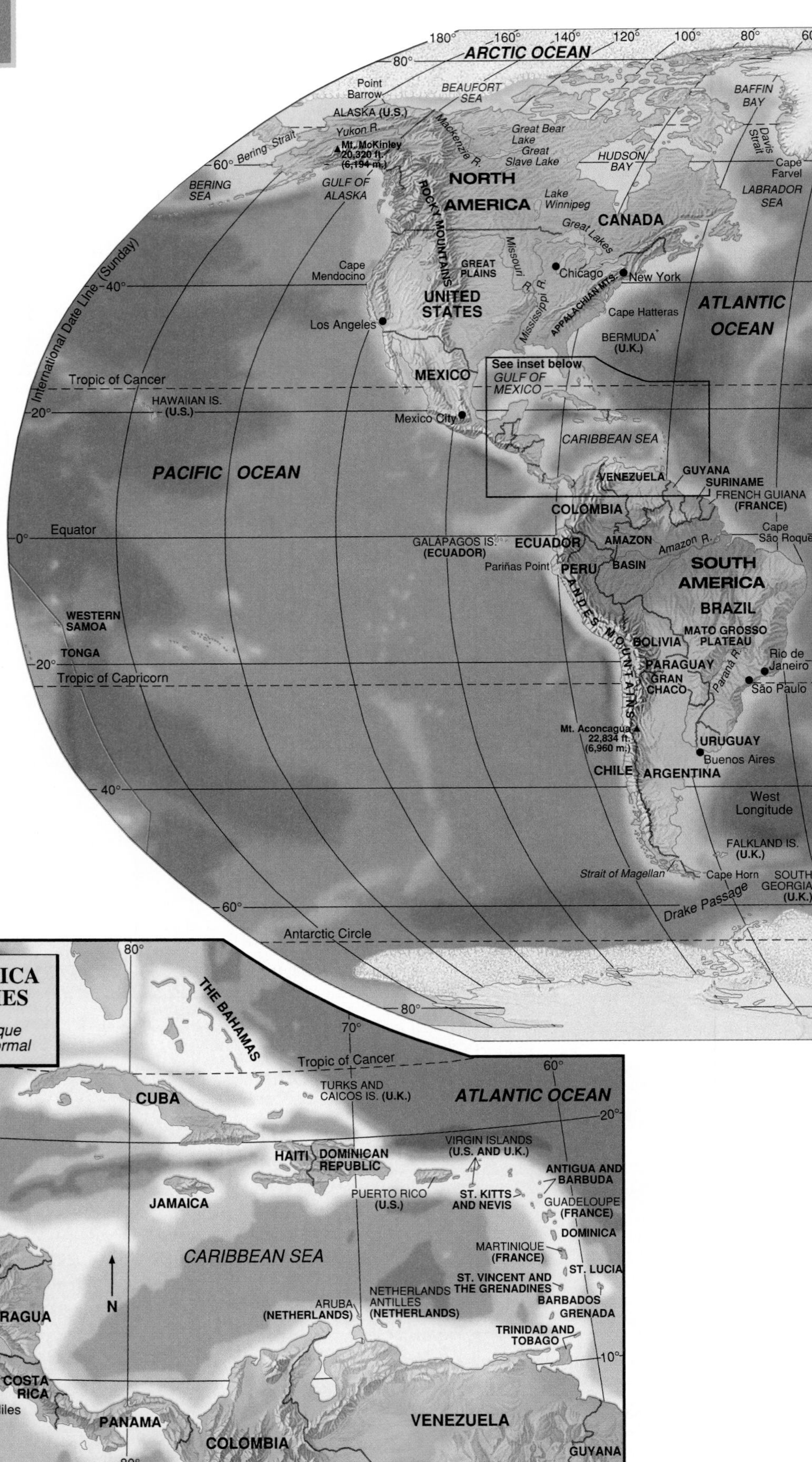

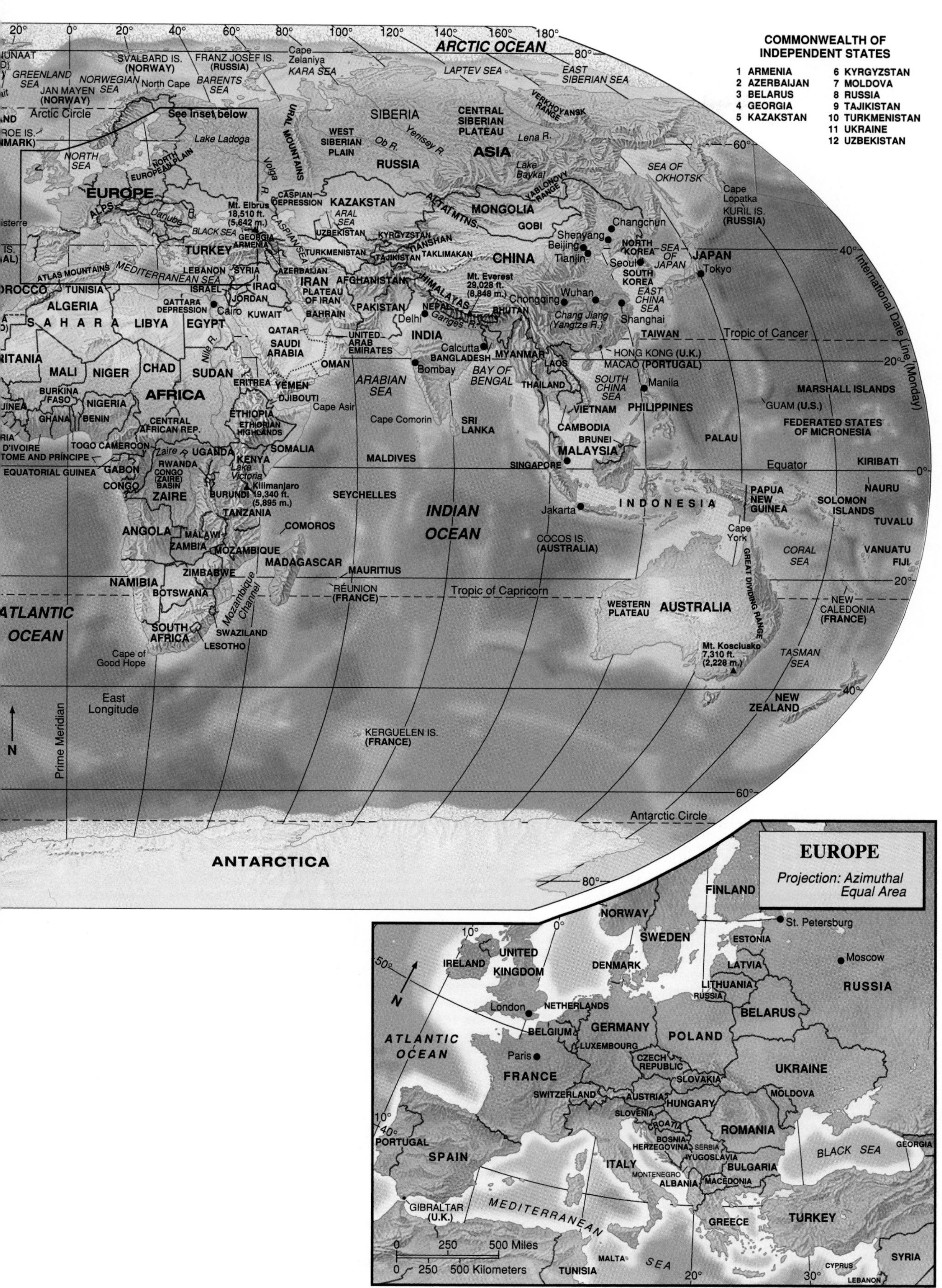

COMMONWEALTH OF INDEPENDENT STATES
1 ARMENIA
2 AZERBAIJAN
3 BELARUS
4 GEORGIA
5 KAZAKSTAN
6 KYRGYZSTAN
7 MOLDOVA
8 RUSSIA
9 TAJIKISTAN
10 TURKMENISTAN
11 UKRAINE
12 UZBEKISTAN
ARCTIC OCEAN
LAPTEV SEA
EAST SIBERIAN SEA
KARA SEA
BARENTS SEA
NORWEGIAN SEA
GREENLAND SEA
SVALBARD IS. (NORWAY)
FRANZ JOSEF IS. (RUSSIA)
Cape Zelaniya
JAN MAYEN (NORWAY)
North Cape
Arctic Circle
See Inset below
Lake Ladoga
URAL MOUNTAINS
WEST SIBERIAN PLAIN
SIBERIA
CENTRAL SIBERIAN PLATEAU
VERKHOYANSK RANGE
Ob R.
Yenisey R.
Lena R.
RUSSIA
ASIA
Lake Baykal
YABLONOVY RANGE
SEA OF OKHOTSK
Cape Lopatka
KURIL IS. (RUSSIA)
NORTH SEA
NORTH EUROPEAN PLAIN
EUROPE
ALPS
Danube
Volga R.
CASPIAN DEPRESSION
KAZAKSTAN
ARAL SEA
ALTAI MTNS.
MONGOLIA
GOBI
Changchun
Shenyang
Beijing
Tianjin
NORTH KOREA
SEA OF JAPAN
JAPAN
Tokyo
Seoul
SOUTH KOREA
Mt. Elbrus 18,510 ft. (5,642 m.)
BLACK SEA
GEORGIA
ARMENIA
CASPIAN SEA
UZBEKISTAN
KYRGYZSTAN
TURKMENISTAN
TIAN SHAN
TAJIKISTAN
TAKLIMAKAN
CHINA
TURKEY
ATLAS MOUNTAINS
MEDITERRANEAN SEA
LEBANON
SYRIA
AZERBAIJAN
ISRAEL
IRAQ
IRAN
PLATEAU OF IRAN
AFGHANISTAN
Mt. Everest 29,028 ft. (8,848 m.)
HIMALAYAS
Wuhan
Chongqing
EAST CHINA SEA
Shanghai
MOROCCO
TUNISIA
ALGERIA
QATTARA DEPRESSION
Cairo
JORDAN
KUWAIT
PAKISTAN
NEPAL
BHUTAN
Delhi
Ganges R.
Chang Jiang (Yangtze R.)
TAIWAN
S A H A R A
LIBYA
EGYPT
QATAR
BAHRAIN
SAUDI ARABIA
UNITED ARAB EMIRATES
INDIA
Calcutta
BANGLADESH
MYANMAR
Tropic of Cancer
HONG KONG (U.K.)
MACAO (PORTUGAL)
MAURITANIA
MALI
NIGER
CHAD
Nile R.
SUDAN
ERITREA
YEMEN
OMAN
Bombay
ARABIAN SEA
BAY OF BENGAL
LAOS
THAILAND
SOUTH CHINA SEA
Manila
MARSHALL ISLANDS
International Date Line (Monday)
BURKINA FASO
NIGERIA
AFRICA
DJIBOUTI
Cape Asir
GUINEA
GHANA
BENIN
CENTRAL AFRICAN REP.
ETHIOPIA
ETHIOPIAN HIGHLANDS
Cape Comorin
SRI LANKA
VIETNAM
PHILIPPINES
GUAM (U.S.)
CAMBODIA
BRUNEI
FEDERATED STATES OF MICRONESIA
PALAU
CÔTE D'IVOIRE
TOGO
CAMEROON
SÃO TOME AND PRÍNCIPE
Zaire R.
UGANDA
SOMALIA
KENYA
MALDIVES
MALAYSIA
SINGAPORE
Equator
KIRIBATI
EQUATORIAL GUINEA
GABON
RWANDA
CONGO
CONGO (ZAIRE) BASIN
Lake Victoria
ZAIRE
BURUNDI
Kilimanjaro 19,340 ft. (5,895 m.)
TANZANIA
SEYCHELLES
INDIAN OCEAN
Jakarta
I N D O N E S I A
PAPUA NEW GUINEA
NAURU
SOLOMON ISLANDS
TUVALU
ANGOLA
MALAWI
ZAMBIA
COMOROS
MOZAMBIQUE
MADAGASCAR
COCOS IS. (AUSTRALIA)
Cape York
CORAL SEA
VANUATU
FIJI
ZIMBABWE
NAMIBIA
BOTSWANA
Mozambique Channel
MAURITIUS
RÉUNION (FRANCE)
Tropic of Capricorn
GREAT DIVIDING RANGE
WESTERN PLATEAU
AUSTRALIA
NEW CALEDONIA (FRANCE)
ATLANTIC OCEAN
SOUTH AFRICA
SWAZILAND
LESOTHO
Cape of Good Hope
Mt. Kosciusko 7,310 ft. (2,228 m.)
TASMAN SEA
NEW ZEALAND
East Longitude
Prime Meridian
N
KERGUELEN IS. (FRANCE)
Antarctic Circle
ANTARCTICA
EUROPE
Projection: Azimuthal Equal Area
FINLAND
NORWAY
St. Petersburg
SWEDEN
ESTONIA
IRELAND
UNITED KINGDOM
DENMARK
LATVIA
Moscow
LITHUANIA
RUSSIA
London
NETHERLANDS
BELARUS
GERMANY
POLAND
BELGIUM
LUXEMBOURG
ATLANTIC OCEAN
Paris
CZECH REPUBLIC
FRANCE
SLOVAKIA
UKRAINE
SWITZERLAND
AUSTRIA
MOLDOVA
HUNGARY
SLOVENIA
CROATIA
ROMANIA
PORTUGAL
BOSNIA-HERZEGOVINA
SERBIA
YUGOSLAVIA
SPAIN
BLACK SEA
GEORGIA
ITALY
MONTENEGRO
BULGARIA
ALBANIA
MACEDONIA
GIBRALTAR (U.K.)
MEDITERRANEAN SEA
GREECE
TURKEY
0 250 500 Miles
0 250 500 Kilometers
MALTA
TUNISIA
CYPRUS
SYRIA
LEBANON

PACIFIC OCEAN
WASHINGTON
OREGON
IDAHO
MONTANA
WYOMING
NORTH DAKOTA
SOUTH DAKOTA
NEBRASKA
KANSAS
OKLAHOMA
TEXAS
NEW MEXICO
COLORADO
UTAH
ARIZONA
NEVADA
CALIFORNIA
MEXICO
Cape Flattery
Bellingham
Juan de Fuca Strait
Puget Sound
Seattle
Tacoma
Olympia
CASCADE RANGE
F.D. Roosevelt Lake
Mt. Rainier 14,410 ft. (4,392 m.)
COLUMBIA PLATEAU
Spokane
Pend Oreille Lake
Flathead Lake
Portland
Columbia River
Lewiston
Salem
Mt. Hood 11,235 ft. (3,424 m.)
Corvallis
Eugene
COAST RANGES
Medford
BITTERROOT RANGE
ROCKY MOUNTAINS
Helena
Butte
Missouri River
Great Falls
Fort Peck Lake
Yellowstone R.
Billings
BIGHORN MTN.
Powder River
Minot
Lake Sakakawea
Bismarck
Lake Oahe
Aberdeen
Pierre
BLACK HILLS
Rapid City
GREAT PLAINS
Borah Peak 12,662 ft. (3,859 m.)
Boise
Grand Teton Peak 13,770 ft. (4,197 m.)
Idaho Falls
Twin Falls
Snake River
Pocatello
Mt. Shasta 14,162 ft. (4,316 m.)
Goose Lake
Eureka
Cape Mendocino
Sacramento
Pyramid Lake
Reno
Lake Tahoe
Carson City
GREAT BASIN
GREAT SALT LAKE DESERT
Great Salt Lake
Ogden
Salt Lake City
Orem
Provo
Utah Lake
WASATCH RANGE
Continental Divide
Casper
North Platte River
Rock Springs
Laramie
Cheyenne
Fort Collins
Greeley
South Platte River
North Platte
Grand Island
Platte
Republican River
Boulder
Denver
Mt. Elbert 14,433 ft. (4,399 m.)
Pikes Peak 14,110 ft. (4,301 m.)
Colorado Springs
Pueblo
Arkansas River
Hutchinson
Green River
Sacramento
San Francisco
Oakland
San Jose
Stockton
SIERRA NEVADA
Mono Lake
San Joaquin R.
Fresno
Mt. Whitney 14,494 ft. (4,418 m.)
Death Valley –282 ft. (–89 m.)
Las Vegas
Lake Mead
Lake Powell
COLORADO PLATEAU
PAINTED DESERT
SANGRE DE CRISTO MTNS.
Bakersfield
Point Conception
MOJAVE DESERT
Colorado River
Grand Canyon
Flagstaff
Santa Fe
Albuquerque
Rio Grande
Canadian River
Amarillo
Red River
Los Angeles
San Bernardino
Riverside
Long Beach
Salton Sea
San Diego
Glendale
Phoenix
Mesa
Gila River
Yuma
Tucson
Las Cruces
El Paso
Roswell
LLANO ESTACADO
Lubbock
Brazos River
Pecos River
EDWARDS PLATEAU
San Antonio
Corpus Christi
GULF OF CALIFORNIA
HAWAII
Kauai Channel
Kailua
Honolulu
PACIFIC OCEAN
Alenuihaha Channel
Mauna Kea 13,796 ft. (4,205 m.)
Hilo
0 100 Miles
0 100 Kilometers
RUSSIA
Arctic Circle
Pt. Barrow
BROOKS RANGE
ALASKA
Bering Strait
SEWARD PEN.
Yukon River
Fairbanks
Tanana River
Mt. McKinley 20,320 ft. (6,194 m.)
ALASKA RANGE
CANADA
Bethel
BERING SEA
Anchorage
Iliamna Lake
BRISTOL BAY
ALASKA PENINSULA
Shelikof Str.
Kodiak
GULF OF ALASKA
Juneau
Sitka
ALEUTIAN ISLANDS
0 250 500 Miles
0 250 500 Kilometers

Appendix H
95° 90° 85° 80° 75° 50° 70° 65°
CANADA
Lake of the Woods
Red Lake
Lake Superior
Duluth
NNESOTA
MICHIGAN
Lake Huron
St. Lawrence River
MAINE
Moosehead Lake
Bangor
Mt. Washington 6,288 ft. (1,905 m.)
Augusta
Lewiston
Portland
Lake Champlain
Burlington
ADIRONDACK MTNS.
Montpelier
N.H.
VT.
Hudson R.
Concord
Manchester
MASS.
Boston
Cape Cod
Worcester
Springfield
Providence
R.I.
Hartford
New Haven
CONN.
Albany
Utica
Syracuse
Lake Ontario
Rochester
NEW YORK
Niagara Falls
Buffalo
Binghamton
WISCONSIN
Minneapolis
St. Paul
Mississippi
Green Bay
Appleton
Rochester
Lake Michigan
Grand Rapids
Flint
Lansing
Detroit
Ann Arbor
Lake Erie
Milwaukee
Madison
Racine
River
Dubuque
Rockford
Erie
Susquehanna River
Newark
Yonkers
New York
Allentown
N.J.
x City
IOWA
Cedar Rapids
Davenport
Des Moines
aha
Council Bluffs
coln
Chicago
Aurora
Joliet
Gary
Hammond
South Bend
Toledo
Cleveland
Akron
Canton
PENNSYLVANIA
Youngstown
Harrisburg
Philadelphia
Trenton
Camden
Wilmington
Pittsburgh
Fort Wayne
ILLINOIS
Peoria
LOWLAND
CENTRAL
Springfield
Decatur
OHIO
Columbus
Muncie
Dayton
Indianapolis
Wheeling
MD.
Dover
Baltimore
DELAWARE BAY
Annapolis
Arlington
Washington
DEL.
D.C.
Parkersburg
WEST VIRGINIA
INDIANA
Cincinnati
River
Charleston
MOUNTAINS
Richmond
Newport News
Norfolk
CHESAPEAKE BAY
ATLANTIC OCEAN
Kansas City
eka
ence
Independence
Jefferson City
Kansas City
Harry S. Truman Res.
East St. Louis
St. Louis
Ohio
Frankfort
Lexington
Louisville
Evansville
Wabash R.
Huntington
Roanoke
VIRGINIA
KENTUCKY
Owensboro
Roanoke River
Durham
Raleigh
MISSOURI
Springfield
OZARK PLATEAU
Cumberland River
Greensboro
Winston-Salem
Cape Hatteras
Mt. Mitchell 6,684 ft. (2,037 m.)
NORTH CAROLINA
Charlotte
APPALACHIAN
Knoxville
Nashville
PLATEAU
TENNESSEE
ulsa
ARKANSAS
R.S. Kerr Res.
Chattanooga
Spartanburg
Greenville
Columbia
Mississippi River
Memphis
North Little Rock
Fort Smith
Little Rock
Lake Eufaula
Hot Springs
Pine Bluff
Huntsville
Tennessee R.
CUMBERLAND
Chattahoochee R.
Atlanta
SOUTH CAROLINA
Augusta
Charleston
Birmingham
GEORGIA
Macon
Tuscaloosa
ALABAMA
Montgomery
Columbus
Savannah
Greenville
Meridian
Jackson
Alabama R.
Albany
PLAIN
COASTAL
Shreveport
LOUISIANA
Toledo Bend Res.
am Rayburn Reservoir
Hattiesburg
MISSISSIPPI
Mobile
Jacksonville
Tallahassee
Pensacola
Lake Pontchartrain
Biloxi
Baton Rouge
Lafayette
Lake Charles
New Orleans
FLORIDA
Houston
Orlando
Cape Canaveral
Tampa
St. Petersburg
Lake Okeechobee
Palm Beach
Miami Beach
Miami
GULF OF MEXICO
N
Cape Sable
Key West
Straits of Florida
THE BAHAMAS
45° 40° 35° 30° 25° 65° 70°
UNITED STATES
National capital
State capital
Major city
International boundary
State boundary
0 150 300 Miles
0 150 300 Kilometers
Projection: Albers Equal Area

Appendix I

Solar System Information

Planet	Mercury	Venus	Earth	Mars	Jupiter	Saturn	Uranus	Neptune	Pluto
Diameter (km)	4878	12 104	12 756	6794	142 796	120 660	51 118	49 528	2290
Diameter (E = 1.0)*	0.38	0.95	1.00	0.53	11.19	9.46	4.01	3.88	0.18
Mass (E = 1.0)*	0.06	0.82	1.00	0.11	317.83	95.15	14.54	17.23	0.002
Density (g/cm^3)	5.42	5.24	5.50	3.94	1.31	0.70	1.30	1.66	2.03
Period of Rotation days hours minutes **R = retrograde**	58 15 28	243 00 14_R	00 23 56	00 24 37	00 09 55	00 10 39	00 17 14_R	00 16 03	06 09 17
Surface gravity (E = 1.0)*	0.38	0.90	1.00	0.38	2.53	1.07	0.92	1.12	0.06
Average distance to sun (AU)	0.387	0.723	1.000	1.524	5.203	9.529	19.191	30.061	39.529
Period of revolution	87.97d	224.70d	365.26d	686.98d	11.86y	29.46y	84.04y	164.79y	248.53y
Eccentricity of orbit	0.206	0.007	0.017	0.093	0.048	0.056	0.046	0.010	0.248
Average orbital speed (km/s)	47.89	35.03	29.79	24.13	13.06	9.64	6.81	5.43	4.74
Number of known satellites	0	0	1	2	16	18	15	8	1
Known rings	0	0	0	0	1	thou-sands	11	4	0

*Earth = 1.0

Appendix J

Table of Relative Humidity

Relative Humidity %										
Dry Bulb Temperature	**Dry Bulb Temperature Minus Wet Bulb Temperature, °C**									
	1	**2**	**3**	**4**	**5**	**6**	**7**	**8**	**9**	**10**
0°C	81	64	46	29	13					
1°C	83	66	49	33	18					
2°C	84	68	52	37	22	7				
3°C	84	69	55	40	25	12				
4°C	85	71	57	43	29	16				
5°C	85	72	58	45	32	20				
6°C	86	73	60	48	35	24	11			
7°C	86	74	61	49	38	26	15			
8°C	87	75	63	51	40	29	19	8		
9°C	87	76	65	53	42	32	21	12		
10°C	88	77	66	55	44	34	24	15	6	
11°C	89	78	67	56	46	36	27	18	9	
12°C	89	78	68	58	48	39	29	21	12	
13°C	89	79	69	59	50	41	32	22	15	7
14°C	90	79	70	60	51	42	34	26	18	10
15°C	90	80	71	61	53	44	36	27	20	13
16°C	90	81	71	63	54	46	38	30	23	15
17°C	90	81	72	64	55	47	40	32	25	18
18°C	91	82	73	65	57	49	41	34	27	20
19°C	91	82	74	65	58	50	43	36	29	22
20°C	91	83	74	66	59	51	44	37	31	24
21°C	91	83	75	67	60	53	46	39	32	26
22°C	92	83	76	68	61	54	47	40	34	28
23°C	92	84	76	69	62	55	48	42	36	30
24°C	92	84	77	69	62	56	49	43	37	31
25°C	92	84	77	70	63	57	50	44	39	33
26°C	92	85	78	71	64	58	51	46	40	34
27°C	92	85	78	71	65	58	52	47	41	36
28°C	93	85	78	72	65	59	53	48	42	37
29°C	93	86	79	72	66	60	54	49	43	38
30°C	93	86	79	73	67	61	55	50	44	39
31°C	93	86	80	73	67	62	56	50	45	40
32°C	93	86	80	74	68	62	57	51	46	41

Appendix K

Star Charts

Shown here are star charts for viewing stars in the Northern Hemisphere during the four different seasons. These charts are drawn from the night sky at about 35° North Latitude, but they can be used for most locations in the Northern Hemisphere. The lines on the charts outline major constellations. The dense band of stars is the Milky Way. To use, hold the chart vertically, with the direction you are facing at the bottom of the map.

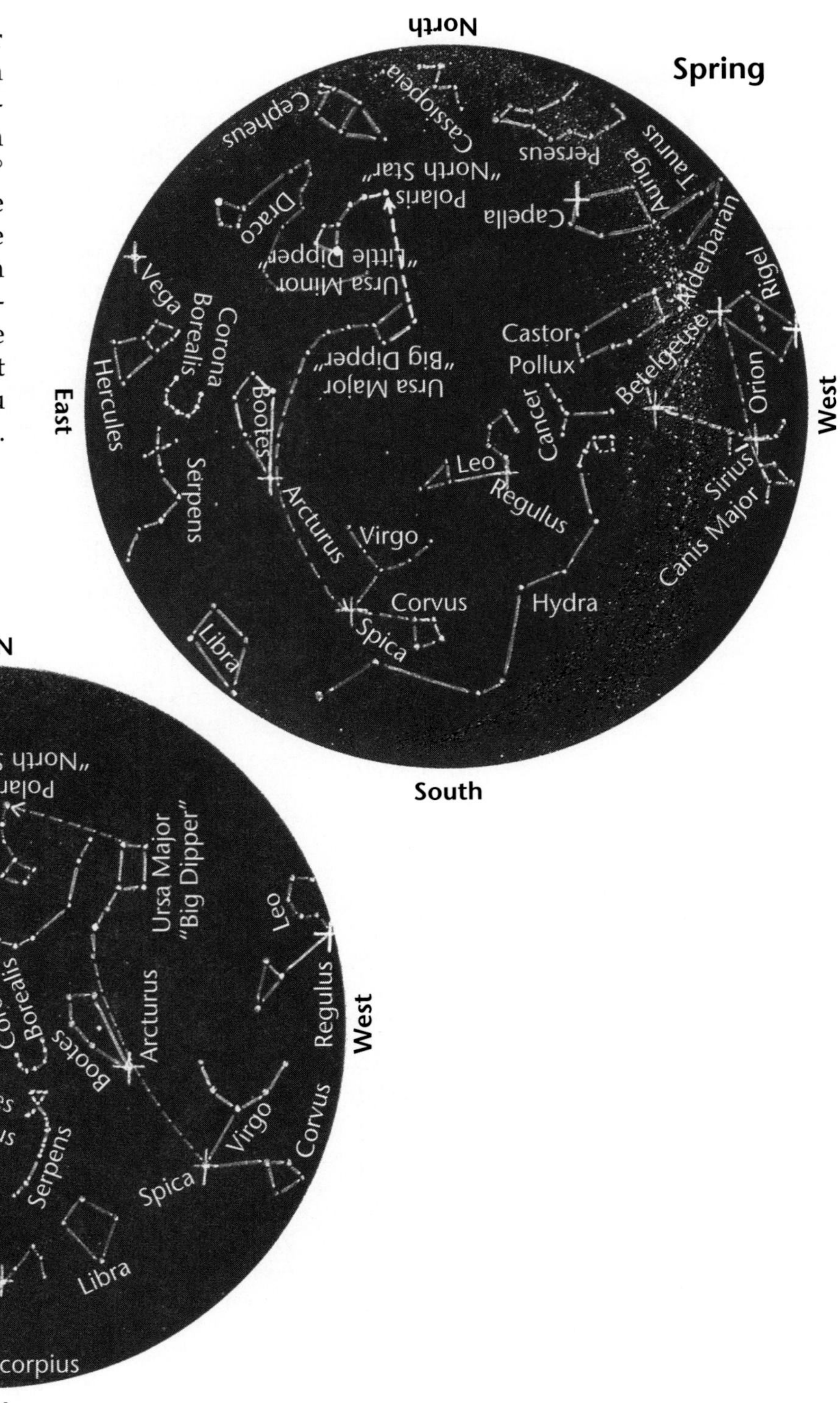

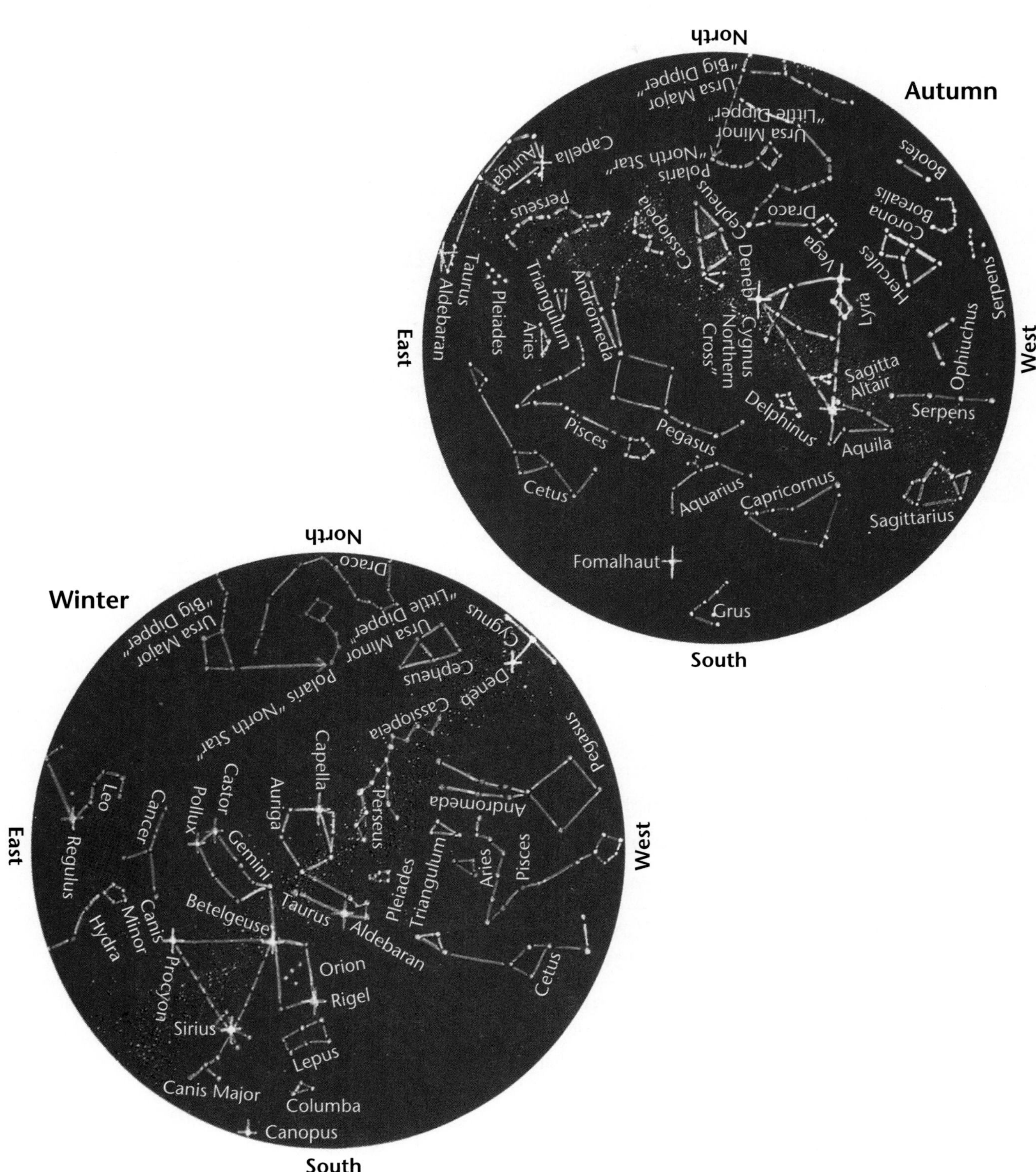

Autumn
North
Ursa Major "Big Dipper"
Ursa Minor "Little Dipper"
Polaris "North Star"
Capella
Auriga
Perseus
Cassiopeia
Cepheus
Draco
Bootes
Corona Borealis
Hercules
Vega
Lyra
Deneb
Cygnus
"Northern Cross"
Serpens
Ophiuchus
Sagitta
Altair
Serpens
Aquila
Delphinus
Pegasus
Andromeda
Triangulum
Aries
Pleiades
Taurus
Aldebaran
Pisces
Cetus
Aquarius
Capricornus
Sagittarius
Fomalhaut
Grus
East
West
South
Winter
North
Draco
Ursa Major "Big Dipper"
Ursa Minor "Little Dipper"
Cygnus
Deneb
Cepheus
Cassiopeia
Polaris "North Star"
Pegasus
Capella
Castor
Auriga
Perseus
Andromeda
Leo
Cancer
Pollux
Gemini
Triangulum
Aries
Pisces
Regulus
Pleiades
Taurus
Betelgeuse
Aldebaran
Canis Minor
Hydra
Procyon
Orion
Rigel
Cetus
Sirius
Lepus
Canis Major
Columba
Canopus
East
West
South

Appendix L

Weather Map Symbols

Sample Plotted Report at Each Station

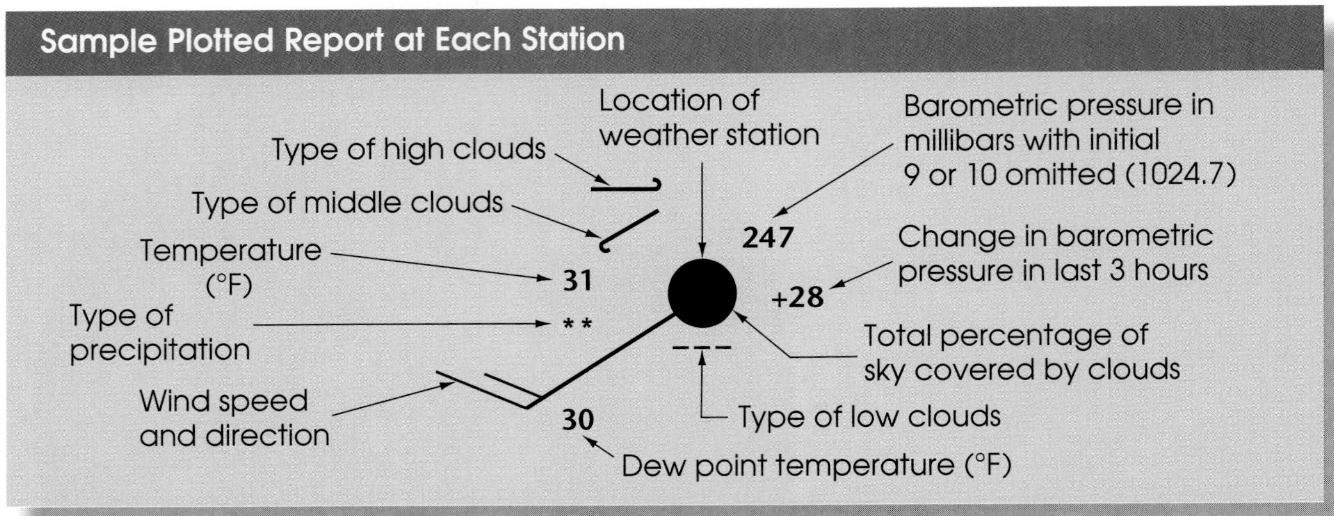

Symbols Used in Plotting Report

Precipitation	Wind speed and direction	Sky coverage	Some types of high clouds
Fog	0 calm	No cover	Scattered cirrus
Snow	1–2 knots	1/10 or less	Dense cirrus in patches
Rain	3–7 knots	2/10 to 3/10	Veil of cirrus covering entire sky
Thunder-storm	8–12 knots	4/10	Cirrus not covering entire sky
Drizzle	13–17 knots	1/2	
Showers	18–22 knots	6/10	
	23–27 knots	7/10	
	48–52 knots	Overcast with openings	
	1 knot = 1.852 km/h	Complete overcast	

Some types of middle clouds	Some types of low clouds	Fronts and pressure systems	
Thin altostratus layer	Cumulus of fair weather	(H) or High (L) or Low	Center of high- or low-pressure system
Thick altostratus layer	Stratocumulus		Cold front
Thin altostratus in patches	Fractocumulus of bad weather		Warm front
Thin altostratus in bands	Stratus of fair weather		Occluded front
			Stationary front

Appendix M

Minerals with Metallic Luster

Mineral (formula)	Color	Streak	Hardness	Specific gravity	Crystal system	Breakage pattern	Uses and other properties
graphite (C)	black to gray	black to gray	1-2	2.3	hexagonal	basal cleavage (scales)	pencil lead, lubricants for locks, rods to control some small nuclear reactions, battery poles
silver (Ag)	silvery white, tarnishes to black	light gray to silver	2.5	10-12	cubic	hackly	coins, fillings for teeth, jewelry, silverplate, wires; malleable and ductile
galena (PbS)	gray	gray to black	2.5	7.5	cubic	cubic cleavage perfect	source of lead, used in pipes, shields for X rays, fishing equipment sinkers
gold (Au)	pale to golden yellow	yellow	2.5-3	19.3	cubic	hackly	jewelry, money, gold leaf, fillings for teeth, medicines; does not tarnish
bornite (Cu_5FeS_4)	bronze, tarnishes to dark blue, purple	gray-black	3	4.9-5.4	tetragonal	uneven fracture	source of copper; called "peacock ore" because of the purple shine when it tarnishes
copper (Cu)	copper red	copper red	3	8.5-9	cubic	hackly	coins, pipes, gutters, wire, cooking utensils, jewelry, decorative plaques; malleable and ductile
chalcopyrite ($CuFeS_2$)	brassy to golden yellow	greenish black	3.5-4	4.2	tetragonal	uneven fracture	main ore of copper
chromite ($FeCr_2O_4$)	black or brown	brown to black	5.5	4.6	cubic	irregular fracture	ore of chromium, stainless steel, metallurgical bricks
pyrrhotite (FeS)	bronze	gray-black	4	4.6	hexagonal	uneven fracture	often found with pentlandite, an ore of nickel; may be magnetic
hematite (specular) (Fe_2O_3)	black or reddish brown	red or reddish brown	6	5.3	hexagonal	irregular fracture	source of iron; roasted in a blast furnace, converted to "pig" iron, made into steel
magnetite (Fe_3O_4)	black	black	6	5.2	cubic	conchoidal fracture	source of iron, naturally magnetic, called lodestone
pyrite (FeS_2)	light, brassy, yellow	greenish black	6.5	5.0	cubic	uneven fracture	source of iron, "fool's gold," alters to limonite

Appendix N

Minerals with Nonmetallic Luster

Mineral (formula)	Color	Streak	Hardness	Specific gravity	Crystal system	Breakage pattern	Uses and other properties
talc ($Mg_3(OH)_2 Si_4O_{10}$)	white, greenish	white	1	2.8	monoclinic	cleavage in one direction	easily cut with fingernail; used for talcum powder; soapstone; is used in paper and for tabletops
bauxite (hydrous aluminum compound)	gray, red, white, brown	gray	1-3	2.0-2.5	—	—	source of aluminum; used in paints, aluminum foil, and airplane parts
kaolinite ($Al_2Si_2O_5 (OH)_4$)	white, red, reddish, brown, black	white	2	2.6	triclinic	basal cleavage	clays; used in ceramics and in china dishes; common in most soils; often microscopic-sized particles
gypsum ($CaSO_4 \cdot 2H_2O$)	colorless, gray, white, brown	white	2	2.3	monoclinic	basal cleavage	used extensively in the preparation of plaster of paris, alabaster, and dry wall for building construction
sphalerite (ZnS)	brown	pale yellow	3.5-4	4	cubic	cleavage in six directions	main ore of zinc; used in paints, dyes, and medicine
sulfur (S)	yellow	yellow to white	2	2.0	ortho-rhombic	conchoidal fracture	used in medicine, fungicides for plants, vulcanization of rubber, production of sulfuric acid
muscovite ($KAl_3Si_3O_{10} (OH)_2$)	white, light gray, yellow, rose, green	colorless	2.5	2.8	monoclinic	basal cleavage	occurs in large flexible plates; used as an insulator in electrical equipment, lubricant
biotite ($K(Mg,Fe)_3 AlSi_3O_{10} (OH)_2$)	black to dark brown	colorless	2.5	2.8-3.4	monoclinic	basal cleavage	occurs in large flexible plates
halite ($NaCl$)	colorless, red, white, blue	colorless	2.5	2.1	cubic	cubic cleavage	salt; very soluble in water; a preservative
calcite ($CaCO_3$)	colorless, white, pale, blue	colorless, white	3	2.7	hexagonal	cleavage in three directions	fizzes when HCl is added; used in cements and other building materials
dolomite ($CaMg (CO_3)_2$)	colorless, white, pink, green, gray, black	white	3.5-4	2.8	hexagonal	cleavage in three directions	concrete and cement; used as an ornamental building stone

Mineral (formula)	Color	Streak	Hardness	Specific gravity	Crystal system	Breakage pattern	Uses and other properties
fluorite (CaF_2)	colorless, white, blue, green, red, yellow, purple	colorless	4	3-3.2	cubic	cleavage	used in the manufacture of optical equipment; glows under ultraviolet light
limonite (hydrous iron oxides)	yellow, brown, black	yellow, brown	5.5	2.74-4.3	—	conchoidal fracture	source of iron; weathers easily, coloring matter of soils
hornblende ($CaNa(Mg,Al,Fe)_5(Al,Si)_2Si_6O_{22}(OH)_2$)	green to black	gray to white	5-6	3.4	monoclinic	cleavage in two directions	will transmit light on thin edges; 6-sided cross section
feldspar (orthoclase) ($KAlSi_3O_8$)	colorless, white to gray, green and yellow	colorless	6	2.5	monoclinic	two cleavage planes meet at 90° angle	insoluble in acids; used in the manufacture of porcelain
feldspar (plagioclase) ($NaAlSi_3O_8$) ($CaAl_2Si_2O_8$)	gray, green, white	colorless	6	2.5	triclinic	two cleavage planes meet at 86° angle	used in ceramics; striations present on some faces
augite ($(Ca,Na)(Mg,Fe,Al)(Al,Si)_2O_6$)	black	colorless	6	3.3	monoclinic	2-directional cleavage	square or 8-sided cross section
olivine ($(Mg,Fe)_2SiO_4$)	olive green	colorless	6.5	3.5	orthorhombic	conchoidal fracture	gemstones, refractory sand
quartz (SiO_2)	colorless, various colors	colorless	7	2.6	hexagonal	conchoidal fracture	used in glass manufacture, electronic equipment, radios, computers, watches, gemstones
garnet ($(Mg,Fe,Ca)_3(Al_2Si_3O_{12})$)	deep yellow-red, green, black	colorless	7.5	3.5	cubic	conchoidal fracture	used in jewelry; also used as an abrasive
topaz ($Al_2SiO_4(F,OH)_2$)	white, pink, yellow, pale blue, colorless	colorless	8	3.5	orthorhombic	basal cleavage	valuable gemstone
corundum (Al_2O_3)	colorless, blue, brown, green, white, pink, red	colorless	9	4.0	hexagonal	fracture	gemstones: ruby is red, sapphire is blue; industrial abrasive

Skill Handbook

Table of Contents

Organizing Information

Thinking Critically

Practicing Scientific Processes

Representing and Applying Data

Organizing Information

Communicating

The communication of ideas is an important part of our everyday lives. Whether reading a book, writing a letter, or watching a television program, people everywhere are expressing opinions and sharing information with one another. Writing in your Science Journal allows you to express your opinions and demonstrate your knowledge of the information presented on a subject. When writing, keep in mind the purpose of the assignment and the audience with which you are communicating.

Examples Science Journal assignments vary greatly. They may ask you to take a viewpoint other than your own; perhaps you will be a scientist, a TV reporter, or a committee member of a local environmental group. Maybe you will be expressing your opinions to a member of Congress, a doctor, or to the editor of your local newspaper, as shown in **Figure 1.** Sometimes, Science Journal writing may allow you to summarize information in the form of an outline, a letter, or in a paragraph.

Figure 1
A Science Journal entry.

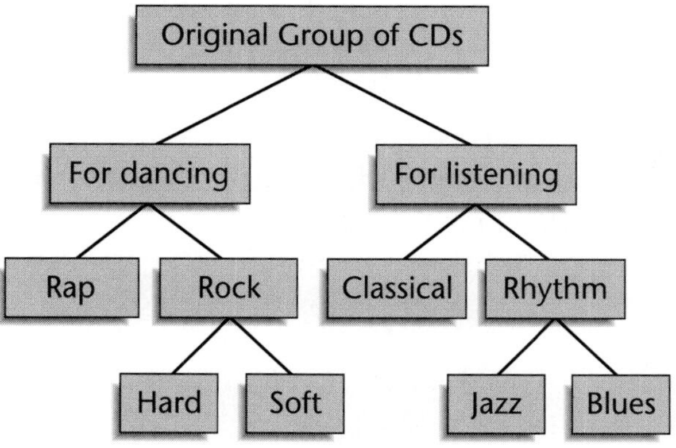

Figure 2
Classifying CDs.

Classifying

You may not realize it, but you make things orderly in the world around you. If you hang your shirts together in the closet or if your favorite CDs are stacked together, you have used the skill of classifying.

Classifying is the process of sorting objects or events into groups based on common features. When classifying, first observe the objects or events to be classified. Then, select one feature that is shared by some members in the group but not by all. Place those members that share that feature into a subgroup. You can classify members into smaller and smaller subgroups based on characteristics.

Remember, when you classify, you are grouping objects or events for a purpose. Keep your purpose in mind as you select the features to form groups and subgroups.

Example How would you classify a collection of CDs? As shown in **Figure 2,** you might classify those you like to dance to in one subgroup and CDs you like to listen to in the next column. The CDs you like to dance to could be subdivided into a rap subgroup and a rock subgroup. Note that for each feature selected, each CD fits into only one subgroup. You would keep selecting features until all the CDs are classified. **Figure 2** shows one possible classification.

Figure 3

A recipe for bread contains sequenced instructions.

Sequencing

A sequence is an arrangement of things or events in a particular order. When you are asked to sequence objects or events within a group, figure out what comes first, then think about what should come second. Continue to choose objects or events until all of the objects you started out with are in order. Then, go back over the sequence to make sure each thing or event in your sequence logically leads to the next.

Example A sequence with which you are most familiar is the use of alphabetical order. Another example of sequence would be the steps in a recipe, as shown in **Figure 3.** Think about baking bread. Steps in the recipe have to be followed in order for the bread to turn out right.

Concept Mapping

If you were taking an automobile trip, you would probably take along a road map. The road map shows your location, your destination, and other places along the way. By looking at the map and finding where you are, you can begin to understand where you are in relation to other locations on the map.

A concept map is similar to a road map. But, a concept map shows relationships among ideas (or concepts) rather than places. A concept map is a diagram that visually shows how concepts are related. Because the concept map shows relationships among ideas, it can make the meanings of ideas and terms clear, and help you understand better what you are studying.

There is usually not one correct way to create a concept map. As you construct one type of map, you may discover other

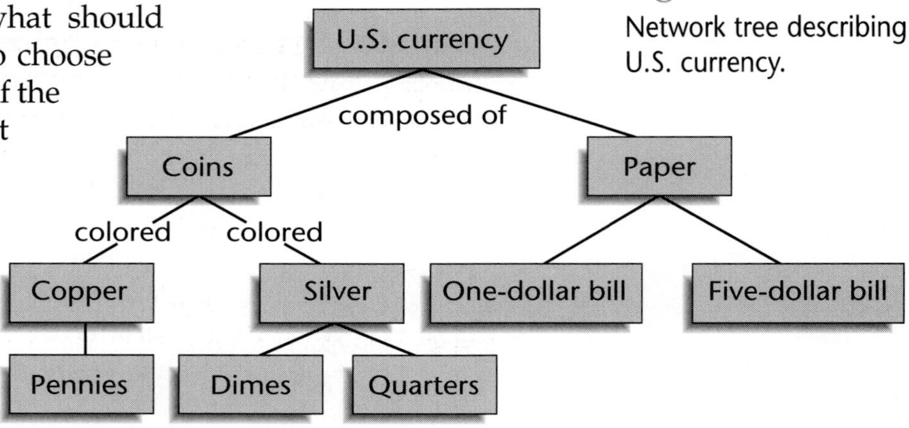

U.S. currency
composed of
Coins
Paper
colored
colored
Copper
Silver
One-dollar bill
Five-dollar bill
Pennies
Dimes
Quarters

Figure 4

Network tree describing U.S. currency.

ways to construct the map that show the relationships between concepts in a better way. If you do discover what you think is a better way to create a concept map, go ahead and use the new one. Overall, concept maps are useful for breaking a big concept down into smaller parts, making learning easier.

Examples

Network Tree Look at the concept map about U.S. currency in **Figure 4.** This is called a network tree. Notice how some words are in rectangles while others are written across connecting lines. The words inside the rectangles are science concepts. The lines in the map show related concepts. The words written on the lines describe the relationships between concepts.

When you are asked to construct a network tree, write down the topic and list the major concepts related to that topic on a piece of paper. Then look at your list and begin to put them in order from general to specific. Branch the related concepts from the major concept and describe the relationships on the lines. Continue to write the more specific concepts. Write the relationships between the concepts on the lines until all concepts are mapped. Examine the concept map for relationships that cross branches, and add them to the concept map.

Events Chain An events chain is another type of concept map. An events chain map, such as the one describing a typical morning routine in **Figure 5,** is used to describe ideas in order. In science, an events chain can be used to describe a sequence of events, the steps in a procedure, or the stages of a process.

When making an events chain, first find the one event that starts the chain. This event is called the initiating event. Then, find the next event in the chain and continue until you reach an outcome. Suppose you are asked to describe what happens when your alarm rings. An events chain map describing the steps might look like **Figure 5.** Notice that connecting words are not necessary in an events chain.

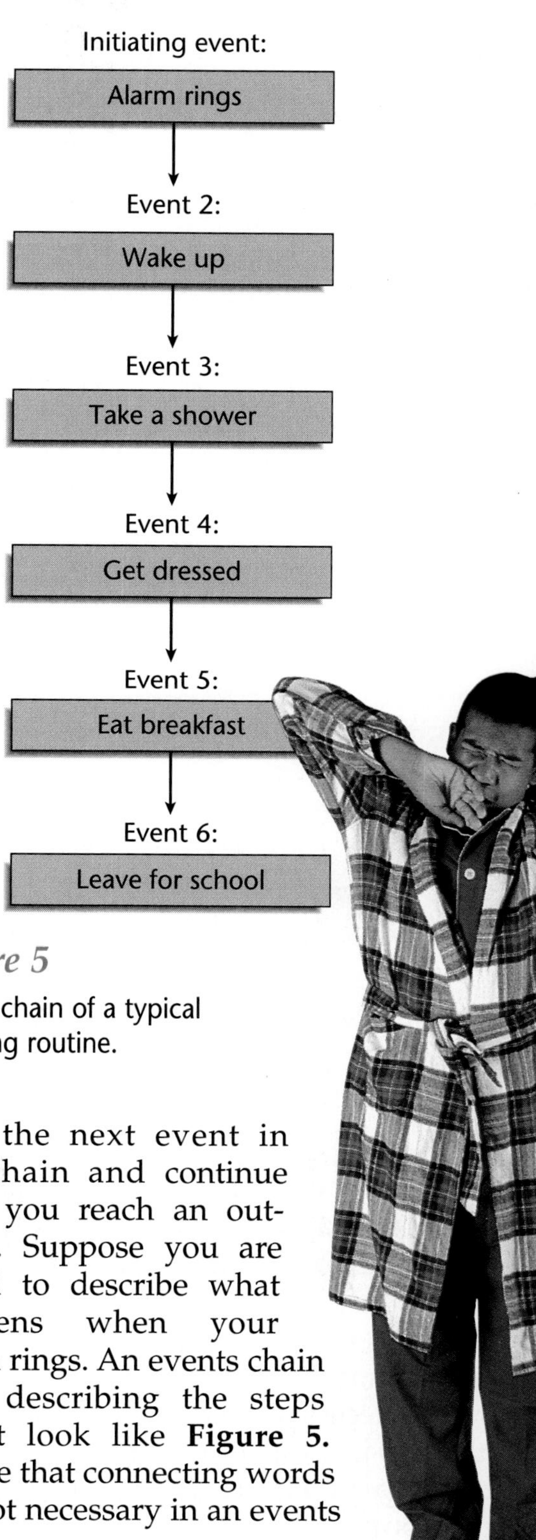

Figure 5
Events chain of a typical morning routine.

Cycle Map A cycle concept map is a special type of events chain map. In a cycle concept map, the series of events does not produce a final outcome. Instead, the last event in the chain relates back to the initiating event.

As in the events chain map, you first decide on an initiating event and then list each event in order. Because there is no outcome and the last event relates back to the initiating event, the cycle repeats itself. Look at the cycle map describing the relationship between day and night in **Figure 6.**

Figure 6

Cycle map of day and night.

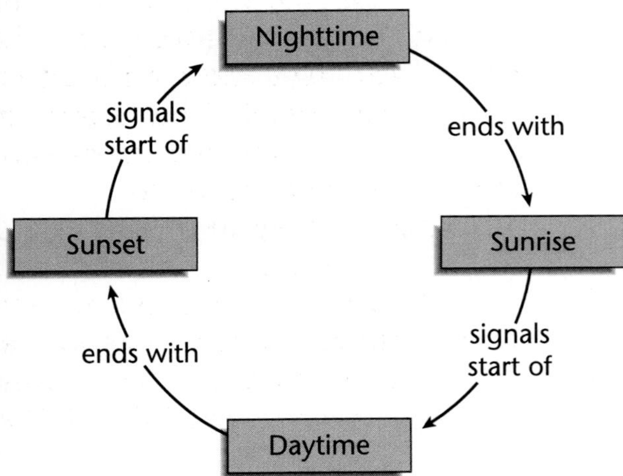

Spider Map A fourth type of concept map is the spider map. This is a map that you can use for brainstorming. Once you have a central idea, you may find you have a jumble of ideas that relate to it, but are not necessarily clearly related to each other. As illustrated by the homework spider map in **Figure 7,** by writing these ideas outside the main concept, you may begin to separate and group unrelated terms so that they become more useful.

Figure 7

Spider map about homework.

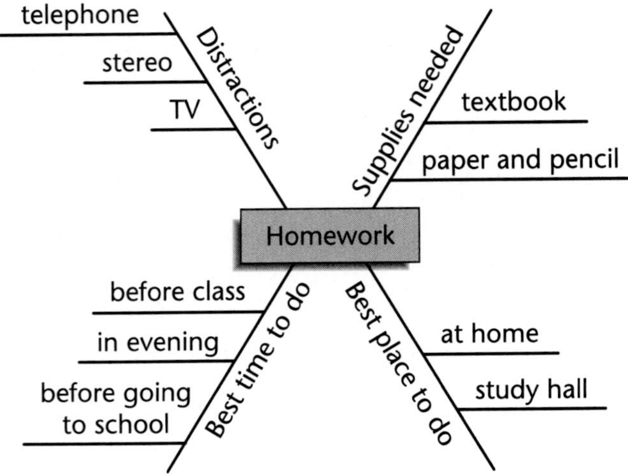

Making and Using Tables

Browse through your textbook and you will notice tables in the text and in the activities. In a table, data or information is arranged in a way that makes it easier for you to understand. Activity tables help organize the data you collect during an activity so that results can be interpreted more easily.

Examples Most tables have a title. At a glance, the title tells you what the table is about. A table is divided into columns and rows. The first column lists items to be compared. In **Figure 8,** the collection of recyclable materials is being compared in a table. The row across the top lists the specific characteristics being compared. Within the grid of the table, the collected data are recorded.

What is the title of the table in **Figure 8?** The title is "Recycled Materials." What is being compared? The different materials being recycled and on which days they are recycled.

Making Tables To make a table, list the items to be compared down in columns

Figure 8

Table of recycled materials.

Recycled Materials			
Day of Week	Paper (kg)	Aluminum (kg)	Plastic (kg)
Mon.	4.0	2.0	0.5
Wed.	3.5	1.5	0.5
Fri.	3.0	1.0	1.5

and the characteristics to be compared across in rows. The table in **Figure 8** compares the mass of recycled materials collected by a class. On Monday, students turned in 4.0 kg of paper, 2.0 kg of aluminum, and 0.5 kg of plastic. On Wednesday, they turned in 3.5 kg of paper, 1.5 kg of aluminum, and 0.5 kg of plastic. On Friday, the totals were 3.0 kg of paper, 1.0 kg of aluminum, and 1.5 kg of plastic.

Using Tables How much plastic, in kilograms, is being recycled on Wednesday? Locate the column labeled "Plastic (kg)" and the row "Wed." The data in the box where the column and row intersect are the answer. Did you answer "0.5"? How much aluminum, in kilograms, is being recycled on Friday? If you answered "1.0," you understand how to use the parts of the table.

Making and Using Graphs

After scientists organize data in tables, they may display the data in a graph. A graph is a diagram that shows the relationship of one variable to another. A graph makes interpretation and analysis of data easier. There are three basic types of graphs used in science—the line graph, the bar graph, and the circle graph.

Examples

Line Graphs A line graph is used to show the relationship between two variables. The variables being compared go on two axes of the graph. The independent variable always goes on the horizontal axis, called the x-axis. The dependent variable always goes on the vertical axis, called the y-axis.

Suppose your class started to record the amount of materials they collected in one week for their school to recycle. The collected information is shown in **Figure 9.**

You could make a graph of the materials collected over the three days of the school week. The three weekdays are the independent variables and are placed on the x-axis of your graph. The amount of materials collected is the dependent variable and would go on the y-axis.

After drawing your axes, label each with a scale. The x-axis lists the three weekdays. To make a scale of the amount of materials collected on the y-axis, look at the data values. Because the lowest amount collected was 1.0 and the highest

Figure 9

Amount of recyclable materials collected during one week.

Materials Collected During Week		
Day of Week	Paper (kg)	Aluminum (kg)
Mon.	5.0	4.0
Wed.	4.0	1.0
Fri.	2.5	2.0

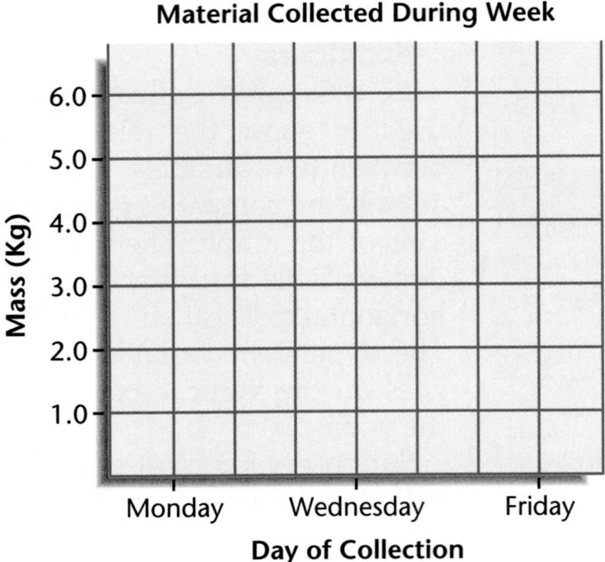

Figure 10

Graph outline for material collected during week.

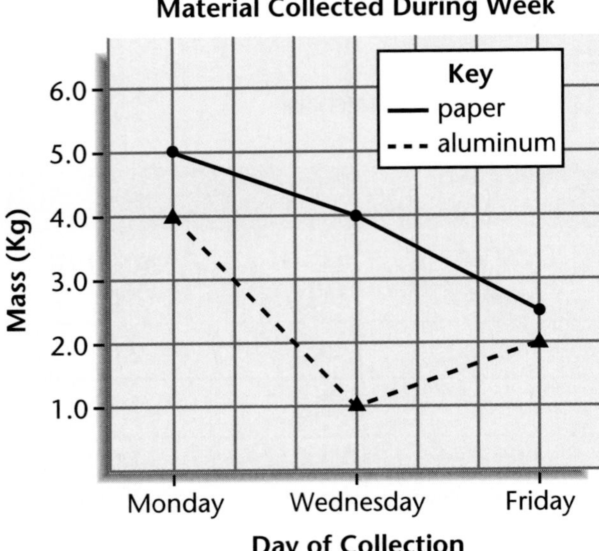

Figure 11

Line graph of materials collected during week.

was 5.0, you will have to start numbering at least at 1.0 and go through 5.0. You decide to start numbering at 0 and number by ones through 6.0, as shown in **Figure 10.**

Next, plot the data points for collected paper. The first pair of data you want to plot is Monday and 5.0 kg paper. Locate "Monday" on the x-axis and locate "5.0" on the y-axis. Where an imaginary vertical line from the x-axis and an imaginary horizontal line from the y-axis would meet, place the first data point. Place the other data points the same way. After all the points are plotted, connect them with the best smooth curve. Repeat this procedure for the data points for aluminum. Use continuous and dashed lines to distinguish the two line graphs. The resulting graph should look like **Figure 11.**

Bar Graphs Bar graphs are similar to line graphs. They compare data that do not continuously change. In a bar graph, vertical bars show the relationships among data.

To make a bar graph, set up the x-axis and y-axis as you did for the line graph. The data are plotted by drawing vertical bars from the x-axis up to a point where the y-axis would meet the bar if it were extended.

Look at the bar graph in **Figure 12** comparing the mass of aluminum collected over three weekdays. The x-axis is the days on which the aluminum was collected. The y-axis is the mass of aluminum collected, in kilograms.

Circle Graphs A circle graph uses a circle divided into sections to display data. Each section represents part of the whole. All the sections together equal 100 percent.

Suppose you wanted to make a circle graph to show the number of seeds that germinated in a package. You would count the total number of seeds. You find that there are 143 seeds in the package. This represents 100 percent, the whole circle.

You plant the seeds, and 129 seeds germinate. The seeds that germinated will make up one section of the circle graph, and the seeds that did not germinate will make up the remaining section.

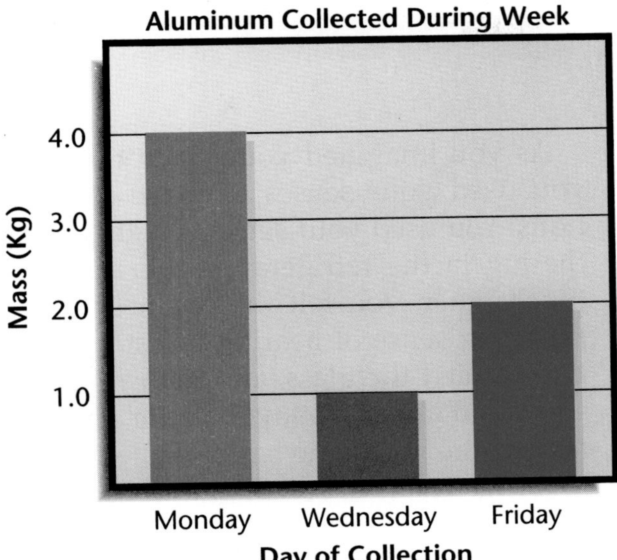

Figure 12

Bar graph of aluminum collected during week.

To find out how much of the circle each section should take, divide the number of seeds in each section by the total number of seeds. Then multiply your answer by 360, the number of degrees in a circle, and round to the nearest whole number. The section of the circle graph in degrees that represents the seeds germinated is figured below.

$$\frac{129}{143} \times 360 = 324.75 \text{ or } 325 \text{ degrees (or } 325^\circ)$$

Plot this group on the circle graph using a compass and a protractor. Use the compass to draw a circle. It will be easier to measure the part of the circle representing the nongerminating seeds, so subtract 325° from 360° to get 35°. Draw a straight line from the center to the edge of the circle. Place your protractor on this line and use it to mark a point at 325°. Use this point to draw a straight line from the center of the circle to the edge. This is the section for the group of seeds that did not germinate. The other section represents the group of 129 seeds that did germinate. Label the sections of your graph and title the graph as shown in **Figure 13.**

Figure 13

Circle graph of germinated seeds.

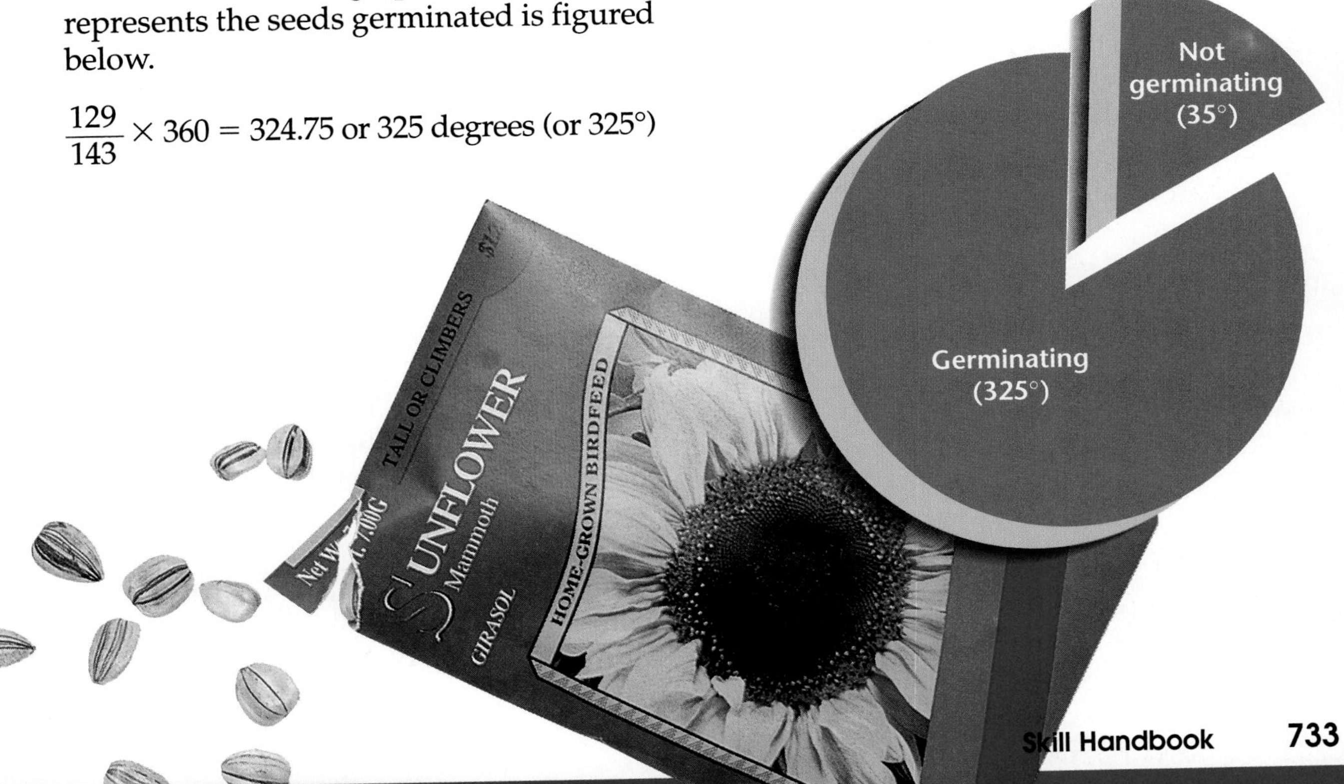

Thinking Critically

Observing and Inferring

Observing Scientists try to make careful and accurate observations. When possible, they use instruments such as microscopes, thermometers, and balances to make observations. Measurements with a balance or thermometer provide numerical data that can be checked and repeated.

When you make observations in science, you'll find it helpful to examine the entire object or situation first. Then, look carefully for details. Write down everything you observe.

Example Imagine that you have just finished a volleyball game. At home, you open the refrigerator and see a jug of orange juice on the back of the top shelf. The jug, shown in **Figure 14,** feels cold as you grasp it. Then you drink the juice, smell the oranges, and enjoy the tart taste in your mouth.

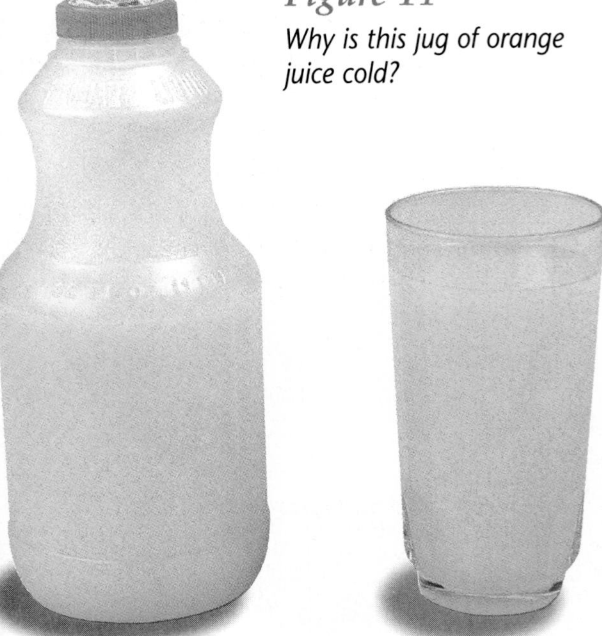

Figure 14
Why is this jug of orange juice cold?

As you imagined yourself in the story, you used your senses to make observations. You used your sense of sight to find the jug in the refrigerator, your sense of touch when you felt the coldness of the jug, your sense of hearing to listen as the liquid filled the glass, and your senses of smell and taste to enjoy the odor and tartness of the juice. The basis of all scientific investigation is observation.

Inferring Scientists often make inferences based on their observations. An inference is an attempt to explain or interpret observations or to say what caused what you observed.

When making an inference, be certain to use accurate data and observations. Analyze all of the data that you've collected. Then, based on everything you know, explain or interpret what you've observed.

Example When you drank a glass of orange juice after the volleyball game, you observed that the orange juice was cold as well as refreshing. You might infer that the juice was cold because it had been made much earlier in the day and had been kept in the refrigerator, or you might infer that it had just been made using both cold water and ice. The only way to be sure which inference is correct is to investigate further.

Comparing and Contrasting

Observations can be analyzed by noting the similarities and differences between two or more objects or events that you observe. When you look at objects or events to see how they are similar, you are comparing them. Contrasting is looking for differences in similar objects or events.

Figure 15
Table comparing the nutritional value of *Candy A* and *Candy B.*

Nutritional Value	Candy A	Candy B
Serving size	103 g	105 g
Calories	220	160
Total Fat	10 g	10 g
Protein	2.5 g	2.6 g
Total Carbohydrate	30 g	15 g

Example Suppose you were asked to compare and contrast the nutritional value of two candy bars, *Candy A* and *Candy B.* You would start by looking at what is known about these candy bars. Arrange this information in a table, like the one in **Figure 15.**

Similarities you might point out are that both candy bars have similar serving sizes, amounts of total fat, and protein. Differences include *Candy A* having a higher Calorie value and containing more total carbohydrates than *Candy B.*

Recognizing Cause and Effect

Have you ever watched something happen and then made suggestions about why it happened? If so, you have observed an effect and inferred a cause. The event is an effect, and the reason for the event is the cause.

Example Suppose that every time your teacher fed the fish in a classroom aquarium, she or he tapped the food container on the edge of the aquarium. Then, one day your teacher just happened to tap the edge of the aquarium with a pencil while making a point. You observed the fish swim to the surface of the aquarium to feed, as shown in **Figure 16.** What is the effect, and what would you infer to be the cause? The effect is the fish swimming to the surface of the aquarium. You might infer the cause to be the teacher tapping on the edge of the aquarium. In determining cause and effect, you have made a logical inference based on your observations.

Perhaps the fish swam to the surface because they reacted to the teacher's waving hand or for some other reason. When scientists are unsure of the cause of a certain event, they design controlled experiments to determine what causes the event. Although you have made a logical conclusion about the behavior of the fish, you would have to perform an experiment to be certain that it was the tapping that caused the effect you observed.

Figure 16
What cause and effect situations are occurring in this aquarium?

Practicing Scientific Processes

You might say that the work of a scientist is to solve problems. But when you decide how to dress on a particular day, you are doing problem solving, too. You may observe what the weather looks like through a window. You may go outside and see whether what you are wearing is warm or cool enough.

Scientists use an orderly approach to learn new information and to solve problems. The methods scientists may use include observing to form a hypothesis, designing an experiment to test a hypothesis, separating and controlling variables, and interpreting data.

Forming Operational Definitions

Operational definitions define an object by showing how it functions, works, or behaves. Such definitions are written in terms of how an object works or how it can be used; that is, what is its job or purpose?

Figure 17

What observations can be made about this dog?

Example Some operational definitions explain how an object can be used.

- A ruler is a tool that measures the size of an object.
- An automobile can move things from one place to another.

Or such a definition may explain how an object works.

- A ruler contains a series of marks that can be used as a standard when measuring.
- An automobile is a vehicle that can move from place to place.

Forming a Hypothesis

Observations You observe all the time. Scientists try to observe as much as possible about the things and events they study so they know that what they say about their observations is reliable.

Some observations describe something using only words. These observations are called qualitative observations. Other observations describe how much of something there is. These are quantitative observations and use numbers, as well as words, in the description. Tools or equipment are used to measure the characteristic being described.

Example If you were making qualitative observations of the dog in **Figure 17,** you might use words such as *furry, yellow,* and *short-haired.* Quantitative observations of this dog might include a mass of 14 kg, a height of 46 cm, ear length of 10 cm, and an age of 150 days.

Hypotheses Hypotheses are tested to help explain observations that have been made. They are often stated as *if* and *then* statements.

Examples Suppose you want to make a perfect score on a spelling test. Begin by thinking of several ways to accomplish this. Base these possibilities on past observations. If you put each of these possibilities into sentence form, using the words *if* and *then*, you can form a hypothesis. All of the following are hypotheses you might consider to explain how you could score 100 percent on your test:

If the test is easy, then I will get a perfect score.

If I am intelligent, then I will get a perfect score.

If I study hard, then I will get a perfect score.

Perhaps a scientist has observed that plants that receive fertilizer grow taller than plants that do not. A scientist may form a hypothesis that says: If plants are fertilized, then their growth will increase.

Designing an Experiment to Test a Hypothesis

In order to test a hypothesis, it's best to write out a procedure. A procedure is the plan that you follow in your experiment. A procedure tells you what materials to use and how to use them. After following the procedure, data are generated. From this generated data, you can then draw a conclusion and make a statement about your results.

If the conclusion you draw from the data supports your hypothesis, then you can say that your hypothesis is reliable. *Reliable* means that you can trust your conclusion. If it did not support your hypothesis, then you would have to make new observations and state a new hypothesis—just make sure that it is one that you can test.

Example Super premium gasoline costs more than regular gasoline. Does super premium gasoline increase the efficiency or fuel mileage of your family car? Let's figure out how to conduct an experiment to test the hypothesis, "*if* premium gas is more efficient, *then* it should increase the fuel mileage of our family car." Then a procedure similar to **Figure 18** must be written to generate data presented in **Figure 19** on page 738.

These data show that premium gasoline is less efficient than regular gasoline. It took more gasoline to travel one mile (0.064) using premium gasoline than it does to travel one mile using regular gasoline (0.059). This conclusion does not support the original hypothesis made.

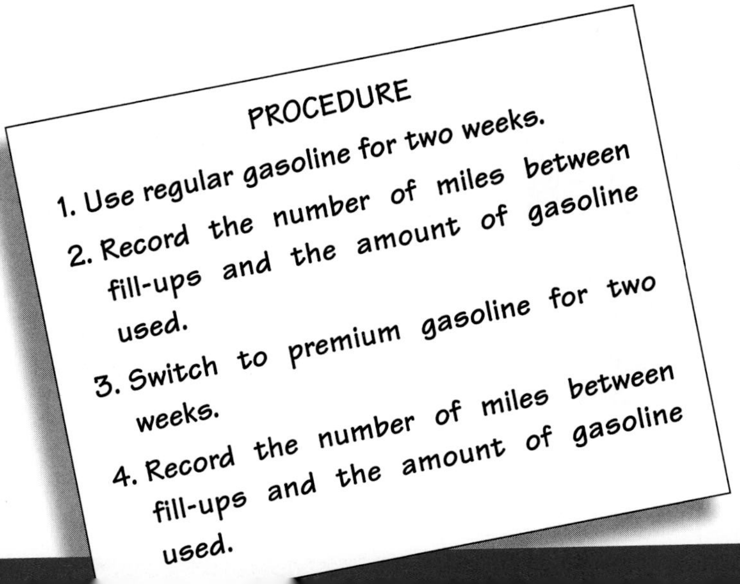

PROCEDURE

1. Use regular gasoline for two weeks.
2. Record the number of miles between fill-ups and the amount of gasoline used.
3. Switch to premium gasoline for two weeks.
4. Record the number of miles between fill-ups and the amount of gasoline used.

Figure 18
Possible procedural steps.

Figure 19

Data generated from procedure steps.

	Miles traveled	Gallons used	Gallons per mile
Regular gasoline	762	45.34	0.059
Premium gasoline	661	42.30	0.064

Separating and Controlling Variables

In any experiment, it is important to keep everything the same except for the item you are testing. The one factor that you change is called the *independent variable.* The factor that changes as a result of the independent variable is called the *dependent variable.* Always make sure that there is only one independent variable. If you allow more than one, you will not know what causes the changes you observe in the independent variable. Many experiments have *controls*—a treatment or an experiment that you can compare with the results of your test groups.

Example In the experiment with the gasoline, you made everything the same except the type of gasoline being used. The driver, the type of automobile, and the weather conditions should remain the same throughout. The gasoline should also be purchased from the same service station. By doing so, you made sure that at the end of the experiment, any differences were the result of the type of fuel being used—regular or premium. The type of gasoline was the *independent factor* and the gas mileage achieved was the *dependent factor.* The use of regular gasoline was the *control.*

Interpreting Data

The word *interpret* means "to explain the meaning of something." Look at the problem originally being explored in the gasoline experiment and find out what the data show. Identify the control group and the test group so you can see whether or not the variable has had an effect. Then you need to check differences between the control and test groups.

Figure 20

Which gasoline type is most efficient?

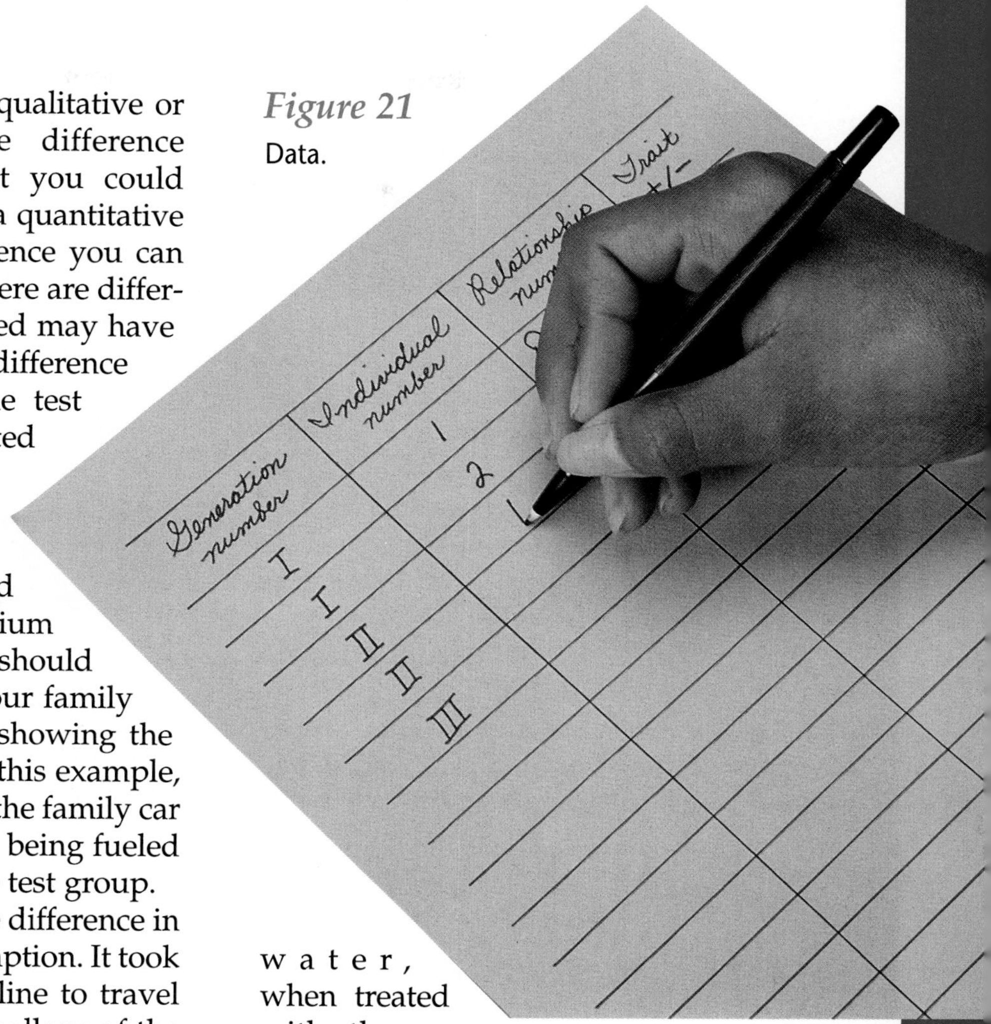

These differences may be qualitative or quantitative. A qualitative difference would be a difference that you could observe and describe, while a quantitative difference would be a difference you can measure using numbers. If there are differences, the variable being tested may have had an effect. If there is no difference between the control and the test groups, the variable being tested apparently has had no effect.

Example Perhaps you are looking at a table from an experiment designed to test the hypothesis: If premium gas is more efficient, then it should increase the fuel mileage of our family car. Look back at **Figure 19** showing the results of this experiment. In this example, the use of regular gasoline in the family car was the control, while the car being fueled by premium gasoline was the test group.

Data showed a quantitative difference in efficiency for gasoline consumption. It took 0.059 gallons of regular gasoline to travel one mile, while it took 0.064 gallons of the premium gasoline to travel the same distance. The regular gasoline was more efficient; it increased the fuel mileage of the family car.

What are data? In the experiment described on these pages, measurements were taken so that at the end of the experiment, you had something concrete to interpret. You had numbers to work with. Not every experiment that you do will give you data in the form of numbers. Sometimes, data will be in the form of a description. At the end of a chemistry experiment, you might have noted that one solution turned yellow when treated with a particular chemical, and another remained clear, like water, when treated with the same chemical. Data, therefore, are stated in different forms for different types of scientific experiments.

Figure 21
Data.

Are all experiments alike? Keep in mind as you perform experiments in science that not every experiment makes use of all of the parts that have been described on these pages. For some, it may be difficult to design an experiment that will always have a control. Other experiments are complex enough that it may be hard to have only one dependent variable. Real scientists encounter many variations in the methods that they use when they perform experiments. The skills in this handbook are here for you to use and practice. In real situations, their uses will vary.

Representing and Applying Data

Interpreting Scientific Illustrations

As you read a science textbook, you will see many drawings, diagrams, and photographs. Illustrations help you to understand what you read. Some illustrations are included to help you understand an idea that you can't see easily by yourself. For instance, we can't see atoms, but we can look at a diagram of an atom and that helps us to understand some things about atoms. Seeing something often helps you remember more easily. Illustrations also provide examples that clarify difficult concepts or give additional information about the topic you are studying. Maps, for example, help you to locate places that may be described in the text.

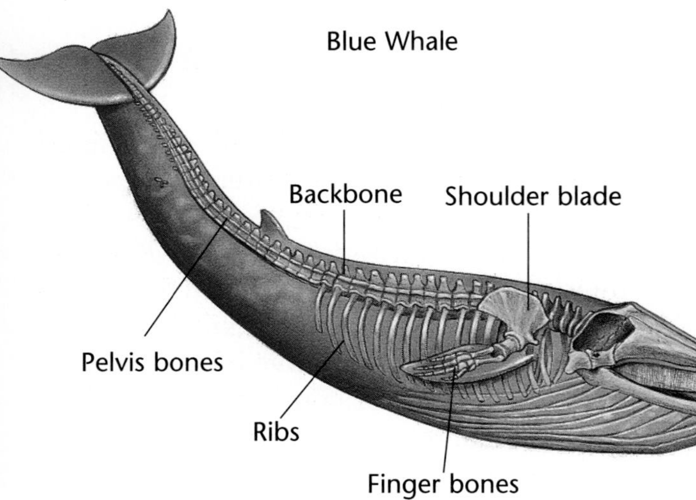

Figure 22

A labeled diagram of a blue whale.

Figure 23

If an inanimate object had dorsal and ventral surfaces, its various sides would be marked for orientation like the one in this drawing.

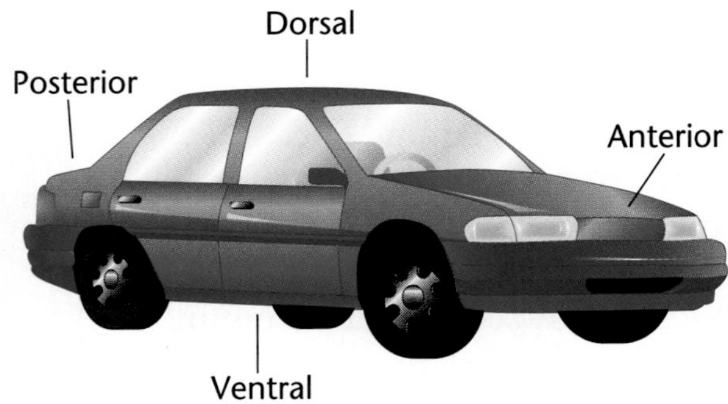

Examples

Captions and Labels Most illustrations have captions. A caption is a comment that identifies or explains the illustration. Diagrams, such as **Figure 22,** often have labels that identify parts of the organism or the order of steps in a process.

Learning with Illustrations An illustration of an organism shows that organism from a particular view or orientation. In order to understand the illustration, you may need to identify the front (anterior) end, tail (posterior) end, the underside (ventral), and the back (dorsal) side of the object shown in **Figure 23.**

You might also check for symmetry. The shark in **Figure 24** has bilateral symmetry. This means that drawing an imaginary line through the center of the animal from the anterior to posterior end forms two mirror images.

Radial symmetry is the arrangement of similar parts around a central point. An

Figure 24

A shark (A) illustrating bilateral symmetry and a pear (B) illustrating a longitudinal section and a cross section.

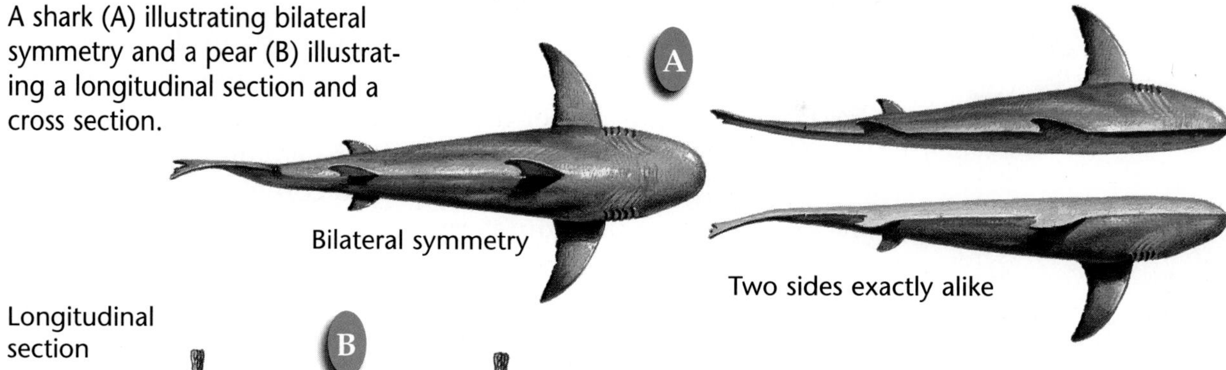

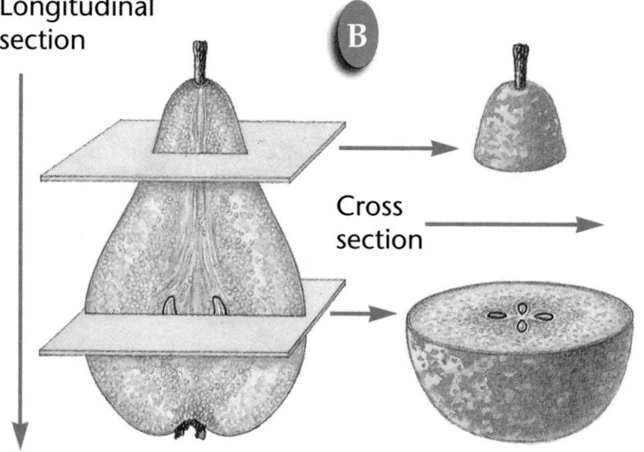

object or organism such as a hydra can be divided anywhere through the center into similar parts.

Some organisms and objects cannot be divided into two similar parts. If an organism or object cannot be divided, it is asymmetrical. Regardless of how you try to divide a natural sponge, you cannot divide it into two parts that look alike.

Some illustrations enable you to see the inside of an organism or object. These illustrations are called sections. **Figure 24** also illustrates some common sections.

Look at all illustrations carefully. Read captions and labels so that you understand exactly what the illustration is showing you.

Making Models

Have you ever worked on a model car or plane or rocket? These models look, and sometimes work, much like the real thing, but they are often on a different scale than the real thing. In science, models are used to help simplify large or small processes or structures that otherwise would be difficult to see and understand. Your understanding of a structure or process is enhanced when you work with materials to make a model that shows the basic features of the structure or process.

Example In order to make a model, you first have to get a basic idea about the structure or process involved. You decide to make a model to show the differences in size of arteries, veins, and capillaries. First, read about these structures. All three are hollow tubes. Arteries are round and thick. Veins are flat and have thinner walls than arteries. Capillaries are small.

Now, decide what you can use for your model. Common materials are often best and cheapest to work with when making models. As illustrated in **Figure 25** on page 742, different kinds and sizes of pasta might work for these models. Different sizes of rubber tubing might do just as well. Cut and glue the different noodles or tubing onto thick paper so the openings can be seen. Then label each. Now you have a simple, easy-to-understand model showing the differences in size of arteries, veins, and capillaries.

What other scientific ideas might a model help you to understand? A model

Figure 25

Different types of pasta may be used to model blood vessels.

of a molecule can be made from gumdrops (using different colors for the different elements present) and toothpicks (to show different chemical bonds). A working model of a volcano can be made from clay, a small amount of baking soda, vinegar, and a bottle cap. Other models can be devised on a computer. Some models are mathematical and are represented by equations.

Measuring in SI

The metric system is a system of measurement developed by a group of scientists in 1795. It helps scientists avoid problems by providing standard measurements that all scientists around the world can understand. A modern form of the metric system, called the International System, or SI, was adopted for worldwide use in 1960.

The metric system is convenient because unit sizes vary by multiples of 10. When changing from smaller units to larger units, divide by 10. When changing from larger units to smaller, multiply by 10. For example, to convert millimeters to centimeters, divide the millimeters by 10. To convert 30 millimeters to centimeters, divide 30 by 10 (30 millimeters equal 3 centimeters).

Prefixes are used to name units. Look at **Figure 26** for some common metric prefixes and their meanings. Do you see how the prefix *kilo-* attached to the unit *gram* is *kilogram,* or 1000 grams? The prefix *deci-* attached to the unit *meter* is *decimeter,* or one-tenth (0.1) of a meter.

Examples

Length You have probably measured lengths or distances many times. The meter is the SI unit used to measure length. A baseball bat is about one meter long. When measuring smaller lengths, the meter is divided into smaller units called centimeters and millimeters. A centimeter is one-hundredth (0.01) of a meter, which is about the size of the width

Figure 26

Common metric prefixes.

Metric Prefixes

Prefix	Symbol	Meaning	
kilo-	k	1000	thousand
hecto-	h	200	hundred
deka-	da	10	ten
deci-	d	0.1	tenth
centi-	c	0.01	hundredth
milli-	m	0.001	thousandth

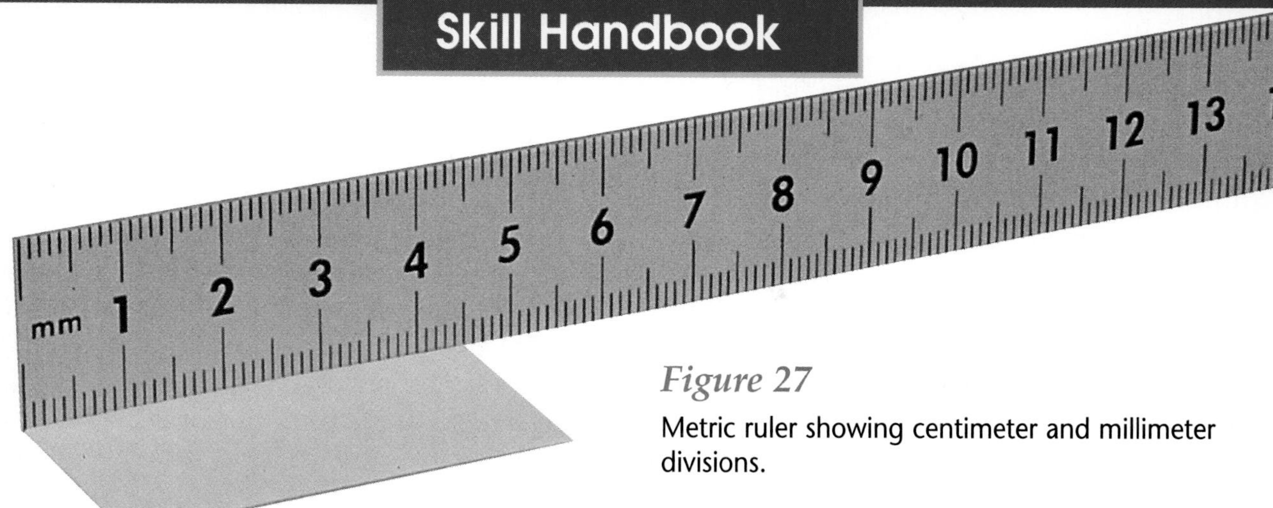

Figure 27

Metric ruler showing centimeter and millimeter divisions.

of the fingernail on your ring finger. A millimeter is one-thousandth of a meter (0.001), about the thickness of a dime.

Most metric rulers have lines indicating centimeters and millimeters, as shown in **Figure 27.** The centimeter lines are the longer, numbered lines; the shorter lines are millimeter lines. When using a metric ruler, line up the 0-centimeter mark with the end of the object being measured, and read the number of the unit where the object ends, in this instance 4.5 cm.

Surface Area Units of length are also used to measure surface area. The standard unit of area is the square meter (m^2). A square that's one meter long on each side has a surface area of one square meter. Similarly, a square centimeter, (cm^2), shown in **Figure 28,** is one centimeter long on each side. The surface area of an object is determined by multiplying the length times the width.

Volume The volume of a rectangular solid is also calculated using units of length. The cubic meter (m^3) is the standard SI unit of volume. A cubic meter is a cube one meter on each side. You can determine the volume of rectangular solids by multiplying length times width times height.

Liquid Volume During science activities, you will measure liquids using beakers and graduated cylinders marked in milliliters, as illustrated in **Figure 29.** A graduated cylinder is a cylindrical container marked with lines from bottom to top.

Liquid volume is measured using a unit called a liter. A liter has the volume of 1000 cubic centimeters. Because the prefix *milli-* means thousandth (0.001), a milliliter equals one cubic centimeter. One

Figure 28

A square centimeter.

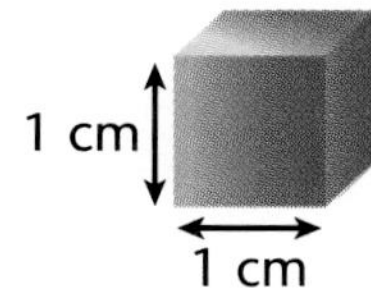

Figure 29

A volume of 79 mL is measured by reading at the lowest point of the curve.

milliliter of liquid would completely fill a cube measuring one centimeter on each side.

Mass Scientists use balances to find the mass of objects in grams. You will use a beam balance similar to **Figure 30.** Notice that on one side of the balance is a pan and on the other side is a set of beams. Each beam has an object of a known mass called a *rider* that slides on the beam.

Before you find the mass of an object, set the balance to zero by sliding all the riders back to the zero point. Check the pointer on the right to make sure it swings an equal distance above and below the zero point on the scale. If the swing is unequal, find and turn the adjusting screw until you have an equal swing.

Place an object on the pan. Slide the rider with the largest mass along its beam until the pointer drops below zero. Then move it back one notch. Repeat the process on each beam until the pointer swings an equal distance above and below the zero point. Add the masses on each beam to find the mass of the object.

You should never place a hot object or pour chemicals directly onto the pan. Instead, find the mass of a clean beaker or a glass jar. Place the dry or liquid chemicals in the container. Then find the combined mass of the container and the chemicals. Calculate the mass of the chemicals by subtracting the mass of the empty container from the combined mass.

Predicting

When you apply a hypothesis, or general explanation, to a specific situation, you predict something about that situation. First, you must identify which hypothesis fits the situation you are considering.

Examples People use prediction to make everyday decisions. Based on previous observations and experiences, you may form a hypothesis that if it is wintertime, then temperatures will be lower. From past experience in your area, temperatures are lowest in February. You may then use this hypothesis to predict specific temperatures and weather for the month of February in advance. Someone could use these predictions to plan to set aside more money for heating bills during that month.

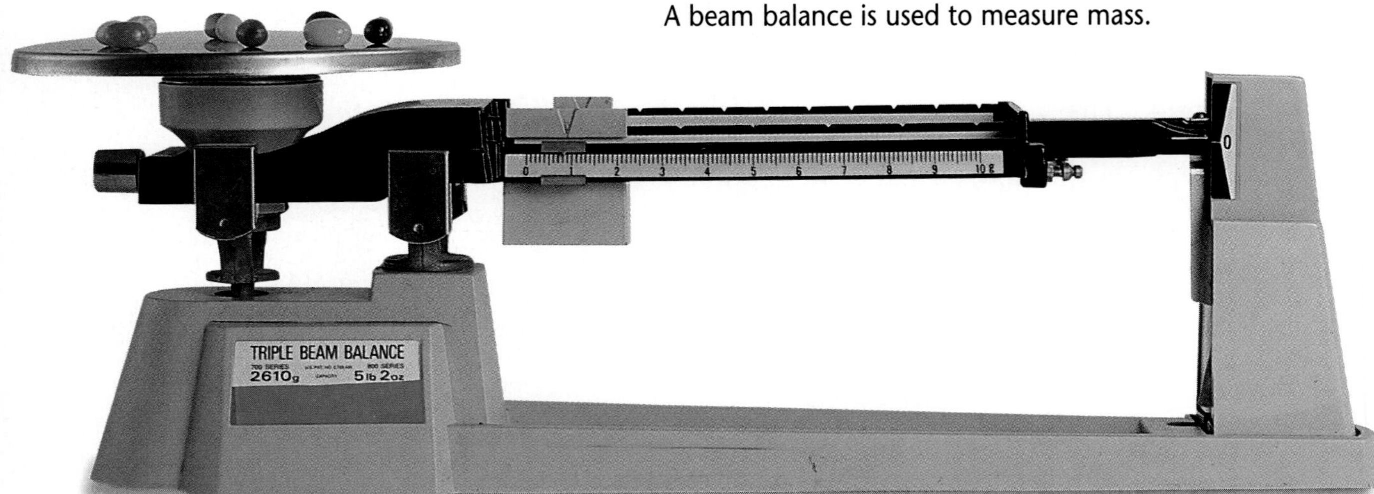

Figure 30
A beam balance is used to measure mass.

Using Numbers

When working with large populations of organisms, scientists usually cannot observe or study every organism in the population. Instead, they use a sample or a portion of the population. To sample is to take a small representative portion of organisms of a population for research. By making careful observations or manipulating variables within a portion of a group, information is discovered and conclusions are drawn that might then be applied to the whole population.

Scientific work also involves estimating. To estimate is to make a judgment about the size of something or the number of something without actually measuring or counting every member of a population.

Examples Suppose you are trying to determine the effect of a specific nutrient on the growth of black-eyed Susans. It would be impossible to test the entire population of black-eyed Susans, so you would select part of the population for your experiment. Through careful experimentation and observation on a sample of the population, you could generalize the effect of the chemical on the entire population.

Here is a more familiar example. Have you ever tried to guess how many beans were in a sealed jar? If you did, you were estimating. What if you knew the jar of beans held one liter (1000 mL)? If you knew that 30 beans would fit in a 100-milliliter jar, how many beans would you estimate to be in the one-liter jar? If you said about 300 beans, your estimate would be close to the actual number of beans. Can you estimate how many jelly beans are on the cookie sheet in **Figure 31?**

Scientists use a similar process to estimate populations of organisms from bacteria to buffalo. Scientists count the actual number of organisms in a small sample and then estimate the number of organisms in a larger area. For example, if a scientist wanted to count the number of bacterial colonies in a petri dish, a microscope could be used to count the number of organisms in a one-square-centimeter sample. To determine the total population of the culture, the number of organisms in the square-centimeter sample is multiplied by the total number of square centimeters in the culture.

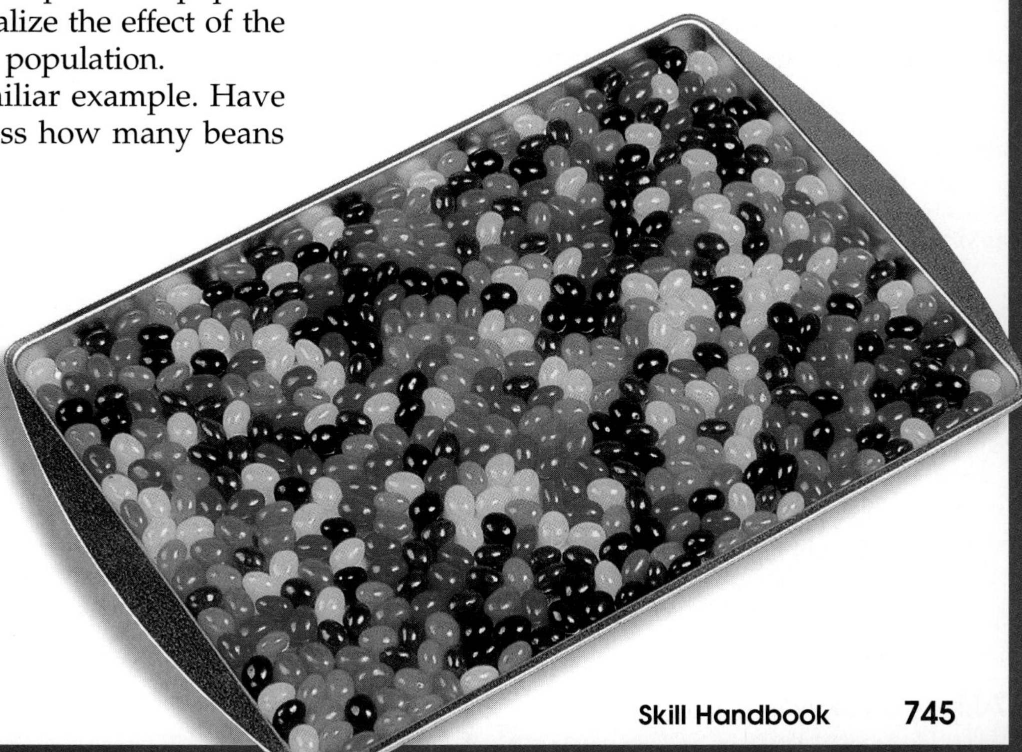

Figure 31

Sampling a group of jelly beans allows for an estimation of the total number of jelly beans in the group.

Glossary

This glossary defines each key term that appears in **bold type** in the text. It also shows the page number where you can find the word used. Some other terms that you may need to look up are included as well. We also show how to pronounce some of the words. A key to pronunciation is in the table below.

Pronunciation Key

a . . . b**a**ck (bak)	oh . . . g**o** (goh)	sh . . . **sh**elf (shelf)
ay . . . d**ay** (day)	aw . . . s**o**ft (sawft)	ch . . . na**t**ure (nay chur)
ah . . . f**a**ther (fahth ur)	or . . . **or**bit (or but)	g . . . **g**ift (gihft)
ow . . . fl**ow**er (flow ur)	oy . . . c**oi**n (coyn)	j . . . **g**em (jem)
ar . . . c**ar** (car)	oo . . . f**oo**t (foot)	ing . . . s**ing** (sing)
e . . . l**e**ss (les)	ew . . . f**oo**d (fewd)	zh . . . vi**s**ion (vihzh un)
ee . . . l**ea**f (leef)	yoo . . . p**u**re (pyoor)	k . . . **c**a**k**e (kayk)
ih . . . tr**i**p (trihp)	yew . . . f**ew** (fyew)	s . . . **s**eed, **c**ent (seed, sent)
i (i + con + e) . . . **i**dea (i dee uh), l**i**fe (life)	uh . . . comm**a** (cahm uh)	z . . . **z**one, rai**s**e (zohn, rayz)
	u (+ con) . . . flow**er** (flow ur)	

abrasion: a type of erosion caused by wearing or scraping away by sand grains or other particles striking other sand grains and rocks, breaking off small fragments. The particles can be transported by wind, water, ice, or gravity. (Chap. 7, p. 188)

absolute dating: determining the age, in years, of rocks or other objects, using the radioactive decay of their atoms. (Chap. 12, p. 344)

absolute magnitude: a measure of the amount of light a star or other space object actually gives off. (Chap. 24, p. 674)

abyssal (a BIHS uhl) **plain:** the flat seafloor in the deep ocean, formed by the deposition of sediment by ocean currents. (Chap. 18, p. 505)

acid: any substance with a pH lower than seven; the lower the pH number, the greater the acidity. (Chap. 20, p. 558)

acid rain: acidic rain, snow, sleet, or hail that can form when sulfur dioxide (from coal-burning power plants) and nitrogen oxide (from car exhausts) combine with moisture in the air. (Chap. 20, p. 558)

adaptation: any structural or behavioral characteristic organisms have developed over time that helps them survive in a particular environment. (Chap. 16, p. 459)

air mass: a large body of air that has the same properties as the surface over which it formed. (Chap. 15, p. 432)

alluvial fan: a triangular deposit of sediment that forms when water rushing down a slope loses its energy and abruptly slows at the bottom, depositing its sediment load. (Chap. 8, p. 213)

amphibians (am FIHB ee unz)**:** vertebrate animals that breathe air and live on land but must return to water to reproduce. (Chap. 13, p. 370)

angular unconformity: a type of unconformity where older, tilted rock layers meet younger, horizontal rock layers; this indicates that layers are missing and there is a gap in the time record.

apparent magnitude: a measure of light received on Earth from a star or other space object. (Chap. 24, p. 674)

aquifer: a layer of permeable rock that has connecting pores and transmits water freely. (Chap. 8, p. 215)

arête (uh RAYT): a sharp-edged mountain ridge carved by glaciers.

artesian (ahr TEE zhun) **well:** a well in which water under natural pressure rises to the surface without being pumped. (Chap. 8, p. 216)

asteroid: a piece of rock, smaller than a planet, that orbits the sun; most are between the orbits of Mars and Jupiter in an area called the asteroid belt. (Chap. 23, p. 665) asteroid belt: the area between the orbits of Mars and Jupiter where most asteroids are found.

asthenosphere (as THEN uh sfihr)**:** the plasticlike layer below the lithosphere in Earth's mantle. (Chap. 11, p. 304)

astronomical unit: the average distance from Earth to the sun (150 million km), used for measuring distances to objects in the solar system. (Chap. 23, p. 648)

astronomy: the study of objects in space, including stars, planets, comets, and their origins. (Chap. 1, p. 8)

atomic number: equals the number of protons in the nucleus of an atom. (Chap. 2, p. 36)

atoms: the smallest particles of an element that still have all the properties of that element; the building blocks of matter. (Chap. 2, p. 32)

Australopithecus: the earliest animals that have distinct humanlike characteristics; now extinct.

axis: an imaginary line around which an object spins; for example, Earth spins around its axis. (Chap. 22, p. 615)

barred spiral galaxy: a galaxy having two spiral arms extending from a large bar.

barrier island: a temporary sand deposit that parallels the shore but is separated from the mainland by water. (Chap. 8, p. 226)

basaltic: dark-colored, dense igneous rocks that form from magma rich in iron and magnesium. (Chap. 4, p. 93)

base: any substance with a pH above seven; the higher the pH number, the more basic the solution. (Chap. 20, p. 558)

basin: a low area on Earth's surface that contains an ocean. (Chap. 17, p. 479)

batholiths: the largest intrusive igneous rock bodies that form when magma cools and solidifies underground and stops rising to the surface. (Chap. 10, p. 283)

beach: a deposit of sediments, such as rock fragments or seashell fragments, that runs parallel to a shore. (Chap. 8, p. 224)

benthos (BEN thohs)**:** plants and animals that live near and on the ocean bottom, such as kelp, corals, snails, sea urchins, and bottom-dwelling fish. (Chap. 18, p. 514)

big bang theory: the leading theory about the formation of the universe, which states that the universe began expanding out of a huge explosion about 15 billion years ago. (Chap. 24, p. 696)

binary system: a system of two stars orbiting one another. (Chap. 24, p. 681)

black hole: the final stage in the life cycle of some massive stars; the remnant of a star that is so dense that not even light can escape its gravity field. (Chap.24, p. 688)

brachiopods (BRAY kee uh pahdz): fan-shaped marine invertebrate animals commonly found as fossils in Paleozoic Era rocks.

breaker: an ocean wave that collapses and tumbles forward as it approaches the shore because the top is moving faster than the bottom. (Chap. 17, p. 491)

buoyancy (BOY un see): the lifting effect on an object immersed in water.

caldera (kal DARE uh)**:** the large opening formed at the top of a volcano when a crater collapses into the vent following an eruption. (Chap. 10, p. 285)

carbonaceous film: a fossil impression in a rock, consisting only of a thin carbon residue that forms an outline of the original organism. (Chap. 12, p. 328)

carbonic acid: a weak acid that forms when water mixes with carbon dioxide from air.

carrying capacity: the maximum number of individuals of a particular species that an environment can support. (Chap. 19, p. 532)

cast: a type of fossil formed when an earlier fossil in the rock is dissolved away, leaving behind the impression of that fossil (a mold), and new sediments or mineral crystals fill the mold. (Chap. 12, p. 329)

cave: a large underground opening formed when groundwater gradually dissolves limestone. (Chap. 8, p. 217)

cementation: a sedimentary rock-forming process in which large sediments are glued together by minerals deposited between the sediments. (Chap. 4, p. 102)

Cenozoic (sen uh ZOH ihk) **Era:** the most recent era of Earth's geologic history; began about 66 million years ago when dinosaurs became extinct. (Chap. 13, p. 379)

chemical properties: characteristics of an element or compound that determine how it will react with other elements or compounds. (Chap. 2, p. 39)

chemical weathering: the breaking up of rocks due to a change in their chemical composition. (Chap. 6, p. 151)

chemosynthesis (kee moh SIHN thuh sihs)**:** the process used by bacteria along mid-ocean ridges to produce food and oxygen by using dissolved sulfur compounds from magma. (Chap. 18, p. 512)

chlorofluorocarbons: a group of chemicals used as refrigerants and aerosol spray propellants. Chlorofluorocarbons destroy ozone molecules. (Chap. 14, p. 401)

chromosphere: the intermediate layer of the three layers of the sun's atmosphere, lying between the photosphere and the corona. (Chap. 24, p. 679)

cinder cone: a type of volcano in which tephra (cinders) piles up into a steep-sided cone. (Chap. 10, p. 279)

circumpolar constellation: a constellation that appears to circle Polaris as Earth rotates.

cirque (SURK): a bowl-shaped basin carved by erosion at the head of a valley glacier.

clay: sediment particle less than 0.004 mm in size.

Clean Air Act: this 1990 U.S. law sets a maximum level for major air pollutants. (Chap. 20, p. 561)

Clean Water Act: this 1987 U.S. law gives money to the states for building sewage and wastewater-treatment facilities and for controlling runoff from streets, farms, and mines. (Chap. 20, p. 571)

cleavage: the physical property of a mineral that causes it to break along smooth, flat surfaces. (Chap. 3, p. 70)

climate: the pattern of weather in a particular area over a period of many years; affects the rate and type of weathering. (Chap. 6, p. 154)

cloud seeding: the addition of silver iodide or dry ice particles to clouds to produce a change in weather conditions. (Chap. 15, p. 444)

coal: a sedimentary rock formed from compacted, decayed plants; burned as a fossil fuel.

coastal plain: a landform that is a broad, flat area along a coastline; also called a lowland.

cogeneration: process in which a power plant uses both the electrical and thermal energy produced by the plant. (Chap. 4, p. 109)

cold front: the boundary that develops when a cold air mass pushes under a warm air mass.

comet: a mass of frozen gases, dust, and rock particles that often develops a bright tail when it passes near the sun and is acted on by solar winds. (Chap. 23, p. 663)

compaction: a sedimentary rock-forming process that occurs when layers of small sediments become compressed by the weight of layers above them. (Chap. 4, p. 102)

composite volcano: a type of volcano built of silica-rich lava and tephra layers accumulated from repeated alternating cycles of tephra eruptions and lava eruptions. (Chap. 10, p. 281)

composting: the piling up of grass clippings, dead leaves, and other organic matter so they can gradually decompose. (Chap. 19, p. 545)

compound: a type of matter containing two or more chemically combined elements and having physical and chemical properties different from each of the elements in it. (Chap. 2, p. 38)

compression: squeezing forces that compress rocks together at convergent plate boundaries, causing them to deform, fold, and sometimes break.

conduction: the transfer of heat that occurs when molecules collide. (Chap. 14, p. 405)

conic projection: a map projection made by projecting points and lines from a globe onto a cone; it produces accurate maps of areas smaller than the whole Earth, such as a nation or state. (Chap. 5, p. 131)

conservation: the careful use of resources to avoid wasting them and damaging the environment; includes reusing and recycling resources. (Chap. 19, p. 544)

constellation: a grouping of stars that has a shape resembling an animal, mythological character, or other familiar object and thus is named for it. (Chap. 24, p. 672)

container law: a law requiring you to pay a deposit each time you buy a container of beverage.

continental drift: a hypothesis proposed by Alfred Wegener, which states that continents have moved horizontally around the globe, over time, to reach their current locations. (Chap. 11, p. 294)

continental glacier: a glacier of considerable thickness that covers a vast area; existing now only in Greenland and Antarctica.

continental shelf: the gradually sloping end of every continent that extends out under the ocean. (Chap. 18, p. 504)

continental slope: a part of the continental shelf that dips steeply down to the ocean floor. (Chap. 18, p. 505)

contour interval: the difference in elevation between two side-by-side contour lines on a topographic map. (Chap. 5, p. 132)

contour line: a line on a topographic map that connects points of equal elevation. (Chap. 5, p. 132)

control: in an experiment, the standard used for comparison. (Chap. 1, p. 17)

convection: the transfer of heat by a flow of a heated material; occurs in gases or liquids. (Chap. 14, p. 405)

convection current: the driving force of plate tectonics in which hot, plasticlike material from the mantle rises to the lithosphere, moves horizontally, cools, and sinks back to the mantle. (Chap. 11, p. 310)

convergent boundary: in plate tectonics, the boundary between two plates that are converging, or moving toward each other. (Chap. 11, p. 307)

Copernicus, Nicholas: Polish astronomer (1473-1543) who hypothesized a sun-centered solar system.

Coriolis (kohr ee OH lus) **effect:** the effect of Earth's rotation on the movement of air masses; changes the direction of the air flow. (Chap. 14, p. 411)

corona: the outermost and largest of the three layers of the sun's atmosphere, extending from the chromosphere outward millions of kilometers into space. (Chap. 24, p. 679)

crater: the steep-walled depression at the top of a volcanic vent. (Chap. 10, p. 268)

creep: a type of mass movement in which sediments move down a hill slowly, sometimes causing posts and trees to lean. (Chap. 7, p. 175)

crest: the highest point of a wave. (Chap. 17, p. 490)

crinoids (KRI noyds): marine invertebrate animals that resemble plants, commonly found as fossils in Paleozoic Era rocks and still a living animal group.

crust: the outermost layer of Earth, varying in thickness from more than 60 km to less than 5 km. (Chap. 9, p. 249)

crystal: a solid having a distinctive shape because its atoms are arranged in repeating patterns. (Chap. 3, p. 63)

crystal system: the pattern that atoms form in a crystal.

cyanobacteria: blue-green bacteria thought to be one of the earliest life-forms on Earth. (Chap. 13, p. 367)

deflation: wind erosion that removes loose, fine-grained sediments such as clay or silt, and leaves behind coarser material. (Chap. 7, p. 188)

deforestation: the removal of forests, mostly to clear land for farming, construction, mining, and drilling for oil. (Chap. 16, p. 470)

delta: a triangular deposit of sediment that forms when a stream or a river slows as it empties into an ocean, gulf, or lake. (Chap. 8, p. 213)

density: how highly packed a substance's molecules are; the mass of an object divided by its volume, in g/cm^3. (Chap. 2, p. 44)

density current: an ocean current that occurs when more dense seawater sinks under an area of less dense seawater. (Chap. 17, p. 487)

deposition (dep uh ZIHSH un): the final step in an erosional process, in which sediments are dropped by running water, wind, gravity, or glaciers as their energy of motion decreases. (Chap. 7, p. 173)

desertification: the formation of a desert. (Chap. 6, p. 165)

dew point: the temperature at which air is saturated with water and condensation begins. (Chap. 15, p. 424)

dike: an intrusive igneous rock body formed when magma is squeezed into a vertical crack that cuts across rock layers and solidifies underground. (Chap. 10, p. 284)

disconformity: a type of unconformity in which the top rock layer is eroded before the next layer can be deposited, causing a gap in the rock record.

divergent boundary: in plate tectonics, the boundary between two plates that are diverging, or moving away from each other. (Chap. 11, p. 305)

Doppler shift: the change in wavelength that occurs in any kind of wave energy (light, radio, sound) as the source of the energy moves toward you (the wavelength shortens) or away from you (the wavelength lengthens).

drainage basin: the land area drained by a river system. (Chap. 8, p. 207)

dust bowl: name given to the Great Plains area of the United States in the 1930s, when it was struck by devastating drought and dust storms.

Earth: in our solar system, the third planet from the sun; the only planet known to support life; appears blue from space because more than 70 percent of its surface is covered by liquid water. (Chap. 23, p. 648)

Earth science: the study of Earth and space; includes geology, meteorology, astronomy, and oceanography. (Chap. 1, p. 8)

earthquake: the movement of the ground, caused by waves from energy released as rocks move along faults. (Chap. 9, p. 237)

El Niño: a climatic event that occurs when trade winds weaken west of Peru and whose effects can be felt worldwide. (Chap. 16, p. 463)

electromagnetic spectrum: the classification of electromagnetic radiation (including radio waves, visible light, X rays, and gamma rays) according to their wavelengths. (Chap. 21, p. 585)

electron: one of the three subatomic particles; moves around the nucleus of an atom; has a negative electric charge. (Chap. 2, p. 35)

element: a form of matter that contains only one kind of atom; cannot be broken down into simpler substances. (Chap. 2, p. 33)

ellipse: an elongated, closed curve; the shape of Earth's orbit. (Chap. 22, p. 617)

elliptical galaxy: the most common type of galaxy; shaped like a large, three-dimensional ellipse.

endangered: describes a species that has only a relatively small number of individuals living and thus is in danger of dying out. (Chap. 13, p. 364)

epicenter: the point on Earth's surface directly above an earthquake's focus. (Chap. 9, p. 244)

epochs: subdivisions of periods on the geologic time scale. (Chap. 13, p. 359)

equator: an imaginary line, at 0° latitude, that circles Earth exactly halfway between the North and South Poles; separates Earth into the northern hemisphere and the southern hemisphere. (Chap. 5, p. 126)

equinox: the two times each year that the sun is directly above Earth's equator and the day and night are of equal length all over the world; the start of spring and fall. (Chap. 22, p. 619)

eras: the four largest subdivisions of the geologic time scale—Precambrian, Paleozoic, Mesozoic, Cenozoic—based on differences in life-forms. (Chap. 13, p. 357)

erosion (ih ROH zhun)**:** the process that wears away surface materials and moves them from one location to another, usually by gravity, glaciers, wind, or water. (Chap. 7, p. 172)

esker: a winding ridge of sand and gravel formed by streams flowing beneath a glacier.

extinct: describes a species that no longer has any living members anywhere on Earth—for example, dinosaurs. (Chap. 12, p. 334)

extrusive igneous rocks: igneous rocks that form when magma extrudes onto Earth's surface and cools as lava; have a fine-grained texture. (Chap. 4, p. 92)

fault-block mountains: jagged mountains formed from huge, tilted blocks of rock that are separated from surrounding rock by faults. (Chap. 5, p. 124)

faults: surfaces along which rocks break and move; rocks on either side of a fault move in different directions relative to the fault surface. (Chap. 9, p. 236)

first quarter: the moon phase halfway between new moon and full moon, when half of the side facing Earth is lighted. (Chap. 22, p. 624)

fission: the splitting of the nuclei of atoms of heavy elements to release energy. (Chap. 2, p. 54)

floodplain: the broad, flat valley floor carved by a meandering stream and often covered with water when the stream floods. (Chap. 8, p. 209)

focus: the point in Earth's interior where earthquake energy is released. (Chap. 9, p. 241)

fog: a stratus cloud that forms on or near the ground when air is cooled to its dew point and condenses. (Chap. 15, p. 425)

folded mountains: mountains created when rock layers are squeezed from opposite sides, causing them to buckle and fold. (Chap. 5, p. 123)

foliated: a type of metamorphic rock created when mineral grains flatten and line up in parallel bands. (Chap. 4, p. 99)

fossils: the remains, imprints, or traces of once-living organisms, usually preserved in rock, that tell us when, where, and how those organisms lived. (Chap. 12, p. 327)

fracture: the physical property of a mineral that causes it to break with rough or jagged edges. (Chap. 3, p. 70)

front: in weather systems, the boundary between two air masses. (Chap. 15, p. 433)

full moon: the moon phase when the side facing Earth is completely lighted because Earth is between the sun and the moon. (Chap. 22, p. 624)

fusion: the process that powers our sun and the other stars; hydrogen fusion occurs when great temperatures and pressures fuse hydrogen atoms to form helium atoms and energy is released.

galaxy: a massive grouping of stars, gas, and dust in space, held together by gravity; can be elliptical, spiral, or irregular. (Chap. 24, p. 691)

Galilei, Galileo: Italian astronomer (1564-1642) who supported a sun-centered solar system by discovering that Venus has phases similar to the moon.

gamma rays: electromagnetic waves having short wavelengths and high energy.

Ganymede: one of Jupiter's four largest moons; the largest satellite in the solar system.

gasohol: a biomass fuel that is about 90 percent gasoline and about ten percent alcohol; used in cars and trucks.

gem: a valuable mineral highly prized because it is rare and beautiful. (Chap. 3, p. 74)

geologic time scale: a record of Earth's history, beginning 4.6 billion years ago, that shows events, time units, and ages. (Chap. 13, p. 357)

geology: the study of Earth and its matter, processes, and history. (Chap. 1, p. 8)

geothermal energy: thermal energy from magma bodies inside Earth that can be used to produce electricity with very little environmental pollution. (Chap. 10, p. 274)

geyser: a hot spring of groundwater that erupts periodically, shooting water and steam into the air. (Chap. 8, p. 217)

giant: a late stage in a star's life cycle in which the hydrogen in the core has contracted and grown hotter, causing its outer layers to expand. (Chap. 24, p. 687)

glacier: a mass of snow and ice that moves slowly downhill due to its weight. (Chap. 7, p. 180)

global warming: the rise in global temperatures due to an increased greenhouse effect. (Chap. 16, p. 467)

Glossopteris (glah SAHP tuhr ihs): a fossil fern providing evidence that continents once were joined.

granitic: light-colored igneous rocks formed from magma rich in silicon and oxygen. (Chap. 4, p. 93)

gravity: an attractive force that exists between all objects. (Chap. 1, p. 20)

Great Red Spot: a giant, swirling, high-pressure gas storm in Jupiter's atmosphere. (Chap. 23, p. 654)

greenhouse effect: natural heating caused by atmospheric gases trapping heat at Earth's surface. (Chap. 16, p. 467)

groundwater: water that soaks into the ground and collects in the pore spaces between particles of rock and soil. (Chap. 8, p. 214)

Gulf Stream: an ocean current that flows out of the Gulf of Mexico, northward along the east coast of the United States, and then toward Europe.

gully erosion: a type of surface water erosion due to runoff, in which water swiftly running down a slope creates large channels in the soil or rock. (Chap. 8, p. 204)

habitat: any place where organisms live, grow, and interact with one another and with the environment. (Chap. 13, p. 364)

hachures (ha SHOORZ): lines drawn at right angles to contour lines on a topographic map; they indicate depressions.

half-life: the time it takes for half of the atoms of an isotope to decay. (Chap. 12, p. 345)

hardness: a measure of how easily a mineral can be scratched. (Chap. 3, p. 68)

hazardous waste: waste that is dangerous to organisms because it is poisonous, cancer causing, or radioactive. (Chap. 19, p. 541)

Hess, Harry: a Princeton University scientist who proposed the theory of seafloor spreading in the 1960s.

hibernation: a period of inactivity during which an animal's body temperature drops and its body processes slow down. (Chap. 16, p. 460)

high-pressure system: an air mass with densely packed air molecules where cold air descends and rotates clockwise (in the northern hemisphere).

Homo erectus: an ancestor of modern *Homo sapiens,* this extinct primate lived in Africa and Asia from 1.6 million to 250 000 years ago.

Homo habilis: the earliest species to have fully human characteristics, this extinct primate evolved from *Austraulopithecus,* lived in Africa 1.5 to 2 million years ago, and regularly used tools.

horizon: a soil layer; most areas of Earth have three, called the A horizon (topsoil), B horizon, and C horizon. (Chap. 6, p. 158)

horn: a sharp mountain peak pointed like an animal's horn, carved by glaciers.

hot dry rock (HDR): a new technology in which heat from Earth's internal hot dry rock material is used to generate energy. (Chap. 10, p. 275)

hot spots: areas in Earth's mantle that are hotter than the neighboring areas, forming melted rock that rises toward the crust. (Chap. 10, p. 273)

hot spring: a spring of heated groundwater, caused when the water is warmed by rocks that come into contact with molten material under Earth's surface. (Chap. 8, p. 216)

humidity: the amount of water vapor held in the air. (Chap. 15, p. 423)

humus: dark-colored organic matter found in soil; made of decayed plants and animals. (Chap. 6, p. 157)

hurricane: a large, swirling, low-pressure system that forms over tropical oceans; winds must reach at least 120 km/hr. (Chap. 15, p. 438)

hydrosphere: all the water that occurs at Earth's surface. (Chap. 14, p. 406)

hypothesis: a prediction about a problem that can be tested; may be used to form a theory. (Chap. 1, p. 16)

ice wedging: the breaking of rocks when water in cracks freezes and expands; a type of mechanical weathering. (Chap. 6, p. 150)

igneous rock: rock formed by the cooling and hardening of molten material from a volcano or from deep inside Earth. (Chap. 4, p. 91)

impermeable: rock or soil that has few pores or small pores, preventing water from passing through. (Chap. 8, p. 215)

index fossil: a fossil of a species that was abundant, existed briefly, and was widespread geographically; used in determining the relative ages of rock layers. (Chap. 12, p. 331)

infrared waves: electromagnetic waves that are the heat waves that we feel.

inner core: the dense, solid center of Earth, formed mostly of iron and nickel. (Chap. 9, p. 249)

inner planets: the four solid, rocky planets closest to the sun—Mercury, Venus, Earth, and Mars. (Chap. 23, p. 643)

International Date Line: the 180° meridian, on the other side of Earth from the prime meridian; an imaginary line in the Pacific Ocean where we change calendar days. (Chap. 5, p. 129)

International System of Units (SI): a modern version of the metric system that is used by most people around the world. (Chap. 1, p. 19)

intrusive igneous rocks: igneous rocks formed when magma cools below Earth's surface; generally have large mineral grains. (Chap. 4, p. 92)

invertebrates: animals without backbones.

ion: an atom with an electrical charge. (Chap. 2, p. 40)

ionosphere: a layer of Earth's atmosphere, where ions and free electrons absorb AM radio waves during the day and reflect them back toward Earth at night. (Chap. 14, p. 395)

irregular galaxies: galaxies having irregular shapes.

isobar: on a weather map, a line connecting points of equal atmospheric pressure. (Chap. 15, p. 441)

isotherm: on a weather map, a line connecting points of equal temperature. (Chap. 15, p. 441)

isotopes: atoms of the same element that have different numbers of neutrons in their nuclei, but the same number of protons. (Chap. 2, p. 36)

jet stream: narrow wind belts occurring near the top of the troposphere. (Chap. 14, p. 414)

Jupiter: in our solar system, the fifth planet from the sun; it is the largest planet, mostly gas and liquid, and has continuous storms of high-pressure gas. (Chap. 23, p. 654)

L

land breeze: wind blowing from land to sea at night because the land cools faster and cool air over the land moves out over the sea. (Chap. 14, p. 415)

landfill: an area of land where waste is deposited. (Chap. 19, p. 541)

landforms: features that make up the shape of the land at Earth's surface, such as plains, plateaus, and mountains.

Landsat Satellite: satellite that collects information about Earth's surface by using a mirror to detect different wavelengths of reflected or emitted energy. (Chap. 5, p. 138)

latitude: a distance north or south of the equator, expressed in degrees. (Chap. 5, p. 127)

lava: molten rock from a volcano flowing onto Earth's surface. (Chap. 4, p. 91)

law: a scientific rule of nature that describes the behavior of something in nature, but doesn't explain why something will happen in a given situation. (Chap. 1, p. 18)

leaching: occurs when soil materials dissolved in water are carried down through soil layers. (Chap. 6, p. 159)

legend: a list of symbols used on a map that explains their meaning.

light pollution: the glow in the night sky caused by urban lights, suburban lights, and rural security lights. (Chap. 21, p. 591)

light-year: a unit used to measure distance in space; the distance that light travels in one year (about 9.5 trillion km). (Chap. 24, p. 675)

lithosphere (LITH uh sfihr)**:** the rigid, outermost layer of Earth, about 100 km thick, composed of the crust and part of the mantle. (Chap. 11, p. 304)

Local Group: the cluster of about 25 galaxies that includes our galaxy (the Milky Way).

loess (LES)**:** a thick deposit of fine, wind-eroded sediments. (Chap. 7, p. 191)

longitude: a distance east or west of the prime meridian, expressed in degrees. (Chap. 5, p. 127)

longshore current: an ocean current that runs parallel to the shore, caused by waves colliding with the shore at slight angles. (Chap. 8, p. 223)

low-pressure system: in weather systems, an area where warm air rises and rotates counterclockwise (in the northern hemisphere).

lunar eclipse: an eclipse that occurs when Earth passes between the sun and moon, and Earth's shadow falls on the moon, preventing sunlight from reaching all or part of the moon. (Chap. 22, p. 627)

luster: the physical property of a mineral that describes how light is reflected from its surface; is defined as either metallic or nonmetallic. (Chap. 3, p. 69)

M

magma: hot, melted rock material beneath Earth's surface. (Chap. 3, p. 64)

magnetometer: a sensitive instrument that records magnetic data and is used to study Earth's magnetic field. (Chap. 11, p. 300)

magnitude: in earthquakes studies, a measure of the energy released by an earthquake; the Richter scale is used to describe earthquake magnitude. (Chap. 9, p. 253)

main sequence: on a Hertzsprung-Russell diagram, the diagonal band that includes 90 percent of all stars; shows relationships among a star's color, brightness, and temperature. (Chap. 24, p. 684)

mammals: warm-blooded vertebrates.

mantle: the thickest layer inside Earth; it lies between the outer core and the crust and is described as plasticlike; formed mostly of silicon, oxygen, magnesium, and iron. (Chap. 9, p. 249)

map legend: the key on most maps that is used to explain what the symbols on the map mean. (Chap. 5, p. 134)

map scale: the relationship between distances drawn on a map and actual distances on Earth's surface. (Chap. 5, p. 135)

maria: dark, flat regions of ancient lava on the moon; viewed from Earth, they resemble oceans, the Latin word for which is *maria.* (Chap. 22, p. 628)

market: the people or businesses that want to purchase a product.

Mars: in our solar system, the fourth planet from the sun; known as the red planet due to the iron oxide content of its weathered rocks. (Chap. 23, p. 648)

mascon: the concentration of mass on the moon located under an impact basin. (Chap. 22, p. 633)

mass: a measure of the amount of matter in an object; SI unit is the kilogram. (Chap. 1, p. 20)

mass movement: a type of erosion in which gravity causes loose materials to move downslope; can occur slowly or quickly. (Chap. 7, p. 173)

mass number: the sum of the number of protons and neutrons in an atom's nucleus. (Chap. 2, p. 36)

matter: anything that takes up space and has mass; the characteristics of matter are determined by its atoms. (Chap. 2, p. 32)

meander: a curve in a mature stream. (Chap. 8, p. 208)

mechanical weathering: the breaking apart of rocks without changing their chemical composition; for example, by plant roots and ice. (Chap. 6, p. 149)

Mercator (mur KAYT ur) **projection:** a map-projection method using parallel longitude lines; continent shapes are accurate, but their areas are distorted. (Chap. 5, p. 130)

Mercury: in our solar system, the first planet from the sun; the second-smallest planet that has a cratered surface like our moon and cliffs as high as 3 km. (Chap. 23, p. 646)

Mesosaurus (mes oh SAR uhs): a fossil reptile found both in South America and Africa, providing evidence that these continents once were joined.

Mesozoic (mez uh ZOH ihk) **Era:** the middle era of Earth's geologic history; began about 245 million years ago; reptiles and gymnosperms were the dominant land life-forms. (Chap. 13, p. 374)

metallic luster: the physical property of any mineral that has a shiny appearance resembling metal.

metamorphic rock: rock formed from sedimentary, igneous, or other metamorphic rock due to increases in heat or pressure. (Chap. 4, p. 97)

meteor: a meteoroid that enters Earth's atmosphere and burns up as it falls; also called a shooting star. (Chap. 23, p. 664)

meteorite: a meteor that does not completely burn up in Earth's atmosphere and strikes Earth's surface. (Chap. 23, p. 665)

meteorologist: a scientist who studies weather conditions using radar, satellites, and other instruments to make weather maps and forecasts. (Chap. 15, p. 440)

meteorology: the study of Earth's weather and the forces that cause it. (Chap. 1, p. 8)

microwaves: electromagnetic waves that are shorter than radio waves but longer than light waves; we use them for radar and transmitting voices, music, video, and data.

mid-ocean ridge: the place where new ocean floor forms; resembles an underwater mountain range; formed when forces within Earth spread the seafloor apart. (Chap. 18, p. 506)

mineral: a naturally occurring, inorganic solid with a distinct internal structure and chemical composition. (Chap. 3, p. 62)

mixture: a combination of different substances in which each of the components keeps its own physical and chemical properties despite being mixed. (Chap. 2, p. 40)

Moho discontinuity: the boundary between Earth's crust and the mantle; seismic waves travel faster below the Moho and slower above it. (Chap. 9, p. 249)

Mohs Scale of Hardness: a list of common minerals and their hardnesses, developed by German mineralogist Friedrich Mohs.

mold: a cavity in a rock that has the shape of a fossil that was trapped there; water dissolved the fossil away, leaving its imprint. (Chap. 12, p. 329)

molecule: the smallest particle of a compound that still keeps all the properties of the compound; it is formed by the combining of two or more atoms. (Chap. 2, p. 38)

moon phases: the change in appearance of the moon as it orbits Earth every 29½ days, depending on the relative positions of the moon, Earth, and the sun; for example, full moon and new moon. (Chap. 22, p. 624)

moraine: a ridge of unsorted rock and soil bulldozed ahead and to the sides of a glacier; left behind when the glacier melts. (Chap. 7, p. 184)

natural gas: a mixture of gases formed as ancient plants and animals decayed; burned as a fossil fuel.

natural selection: the process by which organisms with traits best suited to an environment survive and reproduce, while others die out because they lack those traits. (Chap. 13, p. 360)

neap tide: a tide that is lower than normal because the sun, Earth, and moon form a right angle.

nebula: a large cloud of gas and dust in space that may be the beginning of a star. (Chap. 24, p. 686)

nekton (NEK tuhn)**:** marine animals that swim and determine their own course of movement rather than drift with currents; for example, shrimp and whales. (Chap. 18, p. 513)

Neptune: in our solar system, usually the eighth planet from the sun; it is large, gaseous, and has stormlike features. (Chap. 23, p. 658)

neutron: one of the two particles that make up the nucleus of an atom; it has no electric charge. (Chap. 2, p. 35)

neutron star: the final stage in the life cycle of some stars, where the core of a supernova collapses, shrinks to about 10-15 km, and becomes so dense that only neutrons can exist there. (Chap. 24, p. 688)

new moon: the moon phase when the side facing Earth is completely dark and cannot be seen because the moon is between Earth and the sun. (Chap. 22, p. 624)

nonfoliated: a type of metamorphic rock created when mineral grains change, grow, and rearrange, but don't form bands. (Chap. 4, p. 99)

nonmetallic luster: a physical property of a mineral that has a dull appearance and does not resemble a metal.

normal fault: a pull-apart (tension) fracture in rocks, where rocks that are above the fault surface drop downward in relation to rocks that are below the fault surface like this:
<—normal/fault—>. (Chap. 9, p. 238)

nuclear energy: an alternate energy source produced from atomic reactions. (Chap. 2, p. 54)

nuclear reactor: a device in which uranium atoms fission to release energy, used to generate electricity.

nuclear waste: radioactive waste material produced by nuclear power plants; must be safely stored to prevent it from entering the environment. (Chap. 2, p. 55)

observatory: a building that contains an optical telescope for observing objects in space. (Chap. 21, p. 586)

occluded front: in weather systems, the boundary that results when two cool air masses merge and force warmer air to rise between them.

oceanography: the study of Earth's oceans, their processes, and life within them. (Chap. 1, p. 9)

oil: a liquid formed as ancient plants and animals decay; burned as a fossil fuel and used to make lubricants and plastics.

old stream: a stream that flows slowly down a gradual slope through a broad floodplain that it has made; often meandering.

Olympus Mons: the largest known volcano in the solar system—on Mars.

Oort Cloud: a cloud of comets surrounding the solar system beyond Pluto's orbit; it may be the source of most comets. (Chap. 23, p. 663)

orbit: the curved path followed by a satellite as it travels around a star, planet, or other object. (Chap. 21, p. 594)

ore: a mineral containing a useful substance, such as a metal, that can be mined at a profit. (Chap. 3, p. 75)

organic evolution: the gradual change in life-forms over time. (Chap. 13, p. 359)

organic matter: any material that originated as plant or animal tissue; decaying animals or plants that become sediment and a part of soils.

outer core: the liquid layer of Earth's core that surrounds the solid inner core and is comprised of iron and nickel. (Chap. 9, p. 249)

outer planets: the five planets farthest from the sun—Jupiter, Saturn, Uranus, Neptune, and Pluto. (Chap. 23, p. 643)

outwash: stratified material washed out from a glacier by meltwater.

overgrazing: occurs when too many animals graze too small an area and eat all the vegetation from the land.

oxidation: chemical weathering that occurs when a substance is exposed to oxygen and water. (Chap. 6, p. 154)

ozone layer: a layer of the stratosphere with a high concentration of ozone; protects living things by absorbing ultraviolet radiation from the sun. (Chap. 14, p. 400)

Pacific Ring of Fire: the area around the Pacific Plate where volcanoes and earthquakes are common due to tectonic movement. (Chap. 10, p. 272)

Paleozoic (pay lee uh ZOH ihk) **Era:** the second-oldest division of geologic time; began about 544 million years ago, when organisms developed hard parts. (Chap. 13, p. 367)

Pangaea (pan JEE uh)**:** the name Alfred Wegener gave to the large landmass, made up of all continents, that he believed existed before it broke apart to form the present continents. (Chap. 11, p. 294)

parallax: the apparent shift in position of an object when viewed from two different points, such as your left eye and right eye. (Chap. 24, p. 675)

pebble: a sediment particle measuring 2.0 mm to 64 mm in size.

penumbra: during an eclipse, the highest outer portion of the shadow.

periods: subdivisions of eras on the geologic time scale, based on life-forms and geologic events. (Chap. 13, p. 357)

permeable: describes rock or soil that has connecting pores that allow water to pass through easily. (Chap. 8, p. 215)

petrified remains: plant or animal remains that have been petrified, or "turned to rock"; this happens when minerals carried in groundwater replace the original materials. (Chap. 12, p. 328)

petroleum: a naturally occurring liquid formed over millions of years from organisms; is refined into fuel such as gasoline.

pH scale: a logarithmic scale used to describe how acidic or how basic a solution is; an abbreviation of *p*otential of *H*ydrogen. (Chap. 20, p. 558)

photochemical smog: a brown-colored air pollution that forms when sunlight chemically changes the pollutants released into the air by burning fossil fuels. (Chap. 20, p. 556)

photosphere: the innermost (lowest) of the three layers of sun's atmosphere; radiates the light we see; often called the surface of the sun. (Chap. 24, p. 679)

photosynthesis (foh toh SIHN thuh sihs)**:** the process that plants use to make food, using light energy, carbon dioxide, and water. (Chap. 18, p. 510)

physical properties: characteristics of an element or a compound that affect weight, color, density, etc., but don't affect how it will react with other elements or compounds. (Chap. 2, p. 44)

plain: a landform that is a large, relatively flat area; interior plains and coastal plains make up one-half the land area in the United States. (Chap. 5, p. 120)

plankton (PLANK tuhn)**:** tiny marine plants and animals that drift with the ocean currents. (Chap. 18, p. 513)

plate: in plate tectonics, a section of Earth's lithosphere (crust and upper mantle) that moves around on the mantle. (Chap. 11, p. 304)

plate tectonics: the theory that Earth's crust and upper mantle (lithosphere) are broken into sections, called plates, that slowly move around on the mantle. (Chap. 11, p. 304)

plateaus: landforms created next to mountains, when forces within Earth raised high, relatively flat areas of nearly horizontal rocks. (Chap. 5, p. 122)

plucking: a type of glacial erosion in which rock fragments are loosened, broken off, and carried away by the freezing of water in rock cracks. (Chap. 7, p. 182)

Pluto: in our solar system, usually the ninth and last planet from the sun; it is the smallest planet and the only outer planet with a solid, icy rock surface. (Chap. 23, p. 658)

polar zones: the zones that receive solar radiation at an angle and are the coldest areas on Earth; extend from the poles to 66½° north and south latitudes. (Chap. 16, p. 452)

pollutant: a substance that interferes with biochemical processes and causes harmful change to organisms; produced both by human activities and natural processes. (Chap. 18, p. 519)

pollution: the introduction of harmful substances into an environment. (Chap. 18, p. 519)

population: the total number of individuals of a particular species that exists in a specific area. (Chap. 19, p. 533)

population explosion: a large increase in the population of a species, due to a rapid increase in the birthrate, or a sharply reduced death rate, or both. (Chap. 19, p. 533)

Precambrian (pree KAM bree un) **time:** the oldest and longest division of geologic time, including about 90 percent of Earth's history; spans 4.6 billion to about 544 million years ago. (Chap. 13, p. 366)

precipitation: water or ice that condenses in the air and falls to the ground as rain, snow, sleet, or hail, depending on the air temperature. (Chap. 15, p. 430)

primary waves: waves of energy, released during an earthquake, that travel through Earth by causing particles in rocks to compress and stretch apart in the direction of the wave. (Chap. 9, p. 244)

prime meridian: an imaginary line running from the North Pole to the South Pole, passing through Greenwich, England; the 0° reference line for longitude. (Chap. 5, p. 127)

principle of superposition: states that in an undisturbed layer of rock, older rocks lie at the bottom and the rocks become younger toward the top. (Chap. 12, p. 337)

Project Apollo: a project of the U.S. space program in which astronauts first traveled to the moon in the spacecraft *Apollo 11*. (Chap. 21, p. 598)

Project Gemini: an early project of the U.S. space program in which two crewed *Gemini* spacecraft successfully linked in orbit. (Chap. 21, p. 598)

Project Mercury: an early project of the U.S. space program in which a piloted spacecraft orbited Earth and returned safely. (Chap. 21, p. 597)

proton: one of the two particles that make up the nucleus of an atom; it has a positive electric charge. (Chap. 2, p. 35)

psychrometer (si KRAH muh tur): a device used by meteorologists to measure relative humidity.

radiation: the transfer of energy through matter or space by electromagnetic waves. (Chap. 14, p. 403)

radio telescope: an instrument that uses a large curved dish to reflect radio waves from space to a focal point; used to study space objects and map the universe. (Chap. 21, p. 587)

radio waves: electromagnetic waves having long wavelengths; we use them to transmit voices, music, video, and data over distances.

radioactive decay: the decay of an atom of one element to form another element, occurring when an alpha particle or beta particle is expelled from the original atom. (Chap. 12, p. 344)

radiometric dating: a dating method that uses the rate of decay of radioactive isotopes in rocks and measures the amount of parent and daughter materials to determine the absolute age of the rock. (Chap. 12, p. 345)

recyclable: items that can be processed and used again. (Chap. 19, p. 546)

red shift: a kind of Doppler shift in which light from a star that is moving away from us has its wavelength shifted toward the red end of the spectrum.

reef: on the continental shelf, a large, rigid underwater colony of coral animals that have become cemented together. (Chap. 18, p. 518)

reflecting telescope: an optical magnifying instrument in which light from an object strikes a concave mirror, which then reflects the light to form an image at the focal point. (Chap. 21, p. 586)

reforestation: the planting of a large number of trees. (Chap. 16, p. 471)

refracting telescope: an optical magnifying instrument in which light from an object passes through a convex lens and is bent to form an image at the focal point. (Chap. 21, p. 586)

relative dating: determining the order of events and the relative age of rocks (older or younger) by examining the positions of rocks in layers. (Chap.12, p. 338)

relative humidity: the measure of the amount of water vapor in the air, compared to the maximum it can hold at a specific temperature; it varies with temperature and is between zero percent and 100 percent. (Chap. 15, p. 424)

reptiles: scaly skinned, vertebrate animals that evolved from the same ancestors as amphibians but do not return to water to reproduce. (Chap. 13, p. 370)

reverse fault: a compression fracture in rocks, where rocks that are above the fault surface are forced up over rocks that are below the fault surface, like this: —>reverse/fault<—. (Chap. 9, p. 238)

revolution: the orbiting of one object around another, like Earth's yearly orbit around the sun. (Chap. 22, p. 617)

Richter (RIHK tur) scale: describes how much energy is released by an earthquake.

rill erosion: a type of surface water erosion due to runoff, in which water swiftly running down the slope creates small channels in the soil; these channels can enlarge into gullies. (Chap. 8, p. 204)

Robinson projection: a map-projection method using curved longitude lines; continent shapes and land areas are accurate with little distortion. (Chap. 5, p. 131)

rock: Earth material made of a mixture of one or more minerals, glass, mineraloids, or organic matter. (Chap. 4, p. 87)

rock cycle: the processes by which, over many years, Earth materials form and change back and forth among igneous rocks, sedimentary rocks, and metamorphic rocks. (Chap. 4, p. 87)

rock-forming minerals: a group of minerals that make up most of the rocks in Earth's crust.

rotation: the spinning of an object around its axis; causes day and night to occur on Earth. (Chap. 22, p. 615)

runoff: water that neither soaks into the ground nor evaporates but instead flows across Earth's surface and eventually into streams, lakes, or oceans. (Chap. 8, p. 202)

Safe Drinking Water Act: this 1986 law sets safety standards for drinking water in the United States. (Chap. 20, p. 570)

salinity: a measure of the amount of solids (mostly salts) dissolved in seawater; usually measured in grams of salt per kilogram of ocean water. (Chap. 17, p. 480)

sand: a sediment particle measuring 0.06 mm to 2.0 mm in size.

sanitary landfill: a waste-disposal area that is excavated; lined with plastic, concrete, or clay; and filled with layers of waste and dirt. (Chap. 19, p. 541)

satellite: any object that revolves around another object; planets and human-made satellites are examples. (Chap. 21, p. 594)

saturated (SACH uh rayt id)**:** condition when air contains as much moisture as possible at a specific temperature. (Chap. 15, p. 424)

Saturn: in our solar system, the sixth planet from the sun; it is the second-largest planet, is mostly gas and liquid, and has prominent rings. (Chap. 23, p. 656)

science: the process of observing, explaining, and understanding things in our world; means "having knowledge"; divided into four general areas: chemistry, physics, life science, and Earth science. (Chap. 1, p. 6)

scientific method: a series of problem-solving procedures used by scientists; may include the following basic steps: determining the problem, making a hypothesis, testing the hypothesis, analyzing the results, and drawing conclusions. (Chap. 1, p. 16)

scrubber: a device that "scrubs" the smoke from coal-burning power plants with basic compounds to increase the pH to a safe level. (Chap. 20, p. 563)

sea breeze: wind blowing from sea to land during the day when the sun warms the land faster, and cool air from above the water forces the warm air above the land to rise. (Chap. 14, p. 414)

seafloor spreading: the theory that magma from Earth's mantle rises to the surface at mid-ocean ridges and cools to form new seafloor, which new magma slowly pushes away from the ridge. (Chap. 11, p. 299)

seamount: an elevated part of the ocean floor, possibly an underwater volcano.

season: a regular, short-term period of change in the climate of an area due to changes in the amount of solar radiation the area receives. (Chap. 16, p. 462)

secondary waves: waves of energy, released during an earthquake, that travel through Earth by causing particles in rocks to move at right angles to the direction of the wave. (Chap. 9, p. 244)

sedimentary rock: rock formed when fragments of rocks, minerals, and/or organic matter are compacted or cemented together or precipitate out of solution. (Chap. 4, p. 101)

sediments: loose materials such as rock fragments, mineral grains, and bits of plants and animals that have been transported by wind, water, or glaciers. (Chap. 4, p. 101)

seismic waves: in an earthquake, the energy waves that move outward from the earthquake focus and make the ground quake. (Chap. 9, p. 241)

seismograph: an instrument used by seismologists to record primary, secondary, and surface waves from earthquakes. (Chap. 9, p. 252)

seismologist: a scientist who studies earthquakes and seismic waves. (Chap. 9, p. 252)

shadow zone: the area where seismic waves cannot reach because Earth's liquid outer core bends primary waves and stops secondary waves.

shearing forces: along strike-slip faults, forces that push on rocks from various directions, causing them to twist and break.

sheet erosion: a type of surface water erosion due to runoff, in which rainwater flowing over a gentle slope slowly moves sediments from the entire surface. (Chap. 8, p. 205)

shield volcano: a broad volcano with gently sloping sides, built by quiet eruptions of fluid basaltic lava, which spreads out in flat layers; example: the Hawaiian Islands. (Chap. 10, p. 279)

silicates: minerals containing silicon and oxygen, usually with other elements; the largest group of minerals. (Chap. 3, p. 66)

sill: an intrusive igneous rock body formed when magma is squeezed into a horizontal crack between rock layers and solidifies underground. (Chap. 10, p. 284)

silt: a sediment particle measuring 0.004 mm to 0.06 mm in size.

sinkhole: a depression in the ground caused when groundwater dissolves limestone beneath the hole, causing the ground to collapse.

slump: a type of mass movement in which one large mass of loose material or rock layers moves downhill, leaving a curved scar. (Chap. 7, p. 174)

smog: air pollution seen around cities, resulting from burning fossil fuels.

soil: a mixture of weathered rock, decaying organic matter (plants and animals), mineral fragments, water, and air. (Chap. 6, p. 156)

soil profile: a vertical section of soil layers (horizons). (Chap. 6, p. 158)

solar eclipse: an eclipse that occurs when the moon passes directly between Earth and the sun, so that the moon casts a shadow on part of Earth and blocks sunlight from reaching Earth. (Chap. 22, p. 626)

solar flare: an intense bright spot in the sun's chromosphere, associated with sunspots and radio interference on Earth.

solar system: the system of nine planets and many other objects that orbit our sun; may have been formed about 5 billion years ago from a cloud of ice, gas, and dust. (Chap. 23, p. 641)

solstice: the two times each year that Earth's tilt makes the sun reach its greatest angle north or south of the equator, marking the start of summer or winter. (Chap. 22, p. 622)

solution: a kind of mixture formed when one substance is dissolved in another. (Chap. 2, p. 41)

sonar: the use of sound-wave echoes to detect the size and shape of structures found underwater. (Chap. 5, p. 139)

space probe: an instrument that travels out into the solar system, gathers information, and transmits it back to Earth. (Chap. 21, p. 595)

space shuttle: a reusable spacecraft that transports astronauts, satellites, and other material to and from space. (Chap. 21, p. 604)

space station: a facility in space with living quarters, work and exercise space, and the support systems needed for people to live and work. (Chap. 21, p. 605)

species: a group of individuals that normally breed only among themselves. (Chap. 13, p. 359)

sphere: a round, three-dimensional object whose surface at all points is the same distance from its center; Earth is a sphere that is slightly flattened at its poles. (Chap. 22, p. 614)

spiral galaxy: a galaxy having spiral arms.

spring: the point at which the water table meets Earth's surface, causing water to flow from the ground. (Chap. 8, p. 216)

spring tide: a tide level greater than normal because the moon, Earth, and sun are aligned.

stalactite: a icicle-like deposit of calcite hanging from the ceiling of a cave.

station model: in weather forecasting, a group of meteorological symbols on a weather map that depict weather information at one specific location. (Chap. 15, p. 441)

stationary front: in weather systems, a warm front or cold front that has stopped moving.

streak: the color of a mineral when it is powdered; usually observed by rubbing the mineral on a ceramic streak plate. (Chap. 3, p. 69)

striations (stri AY shunz): long, parallel scars in rocks, caused by rock fragments being dragged across them by a glacier.

strike-slip fault: a break in rocks due to shearing forces, where rocks on either side of the fault move past each other without much upward or downward movement. (Chap. 9, p. 238)

subduction zone: in plate tectonics, the area where an ocean-floor plate collides with a continental plate, and the denser ocean plate sinks under the less dense continental plate. (Chap. 11, p. 307)

sulfurous smog: a gray-colored air pollution created when power plants and home furnaces burn fossil fuels, releasing sulfur compounds and smoke particles into the air. (Chap. 20, p. 557)

sunspot: a dark-appearing spot on the sun's surface that shows up because it is cooler than the surrounding areas; caused by the sun's intense magnetic field. (Chap. 24, p. 680)

supergiant: a late stage in the life cycle of a large star, when the core reaches high temperatures, heavy elements form by fusion, and the star's outer layers expand. (Chap. 24, p. 688)

supernova: a late stage in the life cycle of some stars, when the core collapses, causing the outer portion to explode.

surface current: an ocean current powered by wind, found in the upper few hundred meters of seawater; moves water horizontally and parallel to Earth's surface. (Chap. 17, p. 484)

surface waves: waves of energy, released during an earthquake, that reach Earth's surface and travel outward from the epicenter in all directions; travel through Earth by giving rock particles an elliptical and side-to-side motion. (Chap. 9, p. 244)

technology: the application of scientific discoveries; can both contribute to problems and solve problems. (Chap. 1, p. 10)

temperate zones: the two zones of moderate, seasonal weather that exist between the tropics and the polar regions. (Chap. 16, p. 452)

tension: stretching forces that can be strong enough to pull rocks apart at divergent plate boundaries.

tephra: lava that is blasted into the air by violent volcanic eruptions and solidifies as it falls to the ground as ash, cinders, and volcanic bombs. (Chap. 10, p. 279)

theory: an explanation backed by results from repeated tests, experiments, or observations. (Chap. 1, p. 18)

third quarter: the moon phase halfway between full moon and new moon, when half of the side facing Earth is lighted. (Chap. 22, p. 625)

tidal energy: electricity generated by the ocean tides.

tidal range: the vertical distance between the level of the ocean at high tide and low tide. (Chap. 17, p. 492)

tide: the periodic rise and fall of the surface level of the oceans; caused by a giant wave formed by the gravitational attraction between the sun, moon, and Earth. (Chap. 17, p. 492)

till: an unsorted mixture of boulders, sand, silt, and clay deposited by a glacier. (Chap. 7, p. 184)

Titan: Saturn's largest moon.

titanium: durable, lightweight metal obtained from minerals such as ilmenite or rutile. (Chap. 3, p. 78)

Topex-Poseidon Satellite: satellite that collects information about Earth's oceans by using radar. (Chap. 5, p. 138)

topographic map: a map that uses contour lines to show changes in elevation at Earth's surface; shows natural features such as lakes and cultural features such as cities and dams. (Chap. 5, p. 132)

topsoil: the top layer of soil, also called A horizon; usually contains humus and is dark in color.

tornado: a violent, whirling, funnel-shaped windstorm, up to 500 km/hour, that moves in a narrow path over land. (Chap. 15, p. 436)

trace fossils: footprints, worm holes, burrows, and other traces of animal activity preserved in rock.

transform fault: in plate tectonics, a boundary between two plates that are sliding horizontally past one another. (Chap. 11, p. 309)

trench: a long, narrow, steep-sided depression in the ocean floor formed where one crustal plate is forced beneath another. (Chap. 18, p. 507)

trilobites (TRI luh bites): shield-shaped marine invertebrate animals commonly found as fossils in Paleozoic Era rocks; they are now extinct.

Triton: Neptune's largest moon.

tropics: the region that receives the most solar radiation; extends between latitudes 23½° north and 23½° south of the equator. (Chap. 16, p. 452)

troposphere: the layer of Earth's atmosphere closest to the ground; contains clouds, smog, weather, and 75 percent of atmospheric gases. (Chap. 14, p. 394)

trough: the lowest point of a wave. (Chap. 17, p. 490)

tsunami (soo NAHM ee): an ocean wave (seismic sea wave) that begins over an earthquake focus and can reach 30 m high. (Chap. 9, p. 255)

ultraviolet radiation: a type of energy that comes to Earth from the sun and is mostly absorbed by the ozone layer; rays can damage skin and cause cancer and other health problems. (Chap. 14, p. 400)

ultraviolet waves: electromagnetic waves that are a little shorter than visible light waves; they cause sunburn and skin cancer.

umbra: during an eclipse, the darker central portion of the shadow.

unconformities: gaps in a sequence of rock layers (the rock record), resulting from erosion or a lack of new deposition, or both. (Chap. 12, p. 339)

uniformitarianism: a basic principle of geology stating that Earth processes occurring today are similar to those that occurred in the past. (Chap. 12, p. 347)

upwarped mountains: mountains formed when Earth's crust is pushed up and eroded, forming sharp peaks and ridges. (Chap. 5, p. 124)

upwelling: the rising of cold, nutrient-rich water from deep in the ocean to the surface; occurs when surface water is pushed away by winds. (Chap. 17, p. 487)

Uranus: in our solar system, the seventh planet from the sun; it is large, gaseous, and is the only planet that lays on its side in orbit. (Chap. 23, p. 657)

Valles Marineris: a huge rift on Mars that is more than 4000 km long.

valley glacier: the most common type of glacier, occurs in mountain valleys where the temperatures are low enough to allow snow to accumulate faster than it can melt.

variable: in an experiment, the factor that can be changed to see what will happen. (Chap. 1, p. 17)

vent: in volcanic regions, an opening in Earth's surface through which can flow lava, ash, cinders, smoke, and steam. (Chap. 10, p. 268)

Venus: in our solar system, the second planet from the sun; it is similar in size and mass to Earth, is blanketed by a dense atmosphere of carbon dioxide and sulfuric acid, and is hot. (Chap. 23, p. 647)

vertebrates: animals with backbones; evolved during Ordovician Period.

visible light: electromagnetic waves having short wavelengths; the only part of the electromagnetic spectrum that we can see.

volcanic mountains: mountains created when magma within Earth escapes to the surface, building cones of lava and ash. (Chap. 5, p. 125)

volcanic neck: the solid igneous core of a vent that remains after the outer layers of lava and tephra have been eroded away from an extinct volcano. (Chap. 10, p. 284)

volcano: a vent in Earth's surface that often forms a mountain built of lava and volcanic ash, which erupts and builds up. (Chap. 10, p. 266)

volume: the amount of space occupied by an object; SI unit is the cubic meter (m^3).

waning: describes the moon following a full moon, as its visible lighted area grows smaller during the lunar cycle. (Chap. 22, p. 625)

waning crescent: the shrinking slice of lighted moon when the visible lighted area is decreasing from third quarter to new moon.

waning gibbous: the shrinking area of moon as the visible lighted area is decreasing from full moon to third quarter.

warm front: the moving boundary that develops when a warm air mass meets a cold air mass.

waste coal: large piles of poor-quality coal lying near abandoned coal mines; can generate acid runoff. (Chap. 4, p. 108)

water cycle: the continuous movement of water between Earth's surface and the atmosphere through evaporation, condensation, and precipitation. (Chap. 14, p. 406)

water table: the upper surface of the zone of saturation (the area where all of the pores in the rock are completely filled with water). (Chap. 8, p. 215)

wave: a movement in which ocean water rises and falls as energy passes through; crest, trough, wavelength, and wave height are parts of a wave. (Chap. 17, p. 490)

waxing: describes the moon shortly after a new moon, as its visible lighted area grows larger during the lunar cycle. (Chap. 22, p. 624)

waxing crescent: the growing slice of moon when the visible lighted area is increasing from new moon to first quarter.

waxing gibbous: the growing area of visible lighted moon as the lighted area is increasing from first quarter to full moon.

weather: the behavior of the atmosphere—wind, temperature, pressure, precipitation—at a particular place and time. (Chap. 15, p. 422)

weathering: the breaking of rocks into smaller pieces, either mechanically or chemically. (Chap. 6, p. 148)

Wegener, Alfred: a German scientist who proposed the idea of continental drift in 1912.

weight: a measure of the force of gravity on an object. (Chap. 1, p. 20)

white dwarf: a late stage in a star's life cycle when its core runs out of fuel (helium) and its unstable outer layers escape into space, leaving the white-hot, dense core. (Chap. 24, p. 688)

young stream: a stream that flows swiftly down a steep slope or a valley with steep sides, causing rapid erosion.

zone of saturation: an area where all the pores in the rock are completely filled with water, usually near the ground surface. (Chap. 8, p. 215)

Glossary/Glosario

This glossary defines each key term that appears in **bold type** in the text. It also shows the page number where you can find the word used.

abrasion/abrasión: tipo de erosión que ocurre cuando los sedimentos soplados y acarreados por el viento golpean las rocas. (Cap. 7, pág. 188)

absolute dating/datación absoluta: método que utilizan los geólogos para determinar la edad, en años, de una roca u otro objeto. (Cap. 12, pág. 344)

absolute magnitude/magnitud absoluta: medida de la cantidad de luz que una estrella emite verdaderamente. (Cap. 24, pág. 674)

abyssal plain/llanura abisal: suelo oceánico llano. (Cap. 18, pág. 505)

acid/ácido: sustancia que contiene un pH menor que siete. (Cap. 20, pág. 558)

acid rain/lluvia ácida: tipo de precipitación más ácida de lo normal. (Cap. 20, pág. 558)

adaptation/adaptación: cualquier característica que los organismos han desarrollado a lo largo del tiempo para garantizar su sobrevivencia. (Cap. 16, pág. 459)

air mass/masa de aire: gran extensión de aire que posee las mismas propiedades que la superficie sobre la cual se desarrolla. (Cap. 15, pág. 432)

alluvial fan/abanico aluvial: depósito en forma de triángulo que se forma cuando corrientes de agua que se mueven cuesta abajo llegan a superficies llanas, pierden su energía de movimiento y depositan sus sedimentos. (Cap. 8, pág. 213)

amphibian/anfibio: animal que vive en tierra y respira aire, pero que debe regresar al agua con el fin de reproducirse. (Cap. 13, pág. 370)

apparent magnitude/magnitud aparente: medida de la cantidad de luz de una estrella que llega a la Tierra. (Cap. 24, pág. 674)

aquifer/acuífero: capa de roca permeable que deja pasar el agua libremente. (Cap. 8, pág. 215)

artesian well/pozo artesiano: pozo en el cual el agua bajo presión se eleva a la superficie. (Cap. 8, pág. 216)

asteroid/asteroide: fragmento rocoso semejante al material que posteriormente formó los planetas. (Cap. 23, pág. 665)

asthenosphere/astenosfera: capa tipo plástico situada debajo de la litosfera. (Cap. 11, pág. 304)

astronomical unit/unidad astronómica: medida que se usa para medir distancias hacia los objetos en el sistema solar; corresponde a 150 millones de kilómetros, la cual es la distancia promedio entre la Tierra y el sol. (Cap. 23, pág. 648)

astronomy/astronomía: es el estudio de los objetos del espacio, incluidos las estrellas, los planetas y los cometas. (Cap. 1, pág. 8)

atom/átomo: elemento constitutivo de la materia. (Cap. 2, pág. 32)

atomic number/número atómico: equivale al número de protones en el núcleo de un átomo. (Cap. 2, pág. 36)

axis/eje: línea imaginaria alrededor de la cual gira la Tierra. (Cap. 22, pág. 615)

barrier island/crestas prelitorales: depósitos de arena paralelos a la costa pero separados del continente. (Cap. 8, pág. 226)

basaltic/basáltica: roca ígnea densa, pesada y de color oscuro que se forma a partir del magma basáltico. (Cap. 4, pág. 93)

base/base: sustancia que contiene un pH mayor que siete. (Cap. 20, pág. 558)

basin/cuenca: área baja sobre la cual se formaron los océanos. (Cap. 17, pág. 479)

batholith/batolito: las masas más grandes de rocas ígneas intrusivas. (Cap. 10, pág. 283)

beach/playa: depósito de sedimentos que corre paralelo a la costa. (Cap. 8, pág. 224)

benthos/bentos: organismos vegetales y animales que habitan los niveles más profundos del océano. (Cap. 18, pág. 514)

big bang theory/teoría de la gran explosión: teoría que dice que hace unos 15 billones de años, el universo comenzó a expandirse después de una enorme explosión. Pruebas recientes indican una fecha más reciente entre 8 a 10 billones de años. (Cap. 24, pág. 696)

binary system/sistema binario: sistema en el cual dos estrellas giran alrededor de la una a la otra. (Cap. 24, pág. 681)

black hole/agujero negro: núcleo restante de una estrella de neutrones, el cual es tan denso y masivo que nada puede escapar de su campo de gravedad, ni siquiera la luz. (Cap. 24, pág. 688)

breaker/cachón: ola de mar que rompe en la playa. (Cap. 17, pág. 491)

caldera/caldera: gran depresión que resulta cuando la cima de un volcán se hunde en la cámara magmática parcialmente vacía. (Cap. 10, pág. 285)

carbonaceous film/película carbonácea: tipo de fósil producido por una película fina de residuo carbonoso la cual forma un bosquejo del organismo original. (Cap. 12, pág. 328)

carrying capacity/capacidad de carga: número máximo de individuos de una especie en particular que puede sustentar el planeta. (Cap. 19, pág. 532)

cast/impresión fósil: se forma cuando otros sedimentos llenan el molde, se endurecen y forman una roca produciendo una impresión del objeto original. (Cap. 12, pág. 329)

cave/caverna: abertura subterránea; se forma cuando el agua subterránea ácida corre por las resquebrajaduras naturales de la piedra caliza disolviéndola y formando, paulatinamente, una caverna. (Cap. 8, pág. 217)

cementation/cementación: proceso de formación de roca sólida a partir de la disolución de cementos naturales en la tierra y las rocas, los cuales se depositan alrededor de los sedimentos y los unen. (Cap. 4, pág. 102)

Cenozoic Era/ Era cenozoica: era de la vida reciente que comenzó hace aproximadamente 66 millones de años. (Cap. 13, pág. 379)

chemical property/propiedad química: describe la manera en que una sustancia cambia cuando reacciona con otras sustancias. (Cap. 2, pág. 39)

chemical weathering/meteorización química: meteorización que se presenta cuando el agua, el aire y otras sustancias reaccionan con los minerales presentes en las rocas; la meteorización química cambia la composición química de las rocas. (Cap. 6, pág. 151)

chemosynthesis/quimiosíntesis: proceso de fabricación de alimentos que no depende de la luz solar y el cual llevan a cabo las bacterias en las dorsales oceánicas. (Cap. 18, pág. 512)

chlorofluorocarbons/clorofluorocarburos: compuestos químicos que se usan para elaborar refrigerantes, rociadores aerosoles y productos de espuma; su uso puede llegar a destruir la capa de ozono. (Cap. 14, pág. 401)

chromosphere/cromosfera: capa que se encuentra encima de la fotosfera y que se extiende hacia arriba unos 2000 km. (Cap. 24, pág. 679)

cinder cone/cono de carbonilla: volcán de lados empinados y ligeramente consolidado que se forma cuando la tefrita llega al suelo. (Cap. 10, pág. 279)

Clean Air Act/Ley para el Control de la Contaminación del Aire: Ley promulgada en 1990 para combatir los problemas del smog, los clorofluorocarburos, las macropartículas y la lluvia ácida. Regula la industria automotriz, las tecnologías del carbón y otras industrias. (Cap. 20, pág. 561)

Clean Water Act/Ley para el Control de la Contaminación del Agua: ley promulgada en 1987 que otorga fondos a los estados para que construyan instalaciones de tratamiento de aguas negras y residuales. (Cap. 20, pág. 571)

cleavage/crucero: propiedad de un mineral que lo hace partirse a lo largo de superficies suaves y planas. (Cap. 3, pág. 70)

climate/clima: patrón de tiempo que ocurre en una región en particular a lo largo de muchos años. (Cap. 6, pág. 154)

cloud seeding/lluvia artificial: bombardeo de las nubes con partículas de hielo seco o con yoduro de plata como núcleos para las moléculas de agua con el objeto de que se agrupen, se congelen y formen precipitación. (Cap. 15, pág. 444)

cogeneration/cogeneración: proceso en el cual tanto la energía eléctrica como la térmica son aprovechadas por la central que las produce. (Cap. 4, pág. 109)

comet/cometa: astro compuesto de polvo y partículas rocosas mezclados con agua congelada, metano y amoníaco. (Cap. 23, pág. 663)

compaction/compactación: proceso formador de roca sólida al presionar las capas superiores de sedimento sobre las inferiores. (Cap. 4, pág. 102)

composite volcano/volcán compuesto: volcán que se forma del continuo y alternado ciclo de erupción de lava y tefrita. (Cap. 10, pág. 281)

composting/abono orgánico: desechos vegetales que se apilan en el patio y que se dejan decomponer paulatinamente mediante la acción del calor y de las bacterias. (Cap. 19, pág. 545)

compound/compuesto: tipo de materia que posee propiedades diferentes de las propiedades de cada uno de los elementos que existen en él. (Cap. 2, pág. 38)

conduction/conducción: transferencia de energía que ocurre cuando las moléculas chocan entre sí. (Cap. 14, pág. 405)

conic projection/proyección cónica: se usa para producir mapas de áreas pequeñas. Se hacen proyectando puntos y líneas desde un globo a un cono. (Cap. 5, pág. 131)

conservation/conservación: uso cuidadoso de los recursos, lo cual disminuye el daño al ambiente. (Cap. 19, pág. 544)

constellation/constelación: un grupo de estrellas que presenta una configuración que parece un animal, un personaje mitológico o cualquier otro objeto, del cual obtiene su nombre. (Cap. 24, pág. 672)

continental drift/deriva continental: teoría que dice que los continentes se han movido horizontalmente a sus posiciones actuales. (Cap. 11, pág. 294)

continental shelf/plataforma continental: extremo, inclinado paulatinamente, de un continente que se extiende bajo el océano. (Cap. 18, pág. 504)

continental slope/talud continental: extremo de la plataforma continental y más empinado y que esta que se extiende desde su borde hacia abajo del suelo oceánico. (Cap. 18, pág. 505)

contour interval/intervalo entre curvas de nivel: diferencia de elevación entre dos curvas de nivel consecutivas. (Cap. 5, pág. 132)

contour line/curva de nivel: línea de un mapa que conecta puntos de igual elevación. (Cap. 5, pág. 132)

control/control: estándar de comparación. (Cap. 1, pág. 17)

convection/convección: transferencia de calor por medio de un flujo de material calentado. (Cap. 14, pág. 405)

convection current/corriente de convección: ciclo completo de calentamiento, ascenso, enfriamiento y hundimiento. (Cap. 11, pág. 310)

convergent boundary/límite convergente: lugar donde chocan dos placas. (Cap. 11, pág. 307)

Corioles effect/efecto de Coriolis: efecto producido por el movimiento de rotación de la Tierra, que hace que el viento que sopla en dirección sur se desvíe hacia el oeste, en el hemisferio norte. (Cap. 14, pág. 411)

corona/corona: la capa más grande de la atmósfera solar, la cual se extiende millones de kilómetros en el espacio. (Cap. 24, pág. 679)

crater/cráter: depresión empinada y más o menos circular alrededor de la chimenea de un volcán. (Cap. 10, pág. 268)

creep/corrimiento: movimiento paulatino de material rocoso suelto debido a congelación, humedecimiento, secado, etc. (Cap. 7, pág. 175)

crest/cresta: el punto más alto de una ola. (Cap. 17, pág. 490)

crust/corteza: la capa más externa de la Tierra y separada del manto por la discontinuidad de Moho. (Cap. 9, pág. 249)

crystal/cristal: sólido cuyos átomos están ordenados en patrones repetitivos. (Cap. 3, pág. 63)

cyanobacteria/cianobacterias: bacterias azul verdosas que se piensa que son una de las primeras formas de vida sobre la Tierra. (Cap. 13, pág. 367)

deflation/deflacción: erosión causada cuando el viento sopla los sedimentos sueltos extrayendo pequeñas partículas tales como la arcilla, el cieno y la arena, y dejando atrás materiales más gruesos. (Cap. 7, pág. 188)

deforestation/desforestación: extracción masiva de árboles. (Cap. 16, pág. 470)

delta/delta: desembocadura en un océano, golfo o en un lago de los sedimentos arrastrados por un río. (Cap. 8, pág. 213)

density/densidad: medida de la masa de un objeto dividida entre su volumen. (Cap. 2, pág. 44)

density current/corriente de densidad: corriente que ocurre cuando el agua marina más densa se hunde bajo agua marina menos densa, mezclando así el agua oceánica de la superficie. (Cap. 17, pág. 487)

deposition/depositación: etapa final de un proceso de erosión, en la cual los agentes erosivos depositan los sedimentos y las rocas cuando disminuye su energía de movimiento. (Cap. 7, pág. 173)

desertification/desertificación: formación de desiertos. (Cap. 6, pág. 165)

dew point/punto de rocío: temperatura a la cual el aire está saturado y se condensa. (Cap. 15, pág. 424)

dike/dique: magma que ha sido apretujado formando un corte vertical, el cual atraviesa capas rocosas y se endurece. (Cap. 10, pág. 284)

divergent boundary/límite divergente: zona situada entre dos placas que se alejan una de la otra. (Cap. 11, pág. 305)

drainage basin/cuenca hidrográfica: extensión territorial de donde obtiene agua una corriente de agua. (Cap. 8, pág. 207)

E

Earth/la Tierra: el tercer planeta a partir del sol. (Cap. 23, pág. 648)

Earth science/ciencia terrestre: es el estudio de la Tierra y el espacio. (Cap. 1, pág. 8)

earthquake/terremoto: vibraciones producida por las rocas que se rompen. (Cap. 9, pág. 237)

El Niño/El Niño: evento climático que comienza en las aguas tropicales del Océano Pacífico y que inicia cambios en la atmósfera. (Cap. 16, pág. 463)

electromagnetic spectrum/espectro electromagnético: arreglo de radiación electromagnética de acuerdo con sus longitudes de onda. (Cap. 21, pág. 585)

electron/electrón: partícula cargada negativamente. (Cap. 2, pág. 35)

element/elemento: sustancia que contiene solo un tipo de átomo. (Cap. 2, pág. 33)

ellipse/elipse: curva cerrada alargada. (Cap. 22, pág. 617)

endangered/en peligro de extinción: una especie cuyo número de miembros vivos es muy pequeño. (Cap. 13, pág. 364)

epicenter/epicentro: punto en la superficie terrestre directamente encima del foco de un terremoto. (Cap. 9, pág. 244)

epoch/época: subdivisión de un período. (Cap. 13, pág. 359)

equator/ecuador: línea imaginaria que circunda la Tierra exactamente equidistante entre los Polos Norte y Sur. El ecuador separa la Tierra en dos mitades llamadas hemisferio norte y hemisferio sur. (Cap. 5, pág. 126)

equinox/equinoccio: momento del año cuando el sol está directamente encima del ecuador lo cual resulta en que los días son iguales a las noches. (Cap. 22, pág. 619)

era/era: subdivisión principal de la escala del tiempo geológico basada en diferencias de biotipos. (Cap. 13, pág. 357)

erosion/erosión: proceso que desgasta materiales en la superficie y los mueve de un lugar a otro. (Cap. 7, pág. 172)

extinct/extinto: especie que ya no tiene ningún miembro vivo. (Cap. 12, pág. 334)

extrusive/extrusiva: roca ígnea que se forma cuando la lava se enfría en la superficie terrestre. (Cap. 4, pág. 92)

fault/falla: superficie a lo largo de la cual se mueven las rocas al exceder su límite de elasticidad y romperse. (Cap. 9, pág. 236)

fault-block mountain/montaña de bloques de fallas: montaña formada por inmensos bloques rocosos inclinados que están separados de rocas circundantes por fallas. (Cap. 5, pág. 124)

first quarter/cuarto creciente: fase lunar cuando, desde la Tierra, se puede observar la mitad de su faz iluminada o un cuarto de la superficie lunar. (Cap. 22, pág. 624)

fission/fisión: es la separación de los núcleos de los átomos en elementos pesados, tales como el uranio. (Cap. 2, pág. 54)

floodplain/llanura aluvial: valle ancho y llano tallado por una corriente de agua serpenteante. (Cap. 8, pág. 209)

focus/foco: punto en el interior de la Tierra donde ocurre la liberación de energía de un terremoto. (Cap. 9, pág. 241)

fog/neblina: nube estrato que se forma cuando el aire se enfría a su punto de rocío y se condensa cerca del suelo. (Cap. 15, pág. 425)

folded mountain/montaña plegada: tipo de montaña que se ha formado al encorvarse y plegarse las capas rocosas cuando estas últimas han sido apretadas desde lados opuestos. (Cap. 5, pág. 123)

foliated/foliada: textura que presenta una roca metamórfica cuando sus granos minerales se aplanan y se alinean en bandas paralelas. (Cap. 4, pág. 99)

fossil/fósil: resto, impresión o huella de organismos que una vez estuvieron vivos, conservado en las rocas. (Cap. 12, pág. 327)

fracture/fractura: propiedad de un mineral que lo hace partirse en puntas ásperas o dentadas. (Cap. 3, pág. 70)

front/frente: límite entre dos masas de aire. (Cap. 15, pág. 433)

full moon/luna llena o plenilunio: fase durante la cual está iluminada la mitad de la superficie lunar que da a la Tierra. (Cap. 22, pág. 624)

galaxy/galaxia: grupo inmenso de estrellas, gas y polvo que se mantiene unido gracias a la gravedad. (Cap. 24, pág. 691)

gem/gema: piedra preciosa muy preciada debido a su rareza y belleza. (Cap. 3, pág. 74)

geologic time scale/escala del tiempo geológico: división de la historia de la Tierra en unidades más pequeñas. (Cap. 13, pág. 357)

geology/geología: es el estudio de la Tierra, su materia y los procesos que la forman y la cambian. (Cap. 1, pág. 8)

geothermal energy/energía geotérmica: tipo de energía térmica en las masas de magma dentro de la Tierra que se utiliza para generar electricidad. (Cap. 10, pág. 274)

geyser/géiser: fuente termal que hace erupción intermitentemente disparando agua y vapor en el aire. (Cap. 8, pág. 217)

giant/gigante roja: etapa en la formación de una estrella en la cual su núcleo se contrae y las temperaturas internas de la estrella aumentan, haciendo que sus capas externas se expandan. (Cap. 24, pág. 687)

glacier/glaciar: masa móvil de hielo y nieve. (Cap. 7, pág. 180)

global warming/calentamiento global: aumento de las temperaturas globales. (Cap. 16, pág. 467)

granitic/granítica: roca ígnea de color claro que posee una menor densidad que la roca basáltica. (Cap. 4, pág. 93)

gravity/gravedad: fuerza de atracción que existe entre todos los objetos. (Cap. 1, pág. 20)

Great Red Spot/la Gran Mancha Roja: espectacular tormenta de gas turbulento y de alta presión que se produce continuamente en Júpiter. (Cap. 23, pág. 654)

greenhouse effect/efecto de invernadero: calentamiento natural causado por gases en la atmósfera que atrapan el calor de la superficie terrestre. (Cap. 16, pág. 467)

groundwater/agua subterránea: agua que se filtra en el suelo y que se junta en los poros subterráneos. (Cap. 8, pág. 214)

gully erosion/erosión en barrancos: surco o zanja que se ensancha y profundiza formando un barranco. (Cap. 8, pág. 204)

habitat/hábitat: lugar donde los organismos viven, crecen e interactúan entre sí y con el ambiente. (Cap. 13, pág. 364)

half-life/media vida: el tiempo que se demora la mitad de los átomos de un isótopo para desintegrarse. (Cap. 12, pág. 345)

hardness/dureza: medida del grado de facilidad que posee un mineral para rayarse. (Cap. 3, pág. 68)

hazardous waste/desecho peligroso: desecho venenoso, cancerígeno o radiactivo. (Cap. 19, pág. 541)

hibernation/hibernación: período de inactividad durante el invierno. (Cap. 16, pág. 460)

horizon/horizonte: nombre que recibe cada capa del perfil del suelo. (Cap. 6, pág. 158)

hot dry rock (HDR)/roca seca caliente: tecnología que produce energía geotérmica sin necesitar la presencia del magma. (Cap. 10, pág. 275)

hot spot/punto cálido: área del manto que posee una mayor temperatura que otras áreas. (Cap. 10, pág. 273)

hot spring/aguas o baños termales: agua subterránea que se sale del suelo a una temperatura elevada. (Cap. 8, pág. 216)

humidity/humedad: cantidad de vapor de agua presente en el aire. (Cap. 15, pág. 423)

humus/humus: materia de color oscuro que se forma cuando se descompone la materia orgánica del suelo. (Cap. 6, pág. 157)

hurricane/huracán: sistema de baja presión, de gran alcance y turbulento, que se forma sobre los océanos tropicales. (Cap. 15, pág. 438)

hydrosphere/hidrosfera: toda el agua que ocurre en la superficie de la Tierra. (Cap. 14, pág. 406)

hypothesis/hipótesis: predicción acerca de un problema que puede probarse. (Cap. 1, pág. 16)

ice wedging/grietas debido al hielo: meteorización mecánica que se produce cuando el agua que se congela en las resquebrajaduras de las rocas se expande y la presión que se acumula rompe las rocas. (Cap. 6, pág. 150)

igneous rock/roca ígnea: roca que se forma cuando el material derretido de un volcán o de las profundidades de la Tierra se enfría. (Cap. 4, pág. 91)

impermeable/impermeable: que no se deja atravesar por el agua. (Cap. 8, pág. 215)

index fossil/fósil guía: proviene de especies que existieron abundantemente en la Tierra durante cortos períodos de tiempo y que se encontraban muy extendidas geográficamente. (Cap. 12, pág. 331)

inner core/núcleo interno: núcleo sólido, muy denso en el mismo centro de la Tierra compuesto principalmente de hierro y níquel. (Cap. 9, pág. 249)

inner planet/planeta interior: planeta sólido rocoso situado más cerca del sol; los planetas interiores son Mercurio, Venus, la Tierra y Marte. (Cap. 23, pág. 643)

International Date Line/Línea Internacional de cambio de fecha: es el meridiano de 180°, es decir, la línea de longitud en el lado opuesto de la Tierra a partir del Primer Meridiano donde las líneas de longitud este se encuentran con las líneas de longitud oeste. Este meridiano es la línea de transición para los días del calendario. (Cap. 5, pág. 129)

International System of Units (SI)/Sistema Internacional de Unidades (SI): versión moderna del sistema métrico. (Cap. 1, pág. 19)

intrusive/intrusiva: roca ígnea que se forma debajo de la superficie terrestre. (Cap. 4, pág. 92)

ion/ion: átomo cargado eléctricamente. (Cap. 2, pág. 40)

ionosphere/ionosfera: capa importante de la termosfera y que es una capa de partículas cargadas eléctricamente. (Cap. 14, pág. 395)

isobar/isobara: línea que conecta puntos de igual presión atmosférica. (Cap. 15, pág. 441)

isotherm/isoterma: línea que conecta puntos de igual temperatura. (Cap. 15, pág. 441)

isotope/isótopo: átomo de un mismo elemento que posee diferentes números de neutrones en su núcleo. (Cap. 2, pág. 36)

jet stream/corriente de chorro: banda estrecha de viento fuerte que sopla cerca de la parte superior de la troposfera. (Cap. 14, pág. 414)

Jupiter/Júpiter: el planeta más grande del sistema solar y está ubicado en quinto lugar a partir del sol. (Cap. 23, pág. 654)

land breeze/brisa terrestre: viento suave que sopla desde la tierra hacia el mar durante la noche, producido por corrientes de convección. (Cap. 14, pág. 415)

landfill/vertedero: área donde se depositan los desechos. (Cap. 19, pág. 541)

Landsat Satellite/satélite Landsat: satélite que detecta diferentes longitudes de onda de energía que se reflejan o emiten desde la superficie terrestre. (Cap. 5, pág. 138)

latitude/latitud: se refiere a la distancia, en grados, ya sea al norte o al sur del ecuador. (Cap. 5, pág. 127)

lava/lava: magma que llega a la superficie terrestre. (Cap. 4, pág. 91)

law/ley: regla de la naturaleza que describe la conducta de algo en ella. (Cap. 1, pág. 18)

leaching/lixiviación: extracción de materiales del suelo al ser disueltos en agua. (Cap. 6, pág. 159)

light pollution/contaminación por luz: brillo en el cielo provocado por las luces de las ciudades y el cual interfiere con la observación de las estrellas opacas. (Cap. 21, pág. 591)

light-year/año-luz: distancia que cual viaja la luz en un año. (Cap. 24, pág. 675)

lithosphere/litosfera: nombre que reciben la corteza y una parte del manto superior terrestres. (Cap. 11, pág. 304)

loess/loess: limo de grano fino sin estratificaciones. (Cap. 7, pág. 191)

longitude/longitud: se refiere a la distancia en grados al este o al oeste del primer meridiano. (Cap. 5, pág. 127)

longshore current/corriente costera: corriente que corre paralela a la costa. (Cap. 8, pág. 223)

lunar eclipse/eclipse lunar: ocurre cuando la sombra de la Tierra cae sobre la luna. (Cap. 22, pág. 627)

luster/lustre: describe la forma en que la luz se refleja desde la superficie de un mineral; el lustre puede ser metálico o no metálico. (Cap. 3, pág. 69)

magma/magma: materia rocosa derretida y de alta temperatura. (Cap. 3, pág. 64)

magnetometer/magnetómetro: instrumento sensible que registra datos magnéticos. (Cap. 11, pág. 300)

magnitude/magnitud: medida de la energía liberada en un terremoto. (Cap. 9, pág. 253)

main sequence/secuencia principal: banda diagonal de estrellas que corre desde la parte izquierda superior (la cual contiene las estrellas calientes, azules y brillantes) hasta la parte derecha inferior (la cual contiene las estrellas frías, rojas y tenues) del diagrama H-R. (Cap. 24, pág. 684)

mantle/manto: la capa más grande de la Tierra ubicada directamente encima del núcleo externo. (Cap. 9, pág. 249)

map legend/leyenda de un mapa: explica el significado de los símbolos de un mapa. (Cap. 5, pág. 134)

map scale/escala de un mapa: relación entre las distancias en el mapa y las distancias verdaderas en la superficie terrestre. (Cap. 5, pág. 135)

maria/maria: regiones oscuras y relativamente planas de la superficie lunar. (Cap. 22, pág. 628)

Mars/Marte: el cuarto planeta a partir del sol. (Cap. 23, pág. 648)

mascon/concentración de masa: concentración de masa ubicada debajo de las cuencas de impacto en la Luna. (Cap. 22, pág. 633)

mass/masa: medida de la cantidad de materia que posee un objeto. (Cap. 1, pág. 20)

mass movement/movimiento de masa: movimiento de materiales cuesta abajo debido a la gravedad. (Cap. 7, pág. 173)

mass number/número de masa: número igual al número de protones y neutrones que componen el núcleo de un átomo. (Cap. 2, pág. 36)

matter/materia: cualquier cosa que ocupa espacio y posee masa. (Cap. 2, pág. 32)

meander/meandro: curva que se forma en una corriente de agua cuando el agua que se mueve rápidamente erosiona el costado de la corriente donde ésta es más fuerte. (Cap. 8, pág. 208)

mechanical weathering/meteorización mecánica: proceso que parte las rocas pero sin cambiar su composición química. (Cap. 6, pág. 149)

Mercator projection/proyección de Mercator: tipo de mapa que muestra las formas correctas de los continentes pero sus superficies están distorsionadas. (Cap. 5, pág. 130)

Mercury/Mercurio: el planeta más cercano al sol. (Cap. 23, pág. 646)

Mesozoic Era/Era mesozoica: era de la vida media que comenzó hace cerca de 245 millones de años. (Cap. 13, pág. 374)

metamorphic rock/roca metamórfica: roca que ha cambiado debido a un aumento en la temperatura y en la presión o que sufre cambios en su composición. (Cap. 4, pág. 97)

meteor/meteoro: meteoroide que se quema en la atmósfera terrestre. (Cap. 23, pág. 664)

meteorite/meteorito: meteoroide que cae sobre la superficie terrestre. (Cap. 23, pág. 665)

meteorologist/meteorólogo: especialista que estudia el tiempo. (Cap. 15, pág. 440)

meteorology/meteorología: estudio del tiempo y las fuerzas y procesos que lo causan. (Cap. 1, pág. 8)

mid-ocean ridge/dorsal oceánica: lugar donde se forma nuevo suelo oceánico. (Cap. 18, pág. 506)

mineral/mineral: sólido inorgánico que ocurre de forma natural y que posee una estructura y una composición definidas. (Cap. 3, pág. 62)

mixture/mezcla: incorporación de varias sustancias entre sí pero en la cual cada una retiene sus propias propiedades. (Cap. 2, pág. 40)

Moho discontinuity/discontinuidad de Moho: límite entre la corteza y el manto. (Cap. 9, pág. 249)

mold/molde: hueco que se forma en roca que contiene fósiles de conchas cuando los poros rocosos permiten que el agua y el aire lleguen hasta la concha o su parte endurecida, permitiendo su desintegración. (Cap. 12, pág. 329)

molecule/molécula: combinación de átomos mediante el compartimiento de electrones en la porción más externa de sus nubes electrónicas. (Cap. 2, pág. 38)

moon phase/fase lunar: apariencia cambiante de la luna vista desde la Tierra. (Cap. 22, pág. 624)

moraine/morena frontal o terminal: tipo de depósito formado por la acumulación de tierra, piedras y otros materiales que arrastra un glaciar; a diferencia de la morena, la morena frontal o terminal no abarca una área muy amplia. (Cap. 7, pág. 184)

natural selection/selección natural: teoría que dice que los organismos que poseen rasgos adaptables a cierto entorno tienen una mejor probabilidad de sobrevivir y de reproducirse que aquellos organismos cuyos rasgos no son apropiados a ese ambiente. (Cap. 13, pág. 360)

nebula/nebulosa: nube extensa de gas y polvo que corresponde a la etapa inicial de formación de una estrella. (Cap. 24, pág. 686)

nekton/necton: animales que nadan libremente en lugar de estar a la deriva con las corrientes. (Cap. 18, pág. 513)

Neptune/Neptuno: planeta grande y gaseoso descubierto en 1846; octavo planeta a partir del sol. (Cap. 23, pág. 658)

neutron/neutrón: partícula que no posee ninguna carga eléctrica. (Cap. 2, pág. 35)

neutron star/estrella de neutrones: la etapa de una supernova cuando el núcleo denso y colapsado de la estrella se encoge hasta unos 10-15 km y solo pueden existir neutrones en él. (Cap. 24, pág. 688)

new moon/luna nueva: ocurre cuando la cara iluminada de la luna mira hacia el Sol y la cara oscura mira hacia la Tierra. (Cap. 22, pág. 624)

nonfoliated/no foliada: textura que presentan los granos minerales de las rocas metamórficas cuando cambian, crecen y se reordenan pero no ocurren bandas. (Cap. 4, pág. 99)

normal fault/falla normal: falla que se forma cuando las rocas sobre la superficie de la falla se mueven hacia abajo en relación con las rocas debajo de la superficie. (Cap. 9, pág. 238)

nuclear energy/energía nuclear: fuente energética alternativa producida a partir de reacciones atómicas. (Cap. 2, pág. 54)

nuclear waste/desechos nucleares: desperdicios que produce la energía nuclear. (Cap. 2, pág. 55)

observatory/observatorio: edificio que alberga la mayoría de los telescopios ópticos usados por astrónomos profesionales. (Cap. 21, pág. 586)

oceanography/oceanografía: es el estudio de los océanos terrestres. (Cap. 1, pág. 9)

Oort Cloud/Nube Oort: nube que, según el astrónomo holandés Jan Oort, está ubicada más allá de la órbita de Plutón y la cual rodea completamente el sistema solar; esta nube albergaría una gran colección de cometas. (Cap. 23, pág. 663)

orbit/órbita: trayectoria curva alrededor de la Tierra que efectúa un satélite. (Cap. 21, pág. 594)

ore/mena: mineral que contiene sustancias útiles que pueden minarse para obtener una ganancia. (Cap. 3, pág. 75)

organic evolution/evolución orgánica: cambio de los biotipos a través del tiempo. (Cap. 13, pág. 359)

outer core/núcleo externo: núcleo líquido ubicado directamente encima del núcleo interno sólido; también compuesto de hierro y níquel. (Cap. 9, pág. 249)

outer planet/ planeta exterior: planeta más alejado del sol; los planetas exteriores son Júpiter, Neptuno, Saturno, Uranio y Plutón; con excepción de Plutón, que es rocoso, los otros planetas exteriores están formados principalmente por elementos más livianos tales como hidrógeno y helio. (Cap. 23, pág. 643)

oxidation/oxidación: combinación con el oxígeno; ocurre cuando, por ejemplo, el hierro es expuesto al oxígeno y al agua. (Cap. 6, pág. 154)

ozone layer/capa de ozono: capa de la estratosfera con una alta concentración de ozono. (Cap. 14, pág. 400)

Pacific Ring of Fire/Cinturón de Fuego del Pacífico: región alrededor de la Placa del Pacífico donde son comunes los terremotos y los volcanes. (Cap. 10, pág. 272)

Paleozoic Era/Era paleozoica: era geológica que marca el momento cuando los organismos desarrollaron partes duras. (Cap. 13, pág. 367)

Pangaea/Pangaea: inmensa extensión territorial que, según Wegener, una vez conectó a todos los continentes y que se separó hace unos 200 millones de años. (Cap. 11, pág. 294)

parallax/paralaje: cambio aparente en la posición de un objeto cuando uno lo observa desde dos posiciones diferentes. (Cap. 24, pág. 675)

period/período: subdivisión de una era. (Cap. 13, pág. 357)

permeable/permeable: que se deja atravesar por el agua. (Cap. 8, pág. 215)

petrified remains/restos petrificados: restos duros y de consistencia parecida a las rocas, en los cuales algunos o todos los materiales originales han sido reemplazados por minerales. (Cap. 12, pág. 328)

pH scale/escala de pH: escala que describe la acidez o basicidad de una solución. (Cap. 20, pág. 558)

photochemical smog/smog fotoquímico: smog color café que se forma cuando la luz solar reacciona con los contaminantes del aire. (Cap. 20, pág. 556)

photosphere/fotosfera: capa más baja de la atmósfera del sol y desde la cual se emite la luz solar. (Cap. 24, pág. 679)

photosynthesis/fotosíntesis: proceso de elaboración de alimento. (Cap. 18, pág. 510)

physical property/propiedad física: propiedad que puede observarse sin cambiar una sustancia en una nueva sustancia. (Cap. 2, pág. 44)

plain/llanura: extensa superficie de terreno relativamente llano. (Cap. 5, pág. 120)

plankton/plancton: algas marinas y animales minúsculos a la deriva con las corrientes oceánicas. (Cap. 18, pág. 513)

plate/placa: sección de la corteza terrestre que se mueve alrededor del manto. (Cap. 11, pág. 304)

plate tectonics/tectónica de placas: teoría que dice que la corteza y el manto superior de la Tierra están separados en secciones. (Cap. 11, pág. 304)

plateau/meseta: parte llana y extensa situada en lugares elevados de terreno. (Cap. 5, pág. 122)

plucking/ablación: cuando un glaciar arranca o desgarra bloques de rocas mediante la acción del agrietamiento debido al hielo. (Cap. 7, pág. 182)

Pluto/Plutón: el planeta más pequeño del sistema solar. (Cap. 23, pág. 658)

polar zone/zona polar: zona que se extiende desde los polos a las latitudes 66° norte y sur. (Cap. 16, pág. 452)

pollutant/contaminante: sustancia que causa daño a los organismos al interferir con procesos bioquímicos. (Cap. 18, pág. 519)

pollution/contaminación: introducción, por parte de los seres humanos, de productos residuales peligrosos, químicos y sustancias en el ambiente. (Cap. 18, pág. 619)

population/población: número total de individuos de una especie en particular en un área particular. (Cap. 19, pág. 533)

population explosion/explosión demográfica: rápido crecimiento de la población humana. (Cap. 19, pág. 533)

Precambrian time/era precámbrica: la era geológica más larga en la historia de la Tierra. (Cap. 13, pág. 366)

precipitation/precipitación: agua que cae de las nubes sea cual sea su forma. (Cap. 15, pág. 430)

primary wave/onda primaria: onda que se mueve a través de la Tierra haciendo que las partículas en las rocas se muevan oscilatoriamente en la misma dirección de la onda. (Cap. 9, pág. 244)

prime meridian/primer meridiano: punto de referencia para medir los grados de longitud de este a oeste y representa longitud 0º. (Cap. 5, pág. 127)

principle of superposition/principio de sobreposición: principio que dice que en una capa rocosa inalterada, las rocas más antiguas se encuentran en las capas inferiores y las más recientes en las capas superiores. (Cap. 12, pág. 337)

Project Apollo/Proyecto Apolo: etapa final del programa americano de llegar a la luna. (Cap. 21, pág. 598)

Project Gemini/Proyecto Géminis: segunda etapa en la meta de llegar a la luna. (Cap. 21, pág. 598)

Project Mercury/Proyecto Mercurio: proyecto que inició el programa americano de llegar a la luna. (Cap. 21, pág. 597)

proton/protón: partícula que posee una carga eléctrica positiva. (Cap. 2, pág. 35)

radiation/radiación: transferencia de energía por medio de ondas electromagnéticas. (Cap. 14, pág. 403)

radio telescope/radiotelescopio: tipo de telescopio que se usa para estudiar ondas radiales que viajan a través del espacio. (Cap. 21, pág. 587)

radioactive decay/desintegración radiactiva: formación de un nuevo elemento al cambiar el número de protones que posee un átomo. (Cap. 12, pág. 344)

radiometric dating/datación radiométrica: proceso que se usa para calcular la edad absoluta de una roca al medir las cantidades de material original y de los productos de desintegración que hay en la roca, conociendo el período de media vida del material original. (Cap. 12, pág. 345)

recyclable/reciclable: que se puede procesar y volver a usar. (Cap. 19, pág. 546)

reef/arrecife: estructura rígida y resistente a las olas que se desarrolla a partir de los materiales óseos y del carbonato cálcico. (Cap. 18, pág. 518)

reflecting telescope/telescopio reflector: telescopio que usa un espejo como objetivo para enfocar la luz proveniente del objeto que se observa. (Cap. 21, pág. 586)

reforestation/reforestación: plantación de árboles. (Cap. 16, pág. 471)

refracting telescope/telescopio refractor: uno en el cual la luz proveniente de un objeto atraviesa una lenta convexa y es doblada para formar una imagen en el punto focal. (Cap. 21, pág. 586)

relative dating/datación relativa: se utiliza en geología para determinar el orden de los sucesos y la edad relativa de las rocas al examinar sus posiciones en una secuencia. (Cap. 12, pág. 338)

relative humidity/humedad relativa: medida de la cantidad de vapor de agua que el aire puede contener a una temperatura específica. (Cap. 15, pág. 424)

reptile/reptil: especie que evolucionó de los anfibios; a diferencia de estos, los reptiles no necesitan regresar al agua para reproducirse. (Cap. 13, pág. 370)

reverse fault/falla invertida: falla en que las rocas sobre la superficie son forzadas hacia arriba y sobre las rocas debajo de la superficie de la falla. (Cap. 9, pág. 238)

revolution/revolución: órbita anual de la Tierra alrededor del sol, la cual dura un año. (Cap. 22, pág. 617)

rill erosion / erosión en regueras: tipo de erosión que comienza cuando se forma un arroyuelo durante una lluvia intensa. Al correr el arroyuelo, este posee suficiente energía como para arrastrar consigo plantas y tierra. (Cap. 8, pág. 204)

Robinson projection/proyección de Robinson: mapa que muestra las formas correctas de los continentes y extensiones territoriales precisas. (Cap. 5, pág. 131)

rock/roca: mezcla de minerales, mineraloides, vidrio o de materia orgánica. (Cap. 4, pág. 87)

rock cycle/ciclo de las rocas: ciclo por el cual las rocas cambian debido a procesos tales como la meteorización, la erosión, la compactación, la cementación, la fusión y el enfriamiento. (Cap. 4, pág. 87)

rotation/rotación: movimiento de la Tierra alrededor de su eje, el cual causa el día y la noche. (Cap. 22, pág. 615)

runoff/agua de desagüe: agua que no se filtra en el suelo o que no se evapora y, por consiguiente, corre sobre la superficie de la Tierra. (Cap. 8, pág. 202)

Safe Drinking Water Act/Ley sobre la Seguridad del Agua Potable: ley promulgada en 1986 para asegurar que el agua potable del país es segura de beber. (Cap. 20, pág. 570)

salinity/salinidad: medida de la cantidad de sólidos disueltos en el agua marina. Por lo general, se mide en gramos de sal por kilogramo. (Cap. 17, pág. 480)

sanitary landfill/vertedero controlado: terraplén donde los depósitos de basura diarios se cubren con tierra para evitar que el viento se los lleve y disminuir también el olor producido por los desechos en descomposición. (Cap. 19, pág. 541)

satellite/satélite: cualquier objeto que gira alrededor de otro objeto. (Cap. 21, pág. 594)

saturated/saturado: se dice que el aire está saturado cuando contiene tanta humedad como le es posible, a una temperatura específica. (Cap. 15, pág. 424)

Saturn/Saturno: conocido como el planeta anular es el sexto planeta a partir del sol. (Cap. 23, pág. 656)

science/ciencia: proceso de observación y estudio de las cosas de nuestro mundo. (Cap. 1, pág. 6)

scientific method/método científico: serie de pasos planificados que usan los científicos para resolver problemas. (Cap. 1, pág. 16)

scrubber/depurador: dispositivo que rocía los gases de escape de las centrales eléctricas con compuestos básicos, aumentándoles el pH a un nivel seguro. (Cap. 20, pág. 563)

sea breeze/brisa marina: viento suave que sopla desde el mar hacia la tierra, producido por corrientes de convección durante el día. (Cap. 14, pág. 414)

seafloor spreading/expansión del suelo marino: teoría famosa y aceptada de Harry Hess que propuso que el material caliente y menos denso del manto es forzado hacia la superficie en una dorsal o cordillera medioceánica, donde se enfría formando el nuevo suelo marino. Este nuevo suelo marino es empujado en ambas direcciones por el nuevo magma. (Cap. 11, pág. 299)

season/estación: período climático de corto plazo causado por una variación regular en la luz del día, la temperatura y los patrones climáticos. (Cap. 16, pág. 462)

secondary wave/onda secundaria: onda que se mueve a través de la Tierra haciendo que las partículas en las rocas se muevan en un ángulo recto a la dirección de la onda. (Cap. 9, pág. 244)

sediment/sedimento: material suelto tal como los fragmentos rocosos, granos minerales y trocitos de restos vegetales y animales, que han sido transportados. (Cap. 4, pág. 101)

sedimentary rock/roca sedimentaria: roca que se forma cuando los sedimentos son presionados o cementados o cuando estos se precipitan de una solución. (Cap. 4, pág. 101)

seismic wave/onda sísmica: ondas generadas por un terremoto. (Cap. 9, pág. 241)

seismograph/sismógrafo: instrumento que mide las ondas primarias, secundarias y de superficie que producen los terremotos. (Cap. 9, pág. 252)

seismologist/sismólogo: científico que estudia los terremotos y las ondas sísmicas. (Cap. 9, pág. 252)

sheet erosion/erosión laminar o en capas: tipo de erosión que ocurre cuando el agua de lluvia corre cuesta abajo arrastrando consigo sedimentos. (Cap. 8, pág. 205)

shield volcano/volcán de escudo: volcán amplio con suaves pendientes formado por la acumulación de capas llanas de lava basáltica. (Cap. 10, pág. 279)

silicate/silicato: mineral que contiene silicio y oxígeno y, por lo general, uno o más elementos. (Cap. 3, pág. 66)

sill/intrusión: magma que ha sido apretujado formando una resquebrajadura horizontal entre capas rocosas y que se endurece. (Cap. 10, pág. 284)

slump/derrumbe: movimiento de masa que ocurre cuando materiales sueltos o capas rocosas se deslizan por una cuesta. (Cap. 7, pág. 174)

soil/suelo: mezcla de roca meteorizada, materia orgánica, fragmentos minerales, agua y aire. (Cap. 6, pág. 156)

soil profile/perfil del suelo: capas diferentes de suelo. (Cap. 6, pág. 158)

solar eclipse/eclipse solar: ocurre cuando la luna se mueve directamente entre el sol y la Tierra y proyecta una sombra sobre parte de la Tierra. (Cap. 22, pág. 626)

solar system/sistema solar: grupo de nueve planetas y muchos objetos más pequeños que giran alrededor del sol. (Cap. 23, pág. 641)

solstice/solsticio: tiempo en que el sol alcanza su mayor distancia al norte o al sur del ecuador. (Cap. 22, pág. 622)

solution/solución: mezcla en la cual una sustancia se disuelve en otra, como por ejemplo el agua salada. (Cap. 2, pág. 41)

sonar/sonar: equipo que usa las ondas sonoras para detectar estructuras en el fondo oceánico. (Cap. 5, pág. 139)

space probe/sonda espacial: instrumento que reúne información y que la envía de vuelta a la Tierra. (Cap. 21, pág. 595)

space shuttle/transbordador espacial: nave espacial que transporta a astronautas, satélites y otros materiales hacia y desde el espacio y la cual se puede volver a usar. (Cap. 21, pág. 604)

space station/estación espacial: estación que posee viviendas, trabajo y espacio para trabajar, y todo el equipo y sistemas auxiliares que necesitan los seres humanos para vivir y trabajar en el espacio. (Cap. 21, pág. 605)

species/especie: grupo de organismos que se reproducen normalmente solo entre ellos mismos. (Cap. 13, pág. 359)

sphere/esfera: objeto redondo tridimensional cuya superficie en cualquiera de sus puntos está a la misma distancia de su centro. (Cap. 22, pág. 614)

spring/manantial: agua que aflora en lugares donde la superficie terrestre se junta con la capa freática. (Cap. 8, pág. 216)

station model/código meteorológico: información meteorológica recogida por el Servicio Meteorológico y registrada en mapas con una combinación de símbolos. (Cap. 15, pág. 441)

streak/veta: es el color del mineral al partirlo y pulverizarlo. (Cap. 3, pág. 69)

strike-slip fault/falla transformante: falla en la cual las rocas en cualquiera de los dos lados de la superficie de la falla se alejan unas de otras sin mucho movimiento ascendente o descendente. (Cap. 9, pág. 238)

subduction zone/zona de subducción: área donde una placa oceánica desciende hacia el manto superior. (Cap. 11, pág. 307)

sulfurous smog/smog sulfuroso: tipo de smog que se forma cuando los combustibles fósiles se queman en centrales eléctricas y en las calderas caseras. (Cap. 20, pág. 557)

sunspot/mancha solar: área de la superficie solar que se ve oscura porque es más fría que las áreas que la rodean. (Cap. 24, pág. 680)

supergiant/supergigante: etapa en la formación de una estrella en la cual se forman elementos cada vez más pesados por medio de la fusión, haciendo que a la larga, se forme hierro en su núcleo. (Cap. 24, pág. 688)

surface current/corriente de superficie: corriente que mueve el agua horizontalmente paralela a la superficie terrestre. Estas corrientes son accionadas por el viento. (Cap. 17, pág. 484)

surface wave/onda de superficie: onda que viaja hacia afuera del epicentro; generada por la energía que llega hasta la superficie terrestre y que se mueve al darles un movimiento elíptico a las partículas como también un movimiento oscilatorio. (Cap. 9, pág. 244)

technology/tecnología: es el uso de los descubrimientos científicos. (Cap. 1, pág. 10)

temperate zone/zona templada: zona ubicada entre los trópicos y las zonas polares. (Cap. 16, pág. 452)

tephra/tefrita: material volcánico de diferentes tamaños causado por el enfriamiento y endurecimiento de la lava. (Cap. 10, pág. 279)

theory/teoría: explicación respaldada por resultados obtenidos de pruebas o experimentos repetidos. (Cap. 1, pág. 18)

third quarter/cuarto menguante: cuando se ve solo la mitad de la faz iluminada de la luna. (Cap. 22, pág. 625)

tidal range/alcance de la marea: la diferencia entre el nivel del océano durante la marea alta y la marea baja. (Cap. 17, pág. 492)

tide/marea: ascenso y descenso del nivel del mar. (Cap. 17, pág. 492)

till/morena: acumulación no estratificada de pedrones, arena. arcilla y cieno arrastrados por un glaciar. (Cap. 7, pág. 184)

titanium/titanio: metal liviano y duradero que se deriva de minerales tales como la ilmenita y el rutilo. (Cap. 3, pág. 78)

Topex-Poseidon Satellite/Satélite Topex-Poseidon: (Topex significa experimento topográfico) satélite que usa un radar para calcular la distancia que hay hasta la superficie del océano. (Cap. 5, pág. 138)

topographic map/mapa topográfico: muestra los cambios en elevación de la superficie terrestre. (Cap. 5, pág. 132)

tornado/tornado: viento violento y arremolinado que se mueve sobre una estrecha trayectoria sobre la tierra, por lo general en dirección suroeste-noroeste. (Cap. 15, pág. 436)

transform fault/falla de transformación: ocurre cuando dos placas se deslizan y se alejan una de la otra moviéndose ya sea en direcciones opuestas o en la misma dirección a distinta velocidad. (Cap. 11, pág. 309)

trench/fosa submarina: depresión larga, estrecha y con laderas empinadas en el suelo oceánico donde una placa de la corteza terrestre es forzada a penetrar debajo de otra. (Cap. 18, pág. 507)

tropics/trópicos: la región entre las latitudes 23° norte y 23° sur. (Cap. 16, pág. 452)

troposphere/troposfera: la capa de la atmósfera terrestre que se encuentra más cerca del suelo; contiene el 75 por ciento de los gases atmosféricos como también polvo, hielo y agua líquida. (Cap. 14, pág. 394)

trough/seno: el punto más bajo de una ola. (Cap. 17, pág. 490)

tsunamis/tsunamis: ondas marinas sísmicas generadas por terremotos. (Cap. 9, pág. 255)

ultraviolet radiation/radiación ultravioleta: uno de los numerosos tipos de energía que llega a la Tierra proveniente del sol. (Cap. 14, pág. 400)

unconformities/discordancias: brechas entre las capas rocosas. (Cap. 12, pág. 339)

uniformitarianism/uniformitarianismo: principio que dice que los procesos terrestres que ocurren actualmente son similares a los que ocurrieron en el pasado. (Cap. 12, pág. 347)

upwarped mountain/montaña plegada anticlinal: montaña que se formó cuando la corteza terrestre fue empujada hacia arriba por fuerzas del interior de la Tierra. (Cap. 5, pág. 124)

upwelling/corriente de aguas resurgentes: circulación en el océano que lleva agua profunda y fría hacia la superficie. (Cap. 17, pág. 487)

Uranus/Uranio: el séptimo planeta a partir del sol, el cual no fue descubierto hasta 1781. (Cap. 23, pág. 657)

variable/variable: factor que se cambia en un experimento. (Cap. 1, pág. 17)

vent/chimenea: abertura por la cual fluye el magma que llega a la superficie terrestre. (Cap. 10, pág. 268)

Venus/Venus: el segundo planeta a partir del sol. (Cap. 23, pág. 647)

volcanic mountain/montaña volcánica: montaña que se forma cuando el material derretido mana hacia la superficie terrestre a través de un área debilitada de la corteza y forma una estructura en forma de cono. (Cap. 5, pág. 125)

volcanic neck/cuello volcánico: núcleo ígneo sólido que queda después de que el cono de un volcán se erosiona cuando un volcán deja de hacer erupción. (Cap. 10, pág. 284)

volcano/volcán: chimenea en la superficie que forma una montaña cuando se arrojan y se acumulan capas de lava y cenizas volcánicas desde el interior de la Tierra. (Cap. 10, pág. 266)

waning/octante menguante: cuando la cantidad de la faz iluminada de la luna comienza a disminuir. (Cap. 22, pág. 625)

waste coal/carbón residual: subproducto de la minería del carbón; carbón de mala calidad. (Cap. 4, pág. 108)

water cycle/ciclo del agua: el constante movimiento del agua entre la atmósfera y la Tierra. (Cap. 14, pág. 406)

water table/nivel hidrostático o capa freática: superficie superior de la zona de saturación. (Cap. 8, pág. 215)

wave/ola: movimiento en el cual el agua sube y baja alternadamente a medida que la energía pasa por ella. (Cap. 17, pág. 490)

waxing/octante creciente: cuando se hace cada vez más visible la cara iluminada de la luna. (Cap. 22, pág. 624)

weather/tiempo: término que se refiere al estado actual de la atmósfera y que describe sus condiciones actuales. (Cap. 15, pág. 422)

weathering/meteorización: proceso que parte las rocas en fragmentos más y más pequeños. (Cap. 6, pág. 148)

weight/peso: medida de la fuerza gravitacional ejercida sobre una masa. (Cap. 1, pág. 20)

white dwarf/enana blanca: etapa en la formación de una estrella en la cual su núcleo se contrae bajo la fuerza de la gravedad al agotar su combustible y las capas externas de la estrella se escapan hacia el espacio. (Cap. 24, pág. 688)

zone of saturation/zona de saturación: área donde todos los poros de una roca permeable están llenos de agua. (Cap. 8, pág. 215)

Index

The Index for *Glencoe Earth Science* will help you locate major topics in the book quickly and easily. Each entry in the Index is followed by the numbers of the pages on which the entry is discussed. A page number given in **boldface type** indicates the page on which that entry is defined. A page number given in *italic type* indicates a page on which the entry is used in an illustration or photograph. The abbreviation *act.* indicates a page on which the entry is used in an activity.

D

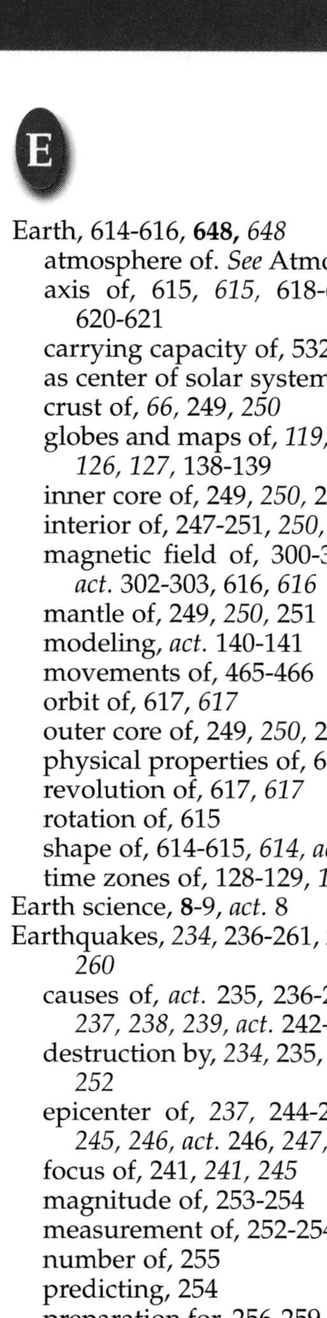

E

F

M

N

O

T

Art Credits

29, Preface, Inc.; **32,** John Edwards; **35,** Kristen Kest; **35, 36,** Jim Shough; **37,** Preface, Inc.; **39, 46, 47,** Jim Shough; **59,** Preface, Inc.; **64,** Tom Kennedy; **66,** Dartmouth Publishing, Inc.; **83,** Preface, Inc.; **87,** Dartmouth Publishing, Inc.; **90-1, 97, 102,** John Edwards; **113,** Preface, Inc.; **120,** John Edwards; **121,** Tom Kennedy; **123, 124, 125,** John Edwards; **126, 127,** Precision Graphics; **129, 130, 131,** Dartmouth Publishing, Inc.; **134,** Preface, Inc.; **134,** John Edwards; **135,** USGS; **138,** John Edwards; **139,** Edwin Huff; **140,** USGS; **145,** Preface, Inc.; **148,** Tom Kennedy; **149,** John Edwards; **150,** Rolin Graphics; **156-7, 158, 162,** John Edwards; **163,** Dartmouth Publishing, Inc.; **169,** Preface Inc.; **174, 175,** John Edwards; **180,** Tom Kennedy; **182-3, 184,** John Edwards; **188,** Preface, Inc.; **195,** Dartmouth Publishing, Inc.; **199,** Tom Kennedy; **199,** Preface, Inc.; **205,** Edwin Huff; **206,** Thomas Gagliano; **208-9,** John Edwards; **214,** Rolin Graphics; **215,** Leon Bishop; **216,** Thomas Gagliano; **217,** Leon Bishop; **222,** Chris Forsey/Morgan-Cain and Associates; **225,** Rolin Graphics; **225, 229,** Preface, Inc.; **229, 237,** Tom Kennedy; **238, 239, 244-5,** Chris Forsey/Morgan-Cain and Associates; **246,** Preface, Inc.; **247,** Tom Kennedy; **250,** Edwin Huff; **250-1,** Gary Hinks/Publisher's Graphics; **256,** Deborah Morse/Morgan-Cain and Associates; **256,** George Bucktell; **257,** Tom Kennedy; **259,** George Bucktell; **260,** Dartmouth Publishing, Inc.; **263,** Preface, Inc.; **268,** Tom Kennedy; **271,** Preface, Inc.; **272,** John Edwards; **272,** Gary Hinks/Publisher's Graphics; **274,** John Edwards; **275,** Dartmouth Publishing, Inc.; **278, 279,** Gary Hinks/Publisher's Graphics; **282,** Preface, Inc.; **283,** Gary Hinks/Publisher's Graphics; **287,** Preface, Inc.; **286,** Gary Hinks/ Publisher's Graphics; **291,** Preface, Inc.; **294-5,** Chris Forsey/Morgan-Cain and Associates; **295,** John Edwards; **297,** Precision Graphics; **298,** John Edwards; **301, 304,** Leon Bishop; **305,** Tom Kennedy; **308-9,** Gary Hinks/Publisher's Graphics; **311,** David De Gasperis; **311,** Edwin Huff; **312, 315,** John Edwards; **319,** Preface, Inc.; **326,** John Edwards; **329,** Rolin Graphics; **331,** Preface, Inc.; **331,** Jim Shough; **333,** Leon Bishop; **334, 338-9, 340,** John Edwards; **341, 342,** Dartmouth Publishing; **343, 345,** Tom Kennedy; **346,** Dartmouth Publishing, Inc.; **353,** John Edwards; **356-7,** Laurie O'Keefe; **358,** Dartmouth Publishing, Inc.; **362,** Laurie O'Keefe; **363,** Thomas Gagliano; **363,** Rolin Graphics; **366,** Barbara Hoopes; **367, 371,** Laurie O'Keefe; **371,** Dartmouth Publishing, Inc.; **374,** Precision Graphics; **376,** Laurie O'Keefe; **376, 377, 379,** Sarah Woodward; **381,** Felipe Passalacqua; **385,** Tom Kennedy; **392,** Jim Shough; **394,** Tom Kennedy; **395,** Edwin Huff; **397,** John Edwards; **398,** Tom Kennedy; **401,** Thomas Gagliano; **402, 404, 405,** John Edwards; **408,** Leon Bishop; **410,** Laurie O'Keefe; **411,** Dartmouth Publishing, Inc.; **411,** Edwin Huff; **412,** Bruce Sanders; **415, 416, 418,** John Edwards; **419,** Preface, Inc.; **423,** Dartmouth Publishing, Inc.; **425, 426, 428,** Tom Kennedy; **427,** Precision Graphics, Inc.; **432,** Tom Kennedy; **434,** John Edwards; **436,** Barbara Hoopes; **438,** John Edwards; **440,** Dartmouth Publishing, Inc.; **440,** Tom Kennedy; **443,** Ortelius Design; **449,** Preface, Inc.; **449,** Jim Shough; **452,** Leon Bishop; **453,** Tom Kennedy; **455,** John Edwards; **459,** Dartmouth Publishing, Inc.; **460,** Kristen Kest; **462,** John Edwards; **464,** Rolin Graphics; **466, 467,** John Edwards; **474,** Rolin Graphics; **475,** Preface, Inc.; **479,** John Edwards; **481,** Dartmouth Publishing, Inc.; **484,** Precision Graphics; **487, 490, 491, 492, 495, 496,** John Edwards; **498,** Dartmouth Publishing, Inc.; **506, 507,** John Edwards; **511,** Laurie O'Keefe; **519,** John Edwards; **533, 536,** Dartmouth Publishing, Inc.; **540,** Tom Kennedy; **541,** John Edwards; **542,** Dartmouth Publishing, Inc.; **546,** Tom Kennedy; **553,** Dartmouth Publishing, Inc.; **557,** Thomas Gagliano; **559,** John Edwards; **560,** Thomas Gagliano; **562,** Tom Kennedy; **563, 569,** John Edwards; **572,** Thomas Gagliano; **574,** Dartmouth Publishing, Inc.; **574, 577,** Preface, Inc.; **584-5,** Dartmouth Publishing, Inc.; **586,** George Bucktell; **594,** John Edwards; **595,** Mike Eddy; **596-7,** John Edwards; **614,** Thomas Gagliano; **615,** John Edwards; **616,** Bruce Sanders; **617, 619, 623,** John Edwards; **624,** Tim Hayes; **625, 626, 627, 630-1,** John Edwards; **637,** Preface, Inc.; **640-1, 642-3,** John Edwards; **645,** Dartmouth Publishing, Inc.; **648, 656, 657,** John Edwards; **662,** Mike Eddy; **662, 664, 666,** John Edwards; **672,** Precision Graphics; **673, 675, 679, 684, 685, 686, 687, 691, 693,** John Edwards; **694-5,** Bruce Sanders; **695,** Jim Shough; **696,** John Edwards; **712-3,** Glencoe; **714-5, 716-7,** Ortelius Design; **720-1,** Mark Lakin

Photo Credits

Cover, Jack Dykinga; **Connect to Chemistry** (flask)K&C Fischer; **vi** (t)Jack Dykinga; (l)Mark Burnett; **vii** (t)Doug Martin, (b)Peter French/DRK Photo; **viii** (t)Wolfgang Kaehler, (c)Jerald P. Fish/Tony Stone Images, (b)Jeff Gnass; **ix** (t)Brad Lewis/Gamma-Liaison, (b)file photo; **x** (t)Phil Degginger/Color-Pic, (b)John Cancalosi/Peter Arnold, Inc.; **xi** Michael Melford/The Image Bank; **xii** Michio Hosino/Minden Pictures; **xiii** Norbert Wu; **xiv xv** NASA; **xvi** Matt Meadows; **xvii** (t)KS Studios, (b)Matt Meadows; **xviii** Animals Animals/Patti Murray; **xix** Doug Martin; **xx** (t)Aaron Haupt, (b)Doug Martin; **xxi** (t)Matt Meadows, (b)Scripps Institution of Oceanography; **xxii** Matt Meadows; **xxiii** (t)Chip Clark, (bl)Bob Kalmbach/University of Michigan, (bc)Norbert Wu, (br)courtesy Leigh Standingbear; **xxiv** (t)courtesy A. Kahn, (others)NASA; **xxv** (t)NASA, (b)Johnny Johnson/Tony Stone Images; **2-3** Bruce Iverson; **3** NASA; **4-5** John Barstow; **5** StudiOhio; **6** (l)Jack Dykinga, (c)Steve Kaufman/DRK Photo, (r)NASA; **7** (tl)Hickson-Bender Photography, (tr)NASA/Photri, (bl)Rod Planck/Tony Stone Images, (br)John Cancalosi/Peter Arnold, Inc.; **8** Jean Miele/The Stock Market; **9** Kitt Peak National Observatory; **10** Telegraph Colour Library/FPG International; **11** NASA; **12** (l)file photo, (r)Bob Daemmrich/Stock Boston; **13** Scripps Institution of Oceanography; **14** UPI/Bettmann; **17** Peter Pearson/Tony Stone Images; **18** Richard Hutchings; **19** (clockwise from top) Chip Clark, Rick Weber, Chip Clark, Aaron Haupt, National Institute of Standards and Technology; **21** (t)StudiOhio, (bl)Ramey/Stock Boston, (br)Chip Clark; **22 23** StudiOhio; **24** (t)Jamie Mosberg/National Geographic Society, (b)Matt Meadows; **25** Doug Martin; **26** Jack Dykinga; **29** Latent Image; **30-31** Francois Gohier/Photo Researchers; **31 33** Matt Meadows; **34** (tl)Dr. E.R. Degginger/Color-Pic, (tlc)Doug Martin, (trc) Martin Land/Science Photo Library/Photo Researchers, (tr)Doug Martin, (bl)Dale O'Dell/The Stock Market, (blc) Doug Martin, (brc)David Stoecklein/The Stock Market, (br)Bruce Iverson; **34-35** Matt Meadows; **35** R.C. Hermes/Photo Researchers; **36** StudiOhio; **37** Jeffery Newbury/(1995) The Walt Disney Co. Reprinted with permission of *Discover* magazine; **38** Inga Spence/Tom Stack & Associates; **39** courtesy IBM Corporation; **40** Matt Meadows; **41** Doug Martin; **42** Carol Sailors; **43 44** Matt Meadows; **45** Vanessa Vick/Photo Researchers; **46** (t)Craig Kramer, (b)Ron Whitby/FPG International; **47** (t)Peter Gridley/FPG International, (b)Warren Faidley/Weatherstock; **48** Bob Kalmbach/University of Michigan; **49** Cameramann/The Image Works; **50** David R. Frazier Photolibrary; **51** NASA; **52 53** Matt Meadows; **54** Dr. E.R. Degginger/Color-Pic; **55 56** DOE; **58** Hickson-Bender; **60** Matt Meadows; **60-61** Jack Dykinga; **62** KS Studios; **63 64** Doug Martin; **65** Matt Meadows; **67 68 69** Doug Martin; **70** (t)Doug Martin, (b)Mark Burnett; **71** Tim Courlas; **72** Matt Meadows; **73** (tr)Chas Krider, (others)Doug Martin; **74** (opal) D.C.H. Plowes, (turquoise)Breck P. Kent/JLM Visuals, (others)Doug Martin; **75** (t)David Austen/Tony Stone Images, (bl)StudiOhio, (br)Doug Martin; **76** Dr. Jeremy Burgess/Science Photo Library/Photo Researchers; **77** Calvin Larsen/Photo Researchers; **78** Jim Cummins/FPG International; **79** (t)Chris Sorensen, (b)P. Peterson/Custom Medical Stock Photo; **80** courtesy A. Kahn; **82** Francois Gohier/Photo Researchers; **83** KS Studios; **84-85** Jack Dykinga; **85** Matt Meadows; **86** (tl)Dr. E.R. Degginger/Color-Pic, (tlc)Richard P. Jacobs/JLM Visuals, (trc)Doug Martin, (tr)Dr. E.R. Degginger/Color-Pic, (b)Doug Martin; **87** (t)Brent Turner/BLT Productions, (b)Doug Martin; **88** (l)Bob Daemmrich, (r)Doug Martin; **89** Jim Brandenburg/Minden Pictures; **90** (t)William E. Ferguson, (b)Doug Martin; **91** Dr. E.R. Degginger/Color-Pic; **92** (rocks)Doug Martin, (other)Douglas Peebles/Westlight; **93** Art Wolfe; **94** KS Studios; **96** Alberto Garcia/Saba Press Photos; **97** (l)Doug Martin, (r)Doug Martin; **98** Mark Burnett; **99** (l)Dr. E.R. Degginger/Color-Pic, (r)William E. Ferguson; **100** Nimatallah/Art Resource, NY; **101** (l)Aaron Haupt, (lc)Doug Martin, (rc)Brent Turner/BLT Productions, (r)Richard P. Jacobs/JLM Visuals; **103** G.R. Roberts; **104** (l)Cliff Leight, (r)StudiOhio; **105** Jack Dykinga; **106** James Westwater; **107** (l)Lynn McLaren/Photo Researchers, (r)StudiOhio; **108** Alan Oddie/PhotoEdit; **109** Phil Degginger/

Photo Credits

Color-Pic; **112** Ward's Natural Science Establishment; **114** (t)David M. Dennis, (c)Brent Turner/BLT Productions, (b)Chas Krider; **115** (t)Doug Martin, (b)Chip Clark; **116-117** Jack Dykinga; **117** Rich Buzzelli/Tom Stack & Associates; **118-119** World Perspectives; **119** StudiOhio; **122** Tom Till; **123** Peter French/DRK Photo; **124** (t)Phil Lauro/Profiles West, (b)Larry Ulrich/Tony Stone Images **125** Paul Chesley/Tony Stone Images; **128** (l)Matt Meadows, (r)Doug Martin; **132** Jeff Gnass; **133** Geographix, Inc.; **136** Tom Carroll/FPG International; **137** StudiOhio; **141** Matt Meadows; **142** SuperStock; **144-145** Galen Rowell/Mountain Light Photography; **146-147** Wolfgang Kaehler; **147** Matt Meadows; **148** Mark Burnett; **149** StudiOhio; **150** Bud Fowle; **151** Harris Photographic/Tom Stack & Associates; **152 153** Matt Meadows; **154** Carr Clifton/Minden Pictures; **155** (l)Henley & Savage/The Stock Market, (r)Tom Stack/Tom Stack & Associates; **156-157** Matt Meadows; **158** Rich Buzzelli/Tom Stack & Associates; **159** William E. Ferguson; **160** Doug Martin; **161** Marty Cordano/DRK Photo; **164** Lee Battaglia/Colorific!; **165** Larry Lefever/from Grant Heilman; 166 courtesy Susan Colclazer; **168** M. Wendler/Photo Researchers; **170-171** Jack Dykinga; **171** StudiOhio; **172** Rod Planck/Photo Researchers; **173** Jack Stein Grove/Profiles West; **174** Thomas G. Rampton/from Grant Heilman; **176** Stephen C. Porter; **177** Giboux/Gamma-Liaison; **178** Wolfgang Kaehler; **179** Reuters/Bettmann; **181** (l)Wolfgang Kaehler, (r)Cliff Leight; **182** (l)Phil Schermeister/National Geographic Society, (r)Mark Burnett; **183** Cliff Leight; **185** Earth Scenes/John Eastcott & Yva Momatiuk; **186** Matt Meadows; **187** (l)Tom Bean/DRK Photo, (r)Tom Till; **188** Debbie Dean; **189** (tl)Floyd Holdman/The Stock Solution, (tr)Runk/Schoenberger/from Grant Heilman, (b) Earth Scenes/M.J. Coe; **190** Mark Boulton/Photo Researchers; **191** (t)Wolfgang Kaehler, (b)Michael Collier; **192 193** Matt Meadows; **194** Jeff Gnass; **196** Bettmann Archive; **200-201** Jack Dykinga; 201 StudiOhio; **202** (l)T.A. Wiewandt/DRK Photo, (r)Tom Bean; **203** Jim Richardson/Westlight; **204** Holt Studios/Photo Researchers; **205** from Grant Heilman; **206** Earth Scenes/Leen van der Slik; **207** Jerald P. Fish/Tony Stone Images; **208** (t)Elaine Shay, (b)Tom Bean/DRK Photo; **209** Ray Fairbanks/Photo Researchers; **210** Matt Meadows; **212** Comstock; **213** (l)John Shelton, (r)NASA/TSADO/Tom Stack & Associates; **214** Matt Meadows; **217** Susan G. Drinker/The Stock Market; **218** Dave Harris/Tom Stack & Associates; **219** Wolfgang Kaehler; **220** StudiOhio; **221** Jack Dykinga; **222** Darrell Gulin/DRK Photo; **223** Charles Krebs/The Stock Market; **224** (t)Matt Meadows, (b)USDA-ASCS; **226** Lynn M. Stone; **229** Jack Dykinga; **230 231** KS Studios; **232-233** Art Wolfe; **233** Dennis Oda/Photo Resource Hawaii; **234-235** Yoshiaki Nagashima PPS/Black Star; **235** StudiOhio; **236** Matt Meadows; **238** Cliff Leight; **239** (t)Tom Till/DRK Photo, (b)David Parker/Photo Researchers; **240** Jonathan Nourok/PhotoEdit; **241** Ron Haviv/Saba Press Photos; **247** Russell D. Curtis/Photo Researchers; **249** Doug Martin; **252** Les Stone/Sygma; **254** H & M Tavetian; **256** Doug Martin; **258** James L. Stanfield/National Geographic Society; **259** Francois Gohier/Photo Researchers; **262** Archive Photos; **264-265** Alberto Garcia/Saba Press Photos; **265** Doug Martin; **266** Brad Lewis/Gamma-Liaison; **267** Emory Kristof/National Geographic Society; **269** Ministry of Construction for Osumi/Fujifotos/The Image Works; **270** Matt Meadows; **273** Terry Duennebier/ SOEST/University of Hawaii; **276 277** Gary Rosenquist; **278** (t)Earth Scenes/Breck P. Kent, (b)Tom Bean/DRK Photo; **279** (t)Robert Madden/National Geographic Society, (b)Earth Scenes/Michael Fogden; **280** Doug Martin; **281** Patrick Aventurier/Gamma-Liaison; **284** Marc Muench; **285** (tl)David Muench, (tr)David Hosking/Photo Researchers, (b)Steve Kaufman/DRK Photo; **286** Greg Vaughn; **287** Doug Martin; **288** Art Resource, NY; **291** Dave B. Fleetham/Pacific Stock; **292-293** Tom Van Sant/Geosphere Project/Photo Researchers; **293** Matt Meadows; **296** David M. Dennis; **299** Scripps Institution of Oceanography; **300** Morin, Tanimoto, Yuen, Zhang. Graphic provided by Paul Morin; **302** Richard Hutchings; **307** file photo; **310** Tom Bean/DRK Photo; **312** Altitude/Peter Arnold, Inc.; **313** CNES/Photo Researchers; **314** Galen Rowell/Mountain Light Photography; **316** Hilsman Jackson/SMU; **318** John Warden/Tony Stone Images; **320** Doug

Photo Credits

Martin; **321** (t)Al Grillo/Alaska Stock Images, (b)Mark Burnett; **322-323** David Muench; **323** Chip Clark; **324-325** Vaughan Fleming/Photo Researchers; **325** StudiOhio; **326** Louis Psihoyos/Matrix; **327** Phil Degginger/Color-Pic; **328** (t)Francois Gohier/Photo Researchers, (b)Sinclair Stammers/Photo Researchers; **329** Dr. E.R. Degginger/Color-Pic; **330** (t)Vaughan Fleming/Photo Researchers, (cb)Louis Psihoyos/Matrix; **332** Richard T. Nowitz/Photo Researchers; **333** Doug Martin; **334** Dr. E.R. Degginger/Color-Pic; **335** Tony Stone Images; **336** Louis Psihoyos/Matrix; **337** Doug Martin; **338** John Shelton; **340** Jack Dykinga; **341** StudiOhio; **344** Doug Martin; **346** Steve Dunwell/The Image Bank; **348** StudiOhio; **350** courtesy University of Michigan; **353** John Cancalosi/Peter Arnold, Inc.; **354-355,** Micheal Skrepnick; **355** KS Studios; **359** (l)Barbara Gerlach/DRK Photo, (c)Animals Animals/Zig Leszczynski, (r)Jeff Lepore/Photo Researchers; **360** Animals Animals/Breck P. Kent; **364** T.Geer/Peter Arnold, Inc.; **365** Earth Scenes/Doug Wechsler; **367** (l)Fred Bavendam/Peter Arnold, Inc., (r)Eric Grave/Photo Researchers; **368** Aaron Haupt; **369 370** Doug Martin; **371** Chris Jones/The Stock Market; **372** Animals Animals/Zig Leszczynski; **373** Burnstein/Corbis/Bettmann Archive; **376** (l)David M. Dennis/Tom Stack & Associates, (r)Michael Collier; **378** Animals Animals/Patti Murray; **380** KS Studios; **382** Dave Watts/Tom Stack & Associates; **386** (t)David M. Dennis, (b)KS Studios; **387** (br)Mark Burnett, (others)David M. Dennis; **388-389** Norbert Wu; **389** NASA/TSADO/Tom Stack & Associates; **390-391** Michael Hildreth/Fascination Photos; **391** Mark Burnett; **393** Rob Badger Photography; **395** David Lawrence/The Stock Market; **396** Frank Rossotto/The Stock Market; **399** Matt Meadows; **400** NASA; **404** Richard Weiss/Peter Arnold, Inc.; **407** T.A. Wiewandt/DRK Photo; **408** KS Studios; **413** Mike & Anne Adair; **414** Michael Melford/The Image Bank; **419** Bill Gillingham/The Stock Solution; **420-421** Warren Faidley/Weatherstock; **421** Mary Lou Uttermohlen; **422** Don & Pat Valenti/DRK Photo; **424** Mary Lou Uttermohlen; **427** (l)Roy Morsch/The Stock Market, (lc)Jose L. Pelaez, (rc)Mark McDermott/Tony Stone Images; (r)EPI Nancy Adams/Tom Stack & Associates; **428** Doug Martin; **429** Earth Scenes/Breck P. Kent; **430** (t)Charles O'Rear/Westlight, (b)Doug Martin; **431** NCAR/Tom Stack & Associates; **435** Roy Johnson/Tom Stack & Associates; **436** Merrilee Thomas/Tom Stack & Associates; **437** Howard Bluestein/Photo Researchers; **439** Gary Williams/Gamma-Liaison; **441** Greg Stumpf; **442** KS Studios; **444** Lowell J. Georgia/Photo Researchers; **446** UPI/Bettmann; **448** Earth Scenes/Phil Degginger; **450-451** Tom Till; **451** Matt Meadows; **456** Doug Martin; **458** (t)Jeff Gnass, (b)David Muench; **460** (l)Tom Till, (r)Johnny Johnson/Tony Stone Images; **461** (l)David R. Frazier Photolibrary, (r)Michio Hoshino/Minden Pictures; **463** (l)Philippe Mazellier/National Geographic Society, (r)Robert Vink/National Geographic Society; **465** Galen Rowell/Mountain Light Photography; **468** Matt Meadows; **469** Bob Cranston/Mo Yung Productions/Norbert Wu; **470** Cliff Leight; **471** Thomas Braise/The Stock Market; **472** courtesy Ellen Mosley Thompson; **475** Darrell Gulin/DRK Photo; **476-477** H. Richard Johnston/Tony Stone Images; **477** Matt Meadows; **478** (l)Mark A. Leman/Tony Stone Images, (r)Zoltan Gaal/Photo Researchers; **480** Haramaty Mula/Phototake; **482 483** Matt Meadows; **485** NOAA; **486** Raven/Explorer/Photo Researchers; **488** Matt Meadows; **489** John Nordquist; **491** Matt Meadows; **492** Vince Cavataio/Pacific Stock; **494** (l)Porterfield-Chickering/Photo Researchers, (r)C. Del/Photo Researchers; **497** Trask/Stock Imagery; **502-503** Norbert Wu; **503** Matt Meadows; **504-505** "The Floor of the Oceans," Bruce C. Heezen and Marie Tharp (1980 by Marie Tharp. Reproduced by permission of Marie Tharp); **508** Institute of Oceanographic Sciences/NERC/Photo Researchers; **510** Norbert Wu; **512** Visuals Unlimited/WHOI/D. Foster; **513** (t)Norbert Wu, (b)Dave B. Fleetham/Tom Stack & Associates; **514** Norbert Wu; **515** (t)Norbert Wu, (b)Urike Welsch/Photo Researchers; **516 517** Matt Meadows; **518** Norbert Wu; **520** Tomas del Amo/Pacific Stock; **521** Randy Brandon/Peter Arnold, Inc.; **522** courtesy Woods Hole Oceanographic Institution; **525** Norbert Wu; **526-527** Mark Steinmetz; **526** StudiOhio; **527** KS Studios; **528-529** Norbert Wu; **529** George H.H. Huey; **530-531** Carr Clifton/Minden Pictures; **531** Mark Burnett; **532** Jon Feingersh/The Stock

Photo Credits

Market; **534 535** Matt Meadows; **537** Richard Hutchings; **538** Michael Ableman; **539** (t)David M. Dennis/Tom Stack & Associates, (b)Larry Lefever/from Grant Heilman; **540** Frans Lanting/ Minden Pictures; **543** (t)Richard Laird/FPG International, (b)Jack Dykinga; **544** Tony Stone Images; **547** (t)Kenji Kerins, (b)Bob Daemmrich/ Stock Boston; **549** Matt Meadows; **550** courtesy Leigh StandingBear; **551** StudiOhio; **552** Terry Donnely/Tom Stack & Associates; **553** Matt Meadows; **554-555** John Hyde/Alaska Stock Images; **555** Tim Courlas; **556** Ken Biggs/The Stock Market; **558** Oliver Strewe/Tony Stone Images; **564** Peter Menzel/Stock Boston; **565** Joe Baraham/The Stock Market; **566** Doug Martin; **567** file photo; **568** (l)Simon Fraser/Science Photo Library/Photo Researchers, (r)Kenji Kerins; **569** (l)courtesy the city of San Diego, (r)John Eastcott & Yva Momatiuk/DRK Photo; **570** C. Hansen Carroll/FPG International; **571** (l)David Woodfall/Tony Stone Images, (r)Ian Adams; **573** Doug Martin; **576-577** Jose L. Pelaez/The Stock Market; **577** David M. Dennis; **578** KS Studios; **579** (t)Larry Ulrich, (b)Ray Pfortner/Peter Arnold, Inc.; **580-581 581 582-583** NASA; **583** Matt Meadows; **584** (l)Chris Collins/The Stock Market, (r)Mark E. Gibson; **585** (l)Daedalus Enterprises/ Peter Arnold, Inc., (c)Curtis Martin/Stock Imagery, (r)David M. Dennis/Tom Stack & Associates; **586** Doug Martin; **587** Tim Davis/ Photo Researchers; **588** (t)NASA/Gamma-Liaison, (c)Eric Sander/Gamma-Liaison, (b)Eric Sander/ Gamma-Liaison; **590** Matt Meadows; **591** Geoff Chester/International Dark-Sky Association; **592** International Dark-Sky Association; **598** Grafton Marshall Smith/The Stock Market; **599** SuperStock; **600** courtesy Gibor Basri; **601** Ken Frick; **603** Doug Martin; **604-605** Mark M. Lawrence/The Stock Market; **606 607 608** NASA; **611** NASA/Sygma; **612-613** Wendy Shattil & Bob Rozinksi/Tom Stack & Associates; **613** Matt Meadows; **616** Skip Comer; **618** Alexander Tsiaras/Photo Researchers; **620 621** Matt Meadows; **626** Bruce Herman/Tony Stone Images; **627** Jerry Schad/Photo Researchers; **628** NASA-TSADO/Tom Stack & Associates; **632** NASA; **633** Zuber et al/Johns Hopkins University/NASA/ Photo Researchers; **636** NASA; **638-639** David P. Anderson, SMU/NASA/Photo Researchers; **639** Matt Meadows; 640 Scala/Art Resource, NY; **644** Tom McHugh/Photo Researchers; **646** (l)JPL/ TSADO/Tom Stack & Associates, (r)NASA; **647** JPL/TSADO/Tom Stack & Associates; **648** NASA; **649** (t)NASA/Gamma-Liaison, (b)NASA/ TSADO/Tom Stack & Associates; **650** (l)USGS/ Phototake, (r)USGS/TSADO/Tom Stack & Associates; **651** Jim Stratford/Black Star; **652** Peter Menzel/Stock Boston; **653** Bob Daemmrich; **654** (t)NASA/Mark Marten/Photo Researchers, (b)NASA; **655** (moons)JPL, (b)NASA/Gamma-Liaison; **657** NASA/JPL/Tom Stack & Associates; **658** (t)NASA/Photo Researchers, (b)NASA; **659** NASA; **663** Ronald E. Royer/Photo Researchers; **664** (l)Tom Till/Tony Stone Images, (r)NASA; **665** NASA; **669** (l)NASA, (r)Larry Ulrich; **670-671** Bill & Sally Fletcher/Tom Stack & Associates; **671** Matt Meadows; **673** Bill & Sally Fletcher/Tom Stack & Associates; **674** Tim Courlas; **676** Matt Meadows; **678** David Parker/Science Photo Library/Photo Researchers; **680** (t)John Bova/Photo Researchers, (b)Hale Observatories/Science Photo Library/ Photo Researchers; **681** (t)NASA, (b)Chris Butler/Science Photo Library/Photo Researchers; **682** California Institute of Technology and Carnegie Institute of Washington; **683** Tim Courlas; **688** NASA; **689** (t)NASA, (b)Joe Stancampiano & Karl Luttrell/IMB Collaboration; **690** NASA; **692** (l)USNO/TSADO/Tom Stack & Associates, (r)David F. Malin; **693** Royal Observatory, Edinburgh/Science Photo Library/ Photo Researchers; **698** Robert Frerck/The Stock Market; **699** JPL; **700** NASA; **703** John Mead/Photo Researchers; **704** Pat Rawlings/NASA; **705** Pat Rawlings; **727** Thomas Veneklasen; **728** Franklin Over; **729** Dominic Oldershaw; **733** Jeff Smith/ Fotosmith; **734 735** Dominic Oldershaw; **736** Phil Degginger/Color-Pic; **738** Timothy Fuller; **739** Dominic Oldershaw; **742** Thomas Veneklasen; **743** Dominic Oldershaw; **744** Jeff Smith/Fotosmith; **745** Dominic Oldershaw